Cement Chemistry

emerald PUBLISHING
ice Publishing

Cement Chemistry

Calcium silicates and anhydrous Portland cement

Third edition

Ian G Richardson and HFW Taylor

Published by Emerald Publishing Limited, Floor 5,
Northspring, 21-23 Wellington Street, Leeds LS1 4DL.

ICE Publishing is an imprint of Emerald Publishing Limited

Other ICE Publishing titles:
ICE Handbook of Concrete Durability, Second edition
Edited by Marios Soutsos. ISBN 978-0-72776-375-4
Structural Design of Buildings: Elemental Design
Edited by Feng Fu and David Richardson. ISBN 978-1-83549-573-5
Fibre-reinforced Concretes for High-performance Structures: Building a more sustainable future
Andreas Lampropoulos. ISBN 978-0-72776-556-7

A catalogue record for this book is available from the British Library

ISBN 978-1-83549-777-7

Cover photo: courtesy of Ian G Richardson

Commissioning Editor: Michael Fenton
Content Development Editor: Cathy Sellars
Production Editor: Emma Sudderick and Benn Linfield

Typeset by KnowledgeWorks Global Limited
Index created by Ian Richardson and Lyn Nesbitt-Smith

Contents

Preface vii
About the authors ix

01 .... **Portland cement and its major constituent phases** **1**
1.1. Introduction 1
1.2. Alite 7
1.3. Belite 20
1.4. Aluminate 31
1.5. Ferrite 38

02 .... **High-temperature chemistry** **57**
2.1. Introduction 57
2.2. Systems containing CaO with SiO_2 or Al_2O_3 or both 57
2.3. Systems containing Fe_2O_3 65
2.4. Systems containing MgO or FeO 70
2.5. Systems containing alkalis or SO_3 or both 74
2.6. Systems with other components 77
2.7. Laboratory preparation of high-temperature phases 79

03 .... **The chemistry of Portland cement manufacture** **89**
3.1. General considerations 89
3.2. Raw materials and manufacturing processes 95
3.3. Reactions below about 1300 °C 100
3.4. Reactions at 1300–1450 °C 105
3.5. Reactions during cooling, grinding or storage 111
3.6. Effects of minor components 117

04 .... **Properties of Portland clinker and cement** **127**
4.1 Macroscopic and surface properties 127
4.2. Light microscopy 132
4.3. Scanning electron microscopy, X-ray diffraction and other techniques 135
4.4. Calculation of quantitative phase composition from bulk chemical analysis 140
4.5. Physical methods for determining quantitative phase composition 142
4.6. Reactivities of clinker phases 147

05 .... **Calcium silicate hydrates: natural minerals** **159**
5.1. Introduction 159
5.2. Natural minerals 159
5.3. The tobermorite phases 160
5.4. Jennite and metajennite 223

5.5. Tacharanite and fukalite 229
5.6. Plombièrite 230
5.7. General aspects of the crystal chemistry of calcium silicate hydrates and related phases 233

06 .... **Calcium (alumino)silicate hydrates: synthetic preparations** **251**
6.1. Introduction 251
6.2. Room temperature preparations with Ca/Si ratios of 0.7–1.5 251
6.3. Room temperature preparations with Ca/Si ratios greater than 1.5 271
6.4. Preparations containing monomeric or high proportions of dimeric silicate ion 282
6.5. Preparations made at elevated temperatures 283
6.6. Tobermorite-based structural models for C–S–H(I) 288
6.7. Thermal behaviour and water content of C–S–H(I) 316
6.8. Preparations containing monomer or high proportions of dimer 322
6.9. Ca–OH groups in C–S–H preparations with Ca/Si ratios >1.3–1.5 322
6.10. Structural substitution and ion exchange 324
6.11. Relation of synthetic C–S–H to the C–S–H in cement pastes 334

07 .... **Hydration of the calcium silicate phases** **371**
7.1. Introduction 371
7.2. Composition, density and other data for C–S–H 377
7.3. Microstructure, microanalysis and electron diffraction 387
7.4. Silicate anion structure 400
7.5. Nanostructures of C–S–H and related materials 407
7.6. Equilibria 412
7.7. Kinetics and mechanisms 421

Index **453**

Preface

Hal Taylor wanted his book *Cement Chemistry* to serve both as an introduction for those with no previous specialist knowledge of cements and as a guide to further research, and so he dealt with recent research but placed the development of the subject in a historical perspective. In retaining the broad scope and ambition of his book, it has been necessary to split the revised third edition into two volumes. This first volume updates second edition chapters on the chemistry and properties of anhydrous Portland cement and its constituent phases, and the hydration of calcium silicates. Highlights include:

- new evidence about the invention of Portland cement
- updated and enhanced treatment of crystal structure data, including the incorporation of space group numbers and extensive use of crystal-chemical information
- updated Bogue equations and illustrative calculation of enthalpy change on formation of Portland cement clinker
- expanded coverage of hydration of calcium silicates, including critical discussion of a wide range of data from the literature
- two new chapters about calcium (alumino) silicate hydrates: natural minerals and synthetic preparations

Hal was a Visiting Professor with me at the University of Leeds in the years 2000-2002, and during that time he reviewed research concerning natural and synthetic calcium silicate hydrates. The work that he produced was a good starting point for two new chapters in the revision of his book, and so we have both contributed to all seven chapters of this volume.

I have attempted to make the third edition of *Cement Chemistry* more accessible as an introduction for those entering the subject for the first time, for example, by setting out the calculations behind some results so that they are more easily understood. But the book has also been very extensively revised and updated to reflect advances in the field, and so there is a lot of material that will only be immediately accessible to those who are already experienced in the subject. In the preface to the second edition, written in 1997, Hal noted that the literature of cement chemistry is voluminous, and that there had been a significant increase in the seven years since publication of the first edition. There has of course been an enormous increase since that time, and so coverage in this edition has again of necessity been selective, but – as then – it is hoped that the most important contributions have been covered. Research on cements is generally very time limited, which constrains the number and length of experiments – for example, a typical PhD project might involve data collection over a period of only 30 months or so. As a consequence, throughout both volumes, I have combined data sets from multiple studies – often conducted across many decades – in order to gain a better understanding of important topics. The result is a very large number of figures and tables, and I hope that the effort expended in creating and discussing them will be considered to have been worthwhile.

Acknowledgements

I have worked on material for this revision (both volumes) over an extended period of time between my work commitments, and I must thank a number of people for their extraordinary patience and support: the team at ICE Publishing / Emerald Publishing, especially Michael Fenton, my contact since 2016; some of the Heads of the School of Civil Engineering at the University of Leeds (in particular Prof Nigel Smith); and most importantly my children, Ella and Charlie, and my wife, Andrea. The publisher's requirement for the use of full references was problematic: anyone who has used the second edition will know that the references are presented in a truncated format and my decision to retain so many of them led to a rather time-consuming effort to obtain the missing information. I am most grateful to my daughter who very ably assisted in that endeavour, and also for transcribing many of the tables from the second edition, which were only available as images. I thank my former PhD students for generating some excellent data, some of which is used in a number of figures, the captions of which include the individual acknowledgement.
I have gained inspiration in cement chemistry from a number of people over the past 40 years: in addition to Hal Taylor, the most notable are Dr GW Groves, Prof CM Dobson, Prof HM Jennings, Prof RJ Kirkpatrick, Prof JF Young, Dr EM Gartner, Dr CR Wilding, Prof JG Cabrera, Dr DD Double, Dr SA Rodger, and Prof JH Sharp.

Ian G Richardson
Leeds
March 2025

About the authors

Harry FW Taylor received a PhD in 1947 for work on the physical chemistry of DNA (under JM Gulland and DO Jordan). This was followed by a Research Fellowship at Birkbeck College (1948–1953, under JD Bernal and JW Jeffery) on the hydrothermal chemistry of silicates that led to his lifelong research interest in the chemistry of cement. His early work laid the foundations for our understanding of the nanostructure of C–S–H, the principal product formed when cement is mixed with water, and the one mainly responsible for its hardening. Subsequent studies took him into many additional aspects of the chemistry and materials science of cement and concrete. He was Professor of Inorganic Chemistry at the University of Aberdeen from 1965 until his retirement in 1983 and was a Visiting Professor at Imperial College London (1989–1994) and at the University of Leeds (2000–2002). His work was recognised by Fellowships and by many other honours and awards.

Ian G Richardson received a DPhil from the University of Oxford in 1991 under GW Groves and was subsequently his assistant. He moved to the University of Leeds in 1995 where he is Professor of Civil Engineering Materials. He is a materials scientist whose main research interests concern the microstructure, chemistry and properties of cementitious materials, and the crystal chemistry of calcium silicate hydrates, and various oxides, hydroxides and carbonates. He is perhaps best known for his work on the composition, morphology and nanostructure of the C–(A–)S–H in cements that contain a wide range of supplementary cementitious materials. He has given numerous invited contributions at conferences or congresses and is a Fellow of the Institute of Materials, Minerals & Mining.

The authors in 1996 (reproduced by permission of Achber Studio, Laconia NH, USA)

Ian G Richardson and HFW Taylor
ISBN 978-1-83549-777-7
https://doi.org/10.1108/978-1-83549-776-020253001

Chapter 1
Portland cement and its major constituent phases

1.1. Introduction
1.1.1 Portland cement: general

Portland cement is made by heating a mixture of limestone and clay, or other materials of similar bulk composition and sufficient reactivity, ultimately to a temperature of about 1450 °C. Partial fusion occurs and nodules of clinker are produced. The clinker is mixed with a few per cent of calcium sulfate and finely ground to make the cement. The calcium sulfate controls the rate of set and influences the rate of strength development. It is commonly described as gypsum, but it may be replaced partly or wholly by other forms of calcium sulfate. Some specifications allow the addition of other materials at the grinding stage. The clinker typically has a composition in the region of 67 % CaO, 22 % SiO_2, 5 % Al_2O_3, 3 % Fe_2O_3 and 3 % other components, and normally contains four major phases, called alite, belite, aluminate and ferrite. Several other phases, such as alkali sulfates and calcium oxide, are normally present in minor amounts. Hardening results from reactions between the major phases and water.

Alite is the most important constituent of all normal Portland cement clinkers, of which it constitutes 50–70 %. It is tricalcium silicate (Ca_3SiO_5) modified in composition and crystal structure by ionic substitutions. It reacts relatively quickly with water, and in normal Portland cements is the most important of the constituent phases for strength development; at ages up to 28 days, it is by far the most important.

Belite constitutes 15–30 % of normal Portland cement clinkers. It is dicalcium silicate (Ca_2SiO_4) modified by ionic substitutions and normally present wholly or largely as the β polymorph. It reacts slowly with water, thus contributing little to the strength during the first 28 days, but substantially to the further increase in strength that occurs at later ages. By one year, the strengths obtainable from pure alite and pure belite are about the same under comparable conditions.

Aluminate constitutes 5–10 % of most normal Portland cement clinkers. It is tricalcium aluminate ($Ca_3Al_2O_6$), substantially modified in composition and sometimes also in structure by ionic substitutions. It reacts rapidly with water and can cause undesirably rapid setting unless a set-controlling agent, usually gypsum, is added.

Ferrite makes up 5–15 % of normal Portland cement clinkers. It is tetra-calcium aluminoferrite (Ca_2AlFeO_5), substantially modified in composition by variation in Al/Fe ratio and ionic substitutions. The rate at which it reacts with water appears to be somewhat variable, perhaps due to differences in composition or other characteristics, but in general is high initially and low or very low at later ages.

1.1.2 The invention of Portland cement

Portland cement has been central to the development of the modern world, with billions of tonnes now used annually (CEMBUREAU, 2018; US Geological Survey, 2019). It is perhaps surprising therefore that its early history is shrouded in mystery, and that there is no consensus about the identity of the inventor. While Joseph Aspdin, a bricklayer from Hunslet, Leeds, UK was granted a patent in 1824 for a material that he called Portland cement (Aspdin, 1824), conclusions differ considerably on how similar Aspdin's material would have been to the Portland cement that is in use today. For example, Blezard (1998) believed that Joseph had produced '*Nothing more than a hydraulic lime*', whereas Skempton (1962) stated that '*The truth would seem to be that the "break-through"* [of clinkering] *had been made by 1843 at the latest, and more probably a good deal earlier by the elder Aspdin* [i.e. by Joseph rather than his son William].' Microstructural characterisation of a sample of render from a former public house that is listed as '*...the only surviving example of a building covered with Joseph Aspdin's patent Portland cement*' (The Wakefield Arms: Historic England, 1992) combined with historical research has shown that the render, which was dated with high confidence to 1840 or early 1841, was certainly made using a Portland cement because it contained a lot of relicts of alite (Richardson and Zhu, 2022) and so Skempton was correct. The common belief that Joseph used a lime kiln is due to the vagueness of the description of his process in the patent (including the description of the kiln as being '*similar to a lime kiln*'), and also perhaps because of the regular publication of a painting that purports to depict his first cement works in Wakefield (e.g. in Anon, 1999; Beadle, 1977; Blezard, 1998; Stanley, 1979; Trout, 2017). Regarding the latter, it is most unlikely that the picture is genuinely of the Wakefield cement works. This is because it shows two lime kilns that are surrounded by trees and open ground, whereas extant plans of the works show a compound with buildings, a perimeter wall and one kiln, as shown in Figure 1.1(a) (WYAS JG001578, 1837).

The cement works shown in Figure 1.1(a) will be referred to as Aspdin Works 1. If the painting was really commissioned by the Aspdin family as claimed (e.g. by Stanley and Bond, 1999), then it is possible that it was part of a deliberate attempt to mislead competitors (Bogue, 1947; Davis, 1924; Steinour, 1960). Bogue (1947), for example, speculated '*...on the possibility that Aspdin may not have revealed in his patent certain features of his operation, and that he may have used much higher temperatures than are indicated.*' Halstead (1961) noted that a glass works existed in Hunslet when Joseph Aspdin lived there, and that '*a glass kiln would be more likely to reach the temperatures required for successful burning of Portland cement than would the lime-burning kilns used by his competitors*'. The glass works in Hunslet consisted of two parts: one for making crown glass and the other for flint glass. The works were owned by two brothers, John and Joshua Bower (and their partners). While referred to as a carpenter/joiner in early documents, John Bower had been a manufacturer of sulfuric and nitric acids at a site on Jack Lane in Hunslet since at least 1795 (WYAS Wakefield Deed 1795/DR/548/575; WYAS Calderdale WYC:1089/226), and he was as a consequence often described as an '*Oil of Vitriol and Aqua Fortis Maker*' (e.g. WYAS Wakefield Deeds 1795/DR/548/575 and 1814/GC/318/355) or chemist (e.g. in a newspaper announcement of his eldest daughter's marriage; *Leeds Mercury*, 1813). Abutting this site on the south were houses built by Joseph Aspdin's father, Thomas, who was a bricklayer, and who had been John Bower's neighbour since at least 1792 (WYAS Wakefield Deeds 1792/DH/716/925, 1792/DH/717/927, 1794/DO/51/67). These houses – which are referred to as '*Aspdin Row*' on the 1850 Town Plan of Leeds (Ordnance Survey, 1847) – were left by Thomas to his wife Mary (and on her death, to their

Figure 1.1(a) Plan of the world's first dedicated Portland cement works at Kirkgate Wakefield, West Yorkshire, England (reproduced by permission of WYAS, Wakefield: JG001578/1837 plan)

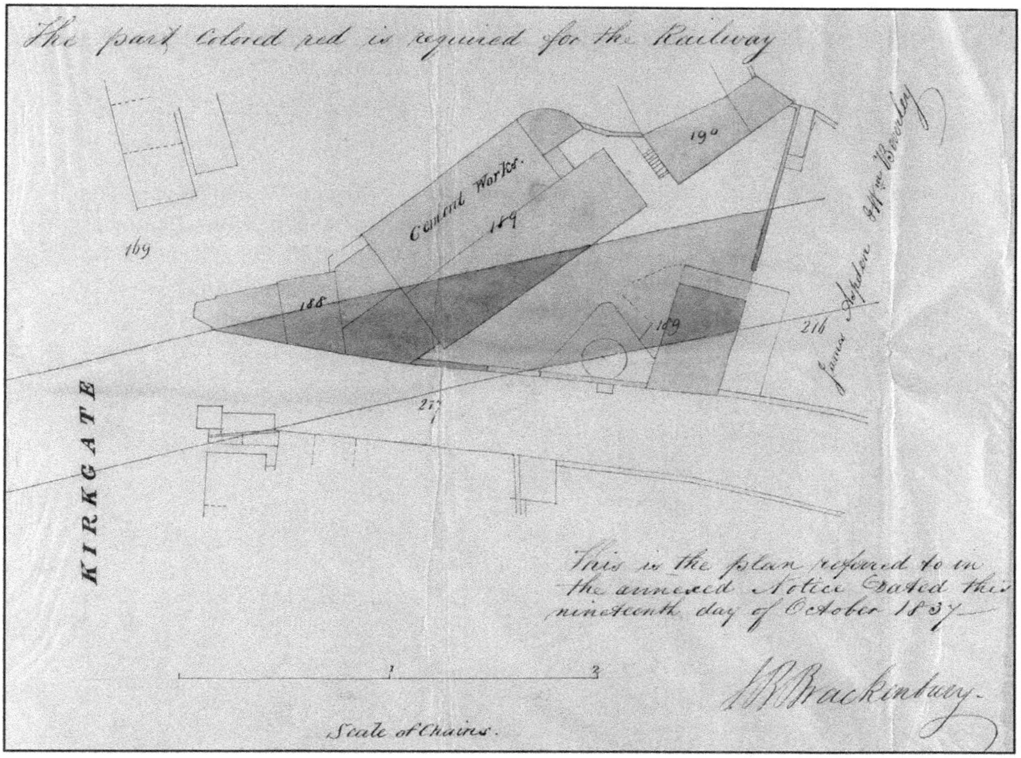

It was drawn to show the part of the cement works that was required for the new railway. The text states that the plan was referred to in a '*Notice Dated this nineteenth day of October 1837*'. It is based on the 1835 survey plan for the Manchester and Leeds Railway (WYAS QE20/1/1835/10). The book of reference that accompanies the survey plan indicates that plots 188, 189 and 190 were all owned by James and Rachel Strafford, which they inherited in 1832 from their father, George Strafford snr. (Exchequer Court of York Wills, 1832), who was a timber merchant (e.g. WYAS Wakefield Deed 1811/FP/84/82) and builder (e.g. WYAS Wakefield Deeds 1811/FP/52/56 and 1831/LD/364/256). Plots 189 and 190 were leased by Joseph Aspdin's eldest son, James, and their partner William Beverley, who also owned plot 216 (as indicated on the plan). Plot 216 consisted of '*Cement Pits and Garden*' and plot 189 consisted of a '*Cement Manufactory, Engine, Engine-house, and Burning-kiln*'. All of the buildings on the plan that are shaded grey are present on Walker's Wakefield town plan of 1823 (WYAS C509) and so were part of George Strafford's '*Sawing Mill*' with '*Engine House*' (WYAS Wakefield Deeds 1814/FX/447/518 and 1816/GK/325/340); the unshaded parts constitute the '*Burning-kiln*' that was built after 1823. The diameter of the circle (i.e. the chimney of the kiln) is about 14′ ≈ 4·3 m.

children), and John Bower was one of three witnesses to the will, which is dated 14 August 1800 (Exchequer Court of York Wills, 1802). Thomas Aspdin and his five sons were all bricklayers, and so it seems entirely plausible that they may have been involved in the construction of properties owned by their close neighbours John and Joshua Bower (including the furnaces of the glass works), and that John may have supported Joseph with his experiments to develop a new cement. It is notable that the diameter of the '*Burning-kiln*' on Figure 1.1(a) is about the same as John Bower's first glass house (WYAS Leeds WYL160/14/113 and 114; probably his first), and it is also consistent with the maximum diameter of the bottle-shaped cement kilns in 19th century England (Butler, 1913; Eckel, 1905). It is relevant to note that John and Joshua Bower (and partners) were manufacturing glass in Hunslet some time earlier than November 1805 (WYAS Wakefield Deed 1805/EX/82/117) and that Joseph's younger son, William, claimed that his father had '*first introduced*' his cement in 1813 (Aspdin, 1848).

The compulsory purchase of Aspdin Works 1 by the Manchester and Leeds Railway Company in 1839 (WYAS Wakefield QE21/49) resulted in the construction of Aspdin Works 2 only a short distance from Works 1. Works 2 is present on a scale plan of Wakefield Kirkgate Railway Station that was prepared in March 1851 (National Archives, 1851). Richardson and Zhu (2022) and Richardson *et al.* (2024) showed that the diameter of the base of the kiln of Works 2 is $> 30'$ (or $> 9{\cdot}1$ m), which is much larger than the kiln of Works 1 (Figure 1.1(a)). Comparison of the contemporary images of Aspdin Works 2 that were drawn by Kilby (1843) and Tait (1845) with Figure 1.1(b) reveals that it is very similar in appearance to the cones and associated buildings of John Bower's glass houses in Hunslet in 1829. The cement in the render of the Wakefield Arms public house mentioned above would have been produced at Aspdin Works 2.

Little appears to be known about Joseph Aspdin's education, but it is perhaps significant that another of the witnesses to his father's will was John Ryley (1747–1815) who was a schoolmaster (WYAS Wakefield Deeds 1811/FO/58/77, 1814/FX/487/560, 1814/GA/373/455). He was highly

Figure 1.1(b) Billhead (engraved) showing John Bower's flint glass works in Hunslet, Leeds, dated 11 February 1829 (reproduced by permission of WYAS, Leeds: WYL33/A/33)

The diameter at the base of the glasshouse cone that is at the centre of the image is 49′ (14·9 m) (measured using the scale bar on the Ordnance Survey (1850) Town plan of Leeds sheet 22, surveyed in 1847). The appearance of the cone and adjacent buildings is strikingly similar to contemporary images of Aspdin Works 2 drawn by Kilby (1843) and Tait (1845).

regarded as a teacher of mathematics in Leeds (Nichols *et al.*, 1818), author of *The Leeds Guide* (Ryley, 1806) and the first editor of *The Leeds Correspondent*. While a Master at the charity school in Leeds from 1789, he also provided '*private tuition at the residence of different gentlemen in the town, or by sets of pupils who attended twice or thrice in each week at his own house, either early in the morning, at the hour of noon, or in the evening*' (Nichols *et al.*, 1818). It is plausible that John Ryley taught the children of Thomas Aspdin, including his eldest son, Joseph, who would have been 11 years old in 1789. Despite the assignment by some (e.g. Barfoot, 1974) of low social status to the occupation of bricklayer, and the modest average annual wage of bricklayers in Leeds in the early 19th century (Statistical Committee of the Town Council, 1840), it is quite possible that Thomas could afford to engage John Ryley as a tutor for his children: he left his wife 12 houses when he died in 1800 and he had at least one very wealthy relative, his brother-in-law Obadiah Willans (1745–1833), cloth manufacturer, who bequeathed an annuity to his sister (i.e. Joseph Aspdin's mother) in his 1831 will (Exchequer Court of York Wills, 1834). The houses would have provided Thomas with additional income from rent payments. Interestingly, one of Mary Aspdin's tenants in 1823/4 was John Kilner (WYAS Leeds LO/HU/5(1823–27)), the progenitor of the company where the famous Kilner® jar was invented. It is highly likely that he learned to make glass whilst working for John Bower because of the location of his abode, and the fact that baptism records show that all four of his sons were born in Hunslet and that his occupation is given as labourer for the eldest (Caleb, born Jan. 1818) but glass maker for the other three (George, born 11 Dec. 1823; William, born 20 Jan. 1826; John, born 25 Jan. 1828).

1.1.3 Types of Portland cement

The great majority of Portland cements made throughout the world are designed for general constructional use. The standard specifications with which such cements must comply are similar, but not identical, in all countries and various names are used to define the material, such as Class 42·5 CEM I Portland cement in current European and British standards (42·5 is the minimum 28-day compressive strength in MPa), Types I and II Portland cement in the ASTM (American Society for Testing and Materials) specifications used in the USA, or ordinary Portland cement (OPC) in former British standards. Throughout this book, the term 'ordinary' Portland cements is used to distinguish such general-purpose cements from other types of Portland cement, which are made for special purposes.

Standard specifications are, in general, based partly on chemical composition or physical properties, such as specific surface area, and partly on performance tests, such as setting time or compressive strength developed under standard conditions. The content of MgO* is usually limited to 4–5%, because quantities of this component in excess of about 2% can occur as periclase (magnesium oxide), which through slow reaction with water can cause destructive expansion of hardened concrete. Free lime (calcium oxide) can behave similarly. Excessive contents of SO_3 can also cause expansion, and upper limits, typically 3·5% for ordinary Portland cements, are usually imposed. Alkalis (K_2O and Na_2O) can undergo expansive reactions with certain aggregates, and some specifications limit the content– for example, to 0·6% equivalent Na_2O ($Na_2O + 0{\cdot}66\ K_2O$). Other upper limits of composition widely used in specifications relate to matter insoluble in dilute

* Confusion can arise because the names or formulae of compounds can be used to denote either phases or components; this applies especially to CaO and MgO. Here and elsewhere, chemical or mineral names of oxides (e.g. calcium oxide, magnesium oxide, lime, periclase) will generally be used for phases, and formulae (e.g. CaO, MgO) for components. Mineral names or prefixed formulae (e.g. α-Al_2O_3) are never used for components.

acid, and loss on ignition. Many other minor components are limited in content by their effects on the manufacturing process, or the properties, or both, and in some cases the limits are defined in specifications.

Rapid-hardening Portland cements have been produced in various ways, such as varying the composition to increase the alite content, finer grinding of the clinker, and improvements in the manufacturing process – for example, finer grinding or better mixing of the raw materials. The alite contents of Portland cements have increased steadily over the time that they have been produced, and many cements that would be considered ordinary today would have been described as rapid-hardening only a few decades ago. In the ASTM specifications, rapid-hardening Portland cements are called high early strength or Type III cements. For both ordinary and rapid-hardening cements, both lower and upper limits may be imposed on strengths at 28 days, upper limits being a safeguard against poor durability resulting from the use of inadequate cement contents in concrete.

Destructive expansion from reaction with sulfates can occur not only if the latter are present in excessive proportion in the cement, but also from attack on concrete by sulfate solutions. The reaction involves the Al_2O_3-containing phases in the hardened cement; in sulfate-resisting Portland cements its effects are reduced by decreasing the proportion of the aluminate phase, sometimes to zero. This is achieved by decreasing the ratio of Al_2O_3 to Fe_2O_3 in the raw materials.

White Portland cements are made by increasing the ratio of Al_2O_3 to Fe_2O_3, and thus represent the opposite extreme in composition to sulfate-resisting Portland cements. The normal, dark colour of Portland cement is due to the ferrite, formation of which in a white cement must thus be avoided. It is impracticable to employ raw materials that are completely free from Fe_2O_3 and other components, such as Mn_2O_3, that contribute to the colour. The effects of these components are therefore usually minimised by producing the clinker under slightly reducing conditions and by rapid quenching. In addition to alite, belite and aluminate, some glass may be formed.

The reaction of Portland cement with water is exothermic, and while this can be an advantage under some conditions because it accelerates hardening, it is a disadvantage under others, such as in the construction of large dams or in the lining of oil wells, when a cement slurry has to be pumped over a large distance under pressure and sometimes at a high temperature. Slower heat evolution can be achieved by coarser grinding, and decreased total heat evolution by lowering the contents of alite and aluminate. The ASTM specifications include definitions of a Type II or 'moderate heat of hydration' cement, and a more extreme Type IV or 'low heat' cement. The Type II cement is also suitable for conditions exposed to moderate sulfate attack and is widely used in general construction work. Heat evolution can also be decreased by partially replacing the cement by fly ash (pulverised fuel ash; PFA) or other materials, and this is a widely used solution.

1.1.4 Cement chemical nomenclature and other abbreviations

Chemical formulae in cement chemistry are often expressed as sums of oxides; thus tricalcium silicate, Ca_3SiO_5, can be written as $3CaO \cdot SiO_2$. This does not imply that the constituent oxides have any separate existence within the structure. It is usual to abbreviate the formulae of the commoner oxides to single letters, such as C for CaO or S for SiO_2, Ca_3SiO_5 thus becoming C_3S. This system is often combined with orthodox chemical notation within a chemical equation, for example

$$3CaO + SiO_2 \rightarrow C_3S \quad (1.1)$$

or even within a single formula, as in $C_{11}A_7{\cdot}CaF_2$ for $Ca_{12}Al_{14}O_{32}F_2$. The abbreviations most widely used are as follows.

$$\begin{array}{llll} C = CaO & S = SiO_2 & A = Al_2O_3 & F = Fe_2O_3 \\ M = MgO & K = K_2O & \bar{S} = SO_3 & N = Na_2O \\ T = TiO_2 & P = P_2O_5 & H = H_2O & \bar{C} = CO_2 \end{array}$$

The formulae of the simple oxide phases (e.g. CaO) are usually written in full. Other abbreviations and units used in this book are as follows.

1.1.4.1 Techniques

BEI = backscattered electron imaging; BSE = backscattered electron; DTA = differential thermal analysis; EPMA = electron probe microanalysis; ESCA = electron spectroscopy for chemical analysis; GLC = gas–liquid chromatography; GPC = gel permeation chromatography; IR = infrared. MIP = mercury intrusion porosimetry; NMR = nuclear magnetic resonance; QXDA = quantitative X-ray diffraction analysis; SEM = scanning electron microscop(e,y); STEM = scanning transmission electron microscop(e,y); TEM = transmission electron microscop(e,y); TG = thermogravimetry; TMS = trimethylsilyl(ation); XPS = X-ray photoelectron spectroscopy; XRD = X-ray diffraction; XRF = X-ray fluorescence.

1.1.4.2 Materials

C–S–H = poorly crystalline or amorphous calcium silicate hydrate of unspecified composition; GGBS = ground granulated blast-furnace slag; HCP = hardened cement paste; MK = metakaolin; PFA = pulverised fuel ash (flyash).

1.1.4.3 Properties or reactions

AR = alumina ratio (alumina modulus); C_x = analytical (total) concentration of x, irrespective of species; LSF = lime saturation factor; Na_2O_e = equivalent Na_2O (mass % $Na_2O + 0{\cdot}66\,K_2O$); RH = relative humidity; SR = silica ratio (silica modulus); $[x]$ = concentration of species x; $\{x\}$ = activity of species x; (+)2V, (–)2V, optic sign and optic axial angle.

1.1.4.4 Pressure units

$1\,MPa = 1\,N\,mm^{-2} = 10\,bar = 9{\cdot}87\,atm = 7500\,torr = 145{\cdot}0\,lb\,in^{-2} = 10{\cdot}198\,kg\,cm^{-2}$.

1.2. Alite

1.2.1 Polymorphism and crystal structure

On being heated, pure C_3S undergoes a series of reversible phase transitions, which have been detected by a combination of DTA, high-temperature XRD and high-temperature light microscopy (Bigaré *et al.*, 1967; Maki, 1979, 1986, 1994; Maki and Chromý, 1978a, 1978b; Maki and Kato, 1982; Guinier and Regourd, 1969; Regourd, 1967; Regourd and Guinier, 1976; Yamaguchi and Takagi, 1969).

$$T_1 \overset{620\,°C}{\rightleftarrows} T_2 \overset{920\,°C}{\rightleftarrows} T_3 \overset{980\,°C}{\rightleftarrows} M_1 \overset{990\,°C}{\rightleftarrows} M_2 \overset{1060\,°C}{\rightleftarrows} M_3 \overset{1070\,°C}{\rightleftarrows} R$$

where T = triclinic; M = monoclinic; R = rhombohedral.

The pure compound, when cooled to room temperature, is thus T_1. In production clinkers, due to the incorporation of substituent ions, the form present at room temperature normally approximates to M_1 or M_3 or a mixture of these; rarely, T_2 is found (Maki, 1979, 1986, 1994; Maki and Chromý, 1978a, 1978b; Maki and Kato, 1982; Timashev, 1980). There has been some uncertainty as to the number and nomenclature of these polymorphs; reported M_{1b} and M_{2b} forms appear to be identical with M_3, leaving reported M_{1a} and M_{2a} forms to be called simply M_1 and M_2, respectively (Maki, 1986; Maki and Kato, 1982).

Jeffery (1952) made the first determination of the crystal structure. He showed that the forms now known as R, T_1 and M_3 had closely similar structures, and determined the approximate or pseudostructure common to all three; it was built from Ca^{2+}, SiO_4^{4-} and O^{2-} ions, the last being bonded only to six Ca^{2+} ions, as in CaO. Later, more exact determinations were reported for T_1 (Golovastikov *et al.*, 1975), M_3 stabilised by Mg^{2+} (Nishi *et al.*, 1985), M_3 with increased disorder isolated from a works clinker (Mumme, 1995), R at 1200 °C (Nishi and Takéuchi, 1984) and R stabilised with Sr^{2+} (Il'inets *et al.*, 1985). Figure 1.2 shows the structure of the R form. The known structures are all closely similar as regards the positions of the Ca^{2+} and O^{2-} ions and of the Si atoms, but differ markedly in the orientations of the SiO_4^{4-} tetrahedra, which show varying degrees of disorder. Table 1.1 gives crystal data for the C_3S polymorphs that have been obtained using both single-crystal and powder methods. The unit cells of the T_1, M_3 and R forms are superficially somewhat different, but all are geometrically related; transformation matrices have been given (Hudson and Groves, 1982; Ilyinets *et al.*, 1986), and structural studies reviewed in detail by Dunstetter *et al.* (2006) who proposed a unified analysis of T_1, M_1, M_3 and R polymorphs.

The structural differences between the polymorphs affect the coordination of the Ca^{2+} ions and the O atoms of the SiO_4^{4-} tetrahedra, with the coordination of a given Ca site sometimes varying because of the orientational disorder of the SiO_4^{4-} tetrahedra. The number of crystallographically distinct Ca sites varies between polymorphs, with 3, 6 and 29 in the asymmetric units for the R (Il'inets *et al.*, 1985; Sr stabilised), M_3 (Mumme, 1995), and T_1 (Golovastikov *et al.*, 1975) polymorphs, respectively. In the unit cell of the M_3 polymorph, all 18 of the Ca sites are in six-fold coordination (Mumme, 1995), and so the mean coordination number ($\overline{CN^{Ca}}$) is of course 6·00. In the T_1 polymorph (Golovastikov *et al.*, 1975), 42 of the 54 Ca sites are coordinated to six O, with the remaining 12 Ca coordinated to seven, and so $\overline{CN^{Ca}} = 6{\cdot}22$. In the R polymorph (Il'inets *et al.*, 1985; Sr stabilised), 18 of the 27 Ca sites are coordinated to six O, with the remaining nine Ca coordinated to seven, and so $\overline{CN^{Ca}} = 6{\cdot}33$. In relation to reactivity towards water, the coordination of the oxygen atoms is possibly more important than that of Ca. In all three of these structures, the proportion of O that are O^{2-} ions bonded only to six Ca^{2+} ions is 20 % (see Table 4.4 in Chapter 4), but the mean oxygen coordination number ($\overline{CN^{O}}$) increases with that of calcium ($\overline{CN^{Ca}}$), with $\overline{CN^{O}} = 4{\cdot}4$, 4·53 and 4·6 for the M_3, T_1 and R polymorphs, respectively, and thus $\overline{CN^{O}} = 0{\cdot}6039\left(\overline{CN^{Ca}}\right) + 0{\cdot}7759$.

The silicate tetrahedra do not share O atoms, and so are monomers, or Q^0 using the Q^n notation (see §5.3.1.7). There is some variation in the average Si–O distances of the silicate tetrahedra, with those for the R (Il'inets *et al.*, 1985) and M_3 (Mumme, 1995) polymorphs being rather short when compared with the expected values: i.e. d(Si–O) = 1·607 Å (R) and 1·595 Å (M_3) compared with 1·639 Å from Equation 5.32 in Chapter 5, or with the values calculated using the equations given in Baur (1978) and Brown and Gibbs (1969). The value for the T_1 polymorph (1·633 Å; Golovastikov *et al.*, 1975) is close to the expected value. While Jeffery (1950, 1952) used arguments about the

Figure 1.2 Crystal structure of the R-modification of C_3S, based on the results of Nishi and Takéuchi (1984)

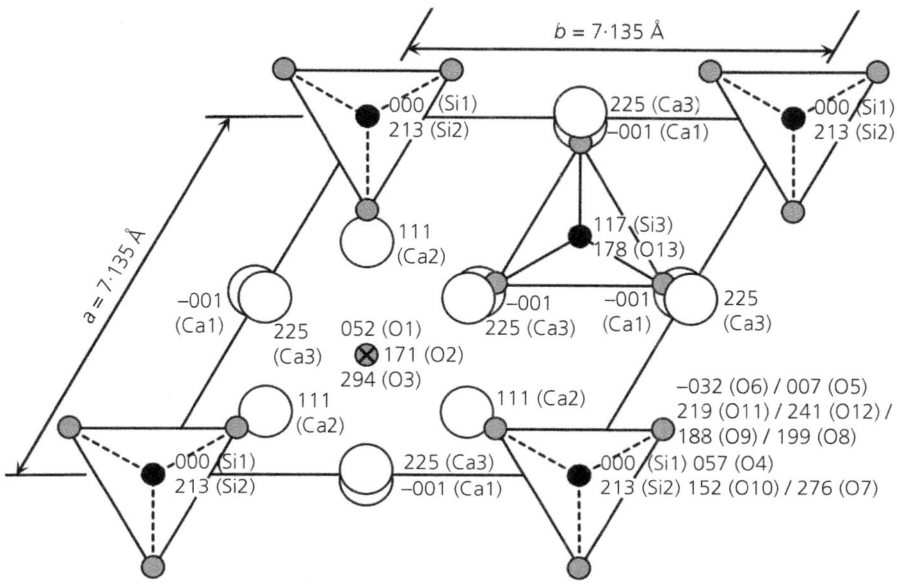

The crystal structure shows Ca atoms (large open circles), Si atoms (small black circles), oxide ions (small grey circle with ×) and oxygen tetrahedra (triangles with grey circles at the vertices). Heights of atoms are in thousandths of the cell height ($c = 25{\cdot}586$ Å), slashes denoting statistical alternatives. Oxygen tetrahedra are shown in averaged orientations; in reality, each is tilted around its enclosed Si atom statistically in any one of three directions so as to preserve the threefold symmetry. All tetrahedra point upwards except those that surround the Si atoms at height 213, of which 70 % point downwards and 30 % upwards. Only the bottom third of the cell is shown, the middle and top thirds being derivable from it by translations of $\frac{1}{3}, \frac{2}{3}, \frac{1}{3}$ and $\frac{2}{3}, \frac{1}{3}, \frac{2}{3}$ parallel to the *a*, *b* and *c* axes, respectively.

packing of spherical ions to assist in his determination of the pseudo-structure of C_3S – and ionic radii are of course used extensively in crystal chemical studies – the evidence from these three structure determinations suggests that their utility is limited in the case of C_3S because the values of the effective ionic radii of Ca and O ions that are required to account for the observed distances are quite different each case.

1.2.2 Tricalcium silicate solid solutions

Hahn *et al.* (1969) reported a detailed study of the limits of substitution in C_3S of MgO, Al_2O_3 and Fe_2O_3, alone and in combination. They concluded that Ca^{2+} could be partly replaced by Mg^{2+} and that both Ca^{2+} and Si^{4+} could be partly replaced by Al^{3+} or Fe^{3+}. They found the maximum content of MgO to be 2·0% at 1550 °C, falling to 1·5% at 1420 °C. The limits for Al_2O_3 and Fe_2O_3 were 1·0% and 1·1%, respectively, but Al^{3+} and Fe^{3+} competed for sites, and the limit for each

Table 1.1 Crystal data for the tricalcium silicate (C_3S) polymorphs

Polymorph	Space group (No.)	Z	Cell parameters (a, b, c: Å)						D_x g cm^{-3}	MV cm^3 mol^{-1}	Exp.[a]	Crystal structure reference / comments
			a	b	c	α	β	γ				
R	*R* 3 *m* (160)	9	7·078		24·94			120°	3·153	72·40	S	O'Daniel and Hellner (1950)[b]
R pseudostructure[c]	*R* 3 *m* (160)	9	7·0		25·0			120°	3·216	70·99	S	Jeffery (1952); $Ca_3Mg_{0·06}Al_{0·11}Si_{0·89}O_5$
R	*R* 3 *m* (160)	9	7·135		25·586			120°	3·021	75·48	S	Nishi and Takéuchi (1984); 1200 °C; $Ca_{2·98}Al_{0·04}Si_{0·98}O_5$
R	*R* 3 *m* (160)	9	7·0567		24·974			120°	3·168	72·07	S	Il'inets *et al.* (1985); stabilised with Sr
M_3	*C* 1 *m 1* (8)	36	*33·08	7·07	18·56		94·17°		3·170	72·42	S	Jeffery (1952); $Ca_3Mg_{0·06}Al_{0·11}Si_{0·89}O_5$
M_3	*C* 1 *m 1* (8)	36 6	*33·083 12·242	7·027 7·027	18·499 9·250		94·12° 116·04°		3·158 3·124	71·75 72·54	S	Nishi *et al.* (1985); $Ca_{2·89}Mg_{0·11}SiO_5$
M_3	*C* 1 *m 1* (8)	36	*33·1078	7·0355	18·5211		94·137°		3·154	71·98	P	de la Torre *et al.* (2002); $Ca_{2·93}Mg_{0·07}Al_{0·02}Si_{0·98}O_{4·99}$
M_3	*C* 1 *m 1* (8)	6	12·235	7·073	9·298		116·31°		3·154	72·39	S	Mumme *et al.* (1995); ex clinker; less ordered

M_1	*P* 1 *m* 1	18	*27·8744	7·0591	12·2578		116·03°		3·149	72·50	P	de Noirfontaine *et al.* (2012); $Ca_{2·85}Mg_{0·06}Fe_{0·03}Al_{0·05}Si_{0·92}S_{0·04}O_5$
	(7)	6	9·2952	7·0792	12·2214		116·08		3·149	72·50		de Noirfontaine *et al.* (2006); clinker
T_3	$P\bar{1}$ (2)	18	11·6389	14·1716	13·6434	104·982	94·622	90·107	3·142	72·47	P	de la Torre *et al.* (2008); $Ca_{2·96}Mg_{0·03}Al_{0·02}Si_{0·99}O_5$
T_2	$P\bar{1}$ (2)	18	11·7416	14·2785	13·7732	105·129	94·415	89·889	3·071	74·34	P	Peterson *et al.* (2004); pure C_3S; 682 °C
T_1	$P\bar{1}$ (2)	18	11·67	14·24	13·72	105·5°	94·33°	90·0°	3·116	73·28	S	Golovastikov *et al.* (1975)

MV = molar volume; * = supercell.

[a]Experiment: S = single-crystal; P = powder.

[b]The structure is wrong: it involves Si_3O_9 rings rather than separate SiO_4 tetrahedra.

[c]Average rhombohedral pseudostructure considered a good approximation to the true structures of triclinic, monoclinic and rhombohedral polymorphs.

was lowered if the other was present. A subsequent study using ^{27}Al NMR indicated that the Al was present almost entirely in tetrahedral sites, thus excluding substitution of Al^{3+} for Ca^{2+} (Skibsted *et al.*, 1994). Woermann *et al.* (1979) found a limit of 1·4% for either Na_2O or K_2O at 1500 °C.

Hahn *et al.* (1969) found that incorporation of Mg^{2+}, Al^{3+} or Fe^{3+} in sufficient quantity caused the higher temperature polymorphs to persist on cooling to room temperature. For each combination of substituents studied, increase in the total proportion of substituents led to the persistence of successively higher temperature polymorphs in the sequence $T_1 \rightarrow T_2 \rightarrow M_1$. In the system CaO–MgO–$Al_2O_3$–$SiO_2$, total contents of substituent oxides of about 1% caused T_1 to be replaced by T_2, and ones of about 2% caused T_2 to be replaced by M_1. Slightly higher total substituent contents were needed to effect each of these changes if Fe_2O_3 was present.

1.2.3 Compositions of alites in clinkers

Pure C_3S contains 73·7% of CaO and 26·3% of SiO_2. Alites in clinkers typically contain 3–4% of substituent oxides. Boikova (1986) found a positive correlation between the total percentage of substituent oxides in the alite (I_a) and the percentage of MgO in the clinker (M_c). Her results are approximately fitted by the equation $I_a = 0{\cdot}7M_c + 2{\cdot}1$.

Many workers have analysed alites from ordinary production clinkers by X-ray microanalysis (e.g. Boikova *et al.*, 1981, 1986; Ghose, 1980; Ghose and Barnes, 1979; Harrisson, 1993; Harrisson *et al.*, 1985, 1986; Kristmann, 1978; Le Saoût *et al.*, 2011; Maki *et al.*, 1992; Mumme, 1995; Sarkar and Roy, 1984; Terrier, 1969; Terrier *et al.*, 1968; Uchikawa *et al.*, 1992) or chemical analysis of separated material (Uchikawa *et al.*, 1985; Yamaguchi and Takagi, 1969). Alites in white (Boikova *et al.*, 1986; Le Saoût *et al.*, 2011) and iron-rich (Gollop and Taylor, 1994; Hall and Scrivener, 1997; Le Saoût *et al.*, 2011) clinkers have been analysed. The most important substituents are Mg^{2+}, Al^{3+} and Fe^{3+}, and there is wide agreement that the contents of MgO and possibly also of Fe_2O_3 in the alite increase with those in the clinker. Since MgO replaces CaO, the content of CaO decreases with that of MgO in the clinker. Taking these relationships into account, as well as some less important variations with clinker composition and the errors inherent in either experimental method, the agreement between the various sets of data is generally good.

In the following discussion, symbols of the type O_a and O_c are used to denote the contents of oxide component O in the alite and in the clinker, respectively. For MgO, Kristmann's (1978) data indicate the approximate relations $M_a = 0{\cdot}67 \times M_c$ for $M_c \leq 3{\cdot}0$ and $M_a \approx 2{\cdot}0$ for $M_c > 3{\cdot}0$. Yamaguchi and Takagi (1969), Terrier *et al.* (1968) and Maki *et al.* (1992) found similar correlations, but with slopes of 0·74, 0·45 and 0·58, respectively. For a series of clinkers of low Al/Fe ratio, Hall and Scrivener (1997) found $M_a = 0{\cdot}64 \times M_c - 0{\cdot}23$. For Fe_2O_3, Kristmann's (1978) data indicate the approximate relation $F_a = 0{\cdot}33 \times F_c$ for $F_c < 4\%$. There does not appear to be a significant correlation between A_a and A_c. Boikova (1986) reported ratios of 0·34–0·82 for M_a/M_c, 0·09–0·23 for A_a/A_c and 0·17–0·36 for F_a/F_c. The SO_3 contents of clinker alites are often very low, but values up to 0·9% have been reported for clinkers low in MgO and alkalis (Gartner and Tang, 1987) or high in SO_3 (Moranville-Regourd and Boikova, 1992).

Taken as a whole, the evidence indicates that the significant substitutions in clinker alites are of Na^+, K^+, Mg^{2+} and Fe^{3+} for Ca^{2+} and of Al^{3+}, P^{5+} and S^{6+} for Si^{4+}. Observed contents of K_2O and Na_2O are much below the limits found by Woermann *et al.* (1979). Table 1.2 includes a typical composition for a clinker with 1·65% MgO and 3·1% Fe_2O_3; Table 1.3 includes atomic ratios calculated from it.

Table 1.2 Typical compositions of phases in Portland cement clinkers (mass per cent)

	Na_2O	MgO	Al_2O_3	SiO_2	P_2O_5	SO_3	K_2O	CaO	TiO_2	Mn_2O_3	Fe_2O_3
Alite[a]	0·1	1·1	1·0	25·2	0·1	0·1	0·1	71·6	0·0	0·0	0·7
Belite[a]	0·1	0·5	2·1	31·5	0·1	0·2	0·9	63·5	0·2	0·0	0·9
Aluminate (cubic)[a]	1·0	1·4	31·3	3·7	0·0	0·0	0·7	56·6	0·2	0·0	5·1
Ferrite[a]	0·1	3·0	21·9	3·6	0·0	0·0	0·2	47·5	1·6	0·7	21·4
Aluminate (orthorhombic)[b]	0·6	1·2	28·9	4·3	0·0	0·0	4·0	53·9	0·5	0·0	6·6
Aluminate (low Fe)[c]	0·4	1·0	33·8	4·6	0·0	0·0	0·5	58·1	0·6	0·0	1·0
Ferrite (low Al)[d]	0·4	3·7	16·2	5·0	0·0	0·3	0·2	47·8	0·6	1·0	25·4

[a]Typical values for an ordinary Portland cement clinker with 1·65 % MgO, 3·1 % Fe_2O_3 and molar $SO_3/(K_2O+Na_2O) < 1{\cdot}0$. For clinkers not approximating to these conditions, the compositions of the phases may differ significantly from those given in the table, as explained in the text.
[b]Orthorhombic or pseudotetragonal forms, present in some clinkers high in alkalis. Na/K ratio varies with that of the clinker.
[c]Tentative composition for aluminate phase in white cement clinkers.
[d]Values for the ferrite phase in a typical sulfate-resisting clinker (MgO, 2·1 %; Al_2O_3, 3·8 %; Fe_2O_3, 4·7 %) (Gollop and Taylor, 1994)). Compositions of other such clinkers may vary considerably (see text).

Table 1.3 Atomic ratios for phases in Portland cement clinkers, calculated from the typical compositions in Table 1.2 (see footnotes to Table 1.2)

	K	Na	Ca	Mg	Mn	Ti	Fe	Al	Si	P	S	O
Alite	<1	1	291	6			2	4	96	<1	<1	500
$Ca_{2.91}Na_{0.01}Mg_{0.06}Fe_{0.02}Al_{0.04}Si_{0.96}O_5$			300					100				
Belite	3	1	197	2			2	7	91	<1	<1	400
$Ca_{1.97}Na_{0.01}K_{0.03}Mg_{0.02}Fe_{0.02}Al_{0.07}Si_{0.91}O_4$			203					100				
Aluminate (cubic)	4	9	274	9		1	17	167	17			600
$Ca_{2.74}Na_{0.09}K_{0.04}Mg_{0.09}Ti_{0.01}Fe_{0.17}Al_{1.67}Si_{0.17}O_6$			297					201				
Ferrite	1	1	198	17	2	5	62	100	14			500
$Ca_{1.98}Na_{0.01}K_{0.01}Mg_{0.17}Mn_{0.02}Ti_{0.05}Fe_{0.62}Al_{1.00}Si_{0.14}O_5$			200					200				
Aluminate (orthorhombic)	23	5	265	8		2	23	156	20			600
$Ca_{2.65}Na_{0.05}K_{0.23}Mg_{0.08}Ti_{0.02}Fe_{0.23}Al_{1.56}Si_{0.20}O_6$			303					199				
Aluminate (low Fe)	3	3	276	7		2	3	177	20			600
$Ca_{2.76}Na_{0.03}K_{0.03}Mg_{0.07}Ti_{0.02}Fe_{0.03}Al_{1.77}Si_{0.20}O_6$			291					200				
Ferrite (low Al)	1	3	201	22	3	2	75	75	20		1	500
$Ca_{2.01}Na_{0.03}K_{0.01}Mg_{0.22}Mn_{0.03}Ti_{0.02}Fe_{0.75}Al_{0.75}Si_{0.20}S_{0.01}O_5$			205					198				

1.2.4 Polymorphic modifications of the alites in clinkers

Maki *et al.* (Maki, 1979, 1986; Maki and Chromý, 1978b; Maki and Goto, 1982; Maki and Kato, 1982) studied the crystallisation of alite by high-temperature light microscopy of thin sections together with thermal and X-ray methods. It crystallises from the clinker liquid at about 1450 °C as the R polymorph, which inverts to lower temperature forms on cooling. Due to the lowering of symmetry, each crystal usually contains several optically distinct domains, recognisable in thin section using polarised light. At the higher temperatures of formation or recrystallisation within the normal range, small crystals, relatively high in substituent ions, are formed, which on cooling invert to M_3. Crystals formed at lower temperatures are larger and poorer in substituent ions, and may undergo further transition, partial or complete, to M_1 or, rarely, to T_2.

The MgO and SO_3 contents of the clinker are especially important in determining whether the transformation into M_1 occurs (Figure 1.3; Maki, 1984, 1994; Maki and Goto, 1982; Maki *et al.*, 1992). High ratios of MgO to SO_3 promote nucleation rather than growth, resulting in formation of small, euhedral crystals of M_3. In contrast, low ratios favour rapid, unstable growth, giving rise to large, irregularly shaped crystals, often of M_1. These can be skeletal or dendritic, with relatively high contents of substituents and inclusions both of clinker liquid and of minute crystals of belite and free lime. High ratios of alkalis to SO_3 lessen the tendency to form M_1 as the SO_3 is then largely combined as alkali sulfates. Crystals containing both M_3 and M_1 can occur, and are often zoned, with cores of M_1 and peripheral regions of M_3. This is because the clinker liquid becomes richer in MgO as crystallisation proceeds, causing the material deposited later to persist as M_3. The M_3 to M_1 transformation is also affected by the cooling rate, slow cooling favouring the transformation to M_1. In certain slowly cooled clinkers, further transformation may occur, giving T_2. This is likely to occur only if the alite is particularly low in substituents.

Figure 1.3 Dependence of the polymorphic modification of alite in production clinkers on the MgO and SO_3 contents of the clinker (Maki and Goto, 1982)

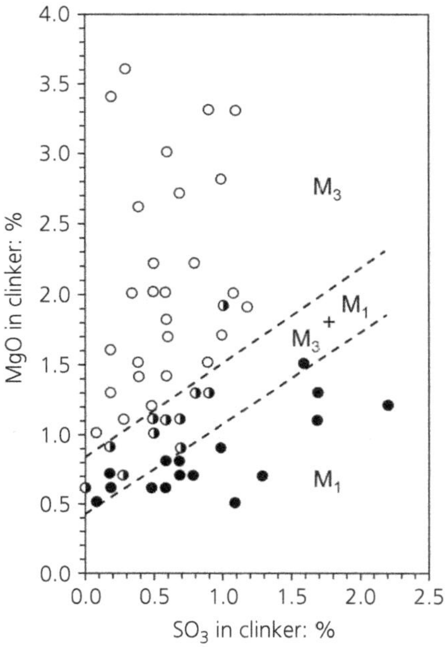

On prolonged heating at high temperatures, the larger alite crystals grow at the expense of smaller ones, and in this process the content of substituent ions decreases, favouring inversion of M_3 to M_1 or T_2. Production clinkers are usually sufficiently high in MgO to prevent this inversion from taking place, even though considerable recrystallisation may occur (Maki and Goto, 1982). Recrystallisation can thus markedly affect the relation between temperature of formation and crystal size.

Guinier and Regourd (1969) concluded from XRD evidence that the R polymorph occurred in certain clinkers. This view has been accepted by many other workers, but Maki and Kato (1982) concluded from thermal and other evidence that the supposed R form was really M_1, which had been misidentified because of the similarity of its XRD powder pattern to that of the R form. They gave two strong reasons in support of this view. Firstly, the clinkers contained less MgO than M_3 alites; for R to persist at room temperature it would have to contain more MgO than M_3 because it is a higher temperature form. Secondly, phase transitions were detected on heating; R, again because it is the high-temperature form, would not undergo these. Sinclair and Groves (1984) also described a case of the misidentification of an alite as rhombohedral; using single-crystal electron diffraction, they found that some laboratory-prepared alites that had been regarded as rhombohedral were really either monoclinic or triclinic.

On present evidence, it is thus probable that the only alite polymorphs found in normal production clinkers approximate to M_3, M_1 or, occasionally, T_2, and that those that have been described as R are really M_1. However, both XRD and thermal evidence show that some alites formed in the presence of fluoride are rhombohedral (Moir, 1983). XRD and optical evidence indicate that the commonest polymorph in production clinkers is almost certainly M_3 or a less ordered form of it. The presence of M_3 in a works clinker has also been demonstrated using single-crystal electron diffraction (Hudson and Groves, 1982).

1.2.5 X-ray powder patterns and densities of tricalcium silicate and alites

Many experimentally determined XRD powder patterns have been reported for C_3S and alite polymorphs, but because of uncertainties in the indexing and sometimes in the identification of the polymorph, these are usefully supplemented by patterns calculated from the crystal structures (e.g. Mumme, 1995). The patterns of the polymorphs are closely alike, and interpretation is complicated in clinkers by the fact that many peaks overlap ones of other phases. The pattern depends not only on the polymorph or mixture of polymorphs present, but also on the nature and amounts of the substituent ions present in each. It is therefore not always easy or even practicable to identify the polymorph or polymorphs in a clinker by the XRD powder pattern alone. The patterns differ principally in the fine structures of certain peaks (Boikova, 1986; Courtial *et al.*, 2003; de la Torre *et al.*, 2008; Guinier and Regourd, 1969; Jeffery, 1952; Le Saoût *et al.*, 2011; Maki, 1986; Maki and Kato, 1982; Regourd and Guinier, 1976). Figure 1.4 shows characteristic portions of the patterns of T_1, M_1, M_3 and R polymorphs. M_1 alite, as present in clinkers, probably always has the maximum MgO content possible for this polymorph, and as such has a unit cell or pseudocell that is geometrically almost hexagonal. This causes the peak at about 51·7° 2θ (Cu Kα radiation) to be almost a singlet; the corresponding peak of the M_3 alite is a well-defined doublet, and those of T_1 C_3S and T_2 alite are triplets (Guinier and Regourd, 1969). There are also differences in the peaks at 32–33° 2θ, but these are less useful because of overlaps with other phases. Because of their high relative intensities, care is needed with these latter peaks to avoid intensity errors from extinction. The X-ray density of T_1 C_3S is $3{\cdot}116\,g\,cm^{-3}$ (Golovastikov *et al.*, 1975); that of M_3 alite, calculated assuming the atomic parameters of Nishi *et al.* (1985) and the composition $(Ca_{0{\cdot}98}Mg_{0{\cdot}01}Al_{0{\cdot}0067}Fe_{0{\cdot}0033})_3(Si_{0{\cdot}97}Al_{0{\cdot}03})O_5$, is $3{\cdot}149\,g\,cm^{-3}$. As an example, the calculated XRD powder pattern for this M_3 alite is given in Table 1.4. This pattern – and others tabulated in this chapter – were calculated as described by Morris *et al.* (1979). The columns headed d (Å), I_{int}, h, k and l give the integrated intensity pattern as defined by those authors – that is, they refer to individual reflections, which can be components of overlaps. I_{int} is the integrated intensity relative to a value of 100 for the strongest. Reflections with $I_{int} < 1{\cdot}5$ are not listed. The column headed I_{pk} gives an indication of relative peak heights assuming typical experimental conditions using a diffractometer, but these depend on resolution and on any variation in cell parameters because these factors affect separations between overlapping peaks. They take into account contributions from reflections too weak to appear in the list of integrated intensities and their spacings sometimes differ significantly from those of the individual components. Peaks of significant intensity can arise from groups of reflections all of which are too weak to be listed individually; the most prominent are noted in footnotes to the tables. The cell parameters and compositions assumed are typical ones for phases in Portland cement clinkers. Reference intensity ratios I_{hkl}/I_c (Hubbard *et al.*, 1976; Morris *et al.*, 1979) are also given. Their use is explained in Chapter 4.

Figure 1.4 Portions of XRD powder patterns (calculated) for four polymorphs of C_3S

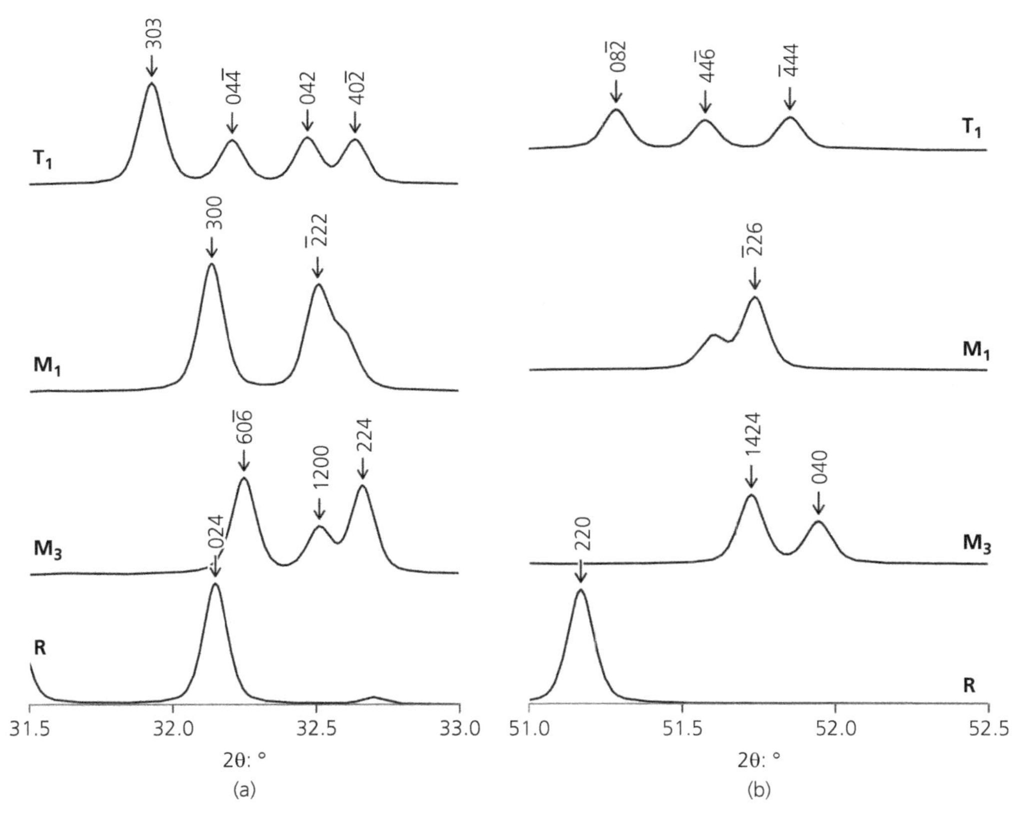

The structures are from Golovastikov *et al.* (1975; T_1), de Noirfontaine *et al.* (2006; M_1 alite), de la Torre *et al.* (2002; M_3 alite), Nishi and Takéuchi (1984; R); lattice parameters are given in Table 1.1. The patterns were calculated using CrystalDiffract® software using monochromatic radiation (Cu, $\lambda = 1{\cdot}5406$ Å), instrument peak width 0·1°, and a pseudo-voigt lineshape (mix parameter = 0·5).

Table 1.4 Calculated XRD powder pattern for M_3 alite (continued on next page)

d (Å)	I_{int}	I_{pk}	h	k	l
5·949	4	11	4	0	2
5·940	7		3	1	0
5·490	2	3	1	1	2
3·877	4	8	$\bar{8}$	0	2
3·869	6		$\bar{1}$	1	4
3·535	5	6	7	1	2
3·527	3		0	2	0
3·180	2	3	$\bar{6}$	0	5
3·042	34	90	10	0	2
3·034	78		4	2	2
2·975	10	27	8	0	4
2·970	20		6	2	0
2·778	100	90	$\bar{6}$	0	6
2·753	43	99	12	0	0
2·745	84		2	2	4
2·697	2	8	$\overline{11}$	1	2
2·694	2		3	1	6
2·692	3		$\bar{4}$	2	4
2·611	41	100	6	0	6
2·609	80		$\bar{8}$	2	2
2·453	5	5	$\bar{6}$	0	7
2·327	6	18	$\overline{14}$	0	2
2·321	16		0	2	6
2·318	2		6	0	7
2·299	2	1	3	3	0
2·287	2	3	$\overline{10}$	2	3
2·186	15	48	4	0	8
2·186	32		$\overline{10}$	2	4
2·186	5		$\overline{13}$	1	4
2·183	6		1	1	8
2·182	6		$\bar{6}$	2	6
2·170	5	17	9	1	6
2·170	6		12	2	0
2·166	6		$\bar{5}$	3	2

Table 1.4 Continued

d (Å)	I_{int}	I_{pk}	*h*	*k*	*l*
2·071	2	2	$\overline{10}$	2	5
1·984	4	11	16	0	2
1·980	4		9	3	0
1·979	7		5	3	4
1·938	7	18	$\overline{16}$	0	4
1·934	17		$\bar{2}$	2	8
1·837	3	5	14	2	3
1·835	2		18	0	0
1·830	5	7	3	3	6
1·825	4	7	$\overline{12}$	2	6
1·824	2		2	0	10
1·768	58	50	14	2	4
1·763	31	52	0	4	0
1·644	2	5	8	0	10
1·644	3		$\overline{13}$	3	4
1·630	14	26	$\overline{18}$	0	6
1·627	27		$\bar{4}$	2	10
1·542	14	16	$\overline{14}$	2	8
1·542	8		0	0	12

The pattern was calculated assuming $a = 33{\cdot}12$ Å, $b = 7{\cdot}054$ Å, $c = 18{\cdot}55$ Å, $\beta = 94{\cdot}17°$; atomic parameters given by Nishi *et al.* (1985); site occupancies $(Ca_{0{\cdot}98}Mg_{0{\cdot}01}Al_{0{\cdot}0067}Fe_{0{\cdot}003})_3(Si_{0{\cdot}97}Al_{0{\cdot}03})O_5$.
Other peaks of significant height: 3·248 Å (2), $\bar{2}$ 2 2 +; 3·211 Å (1), 10 0 1 +; 2·673 Å (3), $\bar{8}$ 2 1; 2·562 Å (2), $\bar{8}$ 0 6 +; 2·504 Å (1), 2 2 5 +; 2·276 Å (2), 13 1 2 +; 2·097 Å (2), $\bar{1}$ 3 4 +; 2·036 Å (1), 16 0 1 +; 1·861 Å (1), $\overline{16}$ 0 5 +; 1·794 Å (2), $\bar{8}$ 2 8 +; 1·714 Å (1), $\bar{7}$ 1 10 +; 1·693 Å (2), $\overline{18}$ 0 5; 1·691 Å (2), 14 2 5 +; 1·594 Å (1), $\overline{13}$ 3 5; 1·561 Å (1), $\overline{18}$ 0 7; $I_{\bar{6}06}/I_c = 0{\cdot}742$.

1.2.6 Optical, thermal and other data

T_1 C_3S is biaxial with (–)2V = 20–60° and $\alpha \wedge c_{hex} = 0 - 15°$ (Ono *et al.*, 1969); the refractive indices are $\alpha = 1{\cdot}7139$, $\gamma = 1{\cdot}7172$ for pure C_3S and $\alpha = 1{\cdot}7158$–$1{\cdot}7197$, $\gamma = 1{\cdot}7220$–$1{\cdot}7238$ for typical clinker alites, all determined using sodium light (Burke, 1954). In thin sections in polarised light, M_1 and M_3 are most readily distinguished by the maximum birefringence, which is 0·005 to 0·006 for M_3 and 0·003 for M_1 (Maki, 1979; 1986; Maki and Chromý, 1978b; Maki and Kato, 1982). Other differences in optical properties include the optic orientation referred to the hexagonal axes of the R form, whose shape persists through the transitions that occur on cooling, and the optic axial angle; Maki and Kato (1982) give details.

It is doubtful whether the polymorphs can be reliably distinguished by DTA, since both the M_1 and the M_3 forms are metastable at low temperatures and transform to T_2 at about 700 °C (Maki, 1986; Maki and Kato, 1982). This effect is normally swamped by the endothermic β to α'_L transition of the belite. Due to the effect of substituents, neither T_3 nor M_2 appears to form on heating. With alites relatively low in substituents, endotherms occur at 790–850 °C, giving M_1, and at 950 °C, giving M_3. Clinkers high in MgO, in which the alite is present entirely as M_3, normally show no thermal effects attributable to the alite until more than 1000 °C, when transition to R occurs.

A substituted C_3S has been found in nature and given the mineral name 'hatrurite' (Gross, 1977). The rock containing it is interesting because it contains other phases present in clinker and appears to have been formed under conditions somewhat similar to those existing in a cement kiln.

1.3. Belite

1.3.1 Polymorphism and crystal structure

Thermal and X-ray evidence (Barnes *et al.*, 1977; Niesel and Thormann, 1967; Regourd and Guinier, 1976) shows that five polymorphs of C_2S exist at ordinary pressures, namely

$$\alpha \overset{1425\,°C}{\rightleftarrows} \alpha'_H \overset{1160\,°C}{\rightleftarrows} \alpha'_L \overset{630\text{–}680\,°C}{\underset{690\,°C}{\rightleftarrows}} \beta \overset{<500\,°C}{\longrightarrow} \gamma$$

$$\gamma \xrightarrow{780\text{–}860\,°C} \alpha'_L$$

where H = high; L = low.

The structures of all are built from Ca^{2+} and SiO_4^{4-} ions. The arrangements of these ions are closely similar in the α, α'_H, α'_L and β polymorphs, but that in γ-C_2S is somewhat different. As with C_3S, the higher temperature polymorphs cannot normally be preserved on cooling to room temperature unless stabilised by substituent ions. γ-C_2S is much less dense than the other polymorphs and this causes crystals or sintered masses of β-C_2S to crack and fall to a more voluminous powder on cooling, a phenomenon known as dusting. If the crystallites of β-C_2S are sufficiently small, the transformation does not occur, even if no stabiliser is present (Yannaquis and Guinier, 1959). In all normal Portland cement clinkers, the belite contains enough stabilising ions to prevent the transformation to γ-C_2S from taking place.

The crystal structures have been studied extensively: data derived from both single-crystal and powder evidence are summarised in Table 1.5. The structures of the α, α'_H, α'_L and β polymorphs all belong to a large family typified by that of glaserite, $K_3Na(SO_4)_2$ (Moore, 1973; Yamnova *et al.*, 2011). Structure determinations have been reported for all four: the references are given in Table 1.5. The space group of the α form may vary according to the temperature and the nature of the stabiliser, if any (Kim *et al.*, 1992). α'_L is in practice a generic term for complex variants of α'_H, as two different superlattices have been found to occur (Regourd *et al.*, 1969; Kim *et al.*, 1992).

Table 1.5 Crystal data for the dicalcium silicate (C_2S) polymorphs (continued on next page)

Polymorph (mineral)	Space group (No.)	*Z*	*a*: Å	*b*: Å	*c*: Å	Angle (°)[a]	D_x: g cm^{-3}	Molar volume cm^3 mol^{-1}	Exp.[b]	Crystal structure reference/comments
γ-C_2S (calico-olivine)	*P b n m* (62; setting **cab**)	4	5·091	11·371	6·782		2·914	59·11	S	Smith *et al.* (1965); synthetic 'pure' C_2S
			5·078	11·225	6·760		2·969	58·01	S	Czaya (1971); synthetic C_2S
			5·081	11·224	6·778		2·960	58·20	S	Udagawa and Urabe (1978); Udagawa *et al.* (1980b); synthesised by flux method using $CaCl_2$ as the flux
			5·0821	11·2237	6·7638		2·965	58·09	P	Mumme *et al.* (1995); synchrotron
			5·0762	11·2136	6·7583		2·974	57·92	P	Mumme *et al.* (1996); synthetic C_2S; neutrons
			5·07389	11·21128	6·75340		2·978	57·84	P	Gobechiya *et al.* (2008); Zadov *et al.* (2008, 2009); natural calico-olivine
β-C_2S (larnite)	*P* 2_1/*n* (14; unique axis *b*, cell choice 2)	4	5·48	6·76	9·28	94·55	3·338	51·59	S	Midgley (1952); stabilised with 0·5 wt.% B_2O_3
			5·506	6·749	9·304	94·62	3·320	51·88	P	Regourd *et al.* (1969); synthetic pure C_2S
			5·502	6·745	9·297	94·59	3·326	51·78	S	Jost *et al.* (1977); synthetic C_2S; grown from $CaCl_2$ melt without additional stabiliser
			5·51	6·76	9·32	94·5	3·306	52·10	S	Udagawa and Urabe (1978)
			5·5041	6·7622	9·3281	94·172	3·304	52·13	P	Tsurumi *et al.* (1994)
			5·5121	6·7575	9·3138	94·581	3·308	52·06	P	Mumme *et al.* (1995); synchrotron; stabilised with 0·5 wt.% Cr_2O_3
			5·50684	6·74839	9·30320	94·590	3·320	51·88	P	Mori *et al.* (2006); synthetic C_2S; neutrons; RT
			5·5051	6·7551	9·3108	94·513	3·311	51·97	S	Yamnova *et al.* (2011); natural larnite; $(Na_{0·002}Ca_{0·998})_2(Al_{0·005}Si_{0·995})O_{3·996}$ trace Mg and P
			5·511	6·670	9·317	94·447	3·351	51·41	P	Chen *et al.* (2015); $(Ca_{1·98}Ce_{0·01}Li_{0·01})SiO_4$

Table 1.5 Continued

Polymorph (mineral)	Space group (No.)	Z	a: Å	b: Å	c: Å	Angle (°)[a]	D_x: g cm^{-3}	Molar volume cm^3 mol^{-1}	Exp.[b]	Crystal structure reference/comments
α'_L-C_2S	*P n a* 2_1 1 2_1 (33 setting **abc**)	12	20·871	9·496	5·600		3·946	55·70		Il'inets *et al.* (1982); for $CaSrSiO_4$
			20·8630	9·5000	5·6005		4·082	55·71	S	Il'inets and Bikbau (1990); for $Ca_{0.84}Sr_{1.16}SiO_4$
			20·5266	9·4963	5·5897		3·150	54·68	P	Mumme *et al.* (1996); neutrons 1060 °C; synthetic C_2S
α'_H-C_2S	*P n m a* (62; setting **abc**)	4	6·748	5·494	9·261		3·332	51·69	P	Yamaguchi *et al.* (1963b); stabilised with 10 % $CaMgSiO_4$ + 4 % K_2O; cooled quickly from 1450 °C
			6·883	5·605	9·543		3·107	55·43	P	Yamaguchi *et al.* (1963a); pure synthetic C_2S; X-rays 1300 °C
			6·7673	5·5191	9·3031		3·292	52·31	P	Mumme *et al.* (1995); synchrotron; stabilised with 5 mol % $Ca_3(PO_4)_2$
			6·8709	5·6010	9·5563		3·111	53·92	P	Mumme *et al.* (1996); neutrons 1250 °C; synthetic C_2S
			6·8432	5·4555	9·2346		3·119	51·69	P	Cuesta *et al.* (2012); for $Ca_{1.85}Na_{0.15}(SiO_4)_{0.85}(BO_3)_{0.15}$

α-C_2S	*P* 6_3/*m m c* (194)	2	5·5790	5·5790	7·15	120	2·968	58·03		Udagawa *et al.* (1977); stabilised with Ba
			5·532	5·532	7·327	120	2·946	58·47	P	Mumme *et al.* (1996); neutron 1545 °C; synthetic C_2S
			5·3936	5·3936	7·1275	120	3·028	54·07	S	Widmer *et al.* (2015); for $Ca_{1·4}Na_{0·6}P_{0·6}Si_{0·4}O_4$
	P 6_3 *m c* (186)	2	5·4700	5·4700	7·23	120	3·053	56·41	S	Eysel and Hahn (1970)
	P $\bar{3}$ *m* 1 (164)	2	5·527	5·527	7·311	120	2·957	58·24	P	Yamaguchi *et al.* (1963a); pure synthetic C_2S; X-rays 1500 °C
			5·526	5·526	7·307	120	2·960	58·19	P	Regourd *et al.* (1969); synthetic C_2S; X-rays 1500 °C

[a]Where not 90°. The angle is β for SG *P* 2_1/*n* and γ for *P* 6_3/*m m c* and *P* 6_3 *m c*.
[b]Experiment: S = single-crystal; P = powder.

The structures of the α'_H, α'_L and β polymorphs are derived from that of α-C_2S by progressive decreases in symmetry, which arise from changes in the orientations of the SiO_4^{4-} tetrahedra and small movements of Ca^{2+} ions. The close similarity of the structures is evident on inspection of Figures 1.5(a)–(c). The quantities of substituent ions needed to stabilise the higher temperature

Figure 1.5 Representations of the crystal structures of four C_2S polymorphs (continued on next page)

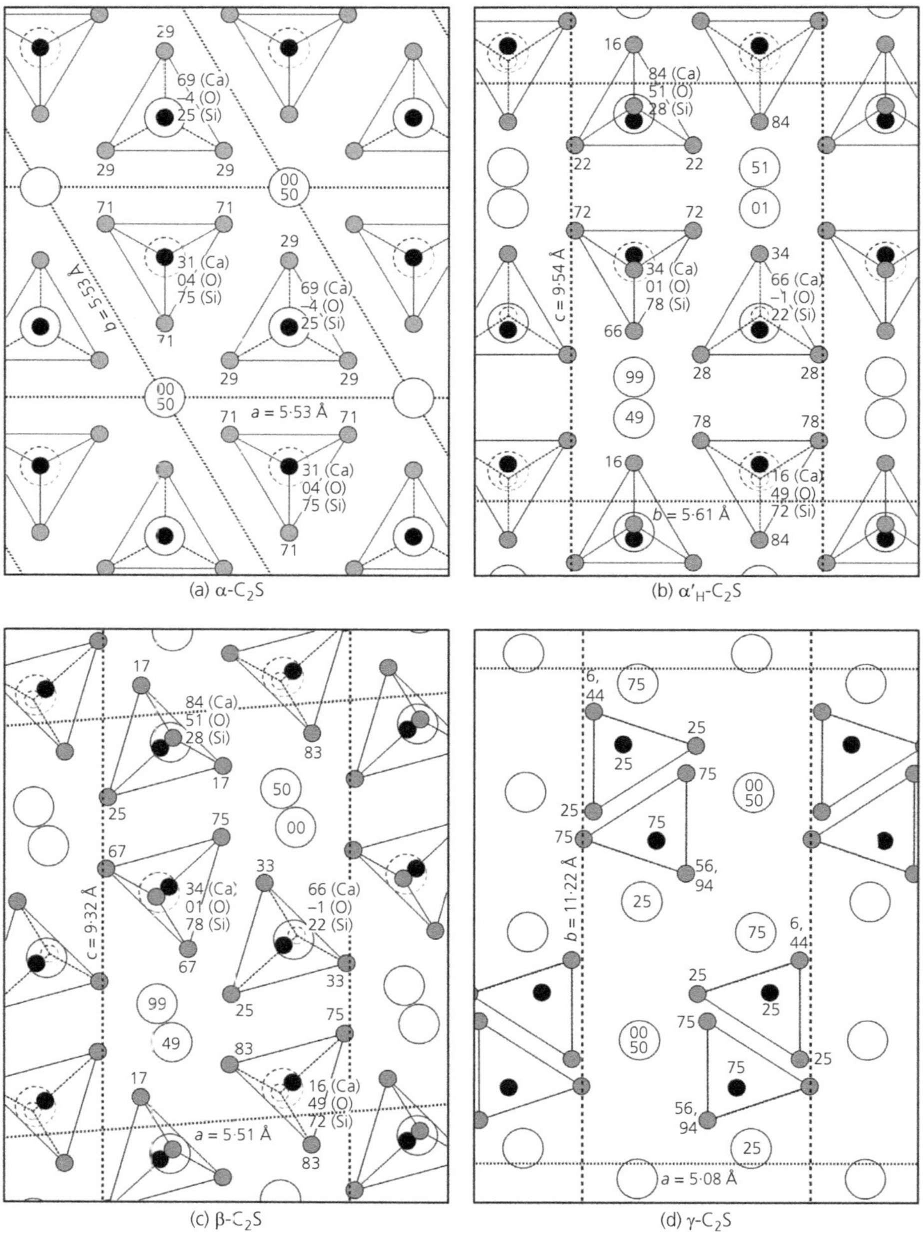

Figure 1.5 Continued

The diagram shows Ca atoms (large open circles), Si atoms (small black circles) and oxygen tetrahedra (triangles with grey circles at the vertices). The diagram for α-C_2S is based on Yamaguchi *et al.* (1963a), Udagawa *et al.* (1977) and Mumme *et al.* (1996); α'_H-C_2S is based on Yamaguchi *et al.* (1963a), Mumme *et al.* (1996) and McGinnety (1972; idealised depiction derived in part from the structure of β-K_2SO_4); β-C_2S is based on Udagawa and Urabe (1978); γ-C_2S is based on Udagawa and Urabe (1978). The four diagrams are drawn to the same scale to facilitate comparison. Heights of atoms are given as hundredths of the cell height, which is: 7·31 Å for α-C_2S; 6·88 Å for α'_H-C_2S; 6·76 Å for β-C_2S; 6·78 Å for γ-C_2S. An unconventional choice of origin has been adopted for γ-C_2S to clarify the relation to the other polymorphs.

polymorphs at room temperature decrease along the sequence from α- to β-C_2S. The situation parallels that existing with the C_3S polymorphs, as described in §1.2.2. Also as in the latter case, and for the same reason, these C_2S polymorphs differ in calcium ion coordination. In β-C_2S, some of the Ca^{2+} ions have seven and others eight oxygen atoms within 2·9 Å (Jost *et al.*, 1977; Mumme *et al.*, 1995). A modulated structure – that is, one in which compositional or structural variations occur with a statistical periodicity that is not a rational multiple of that of the basic structure – has been observed in a polymorph of C_2S (Fukuda and Maki, 1989a).

The crystal structure of γ-C_2S is similar to that of olivine, $(Mg,Fe)_2SiO_4$ (Figure 1.5(d); Mumme *et al.*, 1995; Udagawa *et al.*, 1980b). The calcium is octahedrally coordinated. The unit cell and the arrangement of Ca^{2+} and SiO_4^{4-} ions show some similarities to those of the other polymorphs, but also important differences. It has often been supposed that transformations involving this form entail total reorganisation of the structure, but single-crystal high-temperature XRD evidence indicates that this is not correct; the products are formed topotactically (i.e. in a definite crystallographic orientation relative to the starting material; Figure 1.5), but fragment because of the large volume change (Barbier and Hyde, 1985; Gazzozi and Chiari, 1986; Kim *et al.*, 1992). γ-C_2S scarcely reacts with water at ordinary temperatures.

1.3.2 Polymorphic forms and textures in clinker belites

XRD powder evidence shows that in the majority of clinkers the belite is predominantly or entirely of β-C_2S structure (Guinier and Regourd, 1969; Yamaguchi and Takagi, 1969), although some peaks are broadened (Guinier and Regourd, 1969) and the presence also of both α and α' (presumably α'_L) forms has been reported (Guinier and Regourd, 1969; Ono *et al.*, 1969; Regourd and Guinier, 1976; Yamaguchi and Takagi, 1969). Characterisation of the polymorphic form is rendered difficult by the similarities between their powder patterns (Figure 1.6) and overlaps between the peaks and ones of other phases, especially alite, but has been aided by examination of fractions in which the belite has been concentrated by chemical (Regourd and Guinier, 1976) or heavy liquid (Yamaguchi and Takagi, 1969) separation.

The belite grains in Portland cement clinkers frequently show complex, striated structures. These have been studied over a long period by workers using light microscopy; Yamaguchi and Takagi (1969) and Ono *et al.* (1969), who also used XRD and other methods, gave the first substantially complete interpretations. Fukuda and Maki (1989b; 1993) and Groves (1982) extended these results, using single-crystal electron diffraction and other methods. Transformation mechanisms have been discussed (Kim *et al.*, 1992).

Figure 1.6 Portions of XRD powder patterns (calculated) of (a) α-C_2S at 1545 °C, (b) α'_L-C_2S at 1060 °C and (c) β-C_2S

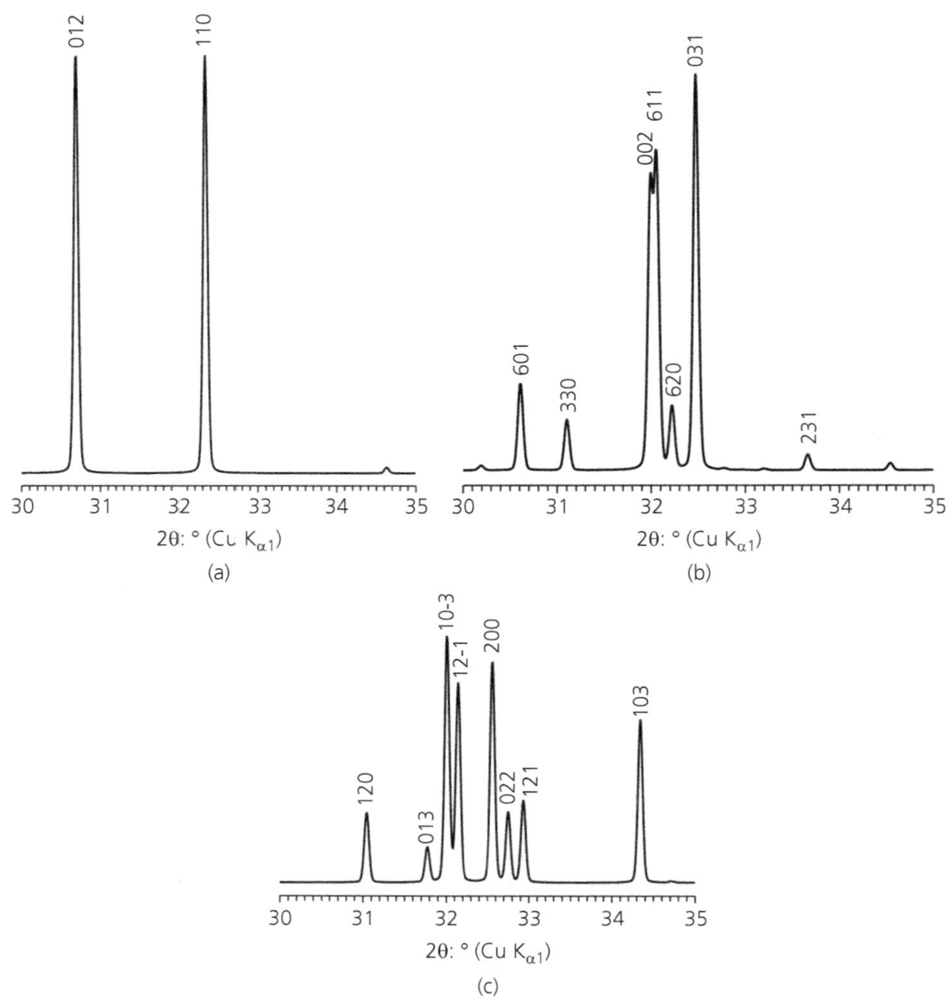

The structures in (a) and (b) are from Mumme *et al.* (1996) and in (c) from Udagawa and Urabe (1978); lattice parameters are given in Table 1.4.

A very common type of belite grain in production clinkers, called Type I belite, is rounded, typically 20 to 40 μm in mean dimension, and shows several sets of parallel striations. Such crystals have formed as the α polymorph and the striations arise from the lowering of symmetry inherent in the α to α'_H and α'_L to β transformations that occur on cooling. Type I belite crystals consist largely or wholly of the β polymorph, although some α may remain. Other phases tend to be exsolved between the lamellae since the lower-temperature polymorphs are less able to accommodate substituent ions.

Type II belite grains are typically irregular in shape and show only one set of striations. They are rare in modern clinkers. The striations arise from the α'_L to β transformation. The restricted number of orientations can arise either because the crystals had formed below the α to α'_H transition temperature or because one of the twin components of the α' modification grew at the expense of the others (Ono *et al.*, 1969). Belite can form in clinkers by additional mechanisms, sometimes at lower temperatures than those discussed above, and may not then show any striations.

Chan *et al.* (1988) showed that material exsolved between lamellae or at grain boundaries in synthetic belites was often compositionally far removed from C_2S, and structurally either amorphous or composed of crystalline phases of entirely different structure, such as C_2AS, C_3A or C_3S_2. They concluded that the possibilities of compositional variation in β-C_2S were considerably narrower than had often been supposed, the grosser deviations from ideal composition being attributable to such effects. Evidence from ^{27}Al NMR is consistent with these results (Skibsted *et al.*, 1994). Chan *et al.* (1998) also concluded that, contrary to some earlier views, β-C_2S cannot be stabilised by excess CaO alone, and that the role of ionic substitution in stabilising forms other than γ to room temperature needed to be reassessed, mechanical constraints associated with exsolution providing a possible alternative explanation.

1.3.3 Compositions of belites in clinkers

Table 1.2 includes an average composition for belites in ordinary production clinkers, based on the results of many studies by X-ray microanalysis (Boikova *et al.*, 1986; Ghose, 1980; Ghose and Barnes, 1979; Harrisson *et al.*, 1985, 1986; Kristmann, 1978; Sarkar and Roy, 1984; Terrier *et al.*, 1968) or chemical analysis of separated material (Daimon *et al.*, 1992; Yamaguchi and Takagi, 1969). As with alite (§1.2.3), most of the results from different laboratories are in relatively close agreement. Belites in white (Boikova *et al.*, 1986; Le Saoût *et al.*, 2011) and iron-rich (Gollop and Taylor, 1994; Hall and Scrivener, 1997; Le Saoût *et al.*, 2011) clinkers have also been analysed. Pure C_2S, for comparison, contains 34·9% of SiO_2 and 65·1% of CaO.

Clinker belites typically contain 4–6% of substituent oxides, of which the chief are usually Al_2O_3 and Fe_2O_3. The Al^{3+} is in tetrahedral sites (Skibsted *et al.*, 1994). With clinkers high in SO_3, the belite tends to be high in that component; coupled substitution of $2Al^{3+} + S^{6+}$ for $3Si^{4+}$ probably occurs (Bonafous *et al.*, 1995). The SO_3 content of the belite ($\bar{S}_b$) is reported to be related to that of the clinker ($\bar{S}_c$) by the relation $\bar{S}_b = 1{\cdot}23\bar{S}_c + 0{\cdot}24$ or, alternatively, $\bar{S}_b = 3{\cdot}36\bar{S}_{c'} + 0{\cdot}55$, where $\bar{S}_{c'}$ excludes SO_3 present in soluble sulfate phases (Hall and Scrivener, 1997). Correlations have been reported to exist between the MgO contents of the belite and the clinker (Kristmann, 1978), and the Al_2O_3/Fe_2O_3 ratios of the belite and the clinker (Fukuda *et al.*, 1994; Hall and Scrivener, 1997), but there is disagreement in the case of Fe_2O_3 (Hall and Scrivener, 1997; Kristmann, 1978). Early reports mention a compound $KC_{23}S_{12}$, but the K_2O content of 3·5% corresponding to this formula is well above those found in clinker belites; electron optical work indicates that the limit of K_2O substitution is about 1·2% (Chan *et al.*, 1988).

Any interpretation of belite analyses is at present tentative because of the uncertainties arising from exsolution effects. From the practical standpoint, it may be necessary to regard the exsolved material as forming part of the belite. Clinker belite compositions seem often to indicate ratios of other atoms to oxygen above the theoretical value of 0·75. As Regourd *et al.* (1969) noted, the C_2S structures are too densely packed for the presence of interstitial ions to appear likely, and it is possible that vacancies occur in the oxygen sites. Table 1.3 includes atomic ratios, with a possible allocation of atoms to sites, corresponding to the composition given in Table 1.2.

1.3.4 Cell parameters, X-ray powder patterns and other data

Regourd *et al.* (1969) synthesised several belites of β-C_2S type with compositions similar to those found in clinkers and determined their cell parameters. With increased substitution the crystallinity tended to decrease, causing broadening of XRD powder peaks. Table 1.6 gives two of their results and one for a belite extracted from a Portland cement clinker. Table 1.7 is a calculated pattern for a typical clinker belite. Assuming $a = 5{\cdot}50$ Å, $b = 6{\cdot}75$ Å, $c = 9{\cdot}33$ Å, $\beta = 94{\cdot}4°$, with the atomic parameters of Jost *et al.* (1977) and the site occupancies $(K_{0{\cdot}01}Na_{0{\cdot}005}Ca_{0{\cdot}975}Mg_{0{\cdot}01})_2(Fe_{0{\cdot}02}Al_{0{\cdot}06}Si_{0{\cdot}90}P_{0{\cdot}01}S_{0{\cdot}01})O_{3{\cdot}96}$, the X-ray density is 3·300 g cm^{-3}. That for pure β-C_2S, based on Jost *et al.*'s (1977) data, is 3·326 g cm^{-3}, and that for pure γ-C_2S, based on data of Udagawa *et al.* (1980a), is 2·960 g cm^{-3}.

Table 1.6 Cell parameters of belites

Composition	*a*: Å	*b*: Å	*c*: Å	β
$Ca_2Fe_{0{\cdot}035}Al_{0{\cdot}035}Si_{0{\cdot}930}O_{3{\cdot}965}$ [a]	5·502	6·750	9·316	94·45°
$Ca_2Fe_{0{\cdot}050}Al_{0{\cdot}050}Si_{0{\cdot}900}O_{3{\cdot}950}$ [a]	5·502	6·753	9·344	94·19°
$Ca_2Mg_{0{\cdot}044}K_{0{\cdot}020}Fe_{0{\cdot}032}Al_{0{\cdot}043}Si_{0{\cdot}933}O_4$ [b]	5·504	6·762	9·328	94·17°

[a]Laboratory preparation (Regourd *et al.*, 1969).
[b]Isolated from a clinker; also, Na, Ti, Mn, P, each ≤ 0·01 (Daimon *et al.*, 1992).

Synthetic β-C_2S is biaxial positive with high 2V and α = 1·717, γ = 1· 735; γ-C_2S is biaxial negative with 2V = 60°, α = 1·642, β = 1·645, γ = 1·654 (Rankin and Wright, 1915). In the Type I belites of clinkers, lamellae of β structure have mean refractive index 1·720 and birefringence 0·015–0·018, while for the intervening material the corresponding values are 1·700–1·710 and 0·000–0·003 (Ono *et al.*, 1969). Ono *et al.* (1969) and Yamaguchi and Takagi (1969) give further data on the optical properties of clinker belites.

Guinier and Regourd (1969) summarised the thermal behaviour of C_2S polymorphs. On heating, β-C_2S shows endotherms beginning at 693 °C and 1450 °C, due respectively to the $\beta \rightarrow \alpha'_L$ and $\alpha'_H \rightarrow \alpha$ transitions; with γ-C_2S, the 693 °C endotherm is replaced by a broad one beginning at about 748 °C due to the transition to α'_L. The curve obtained on cooling α-C_2S from 1500 °C shows exotherms beginning at about 1456 °C, 682 °C and 530 °C, due respectively to the $\alpha \rightarrow \alpha'_H$, $\alpha'_L \rightarrow \beta$ and $\beta \rightarrow \gamma$ transitions. The $\alpha'_H \rightarrow \alpha'_L$ transition gives only a very weak effect at 1160 °C.

β-C_2S, modified by solid solution, occurs in nature as larnite. Early identifications of α′-C_2S with bredigite were incorrect (Biggar, 1971; Moore and Araki, 1976). α-C_2S has been assumed to resemble nagelschmidtite ($Ca_7Si_2P_2O_{16}$), but the unit cell of the latter seems to be more complex (Guinier and Regourd, 1969).

Table 1.7 Calculated XRD powder pattern for a belite (continued on next page)

d: Å	I_{int}	I_{pk}	*h*	*k*	*l*
4·891	3	3	$\bar{1}$	0	1
4·651	10	10	0	0	2
3·830	7	7	0	1	2
3·786	6	6	1	1	1
3·375	8	7	0	2	0
3·237	9	9	$\bar{1}$	1	2
3·173	6	6	0	2	1
3·051	12	11	1	1	2
2·874	26	24	1	2	0
2·818	14	19	0	1	3
2·793	100	100	$\bar{1}$	0	3
2·778	81	94	$\bar{1}$	2	1
2·742	88 }	88	{ 2	0	0
2·732	28 }		{ 0	2	2
2·716	32	38	1	2	1
2·615	69	60	1	0	3
2·540	16	14	2	1	0
2·446	17	16	$\bar{2}$	0	2
2·438	8	16	1	1	3
2·407	13 }	24	{ 2	1	1
2·402	18 }		{ 1	2	2
2·326	3	3	0	0	4
2·299	3	4	$\bar{2}$	1	2
2·287	3 }	28	{ 2	0	2
2·283	31 }		{ 0	2	3
2·199	18	18	0	1	4
2·187	51	45	0	3	1
2·166	19 }	18	{ 2	1	2
2·152	2 }		{ $\bar{1}$	2	3
2·128	8	7	2	2	0
2·102	2	<1	$\bar{2}$	2	1
2·094	8	8	$\bar{1}$	1	4
2·082	5	5	1	3	0
2·048	17	16	2	2	1

Table 1.7 Continued

d: Å	I_{int}	I_{pk}	*h*	*k*	*l*
2·044	3	13	$\bar{1}$	3	1
2·037	10	12	$\bar{2}$	1	3
2·025	11	13	0	3	2
2·019	15	18	1	3	1
1·992	18	16	1	1	4
1·980	31	28	$\bar{2}$	2	2
1·915	9	8	0	2	4
1·900	8	8	2	1	3
1·893	16	15	2	2	2
1·845	6	5	$\bar{2}$	0	4
1·821	4	4	0	3	3
1·805	12	16	$\bar{2}$	2	3
1·804	8		$\bar{1}$	0	5
1·794	13	12	0	1	5
1·764	5	4	3	1	0
1·743	2	2	$\bar{1}$	1	5
1·725	7	7	$\bar{2}$	3	1
1·722	4	7	1	0	5
1·710	3	14	2	0	4
1·708	15		2	2	3
1·695	5	5	2	3	1
1·687	5	5	0	4	0
1·660	2	2	0	4	1
1·630	21	23	$\bar{3}$	0	3
1·629	10		0	2	5
1·619	4	6	$\bar{2}$	2	4
1·613	5	8	1	4	0
1·611	5		3	1	2
1·607	8	11	3	2	0
1·604	2	15	2	3	2
1·602	14		$\bar{3}$	2	1

Table 1.7 Continued

d: Å	I_{int}	I_{pk}	*h*	*k*	*l*
1·595	2	5	$\bar{1}$	4	1
1·591	3		$\bar{1}$	2	5
1·586	6	8	0	4	2
1·585	2		$\bar{3}$	1	3
1·583	5	8	1	4	1
1·574	16	13	$\bar{1}$	3	4
1·555	4	10	$\bar{2}$	1	5
1·552	4		$\bar{3}$	2	2
1·550	6		0	0	6
1·549	5		$\bar{2}$	3	3

The pattern was calculated assuming $a = 5{\cdot}50$ Å, $b = 6{\cdot}75$ Å, $c = 9{\cdot}33$ Å, $\beta = 94{\cdot}40°$; atomic parameters given by Jost *et al.* (1977); site occupancies $(K_{0{\cdot}01}Na_{0{\cdot}005}Ca_{0{\cdot}975}Mg_{0{\cdot}01})_2(Fe_{0{\cdot}02}Al_{0{\cdot}06}Si_{0{\cdot}90}P_{0{\cdot}01}S_{0{\cdot}01})O_{3{\cdot}96}$.
Other peaks of significant height: 1·760 Å (3), $\bar{3}$ 1 1; 1·656 Å (2), $\bar{2}$ 3 2 +; $I_{\bar{1}03} / I_c = 0{\cdot}724$.

1.4. Aluminate

1.4.1 Crystal structure: cubic (C), orthorhombic (O) and monoclinic (M) modifications

Pure C_3A does not exhibit polymorphism. It is cubic, with $a = 15{\cdot}263$ Å, space group $P2_1 / a\bar{3}$ (No. 205) and $Z = 24$; the structure is built from Ca^{2+} ions and rings of six AlO_4 tetrahedra, of formula $Al_6O_{18}^{18-}$ (Mondal and Jeffery, 1975), as illustrated in Figure 1.7. These rings are highly puckered, such that the Al atoms lie near to six of the corners of a cube. The unit cell is composed of 64 (i.e. 4^3) subcells of edge length 3·816 Å. Of these subcells, eight are occupied by Al_6O_{18} rings; the Ca^{2+} ions occur at the body centres of some of the remaining subcells and near the corners of others. The coordination of those near the corners is octahedral, while that of those at the body centres is irregular, with either six or eight oxygen atoms, the latter including three long bonds of ≈2·95 Å and one short bond of ≈2·26 Å, and sharing edges with five Al–O tetrahedra and a corner with a sixth tetrahedron.

C_3A can incorporate Na^+ by substitution of Ca^{2+} with inclusion of a second Na^+ ion in an otherwise vacant site, thus giving solid solutions of general formula $Na_{2x}Ca_{3-x}Al_2O_6$ (Fletcher *et al.*, 1965; Regourd *et al.*, 1973; Regourd and Guinier, 1976; Takéuchi *et al.*, 1980). The substitution occurs without change in structure up to a limit of about 1 % Na_2O ($x \approx 0{\cdot}04$). Higher degrees of substitution lead to a series of variants of the structure (Table 1.8). In the absence of other substituents, the upper limit of Na_2O substitution is 5·7 %.

Figure 1.7 An Al_6O_{18} puckered ring in the structure of C_3A surrounding the hole at $\frac{1}{8}\frac{1}{8}\frac{1}{8}$ viewed along $[\bar{1}\bar{1}\bar{1}]$; based on Figure 4 of Mondal and Jeffery (1975)

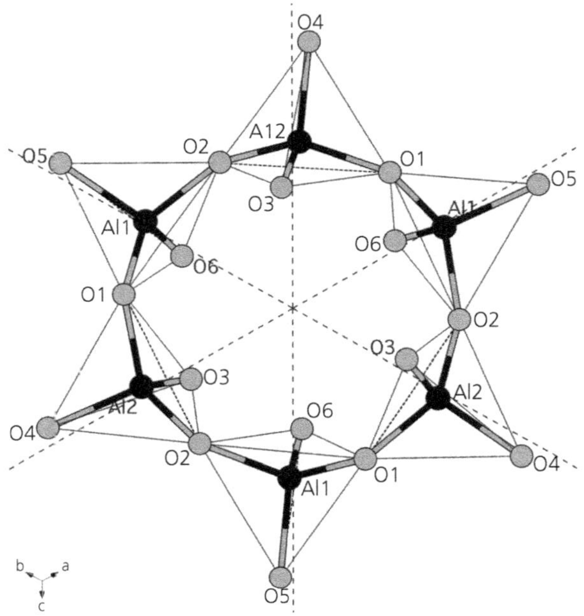

The C_I structure (Table 1.8) is the C_3A structure, with up to about 1% of Na_2O present as described above. The C_{II} structure is a minor variant of it, of lower symmetry (Takéuchi *et al.*, 1980), space group $P\,2_1\,3$ (No. 198); comparison of calculated XRD patterns shows that there is little possibility of distinguishing the two by powder XRD, unless by very precise determinations of the cell parameter, C_{II} with 2·4% Na_2O having a = 15·248 Å (Takéuchi *et al.*, 1980).

Table 1.8 Modifications of the tricalcium aluminate structure, of general formula $Na_{2x}Ca_{3-x}Al_2O_6$ (Takéuchi *et al.*, 1980)

Approximate Na_2O: %	Compositional range: x	Designation	Crystal system	Space group (No.)
0–1·0	0–0·04	C_I	Cubic	$P\,2_1\,/\,a\,\bar{3}$(205)
1·0–2·4	0·04–0·10	C_{II}	Cubic	$P\,2_1\,3$ (198)
2·4–3·7	0·10–0·16	C_{II} + O	-	-
3·7–4·6	0·16–0·20	O	Orthorhombic	$P\,b\,c\,a$ (61)
4·6–5·7	0·20–0·25	M	Monoclinic	$P\,112_1\,/\,a$ (14, unique axis c, cell choice 1)

In C_{II}, and by analogy also in C_I, the additional Na^+ ions are located at the centres of the Al_6O_{18} rings. The O structure (originally called O_I) resembles C_I and C_{II} in having a unit cell that is composed of pseudocubic subcells with edge lengths of approximately 3·8 Å, which contain Ca^{2+} ions and Al_6O_{18} rings, but the arrangement of the rings within the true unit cell is entirely different (Nishi and Takéuchi, 1975; Takéuchi *et al.*, 1980). A preparation of composition $Na_{0\cdot875}Ca_{8\cdot375}Al_{5\cdot175}Fe_{0\cdot450}Si_{0\cdot375}O_{18}$ (3·3 % Na_2O) had a = 10·879 Å, b = 10·845 Å, c = 15·106 Å, space group $Pbca$ (No. 61) $Z = 4$ (Takéuchi *et al.*, 1980). The M structure, which was originally considered to be orthorhombic and called O_{II} (Regourd *et al.*, 1973), is a slightly distorted variant of the O structure with space group $P112_1/a$ (No. 14, unique axis c, cell choice 1), $Z = 4$ (Takéuchi *et al.*, 1980). Preparations with compositions $Na_{1\cdot50}Ca_{8\cdot25}Al_6O_{18}$ (5·7% Na_2O) and $Na_{1\cdot73}Ca_{7\cdot90}Al_{5\cdot5}Si_{0\cdot5}O_{18}$ (6·6 % Na_2O) had parameters: a = 10·877 Å, b = 10·854 Å, c = 15·135 Å, γ = 90·1; and a = 10·844 Å, b = 10·855 Å, c = 15·122 Å, γ = 90·38 (Takéuchi *et al.*, 1980).

Substantial proportions of the Al in these structures can be replaced by other ions, of which Fe^{3+} and Si^{4+} are the most important. Lee *et al.* (1982) found the limits of substitution under equilibrium conditions to be around 2 % for SiO_2 and 3–4 % for Fe_2O_3, but higher degrees of substitution were obtainable under non-equilibrium conditions, such as crystallisation from undercooled melts. If Si^{4+} is present, fewer large cations are needed to maintain charge balance; this extends the solid solution range of the O structure to lower Na_2O contents and that of the M structure to higher ones (Takéuchi *et al.*, 1980). The literature contains many references to a supposed compound NC_8A_3. This formula corresponds to an Na_2O content of 7·6 %, which cannot be attained in C_3A substituted only with Na^+, and it is reasonably certain that this compound does not exist (Fletcher *et al.*, 1965; Maki, 1973; Regourd *et al.*, 1973; Takéuchi *et al.*, 1980); however, if Si is also present, the upper limit of Na substitution is near to this value (Maki, 1973; Takéuchi *et al.*, 1980).

Pollitt and Brown (1969) were unable to prepare an analogue of the orthorhombic phase with K^+ as the sole substituent but obtained evidence that K^+ could stabilise it in clinker, probably because other substituents were also present. Maki (1973) also failed to prepare K forms of the orthorhombic or monoclinic phases under equilibrium conditions, but by moderately rapid cooling of melts he obtained orthorhombic crystals having cell parameters close to those of the corresponding sodium-containing phase. He considered that the presence of Si in the clinker liquid would favour supercooling and thereby also non-equilibrium formation of the orthorhombic or monoclinic phase.

1.4.2 Other modifications

Two further modifications of the C_3A structure have been described. One was obtained as a high-temperature polymorph of the O and M forms and was considered on the basis of powder XRD evidence to be tetragonal (Regourd and Guinier, 1976; Regourd *et al.*, 1973). It appeared to be metastable at room temperature but could be preserved by quenching. Later studies, using thermal analysis, high-temperature light microscopy and high-temperature single-crystal XRD (Maki, 1976a, 1976b; Takéuchi *et al.*, 1980) showed, however, that the O modification was structurally unchanged up to its decomposition temperature and that the M modification was reversibly transformed on heating into the O form. A study of quenched material using powder XRD and single-crystal electron diffraction (Lee and Glasser, 1979) confirmed that the supposed tetragonal form was really orthorhombic, although with a equal to b and thus geometrically tetragonal. The space group was considered to be probably $P\,c\,a\,a$, (No. 54, setting $\mathbf{\bar{c}ba}$) which differs from that ($P\,b\,c\,a$, No. 61) of the normal O form; however, three of the indices assigned to the XRD powder peaks (023, 045 and 047) are incompatible with this space group, although not with $P\,b\,c\,a$. The high-temperature material may thus have the normal O structure and a composition such that the a and b axial lengths are equal to within experimental error. Its persistence to room temperature appears to be favoured by relatively high contents of Na_2O and Fe_2O_3 (Lee *et al.*, 1982).

The other C_3A modification is a disordered, poorly crystalline form, called 'proto-C_3A', which was obtained metastably from simulated clinker liquids, either by rapid cooling or by static crystallisation at low temperatures (Banda and Glasser, 1978). It gave an XRD powder pattern with broadened peaks, corresponding to the systematically strong reflections of cubic C_3A indexable on the subcell with $a \approx 3{\cdot}9$ Å. Analytical electron microscopy of individual crystals showed it to be very high in substituents, one preparation, for example, having an Fe/Al ratio of 0·54 (Han *et al.*, 1981).

1.4.3 Structural modifications of clinker aluminates

Production clinkers have been found to contain cubic or orthorhombic forms of aluminate, alone or in combination. The monoclinic modification has not been observed. The orthorhombic modification is also known as the prismatic, dark interstitial material and is sometimes pseudotetragonal. It can arise only if sufficient alkali is available, but its formation appears to be favoured also by rapid cooling and by bulk compositions potentially able to yield a relatively high proportion of aluminate (Maki, 1973).

Cubic aluminate in clinker is often finely grained and closely admixed with dendritic crystals of ferrite; when it forms larger crystals, these tend to be equidimensional. The XRD powder pattern is characterised by strong, singlet peaks at approximately 33·3°, 47·7° and 59·4° 2θ (Cu Kα radiation; Figure 1.8(a)), whose indices are, respectively, 044, 008 and 448. Patterns obtained either from clinker or from material in which the aluminate and ferrite phases have been concentrated by chemical extraction of the silicates give $a = 15{\cdot}223$–$15{\cdot}255$ Å (Regourd and Guinier, 1976). The slight decrease relative to the value of 15·263 Å for pure C_3A agrees with results for synthetic sodium-substituted preparations (Regourd and Guinier, 1976; Takéuchi *et al.*, 1980). These data are probably equally compatible with the presence of proto-C_3A, but the contents of substituents (§1.4.4) are considerably lower than those found in the latter material. There is probably not enough evidence to show whether the modification is C_I ($P2_1/a\bar{3}$, No. 205) or C_{II} ($P2_13$, No. 198).

The orthorhombic phase is recognisable in the light microscope or SEM by its occurrence as characteristic, lath-shaped crystals, which are often twinned (Maki, 1973). An XRD powder pattern of material from which the silicate phases had been removed showed a splitting of the strong peak at 33·3° 2θ into a strong singlet at approximately 33·2° and a weaker, close doublet at 32·9–33·0° (Figure 1.8(b)). The unit cell parameters were $a = 10{\cdot}874$ Å, $b = 10{\cdot}860$ Å, $c = 15{\cdot}120$ Å (Regourd and Guinier, 1976). Another clinker, in which the aluminate was pseudotetragonal, gave a strong singlet peak at 33·2° and a weaker singlet at 33·0° (Figure 1.8(c)), and the cell parameters were $a = b = 10{\cdot}867$ Å, $c = 15{\cdot}123$ Å.

1.4.4 Compositions of clinker aluminates

Because of the close admixture with other phases, which is often on a scale of 10 μm or less, X-ray microanalysis of the aluminate in clinkers is frequently difficult or unreliable. Data have been reported for cubic, orthorhombic, pseudotetragonal or unspecified forms of the aluminate in ordinary clinkers (Boikova *et al.*, 1981; Harrisson *et al.*, 1985; Kristmann, 1978; Le Saoût *et al.*, 2011; Regourd and Guinier, 1976), for aluminate in high-iron clinkers (Hall and Scrivener, 1997) and

Figure 1.8 Portions of XRD powder patterns of clinkers containing (a) cubic, (b) orthorhombic and (c) pseudotetragonal modifications of the aluminate phase. Peaks marked A and F are due to aluminate and ferrite phases, respectively (after Regourd and Guinier, 1976)

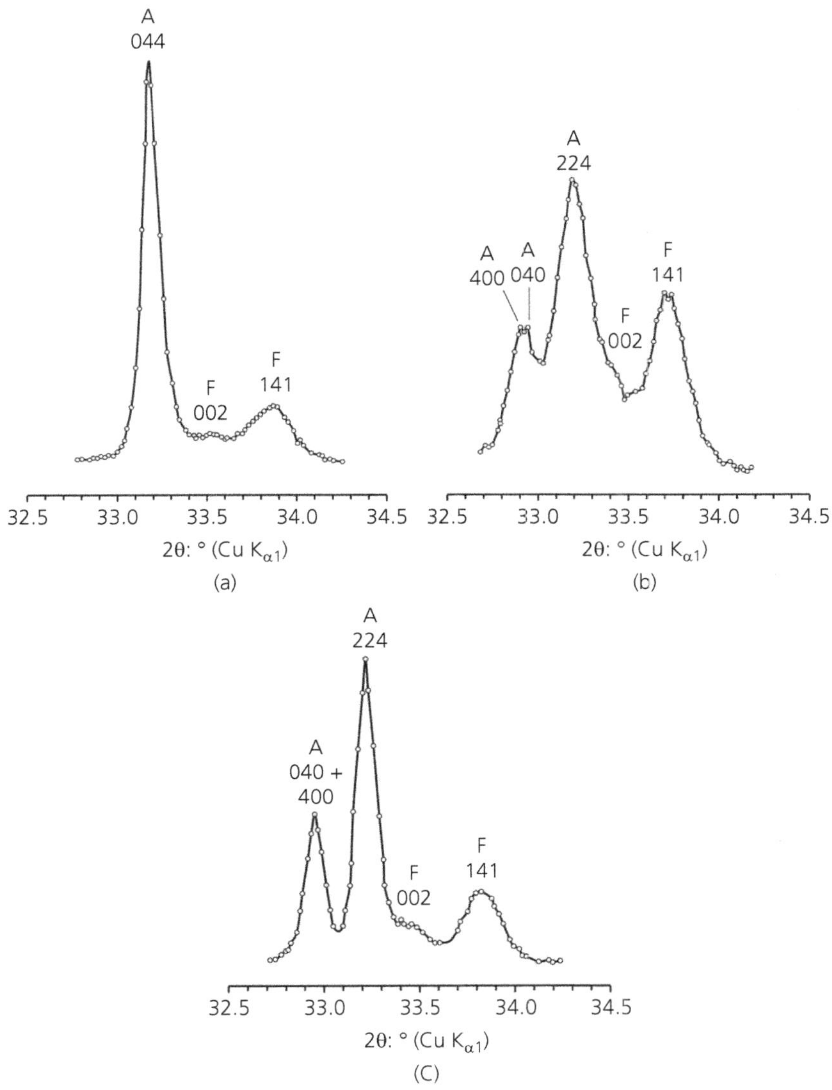

for aluminate (Boikova *et al.*, 1986; Ghose, 1980; Ghose and Barnes, 1979; Sarkar and Roy, 1984) and glass (Boikova *et al.*, 1986) in white cement clinkers. A study of laboratory clinkers indicated that slow cooling decreased Mg^{2+} substitution (Ichikawa *et al.*, 1994). Tables 1.2 and 1.3 include, respectively, average compositions based on these data and suggested site occupancies based on them. The values in both tables take into account both the experimental data and the requirement of reasonable site occupancies.

Pure C_3A contains 62·3% CaO and 37·7% Al_2O_3. Substantial proportions of both Ca and Al are thus replaced, the total content of substituent oxides being typically around 13% for the cubic and up to about 20% for the orthorhombic modification. The content of equivalent Na_2O ($Na_2O + 0{\cdot}66\ K_2O$) appears to be around 1% for the cubic form and 2–4% for the orthorhombic form. None of the analyses indicates alkali contents as high as that required by the formula NC_8A_3, even though considerable amounts of Si are present.

1.4.5 X-ray powder data, densities and optical properties

Calculated XRD powder patterns are given in Tables 1.9 and 1.10 for cubic and orthorhombic clinker aluminates, respectively. The X-ray density of the cubic modification is $3{\cdot}064\,g\,cm^{-3}$, assuming $a = 15{\cdot}240$ Å and composition $(K_{0{\cdot}03}Na_{0{\cdot}06}Ca_{2{\cdot}76}Mg_{0{\cdot}05}Ti_{0{\cdot}01})(Fe_{0{\cdot}22}Al_{1{\cdot}60}Si_{0{\cdot}18})O_6$. That of the orthorhombic modification is $3{\cdot}052\,g\,cm^{-3}$, assuming $a = 10{\cdot}879$ Å, $b = 10{\cdot}845$ Å, $c = 15{\cdot}106$ Å and composition $(Na_{0{\cdot}292}Ca_{2{\cdot}792})(Fe_{0{\cdot}150}Al_{1{\cdot}725}Si_{0{\cdot}125})O_6$. For pure C_3A with $a =$ 15·263 Å, the value is $3{\cdot}029\,g\,cm^{-3}$. Pure C_3A is optically isotropic, with refractive index 1·710 (Rankin and Wright, 1915); iron substitution can raise this value to 1·735 (Hansen *et al.*, 1928). Maki (1974, 1976a) described the morphology and optical properties of the orthorhombic modification, including variations with composition.

Table 1.9 Calculated XRD powder pattern for a cubic aluminate (continued on next page)

d: Å	I_{int}	I_{pk}	*h*	*k*	*l*
6·815	2	2	1	0	2
5·080	3	4	1	2	2
4·595	2	2	1	1	3
4·227	5	6	0	2	3
4·073	5	14	1	2	3
4·073	7		2	1	3
3·810	3	3	0	0	4
3·326	3	3	2	1	4
3·048	2	2	3	0	4
2·830	4	5	0	2	5
2·782	5	8	1	2	5
2·782	2		2	1	5
2·694	100	100	0	4	4
2·410	2	3	0	2	6
2·380	2	3	0	4	5
2·200	7	7	4	4	4
2·037	2	2	2	4	6
1·951	3	3	3	4	6
1·905	31	29	0	0	8
1·890	2	4	1	0	8

Table 1.9 Continued

d: Å	I_{int}	I_{pk}	*h*	*k*	*l*
1·822	2	3	3	5	6
1·555	34	30	4	4	8

The pattern was calculated assuming $a = 15{\cdot}24$Å, space group $Pa\bar{3}$, atomic parameters given by Mondal and Jeffery (1975) and composition $(K_{0{\cdot}03}Na_{0{\cdot}06}Ca_{2{\cdot}76}Mg_{0{\cdot}05}Ti_{0{\cdot}01})(Fe_{0{\cdot}22}Al_{1{\cdot}60}Si_{0{\cdot}18})O_6$.
Other peaks of significant height: 6·215Å (2), 1 1 2; 2·272Å (2), 4 2 5 +; 2·177Å (2), 2 3 6 +; 1·984Å (2), 3 1 7 +; 1·643Å (2), 5 5 6 +; $I_{044}/I_c = 3{\cdot}06$.

Table 1.10 Calculated XRD powder pattern for an orthorhombic aluminate phase (continued on next page)

d: Å	I_{int}	I_{pk}	*h*	*k*	*l*
7·553	2	2	0	0	2
6·846	3	4	1	1	1
5·104	5	6	0	2	1
4·414	5	5	2	0	2
4·211	8	8	1	1	3
3·840	2	2	2	2	0
3·690	2	2	0	2	3
3·423	9	9	2	2	2
3·389	2	3	1	1	4
3·353	4 }	7	{ 3	1	1
3·345	3 }		{ 1	3	1
3·054	6	6	2	2	3
2·812	5	5	1	1	5
2·720	31 }	49	{ 4	0	0
2·711	25 }		{ 0	4	0
2·693	100	100	2	2	4
2·431	3	3	4	2	0
2·392	2	3	1	1	6
2·374	4	4	2	2	5
2·354	2	2	2	3	4
2·310	2	3	2	4	2
2·282	2	2	3	3	3
2·270	2	3	3	1	5

Table 1.10 Continued

d: Å	I_{int}	I_{pk}	*h*	*k*	*l*
2·207	5	9	4	0	4
2·202	6		0	4	4
2·186	2	3	2	4	3
2·030	2	2	1	3	6
1·964	3	3	5	1	3
1·920	32	29	4	4	0
1·905	4	5	4	4	1
1·892	2	14	2	4	5
1·888	15		0	0	8
1·848	2	2	4	0	6
1·749	2	2	4	2	6
1·688	2	1	0	4	7
1·624	3	3	2	6	3
1·565	17	17	6	2	4
1·562	15	21	2	6	4
1·551	9	11	4	0	8
1·549	6		0	4	8

The pattern was calculated assuming *a* = 10·879 Å, *b* = 10·845 Å, *c* = 15·106 Å, space group *P b c a*; composition $(Na_{0\cdot292}Ca_{2\cdot792})(Fe_{0\cdot150}Al_{1\cdot725}Si_{0\cdot125})O_6$; atomic parameters given by Nishi and Takéuchi (1975).
Other peaks of significant height: 5·414Å (2), 0 2 0; 5·388 Å (2), 1 1 2; 2·639 Å (2), 0 2 5; 2·586 Å (2), 3 2 3+; 2·425 Å (3), 3 3 2; 2·047 Å (2), 1 5 2; 2·042 Å (2), 2 4 4; 1·827 Å (2), 1 3 7 +; 1·616 Å (2), 6 1 4; $I_{224}/I_c = 1\cdot94$.

1.5. Ferrite

1.5.1 Crystal structure and composition in the $Ca_2(Al_xFe_{1-x})_2O_5$ series

At ordinary pressures in the absence of oxide components other than CaO, Al_2O_3 and Fe_2O_3, the ferrite phase can be prepared with any composition in the solid solution series $Ca_2(Al_xFe_{1-x})_2O_5$, where $0 < x < \approx 0\cdot7$. The composition C_4AF is only a point in this series, with $x = 0\cdot5$. The end member, C_2A, with $x = 1$, has been prepared, but only at a pressure of 2·5 GPa (Aggarwal *et al.*, 1972; Kahlenberg *et al.*, 2000; Lazic *et al.*, 2008), although it is readily preserved at ambient pressure by quenching. The series with $x < \approx 0\cdot7$ is not quite isostructural, with a change in space group that has been reported at a composition between $x = 0\cdot235$ and 0·33 (Fukuda and Ando, 2002; Redhammer *et al.*, 2004; Smith, 1962). Table 1.11 gives crystal data for various compositions. Care is needed in referring to the unit cell and space group, as some workers reverse the choice of *a* and *c* axes used here, with appropriate changes in space group symbol.

Table 1.11 Crystal data for some calcium aluminoferrites in the series $Ca_2(Al_xFe_{1-x})_2O_5$

x	Unit cell parameters: Å[a]			Unit cell volume: $Å^3$	Space group[b]	X-ray density: $g\,cm^{-3}$	Reference
	a	b	c				
0	5·4153	14·7687	5·5980	448·54	*P n m a*	4·026	Berggren *et al.* (1971)
0·138	5·4071	14·6949	5·5941	444·49	*P n m a*	3·943	Redhammer *et al.* (2004)
0·276	5·3808	14·6131	5·5911	439·63	*I 2 m b*	3·867	Redhammer *et al.* (2004)
0·285	5·380	14·61	5·588	439·23	*I 2 m b*[c]	3·862	Colville and Geller (1972)
0·50	5·3415	14·5065	5·5589	430·74	*I 2 m b*[c]	3·747	Jupe *et al.* (2001)
1	5·2281	14·4686	5·4004	408·50	*I 2 m b*	3·481	Kahlenberg *et al.* (2000)

[a]The unit cell is orthorhombic, with $Z = 4$.
[b]*I 2 m b* is SG No. 46, setting **cab**; *P n m a* is SG No. 62, setting **abc**.
[c]The SG is *I b m 2* in the reference, which is No. 46, setting $\mathbf{ba\bar{c}}$.

Plots of unit cell parameters or XRD powder spacings against x derived from studies of the pure $Ca_2(Al,Fe)_2O_5$ series have generally shown a change in slope that is attributable to the structural change. The value of x at which the change has been observed has varied between studies – for example, 0·235 (Fukuda and Ando, 2002), 0·28 (Redhammer *et al.*, 2004) and 0·33 (Smith, 1962). Data from 16 separate studies are compiled on Figure 1.9, which includes plots for the a, b and c parameters, the unit cell volume and the calculated density. For these combined data, the change in slope is clearly evident for the c parameter (at $x = 0{\cdot}35$), only very slight for the a parameter, and it does not occur for the b parameter. The change is also very slight for the unit cell volume (Figure 1.9(d)), the value of which can as a consequence be used to provide an estimate of x – that is, linear regression analysis gives the equation $x = 12{\cdot}088 - 0{\cdot}0269V$ (where V is in $Å^3$; $r^2 = 0{\cdot}984$). This can be useful for checking the composition of synthetic preparations: for example, the values of the unit cell parameters given in Colville and Geller (1971) and in Sharma *et al.* (2018) are not consistent with the reported $x = 0{\cdot}5$ (Colville and Geller's (1971) data are represented by the filled triangle on Figure 1.9). It is important to emphasise that this is only strictly valid if oxide components other than CaO, Al_2O_3 and Fe_2O_3 are absent. Application to ferrites in clinkers would give inaccurate results because of the effects of other substituents (§1.5.2 and §1.5.3).

Figure 1.9(e) shows that there is a strong linear relationship between X-ray density and composition for $x < 0{\cdot}7$: regression analysis gives the equation D_x ($g\,cm^{-3}$) $= 4{\cdot}0233 - 0{\cdot}5641x$ ($r^2 = 0{\cdot}995$), which actually provides a reasonable prediction of the density of pure C_2A (i.e. $x = 1$; predicted 3·46, actual 3·48 $g\,cm^{-3}$ (Aggarwal *et al.*, 1972; Kahlenberg *et al.*, 2000)).

Figure 1.9 Variations of unit cell parameters, volume and calculated density with *x* for the $Ca_2(Fe_{1-x}Al_x)_2O_5$ solid solution series at ambient temperature (small unfilled circles) (continued on next page)

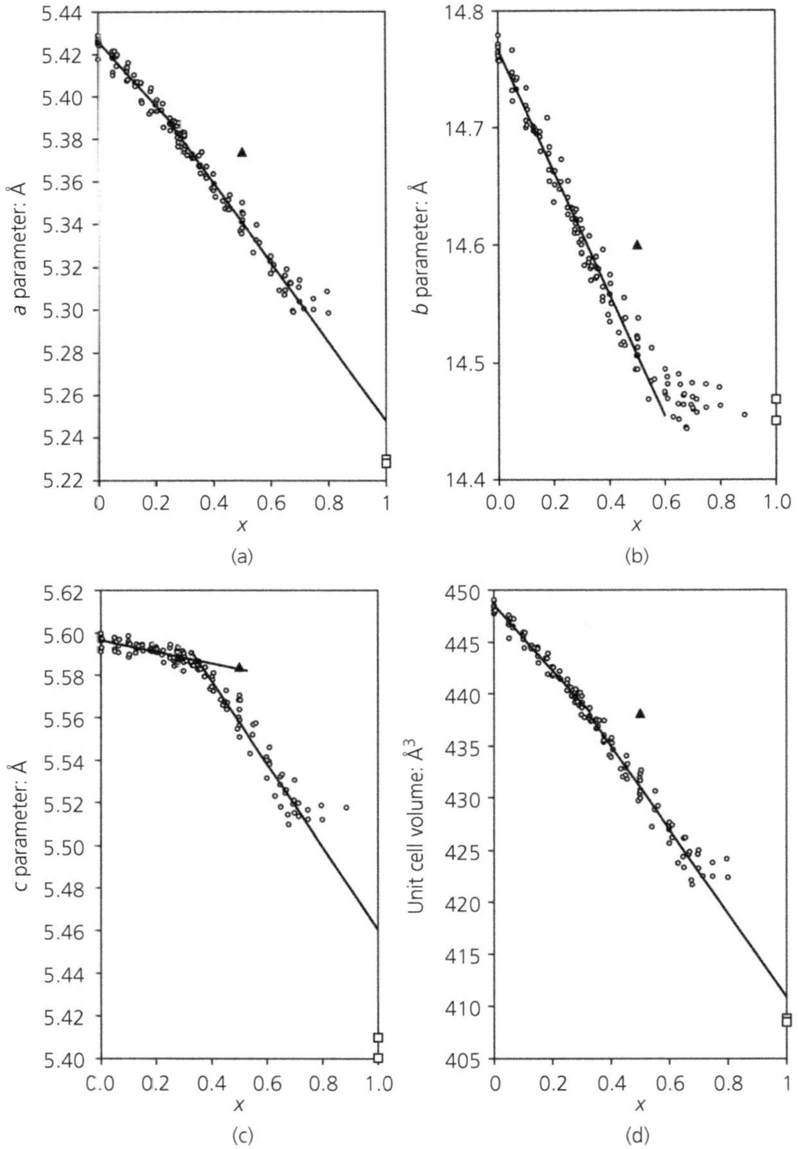

Figure 1.9 Continued

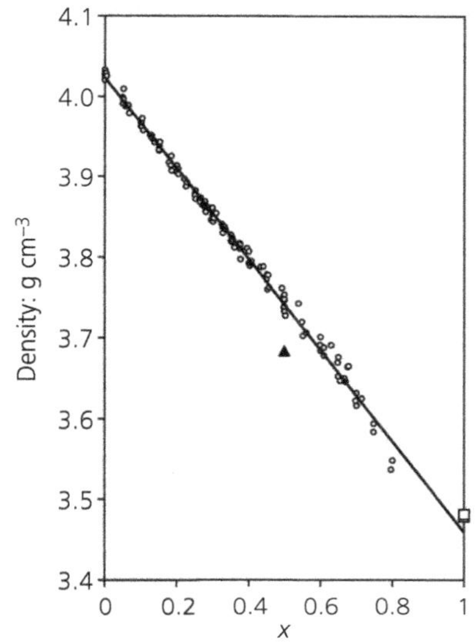

(e)

Data are from: Berggren (1971); Carvalho *et al.* (2008); Colville (1970); Colville and Geller (1972); Fukuda and Ando (2002); Jupe *et al.* (2001); Kahlenberg and Fischer (2000); Newkirk and Thwaite (1958); Phan *et al.* (2018); Redhammer *et al.* (2004); Smith (1962); Vanpeteghem *et al.* (2008); Woermann *et al.* (1965, 1969); Zötzl and Pöllmann (2006). Newkirk and Thwaite's (1958) data are included only on the plots for the *b* and *c* parameters because their values for the *a* parameter are clearly in error, being about 0·06 Å lower than the values of the other workers. The squares are for C_2A (Aggarwal *et al.*, 1972; Kahlenberg *et al.*, 2000) and the filled triangle is Colville and Geller's (1971) C_4AF.

Büssem (1939) determined the approximate crystal structure of C_4AF. Subsequent determinations or refinements have been reported for many preparations, including with $x = 0$ (C_2F) (Berggren, 1971; Bertaut *et al.*, 1959; Colville, 1970), $x = 0{\cdot}285$ and $0{\cdot}36$ (Colville and Geller, 1972) and $x = 0{\cdot}5$ (Colville and Geller, 1971). Figure 1.10 shows the structure for the compositional range with $\approx 0{\cdot}33 < x < \approx 0{\cdot}7$. It is derived from that of perovskite ($CaTiO_3$) by the substitution of Al and Fe for Ti, together with ordered omission of oxygen atoms, which causes one-half of the sheets of octahedra in perovskite to be replaced by chains of tetrahedra. The empirical composition may thus be written $Ca_2M_2T_2O_5$, where M and T represent octahedral and tetrahedral cations, respectively. The structure for compositions with $x < 0{\cdot}33$ differs from that shown in Figure 1.10 in that one-half of the chains of tetrahedra have the opposite polarity.

Each Ca^{2+} ion in C_4AF has seven oxygen neighbours at 2·3 to 2·6 Å (Colville and Geller, 1971). The aluminium and iron atoms are both distributed between octahedral and tetrahedral sites, the fraction of the aluminium entering tetrahedral sites under equilibrium conditions decreasing with temperature. For the three preparations with $x = 0{\cdot}285$, $0{\cdot}36$ and $0{\cdot}5$ on which X-ray structure

Figure 1.10 Crystal structure of C_4AF, based on the results of Colville and Geller (1971)

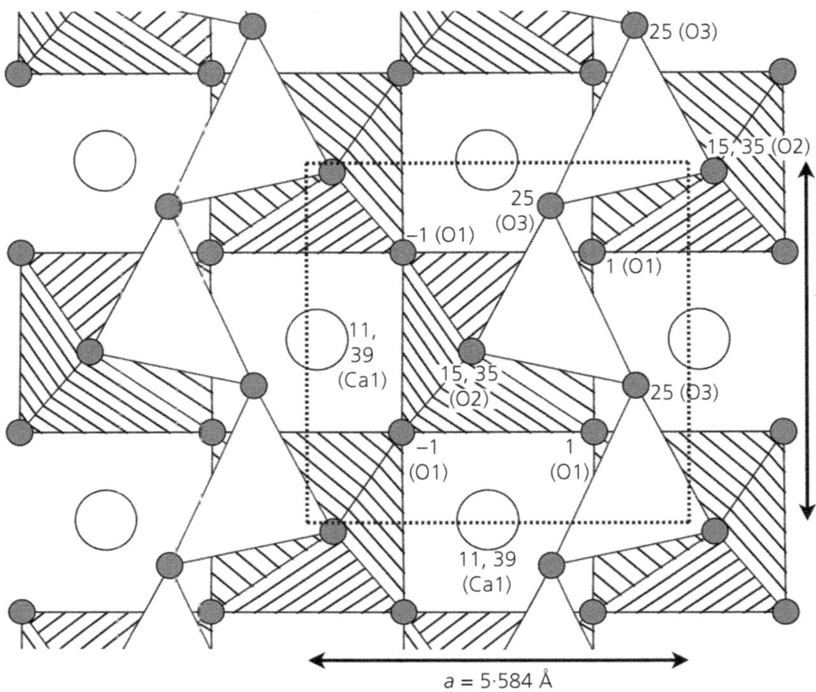

The structure is orthorhombic, space group *I b m* 2 (No. 46, cell choice **ba$\bar{\mathbf{c}}$**). The diagram shows calcium atoms (open circles), (Al,Fe)O_4 tetrahedra (triangles with grey circles at the vertices) and (Fe,Al)O_6 octahedra (hatched squares with grey circles at the vertices). Heights of atoms are given as hundredths of the cell height (b = 14·5 Å). Atoms at heights outside the range of –1 to +39 are not shown, their positions being derivable from those shown by translations of $\frac{1}{2}$ parallel to each of the axes.

determinations were made, 75–76% of the total content of aluminium was found to be in tetrahedral sites. These preparations were shown to have been in equilibrium at about 750 °C (Colville and Geller, 1971); for a C_4AF preparation quenched from 1290 °C, the Mössbauer spectrum indicated that only 68% of the aluminium was in tetrahedral sites (Grant *et al.*, 1968). It was suggested that x does not exceed the observed limit of about 0·7 because at this composition the tetrahedral sites are all occupied by aluminium (Colville and Geller, 1972). There is evidence of clustering of aluminium and iron atoms to an extent, depending on composition and conditions of formation (Zheng and Yang, 1986).

1.5.2 Compositions of clinker ferrites

In many clinkers, the ferrite is closely mixed with aluminate; due to a similarity in cell parameters, oriented intergrowth can occur (Maki, 1974). The close admixture often renders X-ray microanalysis difficult or unreliable. For ordinary Portland cement clinkers, the compositions found in different laboratories are nevertheless remarkably consistent. They show that there is little or no substitution for Ca^{2+}, the substituent ions being contained essentially in the octahedral and tetrahedral sites. Table 1.2 includes an average value based on the results of investigations using X-ray microanalysis (Boikova, 1986; Boikova *et al.*, 1981; Harrisson *et al.*, 1985; Hornain, 1973; Kristmann, 1978; Scrivener and Taylor, 1995a; Uchikawa *et al.*, 1985) or chemical analysis of separated material (Yamaguchi and Takagi, 1969). Table 1.3 includes suggested site occupancies corresponding to these data. Data on the distribution of substituents between octahedral and tetrahedral sites (Zheng and Yang, 1991) are insufficient to justify a distinction between them, as the preferences of some of the cations, especially Mg^{2+}, in this structure are unknown, as is the temperature at which equilibrium is attained. This temperature probably varies between clinkers and may be expected to affect the distribution.

The typical composition differs markedly from that of C_4AF (46·1% CaO, 21·0% Al_2O_3, 32·9% Fe_2O_3). It contains about 10% of substituent oxides and is much lower in Fe_2O_3. It approximates to $Ca_2AlFe_{0\cdot6}Mg_{0\cdot2}Si_{0\cdot15}Ti_{0\cdot05}O_5$, which is derived from that of C_4AF by replacing some of the Fe^{3+} by Mg^{2+} and an equal amount by Si^{4+} and Ti^{4+}. A conclusion (Neubauer *et al.*, 1996) based on Rietveld analysis of the XRD powder pattern that Si^{4+} is not present in solid solution but in closely admixed belite appears incompatible with the observation that incorporation of Si^{4+} markedly affects the colour and electrical conductivity (Ichikawa and Komukai, 1993). A Mössbauer study also provided indirect evidence for silicon substitution (Zheng and Yang, 1991).

Sulfate-resisting Portland cement clinkers, including those made for use in oil-well cements, have relatively high ratios of Fe to Al and, if it is assumed that most of the iron is in the ferrite, the latter cannot have the composition given above. X-ray microanalyses (Bergstrom *et al.*, 1992; Ghose, 1980; Ghose and Barnes, 1979; Gollop and Taylor, 1994; Hall and Scrivener, 1997; Richardson *et al.*, 1993) indicate that, in contrast to the situation in normal Portland cement clinkers, there is substantial replacement of Al^{3+} as well as, or even instead of, Fe^{3+} in the C_4AF composition. Microanalyses show that there is much variation in composition, both between clinkers and within clinkers, especially at the micrometre level that can be studied using TEM (Richardson *et al.*, 1993). The mean contents of MgO and Fe_2O_3 are related to those of the clinker, and the SiO_2 content tends to increase with that of MgO; SO_3 contents of up to 0·9% have been reported (Bergstrom *et al.*, 1992; Hall and Scrivener, 1997). The Al_2O_3/Fe_2O_3 ratio is probably also related to that in the clinker. Tables 1.2 and 1.3 include what is possibly a typical mean composition and atomic ratios calculated from it.

1.5.3 Crystal data and X-ray powder patterns for ferrites containing substituent ions

Mn^{3+} can replace all the Fe^{3+} or up to 60% of the Al^{3+} in C_4AF (Kondo *et al.*, 1978). Yamaguchi and Takagi (1969) summarised data showing that incorporation of Mg^{2+} or Si^{4+} or both causes increases in cell parameters. Boikova (1986) determined the cell parameters of four preparations with compositions similar to those of typical clinker ferrites; the ranges were $a = 5{\cdot}535$–$5{\cdot}554$ Å, $b = 14{\cdot}492$–$14{\cdot}642$ Å, $c = 5{\cdot}288$–$5{\cdot}333$ Å. Marinho and Glasser (1984) found that Ti^{4+} substitution caused stacking changes in C_2F or C_4AF. It had a negligible effect on the cell parameters and the charge was balanced by incorporation of additional oxygen atoms. This last conclusion may be irrelevant to clinker ferrites since Ti^{4+} is a relatively minor substituent and the charge can be balanced in other ways.

Regourd and Guinier (1976) reported unit cell parameters for the ferrite in five clinkers. The ranges observed were $a = 5{\cdot}517$–$5{\cdot}555$ Å, $b = 14{\cdot}55$–$14{\cdot}62$ Å, $c = 5{\cdot}335$–$5{\cdot}350$ Å. Boikova (1986) reported XRD powder spacings for clinker ferrites which indicate similar values. The similarity of these cell parameters to those of the laboratory preparations (Boikova, 1986) mentioned in the previous paragraph supports the results of X-ray microanalyses of clinker ferrites described in §1.5.2. Compared with pure C_4AF, typical clinker ferrites have smaller values of *a* and *c*, but larger values of *b*.

The XRD powder patterns of clinker ferrites are affected by the cooling rate (Ichikawa *et al.*, 1994; Miyazawa and Tomita, 1969; Ono and Nagashima, 1976). The effect is associated with differing uptake of MgO and SiO_2. Extreme values of the cell parameters were $a = 5{\cdot}57$ Å, $b = 14{\cdot}62$ Å, $c = 5{\cdot}32$ Å for a quenched sample and $a = 5{\cdot}43$ Å, $b = 14{\cdot}65$ Å, $c = 5{\cdot}33$ Å for one that was slowly cooled (Ono and Nagashima, 1976). The ferrite in the quenched samples was poorly crystalline, many peaks other than the three most intense (200, 141 and 202) disappearing (Ono and Nagashima, 1976). Broadening of peaks of clinker ferrites (Boikova, 1986; Miyazawa and Tomita, 1969; Regourd and Guinier, 1976) might be caused not only by poor crystallinity, but also by zoning. The cell parameters observed by Boikova (1986) and Regourd and Guinier (1976) are near to those of the quenched sample.

Table 1.12 Calculated XRD powder pattern for a clinker ferrite (continued on next page)

d: Å	I_{int}	I_{pk}	*h*	*k*	*l*
7·321	19	21	0	2	0
5·177	4	4	1	1	0
3·839	2	2	1	0	1
3·661	16	17	1	3	0
3·400	7	7	1	2	1
2·767	35	33	2	0	0
2·664	47	53	0	0	2
2·649	100	100	1	4	1
2·588	14	14	1	5	0
2·422	5	4	2	1	1
2·208	9	9	2	4	0
2·154	7	14	1	3	2
2·154	9		0	4	2
2·059	23	20	1	6	1
1·919	48	41	2	0	2
1·882	3	3	2	5	1
1·857	9	8	2	2	2
1·830	2	20	2	6	0
1·830	21		0	8	0
1·799	3	3	0	6	2
1·726	10	8	3	3	0

Table 1.12 Continued

d: Å	I_{int}	I_{pk}	*h*	*k*	*l*
1·648	2	2	1	2	3
1·592	4	3	2	7	1
1·574	17	15	3	4	1
1·561	3	4	3	5	0
1·535	21	17	1	4	3
1·527	6	6	2	8	0

The pattern was calculated assuming a = 5·535 Å, b = 14·642 Å, c = 5·328 Å, space group *I b m* 2, atomic parameters given for C_4AF by Colville and Geller (1971) and composition $Ca_2[Al_{0·7}Fe_{0·3}](Al_{0·7}Fe_{0·3})O_5$, where square and round brackets denote octahedral and tetrahedral sites, respectively; I_{141}/I_c = 1·79.

A calculated powder pattern for a ferrite that has cell parameters and composition that are similar to those for the material in typical clinkers is given in Table 1.12. The X-ray density, assuming $a = 5·535$ Å, $b = 14·642$ Å, $c = 5·328$ Å and composition $Ca_2AlFe_{0·6}Si_{0·15}Ti_{0·05}O_5$, is 3·570 g cm^{-3}.

1.5.4 Optical, magnetic and other data

Ferrite of or near C_4AF composition and free from substituent ions is yellowish brown in transmitted light. The optical properties of C_4AF are as follows: biaxial, negative, with moderate 2V; $\alpha = 1·96$, $\beta = 2·01$, $\gamma = 2·04$ in lithium light; pleochroic, with γ brown, α yellowish brown; the refractive indices increase with Fe/Al ratio (Hansen *et al.*, 1928). The optic orientation is X = *b*, Y = *c*, Z = *a* (Maki *et al.*, 1995). The pleochroism of ferrite solid solutions has been studied (Maki *et al.*, 1995).

If clinker ferrites are cooled in air, they are commonly almost black. The dark colour is associated with markedly higher electrical conductivity (Ichikawa and Komukai, 1993; Miyazawa and Tomita, 1969). In the absence of other substituents, it arises if the ferrite contains an excess of small dipositive ions (Mg^{2+} or Zn^{2+}) over tetrapositive ions (Si^{4+} or Ti^{4+}) and is cooled in air (Ichikawa and Komukai, 1993). If the substituted ferrite is cooled in N_2 or quenched in water, the dark colour disappears (Miyazawa and Tomita, 1969). Its disappearance seems to be due to the absence of Fe^{4+} rather than the presence of Fe^{2+} (Maki et al., 1995; Scrivener and Taylor, 1995b). An excess of dipositive over tetrapositive substituents when all the iron is present as Fe^{3+} implies the presence of oxygen vacancies; these can be filled during cooling, with concomitant oxidation of a little of the iron to Fe^{4+}. Clinker ferrites can apparently be dark even if there is not an excess of dipositive over tetrapositive substituents. This has yet to be explained.

The ferrite is ferromagnetic, and the more iron-rich members of the $Ca_2(Al_xFe_{1-x})_2O_5$ series have been studied for their magnetic properties, especially in regard to the coordinations of the Fe atoms (Colville, 1970; Geller *et al.*, 1971). Ferrite with a composition near C_4AF has been found in nature. The name brownmillerite, already in use in cement chemistry, was adopted as the mineral name (Gross, 1977). Mitchell *et al.* (2017) recommend that brownmillerites are minerals in the solid solution series $Ca_2(Al_xFe_{1-x})_2O_5$ with $x > 0·28$ that adopt space group *I b m* 2, and that srebrodolskites are minerals in the solid solution series with $x < 0{\cdot}28$ that adopt space group *P n m a*; the value of *x* was taken from Redhammer *et al.* (2004) but – as noted above – there is some variation in the exact value of *x* at which the structural change occurs (e.g. it is 0·33 in Smith, 1962).

REFERENCES

Aggarwal PS, Gard JA, Glasser FP and Biggar GM (1972) Synthesis and properties of dicalcium aluminate, $2CaO.Al_2O_3$. *Cement and Concrete Research* **2**: 291–297.

Anon (1813) Married. *Leeds Mercury*, 3 July.

Anon (1999) Concrete through the ages. British Cement Association, Camberley, UK.

Aspdin J (1824) An improvement in the modes of producing an artificial stone. British Patent 5022.

Aspdin W (1848) Letter to the Editor, 9 Sep. *The Builder* VI [292], p. 443.

Banda RMH and Glasser FP (1978) Crystallization of the molten phase in Portland cement clinker. *Cement and Concrete Research* **8(6)**: 665–670.

Barbier J and Hyde BG (1985) The structures of the polymorphs of dicalcium silicate. Ca_2SiO_4. *Acta Crystallographica* **B41**: 383–390.

Barfoot RJ (1974) Joseph, James and William – the Aspdin jigsaw. *Concrete* **8 [8]**: 18–26.

Barnes P, Fentiman CH and Jeffery JW (1977) Chromium(III) complexes of macrocyclic ligands. I. Crystal structure of trans-bis(O-carbamato)(1,4,8,11-tetraazacyclotetradecane)chromium(II) perchlorate sesquihydrate. *Acta Crystallographica* **A36**: 353–359.

Baur WH (1978) Variation of mean Si–O bond lengths in silicon–oxygen tetrahedra. *Acta Crystallographica Section B Structural Crystallography and Crystal Chemistry* **34(6)**: 1751–1756.

Beadle D (1977) *Portland Cement in the Making*. Cement and Concrete Association, Slough, UK.

Berggren J (1971) Refinement of the crystal structure of dicalcium ferrite, $Ca_2Fe_2O_5$. *Acta Chemica Scandinavica* **25**: 3616–3624.

Bergstrom TB, Hall C and Scrivener KL (1992) Interstitial material in oilwell cements: evidence from X-ray microanalysis. *Advances in Cement Research* **4(16)**: 141–147.

Bertaut EF, Blum P and Sagnieres A (1959) Structure du ferrite bicalcique et de la brownmillerite. *Acta Crystallographica* **12(2)**: 149–159. (In French.)

Bigaré M, Guinier A, Mazières C *et al.* (1967) Polymorphism of tricalcium silicate and its solid solutions. *Journal of the American Ceramic Society* **50(11)**: 609–619.

Biggar GM (1971) Phase relationships of bredigite ($Ca_5MgSi_3O_{12}$) and of the quaternary compound ($Ca_6MgAl_8SiO_{21}$) in the system $CaO{\cdot}MgO{\cdot}Al_2O_3{\cdot}SiO_2$. *Cement and Concrete Research* **1(5)**: 493–513.

Blezard RG (1998) The history of calcareous cements. In *Lea's Chemistry of Cement and Concrete*, 4th edn (Hewlett PC (ed.)). Elsevier, Amsterdam, the Netherlands.

Bogue RH (1947) *The Chemistry of Portland Cement*. Reinhold, New York, NY, USA.

Boikova AI (1986) Chemical composition of raw materials as the main factor responsible for the composition, structure and properties of clinker phases. *Proceedings of the 8th International Congress on the Chemistry of Cement* **1**: 19–34.

Boikova AI, Esayan A and Lazukin V (1981) On the composition of cement minerals. *Proceedings of the 7th International Congress on the Chemistry of Cement* **4**: 183–185.

Boikova AI, Fomicheva OI and Zubekhin AP (1986) Chemical composition of phases relations in unbleached and bleached clinkers. *Proceedings of the 8th International Congress on the Chemistry of Cement* **2**: 243–248.

Bonafous L, Bessada C, Massiot D, Coutures J, Holland BL and Colombet P (1995) ^{29}Si MAS NMR study of dicalcium silicate: the structural influence of sulfate and alumina stabilizers. *Journal of the American Ceramic Society* **78(10)**: 2603–2608.

Brown GE and Gibbs GV (1969) Oxygen coordination and the Si-O bond. *American Mineralogist* **54**: 1528–1539.

Burke E (1954) Discussion comment after 'Jeffery, J.W., "The tricalcium silicate phase", Proceedings of the 3rd International Symposium on the Chemistry of Cement, 30-48', *Proceedings of the 3rd International Symposium on the Chemistry of Cement* 50–51.

Büssem W (1939) X-Rays and cement chemistry. *Proceedings of the International Symposium on the Chemistry of Cement*. Ingeniörsvetenskapsakademien, Stockholm, Sweden, pp. 141–168.

Butler DB (1913) *Portland Cement: Its Manufacture, Testing and Use*, 3rd edn. E & FN Spon, London, UK.

Carvalho MD, Ferreira LP, Waerenborgh JC, Tsipis E, Lopes AB and Godinho M (2008) Structure and magnetic properties of $Ca_2Fe_{1-x}Mn_xAlO_{5+\delta}$. *Journal of Solid State Chemistry* **181(9)**: 2530–2541.

Cembureau (2018) *Activity Report 2017*. Cembureau, The European Cement Association, Brussels, Belgium. https://cembureau.eu/media/1716/activity-report-2017.pdf

Chan CJ, Kriven WM and Young JF (1988) Analytical electron microscopic studies of doped dicalcium silicates. *Journal of the American Ceramic Society* **71(9)**: 713–719.

Chen M, Xia Z, Molokeev MS and Liu Q (2015) Structural phase transformation and luminescent properties of $Ca_{2-x}Sr_xSiO_4$: Ce(3+) orthosilicate phosphors. *Inorganic Chemistry* **54(23)**: 11369–11376.

Colville AA (1970) The crystal structure of $Ca_2Fe_2O_6$ and its relation to the nuclear electric field gradient at the iron sites. *Acta Crystallographica Section B Structural Crystallography and Crystal Chemistry* **26(10)**: 1469–1473.

Colville AA and Geller S (1971) The crystal structure of brownmillerite, Ca_2FeAlO_5. *Acta Crystallographica Section B Structural Crystallography and Crystal Chemistry* **27(12)**: 2311–2315.

Colville AA and Geller S (1972) Crystal structures of $Ca_2Fe_{1.43}Al_{0.57}O_5$ and $Ca_2Fe_{1.28}Al_{0.72}O_5$. *Acta Crystallographica Section B Structural Crystallography and Crystal Chemistry* **28(11)**: 3196–3200.

Courtial M, de Noirfontaine MN, Dunstetter F, Gasecki G and Signes-Frehel M (2003) Polymorphism of tricalcium silicate in Portland cement: a fast visual identification of structure and superstructure. *Powder Diffraction* **18(1)**: 7–15.

Cuesta A, Losilla ER, Aranda MAG, Sanz J and de la Torre ÁG (2012) Reactive belite stabilization mechanisms by boron-bearing dopants. *Cement and Concrete Research* **42(4)**: 598–606.

Czaya R (1971) Refinement of the structure of γ-Ca_2SiO_4. *Acta Crystallographica Section B Structural Crystallography and Crystal Chemistry* **27(4)**: 848–849.

Daimon M, Hirano Y, Tsurumi T and Asaga K (1992) Crystal structure analysis of major constituent phases in ordinary Portland cement. *Proceedings of the 9th International Congress on the Chemistry of Cement* **2**: 17–22.

Davis AC (1924) *A Hundred Years of Portland Cement,1824–1924*. Concrete Publications, London, UK.

de la Torre ÁG, Bruque S, Campo J and Aranda MAG (2002) The superstructure of C_3S from synchrotron and neutron powder diffraction and its role in quantitative phase analyses. *Cement and Concrete Research* **32(9)**: 1347–1356.

de la Torre ÁG, de Vera RN, Cuberos AJM and Aranda MAG (2008) Crystal structure of low magnesium-content alite: Application to Rietveld quantitative phase analysis. *Cement and Concrete Research* **38(11)**: 1261–1269.

de Noirfontaine MN, Dunstetter F, Courtial M, Gasecki G and Signes-Frehel M (2006) Polymorphism of tricalcium silicate, the major compound of Portland cement clinker 2. Modelling alite for Rietveld analysis, an industrial challenge. *Cement and Concrete Research* **36(1)**: 54–64.

de Noirfontaine MN, Courtial M, Dunstetter F, Gasecki G and Signes-Frehel M (2012) Tricalcium silicate Ca_3SiO_5 superstructure analysis: a route towards the structure of the M_1 polymorph. *Zeitschrift für Kristallographie* **227(2)**: 102–112.

Dunstetter F, de Noirfontaine MN and Courtial M (2006) Polymorphism of tricalcium silicate, the major compound of Portland cement clinker 1. Structural data: Review and unified analysis. *Cement and Concrete Research* **36(1)**: 39–53.

Eckel EC (1905) *Cements, Limes, and Plasters; Their Materials, Manufacture, and Properties*. Wiley, New York, NY, USA. NOTE: Eckel gives the main dimensions of the 'ordinary English form' of bottle-shaped kiln as 'perhaps 9 to 12 feet in diameter at its widest portion, 15 to 18 feet from its base to this widest zone, and 25 to 35 feet in total height'.

Exchequer Court of York Wills (1802) Will of Thomas Aspdin of Leeds. Exchequer Court of York Wills, York Diocesan Archive. Borthwick Institute for Archives, University of York, GB 193, ECY Thomas Aspdin of Leeds, Mar. 1802, vol.146, f.96. Probate copy: WYAS Wakefield Deed 1806/EZ/258/385.

Exchequer Court of York Wills (1832) Will of George Strafford of Wakefield. Exchequer Court of York Wills, York Diocesan Archive, Borthwick Institute for Archives, University of York, GB 193, ECY George Strafford of Wakefield, Sept. 1832, vol.186, f.180.

Exchequer Court of York Wills (1834) Will of Obadiah Willans of Leeds. Exchequer Court of York Wills, York Diocesan Archive, Borthwick Institute for Archives, University of York, GB 193, ECY Obadiah Willans of Leeds, Jan. 1834, vol.189, f.49.

Eysel W and Hahn T (1970) Polymorphism and solid solution of Ca_2GeO_4 and Ca_2SiO_4. *Zeitschrift für Kristallographie* **131**: 322–341.

Fletcher KE, Midgley HG and Moore AE (1965) Data on the binary system $3CaO.Al_2O_3$-$Na_2O.8CaO.3\ Al_2O_3$ within the system CaO-Al_2O_3-Na_2O`. *Magazine of Concrete Research* **17**: 171–176.

Fukuda K and Ando H (2002) Determination of the Pcmn/Ibm2 phase boundary at high temperatures in the system $Ca_2Fe_2O_5$-$Ca_2Al_2O_5$. *Journal of the American Ceramic Society* **85(5)**: 1300–1302.

Fukuda K and Maki I (1989a) Transitional phase of Ca_2SiO_4 solid solution with incommensurate superstructure. *Journal of the American Ceramic Society* **72(11)**: 2204–2207.

Fukuda K and Maki I (1989b) Orientation of β-Ca_2SiO_4 solid solution lamellae formed in the host α-phase. *Cement and Concrete Research* **19(6)**: 913–918.

Fukuda K and Maki I (1993) Orientation of coherent interphase boundaries formed by the α to α'_H phase transition in belite crystals. *Cement and Concrete Research* **23(3)**: 599–602.

Fukuda K, Maki I, Ito S, Yoshida H and Kato C (1994) Effect of Al/Fe ratio in belite on the microtexture induced by the remelting reaction. *Journal of the American Ceramic Society* **77(11)**: 3027–3029.

Gartner EM and Tang FJ (1987) Formation and properties of high sulfate Portland cement clinkers. *Il Cemento* **84**: 141–164.

Gazzozi G and Chiari G (1986) Crystallographic relations between the β and γ forms of $(Ca_2,Ca_{1.8}Sr_{0.2})SiO_4$. *Acta Crystallographica* **B42**: 146–151.

Geller S, Grant RW and Gonser U (1971) Crystal chemistry and magnetic structures of substituted $Ca_2[Fe](Fe)O_5$. *Progress in Solid State Chemistry* **5**: 1–26.

Ghose A (1980) PhD thesis, University of London, London, UK.

Ghose A and Barnes P (1979) Electron microprobe analysis of Portland cement clinkers. *Cement and Concrete Research* **9(6)**: 747–755.

Gobechiya ER, Yamnova NA, Zadov AE and Gazeev VM (2008) Calcio-olivine γ-Ca_2SiO_4: I. Rietveld refinement of the crystal structure. *Crystallography Reports* **53(3)**: 404–408.

Gollop RS and Taylor HFW (1994) Microstructural and microanalytical studies of sulfate attack. II. Sulfate-resisting Portland cement: Ferrite composition and hydration chemistry. *Cement and Concrete Research* **24(7)**: 1347–1358.

Golovastikov NI, Matveeva RG and Belov NV (1975) Crystal structure of the tricalcium silicate $3CaO{\cdot}SiO_2{=}C_3S$. *Soviet Physics. Crystallography* **20**: 441–445. (Translated from *Kristallografiya* 20, 721).

Grant RW, Geller S, Wiedersich H, Gonser U and Fullmer LD (1968) Spin orientation and magnetic properties of Ca_2FeAlO_5. *Journal of Applied Physics* **39(2)**: 1122–1123.

Gross S (1977) The mineralogy of the hatrurim formation, Israel. *Geological Survey of Israel* Bulletin no. **70**.

Groves GW (1982) Twinning in β-dicalcium silicate. *Cement and Concrete Research* **12(5)**: 619–624.

Guinier A and Regourd M (1969) Structure of Portland cement minerals. *Proceedings of the 5th International Symposium on the Chemistry of Cement* **1**: 1–32.

Hahn T, Eysel W, Woermann E *et al.* (1969) Crystal chemistry of tricalcium silicate solid solutions *Proceedings of the 5th International Symposium on the Chemistry of Cement* **1**: 61–66. Summarised from: Woermann E, Hahn T and Eysel (1963) Chemical and structural investigations of solid solutions of tricalcium silicate. I. *Zement Kalk Gips* **16**, 370–375; Woermann E, Eysel W and Hahn T (1967) Chemical and structural investigations of solid solutions of tricalcium silicate. II: Phase relations in the systems $CaO\text{-}MgO\text{-}SiO_2$ and $CaO\text{-}Al_2O_3\text{-}SiO_2$. *Zement Kalk Gips* **20**: 385–391; Woermann E, Eysel W and Hahn T (1968) Chemical and structural investigations of solid solutions of tricalcium silicate. III: Combined substitution of MgO and Al_2O_3 in Ca_3SiO_5. *Zement Kalk Gips* **21**: 241–251; Woermann E, Eysel W and Hahn T (1969) Chemical and structural investigations of solid solutions of tricalcium silicate. IV: Combined substitution of Fe_2O_3, Al_2O_3 and MgO in Ca_3SiO_5. *Zement Kalk Gips* **22**: 235; Woermann E, Eysel W and Hahn T (1969) Chemical and structural investigations of solid solutions of tricalcium silicate. V: The alite phase in the quinary system $CaO\text{-}MgO\text{-}Al_2O_3\text{-}Fe_2O_3\text{-}SiO_2$. *Zement Kalk Gips* **22**: 414.

Hall C and Scrivener KL (1998) Oilwell cement clinkers: X-ray microanalysis and phase composition. *Advanced Cement Based Materials* **7(1)**: 28–38.

Halstead PE (1961) The early history of Portland cement. *Transactions of the Newcomen Society* **34(1)**: 37–54.

Han KS, Gard JA and Glasser FP (1981) Compositions of stable and metastable C_3A solid solutions crystallized from simulated clinker melts. *Cement and Concrete Research* **11(1)**: 79–84.

Hansen WC, Brownmiller LT and Bogue RH (1928) Studies on the system calcium oxide-alumina-ferric oxide. *Journal of the American Chemical Society* **50(2)**: 396–406.

Harrisson AM (1993) *Proceedings of the 4th Euroseminar on Microscopy Applied to Building Materials* (Lindqvist JE and Nitz B (eds)), Swedish National Testing and Research Institute. Building and Technology SP Report 1993: 15. Boras, Sweden.

Harrisson AM, Taylor HFW and Winter NB (1985) Electron-optical analyses of the phases in a Portland cement clinker, with some observations on the calculation of quantitative phase composition. *Cement and Concrete Research* **15(5)**: 775–780.

Harrisson AM, Winter NB and Taylor HFW (1986) An examination of some pure and composite Portland cement pastes using scanning electron microscopy with X-ray analytical capability. *Proceedings of the 8th International Symposium on the Chemistry of Cement* **4**: 170–175.

Historic England (1992) The Wakefield Arms, Historic England List Entry 1,245,773. https://historicengland.org.uk/listing/the-list/list-entry/1245773

Hornain H (1973) Electron-probe microanalysis of the ferrite phase in clinker. *Rev. Mater. Constr* **680**: 4–13.

Hubbard CR, Evans EH and Smith DK (1976) The reference intensity ratio, I/I_c, for computer simulated powder patterns. *Journal of Applied Crystallography* **9**: 169–174.

Hudson KE and Groves GW (1982) The structure of alite in Portland cement clinker – TEM evidence. *Cement and Concrete Research* **12(1)**: 61–68.

Ichikawa M and Komukai Y (1993) Effect of burning conditions and minor components on the color of Portland cement clinker. *Cement and Concrete Research* **23(4)**: 933–938.

Ichikawa M, Ikeda S and Komukai Y (1994) Effect of cooling rate and Na_2O content on the character of the interstitial materials in Portland cement clinker. *Cement and Concrete Research* **24(6)**: 1092–1096.

Il'inets AM and Bikbau MY (1990) Structural mechanism of polymorphic transitions of dicalcium silicate, Ca_2SiO_4. Part II: Refinement of crystal structure of high-temperature α'_L modification of dicalcium silicate Ca_2SiO_4. *Soviet Physics – Crystallography* (English Translation) **35**: 54–56.

Il'inets AM, Nevskii NN, Ilyukhin VV, Bikbau MI and Belov NV (1982) Crystal structure of alpha'-modification of dicalcium silicate, $Ca(Ca,Sr)SiO_4$. *Doklady Akademii Nauk SSSR* **267**: 641–644.

Il'inets AM, Malinovskii YA and Nevskii NN (1985) Crystal structure of the rhombohedral modification of tricalcium silicate, Ca_3SiO_5. *Doklady Akademii Nauk SSSR* **281**: 332–336.

Ilyinets AM, Simonov VI and Bikbau MY (1986) New-data concerning atomic structures of Portland cement silicates. *Proceedings of the 8th International Congress on the Chemistry of Cement* **6**: 489–491.

Jeffery JW (1950) *Crystal Structure of Tricalcium Silicate and Cobalt Mercury Thiocyanate*. PhD thesis, University of London, London, UK.

Jeffery JW (1952) The crystal structure of tricalcium silicate. *Acta Crystallographica* **5(1)**: 26–35.

Jost KH, Ziemer B and Seydel R (1977) Redetermination of the structure of β-dicalcium silicate. *Acta Crystallographica Section B Structural Crystallography and Crystal Chemistry* **33(6)**: 1696–1700.

Jupe AC, Cockcroft JK, Barnes P, Colston SL, Sankar G and Hall C (2001) The site occupancy of Mg in the brownmillerite structure and its effect on hydration properties: an X-ray/neutron diffraction and EXAFS study. *Journal of Applied Crystallography* **34(1)**: 55–61.

Kahlenberg V and Fischer RX (2000) Crystal growth and cation distribution in doped dicalcium ferrite $Ca_2(Fe_{1-x}Me_x)_2O_5$ ($Me = Al^{3+}, Ga^{3+}$). *European Journal of Mineralogy* **12(1)**: 129–135.

Kahlenberg V, Fischer RX and Shaw CSJ (2000) Rietveld analysis of dicalcium aluminate ($Ca_2Al_2O_5$) – a new high pressure phase with the brownmillerite-type structure. *American Mineralogist* **85(7-8)**: 1061–1065.

Kilby T (1843) *Scenery in the Vicinity of Wakefield; with a Brief Historical Descriptive account.* Specifically, the lithograph entitled '*Wakefield from near Nelson's foundry*'. Wakefield, UK.

Kim YJ, Nettleship I and Kriven WM (1992) Phase transformations in dicalcium silicate: II, TEM studies of crystallography, microstructure, and mechanisms. *Journal of the American Ceramic Society* **75(9)**: 2407–2419.

Kondo R, Goto S and Fukuhara M (1978) Substitution of aluminum and iron in tetracalcium aluminoferrite by manganese and its effect on hydration. *Review of the 32nd General Meeting of the Cement Association of Japan*, 38–40.

Kristmann M (1978) Portland cement clinker mineralogical and chemical investigations: Part II electron microprobe analysis. *Cement and Concrete Research* **8(1)**: 93–102.

Lazic B, Krüger H, Kahlenberg V, Konzett J and Kaindl R (2008) Incommensurate structure of $Ca_2Al_2O_5$ at high temperatures – structure investigation and Raman spectroscopy. *Acta Crystallographica Section B Structural Science* **64(4)**: 417–425.

Le Saoût G, Kocaba V and Scrivener KL (2011) Application of the Rietveld method to the analysis of anhydrous cement. *Cement and Concrete Research* **41(2)**: 133–148.

Lee FC and Glasser FP (1979) Powder diffraction data for compounds in the series $Na_x(Ca_{3\pm x}Na_x)Al_2O_6$. *Journal of Applied Crystallography* **12(4)**: 407–410.

Lee FC, Banda HM and Glasser FP (1982) Substitution of Na, Fe and Si in tricalcium aluminate and the polymorphism of solid solutions. *Cement and Concrete Research* **12(2)**: 237–246.

Maki I (1973) Nature of the prismatic dark interstitial material in Portland cement clinker. *Cement and Concrete Research* **3(3)**: 295–313.

Maki I (1974) Morphology of the so-called prismatic phase in Portland cement clinker. *Cement and Concrete Research* **4(1)**: 87–97.

Maki I (1976a) Optical properties of the anisotropic C_3A. *Cement and Concrete Research* **6(2)**: 183–192.

Maki I (1976b) Optics and phase relationships of the optically anisotropic C_3A. *Cement and Concrete Research* **6(6)**: 797–802.

Maki I (1979) Phase constitution of the alite in Portland cement clinker from modern manufacturing processes. *Il Cemento* **76**: 167–176.

Maki I (1984) Growth instability of crystalline textures of alite cement clinker. *Il Cemento* **81**: 165–174.

Maki I (1986) Relationship of processing parameters to clinker properties; influence of minor components. *Proceedings of the 8th International Congress on the Chemistry of Cement.* **1**: 34–47.

Maki I (1994) Processing conditions of Portland cement clinker as viewed from the fine textures of the constituent minerals. *Ceramic Transactions* **40**: 3–17.

Maki I and Chromý S (1978a) Microscopic study on the polymorphism of Ca_3SiO_5. *Cement and Concrete Research* **8(4)**: 407–414.

Maki I and Chromý S (1978b) Characterisation of the alite phase in Portland cement clinker by microscopy. *Il Cemento* **75**: 247–252.

Maki I and Goto K (1982) Factors influencing the phase constitution of alite in Portland cement clinker. *Cement and Concrete Research* **12(3)**: 301–308.

Maki I and Kato K (1982) Phase identification of alite in Portland cement clinker. *Cement and Concrete Research* **12(1)**: 93–100.

Maki I, Fukuda K, Yoshida H and Kumaki J (1992) Effect of MgO and SO_3 on the impurity concentration in alite in Portland cement clinker. *Journal of the American Ceramic Society* **75(11)**: 3163–3165.

Maki I, Fukuda K, Oba E, Yoshida H, Mitsumatsu M and Ito S (1995) Anisotropic light absorption of the calcium aluminoferrite phase in Portland cement clinker. *Cement and Concrete Research* **25(4)**: 863–869.

Marinho MB and Glasser FP (1984) Polymorphism and phase changes in the ferrite phase of cements induced by titanium substitution. *Cement and Concrete Research* **14(3)**: 360–368.

McGinnety JA (1972) Redetermination of the structures of potassium sulphate and potassium chromate: the effect of electrostatic crystal forces upon observed bond lengths. *Acta Crystallographica Section B Structural Crystallography and Crystal Chemistry* **28(9)**: 2845–2852.

Midgley CM (1952) The crystal structure of β dicalcium silicate. *Acta Crystallographica* **5(3)**: 307–312.

Mitchell RH, Welch MD and Chakhmouradian AR (2017) Nomenclature of the perovskite supergroup: a hierarchical system of classification based on crystal structure and composition. *Mineralogical Magazine* **81(3)**: 411–461.

Miyazawa K and Tomita K (1969) On the color change of Portland cement. *Proceedings of the 5th International Symposium on the Chemistry of Cement* **1**: 252–261.

Mondal P and Jeffery JW (1975) The crystal structure of tricalcium aluminate. $Ca_3Al_2O_6$. *Acta Crystallographica* **B31**: 689–697.

Moir GK (1983) Improvements in the early strength properties of Portland cement. *Philosophical Transactions of the Royal Society A* **310**: 127–138.

Moore PB (1973) Bracelets and pinwheels: a topological-geometrical approach to the calcium orthosilicate and alkali sulfate structures. *American Mineralogist* **58**: 32–42.

Moore PB and Araki T (1976) The crystal structure of bredigite and the genealogy of some alkaline earth orthosilicates. *American Mineralogist* **61**: 74–87.

Moranville-Regourd M and Boikova AI (1992) Chemistry, structure, properties and quality of clinker. *Proceedings of the 9th International Congress on the Chemistry of Cement* **1**: 3–45.

Mori K, Kiyanagi R, Yonemura M *et al.* (2006) Charge states of Ca atoms in beta-dicalcium silicate. *Journal of Solid State Chemistry* **179(11)**: 3286–3294.

Morris MC, McMurdie HF, Evans EH, Paretzkin B and de Groot JH (1979) *Standard X-ray Diffraction Powder Patterns*. NBS monograph 25 – Section 16. US Department of Commerce, Washington, DC, USA.

Mumme WG (1995) Crystal-structure of tricalcium silicate from a Portland-cement clinker and its application to quantitative XRD analysis. *Neues Jahrbuch für Mineralogie, Monatshefte* **4**: 145–160.

Mumme WG, Hill RJ, Bushnell-Wye G and Segnit ER (1995) Rietveld crystal structure refinements, crystal chemistry and calculated powder diffraction data for the polymorphs of dicalcium silicate and related phases. *Neues Jahrbuch für Mineralogie – Abhandlungen* **169**: 35–68.

Mumme W, Cranswick L and Chakoumakos B (1996) Rietveld crystal structure refinements from high temperature neutron powder diffraction data for polymorphs of dicalcium silicate. *Neues Jahrbuch für Mineralogie – Abhandlungen* **170**: 171–188.

National Archives (1851) *RAIL 795/250/3: Part 3 of RAIL 795/250 – Agreement between L&Y and GNR to joint working of stations at Knottingley and Wakefield. With plans*. National Archives, Surrey, UK.

Neubauer J, Sieber R, Kuzel HJ and Ecker M (1996) Investigations on introducing Si and Mg into brownmillerite – a Rietveld refinement. *Cement and Concrete Research* **26(1)**: 77–82.

Newkirk TF and Thwaite RD (1958) Pseudoternary system calcium oxide-monocalcium aluminate ($CaO{\cdot}Al_2O_3$)–dicalcium ferrite ($2CaO{\cdot}Fe_2O_3$). *Journal of Research of the National Board of Standards* **61**: 233–245.

Nichols J, Ryley J, Gawthorp J and Whitley J (1818) *The Leeds correspondent: a literary, mathematical, and philosophical miscellany. Includes 'A brief memoir of the life of the late Mr. John Ryley, teacher of mathematics in Leeds'*. J. Nichols, Leeds, UK.

Niesel K and Thormann P (1967) The stability fields of dicalcium silicate modifications. *Tonind. Zeit* **91**: 362–369. (In German.)

Nishi F and Takéuchi Y (1975) The $A_{16}O_{18}$ rings of tetrahedra in the structure of $Ca_{8.5}NaAl_6O_{18}$. *Acta Crystallographica* **B31**: 1169–1173.

Nishi F and Takéuchi Y (1984) The rhombohedral structure of tricalcium silicate at 1200°C. *Zeitschrift für Kristallographie* **168(1-4)**: 197–212.

Nishi F, Takéuchi and Y, Iwao (1985) Tricalcium silicate $Ca_3O[SiO_4]$: the monoclinic superstructure. *Zeitschrift für Kristallographie* **172(3-4)**: 297–314.

O'Daniel H and Hellner E (1950) Zur Struktur von $3CaO.SiO_2$. *Neues Jahrbuch für Mineralogie. Monatshefte*: 108–111. (In German.)

Ono M and Nagashima N (1976) Effect of clinker cooling conditions on the strength of cement. *Proceedings of the 6th International Congress on the Chemistry of Cement* **1**: 170–173. (In Russian.)

Ono Y, Kawamura S and Soda Y (1969) Microscopic observations of alite and belite and hydraulic strength of cement. *Proceedings of the 5th International Congress on the Chemistry of Cement* **1**: 275–284.

Ordnance Survey (1850) Sheet 22. Ordnance Survey, UK.

Peterson VK, Hunter BA and Ray A (2004) Tricalcium silicate T_1 and T_2 polymorphic investigations: Rietveld refinement at various temperatures using synchrotron powder diffraction. *Journal of the American Ceramic Society* **87(9)**: 1625–1634.

Phan TL, Tran N, Kim DH *et al.* (2018) Electronic structure and magnetic properties of Al-doped $Ca_2Fe_2O_5$ brownmillerite compounds. *Journal of the American Ceramic Society* **101(5)**: 2181–2189.

Pollitt HWW and Brown AW (1969) The distribution of alkalis in Portland cement clinker. *Proceedings of the 5th International Congress on the Chemistry of Cement* **1**: 322–333.

Rankin GA and Wright FE (1915) The ternary system $CaO\text{-}Al_2O_3\text{-}SiO_2$, with optical study by FE Wright. *American Journal of Science* **s4, 39(229)**: 1–79.

Redhammer G, Tippelt G, Roth G and Amthauer G (2004) Structural variations in the brownmillerite series $Ca_2(Fe_{2-x}Al_x)O_5$: Single-crystal X-ray diffraction at 25°C and high-temperature X-ray powder diffraction (25°C $\leq T \leq$ 1000°C). *American Mineralogist* **89(2-3)**: 405–420.

Regourd M (1967) Polymorphic transformations of tricalcium silicate. *Rev. Mater. Constr* **620**: 167–176.

Regourd M and Guinier A (1976) The crystal chemistry of the constituents of Portland cement clinker. *Proceedings of the 6th International Congress on the Chemistry of Cement*, Principal Paper I–2.

Regourd M, Bigaré M, Forest J and Guinier A (1969) Synthesis and crystallographic investigation of some belites. *Proceedings of the 5th International Symposium on the Chemistry of Cement* **1**: 44–48.

Regourd M, Chromý S, Hjorth L, Mortureux B and Guinier A (1973) Polymorphisme des solutions solides du sodium dans l'aluminate tricalcique. *Journal of Applied Crystallography* **6(5)**: 355–364. (In French.)

Richardson IG and Zhu X (2022) The invention and nature of Joseph Aspdin's patent Portland cement. *Keynote Lecture at the 41st Cement and Concrete Science Conference*, 12 & 13 Sep, Maurice Keyworth Building, University of Leeds, Leeds, UK.

Richardson IG, Hall C and Groves GW (1993) TEM study of the composition of the interstitial phase in an oil-well cement clinker. *Advances in Cement Research* **5(17)**: 15–21.

Richardson IG, Zhu X, Dawson PL, Judkins P and Richardson MJ (2024) The invention and nature of Joseph Aspdin's patent Portland cement. *Keynote Lecture at the 2nd International Conference on Innovation in Low-Carbon Cement and Concrete Technology*, 8–10 Jul, University College London, London, UK.

Ryley J (1806) *The Leeds Guide: Including a Sketch of the Environs and Kirkstall Abbey.* E Baines/R Brown, Leeds, UK.

Sarkar SL and Roy DM (1984) Quantitative phase delineation of four American clinkers by SEM-EDS. *Proceedings of the 6th Annual International Conference on Cement Microscopy*, Albuquerque, NM, USA, pp. 37–46.

Scrivener KL and Taylor HFW (1995a) Clinker nodules with light-coloured centres. *ZKG International* **48**: 34–39.

Scrivener KL and Taylor HFW (1995b) Microstructural and microanalytical study of a strongly reduced Portland cement clinker. *World Cem. Res. Dev* **26(8)**: 34–43.

Sharma G, Kumar D, Tyagi S *et al.* (2018) Spin-lattice coupling and complex thermal expansion in Ca_2FeAlO_5. *Journal of Alloys and Compounds* **732**: 358–362.

Sinclair W and Groves GW (1984) Transmission electron microscopy and X-ray diffraction of doped tricalcium silicate. *Journal of the American Ceramic Society* **67(5)**: 325–330.

Skempton AW (1962) Portland cements, 1843–1887. *Transactions of the Newcomen Society* **35**: 117–152.

Skibsted J, Jakobsen HJ and Hall C (1994) Direct observation of aluminium guest ions in the silicate phases of cement minerals by ^{27}Al MAS NMR spectroscopy. *Journal of the Chemical Society, Faraday Transactions* **90(14)**: 2095–2098.

Smith DK (1962) Crystallographic changes with the substitution of aluminum for iron in dicalcium ferrite. *Acta Crystallographica* **15(11)**: 1146–1152.

Smith DK, Majumdar A and Ordway F (1965) The crystal structure of γ-dicalcium silicate. *Acta Crystallographica* **18(4)**: 787–795.

Stanley CC (1979) *Highlights in the History of Concrete*. Cement and Concrete Association, Slough, UK.

Stanley CC and Bond G (1999) *Concrete Through the Ages*. British Cement Association, Crowthorne, UK.

Statistical Committee of the Town Council (1840) Report upon the condition of the town of Leeds and of its inhabitants. *Journal of the Statistical Society of London* **2**: 397–424 (abridged version of a report written in October 1839).

Steinour HH (1960) Who invented Portland cement? *Journal of the Research and Development Laboratories – Portland Cement Association* **2**: 4–10.

Tait AF (1845) *Views on the Manchester and Leeds Railway, Drawn from Nature and on Stone*. AF Tait, Liverpool, UK.

Takéuchi Y, Nishi F and Maki I (1980) Crystal-chemical characterization of the $3CaO{\cdot}Al_2O_3$–Na_2O solid-solution series. *Zeitschrift für Kristallographie* **152(3-4)**: 259–307.

Terrier P (1969) Contribution of analysis by means of an electron microprobe to the cement chemistry. *Proceedings of the 5th International Symposium on the Chemistry of Cement* **2**: 278–284.

Terrier P, Hornain H and Socroun G (1968) Applications of electronic microprobe analysis to the study of clinker minerals. *Revue des materiaux de construction et de Travaux Publics* **630**: 109–115. (In French.)

Timashev VV (1980) The kinetics of clinker formation: the structure and composition of clinker and its phases. *Proceedings of the 7th International Congress on the Chemistry of Cement 1, I-3/1-I-3/20.*

Trout EAR (2017) The history of calcareous cements. In *Lea's Chemistry of Cement and Concrete,* 5th edn (Hewlett PC and Liska M (eds)), Elsevier, Amsterdam, the Netherlands, pp. 1–29.

Tsurumi T, Hirano Y, Kato H, Kamiya T and Daimon M (1994) Crystal structure and hydration of belite. *Ceramic Transactions* **40**: 19–25.

Uchikawa H, Ogawa K and Uchida S (1985) Influence of character of clinker on the early hydration process and rheological property of cement paste. *Cement and Concrete Research* **15(4)**: 561–572.

Uchikawa H, Hanehara S and Shirasaka T (1992) Relationship between zonal structure in alite, lamellar texture in belite, and their production process, composition and crystal structure. *Proceedings of the 9th International Congress on the Chemistry of Cement* **2**: 3–9.

Udagawa S and Urabe K (1978) Crystal structure of modifications of calcium silicate (Ca_2SiO_4) and their phase transformation. *Semento Gijutsu Nenpo* **32(38)**: 35–38. (In Japanese.) / The crystal structures of the modifications of dicalcium silicate and their phase transformations. *Review of 32nd General Meeting, Cement Association of Japan,* 31–33. (In English.)

Udagawa S, Urabe K and Yano T (1977) Crystal structure of calcium silicate (α-Ca_2SiO_4). *Semento gijutsu nenpo* **31(29)**: 23–25. (In Japanese.) / The crystal structure of α-dicalcium silicate. *Review of 31st General Meeting, Cement Association of Japan*, 23–25. (In English.)

Udagawa S, Urabe K and Yano T (1980a) Stabilization mechanism of polymorphisms of calcium silicate (Ca_2SiO_4). *Semento Gijutsu Nenpo* **34**: 37, 47–51. (In Japanese.) / *Review of the 34th General Meeting, Cement Association of Japan*, Tokyo (1980), p. 37.

Udagawa S, Urabe K, Natsume M and Yano T (1980b) Refinement of the crystal structure of γ-Ca_2SiO_4. *Cement and Concrete Research* **10(2)**: 139–144.

US Geological Survey (2019) Mineral commodity summaries 2019: *US Geological Survey*, https://doi.org/10.3133/70202434.

Vanpeteghem CB, Angel RJ, Zhao J, Ross NL, Redhammer GJ and Seifert F (2008) The effect of oxygen vacancies and aluminium substitution on the high-pressure properties of brownmillerite-structured $Ca_2Fe_{2-x}Al_xO_5$. *Physics and Chemistry of Minerals* **35(9)**: 493–504.

Widmer R, Gfeller F and Armbruster T (2015) Structural and crystal chemical investigation of intermediate phases in the system Ca_2SiO_4–$Ca_3(PO_4)_2$–$CaNaPO_4$. *Journal of the American Ceramic Society* **98(12)**: 3956–3965.

Woermann E, Hahn T and Eysel W (1965) Solid solution of Mg in $Ca_2Fe_2O_5$ and Ca_2FeAlO_5. *American Ceramic Society Bulletin* **44**: 299.

Woermann E, Eysel W and Hahn T (1969) Polymorphism and solid solution of the ferrite phase. *Proceedings of the 5th International Symposium on the Chemistry of Cement* **1**: 54–60.

Woermann E, Hahn T and Eysel W (1979) The substitution of alkalies in tricalcium silicate. *Cement and Concrete Research* **9(6)**: 701–711.

WYAS (West Yorkshire Archive Services) Calderdale WYC:1089/226. Articles of partnership of John Bower and Company, 28 Feb 1804. WYJS (West Yorkshire Joint Services), Morley, UK.

WYAS Leeds LO/HU/5(1823-27): Hunslet Township Overseers' Records, valuation 1823–1827. WYJS, Morley, UK.

WYAS Leeds WYL33/A/33: Billheads. WYJS, Morley, UK.

WYAS Leeds WYL160/14/113 & 114 Plan of The Old Glass House, Hunslet Moor. WYJS, Morley, UK.

WYAS Wakefield C509: collection of maps that includes Walker's Wakefield town plan, 1823. WYJS, Morley, UK.

WYAS Wakefield Deed 1792/DH/716/925 Wilson & others to Thomas Aspdin. WYJS, Morley, UK.

WYAS Wakefield Deed 1792/DH/717/927 Thomas Aspdin to Nixon. WYJS, Morley, UK.

WYAS Wakefield Deed 1794/DO/51/67 Thomas Aspdin to Nixon. WYJS, Morley, UK.

WYAS Wakefield Deed 1795-DR-548-575 Wilson to Pullan. WYJS, Morley, UK.

WYAS Wakefield Deed 1805/EX/82/117 Aspin to Heaton. WYJS, Morley, UK.

WYAS Wakefield Deed 1811/FO/58/77 Othick to Ryleys. WYJS, Morley, UK.

WYAS Wakefield Deed 1811/FP/52/56 Hawke to Townend. WYJS, Morley, UK.

WYAS Wakefield Deed 1811/FP/84/82 Strafford to Machin. WYJS, Morley, UK.

WYAS Wakefield Deed 1814/FX/447/518 Strafford to Moxon. WYJS, Morley, UK.

WYAS Wakefield Deed 1814/FX/487/560 Bischoff to Ryley. WYJS, Morley, UK.

WYAS Wakefield Deed 1814/GA/373/455 Hall to Ryley. WYJS, Morley, UK.

WYAS Wakefield Deed 1814/GC/318/355 Pullan to Bower. WYJS, Morley, UK.

WYAS Wakefield Deed 1816/GK/325/340 Strafford to Drake. WYJS, Morley, UK.

WYAS Wakefield Deed 1831/LD/364/256 Strafford to Charlesworth. WYJS, Morley, UK.

WYAS Wakefield JG001578/1837 plan: Aspdin (cement) box. WYJS, Morley, UK.

WYAS Wakefield QE20/1/1835/10: Plan of an intended railway to be called the Manchester and Leeds Railway from Manchester in the county palatine of Lancaster to Leeds in the West Riding of the county of York. WYJS, Morley, UK.

WYAS Wakefield QE21/49 Yorkshire Inquisition – Manchester and Leeds Railway Co. and Joseph Aspden [sic] and William Beverley, Wakefield. WYJS, Morley, UK.

Yamaguchi G and Takagi S (1969) The analysis of Portland cement clinker. *Proceedings of the 5th International Symposium on the Chemistry of Cement* **1**: 181–218.

Yamaguchi G, Ono Y, Kawamura S and Soda Y (1963a) Differential thermal analysis and high temperature powder X-ray diffraction of $2CaO{\cdot}SiO_2$. *Journal of the Ceramic Association, Japan* **71(805)**: 105–108.

Yamaguchi G, Ono Y, Kawamura S and Soda Y (1963b) Synthesis of the modification of Ca_2SiO_4 and the determination of their powder X-ray diffraction patterns. *Journal of the Ceramic Association, Japan* **71(806)**: 63–68.

Yamnova NA, Zubkova NV, Eremin NN, Zadov AE and Gazeev VM (2011) Crystal structure of larnite β-Ca_2SiO_4 and specific features of polymorphic transitions in dicalcium orthosilicate. *Crystallography Reports* **56(2)**: 210–220.

Yannaquis N and Guinier A (1959) La transition polymorphique β-γ de l'orthosilicate de calcium. *Bulletin de la Société Française de Minéralogie et de Cristallographie* **82(4)**: 126–136. (In French.)

Zadov AE, Gazeev VM, Pertsev NN *et al.* (2008) Discovery and investigation of a natural analog of calcio-olivine (γ-Ca_2SiO_4). *Doklady Earth Sciences* **423(2)**: 1431–1434.

Zadov AE, Gazeev VM, Pertsev NN, Gobechiya ER, Yamnova NA and Chukanov NV (2009) Calcioolivine, γ-Ca_2SiO_4, an old and new mineral species. *Geology of Ore Deposits* **51**: 1–9.

Zheng Y and Yang N (1986) Mössbauer study of $C_2A_xF_{(1-x)}$ solid solution. *Proceedings of the 8th International Congress on the Chemistry of Cement* **2**: 293–299.

Zheng Y and Yang N (1991) A comparative study of the ferrite phase in high-iron cements with the pure $C_2A_xF_{1-x}$ by Mössbauer spectroscopy. *Cement and Concrete Research* **21**: 31–37.

Zötzl M and Pöllmann H (2006) Stability and properties of brownmillerites $Ca_2(Al,Mn,Fe)_2O_5$ and perovskites $Ca(Mn,Fe)O_{3-x}$ in the system Ca_2. Fe_2O_5–'$Ca_2Mn_2O_5$'–'$Ca_2Al_2O_5$'. *Journal of the American Ceramic Society* **89**: 3491–3497.

Ian G Richardson and HFW Taylor
ISBN 978-1-83549-777-7
https://doi.org/10.1108/978-1-83549-776-020253002

Chapter 2
High-temperature chemistry

2.1. Introduction

A knowledge of the relevant high-temperature phase equilibria is necessary for understanding the factors that govern acceptable bulk compositions for Portland cement clinker, the conditions under which the latter can be manufactured and the phase composition and microstructure of the resulting material. This chapter deals with these equilibria and with the phases to which they relate, with the exception of the major clinker phases, which were described in Chapter 1. Some anhydrous phases primarily of interest in relation to other types of cements are also considered here. Principles underlying the preparation of anhydrous silicate, aluminate and other high-temperature phases are outlined.

2.2. Systems containing CaO with SiO_2 or Al_2O_3 or both

2.2.1 The CaO–SiO_2 system

Figure 2.1 shows the phase diagram. For clarity, the polymorphism of C_3S and the distinction between α'_H- and α'_L-C_2S are omitted. Calcium oxide (CaO) has the sodium chloride structure, in which all ions are octahedrally coordinated; the unit cell is cubic, space group $Fm\bar{3}m$ (No. 225), with $a = 4{\cdot}8091 \pm 0{\cdot}0051$ Å, $Z = 4$, $D_x = 3{\cdot}349\,\mathrm{g\,cm^{-3}}$ and molar volume $= 16{\cdot}75\,\mathrm{cm^3\,mol^{-1}}$. The value for the a parameter is the average of nine determinations $\pm$ 1 standard deviation (Ganguly *et al.*, 2000; Huang *et al.*, 1994; Jay and Andrews, 1945; Natta and Passerini, 1929; Primak *et al.*, 1948; Shen *et al.*, 2001; Smith and Leider, 1968; Swanson and Tatge, 1953; Verbraeken *et al.*, 2009). The refractive index is 1·837 (Winchell and Winchell, 1964).

On equilibrium cooling below 1250 °C, C_3S decomposes to give CaO and α'_H-C_2S; this process, which is always slow and below 700 °C imperceptible, is considered in §3.5.5. β-CS, α-CS and C_3S_2 are commonly known by their mineral names of wollastonite, pseudowollastonite and rankinite, respectively; none reacts significantly with water at ordinary temperatures. The crystal structures of β-CS (Ohashi, 1984) and α-CS (Yamanaka and Mori, 1981) differ markedly from each other and the two polymorphs are easily distinguishable by powder XRD. Both exhibit polytypism (β-CS (Henmi *et al.*, 1978, 1983); α-CS (Yamanaka and Mori, 1981)); the polytypes of a given polymorph are barely distinguishable by this method. The name 'parawollastonite' has been used for the 2 M polytype of β-CS. The β-α transition is reversible but slow in the α to β direction, rendering α-CS easily preservable by quenching. The structure of rankinite is known (Saburi *et al.*, 1976); a polymorph, kilchoanite, is known only as a natural mineral and as a product of hydrothermal reactions. Rankinite is monoclinic, space group $P\,2_1\,/\,a$ (No. 14, unique axis b, cell choice 3), with $a = 10{\cdot}557$ Å, $b = 8{\cdot}885$ Å, $c = 7{\cdot}858$ Å, $\beta = 119{\cdot}586°$, $Z = 4$, $D_x = 2{\cdot}989\,\mathrm{g\,cm^{-3}}$ and molar volume $= 96{\cdot}50\,\mathrm{cm^3\,mol^{-1}}$ (Saburi *et al.*, 1976). Some compositional, crystal structure and crystal chemical data for phases in the wollastonite and γ-C_2S (which includes kilchoanite) groups are given in Chapter 5 in Tables 5.1(a)/5.20(a) and 5.1(e)/5.20(e), respectively. The polymorphs of silica relevant to cement chemistry are briefly considered in §3.3.2.

Figure 2.1 The CaO–SiO$_2$ system. After Day *et al.* (1906), with later modifications (Gutt, 1968; Ol'shanskii, 1951; Rankin and Wright, 1915; Welch and Gutt, 1959)

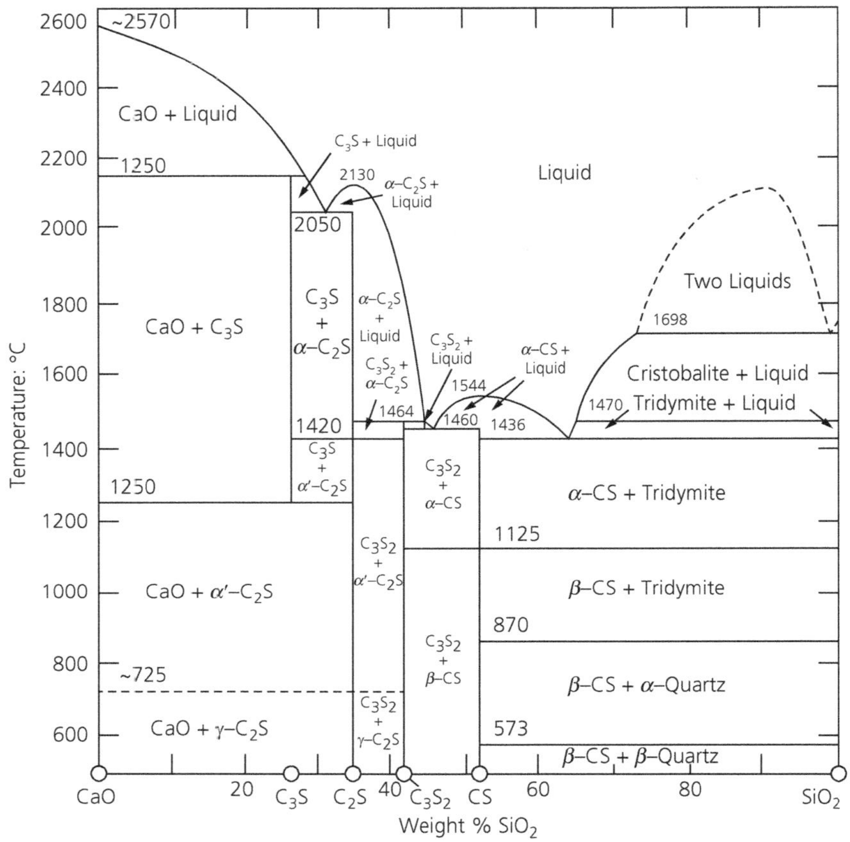

2.2.2 The CaO–Al$_2$O$_3$ system

There have been many phase equilibrium studies on this system, and differences of opinion probably still exist on several points. In part, these differences arise because variations in the humidity and oxygen content of the furnace atmosphere markedly affect the phase relations for compositions in the region of C$_{12}$A$_7$. The phase diagram in Figure 2.2 appears to provide the best description of the system modified by the presence of small amounts of water and oxygen, and thus represents the behaviour in air of ordinary humidity. With both CA and CA$_2$, opinions have differed as to whether melting is congruent or incongruent; Nurse *et al.* (1965a), who found it to be incongruent in both cases, discussed earlier work. Of the five calcium aluminate phases appearing in Figure 2.2, one, tricalcium aluminate (C$_3$A), is described in §1.4.

Monocalcium aluminate (CA), which is the main constituent of calcium aluminate cements, reacts rapidly with water. It is monoclinic and pseudohexagonal, with $a = 8{\cdot}700$ Å, $b = 8{\cdot}092$ Å, $c = 15{\cdot}191$ Å, $\beta = 90{\cdot}3°$, space group $P\,2_1\,/\,n$ (No. 14, unique axis b; cell choice 2), $Z = 12$,

Figure 2.2 The CaO–Al_2O_3 system modified by the presence of small amounts of H_2O and O_2, and thus representing the behaviour in air of ordinary humidity. After Rankin and Wright (1915), with later modifications (Chatterjee and Zhmoidin, 1972a; Nurse and Welch, 1962; Nurse *et al.*, 1965a)

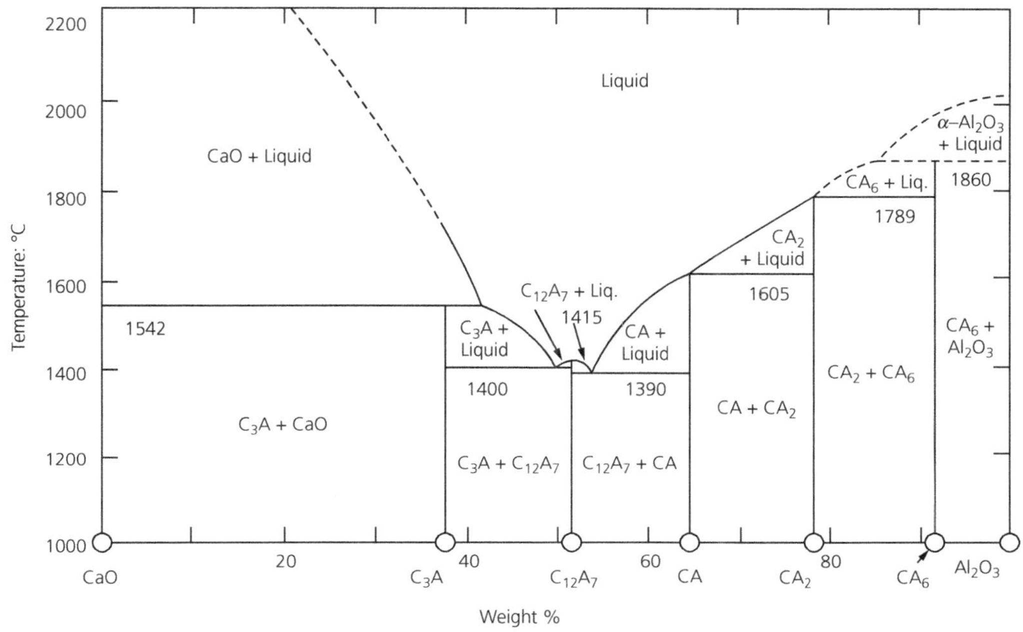

$D_x = 2{\cdot}945\,g\,cm^{-3}$, and has a stuffed tridymite structure, composed of Ca^{2+} ions and an infinite, three-dimensional framework of AlO_4 tetrahedra sharing corners (Hörkner and Müller-Buschbaum, 1976). The asymmetric unit has six Al, three Ca and 12 O sites. In analogy with the Q^n notation for denoting the connectivity of silicate tetrahedra (see §5.3.1.7 and §5.3.5), an aluminate tetrahedron, R, is designated as R^4 if connected by way of oxygen bridges to four other Al atoms; an isolated aluminate tetrahedron would be R^0 and so on (de Jong *et al.*, 1983). The six crystallographically distinct AlO_4 tetrahedra in CA are thus all R^4. Two of the Ca have a distorted octahedral environment ($Ca^{[6]}$, *d*(Ca–O) = 2·29 to 2·72 Å) and share a triangular face with one another. One of these Ca–O polyhedra shares an edge with one AlO_4 tetrahedron and corners with ten other tetrahedra, while the other one shares edges with two tetrahedra and corners with seven. The third Ca is coordinated to nine O atoms, (*d*(Ca–O) = 2·34 to 3·17 Å) and shares edges with all six AlO_4 tetrahedra (i.e. tetrahedra involving sites Al1 to Al6) and corners with a further six (again involving Al1 to Al6). All of the O atoms are coordinated to two Al and to one or two Ca atoms, with the average CN = 3·5. CA crystallises as irregular grains, sometimes prismatic and often twinned, with α = 1·643, β = 1·655, γ = 1·663, (–)2V = 36° (Rankin and Wright, 1915). Formation from $CaCO_3$ and $Al(NO_3)_3$ in the presence of organic reducing agents is reported to proceed through an amorphous material at 500 °C and an orthorhombic polymorph at 850 °C (Bachiorrini, 1985).

Calcium dialuminate (CA_2) occurs in some calcium aluminate cements. It reacts only slowly with water. The formula was at one time wrongly considered to be C_3A_5. CA_2 is monoclinic, with

$a = 12{\cdot}8398$ Å, $b = 8{\cdot}8624$ Å, $c = 5{\cdot}4311$ Å, $\beta = 106{\cdot}83°$, space group $C\ 2/c$ (No. 15), $Z = 4$, $D_x = 2{\cdot}920$ g cm^{-3}, and has a structure based on a framework of R^4 AlO_4 tetrahedra (Boyko and Wisnyl, 1958; Goodwin and Lindop, 1970). The asymmetric unit has just two Al, one Ca and four O sites. The Ca is coordinated to seven O atoms ($Ca^{[7]}$, d(Ca–O) = 2·33 to 2·88 Å, average 2·525 Å; polyhedral volume = 21·124 Å^3), and the Ca–O polyhedron shares edges with two AlO_4 tetrahedra and corners with six others. Twelve of the 28 oxygen atoms in the unit cell are shared between two tetrahedra and one Ca–O polyhedron, eight are shared between two tetrahedra and two Ca–O polyhedra and the remaining eight are shared between three tetrahedra (i.e. a so-called 'oxygen tricluster'; Neuville *et al.*, 2010), and so the average CN = 3·29. CA_2 crystallises as laths or rounded grains, with $\alpha = 1{\cdot}6178$, $\beta = 1{\cdot}6184$, $\gamma = 1{\cdot}6516$, (+)2V = 12° (Boyko and Wisnyl, 1958). CA_2 has been found as a natural mineral (Gross, 1977).

Calcium hexaluminate (CA_6) melts incongruently and does not react with water at ordinary temperatures. Its crystal structure is closely related to those of corundum and 'β-Al_2O_3'; the structure is hexagonal, space group $P\ 6_3\ /\ m\ m\ c$ (No. 194) with $a = 5{\cdot}5587$ Å, $c = 21{\cdot}8929$ Å, $Z = 2$ and $D_x = 3{\cdot}786$ g cm^{-3} (Kato and Saalfeld, 1968; Utsunomiya *et al.*, 1988). The asymmetric unit has five Al, one Ca and five O sites. The Ca is coordinated to 12 O atoms ($Ca^{[12]}$, with d(Ca–O) of 6 × 2·712 and 6 × 2·783 Å, average 2·747 Å; polyhedral volume = 48·534 Å^3). Three of the Al atoms are in octahedral coordination ($Al^{[6]}$), one is tetrahedral ($Al^{[4]}$) and the fifth is at a trigonal bipyramidal site ($Al^{[5]}$), although Utsunomiya *et al.* (1988) showed that the $Al^{[5]}$ probably occupies statistically one of two equivalent sites at room temperature that are 0·338 Å apart. The average CN for all Al atoms in the unit cell is 5·58. The Ca–O_{12} tetradecahedron shares faces with six Al–O_6 octahedra, edges with another six Al–O_6 octahedra and edges with three Al–O_5 bipyramids. Sixteen of the 38 oxygen atoms in the unit cell are shared between three Al–O_6 octahedra and one Al–O_4 tetrahedron, four are shared between three Al–O_6 octahedra and one Al–O_5 bipyramid, 12 are shared between three Al–O_6 octahedra and one Ca–O_{12} tetradecahedron and the remaining six are shared between two Al–O_6 octahedra, one Al–O_5 bipyramid and two Ca–O_{12} tetradecahedra; the average O CN = 4·16.

2.2.3 $C_{12}A_7$ and derived structures

In air of ordinary humidities, a phase of approximate composition $C_{12}A_7$ is readily formed. It reacts rapidly with water and occurs in some calcium aluminate cements. A related phase, $C_{11}A_7{\cdot}CaF_2$, is similarly reactive and occurs in certain special cements. In early work, $C_{12}A_7$ was wrongly assigned the formula C_5A_3 and called 'stable C_5A_3' to distinguish it from another phase of that composition, described below, which was called 'unstable C_5A_3'. $C_{12}A_7$ is cubic, with $a = 11{\cdot}988$ Å (Jeevaratnam *et al.*, 1964), space group $I\bar{4}3d$ (No. 220), $Z = 2$ and $D_x = 2{\cdot}673$ g cm^{-3} (Büssem, 1939; Jeevaratnam *et al.*, 1964). Studies on $C_{12}A_7$ or closely related phases (e.g. Bartl and Scheller, 1970; Boysen *et al.*, 2007; Büssem and Eitel, 1936; Feng *et al.*, 1988; Galuskin *et al.*, 2015a; Palacios *et al.*, 2008; Williams, 1968, 1973) show that the crystal structure of stoichiometric $C_{12}A_7$ consists of a framework of corner-sharing AlO_4 tetrahedra (four in seven are R^3; three in seven are R^4) of empirical composition $[Al_{14}O_{32}]^{22-}$ that includes six 'cages' that are each occupied by two Ca^{2+} ions that are coordinated to seven O atoms ($Ca^{[7]}$, average d(Ca–O) = 2·425 Å; polyhedral volume = 20·198 Å^3). The resulting 2+ excess positive charge is balanced by the partial occupancy of a site that is between the Ca sites and the centre of the cages by a 33rd oxygen, that is easily mobilised between cages (Boysen *et al.*, 2007, 2010); synthetic mayenite, $Ca_{12}Al_{14}O_{32}[\square_5O]$, is thus an anion conductor.

At temperatures above about 930 °C, a reversible equilibrium exists between $C_{12}A_7$ and water vapour in the furnace atmosphere (Jeevaratnam *et al.*, 1964; Nurse *et al.*, 1965b; Nurse and Welch,

1962; Roy and Roy, 1962). Material prepared in ordinary air and quenched from 1360 to 1390 °C is almost anhydrous; on being gradually reheated, it takes up water until at 950 °C a maximum content of about 1·3% is attained, corresponding to the formula $C_{12}A_7H$. On further increase in temperature, the water is lost again, until by the melting point in the region of 1400 °C the material is almost anhydrous. The sorbed water affects the cell parameter and refractive index. There was early disagreement as to the direction of the variations, but it appears established that $C_{12}A_7H$ has the smaller cell parameter and higher refractive index, Jeevaratnam *et al.* (1964) finding $a = 11{\cdot}983$ Å, $n = 1{\cdot}611$ and $D_x = 2{\cdot}680\,g\,cm^{-3}$ for $C_{12}A_7$ and $a = 11{\cdot}977$ Å, $n = 1{\cdot}620$ and $D_x = 2{\cdot}716\,g\,cm^{-3}$ for $C_{12}A_7H$.

Jeevaratnam *et al.* (1964) suggested that $C_{12}A_7$ can take up water because the 12 sites per formula unit that are occupied statistically by one O^{2-} ion can alternatively be occupied by two OH^- ions, $C_{12}A_7H$ thus having the constitution $Ca_6Al_7O_{16}(OH)$. In support of this conclusion, they prepared the halide analogues $C_{11}A_7{\cdot}CaF_2$ and $C_{11}A_7{\cdot}CaCl_2$ of the latter compound. $C_{11}A_7{\cdot}CaF_2$ was found to have a cell parameter of 11·964 Å, giving $D_x = 2{\cdot}732\,g\,cm^{-3}$. It forms a continuous range of solid solutions with $C_{12}A_7$ (Smirnov *et al.*, 1973). Bromide, iodide and sulfide analogues (Timashev, 1980; Zhmoidin and Chatterjee, 1984) and a chloride analogue with partial replacement of Al by Si, balanced by additional occupancy by Cl of the statistically available sites (Feng *et al.*, 1988), have also been described.

Imlach *et al.* (1971) found that they could prepare $C_{12}A_7$ reproducibly by slowly cooling melts in dry oxygen and that samples obtained in this way, or by cooling melts in moist air, contained excess oxygen, which could be detected chemically and was possibly present as peroxide ion. If similar melts were cooled in very dry nitrogen, mixtures consisting of CA, C_3A and, sometimes, C_5A_3 were obtained, with only poor yields of $C_{12}A_7$. These observations have since been confirmed and extended (Borlera and Brisi, 1984; Brisi and Borlera, 1983; Zhmoidin and Chatterjee, 1984, 1985).

Zhmoidin and Chatterjee (1984, 1985) concluded from density, viscosity and mass spectrometric evidence that melts of $C_{12}A_7$ composition contained regions of two kinds, differing in structure and density. The proportion of the less dense regions increased if the melt took up H_2O, O_2, CO, CO_2 or SO_3, or if CaF_2, $CaCl_2$, CaS, $CaSO_4$ or $CaCO_3$ was added; if this proportion was sufficiently high, $C_{12}A_7$ or a derivative was formed on freezing. The proportion of the denser regions in the melt was increased by maintaining a dry and slightly reducing atmosphere and by increase in temperature; if the less dense regions were thus eliminated, the melt yielded a homogeneous glass on quenching or C_5A_3 on slow cooling. The molecules that stabilised the less dense regions acted by surrounding themselves with a particular type of open structure, from which $C_{12}A_7$ readily nucleated; analogies were noted with the roles of hydrated alkali cations in the formation of zeolites and of water or organic molecules in that of compounds of the noble gases. With the exception of water, the molecules thus occluded in crystalline $C_{12}A_7$ could not be removed without destroying the structure. $C_{12}A_7$ melts are considerably denser (2·870–2·910 $g\,cm^{-3}$) than solid $C_{12}A_7$, especially if they have been kept in reducing atmospheres.

The derivatives of $C_{12}A_7$ melt in normal ways, $C_{11}A_7{\cdot}CaF_2$ congruently at 1577 °C and $C_{11}A_7{\cdot}CaS$ incongruently to give CaS and liquid at 1482 °C (Zhmoidin and Chatterjee, 1984). In contrast, $C_{12}A_7$ stabilised only by water shows anomalous behaviour, one study (Nurse *et al.*, 1965b) showing a sharp melting point of 1392 °C over a range of compositions, and another (Smirnov *et al.*, 1973; Zhmoidin and Chatterjee, 1984) melting over a range from 1380 to 1415 °C with evolution of gas and incipient formation of CA, C_3A and C_5A_3. Figure 2.2 is thus approximate in this region.

For practical purposes, the situation in air of ordinary humidities is more important than that in very dry and oxygen-free atmospheres, and the primary phase field of $C_{12}A_7$ will be included in the relevant ternary and quaternary phase diagrams presented below.

Re-examination of the holotype specimen of the natural mineral originally thought to be $C_{12}A_7$ or $C_{12}A_7H$ ($Ca_{12}Al_{14}O_{32}[\square_5 O]$ or $Ca_{12}Al_{14}O_{32}[\square_4(OH)_2]$) and named mayenite (Gross, 1977; Hentschel, 1964) has shown that its correct end-member formula is $Ca_{12}Al_{14}O_{32}[\square_4 Cl_2]$ (Galuskin, 2015a), which as a consequence has been renamed chlormayenite; the name mayenite is reserved for a potential mineral with the end-member composition $Ca_{12}Al_{14}O_{32}[\square_5 O]$ – that is, $C_{12}A_7$.

2.2.4 C_5A_3, C_2A and C_4A_3

The conditions of formation of C_5A_3 are discussed in the preceding section; it is probably an equilibrium phase in the strictly binary system $CaO–Al_2O_3$ but does not form in atmospheres of normal humidity and oxygen content. C_5A_3 is orthorhombic, with $a = 11{\cdot}253$ Å, $b = 10{\cdot}966$ Å, $c = 10{\cdot}290$ Å, space group $C\ m\ c\ 2_1$ (No. 36, setting **abc**), $Z = 4$, $D_x = 3{\cdot}067\,g\,cm^{-3}$ (Vincent and Jeffery, 1978). Its structure is related to that of gehlenite and consists of alternating twisted sheets of distorted $Al–O_4$ tetrahedra and layers of Ca atoms that are all coordinated to six O atoms ($Ca^{[6]}$, average d(Ca–O) $= 2{\cdot}418$ Å; polyhedral volume $= 15{\cdot}076$ Å^3). The tetrahedra form five-membered rings through sharing corners, with two thirds of the tetrahedra R^3 and one third R^4. The density is considerably higher than that of $C_{12}A_7$ and accords with the formation of the compound from the denser regions present in the melt. C_5A_3 reacts rapidly with water (Brisi *et al.*, 1986).

Two other anhydrous calcium aluminates are known. C_2A, a high-pressure phase, is described in §1.5.1. C_4A_3 is formed as the dehydration product of a hydrothermally produced phase, $C_4A_3H_3$. It is cubic with a structure similar to that of sodalite ($Na_4(Al_3Si_3O_{12})Cl$); the constitutional formula is $Ca_4(Al_6O_{12})O$, with $a = 8{\cdot}86$ Å, space group $I\bar{4}3m$ (No. 217), $Z = 2$, $D_x = 2{\cdot}532\,g\,cm^{-3}$ (Ponomarev *et al.*, 1971). $C_4A_3H_3$ is orthorhombic, with $a = 12{\cdot}4289$ Å, $b = 12{\cdot}8111$ Å, $c = 8{\cdot}8867$ Å, space group $A\ b\ m\ a$ (No. 64, setting **cab**), $Z = 4$, $D_x = 2{\cdot}742\,g\,cm^{-3}$ (Peters *et al.*, 2007)

2.2.5 The $CaO–Al_2O_3–SiO_2$ system

The phase diagram was originally determined by Rankin and Wright (1915). Figure 2.3 is based on that of Muan and Osborn (1965), with further amendments mainly following from studies on the bounding $CaO–SiO_2$ and $CaO–Al_2O_3$ systems discussed in the preceding sections. It relates to atmospheric pressure in an atmosphere of normal humidity, the primary phase field of $C_{12}A_7$ thus being shown. Some phases probably form solid solutions within the system, C_3S, for example, being able to accommodate some Al_2O_3; this is not shown.

Two ternary compounds exist stably in the system under these conditions. Gehlenite (C_2AS; $Ca_2Al_2SiO_7$) belongs to the melilite family. In its structure, layers of eight coordinated Ca^{2+} ions alternate with ones of composition $Al_2SiO_7^{4-}$, in which both aluminium and silicon are tetrahedrally coordinated (Louisnathan, 1971). The tetrahedra form five-membered rings through the sharing of corners, with two thirds of the tetrahedra R^3 and one third R^4; half of the R^3 are occupied by Si, and so the Al/Si ratio is 2. Gehlenite forms extensive solid solutions; the solid solution with åkermanite (C_2MS_2; $Ca_2MgSi_2O_7$) is continuous with no compositional breaks between the two end members, although the solution is nonideal (Gemmi *et al.*, 2007; Swainson *et al.*, 1992). In a modified form, it occurs in some calcium aluminate cement clinkers. Gehlenite and åkermanite are tetragonal, space group $P\bar{4}2_1m$ (No. 113), $Z = 2$; synthetic gehlenite has $a = 7{\cdot}677$ Å, $c = 5{\cdot}0594$ Å, $D_x = 3{\cdot}066\,g\,cm^{-3}$ (Kimata and Ii, 1982) and refractive indices $\omega = 1{\cdot}669$, $\varepsilon = 1{\cdot}658$ (Ferguson and Buddington, 1920); and åkermanite has $a = 7{\cdot}835$ Å, $c = 5{\cdot}010$ Å, $D_x = 2{\cdot}940\,g\,cm^{-3}$ (Kimata and

Figure 2.3 The $CaO–Al_2O_3–SiO_2$ system. After Muan and Osborn (1965), with later modifications

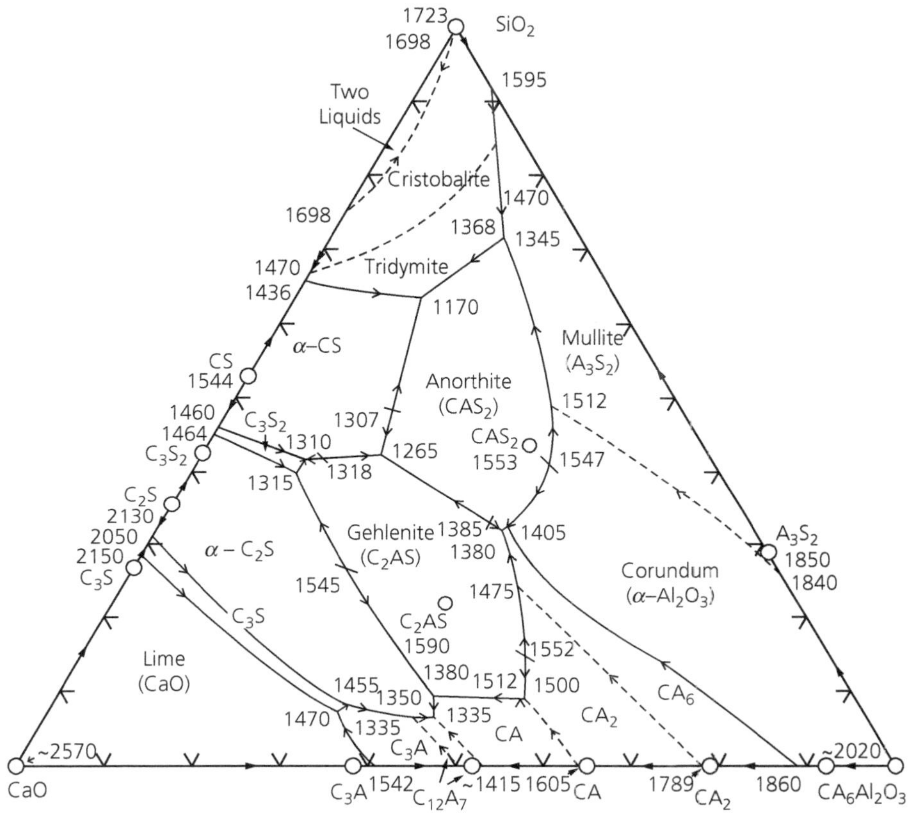

Ii, 1981) and refractive indices $\omega = 1{\cdot}632$, $\varepsilon = 1{\cdot}639$ (Ferguson and Buddington, 1920). The unsubstituted compound does not appear to react with water at ordinary temperatures, but the Al–rich variety present in calcium aluminate cements shows some reactivity at 40 °C (Scrivener and Taylor, 1990). Anorthite (a polymorph of CAS_2; $CaAl_2Si_2O_8$), which is less relevant to cement chemistry, is a triclinic feldspar, space group $P\bar{1}$ (No. 2), $Z = 8$, $a = 8{\cdot}173$ Å, $b = 12{\cdot}869$ Å, $c = 14{\cdot}165$ Å, $\alpha = 93{\cdot}113°$, $\beta = 115{\cdot}913°$, $\gamma = 91{\cdot}261°$, $D_x = 2{\cdot}766\,g\,cm^{-3}$ (Wainwright and Starkey, 1971).

Several other anhydrous calcium aluminosilicates are known, including grossular or garnet (C_3AS_3), which is a high-pressure phase, various dehydration products of zeolites, and various products formed metastably by crystallisation from melts or glasses. Most are too acid in composition to be of clear relevance to cement chemistry, but some of the devitrification products, especially those with compositions near CA and structures similar to those of the feldspathoids nepheline ($Na_3KAl_4Si_4O_{16}$) or kalsilite ($KAlSiO_4$) (Yoshioka, 1970) are of possible interest in relation to the formation of calcium aluminate cements. Nepheline and kalsilite are hexagonal, space group $P\,6_3$ (No. 173) and $Z = 2$, with $a = 9{\cdot}993$ Å, $c = 8{\cdot}374$ Å and $D_x = 2{\cdot}679\,g\,cm^{-3}$ for nepheline (Foreman and Peacor, 1970) and with $a = 5{\cdot}16$ Å, $c = 8{\cdot}69$ Å and $D_x = 2{\cdot}621\,g\,cm^{-3}$ for kalsilite (Perrotta and Smith, 1967).

2.2.6 Clinker formation in the $CaO–Al_2O_3–SiO_2$ system

The $CaO–Al_2O_3–SiO_2$ diagram provides a basis for a preliminary understanding of the chemistry underlying the formation of Portland cement clinker, in which all but the three most important oxide components are ignored. The approximations are least for white cements. In a cement kiln, the maximum temperature reached by the mix, called the clinkering temperature, is commonly 1400–1450 °C; at this temperature, the mix is partly molten. For a mix in the pure $CaO–Al_2O_3–SiO_2$ system, a somewhat higher temperature is required to give a comparable situation. In this discussion, it is assumed that equilibrium is attained at a clinkering temperature of 1500 °C and the $CaO–Al_2O_3–SiO_2$ diagram is used to predict which solid phases will be present for various bulk chemical compositions. The processes by which this equilibrium is approached in cement making, and those taking place during the subsequent cooling, are discussed in Chapter 3.

The compositions in the pure $CaO–Al_2O_3–SiO_2$ system that correspond most nearly to those of typical Portland cement clinkers lie within the ringed area surrounding point P in Figure 2.4. This point lies within the triangle whose apices are the compositions of C_3S, C_2S and point X, which lies at the intersection of the boundary of the C_3S and C_2S primary phase fields with the 1500 °C isotherm. For any bulk composition in this triangle, the equilibrium phases at 1500 °C will be C_3S, C_2S and liquid of composition X. This may be understood by considering what happens if a melt of composition P is cooled. C_3S first crystallises, the liquid composition thus moving away from that of C_3S until the boundary between the primary phase fields of C_3S and C_2S is reached. C_3S and C_2S then both crystallise, and the liquid composition moves along this boundary until the 1500 °C isotherm is reached. For the conditions that exist in cement making, the entire mix is never molten, so that the path to equilibrium differs, but the resulting phase assemblage must be the same. The crystals of C_3S and C_2S, being in contact with the melt, can grow to relatively large dimensions, which are typically around 30 μm. During the subsequent cooling the liquid solidifies, producing the more finely grained interstitial material.

Figure 2.4 Part of the $CaO–Al_2O_3–SiO_2$ system, illustrating the formation of Portland cement clinker; for significance of lettered points and of thin full or broken lines, see text

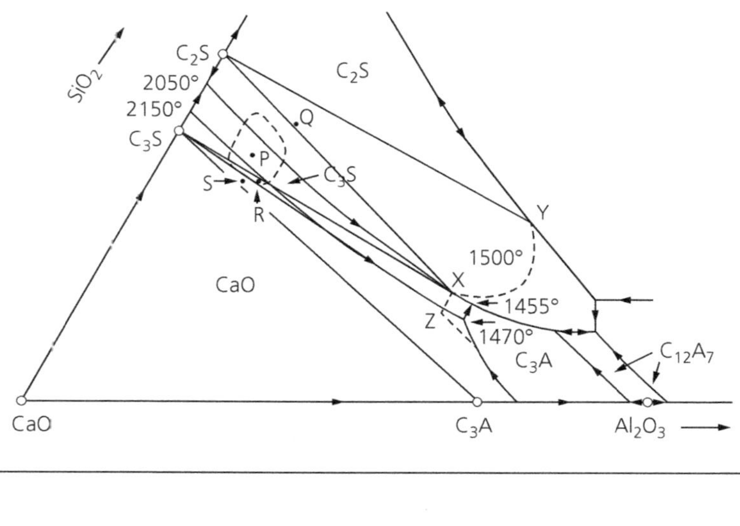

Some other bulk compositions may be considered in the same way. A composition such as Q, lying within the triangle C_2S–X–Y, will at equilibrium at 1500 °C give C_2S and a liquid having a composition on the curve XV. One such as R, in the triangle C_3S–X–Z, will give C_3S and a liquid having a composition on XZ. Any composition lying below the line C_3S–Z, such as S, will give an assemblage that includes calcium oxide as a solid phase.

In making Portland cement clinker, it is essential to avoid the presence of more than minimal amounts of free calcium oxide in the final product, and normally desirable to maximise that of C_3S. If equilibrium were continuously maintained during cooling, apart from decomposition of C_3S into C_2S and calcium oxide at subsolidus temperatures, any bulk composition lying within the triangle C_3S–C_2S–C_3A would yield a final product consisting of these three phases. For bulk compositions below the line C_3S–Z, this would require that the crystals of calcium oxide present at the clinkering temperature be redissolved. This process is slow, and one cannot assume that it will be substantially completed in the conditions existing in the kiln. The line C_3S–Z therefore sets a practical limit to the CaO content of the mix. For the temperature range in which clinkering is practicable, it virtually coincides with a line joining the C_3S composition to the invariant point at 1470 °C involving calcium oxide, C_3S, C_3A and liquid, and that line may therefore be used to define the upper limit of acceptable CaO contents. This approach, modified as discussed in §2.3.3 to allow for the presence of Fe_2O_3, leads to the definition of a quantity called the lime saturation factor (also see §3.1.2), which can be used in practice as an important parameter of the bulk chemical composition.

2.3. Systems containing Fe_2O_3

2.3.1 The $CaO–Al_2O_3–Fe_2O_3$ system

This system includes the $Ca_2(Al_xFe_{1-x})_2O_5$ series of ferrite compositions. The bounding, binary system $CaO–Fe_2O_3$ includes three compounds, namely C_2F, CF and CF_2. C_2F, as an end member of the above series, is discussed in §1.5. The other two compounds are of lesser importance to cement chemistry. Phillips and Muan (1959), in a study of the binary system, found that CF melted incongruently at 1216 °C to give C_2F and liquid, and that CF_2 decomposed at 1155 °C to give CF and hematite (α-Fe_2O_3). In this and other systems containing Fe_2O_3, iron-rich mixes tend to lose oxygen when heated in air above 1200–1300 °C, with consequent replacement of hematite by magnetite (Fe_3O_4). At lower partial pressures of oxygen, such as occur in a cement kiln, oxygen loss is significant even with compositions relatively low in iron.

Figure 2.5 shows part of the ternary system, $CaO–Al_2O_3–Fe_2O_3$. C_3A, $C_{12}A_7$ and CA can all accommodate some Fe^{3+}; for C_3A under equilibrium conditions at 1325 °C, the limit is about 4·5%, expressed as Fe_2O_3 (Majumdar, 1965). The ferrite phase in equilibrium with Fe-substituted C_3A can have compositions with x between 0·48 and 0·7 in the formula $Ca_2(Al_xFe_{1-x})_2O_5$; if CaO is also present, x is fixed at 0·48 – that is, the composition is close to C_4AF. Some reduction of Fe^{3+} to Fe^{2+} occurs when the ferrite phase is prepared from mixes with compositions in the $Ca_2(Al_xFe_{1-x})_2O_5$ series in air; it leads to the formation of minor amounts of other phases, which are not observed when similar experiments are carried out in oxygen (Majumdar, 1965).

Swayze (1946a) noted that equilibrium was often difficult to achieve in this system. One effect was the tendency for crystals of the ferrite phase to be zoned. For bulk compositions in the $Ca_2(Al_xFe_{1-x})_2O_5$ series, the liquid is of higher Al/Fe ratio than the ferrite phase with which it is in equilibrium; the crystals that are initially deposited on cooling such a liquid therefore have a lower Al/Fe ratio than the bulk composition of the mix. On further cooling, the Al/Fe ratio of the material

Figure 2.5 Part of the system $CaO–Al_2O_3–Fe_2O_3$. Hatched lines indicate solid solutions. Ferrite phase compositions are indicated by the quantity x, which relates to the formula $Ca_2(Al_xFe_{1-x})_2O_5$. After Newkirk and Thwaite (1958) and Majumdar (1965)

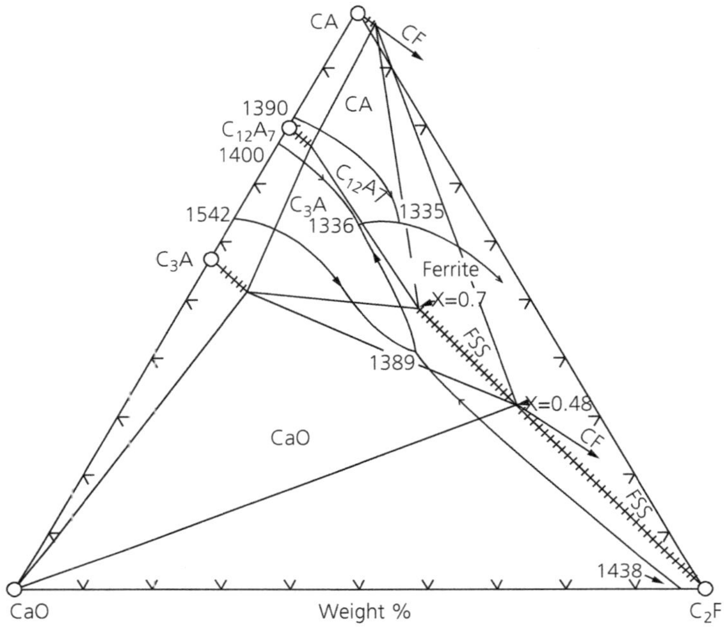

deposited progressively increases. Equilibrium within the ferrite crystals is difficult to attain, causing them to remain zoned, with cores richer in Fe^{3+} and outer regions richer in Al^{3+} than the mean composition.

A second effect was the tendency for protected phases to be formed. If a liquid having a composition somewhat on the CaO-rich side of the boundary between the CaO and C_3A primary phase fields (Figure 2.5) is cooled, calcium oxide is initially deposited and the liquid composition moves away from CaO and towards that boundary. When the latter is reached, and assuming that equilibrium were to be maintained, calcium oxide would redissolve, C_3A would be deposited, and the liquid composition would move along the boundary. In reality, the C_3A quickly surrounds the particles of calcium oxide, which thus becomes a protected phase, effectively removed from the system. This can markedly affect the composition of the ferrite phase which is subsequently formed on further cooling.

2.3.2 The $CaO–Al_2O_3–Fe_2O_3–SiO_2$ system

This system comprises the four major oxide components of Portland cement. Lea and Parker (1934, 1935) made a classic study of the subsystem $CaO–C_2S–C_{12}A_7–C_4AF$, which is the part directly relevant to cement production. Following Lea (1970), the formula $C_{12}A_7$ is substituted in the present description of this and related investigations for C_5A_3, which was used by the original authors in accordance with contemporary opinion on the composition of the phase now regarded as $C_{12}A_7$.

Any errors arising from this compositional difference will be small. Lea and Parker (1934, 1935) assumed the ferrite phase to have the fixed composition C_4AF.

Lea and Parker (1934, 1935) began by studying the bounding ternary subsystems CaO–C_2S–C_4AF and $C_{12}A_7$–C_2S–C_4AF. The first of these, which is the more important, is shown in Figure 2.6, modified in accordance with later work on the CaO–SiO_2 system. In general form, it resembles the lime-rich corner of the CaO–Al_2O_3–SiO_2 system (Figure 2.4), with C_4AF in place of C_3A, the primary phase field of C_3S thus being an elongated area extending away from the CaO–C_2S edge of the diagram.

The quaternary system may be represented on a tetrahedral model, each face of which represents one of the bounding, ternary systems. Within the tetrahedron, primary phase fields and subsolidus compatibility assemblages are represented by volumes, corresponding to the respective areas on the triangular diagram of a ternary system; correspondingly, isotherms are represented by surfaces. Figure 2.7 shows some important features of the CaO–C_2S–$C_{12}A_7$–C_4AF subsystem, as found by Lea and Parker (1934, 1935) and modified to take account of later work (Gutt, 1968; Majumdar, 1965; Newkirk and Thwaite, 1958; Sorrentino and Glasser, 1976; Swayze, 1946a; Welch and Gutt, 1959).

No quaternary compounds exist in this subsystem. The tetrahedron can therefore be divided into three smaller ones representing subsolidus compatibility assemblages for CaO–C_3S–C_3A–C_4AF, C_3S–C_3A–C_2S–C_4AF and $C_{12}A_7$–C_3A–C_2S–C_4AF. For clarity, these are not shown on Figure 2.7. The most important feature of the subsystem is the primary phase volume of C_3S, which is a thin sliver, roughly parallel to the C_2S–$C_{12}A_7$–C_4AF face of the tetrahedron and sandwiched between the larger phase volumes of CaO and C_2S. In accordance with the bounding subsystems

Figure 2.6 The system CaO–C_2S–C_4AF. After Lea and Parker (1934), with later modifications

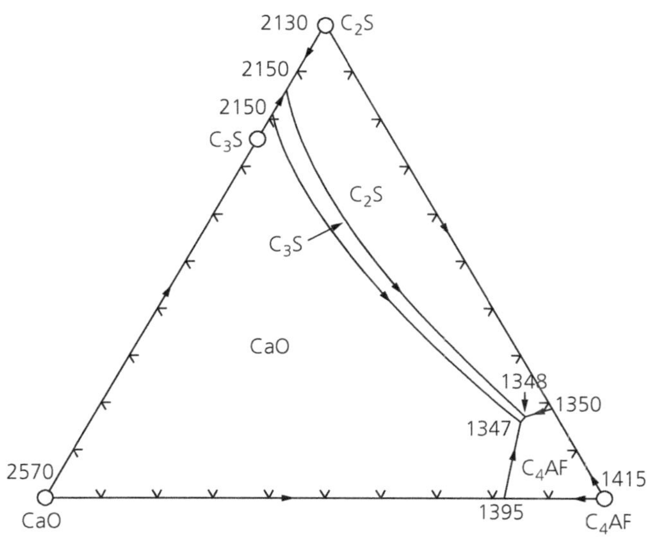

Figure 2.7 The pseudosystem $CaO–C_2S–C_{12}A_7–C_4AF$, showing the primary phase volume of C_3S. For details of invariant points P1–P8, see Table 2.1. After Lea and Parker (1934), with later modifications

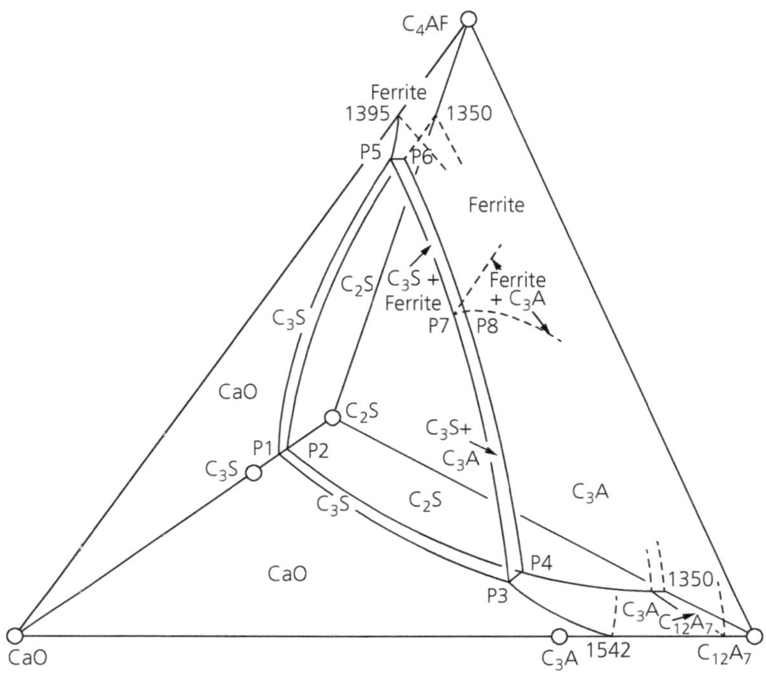

(Figures 2.4 and 2.6), two of its edges lie on the $CaO–C_2S–C_4AF$ and $CaO–C_2S–C_{12}A_7$ faces of the tetrahedron, but it cannot extend to the $CaO–C_4AF–C_{12}A_7$ face. Table 2.1 gives details of the eight invariant points involving C_3S in this subsystem.

Subsequent studies on the $CaO–Al_2O_3–Fe_2O_3$ (Majumdar, 1965; Newkirk and Thwaite, 1958) and $CaO–Al_2O_3–Fe_2O_3–SiO_2$ (Sorrentino and Glasser, 1976; Swayze, 1946a) systems have extended knowledge of the latter to a wider range of compositions, but have only slightly affected conclusions regarding the phase volume of C_3S. All indicate that the phase volume of C_3A is larger, and that of the ferrite phase smaller, than was shown in Lea and Parker's (1934) diagrams, which in this region partly rested on early results for the $CaO–C_{12}A_7–C_4AF$ system. To simplify the diagram, the volumes of phases other than C_3S are indicated only in general terms on Figure 2.7.

2.3.3 Clinker formation in the $CaO–Al_2O_3–Fe_2O_3–SiO_2$ system

Lea and Parker (1935) derived a formula giving the maximum acceptable content of CaO in a Portland cement clinker composition. If equilibrium was maintained during cooling, apart from decomposition of C_3S into C_2S and CaO at subsolidus temperatures, any bulk composition within the tetrahedron $C_3S–C_2S–C_3A–C_4AF$ would yield a clinker consisting of those four phases. However, as in the $CaO–Al_2O_3–SiO_2$ system, it cannot be assumed that any calcium oxide present at equilibrium at the clinkering temperature will subsequently be reabsorbed. Following reasoning similar to that in §2.2.6, calcium oxide will not be present at equilibrium at the clinkering temperature of approximately 1450 °C if the bulk composition lies on the lime-poor side of a plane

Table 2.1 Invariant points involving C_3S in the $CaO–Al_2O_3–Fe_2O_3–SiO_2$ system and in the same modified by 5 % MgO

Point	Solid phases in addition to C_3S	Liquid composition (mass per cent)					x in ferrite phase[a]	Type of point[b]	Temp.: °C	Reference
		CaO	Al_2O_3	Fe_2O_3	SiO_2	MgO				
P1	CaO	71·5	-	-	28·5	-	-	P	2150	Gutt (1968)
P2	C_2S	69·5	-	-	30·5	-	-	E	2050	Gutt (1968)
P3	CaO + C_3A	59·7	32·8	-	7·5	-	-	P	1470	Rankin and Wright (1915)
P4	C_2S + C_3A	58·5	32·9	-	8·6	-	-	P	1455	Rankin and Wright (1915)
P5	CaO + ferrite	52·8	16·2	25·4	5·6	-	0·50	E	1347	Lea and Parker (1934)
P6	C_2S + ferrite	52·4	16·3	25·5	5·8	-	0·50	P	1348	Lea and Parker (1934)
P7	CaO + C_3A + ferrite	55·0	22·7	16·5	5·8	-	0·50	P	1341	Lea and Parker (1934)
P7	CaO + C_3A + ferrite	53·9	21·2	19·1	5·8	-	0·44	P	1342	Swayze (1946a, 1946b)
P7	CaO + C_3A + ferrite + MgO	50·9	22·7	15·8	5·6	<5·0	0·47	P	1305	Swayze (1946a, 1946b)
P8	C_2S + C_3A + ferrite	54·8	22·7	16·5	6·0	-	0·50	?	1338	Lea and Parker (1934)
P8	C_2S + C_3A + ferrite	53·5	22·3	18·2	6·0	-	0·57	P	1338	Swayze (1946a, 1946b)
P8	C_2S + C_3A + ferrite + MgO	50·5	23·9	14·7	5·9	<5·0	0·67	P	1301	Swayze (1946a, 1946b)

[a] x in formula $Ca_2(Al_xFe_{1-x})_2O_5$
[b] E = eutectic; P = peritectic.

defined by the compositions of C_3S, C_4AF and the invariant point P3 (Figure 2.7 and Table 2.1). The oxide ratios by mass for these three compositions are given in Table 2.2. One may then set up three simultaneous equations of the type

$$(SiO_2/CaO)x + (Al_2O_3/CaO)y + (Fe_2O_3/CaO)z = 1{\cdot}0 \qquad (2.1)$$

where chemical formulae represent mass percentages of oxides in the clinker, and solve them to obtain an expression for the content of CaO at any point in the plane, namely

$$CaO = xSiO_2 + yAl_2O_3 + zFe_2O_3 \qquad (2.2)$$

This gives $x = 2{\cdot}80$, $y = 1{\cdot}18$, $z = 0{\cdot}65$.

Table 2.2 Oxide ratios by mass for points on the surface bounding the region in which calcium oxide is an equilibrium phase at 1450 °C

Point on surface	SiO_2/CaO	Al_2O_3/CaO	Fe_2O_3/CaO
C_3S	0·357	0	0
C_4AF	0	0·456	0·713
P3	0·126	0·549	0

The ratio $[CaO/(2{\cdot}80\ SiO_2 + 1{\cdot}18\ Al_2O_3 + 0{\cdot}65\ Fe_2O_3)]$ is called the lime saturation factor (LSF). A mix having an LSF greater than unity will yield free calcium oxide at the clinkering temperature; this phase is liable to persist in the final product, irrespective of the degree of mixing of the raw materials and the time during which the clinkering temperature is maintained. The calculation is approximate because of the neglect of minor oxide components and of ionic substitutions in the solid phases and for other reasons. For values below unity, the LSF provides a measure of the extent to which the maximum attainable content of C_3S is approached. Values of 0·92–0·98 are typical of modern clinkers (§3.1.2).

Lea and Parker (1935) considered that their calculations were valid only for Al_2O_3/Fe_2O_3 mass ratios in the clinker above that in C_4AF (0·64), but this does not appear to be correct (Welch, 1964). The plane within the CaO–C_2S–$C_{12}A_7$–C_4AF tetrahedron defined by the compositions of C_3S, C_4AF and the point P3 passes close to the composition of C_2A, and therefore also to all other compositions in the $Ca_2(Al_xFe_{1-x})_2O_5$ series. Swayze's (1946a) results, discussed in §2.4.2, show that the extension of the CaO–C_3S boundary surface to more iron-rich compositions remains close to this plane, which therefore serves also for Al_2O_3/Fe_2O_3 ratios below that in C_4AF.

2.4. Systems containing MgO or FeO

2.4.1 General

Portland cement raw materials contain small proportions of MgO; as noted in §1.1.3, they must be limited to avoid formation of more than a minor amount of periclase. The iron in Portland cement clinkers is normally present almost entirely as Fe^{3+}, but calcium aluminate cements may contain both Fe^{3+} and Fe^{2+}.

Table 2.3 lists some phases containing MgO that are in varying degrees relevant to cement chemistry. It is not a complete list of phases with essential MgO in the CaO–MgO–Al_2O_3– SiO_2 system. As seen in Chapter 1, some MgO is also taken up by all four of the major clinker phases. Magnesium oxide (periclase), like calcium oxide, has the sodium chloride structure; it is cubic, with $a = 4{\cdot}2135 \pm 0{\cdot}0059$ Å, space group $Fm\bar{3}m$ (No. 225), $Z = 4$, $D_x = 3{\cdot}581\ \mathrm{g\,cm^{-3}}$ and molar volume = $11{\cdot}26\ \mathrm{cm^3\,mol^{-1}}$. The value for the a parameter is the average value from 12 determinations ± 1 standard deviation (Boiocchi *et al.*, 2001; Hazen, 1976; Hirata *et al.*, 1977; Jay and Andrews, 1945; Karen *et al.*, 1999; Kaur *et al.*, 2020; Sasaki *et al.*, 1979; Schimmel *et al.*, 2005; Schmahl and Eikerling, 1968; Smith and Leider, 1968; Taylor, 1984; Tsirelson *et al.*, 1998). The refractive index is 1·7366 (Winchell and Winchell, 1964). FeO (wüstite) has the same structure, but usually contains some Fe^{3+}, balanced by vacancies. It forms extensive solid solutions with MgO, MnO and can also accommodate some CaO (Jay and Andrews, 1945).

Early studies of such systems as MgO–C_2S–$C_{12}A_7$ (Hansen, 1928), CaO–MgO–C_2S–$C_{12}A_7$ (McMurdie and Insley, 1936) and MgO–C_4AF (Insley and McMurdie, 1938) indicated that periclase is the only phase with essential MgO liable to be formed in Portland cement clinkers. A later study on the CaO–MgO–Al_2O_3–SiO_2 system suggested that bredigite and 'phase Q' could possibly occur in rapidly cooled clinkers (Biggar, 1971), but there is no convincing evidence that either of these phases is present in production clinkers.

Allen and Snow (1955) studied equilibria in the low-SiO_2 part of the CaO–'FeO'–SiO_2 system in the presence of metallic iron. The results are relevant to the formation of Portland cement clinkers under reducing conditions.

Table 2.3 Some phases containing Mg relevant to cement chemistry

Name	Formula	Structure type
Periclase	MgO	Sodium chloride
Forsterite	M_2S	Olivine
Monticellite	CMS	Olivine
Merwinite	C_3MS_2	Related to alkali sulfate structures[a]
Bredigite	C_7MS_4	Related to alkali sulfate structures[a]
Åkermanite	C_2MS_2	Melilite (C_2AS etc.)
Enstatite	MS	Pyroxene[b]
Diopside	CMS_2	Pyroxene[b]
Spinel	MA	Spinel
–	C_3A_2M	Related to C_2AS and C_5A_3[c]
–	C_7A_5M (?)	Related to C_2AS and C_5A_3[c]
'Phase Q'	$C_{20}A_{13}M_3S_3$	Related to C_2AS and C_5A_3[c]

[a]'Phase T', of approximate composition $C_{1{\cdot}7}M_{0{\cdot}3}S$ (Gutt, 1961; Sarkar and Jeffery, 1979; Schlaudt and Roy, 1966; Sharp *et al.*, 1960), is identical with bredigite (Biggar 1971; Moseley and Glasser 1981).
[b]Also, the MS polymorphs clinoenstatite and protoenstatite, which, too, are of pyroxene type.
[c]For references and descriptions, see §2.4.3. Phase Q is of variable composition; the formula given is a median.

2.4.2 Effect of MgO on equilibria in the $CaO–Al_2O_3–Fe_2O_3–SiO_2$ system

Swayze (1946a, 1946b) extended the work of Lea and Parker (1934, 1935) by considering the effects of variable ferrite composition and of the presence of MgO as an oxide component. In the work with MgO, a constant 5% of the latter was present in all the compositions examined. This was considered sufficient to saturate the liquid, as small amounts of periclase were detected in nearly all the products. Figure 2.8 shows the results for the system in the presence of MgO, modified to take account of later work on the subsystems. The primary phase is always periclase, and the system is represented by using a tetrahedron whose edges correspond to a total mass percentage of 95%. Volumes within this tetrahedron relate to the second phase to crystallise.

The most important effects, whether of taking into account variable ferrite composition or the presence of MgO, concern the invariant points for C_3S, C_3A, ferrite phase, liquid and either CaO or C_2S, data for which are given in Table 2.1, where they may be compared with those of Lea and Parker (1934, 1935). In the absence of MgO, the temperatures of the corresponding points agree well between the two studies, but the liquid compositions are slightly different. In the presence of MgO, the temperatures are each lowered by 37 °C and the liquid compositions substantially affected, the Al_2O_3 concentrations being increased and those of CaO and Fe_2O_3 decreased. Swayze (1946a, 1946b) found the ferrite phase in equilibrium with C_3A, C_3S, C_2S and liquid to have a composition slightly more iron-rich than C_4AF if MgO was absent and one close to C_6A_2F if MgO was present.

Figure 2.8 The pseudosystem $CaO–C_2S–C_{12}A_7–C_2F$ modified by the presence of 5% of MgO, showing the phase volume of C_3S and tie lines for the ferrite, the compositional range of which is represented by the hatched line. For details of invariant points P1–P8, see Table 2.1. After Swayze (1946b), with later modifications

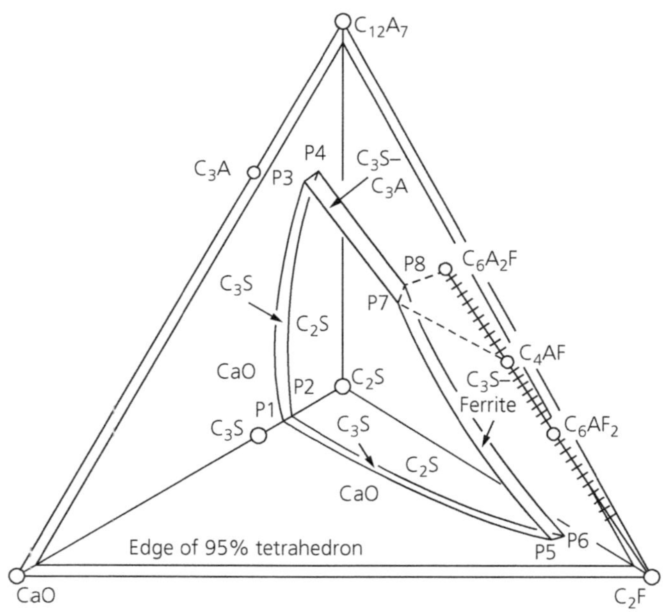

Swayze (1946a, 1946b) pointed out that these results had important consequences for the composition of the ferrite phase in Portland cement clinkers, whose compositions normally include significant amounts of MgO. He concluded that, for Al_2O_3/Fe_2O_3 ratios above 1·6, the ferrite phase composition would be C_6A_2F. Most ordinary Portland cement clinkers have Al_2O_3/Fe_2O_3 ratios in this range. For ratios below 0·84, assuming no free calcium oxide to be present, no C_3A would be formed and the Al_2O_3/Fe_2O_3 ratio in the ferrite phase would be that of the bulk composition. For intermediate values, lack of information about the tie lines for the ferrite phase made it impossible to predict the composition of the ferrite phase. Analyses of the ferrite phase in typical clinkers (§1.5.2) show that these conclusions are essentially correct, although the situation is complicated by the fact that the ferrite phase contains significant proportions of substituent oxide components, especially MgO, SiO_2 and TiO_2.

Swayze (1946a, 1946b) assumed that, for the bulk compositions that he studied, all the MgO was in the liquid phase at equilibrium, apart from the small proportions present as periclase. As noted in the preceding section, all four of the major clinker phases take up significant proportions of MgO. The contents of the latter component in the liquid phase were therefore probably substantially lower than 5%.

2.4.3 Phases structurally related to gehlenite

The phases C_3A_2M, C_7A_5M and 'Q' are structurally related to each other and to C_5A_3 and gehlenite (C_2AS); this is shown by unit cell determinations (Majumdar, 1964) and, for phase Q, by a structure determination (Hanic *et al.*, 1980), which also yielded a probable structure for C_3A_2M. In all the known structures mentioned above, aluminium, magnesium and, where present, silicon are tetrahedrally coordinated. Phase Q was originally assigned the formula C_6A_4MS (Parker, 1954); it has also been suggested that two distinct quaternary phases exist, with compositions $C_{21}A_{14}M_7S$ and $C_{22}A_{13}M_3S_4$ (Glasser and Marr, 1975). The structure determination (Hanic *et al.*, 1980) and an associated phase study (Kaprálik and Hanic, 1980) indicated that there is only one such phase, and that its composition is $Ca_{20}Al_{32-2n}Mg_nSi_nO_{68}$ with n variable between 2·5 and 3·5, a median formula thus being $C_{20}A_{13}M_3S_3$. The unit cell, for $n = 3{\cdot}0$, was found to be orthorhombic with $a = 27{\cdot}638$ Å, $b = 10{\cdot}799$ Å, $c = 5{\cdot}123$ Å, space group $Pmmn$, (No. 59) $Z = 1$ and $D_x = 2{\cdot}985\ \mathrm{g\,cm^{-3}}$.

A phase known as pleochroite, of fibrous or acicular morphology, occurs in some calcium aluminate cements and is phase Q modified by replacement of Mg^{2+} by Fe^{2+} and of Al^{3+} by Fe^{3+} (Glasser and Marr, 1975; Parker, 1954). Unpublished X-ray microanalyses by Conjeaud (1975) as cited by Glasser and Marr (1975) and Midgley (1979) appear compatible with the approximate composition $Ca_{20}Al_{22{\cdot}6}Fe^{3+}_{2{\cdot}4}Mg_{3{\cdot}2}Fe^{2+}_{0{\cdot}3}Si_{3{\cdot}5}O_{68}$. Sourie *et al.* (1994) reported further analyses, most of which are compatible with phase Q, in which the Al^{3+} is partly replaced by Fe^{3+} and the Mg^{2+} is completely replaced by Fe^{2+}, but with values of n ranging from below 1 to more than 4. There were strong compositional variations, both between and within samples. Electron diffraction showed no major abnormalities in the crystals low in silicon, but those high in silicon were ordered intergrowths at a unit-cell level of that phase with a melilite approximating to gehlenite. The results of another EPMA determination (Midgley, 1979) are of uncertain significance because they show improbably large deviations from charge balance.

Several phase equilibrium studies have been made on this part of the $CaO–MgO–Al_2O_3–SiO_2$ system (Biggar, 1971; Glasser and Marr, 1975; Kaprálik and Hanic, 1980; Parker, 1954). Phase Q has a primary phase volume in the system, but there are some uncertainties regarding the stable solid-phase assemblages that include it. C_3A_2M has a primary phase field in the CaO–MgO–Al_2O_3 and CaO–MgO–Al_2O_3–SiO_2 systems (Biggar, 1971; Hallstedt, 1995; Welch, 1961). C_7A_5M is a metastable phase, and the formula given may be only approximate.

2.5. Systems containing alkalis or SO_3 or both

2.5.1 Phases

Portland cement clinkers contain small amounts of alkalis and sulfates derived from the raw materials and fuel. Both alkalis and SO_3 can be present in the major clinker phases but tend to combine preferentially with each other to form alkali or potassium calcium sulfates, and it is necessary to consider these components together. In addition, silicate and aluminate phases containing sulfate can form either as intermediates or in undesirable deposits in cement making; a calcium aluminate sulfate is a major constituent of some expansive and other special cements.

The only phase containing essential alkali but not essential SO_3 known to occur in Portland cement clinkers in more than trace amounts is the orthorhombic aluminate phase described in §1.4. Trace amounts of alkali carbonates (Pollitt and Brown, 1969) or potassium aluminate (Fundal, 1986) have been reported to occur in some clinkers, and some other alkali phases are formed as intermediates or deposits.

Tables 2.4 and 2.5 list the sulfate phases and give crystal data and optical properties, respectively. Arcanite is the polymorph (β) of potassium sulfate stable at room temperature; it transforms reversibly to α-$K\bar{S}$ at 583 °C and melts at 1069 °C (Rowe *et al.*, 1965). Thénardite is the polymorph (V) of sodium sulfate stable in the presence of moisture at room temperature; on heating, it passes through a complex sequence of phase transitions and melts at 883 °C (Eysel, 1973; Rasmussen *et al.*, 1996). Other polymorphs can persist to room temperature. Anhydrite is the polymorph (β) of calcium sulfate stable at room temperature, formed on heating the hydrates or the metastable γ-$C\bar{S}$ ('soluble anhydrite'). A thermal effect at 1195 °C, generally attributed to polymorphic change to an α-$C\bar{S}$, may be associated with onset of ionic rotation (Florke, 1952). On heating in air, decomposition is detectable below 1000 °C (Kaprálik *et al.*, 1986) but slow below 1200 °C; in a closed container, anhydrite melts at 1462 °C (Rowe *et al.*, 1965).

Calcium langbeinite undergoes phase transitions at ≈200 °C and 940 °C and melts incongruently to give calcium sulfate and liquid at 1011 °C (Arceo and Glasser, 1990; Rowe *et al.*, 1965). Speer and Salje (1986) give the lower phase transition temperature as 184 °C, with the structure orthorhombic below that temperature and cubic above it; Speer and Salje (1986) solved both structures, brief details of which are given in Table 2.4. A langbeinite analogue with Ca > Mg has been discovered and named calciolangbeinite (IMA 2011–067; Pekov *et al.*, 2012). The *a* parameter in Pekov *et al.*'s (2012) structure is somewhat shorter than the value given by Speer and Salje (1986) – presumably a result of the substitution of the smaller Mg^{2+} ion for Ca^{2+} – but the structures are otherwise very similar.

Aphthitalite is essentially a solid solution phase of general composition $(K,Na)_2SO_4$. When it is formed stably at room temperature, the composition does not vary much from $K_3N\bar{S}_4$, but at higher temperatures compositions much higher in sodium are possible, up to a limit near $KN_3\bar{S}_4$, and the resulting solid solutions can be quenched to room temperature. Glaserite is a varietal name for aphthitalite of or near $K_3N\bar{S}_4$ composition.

Table 2.4 Crystal data for sulfate phases (Pollitt and Brown, 1969)

Formula	Name	Crystal system[a]	Unit cell parameters: Å			Space group (No.)	*Z*	Molar volume: $cm^3\ mol^{-1}$	D_x: $g\,cm^{-3}$	Reference
			a	*b*	*c*					
K_2SO_4	Arcanite	Or	7·476	10·071	5·763	*P n a m* (62, setting **abc**)	4	65·33	2·668	McGinnety (1972)
$KC_2\bar{S}_3$	Ca langbeinite (below 184 °C)	Or	10·334	10·501	10·186	$P\,2_1\,2_1\,2_1$ (19)	4	166·24	2·686	Speer and Salje (1986)
$KC_2\bar{S}_3$	Ca langbeinite (above 184 °C)	Cu	10·4289	10·4289	10·4289	$P\,2_1\,3$ (198)	4	170·77	2·615	Speer and Salje (1986)
$K_3N\bar{S}_4$	Aphthitalite[b]	Tr	5·680	-	7·309	$P\,\bar{3}\,m\,1$ (164)	½	123·02	2·702	Okada and Ossaka (1980)
Na_2SO_4	Thénardite	Or	5·8596	12·3044	9·8170	*F d d d* (70)	8	53·28	2·666	Nord (1973); Hawthorne and Ferguson (1975)
$CaSO_4$	Anhydrite	Or	7·006	6·998	6·245	*A m m a* (63, setting **cab**)	4	46·10	2·953	Kirfel and Will (1980)
$C_4A_3\bar{S}$	Ye'elimite	Or	13·028	13·037	9·161	*P c c* 2 (27, setting **abc**)	4	234·26	2·605	Calos *et al.* (1995)
$C_5S_2\bar{S}$	Ternesite	Or	6·863	15·387	10·181	*P n m a* (62, setting **abc**)	4	161·87	2·969	Irran *et al.* (1997)

[a]or = orthorhombic, Tr = trigonal ($\gamma = 120°$), Cu = cubic.
[b]Data are for composition given; cell parameters decrease with increasing Na/K ratio.

Table 2.5 Optical properties of sulfate phases

Formula	Name	Refractive indices			Character	2V	Reference
		α	β	γ			
K_2SO_4	Arcanite	1·4935	1·4947	1·4973	Biaxial +	67°	Winchell and Winchell (1964)
$KC_2\bar{S}_3$	Ca langbeinite	1·522	1·526	1·527	Biaxial −	Low	Swanson *et al.* (1969)
$K_3N\bar{S}_4$	Aphthitalite[a]	1·493	-	1·498	Uniaxial +	-	Winchell and Winchell (1964)
Na_2SO_4	Thénardite	1·471	1·477	1·484	Biaxial +	84°	Winchell and Winchell (1964)
$CaSO_4$	Anhydrite	1·5698	1·5754	1·6136	Biaxial +	43°	Winchell and Winchell (1964)
$C_4A_3\bar{S}$	Ye'elimite	-	1·569	-	Isotropic	-	Halstead and Moore (1962)
$C_5S_2\bar{S}$	Ternesite	1·632	1·638	1·640	Biaxial −	60°	Pryce (1972)

[a]Data are for composition given; refractive indices decrease with increasing Na/K ratio (Winchell and Winchell, 1964).

$C_4A_3\bar{S}[Ca_4(Al_6O_{12})(SO_4)]$ is readily formed on heating mixtures of appropriate composition in air at 1350 °C and has been studied extensively (e.g. Andaç and Glasser, 1994; Feng *et al.*, 1991; Halstead and Moore,1962; Hanic *et al.*, 1986; Ikeda *et al.*, 1996; Ragozina, 1957; Zhang *et al.*, 1992). It has been known as 'Klein's compound' and occurs naturally as the mineral ye'elimite. The crystal structure is a variant on that of sodalite $[Na_4(Al_3Si_3O_{12})Cl]$, composed of a three-dimensional framework of AlO_4 tetrahedra sharing corners, with Ca^{2+} and SO_4^{2-} ions in the cavities (Halstead and Moore, 1962). The unit cell is orthorhombic with $a = 13{\cdot}028$ Å, $b = 13{\cdot}037$ Å, $c = 9{\cdot}161$ Å, space group $P\,c\,c\,2$ (No. 27, setting **abc**), $Z = 4$, $D_x = 2{\cdot}605\,g\,cm^{-3}$ (Calos *et al.*, 1995). Cuesta *et al.*'s (2013) work supports the orthorhombic space group $P\,c\,c\,2$, with a $\sqrt{2}a \times \sqrt{2}a \times a$ superstructure based on the cubic sodalite structure. $C_4A_3\bar{S}$ melts at about 1600 °C and reacts readily with water (Halstead and Moore, 1962), forming the basis of calcium sulfoaluminate cements (CSA).

$C_5S_2\bar{S}[Ca_5(SiO_4)_2(SO_4)]$ has sometimes been called sulfospurrite. The name is misleading because the structure is unrelated to that of spurrite $[Ca_5(SiO_4)_2(CO_3)]$ and elemental prefixes conventionally denote similarity not only in formula type, but also in structure. $C_5S_2\bar{S}$ is isostructural with silicocarnotite, $Ca_5(PO_4)_2(SiO_4)$ (Brotherton *et al.*, 1974; Galuskin *et al.*, 2015b). It is obtained on heating mixtures of appropriate composition in air of ordinary humidity (Gutt and Smith, 1967; Pliego-Cuervo and Glasser, 1978), and in this environment it is stable up to 1298 °C (Pliego-Cuervo and Glasser, 1979a). Pliego-Cuervo and Glasser (1978) were unable to prepare it in the absence of water vapour and obtained evidence that it contained a small amount of essential hydroxyl ion. They considered that the atmosphere in a cement kiln was sufficiently moist to stabilise it in the appropriate temperature range, but that this might not hold true for the centres of clinker lumps. It occurs naturally as the mineral ternasite (Irran *et al.*, 1997).

2.5.2 Equilibria

Considering first systems of sulfates alone, phase equilibria have been reported for $K\bar{S}$–$N\bar{S}$ (Eysel, 1973), $K\bar{S}$–$C\bar{S}$ (Arceo and Glasser, 1990; Pliego-Cuervo and Glasser, 1977; Rowe *et al.*, 1965) and $K\bar{S}$–$N\bar{S}$–$C\bar{S}$ (Bellanca, 1942). Melting of $K\bar{S}$–$N\bar{S}$–$C\bar{S}$ mixes begins below 800 °C (Ritzmann, 1971). At high temperatures, $K\bar{S}$ is completely miscible with $N\bar{S}$ (Eysel, 1973) and accommodates up to 20-mole % of $C\bar{S}$ (Arceo and Glasser, 1990) in solid solutions having the α-$K\bar{S}$ structure. It can also dissolve some $CaCO_3$ (Arceo and Glasser, 1990). With increase in temperature above ambient, the compositional range of aphthitalite solid solution rapidly broadens, extending much further in the direction of $N\bar{S}$ and contracting slightly in that of $K\bar{S}$; the maximum Na/K ratio of about 3:1 is attainable at 195 °C (Eysel, 1973). At room temperature, all the phases other than metastable aphthitalites have essentially the fixed compositions shown in Tables 2.4 and 2.5.

A study of the CaO–$K\bar{S}$–$C\bar{S}$ system showed that the sulfate melts dissolved only a little CaO (Pliego-Cuervo and Glasser, 1977). Alkali sulfate melts also show only limited mutual miscibility with oxide melts similar in composition to the main clinker liquid (Pliego-Cuervo and Glasser, 1978, 1979b; Taylor, 1942; Timashev, 1980). For contents of alkali sulfates of 2–3 % in the clinker liquid as a whole, the sulfate liquid is dispersed in the oxide liquid as microregions about 100 nm in diameter (Timashev, 1980) but, with higher contents of alkali sulfates, the two liquids separate on a macroscopic scale (Glasser and Marr, 1980; Timashev, 1980). The partitioning of the components between sulfate melts and simulated clinker liquids at 1350 °C has been studied (Glasser and Marr, 1980). SO_3 is virtually insoluble in the oxide liquid; SiO_2, Al_2O_3 and Fe_2O_3 dissolve only to very minor extents in the sulfate liquid but CaO, Na_2O and K_2O are appreciably soluble in both. For five mixes studied, the ratio of K_2O to Na_2O was between 1·9 and 4·6 times higher in the sulfate than in the oxide liquid.

Gutt and Smith (1968) reviewed earlier work on the effects of SO_3 on the formation of Portland cement clinker and, to augment the available information, studied part of the CaO–Al_2O_3–Fe_2O_3–SiO_2–$CaSO_4$ system at 1400 °C. They concluded that, if Al_2O_3 was present and MgO absent, even small proportions of SO_3 could restrict or prevent formation of C_3S. The effect was lessened, but not eliminated, if MgO was present, and it was noted that other minor components, especially alkalis, might further modify the situation.

Pliego-Cuervo and Glasser (1979a) determined subsolidus phase assemblages in the CaO-rich part of the CaO–Al_2O_3–SiO_2–SO_3 system at 950–1150 °C. Four-phase and three-phase assemblages defining the equilibria between CaO, C_2S, C_3A, $C_{12}A_7$, CA, $C_5S_2\bar{S}$ and $C_4A_3\bar{S}$ were established. This work was extended (Pliego-Cuervo and Glasser, 1979b) to cover parts of the C–A–S–$K\bar{S}$, C–A–$K\bar{S}$–$C\bar{S}$, C–A–S–$K\bar{S}$–$C\bar{S}$ and C–A–F–$K\bar{S}$–$C\bar{S}$ systems, and the role of sulfur-containing species in the solid, liquid and vapour states in the formation of Portland cement clinker was discussed. Neither $C_5S_2\bar{S}$ nor $C_4A_3\bar{S}$ could occur in the final product, but both could play important parts in the kiln reactions.

Kaprálik *et al.* (Kaprálik and Hanic, 1986; Kaprálik *et al.*, 1986) studied equilibria in part of the C–A–S–F–M–$C\bar{S}$–$K\bar{S}$ system at temperatures up to 1300 °C with reference to the formation of clinkers designed to contain $C_4A_3\bar{S}$. Compatible phase assemblages were established. The authors noted that the subsystem C_2S–$C_4A_3\bar{S}$–$C\bar{S}$ included all the main compounds having hydraulic properties that were formed in clinkers of this type.

2.6. Systems with other components

2.6.1 Fluorides and fluorosilicates

Small amounts of F^- may occur in Portland cement clinkers, arising from raw materials or deliberate additions to lower the clinkering temperature. Two ternary phases are known to exist in the CaO–C_2S–CaF_2 system (Bereczky, 1964). One, now known to be of formula $2C_2S{\cdot}CaF_2$ or

$Ca_5(SiO_4)_2F_2$, has a structure analogous to that of chondrodite $[Mg_5(SiO_4)_2(OH,F)_2]$, with layers of γ-C_2S structure alternating with ones containing Ca^{2+} and F^- (Gutt and Osborne, 1966).

The second ternary phase has been described as $3C_3S{\cdot}CaF_2$ (Gutt and Osborne, 1968), $C_{11}S_4{\cdot}CaF_2$ (Tanaka *et al.*, 1969) and $C_{19}S_7{\cdot}2CaF_2$ (Gilioli *et al.*, 1979), but its probable formula is $Ca_{6-x/2}Si_2O_{10-x}F_x$ ($x \approx 1$); the structure is related to that of C_3S with partial replacement of O^{2-} by F^- balanced by omission of Ca^{2+} (Perez-Mendez *et al.*, 1984, 1985). The arrangement of Ca^{2+}, SiO_4^{4-} and O^{2-} or F^- ions differs from that in the C_3S polymorphs, and is the simplest possible in the broader family of structures to which the latter belong (Péréz-Mendez *et al.*, 1984). There is a strongly marked pseudocell, which (for $x = 1{\cdot}0$) is hexagonal with $a = 7{\cdot}099$ Å, $c = 5{\cdot}687$ Å, space group $P\ 6_3/m\ m\ c$ (No. 194) and atomic contents $\frac{1}{2}[C_{10}S_4 \cdot CaF_2]$, but the true cell is triclinic, with $a = 23{\cdot}839$ Å, $b = 7{\cdot}105$ Å, $c = 16{\cdot}755$ Å, $\alpha = 90{\cdot}0°$, $\beta = 117{\cdot}3°$, $\gamma = 98{\cdot}56°$ and atomic contents $5[C_{10}S_4 \cdot CaF_2]$ (Perez-Mendez *et al.*, 1984, 1985). The X-ray density (for $x = 1{\cdot}0$) is $2{\cdot}936\,g\,cm^{-3}$. For brevity, this phase will be called $C_{10}S_4{\cdot}CaF_2$. In the C_2S–CaF_2 system, the two end members form a eutectic at 1110 °C. $2C_2S{\cdot}CaF_2$ can be obtained by solid-state reaction at 950 °C and decomposes at 1040 °C to give C_2S and CaF_2 (Gutt and Osborne, 1968). Two phase equilibrium studies of the CaO–C_2S–CaF_2 system (Gutt and Osborne, 1970; Tanaka *et al.*, 1969) (Figure 2.9) are in good agreement. $C_{10}S_4{\cdot}CaF_2$ melts incongruently at 1170 °C to give C_3S and liquid. In the presence of CaF_2, C_3S can thus be obtained at this temperature through the successive formation and decomposition at lower temperatures of $2C_2S{\cdot}CaF_2$ and $C_{10}S_4{\cdot}CaF_2$ (Bereczky, 1964; Gutt, 1969; Tanaka *et al.*, 1969).

There has been some uncertainty as to the extent to which F^- enters the C_3S structure (as opposed to that of $C_{10}S_4{\cdot}CaF_2$) and also about the effectiveness of C_3S produced in the presence of CaF_2 as a cement. Welch and Gutt (1962) found that addition of CaF_2 to the starting materials caused the polymorph to change from triclinic to monoclinic, and obtained results suggesting that 0·74% of F^- entered the C_3S; CaF_2 accelerated the formation of C_3S but the compressive strengths of mortars made with the latter were reduced. Tanaka *et al.* (1969) found that C_3S formed from $C_{10}S_4{\cdot}CaF_2$ was trigonal (presumably rhombohedral), although it tended to invert to monoclinic on quenching.

Figure 2.9 The CaO–C_2S–CaF_2 system. After Gutt and Osborne (1970)

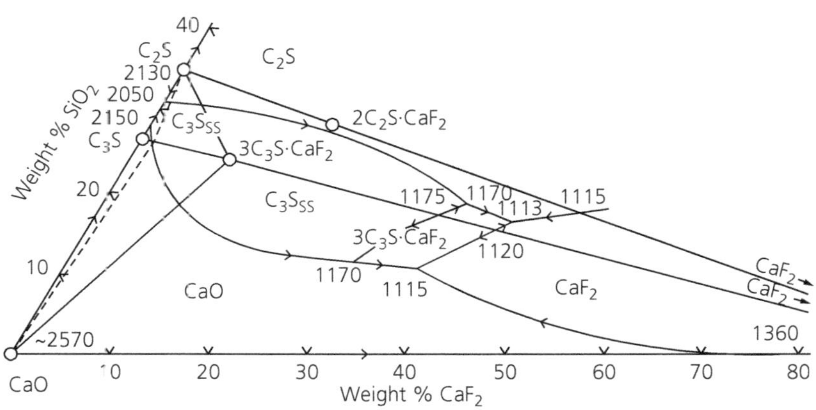

They found that $C_{10}S_4{\cdot}CaF_2$ gave poor strengths and suggested that the poor hydraulic activity of Welch and Gutt's (1962) preparations might have been due to the presence of this phase, formed from the C_3S during cooling. Maki *et al.* (1984) described an alite prepared from a melt containing F^-; it was monoclinic (Maki, 1979) and contained 0·9% F and 3·4% Al_2O_3.

In contrast to some of these results, Moir (1983) described a clinker containing 1·8% Al_2O_3, 2·3% SO_3 and 0·16% F, which gave high early strengths; the C_3S it contained was rhombohedral. Shame and Glasser (1987) found that little F^- could enter the C_3S structure in the absence of Al^{3+}, which permitted coupled substitution to occur, giving material of general formula $Ca_3(Al_xSi_{1-x})(O_{5-x}F_x)$. The upper limit of x was 0·15 (1·2% F). At $x = 0{\cdot}15$, the material was rhombohedral, even if cooled rapidly, and gave high 28-day strengths. In the presence of Al^{3+} and F^-, C_3S could be made by solid-state reaction at temperatures as low as 1025 °C, although reaction was slow below 1050 °C. At higher temperatures, F^- not lost by volatilisation was mainly present in the melt, and thus acted as a flux rather than as a mineraliser.

Another calcium silicate fluoride, cuspidine [$Ca_4(Si_2O_7)F_2$] exists but is not formed under conditions relevant to clinker formation; compositional, crystal structure and crystal chemical data for cuspidine are given in Tables 5.1(f) and 5.20(f). Systems containing fluorosilicate ions [SiF_6^{2-}] and the possible use of calcium fluorosilicate as a flux have been investigated (Smart and Roy, 1979).

In the $CaO–Al_2O_3–CaF_2$ system, the compound $C_{11}A_7{\cdot}CaF_2$ is described in §2.2.3. One other ternary phase, $3CA{\cdot}CaF_2$, is known; it is trigonal, with $a = 17{\cdot}324$ Å, $c = 7{\cdot}000$ Å, space group $R\bar{3}$ (No. 148, hexagonal axes), $Z = 6$, $D_x = 3{\cdot}024\,g\,cm^{-3}$ (Xia *et al.*, 2014) and refractive indices $\omega = 1{\cdot}628$, $\varepsilon = 1{\cdot}618$ (Leary, 1962). Several studies on parts of the system have been reported (Chatterjee and Zhmoidin, 1972a, 1972b; Gutt *et al.*, 1970; Zhmoidin and Chatterjee, 1974).

2.6.2 Carbonates

Spurrite [$Ca_5(SiO_4)_2(CO_3)$] is readily formed under suitably low partial pressures of CO_2; using oxalic acid as a source of CO_2, it has been obtained at 430 °C (Henmi and Henmi, 1978). At sufficiently low CO_2 pressures or sufficiently high temperatures, it decomposes to give C_2S and CaO. Some other syntheses, and conditions under which it is formed or decomposed in the cement kiln, are mentioned in §3.3.3 and §3.3.5. A pressure–temperature equilbrium curve for the reaction of wollastonite and calcite to give spurrite and CO_2 has been reported (Harker and Tuttle, 1956). The crystal structure is known, and the unit cell is monoclinic, with $a = 10{\cdot}484$ Å, $b = 6{\cdot}712$ Å, $c = 14{\cdot}156$ Å, $\beta = 101{\cdot}27°$; space group $P\,2_1/a$ (No. 14, unique axis *b*, cell choice 3), $Z = 4$, $D_x = 3{\cdot}023\,g\,cm^{-3}$ (Grice, 2005; Smith *et al.*, 1960). The refractive indices are: $\alpha = 1{\cdot}638$, $\beta = 1{\cdot}671$, $\gamma = 1{\cdot}676$; $(-)2V = 40°$ (Harker and Tuttle, 1956). Another calcium silicate carbonate, tilleyite [$Ca_5(Si_2O_7)(CO_3)_2$], occurs as a natural mineral and has been synthesised (Henmi and Henmi, 1978), but is not known to form during the manufacture of clinker; compositional, crystal structure and crystal chemical data for tilleyite are given in Tables 5.1(f) and 5.20(f).

2.7. Laboratory preparation of high-temperature phases

In general, high-temperature phases are prepared in the laboratory by heating mixtures of appropriate composition at temperatures for which reference to the phase equilibria shows them to be stable. Platinum apparatus is normally required. Compounds having a lower temperature limit of stability (e.g. C_3S) may have to be cooled rapidly to a temperature at which they are kinetically stable. Some high-temperature polymorphs can be preserved by quenching, but with others it may be necessary to add a stabiliser. For β-C_2S, B_2O_3 (0·1–0·3%) has often been used, but it may be considered preferable to employ a composition closer to that of the material in clinker. As noted in §1.3.1, crystals of β-C_2S below a certain size do not invert to γ-C_2S on cooling. This makes it possible to prepare β-C_2S without any stabiliser by first preparing γ-C_2S and reheating at 1000 °C (Shibata *et al.*, 1984). For the same reason, β-C_2S is the normal product when hydrated calcium silicates of appropriate composition, including the major hydration product of C_3S or Portland cement, are decomposed at 800–1000 °C.

As starting materials for high-temperature preparations, finely ground and well-mixed combinations of crystalline oxides (e.g. quartz) and carbonates (e.g. calcite) have been widely used. Mixtures containing carbonates must initially be heated at a relatively low temperature (900–1000 °C for calcite) to avoid violent decarbonation. Repeated heating and regrinding, followed each time by examination of the product by light microscopy, XRD or other methods, is often necessary to produce an acceptable product. Intermediate products are often formed; for example, in the CaO–SiO_2 system, C_2S is formed initially, and reacts only slowly with any excess of calcium oxide or silica to give other phases. Some phases can be obtained by cooling of melts or devitrification of glasses. Particular care and choice of conditions may be needed if, as with the ferrite phase, protected phases or zoned crystals may be formed. More reactive starting materials can sometimes be used; for example, C_3S of 99% purity has been obtained by a single heating of a mixture of freshly precipitated calcium oxalate and hydrous silica (Odler and Dörr, 1977). It may also be possible first to prepare a hydrated compound and then to heat it; for example, C_3A can be obtained in this way by way of C_3AH_6.

If material can be lost by volatilisation, as with phases containing alkalis, fluoride or sulfate, it may be necessary to use a sealed platinum container and to test its effectiveness by chemically analysing the product. Control of the furnace atmosphere, for example by employing mixtures of CO and CO_2 to buffer the oxygen pressure, may be needed if variable oxidation states are possible, as with iron compounds. Special methods may be needed to make single crystals of sufficient size for X-ray structure determinations or other purposes. These are described in papers on X-ray structure determination. Single crystals of C_3S can be obtained from $CaCl_2$ melts (Nurse, 1949).

REFERENCES

Allen WC and Snow RB (1955) The orthosilicate-iron oxide portion of the system CaO-'FeO'-SiO_2. *Journal of the American Ceramic Society* **38**: 264–272.

Andaç O and Glasser FP (1994) Polymorphism of calcium sulphoaluminate ($Ca_4Al_6O_{16}{\cdot}SO_3$) and its solid solutions. *Advances in Cement Research* **6(22):** 57–60.

Arceo HB and Glasser FP (1990) Fluxing reactions of sulphate and carbonates in cement clinkering. I. Systems $CaSO_4$-K_2SO_4 and K_2SO_4-$CaCO_3$. *Cement and Concrete Research* **20(6)**: 862–868.

Bachiorrini A (1985) Le polymorphisme de l'aluminate monocalcique etudie par spectrometrie infrarouge. *Cement and Concrete Research* **15(1)**: 167–173. (In French.)

Bartl H and Scheller T (1970) Zur Struktur des $12CaO{\cdot}7Al_2O_3$. *Neues Jahrbuch für Mineralogie, Monatshefte*: 547–552. (In German.)

Bellanca A (1942) Aphthitalite in the ternary system K_2SO_4-Na_2SO_4-$CaSO_4$. *Periodico di Mineralogia* **13**: 21–85. (In Italian.)

Bereczky E (1964) Acceleration in the formation and stabilization of tricalcium silicate. II. *Épitöanyag* **16**: 441–448. (In Hungarian.)

Biggar GM (1971) Phase relationships of bredigite ($Ca_5MgSi_3O_{12}$) and of the quaternary compound ($Ca_6MgAl_8SiO_{21}$) in the system $CaO{\cdot}MgO{\cdot}Al_2O_3{\cdot}SiO_2$. *Cement and Concrete Research* **1(5)**: 493–513.

Boiocchi M, Caucia F, Merli M, Prella D and Ungaretti L (2001) Crystal-chemical reasons for the immiscibility of periclase and wuestite under lithospheric P, T conditions. *European Journal of Mineralogy* **13(5)**: 871–881.

Borlera ML and Brisi C (1984) Equilibrium relationships in the zone of the system $CaO-Al_2O_3$ close to 37 % molar of aluminium oxide. *Il Cemento* **81**: 13–20.

Boyko E and Wisnyl LG (1958) The optical properties and structures of $CaO.2Al_2O_3$ and $SrO.2Al_2O_3$. *Acta Crystallographica* **11(6)**: 444–445.

Boysen H, Lerch M, Stys A and Senyshyn A (2007) Structure and oxygen mobility in mayenite ($Ca_{12}Al_{14}O_{33}$): a high-temperature neutron powder diffraction study. *Acta Crystallographica B* **63(5)**: 675–682.

Boysen H, Kaiser-Bischoff I, Lerch M, Berendts S, Hoelzel M and Senyshyn A (2010) Disorder and diffusion in mayenite. *Acta Physica Polonica A* **117(1)**: 38–41.

Brisi C and Borlera ML (1983) Excess oxygen in '$12CaO.7Al_2O_3$' and related phases. *Il Cemento* **80**: 155–164.

Brisi C, Borlera ML, Montanaro L and Negro A (1986) Hydration of $5CaO{\cdot}3Al_2O_3$. *Cement and Concrete Research* **16**: 156–160.

Brotherton PO, Epstein JM, Pryce MW and White AH (1974) Crystal structure of 'calcium sulphosilicate', $Ca_5(SiO_4)_2SO_4$. *Australian Journal of Chemistry* **27(3):** 657–660.

Büssem W (1939) X-rays and cement chemistry. *Proceedings of the 3rd International Symposium on the Chemistry of Cement, Stockholm* (1938) Ingeniörsvetenskapsakademien, Stockholm, Sweden pp. 141–168.

Büssem W and Eitel W (1936) Die Struktur des Pentacalciumtrialuminats. *Zeitschrift für Kristallographie - Crystalline Materials* **95(1–6)**: 175–188. (In German.)

Calos NJ, Kennard CHL, Whittaker AK and Davis RL (1995) Structure of calcium aluminate sulfate $Ca_4A_{16}O_{16}S$. *Journal of Solid State Chemistry* **119(1)**: 1–7.

Chatterjee AK and Zhmoidin GI (1972a) The phase equilibrium diagram of the system $CaO-Al_2O_3-CaF_2$. *Journal of Materials Science* **7(1)**: 93–97.

Chatterjee AK and Zhmoidin GI (1972b) Phase diagram of the $CaO.Al_2O_3-Al_2O_3-CaF_2$ subsystem. *Izvestiya Akademii Nauk SSSR, Neorganicheskie Materialy* **8**: 886–892. (In Russian.)

Cuesta A, de la Torre AG, Losilla ER *et al.* (2013) Structure, atomistic simulations, and phase transition of stoichiometric ye'elimite. *Chemistry of Materials* **25(9)**: 1680–1687.

Day AL, Shepherd ES and Wright FE (1906) Lime-silica series of minerals. *American Journal of Science* **s4 22(130)**: 265–302.

de Jong BHWS, Schramm CM and Parziale VE (1983) Polymerization of silicate and aluminate tetrahedra in glasses, melts, and aqueous solutions. IV. Aluminum coordination in glasses and aqueous solutions and comments on the aluminum avoidance principle. *Geochimica et Cosmochimica Acta* **47(7)**: 1223–1236.

Eysel W (1973) Crystal chemistry of the system $Na_2SO_4-K_2SO_4$. $K_2CrO_4-Na_2CrO_4$ and of the glaserite phase. *American Mineralogist* **58**: 736–747.

Feng QL, Glasser FP, Howie RA and Lachowski E (1988) Chlorosilicate with the $12CaO.7Al_2O_3$ structure and its relationship to garnet. *Acta Crystallographica C* **44(4)**: 589–592.

Feng X, Liao G and Long S (1991) Structure and hydration rate of $3CaO.3Al_2O_3.CaSO_4$. *Il Cemento* **88**: 29–36.

Ferguson JB and Buddington AF (1920) The binary system åkermanite-gehlenite. *American Journal of Science* **s4 50**: 131–140.

Florke H (1952) Kristallographische und röntgenometrische Untersuchungen im System $CaSO_4-CaSO_4{\cdot}2H_2O$. *Neues Jahrbuch für Mineralogie – Abhandlungen* **84**: 189–240. (In German.)

Foreman N and Peacor DR (1970) Refinement of the nepheline structure at several temperatures. *Zeitschrift für Kristallographie* **132(1–6)**: 45–70.

Fundal E (1986) The description of a new potassium-alumina phase in OPC. *Proceedings of the 8th International Congress on the Chemistry of Cement* **2**: 139–145.

Galuskin EV, Galuskina IO, Gfeller F *et al.* (2015b) Silicocarnotite, $Ca_5[(SiO_4)(PO_4)](PO_4)$, a new 'old' mineral from the Negev Desert, Israel, and the ternesite-silicocarnotite solid solution: indicators of high-temperature alteration of pyrometamorphic rocks of the Hatrurim Complex, Southern Levant. *European Journal of Mineralogy* **28(1)**: 105–123.

Galuskin EV, Gfeller F, Galuskina IO, Armbruster T, Bailau R and Sharygin VV (2015a) Mayenite supergroup, part I: Recommended nomenclature. *European Journal of Mineralogy* **27(1)**: 99–111.

Ganguly R, Siruguri V, Gopalakrishnan IK and Yakhmi JV (2000) Stability of the layered $Sr_3Ti_2O_7$ structure in $La_{1.2}(Sr_{1-x}Ca_x)_{1.8}Mn_2O_7$. *Journal of Physics: Condensed Matter* **12**: 1683–1689.

Gemmi M, Merlini M, Cruciani G and Artioli G (2007) Non-ideality and defectivity of the åkermanite-gehlenite solid solution: an X-ray diffraction and TEM study. *American Mineralogist* **92(10)**: 1685–1694.

Gilioli C, Massazza F and Pezzuoli H (1979) Studies on clinker calcium silicates bearing CaF_2 and $CaSO_4$. *Cement and Concrete Research* **9(3)**: 295–302.

Glasser FP and Marr J (1975) Quarternary phases in the system CaO-MgO-Al_2O_3-SiO_2. *Transactions and Journal of the British Ceramic Society* **74**: 113–119.

Glasser FP and Marr J (1980) Sulfates in cement clinkering: Immiscibility between sulfate and oxide melts at 1350°C. *Cement and Concrete Research* **10(6)**: 753–758.

Goodwin DW and Lindop AJ (1970) The crystal structure of $CaO.2Al_2O_3$. *Acta Crystallographica B* **26(9)**: 1230–1235.

Grice JD (2005) The structure of spurrite, tilleyite and scawtite, and relationships to other silicate-carbonate minerals. *The Canadian Mineralogist* **43(5)**: 1489–1500.

Gross S (1977) The mineralogy of the Hatrurim Formation. *Bulletin Geological Survey of Israel* **70**: 82.

Gutt W (1961) A new calcium magnesiosilicate. *Nature* **190(4773)**: 339–340.

Gutt W (1968) In Manufacture of Portland cement from phosphate raw materials. *Building Research Station Paper CP90/68*. HMSO, London, UK, p. 5.

Gutt W (1969) Manufacture of Portland cement from phosphatic raw materials. *Proceedings of the 5th International Symposium on the Chemistry of Cement* **1**: 93–105.

Gutt W and Osborne GJ (1966) The system $2CaO{\cdot}SiO_2$-CaF_2. *Transactions of the British Ceramic Society* **65**: 521–534.

Gutt W and Osborne GJ (1968) The calcium silicofluoride of tentative composition $(3CaO{\cdot}SiO_2)_3{\cdot}CaF_2$. *Transactions of the British Ceramic Society* **67**: 125–133.

Gutt W and Osborne GJ (1970) The system CaO-$2CaO{\cdot}SiO_2$-CaF_2. *Transactions of the British Ceramic Society* **69**: 125–129.

Gutt W and Smith MA (1967) Studies of the sub-system CaO-$CaO{\cdot}SiO_2$-$CaSO_4$. *Transactions of the British Ceramic Society* **66**: 557–567.

Gutt W and Smith MA (1968) Studies of the role of calcium sulphate in the manufacture of Portland cement clinker. *Transactions of the British Ceramic Society* **67**: 487–509.

Gutt W, Chatterjee AK and Zhmoidin GI (1970) The join calcium monoaluminate-calcium fluoride. *Journal of Materials Science* **5(11)**: 960–963.

Hallstedt B (1995) Thermodynamic assessment of the CaO–MgO–Al_2O_3 system. *Journal of the American Ceramic Society* **78(1)**: 193–198.

Halstead PE and Moore AE (1962) The composition and crystallography of an anhydrous calcium aluminosulphate occurring in expanding cement. *Journal of Applied Chemistry* **12(9)**: 413–417.

Hanic F, Handlovič M and Kaprálik I (1980) The structure of a quaternary phase $Ca_{20}Al_{32-2v}Mg_vSi_vO_{68}$. *Acta Crystallographica B* **36(12)**: 2863–2869.

Hanic F, Havlica J, Kapnilik I, Ambruz V, Galikova L and Urbanova O (1986) Crystal chemistry and thermodynamics of the sulfate compounds $Ca_4[Al_6O_{12}](SO_4)$ and $Ca_5(SiO_4)_2(SO_4)$ [calcium aluminate sulfate and calcium silicate sulfate. *Transactions and Journal of the British Ceramic Society* **85**: 52–57.

Hansen WC (1928) Phase equilibria in the system $2CaO{\cdot}SiO_2$–MgO–$5CaO{\cdot}3Al_2O_3$. *Journal of the American Chemical Society* **50(8)**: 2155–2160.

Harker RI and Tuttle OF (1956) Experimental data on the P_{CO2}-T curve for the reaction: calcite + quartz ↔ wollastonite + carbon dioxide. *American Journal of Science* **254(4)**: 239–256.

Hawthorne FC and Ferguson RB (1975) Anhydrous sulphates. I: Refinement of the crystal structure of celestite with an appendix on the structure of thenardite. *Canadian Mineralogist* **13**: 181–187.

Hazen RM (1976) Effects of temperature and pressure on the cell dimension and X-ray temperature factors of periclase. *American Mineralogist* **61**: 266–271.

Henmi C and Henmi K (1978) Synthesis of spurrite and tilleyite at low CO_2 partial pressure. *Mineralogical Journal (Japan)* **9(2)**: 106–110.

Henmi C, Kusachi I, Kawahara A and Henmi K (1978) 7T wollastonite from Fuka, Okayama Prefecture. *Mineralogical Journal (Japan)* **9(3)**: 169–181.

Henmi C, Kawahara A, Henmi K, Kusachi I and Takeuchi Y (1983) The 3T, 4T and 5T polytypes of wollastonite from Kushiro, Hiroshima Prefecture, Japan. *American Mineralogist* **68**: 156–163.

Hentschel G (1964) Mayenit, $12CaO.7Al_2O_3$, und Brownmillerit, 2CaO.(Al, $Fe)_2O_3$, zwei neue Minerale in den Kalksteineinschlüssen der Lava des Ettringer Bellerberges. *Neues Jahrbuch für Mineralogie, Monatshefte*, 22–29. (In German.)

Hirata K, Moriya K and Waseda Y (1977) High temperature thermal expansion of ThO_2, MgO and Y_2O_3 by X-ray diffraction. *Journal of Materials Science* **12(4)**: 838–839.

Hörkner W and Müller-Buschbaum H (1976) Zur Kristallstruktur von $CaAl_2O_4$. *Journal of Inorganic and Nuclear Chemistry* **38(5)**: 983–984. (In German.)

Huang Q, Chmaissem O, Capponi JJ *et al.* (1994) Neutron powder diffraction study of the crystal structure of $HgBa_2Ca_4Cu_5O_{12+\delta}$ at room temperature and at 10 K. *Physica C: Superconductivity* **227(1-2)**: 1–9.

Ikeda K, Kishimoto K and Shima H (1996) Structure refinement of calcium sulfoaluminate, $C_4A_3\overline{S}$ with emphasis to oxygen deficiency. *Cement and Concrete Research* **26(5)**: 743–748.

Imlach JA, Dent Glasser LS and Glasser FP (1971) Excess oxygen and the stability of '$12CaO.7Al_2O_3$'. *Cement and Concrete Research* **1(1)**: 57–61.

Insley H and McMurdie HF (1938) Minor constituents in Portland cement clinker. *Journal of Research of the National Bureau of Standards* **20(2)**: 173–184.

Irran E, Tillmanns E and Hentschel G (1997) Ternesite, $Ca_5(SiO_4)_2SO_4$, a new mineral from the Ettringer Bellerberg/Eifel, Germany. *Mineralogy and Petrology* **60(1–2)**: 121–132.

Jay AH and Andrews KW (1945) Note on oxide systems pertaining to steel-making furnace slags. FeO-MnO, FeO-MgO, CaO-MnO, MgO-MnO. *Journal of the Iron and Steel Institute (London)* **152**: 15–18.

Jeevaratnam J, Glasser FP and Dent Glasser LS (1964) Anion substitution and structure of $12CaO{\cdot}7Al_2O_3$. *Journal of the American Ceramic Society* **47(2)**: 105–106.

Kaprálik I and Hanic F (1980) Studies of the system CaO-Al_2O_3-MgO-SiO_2 in relation to the quarternary phase Q. *Transactions and Journal of the British Ceramic Society* **79**: 128–133.

Kaprálik I and Hanic F (1986) Phase changes in the system CaO-Al_2O_3-SiO_2-Fe_2O_3-MgO-$CaSO_4$-K_2SO_4 in air up to 1300° referred to sulfoaluminate cement clinker. *Transactions and Journal of the British Ceramic Society* **85**: 131–136.

Kaprálik I, Hanic F, Havlica J and Ambruz V (1986) Subsolidus phase relations in the system calcium oxide-alumina-silica-ferric oxide-magnesia-calcium sulfate-potassium sulfate at 950° in air referred to sulfoaluminate cement clinker. *Transactions and Journal of the British Ceramic Society* **85**: 107–110.

Karen P, Kjekshus A, Huang Q and Karen VL (1999) The crystal structure of magnesium dicarbide. *Journal of Alloys and Compounds* **282(1–2)**: 72–75.

Kato K and Saalfeld H (1968) Verfeinerung der Kristallstruktur von $CaO{\cdot}6Al_2O_3$. *Neues Jahrbuch für Mineralogie – Abhandlungen* **109**: 192–200. (In German.)

Kaur P, Kaur A, Singh S, Thakur S and Singh L (2020) Comprehensive analysis of crystal structure, optical and luminescent behavior of Fe doped MgO nanophosphors. *Optik* **219**: 164742.

Kimata M and Ii N (1981) The structural property of synthetic åkermanite, $Ca_2MgSi_2O_7$. *Neues Jahrbuch für Mineralogie, Monatshefte*. 1–10.

Kimata M and Ii N (1982) The structural property of synthetic gehlenite, $Ca_2Al_2SiO_7$. *Neues Jahrbuch für Mineralogie – Abhandlungen* **144(3)**: 254–267.

Kirfel A and Will G (1980) Charge density in anhydrite, $CaSO_4$, from X-ray and neutron diffraction measurements. *Acta Crystallographica B* **36(12)**: 2881–2890.

Lea FM (1970) *The Chemistry of Cement and Concrete*, 3rd edn. Arnold, London, UK.

Lea FM and Parker TW (1934) Investigations of a portion of the quaternary system CaO-Al_2O_3-SiO_2-Fe_2O_3. The quaternary system CaO-$2CaO.SiO_2$-$5CaO._3Al_2O_3$-$4CaO.Al_2O_3.Fe_2O_3$. *Philosophical Transactions of the Royal Society London* **A234**: 1–41.

Lea FM and Parker TW (1935) *Quaternary system CaO-Al_2O_3-SiO_2-Fe_2O_3 in relation to cement technology*. Building Research Board, London, UK, Technical Paper. no. 16.

Leary JK (1962) New compound in the system CaO–AI_2O_3–CaF_2. *Nature* **194(4823)**: 79–80.

Louisnathan SJ (1971) Refinement of the crystal structure of a natural gehlenite, $Ca_2Al(Al,Si)_2O_7$. *The Canadian Mineralogist* **10**: 822–837.

Majumdar AJ (1964) The quarternay phase in high-alumina cement. *Transactions of the British Ceramic Society* **63**: 347–364.

Majumdar AJ (1965) The ferrite phase in cements. *Transactions of the British Ceramic Society* **64**: 105–119.

Maki I (1979) Phase constitution of the alite in Portland cement clinker from modern manufacturing processes. *Il Cemento* **76**: 167–176.

Maki I, Nakagawa K, Hiraiwa K and Nonami T (1984) Characterization of fluorine-bearing alite in rapid-hardening high strength cement clinker manufactured by fusion. *Il Cemento* **81**: 3–10.

McGinnety JA (1972) Redetermination of the structures of potassium sulphate and potassium chromate: the effect of electrostatic crystal forces upon observed bond lengths. *Acta Crystallographica B* **28(9)**: 2845–2852.

McMurdie HF and Insley H (1936) Studies on the quarternary system CaO-MgO-$2CaO.SiO_2$-$5CaO.3Al_2O_3$. *Journal of Research of the National Bureau of Standards* **16**: 467–474.

Mehrotra BN, Hahn T, Eysel W, Ropke H and Illguth A (1978) Crystal chemistry of compounds with thenardite (Na_2SO_4) structure. *Neues Jahrbuch für Mineralogie, Monatshefte*, 408–421.

Midgley HG (1979) The composition of pleochroite in high alumina cement clinker. *Cement and Concrete Research* **9(5)**: 623–630.

Moir GK (1983) Improvements in the early strength properties of Portland cement. *Philosophical Transactions of the Royal Society London* **A310**: 127–138.

Moseley D and Glasser FP (1981) Identity, composition and stability of bredigite and phase T. *Cement and Concrete Research* **11(4)**: 559–565.

Muan A and Osborn EF (1965) *Phase Equilibria among Oxides in Steelmaking*. Addison-Wesley, Reading, MA, USA.

Natta G and Passerini L (1929) Soluzioni solide, isomorfismo e simmorfismo tra gli ossidi dei metalli bivalenti. 1. Sistemi: CaO-CdO, CaO-MnO, CaO-CoO, CaO-NiO, CaO-MgO. *Gazzetta Chimica Italiana* **59**: 129–154. (In Italian.)

Neuville DR, Henderson GS, Cormier L and Massiot D (2010) The structure of crystals, glasses, and melts along the $CaO-Al_2O_3$ join: Results from Raman, Al *L*- and *K*-edge X-ray absorption, and ^{27}Al NMR spectroscopy. *American Mineralogist* **95(10)**: 1580–1589.

Newkirk TF and Thwaite RD (1958) Pseudoternary system calcium oxide-monocalcium aluminate($CaO \cdot Al_2O_3$)-dicalcium ferrite($2CaO \cdot Fe_2O_3$). *Journal of Research of the National Bureau of Standards* **61**: 233–245.

Nord AG (1973) Refinement of the crystal structure of thenardite Na_2SO_4 (V). *Acta Chemica Scandinavica* **27**: 814–822.

Nurse RW (1949) Contribution. In *Summarized Proceedings of Conference on X-Ray Analysis, London, 1948*. Journal of Scientific Instruments (Jeffrey JW, Nelson JB and Nicol ADI (eds)) **26(3)**: 99–105.

Nurse RW and Welch JH (1962) Quoted in Nurse RW (1962) Phase equilibria and constitution of Portland cement clinker, paper II-1. *Proceedings of the Fourth International Symposium on the Chemistry of Cement* **1**: 9–21.

Nurse RW, Welch JH and Majumdar AJ (1965a) The $CaO-Al_2O_3$ system in a moisture-free atmosphere. *Transactions of the British Ceramic Society* **64**: 409–418.

Nurse RW, Welch JH and Majumdar AJ (1965b) The $12CaO \cdot 7Al_2O_3$ system. *Transactions of the British Ceramic Society* **64**: 323–332.

Odler I and Dörr H (1977) Tricalcium silicate formation by solid state reactions. *American Ceramic Society Bulletin* **56**: 1086–1089.

Ohashi Y (1984) Polysynthetically-twinned structures of enstatite and wollastonite. *Physics and Chemistry of Minerals* **10(5)**: 217–229.

Okada K and Ossaka J (1980) Structures of potassium sodium sulphate and tripotassium sodium disulphate. *Acta Crystallographica B* **36(4)**: 919–921.

Ol'shanskii YI (1951) Equilibrium of two immiscible liquids in silicate systems of alkali-earth metals. *Doklady Akademii Nauk SSSR* **76**: 93.

Palacios L, Cabeza A, Bruque S, García-Granda S and Aranda MAG (2008) Structure and electrons in mayenite electrides. *Inorganic Chemistry* **47(7)**: 2661–2667.

Parker TW (1954) The constitution of aluminous cement. *Proceedings of the 3rd International Symposium on the Chemistry of Cement* 485–515.

Pekov IV, Zelenski ME, Zubkova NV *et al.* (2012) Calciolangbeinite, $K_2Ca_2(SO_4)_3$, a new mineral from the Tolbachik volcano, Kamchatka, Russia. *Mineralogical Magazine* **76(3)**: 673–682.

Perez-Mendez M, Howie RA and Glasser FP (1984) Ca_3SiO_5 and its fluorine-stabilized aristotype: Synthesis, stability and postulated structure of $Ca_{6-0.5x}Si_2O_{10-x}F_x$. *Cement and Concrete Research* **14(1)**: 57–63.

Perez-Mendez M, Fayos J, Howie RA, Gard JA and Glasser FP (1985) Calcium fluorosilicates: the $Ca_{6-0.5x}Si_2O_{10-x}F_x$ phase. *Cement and Concrete Research* **15**: 600–604.

Perrotta AJ and Smith JV (1967) The crystal structure of kalsilite $KAlSiO_4$. *Mineralogical Magazine* **35**: 588–595.

Peters L, Knorr K, Evans JSO, Senyshyn A, Rahmoun NS and Depmeier W (2007) Proton positions in and thermal behaviour of the phase $4(CaO).3(Al_2O_3).3(H_2O)$ and its thermal decomposition to $|(OCa_4)_2|[Al_{12}O_{24}]$-SOD, determined by neutron/X-ray powder diffraction and IR spectroscopic investigations. *Zeitschrift für Kristallographie* **222(7)**: 365–375.

Phillips B and Muan A (1959) Phase equilibria in the system CaO-iron oxide-SiO_2, in air. *Journal of the American Ceramic Society* **42**: 413–423.

Pliego-Cuervo YB and Glasser FP (1977) The role of sulphates in cement clinkering reactions: phase formation and melting in the system $CaO-Ca_2SiO_4-CaSO_4-K_2SO_4$. *Cement and Concrete Research* **7(5)**: 477–481.

Pliego-Cuervo YB and Glasser FP (1978) Role of sulphates in cement clinkering: the calcium silicosulphate phase. *Cement and Concrete Research* **8(4)**: 455–459.

Pliego-Cuervo YB and Glasser FP (1979a) The role of sulphates in cement clinkering: subsolidus phase relations in the system $CaO-Al_2O_3-SiO_2-SO_3$. *Cement and Concrete Research* **9**: 51–55.

Pliego-Cuervo YB and Glasser FP (1979b) The role of sulphates in cement clinkering reactions: phase formation in the system $CaO{\cdot}Al_2O_3{\cdot}Fe_2O_3{\cdot}SiO_2{\cdot}CaSO_4{\cdot}K_2SO_4$. *Cement and Concrete Research* **9(5)**: 573–581.

Pollitt HWW and Brown AW (1969) The distribution of alkalis in Portland cement clinker. *Proceedings of the Fifth International Symposium on the Chemistry of Cement* **1**: 322–333.

Ponomarev VI, Kheiker DM and Belov NV (1971) Crystal structure of tetracalcium trialuminate – the aluminate analog of sodalite. *Soviet Physics. Crystallography* **15**: 799–801. (Translated from *Kristallografiya* **15**: 918.)

Primak W, Kaufman H and Ward R (1948) X-ray diffraction studies of systems involved in the preparation of alkaline earth sulfide and selenide phosphors. *Journal of the American Chemical Society* **70(6)**: 2043–2046.

Pryce MW (1972) Calcium sulphosilicate in lime-kiln wall coating. *Mineralogical Magazine* **38(300)**: 968–971.

Raab B and Poellmann H (2011) Heat flow calorimetry and SEM investigations to characterize the hydration at different temperatures of different $12CaO.Al_2O_3$ ($C_{12}A_7$) samples synthesized by solid state reaction, polymer precursor process and glycine nitrate process. *Thermochimica Acta* **513(1–2)**: 106–111.

Rankin GA and Wright FE (1915) The ternary system $CaO-Al_2O_3-SiO_2$, with optical study by F. E. Wright. *American Journal of Science* **s4 39(229)**: 1–79.

Rasmussen SE, Jorgensen JE and Lundtoft B (1996) Structure and phase transitions of Na_2SO_4. *Journal of Applied Crystallography* **29(1)**: 42–47.

Ritzmann H (1971) Cyclic phenomena in rotary kiln systems. *Zement Kalk Gips* **24**: 338.

Rogozina TA (1957) Reaction of calcium sulfate with aluminates at 1200°. *Zhurnal Prikladnoi Khimii* **30**: 1682–1685.

Rowe JJ, Morey GW and Hansen LD (1965) The binary system $K_2SO_4{\cdot}CaSO_4$. *Journal of Inorganic and Nuclear Chemistry* **27**: 53–58.

Roy DM and Roy R (1962) Crystalline solubility and zeolitic behaviour in garnet phases in the system $CaO-Al_2O_3-SiO_2-H_2O$, paper III-S9. *Proceedings of the 4th International Symposium on the Chemistry of Cement* **1**: 307–314.

Saburi S, Kusachi I, Henmi C, Kawahara A, Henmi K and Kawada I (1976) Refinement of the structure of rankinite. *Mineralogical Journal (Japan)* **8(4)**: 240–246.

Sarkar SL and Jeffery JW (1979) A review of the T phase. *Journal of the American Ceramic Society* **62(11–12)**: 630–631.

Sasaki S, Fujino K and Takeuchi Y (1979) X-ray determination of electron-density distributions in oxides, MgO, MnO, CoO, and NiO, and atomic scattering factors of their constituent atoms. *Proceedings of the Japan Academy, Series B* **55(2)**: 43–48.

Schimmel HG, Huot J, Chapon LC, Tichelaar FD and Mulder FM (2005) Hydrogen cycling of niobium and vanadium catalyzed nanostructured magnesium. *Journal of the American Chemical Society* **127(41)**: 14348–14354.

Schlaudt CM and Roy DM (1966) The join $Ca_2SiO_4-CaMgSiO_4$. *Journal of the American Ceramic Society* **49(8)**: 430–432.

Schmahl NG and Eikerling GF (1968) *Über Kryptomodifikationen des Cu(II)-Oxids. Zeitschrift für Physikalische Chemie* **62**: 268–279. (In German.)

Scrivener KL and Taylor HFW (1990) Microstructural development in pastes of a calcium aluminate cement. *Proceedings of International Symposium on Calcium Aluminate Cements* (Mangabhai RJ (ed.)). E & FN Spon, London, UK, pp. 41–51.

Shame EG and Glasser FP (1987) Stable Ca_3SiO_5 solid solutions containing fluorine and aluminium made between 1050 and 1250°C. *Transactions of the British Ceramic Society* **86**: 13–17.

Sharp JD, Johnson W and Andrews W (1960) A visual method of primary slag control for the basic electric-arc furnace. *Journal of the Iron and Steel Institute* **195**: 83–94.

Shen CH, Liu RS, Lin JG and Huang CY (2001) Phase stability study of $La_{1.2}Ca_{1.8}Mn_2O_7$. *Materials Research Bulletin* **36(5–6)**: 1139–1148.

Shibata S, Kishi K, Asaga K, Daimon M and Shrestha PR (1984) Preparation and hydration of β-C_2S without stabilizer. *Cement and Concrete Research* **14(3)**: 323–328.

Smart RM and Roy DM (1979) The system $CaO{\cdot}Al_2O_3{\cdot}Fe_2O_3$ with added fluoride flux. *Cement and Concrete Research* **9(2)**: 269–273.

Smirnov GS, Chatterjee AK and Zhmoidin GI (1973) The phase equilibrium diagram of the ternary subsystem CaO-$CaO{\cdot}Al_2O_3$-$11CaO{\cdot}7Al_2O_3{\cdot}CaF_2$. *Journal of Materials Science* **8(9)**: 1278–1282.

Smith DK and Leider HR (1968) Low-temperature thermal expansion of LiH, MgO and CaO. *Journal of Applied Crystallography* **1(4)**: 246–249.

Smith JV, Karle IL, Hauptman H and Karle J (1960) The crystal structure of spurrite, $Ca_5(SiO_4)_2CO_3$. II. Description of structures. *Acta Crystallographica* **13(6)**: 454–458.

Sorrentino F and Glasser FP (1976) Phase relations in the system CaO-Al_2O_3-SiO_2-Fe-O. Part II. The system CaO-Al_2O_3-Fe_2O_3-SiO_2. *Transactions of the British Ceramic Society* **75**: 95–103.

Sourie A, Glasser FP and Lachowski EE (1994) Microstructural studies on pleochroite. *British Ceramic Transactions* **93**: 41–48.

Speer D and Salje E (1986) Phase transitions in langbeinites I: crystal chemistry and structures of K-double sulfates of the langbeinite type M_2^{++} $K_2(SO_4)_3$, M^{++} = Mg, Ni, Co, Zn, Ca. *Physics and Chemistry of Minerals* **13(1)**: 17–24.

Swainson IP, Dove MT, Schmahl WW and Putnis A (1992) Neutron powder diffraction study of the akermanite-gehlenite solid solution series. *Physics and Chemistry of Minerals* **19(3)**: 185–195.

Swanson HE and Tatge E (1953) Standard X-ray powder diffraction patterns. *National Bureau of Standards Circular* **539**, vol. 1. US Department of Commerce, Washington, DC, USA.

Swanson HE, McMurdie HF, Morris MC and Evans EH (1969) Standard X-ray diffraction patterns. National Bureau of Standards Monograph No. 25, Section 7, US Department of Commerce, Washington, DC, USA.

Swayze MA (1946a) A report on studies of; 1. The ternary system of CaO-C_5A_3-C_2F, 2. The quaternary system CaO-C_5A_3-C_2F-C_2S, 3. The quaternary system as modified by 5% magnesia, part 1. *American Journal of Science* **244(1)**: 1–30.

Swayze MA (1946b) A report on studies of; 1. The ternary system of CaO-C_5A_3-C_2F, 2. The quaternary system CaO-C_5A_3-C_2F-C_2S, 3. The quaternary system as modified by 5% magnesia, part 2. *American Journal of Science* **244(2)**: 65–94.

Tanaka M, Sudoh G and Akaiwa S (1969) New compound $Ca_{12}Si_4O_{19}F_2$ in the system CaO-SiO_2-CaF_2 and the role of CaF_2 in the burning of cement clinker. *Proceedings of the Fifth International Symposium on the Chemistry of Cement* **1**: 122–135.

Taylor D (1984) Thermal expansion data: I. Binary oxides with the sodium chloride and wurtzite structures, MO. *Transactions and Journal of the British Ceramic Society* **83**: 5–9.

Taylor WC (1942) Further phase-equilibrium studies involving the potash compounds of Portland cement. *Journal of Research of the National Bureau of Standards* **29(6)**: 437–451.

Timashev VV (1980) The kinetics of clinker formation: the structure and composition of clinker and its phases. *Proceedings of the 7th International Congress on the Chemistry of Cement. 1, I-3/1-I-3/20.*

Tsirelson VG, Avilov AS, Abramov Y, Belokoneva EL, Kitaneh R and Feil D (1998) X-ray and electron diffraction study of MgO. *Acta Crystallographica B* **54(1)**: 8–17.

Utsunomiya A, Tanaka K, Morikawa H, Marumo F and Kojima H (1988) Structure refinement of $CaO{\cdot}6(Al_2O_3)$. *Journal of Solid State Chemistry* **75**: 197–200.

Verbraeken MC, Suard E and Irvine JTS (2009) Structural and electrical properties of calcium and strontium hydrides. *Journal of Materials Chemistry* **19(18)**: 2766–2770.

Vincent MG and Jeffery JW (1978) The crystal structure of pentacalcium trialuminate, $5CaO.3Al_2O_3$. *Acta Crystallographica B* **34(5)**: 1422–1428.

Wainwright JE and Starkey J (1971) A refinement of the crystal structure of anorthite. *Zeitschrift für Kristallographie* **133**: 75–84.

Welch JH (1961) Ternary compound formation in the system Cao–Al_2O_3–MgO. *Nature* **191(4788)**: 559–560.

Welch JH (1964) Phase equilibria and high-temperature chemistry in the CaO-Al_2O_3-SiO_2 and related systems. In *Chemistry of Cements* (Taylor HFW (ed.)). Academic Press, London, UK, vol. 1, pp. 49–88.

Welch JH and Gutt W (1959) Tricalcium silicate and its stability within the system CaO-SiO_2. *Journal of the American Ceramic Society* **42(1)**: 11–15.

Welch JH and Gutt W (1962) The effect of minor components on the hydraulicity of the calcium silicates, paper II-S1. *Proceedings of the 5th International Symposium on the Chemistry of Cement* **1**: 59–67.

Williams PP (1968) Crystal structure of a fluoride derivative of $12Ca0.7Al_2O_3$. *Journal of the American Ceramic Society* **51(9)**: 531–531.

Williams PP (1973) Refinement of the structure of $11CaO.7Al_2O_3.CaF_2$. *Acta Crystallographica* **B29**: 1550–1551.

Winchell AN and Winchell H (1964) *The Microscopical Characters of Artificial Inorganic Solid Substances*. Academic Press, London, UK.

Xia Z, Molokeev MS, Oreshonkov AS, Atuchin VV, Liu R-S and Dong C (2014) Crystal and local structure refinement in $Ca_2Al_3O_6F$ explored by X-ray diffraction and Raman spectroscopy. *Physical Chemistry Chemical Physics* **16(13)**: 5952–5957.

Yamanaka T and Mori H (1981) The structure and polytypes of α-$CaSiO_3$ (pseudowollastonite). *Acta Crystallographica B* **37(5)**: 1010–1017.

Yoshioka T (1970) A new crystal with kalsilite-type structure on the $CaAl_2O_4$-SiO_2 join. *Bulletin of the Chemical Society of Japan* **43(8)**: 2317–2321.

Zhang P, Chen Y, Shi L, Zhang G, Huang W and Wu J (1992) The crystal structure of C_4A_3S. *Proceedings of the 9th International Congress on the Chemistry of Cement* **3**: 201–208.

Zhmoidin GI and Chatterjee AK (1974) Equilibrium phase diagram of the calcium oxide-calcium oxide-aluminum oxide-calcium fluoride (CaO-CaO.Al_2O_3-CaF_2) subsystem. *Izvestiya akademii nauk SSSR, Neorganicheskie Materialy* **10**: 1846–1851. (In Russian.)

Zhmoidin GI and Chatterjee AK (1984) Conditions and mechanism of interconvertibility of compounds $12CaO.7Al_2O_3$ and $5CaO.3Al_2O_3$. *Cement and Concrete Research* **14(3)**: 386–396.

Zhmoidin GI and Chatterjee AK (1985) Sorption of gases by crystalline and molten $12CaO.7Al_2O_3$. *Cement and Concrete Research* **15(3)**: 442–452.

Ian G Richardson and HFW Taylor
ISBN 978-1-83549-777-7
https://doi.org/10.1108/978-1-83549-776-020253003

Chapter 3
The chemistry of Portland cement manufacture

3.1. General considerations

3.1.1 Summary of the reactions in clinker formation

In the manufacture of Portland cement clinker, the raw materials, typically a limestone and a clay or shale, are intimately mixed and heated, ultimately to a temperature of about 1450 °C. The principal reactions taking place are conveniently divided into three groups, as follows.

1) Reactions below about 1300 °C, of which the most important are (*a*) the decomposition of calcite (calcining), (*b*) the decomposition of clay minerals and (*c*) reaction of calcite or lime formed from it with quartz and clay mineral decomposition products to give belite, aluminate and ferrite. Liquid is formed only to a minor extent at this stage but may have an important effect in promoting the reactions. At the end of this stage, the major phases present are belite, lime, aluminate and ferrite. The last two may not be identical with the corresponding phases in the final product.
2) Reactions at 1300–1450 °C (clinkering): a melt is formed, mainly from the aluminate and ferrite, and by 1450 °C some 20–30 % of the mix is liquid. Much of the belite and nearly all the lime react in the presence of the melt to give alite. The material nodulises, to form the clinker.
3) Reactions during cooling: the liquid crystallises, giving mainly aluminate and ferrite. Polymorphic transitions of the alite and belite occur.

Figure 3.1 shows these changes for a typical clinker. No attempt has been made to show a detailed sequence of phases below 1300 °C, and minor phases, including sulfates, are omitted. Quantitative phase compositions at various stages vary considerably with starting materials and other factors.

3.1.2 Lime saturation factor, silica ratio and alumina ratio

Chemical analyses of cements, clinkers and individual phases are commonly expressed in terms of oxide components, but it is often useful to employ quantities derived from these. In the equations that follow, chemical formulae also denote mass percentages. The following parameters are widely used in the industry (Herfort, 2015).

$$\text{lime saturation factor (LSF)} = \text{CaO}/(2{\cdot}80\ \text{SiO}_2 + 1{\cdot}18\ \text{Al}_2\text{O}_3 + 0{\cdot}65\ \text{Fe}_2\text{O}_3) \quad (3.1)$$

$$\text{silica ratio (SR)} = \text{SiO}_2/(\text{Al}_2\text{O}_3 + \text{Fe}_2\text{O}_3) \quad (3.2)$$

$$\text{alumina ratio (AR)} = \text{Al}_2\text{O}_3/\text{Fe}_2\text{O}_3 \quad (3.3)$$

Figure 3.1 Schematic diagram showing the variations in typical contents of phases during the formation of Portland cement clinker, loosely based on a figure by Wolter (1985)

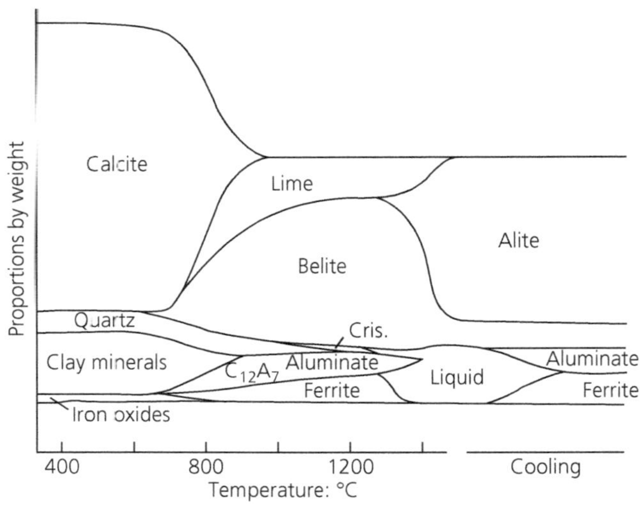

The definition of LSF is theoretically based (§2.3.3), and in the form given above applies to clinkers. It may be applied to cements if 0·7 SO_3 is subtracted from CaO. This is because 1 mole of SO_3 is combined with 1 mole of CaO in $CaSO_4$ and the ratio of the masses of CaO:SO_3 = 56·077/80·063 = 0·700 (Bogue, 1929). The LSF largely governs the ratio of alite to belite and also shows whether the clinker is likely to contain an unacceptable proportion of free lime, a value of 1·0 or above indicating that the latter will be present at equilibrium at the clinkering temperature and thus liable to persist in the product. In practice, values up to 1·02 may be acceptable; typical values for modern clinkers are 0·92–0·98, with convergence on a '*universally acceptable*' value of about 0·96 (Herfort *et al.*, 2010). Other parameters similar in function to LSF are used in some countries. A suggested modification to the LSF definition, intended to allow for Mg substitution in alite, replaces CaO by (CaO + 0·75MgO) for MgO $\leq$2%, or by (CaO + 1·5) for MgO >2% (Spohn *et al.*, 1969).

The SR and AR, also respectively called silica modulus and alumina modulus, are empirically based. For normal types of Portland cement clinker, SR is usually 2·0–3·0 and AR 1·0–4·0, with convergence on values of about 2·6 and 1·5 for SR and AR, respectively (Herfort *et al.*, 2010); these ranges do not apply to special types, such as sulfate-resisting or white cement clinkers. The SR governs the proportion of silicate phases in the clinker. Increase in SR lowers the proportion of liquid at any given temperature in the kiln and thus makes the clinker more difficult to burn. The AR governs the ratio of aluminate to ferrite phases in the clinker, which has important effects on cement properties and also determines the quantity of liquid formed at relatively low temperatures. At 1338 °C, for a given total content of Al_2O_3 and Fe_2O_3, the quantity of liquid passes through a maximum at an AR of 1·38.

The proportions of raw materials are commonly calculated on the basis of the above parameters, most obviously by setting up and solving simultaneous equations; to fix *n* parameters, (*n* + 1) raw materials of appropriate composition are required. It is also necessary to consider the desired or allowable contents of minor components.

3.1.3 The Bogue calculation

In another approach, the quantitative phase composition is estimated using a procedure due to Bogue (1929, 1947), which involves the four main oxides in clinker: CaO, SiO_2, Al_2O_3 and Fe_2O_3. The compositions of the four major phases are assumed to be C_3S, C_2S, C_3A and C_4AF; all of the Fe_2O_3 and some of the Al_2O_3 are present in the C_4AF, and the remaining Al_2O_3 is in the C_3A. Equations 3.4–3.7 give the total mass of each of the four oxides in the clinker where M_{formula} is the formula mass; chemical formulae also denote mass percentages.

$$\mathrm{CaO} = \left(\frac{3 \times M_{\mathrm{CaO}}}{M_{\mathrm{C_3S}}}\right)\mathrm{C_3S} + \left(\frac{2 \times M_{\mathrm{CaO}}}{M_{\mathrm{C_2S}}}\right)\mathrm{C_2S} + \left(\frac{3 \times M_{\mathrm{CaO}}}{M_{\mathrm{C_3A}}}\right)\mathrm{C_3A} + \left(\frac{4 \times M_{\mathrm{CaO}}}{M_{\mathrm{C_4AF}}}\right)\mathrm{C_4AF} \quad (3.4)$$

$$\mathrm{SiO_2} = \left(\frac{M_{\mathrm{SiO_2}}}{M_{\mathrm{C_3S}}}\right)\mathrm{C_3S} + \left(\frac{M_{\mathrm{SiO_2}}}{M_{\mathrm{C_2S}}}\right)\mathrm{C_2S} \quad (3.5)$$

$$\mathrm{Al_2O_3} = \left(\frac{M_{\mathrm{Al_2O_3}}}{M_{\mathrm{C_3A}}}\right)\mathrm{C_3A} + \left(\frac{M_{\mathrm{Al_2O_3}}}{M_{\mathrm{C_4AF}}}\right)\mathrm{C_4AF} \quad (3.6)$$

$$\mathrm{Fe_2O_3} = \left(\frac{M_{\mathrm{Fe_2O_3}}}{M_{\mathrm{C_4AF}}}\right)\mathrm{C_4AF} \quad (3.7)$$

Rearrangement and substitution leads to the following equations (the Bogue equations), in which CaO is assumed to have been corrected for free lime; the formula masses were calculated using atomic masses given by Wieser and Berglund (2009).

$$\mathrm{C_3S} = 4{\cdot}0715\ \mathrm{CaO} - 7{\cdot}5999\ \mathrm{SiO_2} - 6{\cdot}7177\ \mathrm{Al_2O_3} - 1{\cdot}4298\ \mathrm{Fe_2O_3} \quad (3.8)$$

$$\mathrm{C_2S} = 2{\cdot}8666\ \mathrm{SiO_2} - 0{\cdot}7544\ \mathrm{C_3S} \quad (3.9)$$

$$\mathrm{C_3A} = 2{\cdot}6500\ \mathrm{Al_2O_3} - 1{\cdot}6920\ \mathrm{Fe_2O_3} \quad (3.10)$$

$$\mathrm{C_4AF} = 3{\cdot}0432\ \mathrm{Fe_2O_3} \quad (3.11)$$

As with LSF (§3.1.2), the approach is applicable to cements if CaO is further corrected by deducting 0·7 SO_3. Because minor oxide components are ignored, the total for the four main phases plus free lime will not add up to 100%. It is implicit in the approach that all the MgO is assumed to occur as periclase.

In a variant of the Bogue calculation used in US specifications (C150, ASTM, 2024), free lime is not deducted and Al_2O_3 is replaced by Al_2O_3 + P_2O_5 + TiO_2. This latter modification typically raises the calculated C_3A content by some 1·1%.

The results of the Bogue calculation are often called potential phase compositions because when the procedure was devised, it was generally considered that the principal source of error was failure to reach equilibrium during cooling. The results do indeed differ, often markedly, from the true phase compositions, notably in underestimating alite and overestimating belite (§4.5.4), and it is unlikely that equilibrium is maintained during cooling, but the principal direct source of error is that the compositions of the clinker phases differ considerably from those of the pure compounds. One significant compositional error is due to the fact that the ferrite phase has variable Al/Fe ratio (§1.5.1). For example, as noted by Lea and Desch (1956) (and reproduced by Lea, 1970), each 1% of Fe_2O_3 that is present as C_6A_2F instead of C_4AF would result in 1·34 and 1·43% more ferrite and C_3S, respectively, and 1·69 and 1·08% less C_3A and C_2S (the percentages have been recalculated and two of them are very slightly different from those given by Lea (1970), p. 142). The variable Al/Fe ratio of the ferrite is readily incorporated into the three Bogue equations that have Fe_2O_3 terms. As noted, for C_4AF, $x = \frac{1}{2}$ in the formula $Ca_2(Al_xFe_{1-x})_2O_5$ (see §1.5.1) and Equations 3.8, 3.10 and 3.11 can be modified to accommodate different values of x as follows.

The coefficient for Fe_2O_3 in Equation 3.8 has the value of the formula mass for C_3S divided by the formula mass for Fe_2O_3 (i.e. 228·3165 g mol^{-1}/159·6882 g mol^{-1} = 1·4298) but it is negative in the equation because the fraction $\left(\frac{3x-2}{1-x}\right)$ has the value of –1 when $x = \frac{1}{2}$. Equation 3.8 is thus written alternatively as Equation 3.12, which can be used for other values of x, with $x = \frac{2}{3}$ notable because in that case $\left(\frac{3x-2}{1-x}\right) = 0$ and so the Fe_2O_3 term vanishes.

$$C_3S = 4{\cdot}0715\,CaO - 7{\cdot}5999\,SiO_2 - 6{\cdot}7177\,Al_2O_3 + \left(\frac{3x-2}{1-x}\right)1{\cdot}4298\,Fe_2O_3 \qquad (3.12)$$

The situation is similar for Equation 3.10. The coefficient for Fe_2O_3 in Equation 3.10 has the value of the formula mass for C_3A divided by the formula mass for Fe_2O_3 multiplied by –1 (i.e. –(270·1935 g mol^{-1}/159·6882 g mol^{-1}) = –1·6920) but that is only the case because the fraction $\left(\frac{x}{1-x}\right)$ has the value of 1 when $x = \frac{1}{2}$. Equation 3.10 can be written alternatively as Equation 3.13, which can be used for other values of x. The value of $x = \frac{2}{3}$ is again notable because in that case the coefficient for the Fe_2O_3 term is simply multiplied by 2.

$$C_3A = 2{\cdot}6500\,Al_2O_3 - \left(\frac{x}{1-x}\right)1{\cdot}6920\,Fe_2O_3 \qquad (3.13)$$

The coefficient for the Fe_2O_3 term in the last of the Bogue equations, Equation 3.11, has the value of the formula mass for C_4AF divided by the formula mass for Fe_2O_3 (i.e. 485·959 g mol^{-1}/ 159·6882 g mol^{-1} = 3·0432). Alternatively, it is calculated by dividing the formula mass of the ferrite expressed as $Ca_2(Al_xFe_{1-x})_2O_5$ by the formula mass for Fe_2O_3 multiplied by 2 because the fraction $\left(\frac{1}{1-x}\right)$ has the value of 2 when $x = \frac{1}{2}$ (i.e. 2 × (242·9795 g mol^{-1}/159·6882 g mol^{-1}) = 3·0432). Equation 3.11 can therefore be modified to Equation 3.14 for use with different values of x.

$$\text{Ferrite} = \left(\frac{M_{\text{ferrite}}}{(1-x)M_{Fe_2O_3}}\right)Fe_2O_3 \qquad (3.14)$$

where M_{ferrite} is the formula mass (expressed as $Ca_2(Al_xFe_{1-x})_2O_5$) calculated using the value of x that is of interest.

As an example, the Bogue equations modified to involve C_6A_2F instead of C_4AF are

$$C_3S = 4{\cdot}0715\ CaO - 7{\cdot}5999\ SiO_2 - 6{\cdot}7177\ Al_2O_3 \quad (3.15)$$

$$C_2S = 2{\cdot}8666\ SiO_2 - 0{\cdot}7544\ C_3S \quad (3.16)$$

$$C_3A = 2{\cdot}6500\ Al_2O_3 - 3{\cdot}3840\ Fe_2O_3 \quad (3.17)$$

$$C_6A_2F = 4{\cdot}3840\ Fe_2O_3 \quad (3.18)$$

In addition to their use in some specifications, Bogue compositions are used for proportioning by setting up and solving equations to calculate the relative amounts of raw materials needed to obtain given 'potential' contents of C_3S or other phases. They have often been misused for other purposes on the assumption that they are close to the actual phase compositions. Spohn *et al.* (1969) have indicated the dangers of such uncritical use.

3.1.4 Enthalpy changes in clinker formation

The enthalpy change on formation of Portland cement clinker cannot be calculated with high precision, mainly because of uncertainties associated with the clay minerals in the raw material. Table 3.1 gives data for the main thermochemical components of the reaction, almost all of which have been calculated from a self-consistent set of standard enthalpies of formation. The conversion of the clay minerals into oxides is an imaginary reaction, but valid as a component in a Hess's law

Table 3.1 Standard enthalpies of reaction

Reaction	ΔH: kJ	For 1 kg of
$CaCO_3$ (calcite) → $CaO + CO_2$ (g)	+1782	$CaCO_3$
AS_4H (pyrophyllite) → α-Al_2O_3 + $4SiO_2$ (quartz) + H_2O (g)	+224	AS_4H
AS_2H_2 (kaolinite) → α-Al_2O_3 + $2SiO_2$ (quartz) + $2H_2O$ (g)	+538	AS_2H_2
2FeO·OH (goethite) → α-Fe_2O_3 + H_2O (g)	+293	FeO·OH
2CaO + SiO_2 (quartz) → β-C_2S	−734	C_2S
3CaO + SiO_2 (quartz) → C_3S	−495	C_3S
3CaO + α-Al_2O_3 → C_3A	−27	C_3A
6CaO + 2α-Al_2O_3 + α-Fe_2O_3 → C_6A_2F	−157	C_6A_2F
4CaO + α-Al_2O_3 + α-Fe_2O_3 → C_4AF	−105	C_4AF

Values for starting materials and products at 25 °C and 0·101 MPa, calculated from the data of Wagman *et al.* (1982), excepting those for the formation of C_6A_2F (Newman and Hoffman, 1956) and C_4AF (Thorvaldson, 1970). The value for C_4AF is for 20 °C.

Table 3.2 Enthalpy of formation of 1 kg of an idealised Portland cement clinker that has no minor oxide components

Component of reaction	ΔH: kJ
$CaCO_3 \rightarrow CaO + CO_2$ (g)	+2185
AS_4H (pyrophyllite) $\rightarrow \alpha$-$Al_2O_3 + 4SiO_2$ (quartz) + H_2O (g)	+16
AS_2H_2 (kaolinite) $\rightarrow \alpha$-$Al_2O_3 + 2SiO_2$ (quartz) + $2H_2O$ (g)	+44
$2FeO{\cdot}OH$ (goethite) $\rightarrow \alpha$-$Fe_2O_3 + H_2O$ (g)	+11
$3CaO + SiO_2$ (quartz) $\rightarrow C_3S$	−339
$2CaO + SiO_2$ (quartz) $\rightarrow \beta$-C_2S	−97
$3CaO + \alpha$-$Al_2O_3 \rightarrow C_3A$	−1
$6CaO + 2\alpha$-$Al_2O_3 + \alpha$-$Fe_2O_3 \rightarrow C_6A_2F$	−24
Total	+1795

The contributions to ΔH are calculated assuming the starting materials to be pure: calcite, 1·226 kg; quartz, 0·140 kg; pyrophyllite, 0·072 kg; kaolinite, 0·082 kg; goethite, 0·038 kg. The products are calculated to be: C_3S, 0·684 kg; β-C_2S, 0·132 kg; C_3A, 0·023 kg; C_3A_2F, 0·151 kg; free CaO, 0·010 kg; total, 1·000 kg. The oxide composition is: CaO, 0·687 kg; SiO_2, 0·226 kg; Al_2O_3, 0·053 kg; Fe_2O_3, 0·035 kg; LSF = 0·96; AR = 1·52; SR = 2·60. Alternatively, if the ferrite is considered to be C_4AF, then the clinker has: C_3S, 0·635 kg; β-C_2S, 0·169 kg; C_3A, 0·081 kg; C_4AF, 0·105 kg.

calculation. The values for pyrophyllite (AS_4H) and kaolinite (AS_2H_2) can probably be used with sufficient accuracy, on a mass basis, for other 2:1 and 1:1 clay minerals, respectively. The data for the pure compounds may similarly be used for the clinker phases. Table 3.2 illustrates the use of these data for an idealised Portland cement clinker that has no minor oxide components.

The enthalpy of formation of 1 kg of this clinker is 1795 kJ, and the values for LSF, AR and SR are 0.96, 1.5 and 2.6, respectively, which are the same as the '*universally acceptable*' values given by Herfort *et al.* (2010) (§3.1.2). The table includes the masses of the starting materials, the bulk oxide composition and two alternatives for the quantitative phase composition, which were calculated from the bulk composition using Equations 3.9, 3.12, 3.13 and 3.14 (§3.1.3), assuming ferrite with $x = \frac{2}{3}$ (i.e. C_6A_2F) and $x = \frac{1}{2}$ (i.e. C_4AF). The differences in the percentages of the clinker phases calculated for the two values of x are consistent with the numbers discussed in §3.1.3 and noted by Lea and Desch (1956). An experimental method has been described to determine the overall enthalpy change (Castanet and Sorrentino, 1986), and an approximate value can be calculated from the bulk composition using Equation 3.19 (Zur Strassen, 1957; the original equation is in units of $kcal\,kg^{-1}$; chemical formulae denote mass percentages). Substitution of the oxide percentages of Table 3.2 into Equation 3.19 gives $H_{Total} = 1758\,kJ\,kg^{-1}$, which is reasonably close to the value given in the table and above.

$$H_{Total}(kJ\,kg^{-1}) = 9{\cdot}29\,Al_2O_3 + 27{\cdot}11\,MgO + 32{\cdot}05\,CaO - 21{\cdot}41\,SiO_2 - 2{\cdot}47\,Fe_2O_3 \quad (3.19)$$

The overall enthalpy change in forming clinker is dominated by the strongly endothermic decomposition of calcite. The component reactions for the replacement of clay minerals by oxides are endothermic because the heat required for dehydroxylation exceeds that liberated on forming the products. The formation of β-C_2S from lime and quartz is moderately exothermic, but that of C_3S from lime and β-C_2S is endothermic, with $\Delta H = +59\,kJ\,kg^{-1}$. All these calculations refer to reactants and products at 25 °C and 0·1 MPa. The enthalpy changes at the temperatures at which the reactions occur are somewhat different because the specific heats of reactants and products are not the same. The reaction of lime with C_2S giving C_3S changes from endothermic to exothermic at 1430 °C (Jøns, 1980). For the decomposition of calcite at 890 °C, ΔH is $+ 1644\,kJ\,kg^{-1}$ (Lea, 1970).

3.2. Raw materials and manufacturing processes

3.2.1 Raw materials and fuel

These, and manufacturing processes, will be considered only to the extent needed for a basic understanding of their chemistry; fuller accounts are given elsewhere (Duda, 1985; Ghosh, 1983; Kerton and Murray, 1983; Peray, 1986).

The raw mix for making Portland cement clinker is generally obtained by blending a calcareous material, typically limestone, with a smaller amount of an argillaceous one, typically clay or shale. It may be necessary to include minor proportions of one or more corrective constituents, such as iron ore, bauxite or sand, to correct the bulk composition. On the other hand, some argillaceous limestones and marls have compositions near to that required, making it possible to use a blend of closely similar strata from the same quarry.

Limestones vary in physical characteristics from compact rocks of low porosity to friable and highly porous ones, such as chalk, which may contain up to 25 % water. All consist essentially of calcium carbonate, normally in the polymorphic form of calcite. Other naturally occurring forms of $CaCO_3$, such as shell deposits, are sometimes used. Many limestones contain significant amounts of minor components, either as substituents in the calcite or in accessory phases, some of which are deleterious if present in amounts exceeding a few per cent (e.g. MgO, SrO), a few tenths of a per cent (e.g. P_2O_5, CaF_2, alkalis) or even less (some heavy metals).

Suitable shales and clays typically have bulk compositions in the region of 55–60 % SiO_2, 15–25 % Al_2O_3 and 5–10 % Fe_2O_3, with smaller amounts of MgO, alkalis, H_2O and other components. Mineralogically, their main constituents are clay minerals, finely divided quartz and, sometimes, iron oxides. The dominant clay minerals are usually of the illite and kaolinite families, but small amounts of smectites (montmorillonite-type minerals) may also be present. Kaolinite has a 1:1 layer silicate structure and ionic constitution $Al_2(Si_2O_5)(OH)_4$ (in oxide notation the composition is thus AS_2H_2). Illites and smectites have 2:1 layer silicate structures, derived from those of pyrophyllite [$Al_2(Si_2O_5)_2(OH)_2$] (in oxide notation the composition is AS_4H) or talc [$Mg_3(Si_2O_5)_2(OH)_2$] by various ionic substitutions and incorporation of interlayer cations and, in smectites, water molecules. In place of clays or shales, other types of siliceous rocks, such as schists or volcanic rocks of suitable compositions, are sometimes used.

Pulverised coal, oil, natural gas and lignite have all been widely used as fuels but, due to pressing environmental and economic requirements, are increasingly being supplemented by waste materials. In 2014, the most widely used fuels in cement production globally were coal (70 %), oil and gas (24 % combined), and various alternative fuels (6 %) (Fernandez and Leung, 2018). It is possible to use much greater amounts of alternative fuels without affecting the efficiency of the

Figure 3.2 Specific fuel energy consumption (○) in the German cement industry. Energy consumption from fossil (△) and alternative fuels (□), and electrical power consumption (◇). The data are from Figure 4.1 and Tables 4.1 and 4.2 of VDZ (2018)

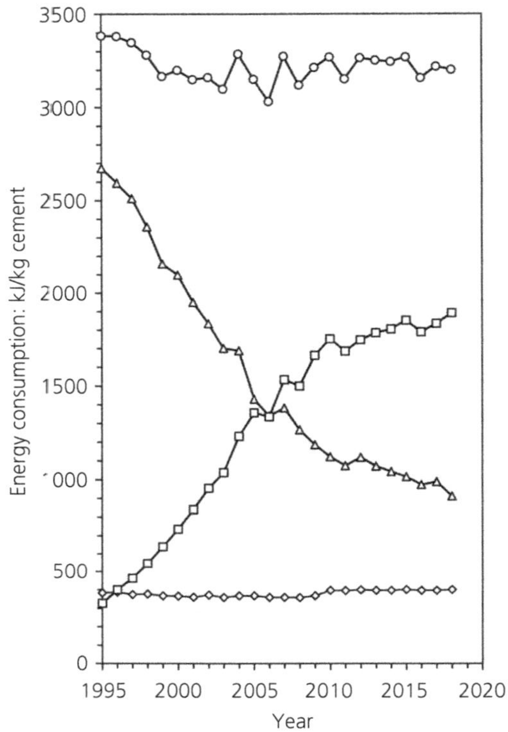

production process; for example, their use has increased steadily in the German cement industry (Figure 3.2), accounting for 67·5% of the total fuel energy demand in 2018 (VDZ, 2018). Worn-out or reject tyres and many kinds of industrial and municipal wastes are used (Kääntee *et al.*, 2004; Rahman *et al.*, 2013; Schneider *et al.*, 2011). In addition to conserving nonrenewable fossil fuels and in some cases lowering CO_2 emissions, the use of waste materials as fuels can provide a means of safe and environmentally acceptable disposal. However, the materials and production process must be monitored more closely because the substitution of fossil fuels with alternatives can influence the properties of the clinker (Schneider *et al.*, 2011). Some waste materials can be used in the raw feed; energy can be saved and CO_2 emission decreased if even a part of the CaO can be provided by a material, such as blast-furnace slag, that does not require decarbonation. Some materials, such as fly ash (pulverised fuel ash (PFA)) can serve as raw materials that also possess some fuel content. The contribution of the fuel to the clinker composition must be taken into account, especially with coal or lignite, which produce significant quantities of ash broadly similar in composition to the argillaceous component. Some fuels also contribute sulfur.

3.2.2 Dry and wet processes; energy requirements

Comminution and mixing of the raw materials may be carried out either dry or wet. The theoretical amount of heat needed to produce 1 kg of clinker from typical raw materials is about 1800 kJ

(§3.1.4). Additional heat is required because heat is retained in the clinker, kiln dust and exit gases after they leave the system, lost from the plant by radiation or convection and, in the wet process, used to evaporate water. The dry process has greater thermal efficiency, largely because there is no added water to be evaporated. The total energy requirement of a cement works also includes electrical energy used in operating the plant, to which grinding of raw material and of clinker make important contributions. This is higher for the dry process than for the wet process, typical values per tonne of cement being 110 kWh (Jankovic *et al.*, 2004; 396 kJ kg^{-1}) and 77 kWh (Davis *et al.*, 1979; 277 kJ kg^{-1}), respectively, including that used for grinding the clinker in each case, but the difference is small compared with that in the amounts of fuel required. The wet process has certain advantages if the raw materials are already soft and moist, but its low fuel efficiency has rendered it obsolescent, and the description in §3.2.3 is of a modern version of the dry process. The total fuel energy consumed in a modern works (dry process with cyclone preheaters) to produce 1 kg of cement is about 1000 kJ higher than the value given above for the theoretical amount of heat needed to produce 1 kg of clinker (i.e. about 1800 kJ), with a further 400 kJ kg^{-1} in electrical power consumed (i.e. about 110 kWh $tonne^{-1}$), giving a total of about 3200 kJ kg^{-1} (VDZ, 2018). This is about half the energy consumption of the older wet process (Barcelo *et al.*, 2014). The approximate efficiency is thus $\frac{1800}{3200} \times 100 = 56\%$. As an example, Figure 3.2 shows the specific energy consumption in the German cement industry over a period of 24 years; the data are from VDZ (2018). The average value for the last five years in Figure 3.2 is about 100 kJ kg^{-1} lower than for the first five years (i.e. 1990s).

3.2.3 The dry process; suspension preheaters and precalciners

The raw materials first pass through a series of crushing, stockpiling, milling and blending stages, which yield an intimately mixed and dry raw meal, of which typically 85 % passes through a 90 μm sieve. With automated, computer-controlled procedures, the LSF, SR and AR can be maintained constant to standard deviations of 1 %, 0·1 and 0·1, respectively.

The raw meal passes through a preheater and frequently also a precalciner before entering a rotary kiln. A preheater is a heat exchanger, usually of a type called a suspension preheater, in which the moving powder is dispersed in a stream of hot gas coming from the kiln. Figure 3.3 shows a common arrangement which employs a series of cyclones. Heat transfer takes place mainly in co-current; the raw material passes through the preheater in less than a minute and leaves it at a temperature of about 800 °C. These conditions are such that about 40 % of the calcite is decarbonated. It is possible to introduce a proportion of the fuel into a preheater, with a corresponding reduction in the quantity fed to the kiln. A precalciner is a furnace chamber introduced into the preheater into which 50–65 % of the total amount of fuel is introduced, often with hot air ducted from the cooler. The fuel in the precalciner is burned at a relatively low temperature; heat transfer to the raw meal, which is almost entirely convective, is very efficient. The material has a residence time in the hottest zone of a few seconds and its exit temperature is about 900 °C; 90–95 % of the calcite is decomposed. Ash from the fuel burned in the precalciner is effectively incorporated into the mix. The series of cyclones in the preheater can have up to six stages, with heat recovery increasing with each additional stage (Habert, 2013); the thermal energy consumption is thus improved by increasing the number of stages and also by the incorporation of a precalciner (Barcelo *et al.*, 2014; Madlool *et al.*, 2013).

Because less heat is supplied to the kiln, precalcination allows the rate at which material can be passed through a kiln of given size to be greatly increased, thus saving on capital cost. Alternatively, the rate of providing heat can be reduced, which lengthens the life of the refractory lining.

Figure 3.3 Schematic diagram of a four-stage cyclone-type suspension preheater. P indicates the position at which a precalciner burner may be incorporated

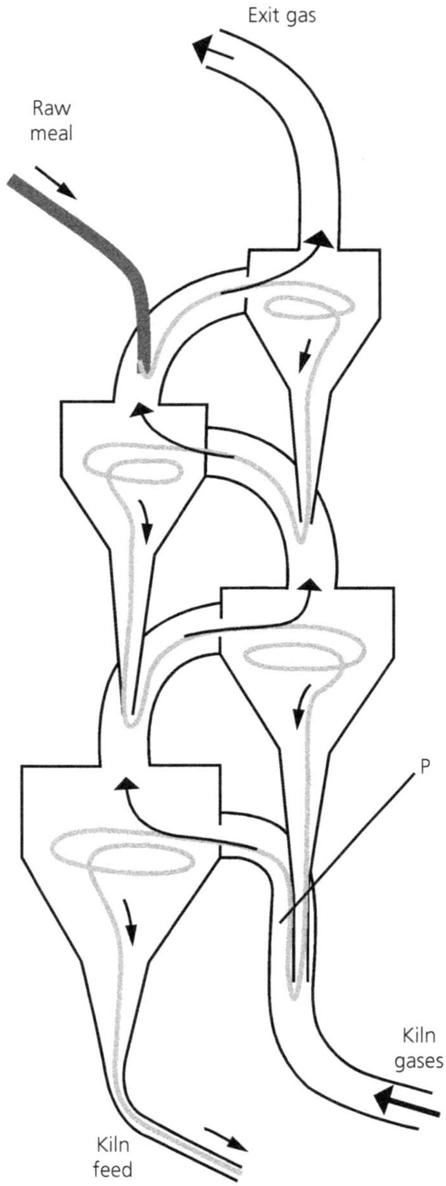

Less NO_x is formed, since much of the fuel is burnt at a low temperature and, with some designs, NO_x formed in the kiln may be reduced to nitrogen. Low-grade fuels can often be used in the precalciner. Precalcination may also make it economically feasible to deal with the problem, discussed in §3.2.5, of formation of deposits in the duct carrying the hot gases from the kiln to the preheater.

3.2.4 The rotary kiln

A rotary kiln is a tube, sloping at 3–4% from the horizontal and rotating at 1–4 rev/min, into which the material enters at the upper end and then slides, rolls or flows counter to the hot gases produced by a flame at the lower or 'front' end. The kiln is typically 35–100 m long and its ratio of length to diameter is 10–17 (diameter = free diameter inside the refractory lining). The maximum material temperature, of about 1450 °C, is reached near the front end of the kiln in the 'burning zone', also called the clinkering or sintering zone, in which the material spends 10–15 minutes. The kiln is lined with refractory bricks, of types that vary along its length in accordance with the varying gas and material temperatures. The bricks become coated with a layer of clinker, which plays an essential part in the insulation and in extending their life. The diameter of the kiln does not exceed 6–7 m (Chatterjee, 2011; FLSmidth, 2014) to maintain reasonable life for the refractory lining (Chatterjee, 2011).

Nodules of clinker, typically 3–20 mm in diameter, are formed in a semisolid state in the burning zone and solidify completely on cooling, which begins in a short cooling zone within the kiln and continues in a cooler. In modern plants, when the nodules leave the kiln, their internal temperatures are around 1350 °C, but their surface temperatures are considerably lower.

Liquid or pulverised solid fuels are blown into the kiln through a nozzle with 'primary' air. Additional 'secondary' air is drawn into the kiln through the clinker cooler. The flame in the rotary kiln must meet several requirements. The clinker must be correctly burned, so as to minimise its content of free lime, with the least expenditure of fuel. The ash from a solid fuel must be uniformly absorbed by the clinker. For normal Portland cements, the conditions must be sufficiently oxidising that the iron is present in the cooled clinker almost entirely as Fe^{3+}; however, for white cements, mildly reducing conditions may be preferable. Proper flame control also extends the life of the refractory lining of the kiln. Computer-aided or fully automated control of kiln operating conditions is increasingly used.

3.2.5 Circulation of volatiles; dust; cooling of clinker

In the hotter parts of the kiln, the K, Na, S, Cl and some other elements provided by the raw material or fuel are partly or wholly volatilised. The fractions that are not, pass into the clinker. The volatilised material is swept back with the hot gases to the cooler parts of the system where, assuming that a suspension preheater or precalciner is used, it is largely reabsorbed, so that cycles are set up. Material that is not reabsorbed may leave the system through the preheater as gas or dust, but some can form solid or semisolid deposits. Deposits formed in this and other ways can seriously restrict the movement of material and gases; in the kiln, they form rings. They can form in the duct carrying the hot gases from the kiln to the preheater. With some raw materials or fuels, it is necessary to divert part of the gas through a bypass to filter out the dust that it carries and to accept the consequent losses of heat and material, but changes in the raw materials or fuel may sometimes be a preferable solution (Davis and Longman, 1984).

Dust from the kiln is largely captured in a suspension preheater; that which is not is removed by an electrostatic precipitator before the gas passes up the chimney. The flow of gas is aided by an exhaust fan. The dust is as far as possible returned to the system by mixing with the raw meal or with the fuel (insufflation), although the extent to which either is possible may be restricted by its content of alkali sulfates or chlorides which, if high, could lead to an unduly high recirculating load, conducive to the formation of deposits or undesirably high concentrations in the clinker.

The clinker cooler is essentially a heat exchanger that extracts heat from the clinker for return to the system; also, a cooled clinker is more readily transported, ground or stored. Rapid cooling from the clinkering temperature down to 1100 °C produces a better-quality clinker (§3.5.5), and the clinker should be effectively air-quenched as soon as it leaves the burning zone. When the clinker enters the grinding mill, its temperature should preferably be below 1100 °C. Current practice favours grate coolers in which the clinker passes over moving grates through which air is blown. More air is needed for cooling than for combustion in the kiln, especially with a fuel-efficient process. Some of the excess ('tertiary air') may be ducted to a precalciner or used to dry the coal and raw materials, but a quantity normally remains for which it has proved difficult to find economically viable uses, and which is therefore cleaned and exhausted; steam raising for electricity generation has been suggested as a possible use.

3.2.6 Other processes for clinker production; clinker grinding

In the wet process, the raw material is fed as a slurry directly to a rotary kiln, which typically has a length to diameter ratio of about 30 and may be up to some 200 m long. In surviving wet process kilns, the slurry typically has a water content of 30–35 %. A system of chains near the back end assists heat transfer. Cycling of volatiles and of dust is much less pronounced than with preheater systems. In the semiwet process, some improvement in fuel efficiency is obtained by pressing out part of the water to produce a cake containing somewhat below 20 % water, which can be fed to the kiln directly or through certain types of preheater. In the semidry or Lepol process, the raw materials are made into nodules with just over 10 % water content which are fed to a moving grate preheater. The fuel efficiency is lower than that obtainable with a suspension preheater. Combined processes have been brought into operation in various countries for making cement and other materials, such as aluminium, iron or sulfuric acid (Goldshtein, 1985). While there are problems associated with such processes, the latter may become increasingly important as classical raw materials become scarcer and environmentally acceptable ways of disposing of industrial wastes more difficult.

To produce Portland cement, the clinker is ground together with gypsum. Portland cement clinker has a long storage life and, while it may be ground immediately, there are often good reasons for grinding it intermittently, and not necessarily in the same plant. Compositional and other variations in the clinker can be much reduced by using storage systems in which blending occurs. Natural gypsum, which is generally used, commonly contains significant proportions of such impurities as anhydrite, quartz, calcite and clay minerals. In some countries, by-product gypsums from various industrial processes are used. A proportion of anhydrite ($C\bar{S}$) may also be added. Some national specifications permit the addition of materials other than calcium sulfate. To make interground composite cements, widely varying proportions of granulated blast-furnace slag, fly ash or other materials are added. Grinding typically requires a power consumption of 35–50 kWh $tonne^{-1}$ (125–180 kJ kg^{-1}), almost all of which is converted into heat, and cooling is required.

3.3. Reactions below about 1300 °C

3.3.1 Decomposition of carbonate minerals

The dissociation pressure of calcite reaches 0·101 kPa (1 atm) at 894 °C (Southard and Royster, 1936) and the decarbonation reaction is highly endothermic (§3.1.4). The rate of decarbonation becomes significant at 500–600 °C if a sufficiently low partial pressure of CO_2 is maintained or if the calcite is intimately mixed with materials, such as quartz or clay mineral decomposition products, that react with the calcium oxide. Even in a precalciner such mixing occurs, aided by agglomeration caused by the presence of low-temperature sulfate melts.

In the absence of other substances, dolomite [$CaMg(CO_3)_2$] begins to decompose rapidly in air at about 750 °C, giving initially periclase and a carbonate of higher Ca/Mg ratio. The decomposition temperature is much affected by the presence of other substances.

The mechanism and kinetics of calcite decomposition have been much studied. The reaction proceeds by the movement inwards from the surface of an interface, behind which the material is converted into lime, thus producing a highly porous pseudomorph. The interface moves at a constant rate, implying that the rate at any instant is proportional to the area of the interface. In principle, the rate is controlled by the slowest of the following five steps

(1) transfer of heat to the exterior surface
(2) transfer of heat from the exterior surface to the interface
(3) chemical reaction at the interface
(4) transfer of carbon dioxide to the exterior surface
(5) transfer of carbon dioxide away from the exterior surface.

Hills (1968) showed that, contrary to some earlier opinions, it was not possible to determine which is the rate-controlling step, either from the observed relationship between the extent of reaction and the time or from the apparent energy of activation, obtained by applying the Arrhenius equation to the rate constant determined at different temperatures. For large particles and low CO_2 pressures, the temperature within the sphere could be as much as 45 °C below that of the surrounding gas, indicating that heat transfer controls the rate. For the conditions existing in a rotary kiln, in which decomposition occurs in a deep, moving mass of material, heat transfer again determines the rate (Bye, 1983). A rotary kiln is well suited to the later stages of clinker production, but much less so for calcination which, if carried out in it, requires some tens of minutes for completion.

Very different conditions exist in a precalciner, where the raw meal is dispersed in hot gas. The reaction still proceeds by the movement of an interface inwards, but the rate is controlled by the chemical reaction, and the temperature within the particle is virtually that of the surrounding gas (Bucchi, 1978; Vosteen, 1974). The rate is much higher than in a rotary kiln and, as seen in §3.2.3, decomposition is normally 90–95 % complete within a few seconds.

3.3.2 Decomposition of clay minerals and formation of products

Laboratory experiments in which single substances or mixtures are heated in stationary containers in air provide a starting point for understanding the reactions that occur in manufacturing clinker, although caution is needed in applying the results because of the important differences in conditions.

The behaviour of clay minerals on heating depends on their structure, composition, crystal size and degree of crystallinity. In general, any interlayer or adsorbed water is lost at 100–250 °C; dehydroxylation begins at 300–400 °C and is rapid by 500–600 °C. Clay minerals in which Al^{3+} is the predominant octahedral cation do not form phases of radically different structure as the immediate products of dehydroxylation, but undergo varying degrees of structural modification and disorganisation due to the replacement of $2OH^-$ by O^{2-}. Kaolinite (AS_2H_2) gives a poorly crystalline product, called metakaolin; with illites, the structural disorganisation is less marked, perhaps because less water has to be lost. Above about 900 °C, all begin to give new, crystalline phases, which are typically Al–Si spinel, cristobalite and, above 1100 °C, mullite.

Quartz undergoes a minor, rapid and reversible phase transition to α-quartz at 573 °C. It is unstable relative to tridymite at 867–1470 °C, and to cristobalite above 1470 °C. These changes are slow in the absence of other materials, but many admixed materials accelerate them and lead to formation not of tridymite, but of cristobalite at temperatures over about 1000 °C.

As noted in §3.3.1, the decarbonation of calcite is greatly favoured by intimate mixing with quartz or clay minerals. Under these conditions, much CO_2 is lost before any free lime can be detected. The formation of C_2S as an early product is well established, but the situation with the calcium aluminate phases is more complex, no one phase being dominant as an initial product (Williamson and Glasser, 1962). The aluminate or aluminosilicate phases most often reported as early or intermediate products in laboratory experiments with pure chemicals or raw meals have been CA, $C_{12}A_7$ and, less frequently, gehlenite (C_2AS) (Chromý, 1976; Cottin *et al.*, 1986; de Keyser, 1951, 1954, 1955; Imlach and Hofmänner, 1976; Lehmann *et al.*, 1964; Ludwig and Ruckensteiner, 1974; Rivera *et al.*, 1987; Weber, 1963; Weyer, 1931). C_3A can form by 850 °C (Cottin *et al.*, 1986; Ludwig and Ruckensteiner, 1974) but seems usually to be a later product. The ferrite phase forms readily and is initially of low Al/Fe ratio (de Keyser, 1954; Rivera *et al.*, 1987).

The reactions are notably accelerated by water vapour (Cottin *et al.*, 1986). In the cement kiln or precalciner agglomeration may influence their course. Glasser (1983, 1986) reviewed kinetic studies. The complexity of the mechanisms has not always been recognised, and it is doubtful whether kinetic data suffice to determine them, but electron microscopy and X-ray microanalysis are beginning to provide reliable information (Rouanet, 1986).

3.3.3 Sampling from cement kilns or preheater outlets

Several investigators have examined samples from rotary kilns during normal operation. In one such study, Weber (1963) obtained samples and measured temperatures, mainly using sampling ports in the kiln walls. In all cases, the raw materials were limestone and marl, and the fuel was pulverised coal. CA and $C_{12}A_7$, but not gehlenite, were found as early products. As in the laboratory studies, much calcite decomposed before any free lime was detected. The free lime content passed through a maximum value of up to 17%, referred to the mass of clinker, about 15 m from the front end of the kiln. At this point, the measured temperature was about 1000 °C. Laboratory studies (Imlach and Hofmänner, 1976) and analyses of kiln input meals from precalciners, described below, show that the maximum content of free lime reached varies greatly with the nature of the raw material.

One may also examine samples from kilns that have been shut down, although uncertainties inevitably exist due to possible phase changes and movement of material during cooling, which is carried out with the kiln still rotating. Moore (1976) reported an XRD study of charge and coating samples taken from numerous points along the lengths of several kilns. Sulfate phases were prominent in both types of sample taken from the cooler parts of the kiln, and comprised anhydrite, $C_4A_3\overline{S}$, $C_5S_2\overline{S}$, calcium langbeinite ($KC_2\overline{S}_3$) and hydroxyl-ellestadite ($C_{10}S_3\overline{S}_3H$). KCl was sometimes found. Spurrite ($C_5S_2\bar{C}$), β-C_2S, $C_{12}A_7$ and ferrite were usually also found before all the calcite had decomposed. Gehlenite was usually found in coating samples at intermediate temperatures, but rarely in the charge; CA was never found. C_3A was a late product. Rarely, ferrite was found as a late product, being replaced earlier by iron-containing spinel, presumably indicating the existence of reducing conditions at the cooler end of the kiln. Moore (1976) concluded that sulfates and chlorides were important mineralisers and that both spurrite and $C_4A_3\bar{S}$ played parts in the normal reaction sequence. Later results probably do not exclude this latter conclusion, but both phases can undoubtedly be undesirable artefacts.

Wolter (1986) determined the phase compositions of kiln inlet meals from about 20 plants using cyclone preheaters, usually with precalciners. XRD showed that the decomposition products of the clay minerals were especially reactive, although some reaction of quartz also occurred. Phases detected, and some notable absences, were as follows

- unreacted phases from the raw meal: calcite, quartz, minor mica and felspars, traces of clay minerals
- product phases: free lime, periclase, poorly crystalline belite (β-C_2S), ferrite, $C_{12}A_7$ assuming that enough Al_2O_3 was present, and often spurrite. Gehlenite was possibly present but, if so, was minor compared with belite. Less basic silicates and aluminates (e.g. CA, CA_2) were absent
- condensed volatiles: arcanite $K\bar{S}$), sylvine (KCl), aphthitalite (rare), NaCl (exceptionally), anhydrite (for high SO_3/alkali ratios, and then less than expected). Absences included thénardite ($N\bar{S}$), calcium langbeinite, alkali carbonates, $C_4A_3\bar{S}$ and $C_5S_2\bar{S}$
- clinker dust and coal ash: alite, fly ash (rare)
- secondary products formed on storage or treatment: calcium hydroxide, hemihydrate.

The quantitative composition varied widely, depending on the degree of calcination, heterogeneity of the raw meal and whether kiln dust or volatiles were recirculated. $C_{12}A_7$ and ferrite formed easily. At high degrees of calcination, virtually all the Al_2O_3 and Fe_2O_3 had reacted, 10–60% of the total SiO_2 content was present as belite or spurrite and the content of free lime was 5–40%. The proportion of the CaO released from the calcite that was combined in other phases was high at low degrees of calcination and decreased as the latter proceeded; at 40–70% calcination, it was 30–75% or, in rare cases, over 90%.

3.3.4 Reaction mechanisms

The mechanisms of the reactions in cement raw meals (Chromý, 1976, 1982a; Imlach and Hofmänner, 1976) and lime–silica composites (Weisweiler *et al.*, 1986) have been studied using light and electron microscopy, XRD, high-temperature XRD and high-temperature light microscopy, EPMA and thermal analysis, and free lime determinations. Below about 1200 °C reactions occur largely in the solid state, the first visible effect in cement raw meals being the appearance of reaction rims around the larger quartz grains, which later become replaced by clusters of belite crystals. Chromý (1976) found that the rims comprised zones of cristobalite, isotropic material of refractive index 1·515–1·525 and (furthest out) belite. By 1200 °C, the belite layer was up to 2–3 μm thick, and the isotropic material, which formed a thicker layer, was tending to break through, forming necks linking adjacent quartz grains. The isotropic material contained Ca and Si, and had Ca/Si $<1{\cdot}0$; it gave a diffuse XRD peak at 3·7 Å and was probably a glass. XRD and optical evidence also showed the presence of wollastonite and a little tridymite; in samples cooled from 1280 °C, both were detectable optically as needles dispersed in the amorphous material and as crystals on the border of the latter with the quartz, respectively. The wollastonite had presumably formed from α-CS on cooling.

At 1200–1300 °C, high-temperature light microscopy showed movement of the silica cores and CS crystals; the amorphous material was softening. Where the belite shell had burst, clusters of belite crystals with a central pore usually replaced the original quartz grain, but where it had not, these clusters only formed at a higher temperature and were more compact, with very little interstitial material.

Maki *et al.* (1995) studied the formation of belite clusters at 1400 °C. Layers of belite and liquid formed within the quartz grains and grew out from them so that the products of neighbouring grains coalesced. Three kinds of belite crystal, differing in origin, texture and composition, were distinguished in the clusters.

Rapid increase in temperature is desirable at temperatures below those at which substantial liquid formation occurs (Bucchi, 1980; Chromý, 1976, 1982a; Glasser, 1986; Sprung, 1980; Wolter, 1985). Most of the belite, and almost all of the other product phases, subsequently either melt or react in the presence of the melt, and there is no merit in promoting crystal growth or removal of imperfections which would impede these processes. Slow heating may also allow the decomposition products of the clay minerals to transform into less reactive phases. It can also lead to the formation of microstructures unfavourable to the later reactions; Chromý (1976) found that it allowed the belite shells around the silica particles to thicken, producing composites slow to react with lime. In contrast, rapid heating increases movement of the liquid phase, when it forms, and thus improves the mixing of the calcareous and siliceous constituents (Chromý, 1982a).

3.3.5 Condensation or reaction of volatiles

Bucchi (1981b) reviewed the cycling of volatiles in kilns and preheaters. Condensation or reaction of species from the gas stream takes place below about 1300 °C. The sulfate and chloride phases produced were noted in §3.3.3. Some of the material thus deposited is liquid, even at quite low temperatures: melting in the Na_2SO_4–K_2SO_4–$CaSO_4$–KCl system begins below 700 °C (Ritzmann, 1971). These low-temperature melts have both good and bad effects. Deposition on the feed can promote reactions, as noted earlier. Deposition in the preheater or kiln can cause serious obstructions, and condensation of volatiles is responsible for the formation of kiln rings and deposits in preheaters at temperatures below 1300 °C (Amafuji and Tsumagari, 1969; Bucchi, 1981b; Chromý and Weber, 1981; Sylla, 1974, 1977). Dust particles become coated with films of liquid condensate that causes them to adhere to obstacles or cooler surfaces. The temperature within a coating or deposit gradually falls as more material is deposited; this allows the formation of compounds within the body of the coating that differ from those present on the hot face.

Spurrite, if formed, reinforces deposits by producing a mass of interlocking crystals; $C_4A_3\overline{S}$ and $C_5S_2\overline{S}$ behave similarly. The kiln atmosphere is high in CO_2, and spurrite is readily formed on heating either clinker (Amafuji and Tsumagari, 1969) or raw meal (Sylla, 1974) in CO_2 at 750–900 °C. It decomposes by 950 °C (Amafuji and Tsumagari, 1969). In SO_2, either raw meal or clinker gives anhydrite at 550–1150 °C and silicosulfate at 1050–1150 °C; in mixed gases, the reactions with SO_2 are favoured over those with CO_2 and, if K_2O is present in the solid, calcium langbeinite is formed at 600–900 °C (Amafuji and Tsumagari, 1969). Formation of spurrite appears to be favoured by the presence of K_2O or chlorides and inhibited by phosphates (Sylla, 1974), which are, however, undesirable constituents of clinker. Fluoride also appears to favour the formation of spurrite and silicosulfate.

Fluorides and compounds of Zn, Cd, Tl and Pb also undergo cycling and can be deposited in preheaters (Kirchner, 1986). Emission of toxic heavy metals must be avoided. Fluoride phases that have been reported in deposits include $KCa_{12}(SO_4)_2(SiO_4)_4O_2F$ (Fayos *et al.*, 1985) and an apatite phase of composition $K_3Ca_2(SO_4)_3F$ (Pérez-Mendez and Fayos, 1986; Triviño Vázquez, 1985).

3.4. Reactions at 1300–1450 °C

3.4.1 Quantity of liquid formed

The processes described in this section occur in the presence of substantial proportions of liquid. They comprise

- melting of the ferrite and aluminate phases and some of the belite
- nodulisation
- reaction of the free lime, unreacted silica and some of the belite to give alite
- polymorphic change of belite to the α form
- recrystallisation and crystal growth of alite and belite
- evaporation of volatiles.

At the clinkering temperature, the principal phases present at equilibrium are alite, belite and liquid. For the pure C_3S–C_2S–C_3A–C_4AF system, this phase assemblage will, except at high Al_2O_3/Fe_2O_3 ratios, be reached by 1400 °C (§2.3.3). Following Lea and Parker (1935), the quantity of liquid formed in a mix of given composition at this temperature may be calculated as follows.

- All the Fe_2O_3 is present in the liquid; hence the percentage of liquid is $100 \times Fe_2O_3/(Fe_2O_3)_l$, where Fe_2O_3 and $(Fe_2O_3)_l$ are the percentages of that component in the mix and in the liquid, respectively.
- The composition of the liquid is governed only by the Al_2O_3/Fe_2O_3 ratio of the mix, being represented by the point on the C_3S–C_2S surface where the 1400 °C isotherm is intersected by a plane corresponding to the value of that ratio. The intersection of the 1400 °C isotherm with the C_3S–C_2S surface runs near and approximately parallel to the boundaries between that surface and the C_3A and C_4AF surfaces. Table 3.3 gives liquid compositions for four values of the Al_2O_3/Fe_2O_3 ratio. The Al_2O_3/Fe_2O_3 ratios and Fe_2O_3 contents in these compositions are related by the empirical equation

  $$Al_2O_3/Fe_2O_3 = -0{\cdot}746 + 33{\cdot}9/(Fe_2O_3)_l \quad (3.20)$$

  which may be rearranged to give

  $$Fe_2O_3/(Fe_2O_3)_l = (Al_2O_3 + 0{\cdot}746Fe_2O_3)/33{\cdot}9 \quad (3.21)$$

  whence the percentage of liquid in the mix is $2{\cdot}95Al_2O_3 + 2{\cdot}20Fe_2O_3$.

- Lea and Parker (1935) corrected for the contributions to the liquid of MgO and alkalis. Assuming that the solubility of MgO in the liquid is 5–6 % and that the percentage of liquid is about 30 %, the contribution from MgO was assumed to be the MgO content of the mix or 2 %, whichever was smaller. The liquid was assumed to contain all the K_2O and Na_2O.

Table 3.3 Compositions of liquids in equilibrium with C_3S and C_2S in the $CaO–Al_2O_3–Fe_2O_3–SiO_2$ system at 1400 °C. From Lea and Parker (1935), with corrections supplied by FM Lea and noted by Bogue (1955)

Mass ratio Al_2O_3/Fe_2O_3	Mass percentage			
	CaO	Al_2O_3	Fe_2O_3	SiO_2
6·06	56·6	30·3	5·0	8·0
2·62	56·4	26·2	10·0	7·4
0·94	55·1	18·8	20·0	6·1
0·64	53·9	15·3	24·0	6·8

Similar calculations were made for other temperatures. At lower temperatures, the relationship differs according to whether the Al_2O_3/Fe_2O_3 ratio is above or below 1·38, the value for the $C_3S–C_2S–C_3A–C_4AF$ invariant point. Formulae for the percentage of liquid for various combinations of temperature and Al_2O_3/Fe_2O_3 ratio are

1450 °C	$3{\cdot}00Al_2O_3 + 2{\cdot}25Fe_2O_3 + MgO^* + K_2O + Na_2O$
1400 °C	$2{\cdot}95Al_2O_3 + 2{\cdot}20Fe_2O_3 + MgO^* + K_2O + Na_2O$
1338 °C ($Al_2O_3/Fe_2O_3 \geq 1{\cdot}38$)	$6{\cdot}10Fe_2O_3 + MgO^* + K_2O + Na_2O$
1338 °C ($Al_2O_3/Fe_2O_3 \leq 1{\cdot}38$)	$8{\cdot}20Al_2O_3 - 5{\cdot}22Fe_2O_3 + MgO^* + K_2O + Na_2O$

where MgO* denotes an upper limit of 2·0% to the term for MgO. Typical percentages of liquid (for a mix with 5·5% Al_2O_3, 3·5% Fe_2O_3, 1·5% MgO and 1·0% K_2O) are thus 24% at 1340 °C, 26% at 1400 °C and 27% at 1450 °C.

Such results, omitting the contributions of MgO and alkalis, are conveniently represented as contours on plots of Al_2O_3 against Fe_2O_3 (Figure 3.4). At 1400 °C, the effects of the two oxides on the quantity of liquid are additive and fairly similar mass for mass. At 1338 °C the effects are not additive, the maximum amount of liquid for a given total content of Al_2O_3 and Fe_2O_3 being obtained at an Al_2O_3/Fe_2O_3 ratio of 1·38. The effects of the minor components on the quantity of liquid formed have not been thoroughly studied. Swayze's (1946b) results indicate that the MgO depresses liquidus temperatures by about 50 °C and increases the Al_2O_3/Fe_2O_3 ratio in the liquid at the $C_3S–C_2S–C_3A–C_4AF$ eutectic to 1·63. Alkalis and sulfate may depress the liquidus temperatures further, but the effects of liquid immiscibility must be taken into consideration.

3.4.2 Burnabilities of raw mixes

The 'burnability' of a mix denotes the ease with which free lime can be reduced to an acceptable value in the kiln. For laboratory use, it has usually been defined either by the quantity of free lime present after a specified regime of heat treatment or by the time of such treatment needed to reduce the free lime content to some specified value, although some investigators have used dynamic heating or other regimes in attempts to simulate kiln conditions more closely. As Bucchi (1980) and Sprung (1980) have noted, none of the measures so far proposed has more than comparative significance as they do not take into account such factors as the atmosphere, movement of material

Figure 3.4 Mass percentages of liquid at 1338 °C (full lines) and 1400 °C (broken lines) in the CaO–Al_2O_3–Fe_2O_3–SiO_2 system (Banda and Glasser, 1978a)

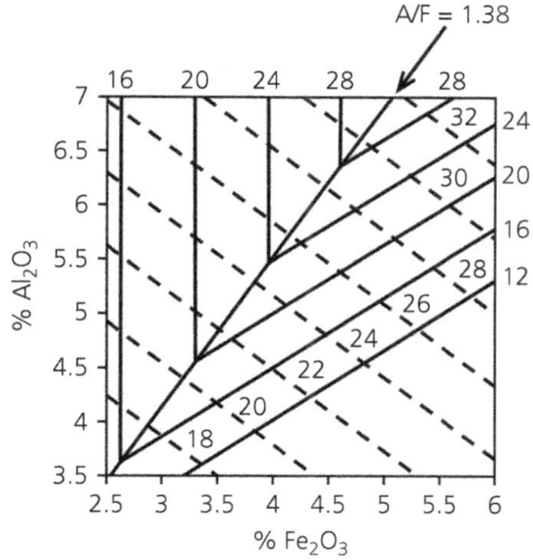

or rate of heat transfer in the kiln. The rate at which the free lime reacts depends not only on characteristics of the raw mix, but also on the effects on the latter of the heat treatment received before entering the clinkering zone and the conditions existing in that zone itself. The measures can strictly be justified only for a given kiln and set of operating conditions.

Many studies have shown that burnability decreases with increasing LSF or increasing SR. Increase in LSF implies more CaO that has to react, while increase in SR implies less liquid at a given temperature. These relationships have been discussed quantitatively (Christensen, 1979a). The AR is relevant at low clinkering temperatures because it then greatly affects the quantity of liquid (§3.4.1) and the temperature at which substantial formation of liquid begins, but at 1400 °C and above its effect is smaller; however, in the extreme case of white cements, the AR is very high and burnability is low. The temperature at which melting begins is also affected by the contents of minor components, such as MgO, and by the addition of fluxes. Burnability is affected not only by the quantity of liquid, but also by its physical properties, some of which are considered in §3.6.

Burnability is affected by the particle size of the raw mix, and especially by the contents of coarse particles (Christensen, 1979b; Heilmann, 1954; Ludwig and Ruckensteiner, 1974). During calcination, large particles of calcite and of siliceous materials are converted into aggregates of lime crystals and clusters of belite crystals respectively. The size distributions of these large particles determine the time required during the last stages of clinkering, when the free lime content is falling to its final value. The fine end of any distribution likely to occur in practice is less important. For assessing burnability at 1400 °C, the proportions of calcite particles larger than 125 μm, and of quartz particles larger than 44 μm, are particularly important (Christensen, 1979b).

Burnability also depends on the nature, microstructures and intimacy of mixing of the raw materials and on the contents of minor components. Silica present as quartz is less reactive than that present in clay minerals, and a limestone high in silica or silicate minerals is likely to be more reactive than one that is nearly pure calcite. Inadequate mixing of the particles of the different constituent rocks present in the raw mix has effects broadly similar to those of too high a proportion of coarse particles. Petersen and Johansen (1979) discussed the mixing of particles from a statistical viewpoint.

Several workers have obtained expressions relating burnability to compositional and other parameters (Christensen, 1979a, 1979b; Chromý, 1982b; Fundal, 1979; Ludwig and Ruckensteiner, 1974). Fundal (1979), defining burnability (C_{1400}) by the percentage of free lime remaining after 30 min at 1400 °C, found that

$$C_{1400} = 0{\cdot}33(\mathrm{LSF} + 5{\cdot}1\mathrm{SR} - 107) + 0{\cdot}93S_{44} + 0{\cdot}56C_{125} + 0{\cdot}2A_{\mathrm{q}} \tag{3.22}$$

where SR denotes silica ratio, S_{44} and C_{125} are the percentages of siliceous particles above 44 μm and of calcareous particles above 125 μm, respectively, and A_{q} is the percentage of acid–insoluble material, assumed to be quartz. This relation was considered to hold for LSF values of 88–100% and silica ratios of 2–6.

3.4.3 Nodulisation

Nodule formation occurs through the sticking together of solid particles by liquid. Petersen (1983a, 1983b, 1986) showed that the nodules grow partly by coalescence and partly by accretion, and developed a theory to predict the size distribution. Timashev (1980) regarded the process as one in which an initially relatively open assemblage of solid particles was compacted through the action of the liquid, initially by rearrangement of particles, and increasingly later by dissolution and crystallisation processes. Nodulisation requires an adequate proportion of liquid, and is favoured by low viscosity and high surface tension of the liquid and by small particle size of the solid (Petersen, 1986; Timashev, 1980). If there is not enough liquid, the outer parts of the nodules become enriched in silicates, some of which form dust (Petersen and Johansen, 1979). Overburning can have a similar effect (Long, 1982a). If the friable skin of large alite crystals remains on leaving the kiln, the clinker has a sparkling appearance. Much is abraded in the kiln, forming a dust that can cause rings or other deposits to form. It can also reaggregate to form porous, alite-rich lumps, which can incorporate small clinker nodules.

Nodulisation, chemical reactions and evaporation of volatiles are interdependent. Alite formation and loss of volatiles affect nodulisation, and the compaction that the latter entails must affect the kinetics of the former processes. Alite formation thus occurs in a changing environment. Initially, the clinker liquid is not a continuous and uniform medium enveloping the grains of lime, belite and other solid phases; rather, it fills separate pores and capillaries, forming thin films on the particles, and there are large local variations in composition (Timashev, 1980). As compaction proceeds, it becomes more continuous and more uniform.

3.4.4 Formation and recrystallisation of alite

In the study described in §3.3.4, Chromý (1976) showed that a melt of low Ca/Si ratio began to form at 1200–1300 °C around the original quartz grains, which were being converted into clusters of belite crystals. From 1300 °C, a liquid also began to form around the grains of lime. It differed in reflectance from that surrounding the silica or belite crystals and was considered to be of higher Ca/Si ratio. Where it bordered on the belite clusters, alite was precipitated, initially as a compact layer. This later recrystallised, allowing the liquids to mix. Alite now formed more rapidly; the belite clusters contracted and the crystals that remained recrystallised.

In further studies, Chromý (1982a, 1982b) and Chromý and Hrabě (1982) distinguished three stages in the reaction at clinkering temperatures. In the first, melt and belite are formed. In the second, belite continues to form; this stage yields a material consisting of isolated clusters of lime crystals having thin layers of alite on their surfaces and dispersed in a matrix of belite crystals and liquid. In the third, the clusters of lime react with the surrounding belite to give alite by a mechanism based on diffusion of CaO through the liquid. The temperature increases rapidly during the first stage, but the subsequent stages are isothermal.

Several workers have discussed the kinetics of the clinker-forming reactions (Butt *et al.*, 1976; Chromý, 1982a, 1982b; Chromý and Hrabě, 1982; Glasser, 1986; Johansen, 1973; Timashev, 1980). Because the material is not uniform, different factors may dominate at different times and in different regions (Timashev, 1980) but the principal one appears to be diffusion of Ca^{2+} through the liquid present between the alite crystals in the layers coating the lime clusters (Chromý and Hrabě, 1982).

From the standpoint of 28-day strength, the optimum size of the alite crystals appears to be about 15 μm, larger crystals containing fewer defects and thus being less reactive (Long, 1982a). Ono (1981) considered that cannibalistic growth of alite crystals was unimportant and that the size depended essentially on the rate of heating to the clinkering temperature; if this was low, the alite formed at relatively low temperatures as large crystals, whereas if it was high, it formed at higher temperatures as smaller crystals. This view has been only partially accepted by other workers, Hofmänner (1975), Long (1982a) and Maki *et al.* (Maki, 1986; Maki and Takahashi, 1984; Maki *et al.*, 1983) all having found that alite crystals grow larger through recrystallisation with increase in either time or temperature in the clinkering zone, and even (Hofmänner, 1975) during cooling. The rate of heating to the clinkering temperature is thus only one factor affecting alite size. In practice, few production clinkers contain alite with an average size as low as 15 μm.

Long (1982a) noted that alite size is indirectly affected by both bulk composition and fineness of the raw feed. Increases in LSF, SR or AR all make a clinker harder to burn, which tends to increase the size, although with increased SR or AR the effect is counteracted by decreases in the quantity and mobility of the liquid, both of which tend to decrease the size. The presence of coarse particles, especially of silica, also demands harder burning and tends to increase the size. Coarse silica particles lead to increased alite size both for this reason and because they yield large, dense clusters of belite crystals which, on subsequent reaction with lime, give large crystals of alite.

3.4.5 Evaporation of volatiles; polymorphic transitions

Miller (1986) summarised factors affecting the evaporation of volatile species in the burning zone. Much of the alkali and SO_3 is often present as sulfate melts. Alkali sulfates volatilise slowly because of their low vapour pressures, because diffusion within the clinker nodules is slow and because the nodules do not spend all their time on the surface of the clinker bed. Evaporation is favoured by increased residence time or temperature, decreased size or increased porosity of the clinker nodules. Low partial oxygen pressure also favours evaporation through decomposition to give SO_2, O_2 and alkali oxides. Chloride favours it through formation of the much more volatile alkali halides. A high partial pressure of H_2O favours it through formation and volatilisation of alkali hydroxides. SO_3 present as anhydrite is more easily lost because decomposition to SO_2 and O_2 occurs, and this is also true of sulfur present in the fuel as FeS_2 or organic compounds. Alkali present in the silicate or aluminate phases also appears to be more easily lost than that present as sulfates.

Figure 3.5 DTA curve for a white cement clinker raw mix, obtained in air at 20 °C min^{-1}. Replotted from Chromý (1976)

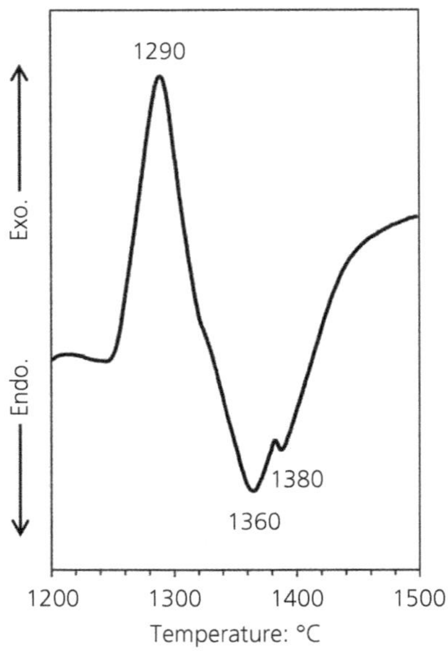

Alite may be expected to form in clinkers as the R polymorph, and belite as α'_L, α'_H or α, according to the temperature. Chromý (1976) and Hofmänner (1975) discussed DTA curves of raw meals (Figure 3.5). The main exotherm is due to belite formation, and the main endotherm to melting. A small superimposed endotherm near 1380 °C was attributed to the $\alpha'_H \rightarrow \alpha$ transition of the belite, at a temperature lowered by solid solution (Chromý, 1976). Curves for different raw meals show considerable variations.

3.4.6 Effects of reducing conditions; colour of clinker

A cement kiln is normally operated with what is commonly described as an oxidising atmosphere, but this term has purely relative significance and the equilibria in the burning zone are such that a significant fraction of the iron is present as Fe^{2+}. This is almost completely converted into Fe^{3+} during cooling. Localised reducing conditions can, however, occur in the burning zone from such causes as deficiency of oxygen in the kiln gas, presence of reducing material in the raw feed or entrapment of solid particles from the fuel. This can have major effects on the constitution of the clinker (Centurione, 1991; Locher, 1980; Long, 1982b, 1983; Scrivener and Taylor, 1995a; Sorrentino, 1994; Sylla, 1981; Woermann, 1962). Decomposition of the alite is promoted, and more aluminate phase is formed at the expense of ferrite. Metallic iron, sulfides (CaS and $KFeS_2$) and wüstite can be produced. Some of the Fe^{2+} that is formed replaces Ca^{2+} in the solid phases; the Fe/Al ratio of the alite increases. For normal cements, the effects on properties are undesirable; strength development is impaired and setting time can be more difficult to control. For white cements, which contain very little Fe, reducing conditions may be preferred because they yield a whiter product. Locher (1980) concluded that the bad effects of reducing conditions were avoided so long as the clinker left the kiln at a temperature of at least 1250 °C and thereafter cooled rapidly in air; however, reoxidation of Fe^{2+} in alite below 1300 °C has been observed to cause the formation of exsolution lamellae consisting of C_2F and belite (Long, 1983).

Portland cement clinker is normally almost black. The colour arises from the ferrite (§1.5.4). It has often been supposed that a brown or yellow colour is a sign of reducing conditions in the kiln but this is not correct. The colour is determined by the extent of oxidation of the ferrite phase during cooling, and thus on the regime of time, temperature and oxygen supply at this stage (Gartner, 1986; Johansen and Jakobsen, 1993; Locher, 1980; Long, 1982b; Scrivener and Taylor, 1995b; Sylla, 1981). Especially in the centres of large nodules, it is possible to have a normal phase assemblage but a light colour. On the other hand, it is possible to have a highly reduced clinker in which oxidation of the ferrite during cooling has been sufficient to produce a dark colour (Scrivener and Taylor, 1995a). Light colours can also arise from underburning.

3.5. Reactions during cooling, grinding or storage

3.5.1 Solidification of the clinker liquid: indications from pure systems

In principle, the clinker liquid can behave in any of three ways during cooling, with intermediate possibilities. In the first, equilibrium between the liquid and the pre-existing solid phases (alite and belite) is continuously maintained. This implies the possibility of material transfer between these solids and the liquid in either direction for as long as any liquid remains. This situation will be called 'equilibrium crystallisation'. In the second, the liquid does not crystallise but forms a glass. In the third, the liquid crystallises independently – that is, without interacting with the solid phases already present. Intermediate modes include, for example, equilibrium crystallisation down to a certain temperature and independent crystallisation below it.

Lea and Parker (1935) found that, in the pure $CaO–Al_2O_3–Fe_2O_3–SiO_2$ system, the compositions of the liquids in equilibrium with C_3S and C_2S at 1400 °C lie within the $C_3S–C_2S–C_3A–C_4AF$ subsystem if the AR is between 0·7 and 1·7; for these compositions, equilibrium crystallisation and independent crystallisation lead to identical quantitative phase compositions. If the AR exceeds 1·7, the liquid composition is deficient in CaO relative to the above subsystem; this results in the formation of a little $C_{12}A_7$ at the expense of C_3A if crystallisation is independent, and in reaction with some of the alite to form belite if it is not. Conversely, for an AR below 0·7, the liquid composition has a surplus of CaO relative to the $C_3S–C_2S–C_3A–C_4AF$ subsystem, resulting in the formation of a little free lime if crystallisation is independent and in reaction with some of the belite to form alite if it is not. Equations were derived for calculating the quantitative phase composition for various cooling conditions (Bogue, 1955; Dahl, as quoted by Bogue, 1939; Lea and Parker, 1935).

3.5.2 Do Portland cement clinkers contain glass or $C_{12}A_7$?

The applicability of these conclusions to production clinkers requires examination. The compositions of the aluminate and ferrite phases differ markedly from C_3A and C_4AF, respectively, (Chapter 1) and, because of the presence of minor components, especially MgO, TiO_2 and alkalis, the compositions of the liquid phase differ from those in the pure system. We shall consider, in turn, first, whether glass is formed and, if so, under what conditions; second, whether either $C_{12}A_7$ or free lime crystallises from the liquid phase under appropriate conditions; and third, whether there is evidence of either equilibrium or independent cooling.

The early literature contains many references to the presence in production clinkers of glass, often in substantial proportions. This view was based partly on observations by light microscopy; however, this method cannot distinguish glass from crystalline solids of the cubic system unless crystals with distinct faces have been formed, nor from crystalline materials of any kind if the crystals are below a certain size. It was also found that if clinkers believed to contain glass were annealed, their heats of solution in an acid medium increased, and this method was used to obtain approximate estimates of the glass content (Lerch and Brownmiller, 1937). This evidence, too, is inconclusive, because the same effect would arise from the presence of small or structurally imperfect crystals.

Quantitative XRD, SEM or X-ray microanalysis do not support the view that modern clinkers contain significant quantities of glass, except in rare cases. Regourd and Guinier (1976) found that a few clinkers gave a strong, diffuse XRD peak at 30–35° 2θ (Cu Kα), indicating the presence of glass or other amorphous material, but concluded that the amount of glassy or microcrystalline phase did not exceed a few per cent. While some studies that used the Rietveld method of quantitative phase analysis of powder XRD data have supported the presence of a significant quantity of amorphous material in Portland cement (e.g. Suherman *et al.*, 2002; Whitfield and Mitchell, 2003), others have provided compelling evidence that such results are artefacts of refinement misfits (Jansen *et al.*, 2011) or of sample preparation procedures (Snellings *et al.*, 2014). For example, Snellings *et al.* (2014) showed that wet-grinding a Portland cement sample to an average particle size around 3.5 μm reduced the content of amorphous material from 20–25 wt.% to an insignificant amount.

An SEM and X-ray microanalytical study showed that white cement clinkers contained glass if quenched in water but not if cooled in air (Boikova *et al.*, 1986). A laboratory study on the cooling of simulated clinker liquids showed that glass, amorphous to X-rays, is formed on quenching small samples in water (Banda and Glasser, 1978b). Somewhat slower cooling, from 1300 to 700 °C in a few seconds, gave 'proto- C_3A', a poorly crystalline, iron-rich form of that phase; still slower cooling (1300–1100 °C in 6 min) gave normally crystalline aluminate and ferrite phases. Material probably identical with proto-C_3A was earlier obtained as a nonequilibrium quench growth (Lerch and Brownmiller, 1937) and appears to form easily. The viscosity of the clinker liquid is near the lower limit for glass formation (Banda and Glasser, 1978b). These observations suggest that glass may occur in clinkers that have been quenched in water but is unlikely to be found with slower methods of cooling.

Neither XRD nor electron or light microscopy indicates that $C_{12}A_7$ is present in any modern clinkers and, when free lime is found, microscopic examination does not suggest that it has formed by precipitation from the liquid.

If clinkers of relatively high AR are cooled slowly, the effect of equilibrium cooling is detectable by light microscopy (Long, 1983). The alite crystals are eroded and a layer of belite forms on their surfaces; 'pinhead' crystals of belite separate from the liquid. At AR values $>2{\cdot}5$, slow cooling can cause a reduction of up to 10% in the concrete strength, attributable to the lowered content of alite. These effects are not observed at the moderately fast cooling rates more usual in practice. This evidence suggests that under normal conditions, crystallisation approximates to independent, and that slower cooling is needed to achieve equilibrium crystallisation. Production clinkers thus behave differently from mixes in the pure, quaternary system, which at AR $>1{\cdot}7$ give $C_{12}A_7$ on independent crystallisation.

3.5.3 Evidence from X-ray microanalysis

X-ray microanalyses by many investigators (Chapter 1) also provide no indication that $C_{12}A_7$ occurs in normal clinkers. Ghose's (1980) results are of particular interest because he did not attempt to distinguish between the individual phases within the interstitial material, of which he thus obtained bulk analyses. Taken in conjunction with the results of other investigations in which individual phases were analysed, they are consistent with the hypothesis that, for AR values between about 1 and 4, the interstitial material consists of aluminate and ferrite phases having the compositions given in Table 1.2, together with 4–10% of belite, the amount of which increases with AR. For lower AR values, the interstitial material consists essentially of ferrite of higher iron contents, while in white cements it consists of aluminate very low in iron, possibly together with glass. In each case, small amounts of silicate phases are probably also present.

The reason why mixes with AR $>1{\cdot}7$ do not yield any $C_{12}A_7$ on independent crystallisation is that the solid phases are not pure C_3A, C_4AF and C_2S. For AR $= 2{\cdot}71$, the quaternary liquid in equilibrium with C_3S, C_2S and C_3A at 1400 °C contains 55·7% CaO, 27·1% Al_2O_3, 10·0% Fe_2O_3 and 7·2% SiO_2 (Swayze, 1946a). This composition can be closely matched by a mixture of aluminate (63%), ferrite (30%) and belite (7%), each having the compositions given in Table 1.2, the bulk composition of this mixture being 54·4% CaO, 26·4% Al_2O_3, 9·7% Fe_2O_3, 5·6% SiO_2 and 1·8% MgO, with <1% each of TiO_2, Mn_2O_3, Na_2O and K_2O. Independent crystallisation can thus yield a mixture of the three phases. The liquid composition cannot be matched by a mixture of pure C_3A, C_4AF and C_2S, which is relatively too high in CaO, so that if no ionic substitutions occurred, some $C_{12}A_7$ would also be formed. A strict comparison would be with the actual composition of the clinker liquid, which is modified by minor components, but lack of adequate data precludes this.

3.5.4 Effects of cooling rate on the aluminate and ferrite phases

It follows from the above discussion that the compositions of the aluminate and ferrite phases in typical clinkers are controlled by the composition of the clinker liquid and the conditions under which the latter crystallises. For clinkers with AR 1–4, their compositions are relatively constant because both phases are present and because cooling rates are normally moderately fast. The compositions must nevertheless depend to some extent on the composition of the raw mix, especially in regard to minor components. The observation that resorption of alite occurs on slow cooling implies that their compositions, or their relative amounts, or both, depend on the cooling rate. They could also be affected by the clinkering temperature and the time during which it is maintained. XRD (Barlow, 1985; Ichikawa *et al.*, 1994; Ono and Nagashima, 1976) and microanalytical (Ichikawa *et al.*, 1994) evidence shows that their compositions are affected by the cooling rate. IR evidence indicates the occurrence of structural and compositional variations in both phases depends on the cooling rate (Vásquez *et al.*, 1984).

The conditions of crystallisation affect the ferrite and aluminate in ways other than in their mean compositions. Slow cooling produces relatively large crystals of each phase, while fast cooling produces close intergrowths; textures are discussed further in §4.2.1. Zoning occurs readily in the ferrite (§2.3.1) and may also occur in the aluminate. The distribution of atoms between octahedral and tetrahedral sites in the ferrite depends on the temperature at which internal equilibrium within the crystal has been achieved (§1.5.1). The degree of crystallinity of both phases appears to vary with cooling rate (Vásquez *et al.*, 1984). All these effects, and perhaps others, may affect the behaviour of the interstitial material on hydration.

3.5.5 Other effects of cooling rate

The behaviour of MgO in clinker formation depends markedly on the cooling rate (Long, 1983). If a clinker is burned at a high temperature (>1500 °C), relatively high contents of MgO can enter the liquid; on rapid cooling, much of it remains in the aluminate and ferrite phases and only a small quantity of periclase separates as small crystals. On slow cooling from high temperatures, only around 1·5% of MgO (referred to the clinker) is taken into solid solution, the excess forming large periclase crystals. At temperatures below 1450 °C, the MgO is less readily taken into solution and clusters of periclase crystals can remain; the crystals grow with time and even with temperature in this range.

The polymorphic transitions of the alite and belite that occur during cooling are discussed in Chapter 1. The belite of normal clinkers contains sufficient proportions of substituent ions to prevent dusting to give the γ form; the mechanical constraint provided by the surrounding material may also contribute. The cooling rate affects the way in which the belite transitions occur, with effects recognisable microscopically (Ono and Nagashima, 1976). Cooling rate is also reported to affect alite crystal size, rapid cooling giving smaller crystals, which are more easily ground (Sprung, 1985).

If a clinker is cooled too slowly in the region from 1250 °C to about 1100 °C, the alite begins to decompose. An intimate mixture of belite and lime forms as pseudomorphs after the alite, and the decomposition is especially likely to occur if the clinker has been fired under reducing conditions (Long, 1983). Mohan and Glasser (1977) reviewed and extended knowledge of the kinetics of this process. The kinetics follow a sigmoidal curve in which an initial induction period merges into an acceleratory phase, followed by a deceleratory one as the process nears completion. These results can be explained by assuming that the rate is controlled by nucleation of the products on the surfaces of the C_3S grains; in accordance with this, the reaction is accelerated and the induction period largely eliminated, if the C_3S is sufficiently intimately mixed with either C_2S or CaO. Some substituent ions affect the rate: Al^{3+} has little effect, Fe^{+} markedly accelerates, Na^{+} slightly retards and Mg^{2+} markedly retards decomposition. Where both Mg^{2+} and Fe^{3+} are present, the accelerating effect of the Fe^{3+} is dominant. The decomposition is also accelerated by water vapour and, very strongly, by contact with sulfate melts; with a $CaSO_4$–Na_2SO_4 melt, decomposition was complete in 2 hours at 1025 °C.

All the effects described above indicate that rapid cooling is desirable: the aluminate phase reacts more slowly with water when finely grained and intimately mixed with ferrite, making it easier to control the setting rate (Sprung, 1985), decrease in alite content either from reactions involving the interstitial material or from decomposition is avoided, a higher MgO content can be tolerated, and the clinker is easier to grind.

3.5.6 Crystallisation of the sulfate phases

Much of the SO_3 is present at the clinkering temperature in a separate liquid phase, immiscible with the main clinker liquid (§2.5.2). The alkali cations are distributed between the two liquids and the alite and belite. During cooling, some redistribution of alkali cations and sulfate ions between the liquids may be expected to occur, the sulfate liquid finally solidifying below 900 °C. The most usual sulfate phases to form are arcanite (K_2SO_4), which can accommodate some Na_2SO_4 or $CaSO_4$ or both in solid solution, and aphthitalite ($K_{4-x}N_x\bar{S}_4$, with x probably usually near 1 but possibly sometimes up to 3). Clinkers relatively high in K_2O and SO_3 may contain calcium langbeinite ($KC_2\bar{S}_3$), and those high in SO_3 (typically above about 2%) may contain anhydrite. Clinkers of unusually high Na/K ratio may contain thénardite (sodium sulfate). Because the sulfate phases are a late product, their crystals are in general to be found between those of the silicate phases but, in clinkers of high SO_3 content, small amounts may also occur as inclusions in those phases.

3.5.7 Quantitative estimation of the distributions of alkalis and SO_3 between phases

Early work, reviewed by Newkirk (1954), indicated that the SO_3 combined as alkali sulfates, the excess of alkalis entering the silicate and aluminate phases, but subsequent work has shown that this is correct only for low ratios of SO_3 to alkalis. Pollitt and Brown (1969) made a systematic study of the distribution of alkalis and SO_3 in a number of works and laboratory-prepared clinkers. Figure 3.6 and the following discussion are based on their conclusions and the tabulated data in their paper. With reservations noted later, the sulfate and alkalis present in alkali sulfates or calcium langbeinite ($KC_2\overline{S}_3$) are rapidly extracted by water and can thus be distinguished from those substituted in the silicate and aluminate phases or, for SO_3, in anhydrite.

The results of such extractions appear to depend significantly on the experimental procedure, and perhaps especially on the length of time during which the sample is in contact with the water. Gartner and Tang (1987) reported results broadly similar to those of Pollitt and Brown (1969) for the alkalis but obtained generally lower results for the fractions of the SO_3 that were rapidly soluble. This may have been because they used a 10 min treatment, whereas Pollitt and Brown (1969) used a method in which the water was drawn through the sample under suction on a filter in a period of

Figure 3.6 (a) Molar fractions of rapidly water-soluble K_2O (open symbols) and Na_2O (filled symbols) plotted against R, the molar ratio of total SO_3 to (total K_2O + total Na_2O) for production clinkers (circles) and laboratory clinkers (squares); (b) molar fractions of rapidly water-soluble K_2O + Na_2O (half-filled symbols) and SO_3 (open symbols) plotted against R for the same samples. Recalculated from the data of Pollitt and Brown (1969)

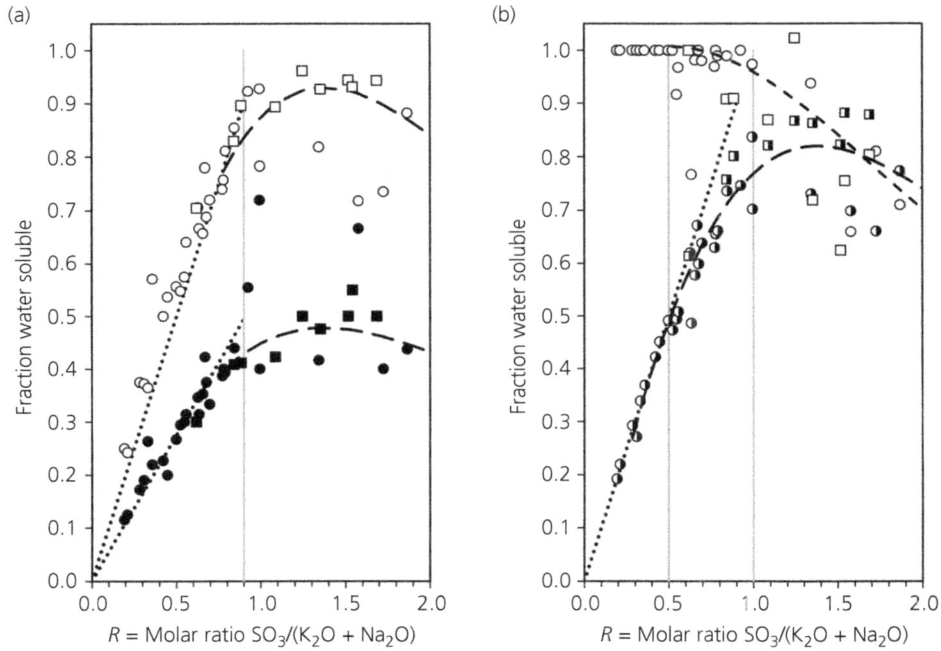

1·5–2·5 min (Moir GK, personal communication with HFWT). A longer time may have allowed some sulfate to be precipitated in ettringite. Results of extraction tests do not necessarily have implications regarding the availability of the alkalis or SO_3 present in these latter phases during hydration, which occurs under different conditions.

Inspection of Figure 3.6 indicates the following points.

- The fractions of the SO_3, K_2O and Na_2O that are rapidly water-soluble will be called $f(SO_3)$, $f(K_2O)$ and $f(Na_2O)$, respectively. They are largely determined by the molar ratio of SO_3 to ($K_2O + Na_2O$) in the clinker, which will be called R. $f(K_2O)$ is always close to twice $f(Na_2O)$ (e.g. compare the dotted lines on Figure 3.6(a)), showing that Na^+ tends more strongly than K^+ to enter the silicate or aluminate phases.
- For values of R up to about 0·5, $f(K_2O)$ and $f(Na_2O)$ increase with R, and $f(SO_3) = 1{\cdot}0$ (Figure 3.6(b)). The rapidly soluble SO_3 is exactly balanced by rapidly soluble alkalis, showing that all the SO_3 is present as alkali sulfates.
- For values of R between about 0·5 and 1·0, $f(SO_3)$ is still usually close to 1.0, but the contents of rapidly soluble alkalis tend to fall below that of the rapidly soluble SO_3. This can be attributed to the formation of calcium langbeinite ($KC_2\overline{S}_3$). $f(K_2O)$ and $f(Na_2O)$ continue to increase with R and reach limits of about 0·9 and 0·45, respectively, at $R \approx 0{\cdot}9$ (indicated on Figure 3.6(a)).
- With increase in R above about 1·0, $f(K_2O)$ and $f(Na_2O)$ tend to level out and then decrease. This might be explained by the presence of small proportions of the alkali sulfates as inclusions within the alite and belite. Alkali sulfates present as inclusions would be less accessible to water than crystals deposited between those of the silicate phases, which probably tend to become broken up and smeared over the surfaces of the cement grains during grinding.
- As R increases above about 1·0, and sometimes at lower values, $f(SO)_3$ also shows an irregular decrease. This could be due to inclusion of alkali sulfate phases within the alite or belite, solid solution of sulfate in the major clinker phases or formation of anhydrite.

Based on these considerations, the probable distribution of alkalis and SO_3 in a clinker can be estimated from the bulk analysis. The contents of rapidly soluble K_2O, Na_2O and SO_3 are first estimated from the smoothed curves in Figure 3.6. For the great majority of clinkers, they can be explained by assuming that any rapidly soluble SO_3 in excess of that equivalent to the rapidly soluble alkalis is present in calcium langbeinite, the rest of the rapidly soluble K_2O and all the rapidly soluble Na_2O then being assigned to alkali sulfates. The relative amounts of rapidly soluble K_2O and Na_2O give an indication of the distribution within the alkali sulfates between arcanite, aphthitalite and, rarely, thénardite, but this can be of doubtful reliability because of uncertainties about the extents of solid solution in the first two of these phases. With a very few clinkers, there is not enough rapidly soluble SO_3 to balance the rapidly soluble alkalis. This has been attributed to the presence of traces of alkali carbonates (Pollitt and Brown, 1969) or of potassium oxide and potassium aluminate (Fundal, 1986). The reverse situation, in which there is not enough K_2O to allow the sulfate balance to be completed through formation of calcium langbeinite, can also occur. Pollitt and Brown's (1969) data show considerable scatter at high SO_3/alkali ratios, and the most likely explanation is perhaps that the smoothed curve in Figure 3.6 does not correctly represent $f(SO)_3$ for the clinkers in question.

The less rapidly soluble alkalis could occur as substituents in the silicate or aluminate phases or in alkali sulfate inclusions in these phases. The less rapidly soluble SO_3 could occur in the same forms or as anhydrite. In neither case is it clearly possible to estimate the distributions between these forms from the bulk analysis alone, although, as noted above, the decreases in $f(K_2O)$ and $f(Na_2O)$ at values of R above about 1·5 suggest that alkali sulfate inclusions are a contributing effect. Tentative calculations based on the magnitudes of these decreases and on the SO_3 contents of the silicate phases suggest that, for clinkers relatively high in alkalis and of the commonly high K_2O/Na_2O ratios, anhydrite is unlikely to be formed unless the SO_3 content of the clinker is greater than about 2·0%. Limited experimental data appear to be largely consistent with this conclusion (Dreizler *et al.*, 1985; Gartner and Tang, 1987; Odler and Zhang, 1996; Pollitt and Brown, 1969).

3.5.8 Changes during grinding or storage

Méric (1980) reviewed the grinding and storage of clinker. Factors influencing the ease with which a clinker can be ground include the size distribution of the nodules, the texture and phase composition, the cooling rate and the conditions under which it has been stored. Grindability increases with alite content and decreases with that of the interstitial phases. Hornain and Regourd (1980) defined a brittleness index, which they showed to have values of 4·7 for alite, 2·9 for aluminate, about 2 for ferrite and 1·8 for belite. They also found that microcracking produced by rapid cooling from 1250 °C increased grindability. Some specifications allow the use of grinding aids. The particle size distributions of cements are discussed in §4.1.

The heat produced on grinding can cause partial conversion of the gypsum into hemihydrate ($2CaSO_4{\cdot}H_2O$) or γ-$CaSO_4$. The extent to which this occurs depends on the temperature reached and the relative humidity within the mill. The conversions affect the rate at which the calcium sulfate dissolves on hydration. Partial conversion to hemihydrate may be desirable as the water present in gypsum can cause the particles of cement to adhere during storage, with formation of lumps (Locher, 1980). To obtain an optimum rate of dissolution on hydration, part of the gypsum may be replaced by anhydrite. Lumpiness on storage can also result from the reaction of gypsum and potassium sulfate to form syngenite ($CK\bar{S}_2H$) (Long, 1983). Maultzsch *et al.* (1980) reported on the effects of prolonged storage; minor amounts of gypsum (from hemihydrate), syngenite, calcite and normal hydration products were present.

3.6. Effects of minor components

3.6.1 General

Minor or trace components derived from raw materials, fuel, refractories or other plant materials, or added deliberately, can affect the reactions of clinker formation, the properties of the product or both. Their effects can be beneficial or harmful. Beneficial effects include acceleration of the clinkering reactions or lowering of the temperature at which they occur, or an increase in the reactivity of the product leading to faster strength development. Harmful effects include decrease in alite content, volatilisation in the kiln with consequent formation of kiln rings or other deposits, decrease in the durability of concrete made with the cement or the introduction of poisonous elements. Some elements have beneficial effects at low concentrations and harmful ones at higher concentrations. Some of the effects on the manufacturing process have been reviewed (Bucchi, 1981a, 1981b; Herfort, 2015; Herfort *et al.*, 2010).

The main clinker liquid can be affected through alteration in the temperature at which it begins to form or in its quantity or properties at a given temperature. A flux is an agent that promotes a reaction by increasing the quantity of liquid at a given temperature; Al_2O_3 and Fe_2O_3 are fluxes for the formation of alite. Butt *et al.* (1976) and Timashev (1980) studied the properties of clinker liquids. They found that the viscosity and surface tension of the liquid are markedly affected by relatively small additions of other ions. Species containing strongly electropositive elements increase the viscosity and ones containing strongly electronegative elements decrease it. This was attributed to changes in the distribution of Al^{3+} and Fe^{3+} between octahedral and tetrahedral coordination. The effects of adding more than one minor component are not additive and can be complex – for example, alkali sulfates decrease the viscosity, but even a small stoichiometric excess of alkali oxides over SO_3 increases it. Many species lower the surface tension because ions that do not easily fit into the structure of the liquid tend to become concentrated near an interface, but the situation may be complicated by liquid immiscibility. Viscosity has a particularly important effect, a low value favouring alite formation by accelerating dissolution of lime and belite and diffusion through the liquid. Decrease in surface tension may be expected to favour wetting of the solids and penetration of the liquid into porous aggregates of lime crystals.

Effects of minor components on the solid phases are of various kinds. A mineraliser is an agent that promotes the formation of a particular solid phase by affecting the equilibria through incorporation in one or more of the solid phases. CaF_2 acts both as a mineraliser and as a flux in promoting the formation of alite. The mechanism of an observed effect is often not clear and, perhaps for this reason, the term 'mineraliser' is often used loosely. Other examples of action of minor components on solid phases include the stabilisation of M_3 C_3S and β-C_2S at room temperature, the modification of the aluminate structure resulting from incorporation of alkali, and the nonformation of $C_{12}A_7$ in situations where it would be formed in the pure $CaO–Al_2O_3–Fe_2O_3–SiO_2$ system. Minor components can affect the properties of a clinker phase in other ways, such as by causing changes in crystal size, morphology or structural perfection, and can also modify the microstructure of the clinker as a whole.

The first 11 minor components of raw mixes, in sequence of decreasing median concentration, are: MgO, K_2O, SO_3, Na_2O, TiO_2, Mn_2O_3, P_2O_5, SrO, fluorides, chlorides and Cr_2O_3 (Bucchi, 1980).

3.6.2 Effects of s-block elements

The effects of K_2O and Na_2O in producing low-temperature melts and volatile species have already been discussed. K_2O and Na_2O lower the temperature of formation of the melt but, in the absence of SO_3, increase its viscosity. If present in sufficient concentration, they lead to the formation of orthorhombic or pseudotetragonal aluminate; this can have undesirable effects on the rheology of fresh concrete. High K_2O contents can lead to syngenite formation and consequent lumpiness on storage. High contents of either K_2O or Na_2O can lead to destructive reactions with certain aggregates. Alkalis increase early strength but often decrease late strength.

MgO also lowers the temperature at which the main melt begins to form and increases the quantity of liquid; it slightly decreases the latter's viscosity (Butt *et al.*, 1976; Timashev, 1980). Its presence is probably the basic cause of the nonformation of $C_{12}A_7$ in normally cooled clinkers of high AR. It stabilises M_3 alite (§1.2.4). It retards the decomposition of alite during cooling. These effects are beneficial, but the formation of periclase, discussed in §3.5.5, limits the permissible content, which most national specifications set at 4–5 %.

Some limestones contain substantial proportions of SrO. Small contents favour alite formation, but phase equilibrium studies show that it increases the range of compositions at which free lime, with SrO in solid solution, is present at the clinkering temperature (Brisi and Appendino, 1965; Gilioli *et al.*, 1972, 1973). This limits the permissible content of SrO in mixes of normal LSFs to about 3 %.

3.6.3 Effects of p- and d-block elements

The behaviour of CaF_2 as a flux and mineraliser is described in §2.6.1. In addition to promoting formation of alite at markedly lower temperatures, CaF_2 accelerates the decomposition of calcite. Volatilisation does not cause serious problems in the kiln, and it has often been added to raw meals, typically in proportions below 0·5%. It retards setting, possibly through formation of $C_{10}S_4{\cdot}CaF_2$ during cooling, and this limits its use. As with admixtures to the raw meal in general, it is also necessary to consider whether the resulting savings outweigh the extra costs. This is most likely to be the case with raw materials that would otherwise require a particularly high burning temperature.

Chloride contents must be kept low to avoid corrosion of steel in reinforced concrete and formation of kiln rings and preheater deposits. Contents below 0·02% are preferred, although higher ones can be acceptable if a sufficient proportion of the kiln gases is by-passed or in less energy-efficient (e.g. wet process) plants.

The effects of SO_3 include formation of alkali sulfates and other low melting phases and volatile species, discussed earlier in this chapter and in §2.5.2. Although much of the SO_3 is present in a separate liquid phase, the main melt begins to form at a lower temperature and its viscosity is reduced, rendering $CaSO_4$ an effective flux. Against this, it leads to the stabilisation of belite and lime at clinkering temperatures and a consequent decrease in content or even disappearance of alite, although this effect is lessened if MgO is present (Gutt and Smith, 1968). Later work showed that high ratios of SO_3 to MgO promote the formation of large, irregularly shaped crystals of M_1 alite, which can engulf other phases (§1.2.4). Sulfate can be removed by increasing the burning temperature, but the inclusions of lime in alite that have formed at lower temperatures are not easily lost (Long, 1982b).

Portland cement clinkers typically contain around 0·2% of P_2O_5; higher contents lead to decreased formation of alite and, if above about 2·5%, to the presence of free lime. An early conclusion that α-C_2S forms a continuous series of solid solutions with a ('super α'-) C_3P above 1450 °C (Nurse *et al.*, 1959) was modified by later work, which showed that a miscibility gap exists at 1500 °C (Gutt, 1963). Solid solutions also exist between other combinations of C_2S and C_3P polymorphs, lowering the temperatures of the C_2S transitions and, for some compositions, stabilising α or α' forms at room temperature. In the CaO–C_2S–C_3P system (Gutt, 1963), C_3S has a primary phase field but, at a clinkering temperature of 1500 °C, mixes with more than a few per cent of P_2O_5 do not yield C_3S. The tolerance for P_2O_5 is increased somewhat if F^- is also present (Gutt, 1969).

The 3d elements from titanium to zinc all lower the viscosity of the clinker liquid in varying degrees, but their effects on clinker formation and properties present a confusing picture. It has been claimed that the content of alite in clinkers decreases with increase in that of TiO_2, but the effect is much less marked if the latter is regarded as replacing SiO_2 (Bucchi, 1980). Chromium can enter the clinker from raw materials or refractories; cements high in it have been reported to give high early strengths but toxicity renders their use impracticable. The enhanced early strength is possibly due to high defect concentrations in the alite (Bucchi, 1980), in which the chromium occurs as Cr^{4+} (Johansen, 1978). Both manganese and titanium occur in their highest concentrations in the ferrite phase and, probably for this reason, yield darker coloured clinkers.

The cycling of zinc, cadmium, thallium, lead, sulfur, fluoride and chloride is noted in §3.2.5 and §3.3.5.

REFERENCES

Amafuji M and Tsumagari A (1969) Formation of double salt in cement burning. *Proceedings of the 5th International Symposium on the Chemistry of Cement* **1**: 136–156.

ASTM (2024) C150/C150M-24: Standard Specification for Portland Cement. ASTM International, West Conshohocken, PA, USA.

Banda RMH and Glasser FP (1978a) Role of iron and aluminum oxides as fluxes during the burning of Portland cement. *Cement and Concrete Research* **8(3)**: 319–324.

Banda RMH and Glasser FP (1978b) Crystallization of the molten phase in Portland cement clinker. *Cement and Concrete Research* **8(6)**: 665–670.

Barcelo L, Kline J, Walenta G and Gartner E (2014) Cement and carbon emissions. *Materials and Structures* **47(6)**: 1055–1065.

Barlow DF (1985) X-ray diffraction studies on the effect of cooling on clinker microstructure. *Proceedings of the Seventh International Conference on Cement Microscopy*: 243–255.

Bogue RH (1929) Calculation of the compounds in Portland cement. *Industrial and Engineering Chemistry, Analytical Edition* **1(4)**: 192–197.

Bogue RH (1939) On Constitution of Portland Cement Clinker. *Proceedings of the Symposium on the Chemistry of Cement*. Ingeniörsvetenskapsakademien, Stockholm, Sweden, pp. 57–98.

Bogue RH (1947) *The Chemistry of Portland Cement*. Reinhold, New York, USA.

Bogue RH (1955) *The Chemistry of Portland Cement*, 2nd edn. Reinhold, New York, USA.

Boikova AI, Fomicheva OI and Zubekhin AP (1986) Chemical composition of phases relations in unbleached and bleached clinkers. *Proceedings of the 8th International Congress on the Chemistry of Cement* **2**: 243–248.

Brisi C and Appendino P (1965) Equilibri allo stato solido nel sistema CaO-SrO-SiO_2. *Annali di Chimica*. **55**: 1213–1222. (In Italian.)

Bucchi R (1978) Calcining systems separate from the suspension preheater cement kilns. *Il Cemento* **75**: 411–440.

Bucchi R (1980) Influence of the nature and preparation of raw materials on the reactivity of raw mix. *Proceedings of the 7th International Congress on the Chemistry of Cement* 1, I-1/3-I-1/43.

Bucchi R (1981a) Features on the role of minor compounds in cement clinker – Part 1. *World Cement Technology* **12**: 210–231.

Bucchi R (1981b) Features on the role of minor compounds in cement clinker – Part 2. *World Cement Technology* **12**: 258–273.

Butt YM, Timashev VV and Osokin AP (1976) Mechanism of clinker formation and modification of its structure. *Proceedings of the 6th International Congress on the Chemistry of Cement* **1**: 132–153.

Bye GC (1983) *Portland Cement. Composition, Production and Properties*. Pergamon Press, Oxford, UK.

Castanet R and Sorrentino FP (1986) Study of clinkering reactions by high temperature isothermal calorimetry. *Proceedings of the 8th International Symposium on the Chemistry of Cement* **2**: 36–39.

Centurione SL (1991) Characterization of reducing environment in Portland clinker by microscopy. *Proceedings of the 13th International Conference on Cement Microscopy*: 120–130.

Chatterjee AK (2011) Chemistry and engineering of the clinkerization process – incremental advances and lack of breakthroughs. *Cement and Concrete Research* **41(7)**: 624–641.

Christensen NH (1979a) Burnability of cement raw mixes at 1400°C I: the effect of the chemical composition. *Cement and Concrete Research* **9(2)**: 219–228.

Christensen NH (1979b) Burnability of cement raw mixes at 1400°C II: the effect of the fineness. *Cement and Concrete Research* **9(3)**: 285–294.

Chromý S (1976) Mechanism of white clinker formation. *Proceedings of the 7th International Congress on the Chemistry of Cement* **3**: 268–273.

Chromý S (1982a) Process of Portland clinker formation, reactivity and burnability of cement raw materials. Part I: Mechanism of Portland clinker formation. *Zement Kalk Gips* **35**: 204.

Chromý S (1982b) Process of Portland clinker formation, reactivity and burnability of cement raw material. *Zement Kalk Gips* **35**: 555 (partial English translation, p. 303).

Chromý S and Hrabě Z (1982) Process of Portland clinker formation, reactivity and burnability of cement raw materials. Part II: Kinetics of clinker formation. *Zement Kalk Gips* **35**: 368 (partial English translation, p. 221).

Chromý S and Weber M (1981) The process of ring formation in rotary cement kilns. *Zement Kalk Gips* **34**: 453 (partial English translation, p. 244).

Cottin B, Rouanet A and Conjeaud M (1986) Mineralogical changes of lime-silica raw materials at 830°C. *Proceedings of the 8th International Congress on the Chemistry of Cement* **2**: 31–35.

Davis P and Longman PA (1984) In *Proceedings of the 19th International Cement Seminar*. Rock Products, Chicago, IL, USA, p. 25.

Davis P, Stringer JA and Watson D (1979) Chapter 7. In *Making the Most of Materials – Cement*. Science Research Council, Swindon, UK.

de Keyser WL (1951) Contribution à l'étude des réactions à l'état solide entre la chaux et l'alumine. *Bulletin des Sociétés Chimiques Belges* **60**: 516–546. (In French.)

de Keyser WL (1954) Réactions à l'état solide dans le système ternaire O_2Si, OCa, O_3Al_2. *Bulletin des Sociétés Chimiques Belges* **63**: 40–58. (In French.)

de Keyser WL (1955) Contribution à l'étude des réactions à l'état solide entre OCa, O_3Fe_2, O_3Al_2. *Bulletin des Sociétés Chimiques Belges* **64**: 395–408. (In French.)

Dreizler IE, Strunge J and Knöfel DFE (1985) Effect of alkalies and sulfur on the formation of clinker phases and on the cement properties. *Proceedings of the Seventh International Conference on Cement Microscopy*, 100–126.

Duda WR (1985) *Cement-Data-Book, 1: International Process Engineering in the Cement Industry*, 3rd edn. Bauverlag, Wiesbaden and Berlin, Germany.

Fayos J, Glasser FP, Howie RA, Lachowski EE and Perez-Mendez M (1985) Structure of dodecacalcium potassium fluoride dioxide tetrasilicate bis(sulphate), $KF.2[Ca_6(SO_4)(SiO_4)_2O]$: a fluorine-containing phase encountered in cement clinker production process. *Acta Crystallographica C* **41(6)**: 814–816.

Fernandez A and Leung Y (2018) Technology Roadmap: Low-carbon transition in the cement industry. *International Energy Agency/World Business Council for Sustainable Development*: 1–61.

FLSmidth (2014) Rotary kilns for cement plants. *Brochure*. www.flsmidth.com.

Fundal E (1979) The burnability of cement raw mixes. *World Cement Technology* **10**: 195–196, 199–200, 203–204.

Fundal E (1986) The description of a new potassium-alumina phase in OPC. *Proceedings of the 8th International Congress on the Chemistry of Cement* **2**: 139–145.

Gartner EM (1986) *Research on the Manufacture and Use of Cements* (Frohnsdorff G (ed.)), Engineering Foundation, New York, NY, USA, p. 1.

Gartner EM and Tang FJ (1987) Formation and properties of high sulfate Portland cement clinkers. *Il Cemento* **84**: 141–164.

Ghose A (1980) PhD thesis. University of London, London, UK.

Ghosh SN (ed.) (1983) *Advances in Cement Technology*. Pergamon, Oxford, UK.

Gilioli C, Massazza F and Pezzuoli M (1972) Equilibrium relations in the solid state in the calcium oxide-aluminum oxide-silicon oxide-sulfur trioxide system. *Il Cemento* **69**: 109–116.

Gilioli C, Massazza F and Pezzuoli M (1973) Solid state relations in the calcium oxide-strontium oxide-aluminum oxide-sulfur trioxide system. *Il Cemento* **70**: 137–148.
Glasser FP (1983) Reactions occurring during cement making. In *Structures and Performance of Cements* (Barnes P (ed.)). Applied Science Publishers, London, UK, pp. 69–108.
Glasser FP (1986) Kinetics of cement-making reactions. *Proceedings of the 8th International Congress on the Chemistry of Cement* **6**: 15–32.
Goldshtein LYA (1985) *Cement Production from Industrial By-Products (Wastes) Utilization.* Stroyizdat, Leningrad, USSR.
Gutt W (1963) High-temperature phase equilibria in the system $2CaO.SiO_2$–$3CaO.P_2O_5$–CaO. *Nature* **197(4863)**: 142–143.
Gutt W (1969) Manufacture of Portland cement from phosphatic raw materials. *Proceedings of the 5th International Symposium on the Chemistry of Cement* **1**: 93–105.
Gutt W and Smith MA (1968) Studies of the role of calcium sulphate in the manufacture of Portland cement clinker. *Transactions of the British Ceramic Society* **67**: 487–509.
Habert G (2013) Environmental impact of Portland cement production. In *Eco-Efficient Concrete* (Pacheco-Torgal F, Jalali S, Labrincha J and John VM (eds)). Woodhead Publishing, Cambridge, UK, pp. 3–25.
Heilmann T (1954) The influence of the fineness of cement raw mixes on their burnability. *Proceedings of the 3rd International Symposium on the Chemistry of Cement*, 711–744.
Herfort D (2015) Chemistry of the clinker production process. *Journal of The Chinese Ceramic Society* **43**: 1314–1323.
Herfort D, Moir GK, Johansen V, Sorrentino F and Bolio Arceo H (2010) The chemistry of Portland cement clinker. *Advances in Cement Research* **22(4)**: 187–194.
Hills AWD (1968) The mechanism of the thermal decomposition of calcium carbonate. *Chemical Engineering Science* **23(4)**: 297–320.
Hofmänner F (1975) *Microstructure of Portland Cement Clinker*. Rheintaler Druckerei, Heerbrugg, Switzerland.
Hornain H and Regourd M (1980) Cracking and grindability of clinker. *Proceedings of the 7th International Congress on the Chemistry of Cement* **2**: 276–281.
Ichikawa M, Ikeda S and Komukai Y (1994) Effect of cooling rate and Na_2O content on the character of the interstitial materials in Portland cement clinker. *Cement and Concrete Research* **24(6)**: 1092–1096.
Imlach JA and Hofmänner F (1976) DTA and optical microscopic study of clinker formation. *Proceedings of the 6th International Congress on the Chemistry of Cement* **1**: 281–285. (In Russian.)
Jankovic A, Valery W and Davis E (2004) Cement grinding optimisation. *Minerals Engineering* **17(11-12)**: 1075–1081.
Jansen D, Stabler C, Goetz-Neunhoeffer F, Dittrich S and Neubauer J (2011) Does ordinary Portland cement contain amorphous phase? A quantitative study using an external standard method. *Powder Diffraction* **26(1)**: 31–38.
Johansen V (1973) Model for reaction between CaO particles and Portland cement clinker. *Journal of the American Ceramic Society* **56(9)**: 450–454.
Johansen V (1978) Absorption spectra of CrO_4^{4-} in Ca_3SiO_5 and $Ca_3Al_2O_6$. *Cement and Concrete Research* **8(2)**: 245–250.
Johansen V and Jakobsen UH (1993) An examination of clinkers with light colored cores. *World Cem. Res. Dev* **24**: 32–37.
Jøns EC (1980) Thermodynamics driving force for formation of tricalcium silicate. *Cement and Concrete Research* **10(1)**: 103–106.

Kääntee U, Zevenhoven R, Backman R and Hupa M (2004) Cement manufacturing using alternative fuels and the advantages of process modelling. *Fuel Processing Technology* **85(4)**: 293–301.

Kerton CP and Murray RJ (1983) Portland cement production. In *Structures and Performance of Cements* (Barnes P (ed.)). Applied Science, London, UK, pp. 205–236.

Kirchner G (1986) Behaviour of heavy metals during cement clinker burning. *Zement Kalk Gips* **39**: 555 (partial English translation, p.368).

Lea FM (1970) *The Chemistry of Cement and Concrete*, 3rd edn. Arnold, London, UK.

Lea FM and Desch CH (1956) *The Chemistry of Cement and Concrete*, 2nd edn. Arnold, London, UK.

Lea FM and Parker TW (1935) *Quaternary system $CaO–Al_2O_3–SiO_2–Fe_2O_3$ in relation to cement technology*. Building Research Board, London, UK, Technical Paper 16.

Lehmann H, Locher FW and Thormann P (1964) The effect of grain size in limestone on the formation of sinter minerals in the temperature range of 850–1450°. *Tonindustrie-Zeitung und Keramische Rundschau* **88**: 489–498. (In German.)

Lerch W and Brownmiller LT (1937) Method for approximating the glass content of Portland cement clinker. *Journal of Research of the National Bureau of Standards* **18(5)**: 609–622.

Locher FW (1980) Influence of burning conditions on clinker characteristics. *World Cement Technology* **11**: 67–73.

Long GR (1982a) Clinker quality characterization by reflected light techniques. *Proceedings of the 4th International Conference on Cement Microscopy*, 92–109.

Long GR (1982b) The effect of burning environment on the microscopic characteristics of Portland cement clinker. *Proceedings of the 4th International Conference on Cement Microscopy*, 128–140.

Long GR (1983) Microstructure and chemistry of unhydrated cements. *Philosophical Transactions of the Royal Society London* **A310**: 43–51.

Ludwig U and Ruckensteiner G (1974) Über die Brennbarkeit von Zementrohmehlen. *Cement and Concrete Research* **4(2)**: 239–246. (In German).

Madlool NA, Saidur R, Rahim NA and Kamalisarvestani M (2013) An overview of energy savings measures for cement industries. *Renewable and Sustainable Energy Reviews* **19**: 18–29.

Maki I (1986) Relationship of processing parameters to clinker properties; influence of minor components. *Proceedings of the 8th International Congress on the Chemistry of Cement* **1**: 34–47.

Maki I and Takahashi S (1984) Redistribution of impurity during recrystallization of alite in Portland cement clinker. *Cement and Concrete Research* **14(3)**: 413–418.

Maki I, Haba H and Takahashi S (1983) Effect of recrystallization on the characters of alite in Portland cement clinker. *Cement and Concrete Research* **13(5)**: 689–695.

Maki I, Fukuda K, Imura T, Yoshida H and Ito S (1995) Formation of belite clusters from quartz grains in Portland cement clinker. *Cement and Concrete Research* **25(4)**: 835–840.

Maultzsch M, Gierloff M and Schimmelwitz P (1980) Aging of cement through long term storage. *Proceedings of the 7th International Congress on the Chemistry of Cement* **2**: 128–133.

Méric JP (1980) Influence of grinding and storage of conditions of clinker. *Proceedings of the 7th International Congress on the Chemistry of Cement* **1**: I-4/1-I-4/16.

Miller FM (1986) *Research on the Manufacture and Use of Cements*. Engineering Foundation, New York, NY, USA, p. 21.

Mohan K and Glasser FP (1977) The thermal decomposition of Ca_3SiO_5 at temperatures below 1250°C: Part. 1. Pure C_3S and the influence of excess CaO and Ca_2SiO_4; part 2. The influence of Mg, Fe, Al, and Na oxides on decomposition; part 3. The influence of water and sulphate on the decomposition. *Cement and Concrete Research* **7(1)**: 1–7, 269–275, 379–383.

Moore AE (1976) The sequence of compound formation in Portland cement rotary kilns. *Cem. Technol* **7**: 85–91, 134–138.

Newkirk TF (1954) The alkali phases in Portland cement clinker. *Proceedings of the 3rd International Symposium on the Chemistry of Cement*, 151–168.

Newman ES and Hoffman R (1956) Heats of formation of hexacalcium dialumino ferrite and dicalcium ferrite. *Journal of Research of the National Bureau of Standards* **56(6)**: 313–318.

Nurse RW, Welch JH and Gutt W (1959) High-temperature phase equilibria in the system dicalcium silicate–tricalcium phosphate. *Journal of the Chemical Society* 1077–1083.

Odler I and Zhang H (1996) Investigations on high SO_3 Portland cement clinkers. *World Cem. Res. Dev* **27**: 73–77.

Ono Y (1981) Microscopical observation of clinker, for the estimation of burning condition, grindability and hydraulic activity. *Proceedings of the 3rd International Conference on Cement Microscopy*, 198–210.

Ono M and Nagashima N (1976) Effect of clinker cooling conditions on the strength of cement. *Proceedings of the 6th International Congress on the Chemistry of Cement* **1**: 170–173. (In Russian.)

Peray K (1986) *The Rotary Cement Kiln*, 2nd edn. Arnold, London, UK.

Péréz-Mendez M and Fayos J (1986) A very modified apatite-type structure $K_3Ca_2(SO_4)_3F$. *Proceedings of the 8th International Congress on the Chemistry of Cement* **2**: 223–227.

Petersen IF (1983a) Isothermal sintering of Portland cement raw mixes. I. *World Cement Technology* **14(5)**: 188–196.

Petersen IF (1983b) Isothermal sintering of Portland cement raw mixes. II. *World Cement Technology* **14(6)**: 220–228.

Petersen IF (1986) The ability of clinker nodule formation. *Proceedings of the 8th International Congress on the Chemistry of Cement* **2**: 21–24.

Petersen IF and Johansen V (1979) Burnability and clinker nodule formation from a statistical point of view. *Cement and Concrete Research* **9(5)**: 631–639.

Pollitt HWW and Brown AW (1969) The distribution of alkalis in Portland cement clinker. *Proceedings on the 5th International Symposium on the Chemistry of Cement* **1**: 322–333.

Rahman A, Rasul MG, Khan MMK and Sharma S (2013) Impact of alternative fuels on the cement manufacturing plant performance – an overview. *Procedia Engineering* **56**: 393–400.

Regourd M and Guinier A (1976) The crystal chemistry of the constituents of Portland cement clinker. *Proceedings of the 6th International Congress on the Chemistry of Cement*, Moscow (1974) 1, Principal Paper I–2, 25–51.

Ritzmann H (1971) Cyclic phenomena in rotary kiln systems. *Zement Kalk Gips* **24**: 338.

Rivera M, Odler I and Abdul-Maula S (1987) Formation of Portland clinker – studies on synthetic raw meals. *Advances in Cement Research* **1(1)**: 52–57.

Rouanet A (1986) Solid-state reactions between lime and alumina. *Proceedings of the 8th International Congress on the Chemistry of Cement* **2**: 25–30.

Schneider M, Romer M, Tschudin M and Bolio H (2011) Sustainable cement production – present and future. *Cement and Concrete Research* **41(7)**: 642–650.

Scrivener KL and Taylor HFW (1995a) Microstructural and microanalytical study of strongly reduced Portland cement clinker. *World Cem. Res. Dev* **26(8)**: 34–43.

Scrivener KL and Taylor HFW (1995b) Clinker nodules with light-coloured centres. *Zement Kalk Gips International* **48**: 34–39.

Snellings R, Bazzoni A and Scrivener KL (2014) The existence of amorphous phase in Portland cements: Physical factors affecting Rietveld quantitative phase analysis. *Cement and Concrete Research* **59**: 139–146.

Sorrentino FP (1994) *Advances in Cement and Concrete* (Grutzeck MW and Sarkar SL (eds.)). American Society of Civil Engineers, New York, NY, USA, p. 65.

Southard JC and Royster PH (1936) The thermal dissociation of calcium carbonate. *The Journal of Physical Chemistry* **40(4)**: 435–438.

Spohn E, Woermann E and Knoefel D (1969) A refinement of the lime standard formula. *Proceedings of the 5th International Symposium on the Chemistry of Cement* **1**: 172–179.

Sprung S (1980) Effect of the burning process on the formation and properties of the clinker. *Proceedings of the 7th International Congress on the Chemistry of Cement* **1**: I-2/1-I-2/15.

Sprung S (1985) Influence of process technology on cement properties. *Zement Kalk Gips* **38**: 577 (partial English translation, p.309).

Suherman PM, van Riessen A, O'Connor B, Li D, Bolton D and Fairhurst H (2002) Determination of amorphous phase levels in Portland cement clinker. *Powder Diffraction* **17(3)**: 178–185.

Swayze MA (1946a) A report on studies of; 1. The ternary system of CaO-C_5A_3-C_2F, 2. The quaternary system CaO-C_5A_3-C_2F-C_2S, 3. The quaternary system as modified by 5% magnesia, part I. *American Journal of Science* **244(1)**: 1–30.

Swayze MA (1946b) A report on studies of; 1. The ternary system of CaO-C_5A_3-C_2F, 2. The quaternary system CaO-C_5A_3-C_2F-C_2S, 3. The quaternary system as modified by 5% magnesia, part II. *American Journal of Science* **244(2)**: 65–94.

Sylla HM (1974) Investigations on the formation of rings in rotary cement kilns. *Zement Kalk Gips* **27**: 499–508.

Sylla HM (1977) Coating formation by salt melts. *Zement Kalk Gips* **30**: 487 (partial English translation, p.264).

Sylla HM (1981) Effect of reductive burning on the properties of cement clinker. *Zement Kalk Gips* **34**: 618 (partial English translation, 1982, 35, p.31).

Thorvaldson T (1970) In Lea FM (1970) *The Chemistry of Cement and Concrete*, 3rd edn. Arnold, London, UK.

Timashev VV (1980) The kinetics of clinker formation: the structure and composition of clinker and its phases. *Proceedings of the 7th International Congress on the Chemistry of Cement*, **1**: *I-3/1-I-3/20*

Triviño Vázquez F (1985) A new compound: K_3Ca_2 (SO_4) F, identified in coatings of heat recover cyclones. *Cement and Concrete Research* **15(4)**: 581–584.

Vásquez T, Blanco-Varela MT and Palomo A (1984) Effect of the cooling rate of clinkers on the formation of the liquid phase. *Proceedings of the British Ceramic Society* **35**: 115–125.

VDZ (Verein Deutscher Zementwerke) (2018) *Environmental data of the German cement industry 2018*. Verein Deutscher Zementwerke e.V., Düsseldorf, Germany.

Vosteen B (1974) Preheating and complete calcination of cement raw meal in a suspension preheater system. *Zement Kalk Gips* **27**: 443.

Wagman DD, Evans WH, Parker VB *et al.* (1982) The NBS tables of chemical thermodynamic properties: Selected values for inorganic and C_1 and C_2 organic substances in SI units. *Journal of Physical and Chemical Reference Data* 11: Suppl.2 American Chemical Society, Washington, DC and American Institute of Physics, New York, NY, USA. An electronic version is available: Reed JJ (1989) (same title). National Institute of Standards and Technology. https://doi.org/10.18434/M32124.

Weber P (1963) *Heat Transfer in Rotary Kilns*. Bauverlag, Wiesbaden and Berlin, Germany.

Weisweiler W, Osen E, Eck J and Höfer H (1986) Kinetic studies in the CaO·SiO_2 system part I. mMechanism and kinetic data of the reactions between CaO– and SiO_2– powder compacts. *Cement and Concrete Research* **16(3)**: 283–295.

Weyer I (1931) Reaction of kaolin and lime in static heating. *Zement* **20**: 560–564, 608–612, 692–694.

Whitfield PS and Mitchell LD (2003) Quantitative Rietveld analysis of the amorphous content in cements and clinkers. *Journal of Materials Science* **38**: 4415–4421.

Wieser ME and Berglund M (2009) Atomic weights of the elements 2007 (IUPAC Technical Report). *Pure and Applied Chemistry* **81(11)**: 2131–2156.

Williamson J and Glasser FP (1962) Reactions in heated lime-alumina mixtures. *Journal of Applied Chemistry* **12(12)**: 535–538.

Woermann E (1962) Decomposition of alite in technical Portland cement clinker. Paper II-S8. *Proceedings on the 4th International Symposium on the Chemistry of Cement* **1**: 119–128.

Wolter A (1985) Influence of the kiln system on the clinker properties. *Zement Kalk Gips* **38**: 612 (partial English translation, p.327).

Wolter A (1986) Phase composition of calcined raw meal. *Proceedings of the 8th International Congress on the Chemistry of Cement* **2**: 89–94.

Zur Strassen H (1957) The theoretical heat requirement for cement burning. *Zement Kalk Gips* **10**: 1–12.

Cement Chemistry, Third edition

Ian G Richardson and HFW Taylor
ISBN 978-1-83549-777-7
https://doi.org/10.1108/978-1-83549-776-020253004

Chapter 4
Properties of Portland clinker and cement

4.1 Macroscopic and surface properties
4.1.1 Unground clinker
This chapter deals with chemical and physical properties other than ones for which the nature of the hydration products must be considered. In general, properties of the whole clinker or cement alone are considered, those of the constituent phases having been dealt with in Chapter 1, but factors affecting the reactivities of these phases are included.

Portland cement clinker emerges from a dry process kiln as rounded pellets or from a wet process kiln as irregularly shaped lumps, in either case typically of 3–20 mm dimensions. The colour, which is normally almost black, is discussed in §3.4.6. The bulk density, determined under standard conditions by rejecting material passing a 5 mm sieve, pouring the remainder into a conical container and weighing, is called the litre weight. This quantity, which is typically 1·25–1·35 $kg\,L^{-1}$, gives some indication of the operating conditions. Its optimum value depends on the composition.

The measured true density of clinker is typically 3·15–3·20 $g\,cm^{-3}$; the value calculated from the X-ray densities and typical proportions of the individual phases is about 3·20 $g\,cm^{-3}$. The difference is probably due mainly to the presence of pores inaccessible to the fluids used in the experimental determination. Using mercury intrusion porosimetry, Butt *et al.* (1974) determined pore size distributions and discussed their relations with burning and cooling conditions and grindability.

4.1.2 Particle size distribution of ground clinker or cement
Portland cement clinker is ground to a powder. The fineness of the powder is an important property because small cement particles react with water more quickly than large particles (they have a higher surface area per unit mass). In the early years of Portland cement manufacture the clinker was ground between revolving burr-stones; specifications in the mid-nineteenth century generally only required cement to be supplied as '*ground extremely fine*' (e.g. Grant, 1866; Redgrave, 1895), which resulted in considerable variation in fineness because each manufacturer would decide what constituted *extremely fine* (Colson, 1875). Specifications consequently began to include a sieving requirement, with a maximum mass of cement that could be retained on a given mesh (Gostling in Bauschinger *et al.*, 1880; Grant, 1880). The large variability in the fineness of cement in the 1870s is evident on comparison of the data of that period in Figure 4.1. These and the other data in Figure 4.1 illustrate the progressive increase in fineness that occurred with time; it is evident that the cement of the nineteenth century was in fact extremely coarse when compared with modern ordinary Portland cements, which typically have very little material coarser than 90 μm (0–4 %), and 7–9 % is finer than 2 μm.

Figure 4.1 Particle size information for Portland cement over time

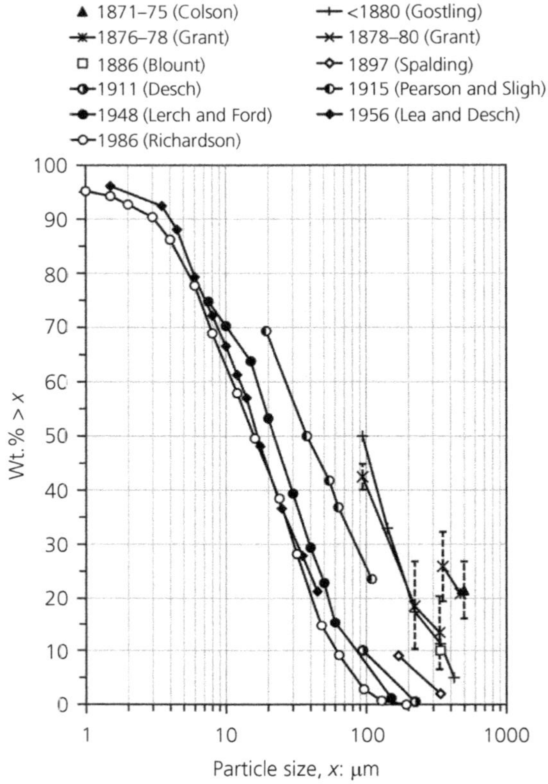

The legend indicates the date that the data were generated (if stated) or when the information was written/published. The references are: 1871–75 (Colson, 1875; sieving; mean ± 1 standard deviation); <1880 (Gostling in Bauschinger *et al.*, 1880; Michaëlis in Bauschinger *et al.*, 1880; Scott and Redgrave, 1880; sieving specifications); 1876–78 and 1878–80 (Grant, 1880; sieving; values are mean ± 1 standard deviation); 1886 (Blount, 1906; sieving); 1897 (Spalding, 1900; sieving); 1911 (Desch, 1911; sieving); 1915 (Pearson and Sligh, 1915; air elutriation device); 1948 (Lerch and Ford, 1948; data extracted from Figure 2.6 of Popovics, 1979 (type I cement)); 1956 (Lea and Desch, 1956; Andreasen sedimentation apparatus); 1986 (Richardson, 1991; laser granulometry). For the purpose of this plot, the apertures of the sieve meshes were assumed to be $\frac{2}{3}$ of the pitch, which was '*the common standard*' in the late nineteenth century (Spalding, 1900). As an example, Colson (1875) reported results for screening cement '*through wire gauze having 1,152 meshes to the square inch*'. The sieve was therefore 34 × 34 meshes per inch, the pitch of the mesh $= \frac{1}{34} \times 25{\cdot}4 = 0{\cdot}747$ mm, and so the aperture $= \frac{2}{3} \times 0{\cdot}747 = 0{\cdot}498$ mm – that is, approximately 500 µm.

Allen (1975, 2003) discussed methods suitable for determining the particle size distribution of cement (PSD). Sieving is suitable only for determining the proportions of particles in ranges above about 45 µm. In airjet sieving, jets speed the passage of material through the sieve and prevent the mesh from becoming blocked. For the entire range of size, the most widely used methods are ones based on either sedimentation in liquids or diffraction of light. In the X-ray sedigraph, the progress of sedimentation is monitored by the absorption of an X-ray beam and the results calculated using Stokes's law. Because highly concentrated suspensions are used, there is some doubt as to whether this law is strictly applicable. An older method employs the Andreason pipette, by which samples are withdrawn from a suspension at a fixed depth after various times so that their solid contents may be determined. In laser granulometry, light from a laser passes through a suspension and the PSD is calculated from the resulting diffraction pattern. Light microscopy can provide information on the distribution of both particle size and particle shape; SEM with image analysis (§4.3.1) can provide additional information. Figure 4.1 includes examples of data obtained using sieving, laser granulometry, Andreasen sedimentation and air elutriation (elutriation is the reverse of sedimentation; see Lea and Desch, 1956, p. 320). Figure 4.2 shows a cumulative PSD for a typical Portland cement ground in an open-circuit mill obtained using an X-ray sedigraph (O; Sumner *et al.*, 1986). With all the methods, complications or uncertainties of interpretation arise due to effects associated with particle shape and with flocculation, so that the results given by different methods do not necessarily agree and usually have relative rather than absolute significance.

Méric (1980) discussed analytical expressions used to represent the PSD of cements. Probably the most successful is the Rosin–Rammler function (Rosin and Rammler, 1933), which may be written in logarithmic form as

$$\ln \ln\left(100/R\right) = n(\ln x - \ln x_0) \tag{4.1}$$

where R is the mass percentage of material of particle size greater than x. The constants x_0 and n describe the distribution; x_0 is a measure of average size, 36·79% (i.e. R = 1/e) of the material being of size greater than x_0; n is a measure of breadth of distribution, increasing as the latter becomes narrower.

For the example shown in Figure 4.2, x_0 is 22·1 μm and n is 1·03. For open-circuit milling, the values of n obtained using an X-ray sedigraph are 0·99–1·07; for closed-circuit milling, in which

Figure 4.2 Rosin–Rammler plot of the particle size distribution of a typical Portland cement

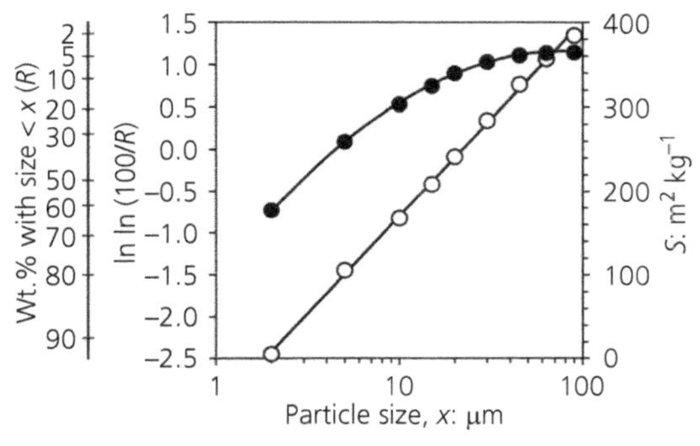

Portland cement (O); S = specific surface area (•; $m^2\,kg^{-1}$) attributable to particles of size smaller than x (μm); open circuit grinding; based on X-ray sedigraph data of Sumner *et al.* (1986).

the fine material is separated and only the coarser material further ground, n can be as high as 1·23–1·28 if the separation is efficient (Sumner *et al.*, 1986). A narrow PSD can also be obtained using a vertical spindle mill. Laser granulometry typically gives values of n lower by 0·15–0·20 (Sumner *et al.*, 1986).

4.1.3 Specific surface area determination

The specific surface area of cement is commonly determined directly by air permeability methods. In the Lea and Nurse method (Lea and Nurse, 1939), a bed of cement of porosity 0·475 is contained in a cell through which a stream of air is passed and steady flow established. The specific surface area is calculated from the density of the cement, the porosity and dimensions of the bed of powder, the pressure difference across the bed and the rate of flow and kinematic viscosity of the air. In the Blaine method (Blaine, 1943), a fixed volume of air passes through the bed at a steadily decreasing rate, which is controlled and measured by the movement of oil in a manometer, the time required being measured. The apparatus is calibrated empirically, most obviously using a cement that has also been examined by the Lea and Nurse method. The two methods give closely similar results. The Blaine method, although not absolute, is simpler to operate and automated variants of it have been devised.

Other methods that have been used to determine specific surface areas of cements include the Wagner turbidimeter (Wagner, 1933) and BET (Brunauer–Emmett–Teller) gas adsorption. The former, as conventionally used, gives very low results because of a false assumption that the mean diameter of the particles smaller than 7·5 μm is 3·8 μm, which is much too high. The BET method gives results two to three times higher than the air permeability methods because it includes internal surfaces present in microcracks or in pores open at only one end.

The specific surface area, like the PSD, is thus a quantity whose value depends on how it is defined and is liable to be affected by any pretreatment or conditions affecting the degree of flocculation. In practice, air permeability methods are widely used. Typical values are 300–350 $m^2 kg^{-1}$ for modern ordinary Portland cements and 400–450 $m^2 kg^{-1}$ for rapid-hardening Portland cements.

From a PSD curve, one may calculate the specific surface area, S

$$S = 6\times10^6 \times F \times \sum f/(d \times \rho) \qquad (4.2)$$

where f is the mass fraction of material consisting of grains assumed to have a diameter or edge length d in μm, ρ is the density in $kg\,m^{-3}$ and the result is in $m^2 kg^{-1}$.

F is an empirical constant which takes into account the differing assumptions regarding surface shape that are made implicitly or explicitly in the determination of PSD and the definition of specific surface area. This formula applies to both cubic and spherical grains. For a calculation starting from a PSD obtained using an X-ray sedigraph to give a specific surface area comparable to that obtained by an air permeability method, F is typically about 1·13. Curve S on Figure 4.2 illustrates the contributions to the calculated specific surface area of particles of differing sizes, assuming that particles smaller than 2 μm have a mean size of 1 μm and that particles that pass a sieve of size a μm but that are retained on a sieve of size b μm have a mean size of $(a + b)/2$ μm. In this case a value of $F = 1{\cdot}18$ is needed to give a match between the calculated and measured specific surface areas – that is, 363 $m^2 kg^{-1}$. Sumner *et al.* (1986) deduced from the application of Equation 4.2 that, for ordinary Portland cement, approximately 40% of the total is provided by particles smaller than 2 μm (it is nearly 50% for the example given in Figure 4.2).

In principle, the specific surface area may also be calculated from the PSD using analytical expressions derived from such functions as the Rosin–Rammler expression. In practice, this is not always satisfactory because the specific surface area is so highly dependent on the bottom end of the distribution.

4.1.4 Particle size distribution, phase composition and cement properties

Because the constituent phases of a cement are not equally easy to grind, different particle size fractions differ in composition. Gypsum and its dehydration products are concentrated in the finer fractions. Osbaeck and Jøns (1980) concluded that each 1% of gypsum contributed about $10\,m^2\,kg^{-1}$ to the specific surface area; in a typical case, some 15% of the total specific surface area is thus due to gypsum. The content of alite decreases, and that of belite increases, with increasing particle size (Gutteridge, 1984; Ritzmann, 1968), the contents of aluminate and ferrite phases being little affected.

The depth of reaction of the cement grains after a given time has been determined by light microscopy; it may be expected to vary with the composition and microstructure of the clinker. Anderegg and Hubbell (1929, 1930) found values of 0·43–0·47 μm at 1 day, 1·7–2·6 μm at 7 days, 3·5–5·4 μm at 28 days and 6·1–8·9 μm at 150 days. Zur Strassen (1959) gave somewhat higher values (e.g. 6–7 μm at 28 days). From the depth of reaction and the PSD curve, and assuming the grains to be spherical, one may calculate the fraction of cement that has reacted after a given time

$$\alpha = \Sigma f \times \left[1 - (1 - 2h/d)^3\right] \tag{4.3}$$

where h is the depth of reaction in μm, f and d have the same meanings as in Equation 4.2 and the quantity in square brackets cannot be less than zero.

For the cement to which the data in Figure 4.1 refer, and using the means of Anderegg and Hubbell's (1929, 1930) values, this gives 0·28 at 1 day, 0·63 at 7 days, 0·80 at 28 days and 0·90 at 150 days. These values are in reasonable agreement with those typically found using XRD. Ritzmann (1968) determined the compressive strengths of pastes of a number of cements of known PSD made by mixing separated fractions. For a given cement and water/cement (w/c) ratio, the strength was linearly related to the fraction of cement reacted, calculated in this way.

At 1 day, the reaction is primarily of the smaller particles, and the fine end of the PSD is therefore the important one for the strength. As the age increases, the coarser part of the distribution becomes progressively more important; by 28 days the particles smaller than about 10 μm have reacted completely, and the content of particles up to about 45 μm in size is of major importance. For a given specific surface area, narrowing the PSD increases the degree of reaction at 28 days and thus also the strength (Kuhlmann *et al.*, 1985; Locher *et al.*, 1973; Sumner *et al.*, 1986). The relation may be quantified using Equation 4.3. Narrowing the PSD also saves energy in grinding (Sumner *et al.*, 1986) but increases the amount of water needed to produce a mix of given workability (Sprung *et al.*, 1985; Sumner *et al.*, 1986); this latter effect can be compensated by adjusting the temperature during milling so as to increase the extent to which the gypsum is dehydrated to hemihydrate or γ-$CaSO_4$ ('soluble anhydrite') (Sumner *et al.*, 1986).

4.1.5 Chemical analysis

Ground clinker or cement may be analysed by standard methods, such as X-ray fluorescence (XRF). Determination of free lime is discussed in §4.3.3. With XRF or other methods in which no fraction of the material is excluded from the analysis, insoluble residue (determined separately) is included in the oxide components and therefore should be omitted from the total. Classical, wet analysis gives inaccurate results for Al_2O_3 unless the effects of P_2O_5 and TiO_2 are allowed for.

Chemical analysis, microscopy, XRD and other methods of examination should be carried out on the same, representative sample of material. A procedure for obtaining such a sample is described in the next section.

4.2. Light microscopy

4.2.1 General

Light microscopy provides much information on clinker microstructure and thereby on the conditions existing at various stages of the manufacturing process. Historically, much the most widely used technique has been the examination of polished and etched sections in reflected light, and the following sections refer to this unless otherwise stated, but additional information has been obtained from the examination of thin sections or powder mounts in transmitted light. Fuller accounts are given elsewhere (Campbell, 1986; Gille *et al.*, 1965; Hofmänner, 1975; Long, 1982a).

It is vital to obtain a representative sample; the following procedure (Long, 1982a) is typical. If a kiln is known to give a reasonably consistent product, samples taken at intervals over a suitable period are aggregated, but if not, they are examined separately. Each sample examined comprises at least 15 kg of clinker and is first examined visually to distinguish kiln deposits or refractories. It is then crushed to below 6 mm and a representative sample obtained – for example, by quartering. The 2–4 mm fraction of this sample is used. Ten to 12 samples of uncrushed nodules, covering the range of nodule sizes, are also examined, together with a sample of the material smaller than 2 mm if much of this is present in the original clinker. The relative amounts of nodules of different sizes and of the material below 2 mm are noted.

For examination in reflected light, the sample is mounted in a plastic resin and is cut and polished using only nonaqueous lubricants. The most generally useful etchant is probably hydrofluoric acid (HF) vapour, which has the merit of not removing alkali sulfates; the slide is inverted over a vessel containing 40% aqueous HF for 10–20 s. Treatment with water followed by 0·25% HNO_3 in ethanol and finally ethanol is also widely used, and many other etchants have been employed for specific purposes. The etchants produce thin films of decomposition products, which yield interference colours when viewed in reflected light. If HF vapour is used properly, the alite is yellowish brown, the belite is blue or red and the aluminate is grey; the ferrite, which is unattacked, is brightly reflective. A good example photomicrograph that demonstrates these colours is Plate 1 in St John *et al.* (1998), which shows crystals of alite and belite embedded in a matrix of aluminate and ferrite phases. Alite crystals are angular, often pseudohexagonal, while belite crystals are rounded and normally striated, as described in §1.3.2. Ideally, the alite crystals are 15–20 μm in average size, although in practice they are usually much larger (e.g. as in St John *et al.*'s (1998) Plate 1, which range between 20 and 80 μm and average about 45 μm), and there should be no clustering of either alite or belite. The average size of the belite crystals is typically 25–40 μm (e.g. they are at the top end of this range in St John *et al.*'s (1998) Plate 1).

A good clinker should have been rapidly cooled and this will cause the ferrite and aluminate phases to be mixed on a scale of a few micrometres, the ferrite often forming dendrites or needles in a matrix of aluminate. In transmitted light, the ferrite is pleochroic. The normal, cubic aluminate tends to fill the spaces between the other phases and thus to exhibit no definite form; it is isotropic in transmitted light. In clinkers with an excess of alkali over SO_3, cubic aluminate may be partly or wholly replaced by the orthorhombic variety. This forms relatively large, prismatic crystals, which occur in a matrix of the other interstitial phases and are birefringent in transmitted light.

There should be little free lime. What there is should occur as rounded grains, typically 10–20 μm in size, and associated with alite and interstitial material. Lime appears cream in sections etched with HF vapour. Its presence may be confirmed by a microchemical test using White's reagent (5 g of phenol in 5 mL of nitrobenzene + 2 drops of water); long, birefringent needles of calcium phenate are formed. The test also responds to CH. Alkali sulfates occur in the clinker pore structure; they are etched black with HF vapour and inhibit the etching of silicate phases with which they are in contact.

4.2.2 Effects of bulk composition, raw feed preparation and ash deposition

Bulk composition affects the relative amounts of phases, the LSF, SR and AR affecting primarily the ratios of alite to belite, silicate phases to interstitial material and aluminate to ferrite phase, respectively. An LSF that is too high, either absolutely or in relation to the burning conditions, also gives rise to an excessive content of free lime. Belite may occur in such clinkers only as inclusions in alite.

Inadequate grinding of the raw meal is shown by the presence of large clusters of free lime or belite, up to several hundred micrometres in size and with defined outlines reflecting those of the original particles of calcareous or siliceous material, respectively. Coarse fragments of dolomitic limestones similarly yield localised concentrations of periclase, usually intimately mixed with lime. Periclase crystals are characteristically angular, up to about 20 μm in size, and are unattacked by normal etchants. Due to their hardness, they stand out in high relief in unetched sections. Less sharply defined segregation indicates inadequate mixing and is typically shown by the presence of regions rich in belite and in alite plus lime, in each case together with interstitial material. In extreme cases, entire nodules can depart seriously from the mean composition.

Incomplete assimilation of coal ash is shown by the existence of large regions high in belite and also containing interstitial phases. The locations of these regions indicate the stage of the burning process during which the ash was deposited; if it was late, the belite-rich material forms coatings on clinker nodules in which smaller nodules can be encapsulated, whereas if it is deposited earlier, it can form part of the kiln coating, which may later break away and form the centres of nodules. If the ash contains unburnt matter, localised reducing conditions can occur. Insufflation of materials into the kiln can produce similar effects (Long, 1982a).

4.2.3 Effects of burning conditions and cooling rate

Underburning is shown by the presence of abundant free lime and a low content of alite, the crystals of which are very small ($<10\,\mu m$). The average size and other characteristics of alite and belite crystals provide much further information about the burning conditions; factors governing them are discussed in Chapters 1 and 3. Ono (1981) considered that alite size was an indication of the rate of heating to the clinkering temperature, which if high led to the formation of smaller crystals, but other studies indicate that many other factors are involved and that overburning, in particular, can cause increase in size through recrystallisation (§3.4.4). Another indication of excessively hard burning is the presence of material high in large alite crystals and low in interstitial material, formed by the withdrawal of the liquid into the centres of the clinker nodules (Long, 1982a).

The effects of slow cooling (§3.5) are readily detectable by light microscopy. The most general are coarsening of the texture of the interstitial material and a change in the belite when viewed in transmitted light from colourless to yellow. The belite crystals may also develop ragged or serrated edges. Slow cooling can also cause resorption of alite, with deposition of small crystals of belite as fringes on the alite and in the body of the interstitial material, increase in alite crystal size and, if it occurs below 1200 °C, decomposition of alite to an intimate mixture of lime and belite. The effects of reducing conditions (§3.4.6) are similarly detectable by light microscopy (Long, 1982b, 1983).

4.2.4 Applications of light microscopic investigations

Light microscopy can be used to determine quantitative phase composition in clinkers, although it will often be impossible to determine aluminate and ferrite separately (§4.5.2). It has proved a highly effective means of finding the causes of unsatisfactory clinker quality or of determining what modifications in composition or plant operation are needed to change the clinker properties in a desired direction. It has also been used to predict strength development. Ono (1981) described results obtained from examinations of powder mounts. The values of four parameters, indicated in Table 4.1, were each estimated on a scale of 1–4 and the strength *R*, in MPa, of a mortar at 28 days then predicted using the regression equation

$$R = 24{\cdot}8 + 0{\cdot}63\mathrm{AS} + 2{\cdot}15\mathrm{AB} + 0{\cdot}39\mathrm{BS} + 2{\cdot}10\mathrm{BC} \tag{4.4}$$

where the four parameters AS, AB, BS and BC were considered to be measures of heating rate, maximum temperature, time at that temperature and cooling rate, respectively.

As noted in the preceding section, the first of these assignments is questionable, but this does not necessarily invalidate the results obtained by the method, which was tested by routine application in many plants of a major cement company over a long period. Some of the other workers using the method have, in contrast, had little success with it (e.g. St John and McGavin, 1981), and the generality of the procedure, or of the coefficients, remains to be established. The constant term in the equation probably subsumes terms relating to composition and cement fineness, which would have to be taken into account if a wider range of samples was considered.

Table 4.1 Parameters of Ono's method (Ono, 1981) for predicting cement strength

AS (alite size: μm)	15–20	20–30	30–40	40–60
AB (alite birefringence)	0·008–0·010	0·006–0·007	0·005–0·006	0·002–0·005
BS (belite size: μm)	25–40	20–25	15–20	5–10
BC (belite colour)	Clear	Faint yellow	Yellow	Amber
Hydraulic activity	Excellent	Good	Average	Poor
Value of parameter	4	3	2	1

4.3. Scanning electron microscopy, X-ray diffraction and other techniques

4.3.1 Scanning electron microscopy

Backscattered electron imaging of polished sections of clinkers in the scanning electron microscope (SEM) provides information essentially similar to that given by light microscopy, with the important addition that phases can be analysed chemically (Scrivener, 1987, 1989; Scrivener and Pratt, 1984). The method can be used to study ground cements as well as clinkers by dispersing the cement in a resin, which is then allowed to harden. The residual, unhydrated cement grains in a hardened cement paste can also be examined.

The polished section is coated with an electrically conducting layer, normally 30 nm of carbon, but not etched. A backscattered electron detector is used which maximises contrast from compositional differences and minimises that from topography. X-ray microanalysis is used to provide quantitative chemical analyses in regions of micrometre dimensions, and X-ray images provide semiquantitative information on the special distributions of individual elements. Techniques are described in standard texts (e.g. Goldstein *et al.*, 1981), and the results of analyses of individual clinker phases are considered in Chapter 1.

Figure 4.3(a) shows a backscattered electron image of a region in a typical clinker. Phases can, in general, be distinguished by their grey levels. The intensity of the electrons backscattered from a particular region of a specimen depends approximately on the mean atomic number of the material of which that region is composed; more precisely, it is represented by the backscattering coefficient η, which may be calculated from the formulae

$$\eta = \Sigma C_i \eta_i \tag{4.5}$$

$$\eta_i = -0\cdot0254 + 0\cdot016 Z_i - 1\cdot86 \times 10^{-4} Z_i^2 + 8\cdot3 \times 10^{-7} Z_i^3 \tag{4.6}$$

where the summation is over elements and C_i is the mass fraction of an element having atomic number Z_i (Reuter, 1972); η is approximately independent of beam energy in the range 10–100 keV.

Figure 4.3 (a) Backscattered electron image of a polished section of a Portland cement clinker, together with X-ray maps for (b) Si, (c), Ca, (d) Al and (e) Fe (continued on next page)

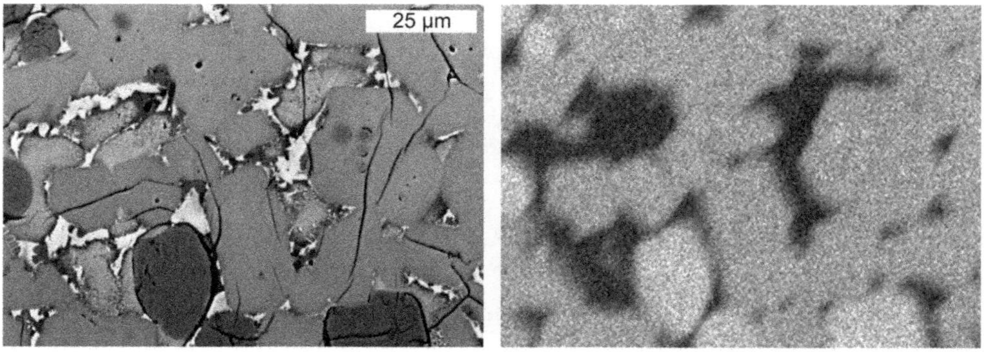

(a) (b)

Figure 4.3 Continued

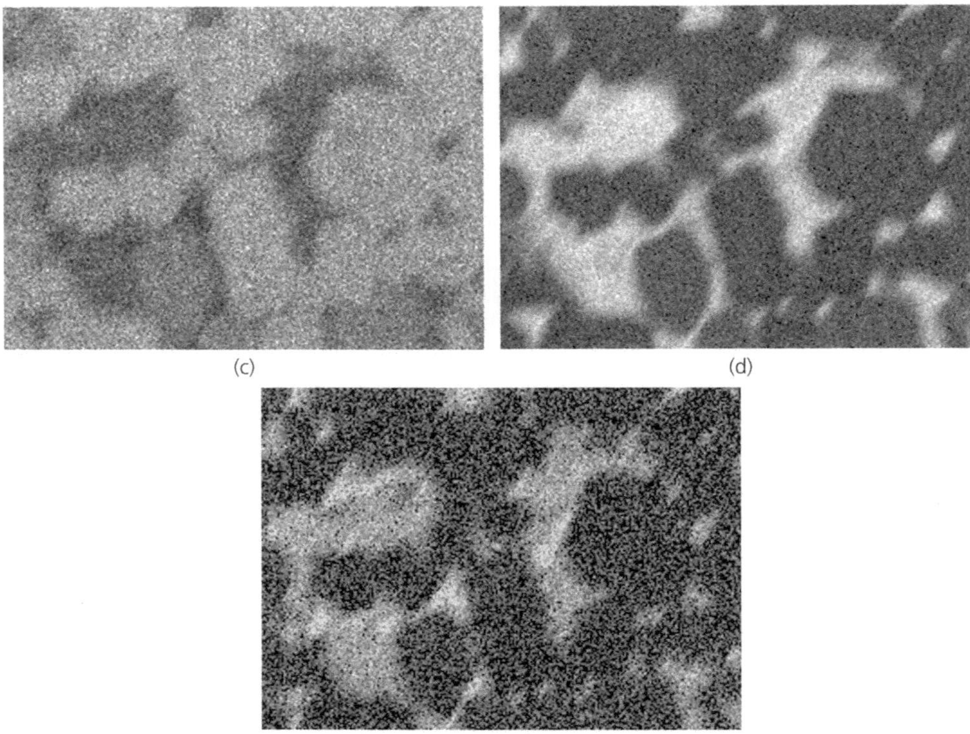
(c) (d) (e)

In (a), the medium-grey regions that constitute most of the micrograph are alite; the belite forms well-defined regions, which are rounded and darker than the alite; the interstitial material consists mainly of ferrite (light) and a mixture of aluminate and ferrite (light grey); cracks and pores (black) are also visible. Data courtesy of Dr R Shirley, University of Leeds. The images were taken using a working distance of 10 mm and an accelerating voltage of 15 kV.

A value for the coefficient is calculated for each phase of interest: for example, pure tricalcium silicate has the formula Ca_3SiO_5 and mass fractions of Ca = 0·5266, Si = 0·1230 and O = 0·3504; the pure-element backscatter coefficients calculated using Equation 4.6 are $\eta_{Ca} = 0{\cdot}2268$, $\eta_{Si} = 0{\cdot}1644$ and $\eta_O = 0{\cdot}0911$, and so using Equation 4.5 $\eta_{C_3S} = (0{\cdot}5266 \times 0{\cdot}2268) + (0{\cdot}1230 \times 0{\cdot}1644) + (0{\cdot}3504 \times 0{\cdot}0911) = 0{\cdot}1716$. The coefficient for pure dicalcium is calculated similarly to 0·1662, and so C_2S will appear dark relative to C_3S because η_{C_2S} is less than η_{C_3S}. The coefficients are affected only slightly by the minor substituents that are present in alite (impure C_3S) and belite (impure C_2S): assuming the typical compositions given in Table 1.2, the values are 0·171 for alite and 0·167 for belite. The values for the cubic aluminate and ferrite are 0·168 and 0·178, respectively. The differences between these values are such that the grey levels of the alite, belite and ferrite phases on micrographs are distinguishable to the eye (e.g. as in Figure 4.3(a)), but those of belite and aluminate are not; however, the phase boundaries are normally visible, presumably due to slight topographical contrast or compositional irregularity. The resolution available is limited by the size of the region from which the backscattered electrons are produced, which is somewhat below 1 μm in each direction. Much higher resolution is obtainable using the secondary electron image, but this depends mainly on topographical contrast and thus gives, at best, poor definition of the phases. The resolution given by the X-ray image is poorer than that given by the backscattered electron image because of the larger volume from which X-rays are generated, which is evident on comparison of the backscattered electron image in Figure 4.3(a) with its associated X-ray images in Figures 4.3(b)–4.3(e). EDX analysis from points in Figure 4.3(a) give compositions for the alite and belite, respectively, as $Ca_{2·94}Mg_{0·04}Fe_{0·02}Al_{0·06}Si_{0·94}O_5$ and $Ca_{1·95}Na_{0·02}Mg_{0·03}Fe_{0·03}Al_{0·06}Si_{0·91}O_4$, which are quite similar to the values given in Table 1.3.

Image analysis makes it possible to quantify images produced by SEM (Bentz *et al.*, 1999; Diamond and Bonen, 1992; Lebiedzik *et al.*, 1981; Richardson, 2002; Scrivener, 1987, 1989; Stutzman, 1991a, 1991b, 2004a; Stutzman *et al.*, 2015). It provides a means of quantitative phase determination and can also provide information on such matters as the distributions of shape, size and surroundings of particles and of phases and pores within them or the relative amounts of each of the phases exposed on the surfaces of polymineralic particles. It is, of course, essential to examine a sufficient number of samples and fields to ensure that the results are representative, but the labour and subjectivity associated with light microscopic studies of a comparable nature are largely eliminated.

Scrivener (1987) showed that it was possible to determine the total contents of silicate phases and interstitial material by image analysis of a combination of backscattered electron images and X-ray maps. She also showed that the proportions of interstitial phases exposed on the surfaces of the grains were significantly greater than those by volume in the bulk material. Overall, fracture during grinding appeared to be predominantly through alite, and to avoid phase boundaries. Subsequent developments in image-analysis systems have made it possible to determine the individual clinker phases quantitatively (Bentz *et al.*, 1999; Diamond and Bonen, 1992; Stutzman, 1991a, 1991b, 2004a; Stutzman *et al.*, 2015, 2016).

4.3.2 X-ray diffraction

The uses of XRD powder diffraction in the study of clinker or anhydrous cement include the qualitative and quantitative (QXDA) determination of phase composition and the determination of polymorphic modification, state of crystallinity and other features of individual phases. In principle, information on compositions of phases is obtainable through cell parameters but, due to the complexity of the problem and lack of adequate reference data, XRD is generally less satisfactory for the clinker phases than X-ray microanalysis. Table 4.2 gives the pattern of a typical Portland cement, with indications of the assignments of peaks to phases. Quantitative determination of phases is discussed in §4.5.3.

Table 4.2 XRD powder pattern of a typical Portland cement (continued on next page)

2θ°	*d*: Å	I_{pk}	Main phases	2θ°	*d*: Å	I_{pk}	Main phases
11·7	7·56	5	gyp	34·4	2·607	83	ali, bel
12·1	7·31	6	fer	36·7	2·449	6	ali
14·9	5·95	6	ali	37·4	2·404	2	bel

Table 4.2 Continued

2θ°	d: Å	I_{pk}	Main phases	2θ°	d: Å	I_{pk}	Main phases
20·7	4·29	7	gyp	38·8	2·321	12	ali
21·9	4·06	2	alu	39·5	2·281	5	bel
23·0	3·867	7	ali	41·3	2·186	41	ali, bel
23·4	3·802	3	bel	41·6	2·171	16	ali, bel
24·4	3·648	3	fer	44·1	2·053	6	fer
25·3	3·520	4	ali	44·5	2·036	3(B)	bel
26·4	3·376	2	bel	44·7	2·027	2	bel
27·6	3·232	2	bel, ali	45·8	1·981	10	ali, bel
28·1	3·175	4	ali, bel	47·0	1·933	11	ali
29·1	3·069	5	gyp	47·4	1·918	8(B)	fer
29·4	3·038	60	ali	47·8	1·903	7	alu
30·1	2·969	19	ali	49·9	1·828	5	ali, fer
31·1	2·876	4	bel, gyp	51·7	1·768	33	ali
32·2	2·780	100	ali, bel, fer	51·8	1·765	35	ali
32·6	2·747	85	ali, bel	56·0	1·642	2	ali
33·2	2·698	40	alu, ali	56·6	1·626	18	ali, bel
33·9	2·644	23	fer	58·7	1·573	3	bel, fer
				59·4	1·555	3	alu
				59·9	1·544	6	ali

Cu Kα radiation; I_{pk} = relative peak height (B = broad); phases are given in decreasing order of their contributions to each peak, ones making only minor contributions being omitted; ali = alite, bel = belite, alu = aluminate, fer = ferrite, gyp = gypsum.

4.3.3 Chemical or physical methods for separation of phases

Partial separations of the clinker phases can be effected by chemical extraction methods, which may be combined with XRD to identify and determine minor constituents. Of several reagents that have been used to dissolve the silicate phases, the most successful has been a solution of salicylic acid in methanol (SAM), introduced by Takashima (1958). This reagent dissolves free lime, alite and belite, at rates decreasing in that order (Hjorth and Lauren, 1971). By adjusting its composition, it is possible either to remove much of the alite, so facilitating investigation of the belite, or all of the alite and belite, facilitating that of the remaining phases. A procedure is as follows (Hjorth and Lauren, 1971).

The clinker is ground in propan-2-ol to an average particle size of 5 μm. After evaporating off the propan-2-ol, the sample (5 g) is treated for 2 h at room temperature with the SAM reagent

(300 mL), with constant stirring. The solution, which is now red, is filtered on a No.4 sintered glass filter and the residue washed with methanol and dried at 90 °C. If the residue is to be examined by QXDA, the internal standard (e.g. Rutile) may be mixed with the sample before the extraction. For removal of alite and belite, the salicylic acid content of the reagent is 20 g per 300 mL. For complete removal of the alite, leaving the maximum quantity of belite undissolved, it is 10–15 g, depending on the alite content; five times the mass of C_3S in the sample, as given by the Bogue calculation, was found to be satisfactory.

A solution of maleic acid in methanol has also been used to dissolve the silicate phases (Tabikh and Weht, 1971). This is quicker than the SAM method, but if the methanol contains water, ettringite may be formed and water-soluble phases, such as K_2SO_4, lost, while if it is anhydrous, gelling occurs; the SAM method is thus preferable (Struble, 1985). Extraction of the silicate phases by the trimethylsilylation method of Tamas *et al.* (1976) also leaves a residue containing the aluminate and ferrite phases.

Gutteridge (1979) found that a solution of sucrose in aqueous potassium hydroxide (KOSH reagent) dissolves the aluminate and ferrite, leaving the silicates and minor phases. The clinker or cement (10 g) is reduced to below 5 μm by grinding in cyclohexane (20 mL) for 40 min in an agate ball mill. After evaporating off the cyclohexane in a stream of N_2, the sample (9 g) is added to the KOSH solution (30 g of KOH and 30 g of sucrose in 300 mL of water) at 95 °C and stirred for 1 min. The solution is then filtered and the residue washed in water (50 mL) followed by methanol (100 mL) and dried at 60 °C. By using a combination of the SAM and KOSH procedures, it is possible to produce a residue consisting largely of belite.

Various solvents may be used to dissolve free lime, which can be determined by titrating the resulting solution with a suitable acid. In one procedure (Javellana and Jawed, 1982), the clinker or cement (1·0 g), ground to about 300 $m^2 kg^{-1}$ (Blaine), is treated with anhydrous ethylene glycol (50 mL) at 80–100 °C for 5 min using a magnetic stirrer. The solution is then filtered under suction and the residue washed at least twice with a further 10–15 mL of hot ethylene glycol. The filtrate and washings are diluted with 25 mL of water and titrated to phenolphthalein with 0·05 *M* HCl. Other procedures employ mixtures of ethanol and glycerol (Lerch and Bogue, 1926) or of acetoacetic ester and butan-2-ol (Franke solvent; Franke, 1941). If properly used, all give similar results. They also dissolve CH and have been used to determine this phase in hydrated cements, but their use for this purpose is questionable as they also appear to attack other hydration products.

Physical methods are of limited effectiveness for separating the phases in cement clinker because of the intimate scale on which the latter are mixed; however, Yamaguchi and Takagi (1969) had some success with the use of dense liquids. Some concentration of the ferrite phase can be effected by magnetic separation (Midgley, 1954; Yamaguchi and Takagi, 1969).

4.3.4 Other methods

Bensted (1980) described the use of TG in the study of clinkers or unhydrated cements. At 4°C min^{-1} in N_2 with cements, losses normally occur at 100–200 °C from gypsum or hemihydrate, 400–500 °C from CH and 500–800 °C from $CaCO_3$. If a normal, open sample cup is used, the losses from conversion of gypsum to hemihydrate and from dehydration of hemihydrate are poorly resolved, making interpretation difficult.

Better separation can be achieved by using a sample cup with a top that is closed except for a narrow exit (Smallwood and Wall, 1981).

Syngenite, if present, decomposes at 250–300 °C. Small amounts of CH can occur due to hydration, and of $CaCO_3$ due to carbonation during milling or storage, impurity in the gypsum or, where specification allows it, deliberate addition. The increasing tendency to allow additions of substances other than calcium sulfate may lead to the presence of other phases detectable by TG. A mass gain at 300–400 °C in air that does not occur in N_2 normally indicates that the clinker was made under reducing conditions but could also conceivably be due to the presence of an excessive quantity of metallic iron incorporated during milling. In this latter case it would, of course, not be shown by the clinker.

Applications of IR and Raman spectroscopy to the study of clinkers and unhydrated cements have been reviewed (Bensted, 1976; Bensted and Varma, 1974; Black, 2009; Potgieter-Vermaak *et al.*, 2006). The laser Raman microprobe, with which regions of micrometre dimensions on a polished surface may be examined, has been used to investigate structure and crystallinity, especially of the alite and belite (Conjeaud and Boyer, 1980).

4.4 Calculation of quantitative phase composition from bulk chemical analysis

4.4.1 General

As seen below in §4.5.4, there is wide agreement that the Bogue calculation gives seriously incorrect results, especially for alite. It has often been stated that the errors arise because equilibrium is not maintained during cooling, but this is only partly correct. Even if equilibrium were maintained, all the major phases would contain significant proportions of substituent ions. The calculation gives incorrect results mainly because the phases do not have the compositions assumed for them, and the absence of equilibrium conditions during cooling merely alters the errors that arise. In addition, the four major components are not completely contained in the four major phases and errors arise because the bulk analysis is not fully corrected for this.

4.4.2 Modified Bogue calculation

The Bogue calculation is a solution of four linear simultaneous equations for four unknowns (Equations 3.8–3.11 in §3.1.3; see also Equations 3.12–3.18 for an extended treatment). Based on an approach due to Yamaguchi and Takagi (1969), it may be modified (Harrisson *et al.*, 1985; Taylor, 1989) by using the best available estimates of the compositions of the four phases instead of those of the pure compounds and by correcting the raw analysis more fully for the presence of major oxide components in phases other than the four major ones. For modern, rapidly cooled clinkers of ordinary Portland cements, the compositions of the phases do not seem to vary excessively, the most important variations in the present context being in the CaO and Fe_2O_3 contents of the alite (Chapter 1). By following these principles, one may expect to obtain results that are substantially more accurate than those given by the orthodox Bogue calculation.

Deductions from the bulk CaO content that may be needed comprise free lime and any CaO present in $CaCO_3$ (equivalent to the CO_2 content), clinker sulfates (calcium langbeinite or anhydrite or both) and, for a cement, added calcium sulfate ('gypsum'). Methods for estimating the amounts of CaO in clinker sulfates are suggested in §3.5.7. To obtain the content of CaO present in calcium sulfate added on grinding of a cement, it is necessary to know either the amount and composition of the added material or the SO_3 content of the clinker.

Assuming that a modern method of analysis (e.g. XRF) has been used, the SiO_2 and Al_2O_3 contents must be corrected for any material present in the insoluble residue. For a clinker, the latter usually consists largely of quartz (SiO_2), but for a cement it is likely to consist largely of impurities in the gypsum; it is often reasonable to assume that the insoluble residue is two-thirds SiO_2 and one-third Al_2O_3. If classical, wet analysis is used, these deductions should not be made, but the apparent content of Al_2O_3 must be corrected for TiO_2 and P_2O_5.

Values are required for the contents of CaO, SiO_2, Al_2O_3 and Fe_2O_3 in each of the four major phases. For a modern, quickly-cooled ordinary Portland cement or clinker, the compositions given in Table 1.2 are probably often a sufficiently good approximation. It may be desirable to modify them in accordance with the bulk analysis, in ways suggested in §1.2.3, §1.3.3, §1.4.4 and §1.5.2 for alite, belite, aluminate and ferrite, respectively, those for the CaO and Fe_2O_3 in the alite being probably the most important. Alternatively, if they are available, compositions determined by X-ray microanalysis of the particular clinker or cement may be used. The situation with other types of Portland cements or clinkers is considered in §4.4.4.

It is then necessary to set up and solve four simultaneous equations, one for each of the four major oxide components. For each equation, there are four terms on the left-hand side, one for each of the four major phases. The coefficient in each term is the mass fraction of the oxide component in the phase, the unknown is the mass percentage of the phase in the clinker or cement, which has to be determined, and the right-hand side is the mass percentage of the oxide component in the material, corrected for any contributions from phases other than the four major ones. As an example, Equation 4.7 might be that for CaO in a case for which the mass fractions of CaO in the alite, belite, aluminate and ferrite were, respectively, 0·710, 0·635, 0·566 and 0·475, the total mass percentage of CaO present in the four major phases was 64·7%, and x_1, x_2, x_3 and x_4 are, respectively, the mass percentages of alite, belite, aluminate and ferrite in the material.

$$0{\cdot}710x_1 + 0{\cdot}635x_2 + 0{\cdot}566x_3 + 0{\cdot}475x_4 = 64{\cdot}7 \qquad (4.7)$$

Similar equations are required for the other three major oxide components and the four equations are then solved to obtain values of x_1, x_2, x_3 and x_4. Any standard method of matrix inversion, such as the Gauss–Jordan method (Norris, 1981), may be used for this purpose.

4.4.3 Mass balance calculations

As a by-product of the modified Bogue calculation described in the preceding section, it is readily possible to calculate a mass-balance table showing how all the oxide components, both major and minor, are distributed among the major and minor phases. This requires values for the mass percentages of all the component oxides in each of the phases. Any MgO in excess of that assigned to the major phases is assumed to occur in periclase. Other constituent phases may include clinker sulfates, gypsum (for a cement), insoluble residue, $CaCO_3$ if present, and any part of the ignition loss not attributable to gypsum or $CaCO_3$. Table 4.3 is such a table for a typical clinker. The compositions assumed for the major phases were approximately those given in Table 1.2, and that of the sulfate was obtained as outlined in §3.5.7. The sulfate is probably largely K_2SO_4 with some Na_2SO_4 and $CaSO_4$ in solid solution, but could include some Ca langbeinite. The residual for each oxide component arises from incorrect assumptions regarding the compositions of phases together with contributions from any phases not included in the calculation.

Table 4.3 Distribution of oxide components among phases in a typical Portland cement clinker, calculated from the bulk chemical analysis

	Na_2O	MgO	Al_2O_3	SiO_2	P_2O_5	SO_3	K_2O	CaO	TiO_2	Mn_2O_3	Fe_2O_3	Total
Alite	0·1	0·7	0·7	16·8	0·1	0·1	0·1	47·7	0·0	0·0	0·5	66·7
Belite	0·0	0·1	0·3	4·1	0·0	0·0	0·1	8·2	0·0	0·0	0·1	13·0
Aluminate	0·1	0·1	2·2	0·3	0·0	0·0	0·0	4·0	0·0	0·0	0·4	7·0
Ferrite	0·0	0·3	2·2	0·4	0·0	0·0	0·0	4·7	0·2	0·1	2·1	9·9
Free lime	0·0	0·0	0·0	0·0	0·0	0·0	0·0	0·9	0·0	0·0	0·0	0·9
Periclase	0·0	0·5	0·0	0·0	0·0	0·0	0·0	0·0	0·0	0·0	0·0	0·5
Sulfates	0·1	0·0	0·0	0·0	0·0	0·9	0·8	0·1	0·0	0·0	0·0	1·9
Insol. Res.	0·0	0·0	0·0	0·1	0·0	0·0	0·0	0·0	0·0	0·0	0·0	0·1
Residual	0·0	0·0	0·0	0·0	0·0	0·0	−0·1	0·0	0·0	0·0	0·0	0·1
Total	0·3	1·6	5·3	21·6	0·2	1·0	1·0	65·6	0·2	0·1	3·1	100·1

Results are given as percentages on the ignited mass. Periclase calculated by difference. Discrepancies in totals arise from rounding.

4.4.4 Limitations and modifications of the modified Bogue calculation

The modified Bogue calculation described above appears to give satisfactory results for most modern Portland cement clinkers or cements having alumina ratios between about 1 and 4, but the following points should be noted.

- If the aluminate phase is orthorhombic, the appropriate composition (Table 1.2) should be used in the equations. If both cubic and orthorhombic forms are present, intermediate values could be used.
- For a sulfate-resisting Portland cement, an appropriate ferrite composition, such as that given in Table 1.2, should be used. If the result for aluminate is negative, this phase is absent, and it may be useful to replace the coefficients for the aluminate composition by ones for a second ferrite composition. By bracketing the ferrite composition in this way, estimates of both the content and the composition of the ferrite can be obtained.
- The results will be less accurate for slowly cooled clinkers, as the compositions of the ferrite and possibly also the aluminate phases may differ significantly from those assumed here. The method is not applicable without major modification to clinkers made under reducing conditions. It is doubtful whether the procedure is applicable to white cements, both for this reason and also because they may contain glass. In addition, the range of applicability of the composition given in Table 1.2 for the aluminate phase in such cements is uncertain.

4.5. Physical methods for determining quantitative phase composition

4.5.1 General

The physical methods that have proved effective for quantitative determination of phases in clinkers are based on light microscopy (§4.2.1), scanning electron microscopy (§4.3.1) and X-ray diffraction (§4.3.2). The last of these, in common with calculation from the bulk chemical analysis (§4.4), is also applicable to cements. Other methods that have been investigated, such as IR spectroscopy, appear to have less potential.

4.5.2 Light microscopy

The phases may be determined quantitatively by point counting (Commission chimique du CETIC, 1978; Hofmänner, 1975; Weigand, 1994). The areas of phases on a polished section of a cement clinker are proportional to the volumes in the material. Random and systematic errors arise from imperfect sampling, counting statistics, the presence of inclusions too small to be observed by this method and incorrect identifications of phases by the operator. Error from counting statistics depends on the number of points, the size of the grid and the grain sizes of the phases; 4000 points normally gives reasonable precision (Hofmänner, 1975). Systematic errors from incorrect identification of phases generally exceed errors from counting statistics and give rise to differences in results between operators (Commission chimique du CETIC, 1978). As an alternative to point counting, image analysis has been used (Märten *et al.*, 1994). With many clinkers, it is impracticable to determine the aluminate and ferrite phases separately by either method. For alite and belite, and for interstitial material as a whole, both are capable of giving highly reliable results, assuming, in the case of the silicate phases, that they do not contain inclusions of other phases too small to be observed.

4.5.3 Quantitative X-ray diffraction analysis (QXDA)

The application of this method to clinkers or cements presents several problems. Alite is normally so much the predominant phase that its pattern tends to swamp those of the other phases. All the stronger peaks of belite are overlapped by ones of alite. Other difficulties in quantitative analysis are caused by variability of patterns of phases due to compositional or polymorphic variation and, especially for the ferrite phase, peak broadening due to compositional zoning or imperfect crystallinity. Other problems in quantitative work are those of QXDA in general; for example, the material must be ground sufficiently finely that primary extinction of the more intense peaks is minimised, and adequate mixing with an internal standard is possible, if one is used, but not so strongly that crystallinity is reduced. Milling to below 5 μm in an agate ball mill with cyclohexane as grinding aid, followed by back loading into the sample holder of a diffractometer, has been found satisfactory for specimen preparation (Gutteridge, 1984).

Struble (1983) summarised early QXDA studies on clinkers and cements. Most investigators used some form of the internal standard method (Klug and Alexander, 1974). Rutile (TiO_2) has been widely used as the internal standard but other substances, such as Si, KBr, CaF_2 and corundum, have also been employed; corundum has the advantage of giving no overlaps with important cement peaks up to 60° 2θ. The method in its conventional form, in which only a *single* peak of the reference standard and of the phase to be determined is used, is moderately effective for alite, using the peak at approximately 51° 2θ, assuming that a means of accommodating the splitting of this peak is found. It is less satisfactory for the aluminate and ferrite, and useless for belite because of the weakness of their peaks and interference from those of alite.

Berger *et al.* (1966) pioneered a least-squares procedure in which the *profile* of a diffractometer trace was matched with a simulated trace summing the contributions from the component phases. In a development of this technique, Gutteridge (1984) collected data over the range of 24–39° 2θ (Cu $K\alpha_1$ radiation) by step counting at 0·05° intervals. Data were collected for the sample and pure phases, including known quantities of an internal standard, and the least-squares calculation carried out. Auxiliary procedures were devised for dealing with background levels, adjusting the peak widths and, to some extent, peak positions in the reference patterns to match those in the clinker, and matching the observed and computed patterns. Minor phases, such as lime, periclase, calcium sulfate phases and mineral additions, could be included in the analysis. Increased computing power allowed later investigators to use wider ranges of 2θ, with consequently improved precision and accuracy (Stutzman, 1992; Taylor and Aldridge, 1993).

The reference standards that are employed in these methods should be as near as possible to the phases in the material as regards composition, polymorphism and degree of crystallinity. This implies both knowledge of the characteristics of the particular clinker under examination and the ability to prepare the necessary specimens or, alternatively, to isolate them physically from the clinker itself, as described in §4.3.3. One solution (Gutteridge, 1984) has been to prepare reference patterns for a wide range of specimens of each phase and to use those that appear to correspond most closely to those present in the clinker under examination. Clinkers for which results had been obtained by point counting have also been used (Knudsen, 1976).

It is also possible to match the observed pattern against a simulated pattern obtained by summing patterns calculated from the known crystal structures. This procedure has been used (Taylor, 1991) to match intensities integrated over selected ranges of 2θ. For each calculated pattern, a '*reference intensity ratio*' (Hubbard *et al.*, 1976) was calculated. This is the integrated intensity of the strongest individual reflection (which may be a component of an overlap) relative to that of the strongest peak of corundum in a 1:1 mixture by mass. The internal standard added to the clinker or cement need not be corundum, provided that its own reference intensity ratio is known.

If calculated patterns and reference intensity ratios are used, the relative proportions of the phases can be determined from the observed diffractometer trace for a clinker or cement to which no internal standard has been added. By normalising the results to some assumed total (e.g. 98% for the four major phases in a clinker or 95% in a cement), it is thus possible to dispense both with reference standards and with an internal standard. Calculated patterns and reference intensity ratios for the four major phases are given in Chapter 1.

The peak overlap problem is tackled in the Rietveld whole-profile method by using a nonlinear least-squares approach to minimise the difference between the intensity at each data point, $2\theta_i$, in an observed pattern (y_i(obs.)) with the intensity at that value of 2θ in a pattern calculated using the crystal structures of the constituent phases (y_i(calc.)), together with a background contribution at that point – that is, the residual difference, $\sum_i \frac{(y_i(\text{obs.})-y_i(\text{calc.}))^2}{y_i(\text{obs.})}$, is minimised. For a sample that contains m crystalline phases, the equation to calculate the intensity at the ith point in the diffraction pattern, $y_i(calc.)$, is Equation 4.8 (Albinati and Willis, 2006; Hill and Howard, 1987)

$$y_i(calc.) = y_i(bkgd.) + \sum_{p=1}^{m}\left[S_p \sum_{k=k_1}^{k_2} m_{kp} L_{kp} \left|F_{kp}\right|^2 \varnothing\left(2\theta_i - 2\theta_{kp}\right) P_{kp}\right] \tag{4.8}$$

where y_i(bkgd.) is the intensity calculated for the background at the ith point; S_p is the scale factor for phase p; m_{kp} is the multiplicity factor for the kth reflexion; L_{kp} is the Lorentz-polarisation factor;

F_{kp} is the structure factor for the kth reflexion; $\varnothing(2\theta_i - 2\theta_{kp})$ is a peak-shape function; P_{kp} concerns any other corrections, including preferred orientation, extinction and so on.

For each phase, the summation is over all overlapping reflexions, k_1 to k_2, that contribute to the ith data point, $2\theta_i$. The structure factors are calculated using Equation 4.9, which necessitates accurate knowledge of the crystal structure of all of the phases, which has posed problems for studies of cement, in particular regarding alite (Courtial *et al.*, 2003; de la Torre *et al.*, 2008; de Noirfontaine *et al.*, 2006, 2012). It is important to stress that the fitting results will not be useful without the availability and selection of appropriate crystal structures. Crystal data for some of the phases that are present in anhydrous Portland cement powder are given in §1.2/Table 1.1 (C_3S polymorphs), §1.3/Table 1.5 (C_2S polymorphs), §1.4 (C_3A), §1.5/Table 1.11 (calcium aluminoferrites in the series $Ca_2(Al_xFe_{1-x})O_5$) and §2.5.1/Table 2.4 (sulfate phases).

$$F_k = \sum_{j=1}^{n} f_j O_j \exp\left[2\pi i\left(hx_j + ky_j + lz_j\right)\right]\exp\left(-B_j\right) \tag{4.9}$$

where x_j, y_j and z_j are the coordinates of the jth atom in the unit cell; O_j is the site occupancy factor of that atom; f_j is the atomic X-ray scattering factor; B_j is the atomic displacement parameter.

On the assumption that all of the crystalline phases in the sample have been included and that there are no amorphous phases (which is valid for Portland cement clinker; see §3.5.2), the mass fraction of each phase can be calculated from its scale factor using the X-ray densities (ρ_p) and unit cell volumes (V_p) using Equation 4.10 (Hill and Howard, 1987).

$$W_p = \frac{s_p \rho_p V_p^2}{\sum_{p=1}^{m} s_p \rho_p V_p^2} \tag{4.10}$$

Equation 4.10 is sometimes written with $\rho_p V_p^2$ replaced by $Z_p M_p V_p$, where M_p is the formula mass of the phase and Z_p is the number of formula units in the unit cell.

The Rietveld whole-profile method has been used widely for the study of cements, but its successful application is not straightforward: detailed reviews, guidance on methodology and examples are available: for example, Roode-Gutzmer and Ballim (2001), Taylor *et al.* (2002), Le Saoût *et al.* (2011), Aranda *et al.* (2012), Snellings (2015) and Stutzman *et al.* (2016). XRD patterns for Portland cement clinker are far too complex to allow refinement of the positions, site occupancies and displacement parameters of the atoms in the crystal structures that are used in the fitting procedure and, while lattice parameters may be allowed to vary to accommodate the effects of substituent ions, any variation must be limited and the resulting structures checked so that bond distances and coordinations remain sensible.

4.5.4 Comparison of results of different methods

From the results of a study in which six clinkers were examined in a number of different laboratories, Aldridge (1982) concluded that light microscopy gave accurate results for the quantitative determination of phases in clinkers, assuming an experienced operator and adequate sampling. Results obtained using QXDA that involved the peak areas/internal standard method varied widely between laboratories and it was concluded that, as carried out in most of them, the method was of low accuracy. One laboratory, however, obtained results that agreed well with those given by microscopy and which were considered to be accurate. In general, the contents of alite found by microscopy were higher than those given by the Bogue calculation.

The results of another cooperative study, in which microscopy and extraction methods were used, showed that the Bogue calculation always gives low results for alite and generally gives high ones for belite when compared with microscopy (Commission chimique du CETIC, 1978). For the aluminate and ferrite phases, agreement was reasonably good. The total contents of silicates, determined by chemical extraction, agreed reasonably well with those found by microscopy.

Several early studies using QXDA gave results in fair agreement with those of the Bogue calculation but, in the light of the above and other recent evidence, are of doubtful validity. Odler *et al.* (1981) found that alite contents found either by microscopy or by QXDA were always greater than those given by the Bogue calculation. Gutteridge (1984) reached the same conclusion for QXDA and also noted that the total amounts of the crystalline phases did not vary significantly from 100%. Kristmann (1977) concluded from a study of 39 clinkers that alite and belite were determined more satisfactorily by microscopy than by QXDA but that aluminate and ferrite, due to their close admixture, were better determined by QXDA or by determining the total content of interstitial material by microscopy and the relative amounts of its constituent phases by QXDA of a residue from SAM extraction (§4.3.3). His results indicated that the values given by the Bogue calculation are low for alite and aluminate, high for belite and reasonably satisfactory for ferrite. The mean discrepancy for alite was 8%.

The modified Bogue calculation described in §4.4.2 was tested using Kristmann's (1977) data (Taylor, 1989). Agreement was substantially better than with the Bogue calculation for alite, belite and aluminate, and about as good for ferrite. For alite and belite, there was no significant bias relative to Kristmann's (1977) results in either direction. For aluminate, there was a possible tendency to give high results; for ferrite, it was not possible to tell whether there was bias, due to inadequacies in the experimental data. Good agreement was also obtained in tests on a wide range of cements using experimental data obtained by Gutteridge (personal communication with HFWT, 1986) using his QXDA method (Gutteridge, 1984); Le Saoût *et al.* (2011) found the same using the Rietveld method. Good agreement has been found between results from the Rietveld method and SEM point counting (Le Saoût *et al.*, 2011) and image analysis (Stutzman, 2004a), and for alite and belite with optical microscopy (Walenta and Füllmann, 2004) and ^{29}Si NMR (Poulsen *et al.*, 2009).

Comparison of the results from round-robin studies involving the Rietveld whole-profile method with earlier ones (described above) has shown an improvement in reproducibility and repeatability, both within and between laboratories, but with alite generally underestimated and belite overestimated (León-Reina *et al.*, 2009; Stutzman, 2004b, 2005).

Taken as a whole, the evidence indicates that all four major phases can be satisfactorily determined by QXDA if an adequate experimental procedure is employed or by calculation. The alite and belite can be determined by microscopy techniques and by ^{29}Si MAS NMR, while the content of interstitial material (total for light optical) can be determined by microscopy. With the Rietveld whole-profile method, the absolute accuracy is probably around 2–3 wt.% for alite and belite, and 1–2 wt.% for aluminate and ferrite.

4.6. Reactivities of clinker phases

4.6.1 Effect of major compositional variation

The ability of a substance to act as a hydraulic cement depends on two groups of factors. First, it must react with water to a sufficient extent and at a sufficient rate; γ-C_2S, which is virtually inert at ordinary temperatures, has no cementing ability. Second, assuming that an appropriate ratio of water to cement is used, the reaction must yield solid products of very low solubility and with a microstructure that gives rise to the requisite mechanical strength, volume stability and other necessary properties. C_3A reacts rapidly and completely with water, but the products that are formed when no other substances are present do not meet these criteria, and its ability to act as a hydraulic cement is very poor. C_3S, in contrast, satisfies both sets of conditions and is a good hydraulic cement.

In this section, only the factors that control reactivity are considered. Any understanding of them must be based on a knowledge of the mechanisms of the reactions with water. It is highly probable that these are in all cases ones in which an essential step is transfer of protons to the solid from the water, which is thus acting as a Brønsted acid (Barret *et al.*, 1983b; Danilov, 1976). As an initial approximation, one may consider the problem purely in terms of the idealised crystal structures. The reactivities of the oxygen atoms towards attack by protons depend on their basicities – that is, the magnitudes of the negative charges localised on them. Any structural feature that draws electrons away from the O atoms renders them less reactive, the basicity thus depending on the electronegativities of the atoms with which they are associated. In this respect, the nearest neighbours are the most important, but the effects of atoms further away are far from negligible. This hypothesis leads to the following predictions, all of which agree with the facts (Taylor, 1990).

- O atoms attached only to atoms of a single element will be less reactive as the electronegativity of that element increases. Thus, C_3S and lime both contain O atoms linked only to Ca (100% for lime and 20% for C_3S; Table 4.4), and are more reactive than β-C_2S, in which all the O atoms are also linked to Si (Table 4.4), while α-Al_2O_3, in which all are linked to Al (Table 4.5), is inert. However, the reactivity of lime decreases if it is strongly burned.
- O atoms forming parts of silicate, aluminate or other anionic groups will be less reactive as the degree of condensation with other such groups increases. β-C_2S, in which the tetrahedra do not share O atoms (i.e. they are Q^0), is more reactive than any of the C_3S_2 or CS polymorphs, in which they do (the asymmetric unit of rankinite (C_3S_2), and has two Si that are both Q^1; kilchoanite (C_3S_2) has three Si, one each of Q^0, Q^1 and Q^2 (Table 5.1(e)); wollastonite-Tc has three Si that are all Q^2 (Table 5.1(a)); wollastonite-2M has three Si that are all Q^2).
- O atoms forming parts of anionic groups similar as regards degree of condensation will be less reactive as the central atoms in the groups become more electronegative. In contrast to the C_3S_2 and CS polymorphs, CA, $C_{12}A_7$ and C_3A are all highly reactive, even though the AlO_4 tetrahedra share corners in all cases.

Other hypotheses based on considerations of crystal structure have been proposed and, in varying degrees, may account for second-order effects. Thus, Jeffery (1952) suggested that irregular coordination of Ca was responsible for the high reactivity of C_3S as compared with γ-C_2S; however, lime reacts with water, but the coordination of its Ca atoms is highly symmetrical. Jost and Ziemer (1984), who noted the weaknesses in this and other earlier hypotheses, considered that high reactivity was associated with the presence of face-sharing polyhedra, but lime contains no such groupings, yet can be highly reactive.

Table 4.4 The environment of oxygen atoms in the unit cells of various calcium silicates

	Phase								
	C	C_3S M_3	C_3S T_1	C_3S R	β-C_2S	γ-C_2S	C_3S_2	CS (2M)	Coordinated to:
$O^{[3]}$								55·6	2 Si, 1 Ca (33·3 %) 1 Si, 2 Ca (22·2 %)
$O^{[4]}$		80	66·7	60	25	100	85·7	44·4	1 Si, 3 Ca*
$O^{[5]}$			13·3	20	75		14·3		1 Si, 4 Ca
$O^{[6]}$	100	20	20·0	20					6 Ca
$\overline{CN_O}$	6·00	4·40	4·53	4·60	4·75	4·00	4·14	3·44	
% O–Si	0	18·2	17·6	17·4	21·1	25·0	27·6	38·7	
% O–Ca	100	81·8	82·4	82·6	78·9	75·0	72·4	61·3	

Values are given for: (i) the percentages of O atoms with different coordination numbers (CN) in each unit cell (rather than in the asymmetric unit i.e. multiplicities were included in the calculations); (ii) the average CN for O in each unit cell ($\overline{CN_O}$); (iii) the percentage of bonds to Si or Ca. The references for the structures are: C_3S M_3 (Mumme, 1995); C_3S T_1 (Golovastikov *et al.*, 1975); C_3S R (Il'inets *et al.*, 1985); β-C_2S (Mumme *et al.*, 1995); γ-C_2S (Mumme *et al.*, 1995); C_3S_2, rankinite (Saburi *et al.*, 1976); CS, wollastonite-2M (Trojer, 1968).
*Except in C_3S_2 where $1/6^{th}$ of these are 2 Si, 2 Ca.

Table 4.5 The environment of oxygen atoms in the unit cells of various calcium aluminates (continued on next page)

	Phase								
	C	C_3A	$C_{12}A_7$	C_5A_3	CA	CA_2	CA_6	A	Coordinated to:
$O^{[3]}$				21·4	50	71·4			3 Al 28·6 % CA_2 2 Al, 1 Ca for CA, C_5A_3 and 42·9 % CA_2
$O^{[4]}$			97·0*	71·4	50	28·6	84·2	100	4 Al for A and 52·6 % CA_6 3 Al, 1 Ca for 31·6 % CA_6 2 Al, 2 Ca for 72·7 % $C_{12}A_7$, 50 % C_5A_3, CA, CA_2 1 Al, 3 Ca for 21·4 % C_5A_3 and 24·2 % $C_{12}A_7$
$O^{[5]}$		100		7·1			15·8		3 Al, 2 Ca for CA_6 2 Al, 3 Ca for C_3A (33·3 %) 1 Al, 4 Ca for C_3A (66·7 %) and C_5A_3
$O^{[6]}$	100								6 Ca

Table 4.5 Continued

	Phase								Coordinated to:
	C	C_3A	$C_{12}A_7$	C_5A_3	CA	CA_2	CA_6	A	
$\overline{CN_O}$	6·00	5·00	4·00	3·86	3·50	3·29	4·16	4·00	
% O–Al	0	26·7	43·8	44·4	42·9	69·6	84·8	100	
% O–Ca	100	73·3	56·3	55·6	57·1	30·4	15·2	0·0	

Values are given for: (i) the percentages of O atoms with different coordination numbers (CN) in each unit cell (rather than in the asymmetric unit i.e. multiplicities were included in the calculations); (ii) the average CN for O in each unit cell ($\overline{CN_O}$); (iii) the percentage of bonds to Al or Ca. The references for the structures are: C_3A (Mondal and Jeffery, 1975); $C_{12}A_7$ (Bartl and Scheller, 1970); C_5A_3 (Vincent and Jeffery, 1978); CA (Hoerkner and Mueller Buschbaum, 1976); CA_2 (Goodwin and Lindop, 1970); CA_6 (Utsunomiya *et al.*, 1988).

*1/33 of O are 'free'; see §2.2.3. This atom has not been considered for the purpose of the $\overline{CN_O}$ or O–Ca bond calculations.

4.6.2 Effects of ionic substitutions, defects and variation in polymorph

The considerations discussed in the preceding section suffice only to explain the effects on reactivity of major variations in composition. They do not, in general, explain the effects of variations in polymorph or introduction of substituent ions. These effects are often complex and, while in some cases the results appear clear cut, there are many others in which they present a confusing and sometimes apparently contradictory picture. The probable reason is that many variables are involved, which are difficult or impossible to control independently. These include, besides chemical composition and polymorphic change, concentrations and types of defects; particle size distribution; textural features such as crystallite size and morphology, mechanical stress, presence of microcracks and the monocrystalline or polycrystalline nature of the grains; influences of reaction products, whether solid or in solution. In a cement as opposed to a pure phase, a foreign ion may not only substitute in a particular phase, but also influence the microstructure of the material in ways unconnected with that substitution – for example, by altering the physical properties of the high-temperature liquid present during its formation. For both microstructural and chemical reasons, the reactivity of a given phase in a cement may not be the same as that observed when it is alone.

Defects play a particularly important role. Sakurai *et al.* (1969) noted that C_3S, alite, C_2F and C_4AF could all take up Cr_2O_3 and that this greatly accelerated reaction at early ages. Electron microscopic studies showed that reaction began at grain boundaries and at points of emergence of screw dislocations, the concentration of which was greatly increased by Cr_2O_3 substitution. It was also shown that the substituted materials were semiconductors and a mechanism of attack based on electron transfer processes was suggested. Fierens *et al.* (Fierens and Verhaegen, 1972; Fierens *et al.*, 1973) also found that the influence of substituent ions on the reactivity of C_3S was due to the presence of defects, which could be studied using thermoluminescence and which could also be introduced by suitable heat treatment.

The complex relations between content of substituent ions, polymorphism, defects and reactivity are well illustrated by the results of Boikova *et al.* (Boikova, 1986; Boikova *et al.*, 1978) on ZnO-substituted C_3S. With increasing ZnO content, triclinic, monoclinic and rhombohedral polymorphs were successively stabilised. Curves in which the extent of reaction after a given time were plotted against ZnO content showed maxima corresponding approximately to the T_2–M_1 and M_2–R transitions, with intermediate minima. As in the work mentioned above, it was shown that these compositions were also ones in which the numbers of defects of a particular type were at a maximum. Reactivity thus depends not so much on the amount of substituent or nature of the polymorph as on the types and concentrations of defects.

Studies on the comparative reactivities of β-, α′- and α-C_2S, reviewed by Skalny and Young (1980), have given contradictory results. As these authors conclude, reactivity probably depends on specimen-specific factors other than the nature of the polymorph and, at the lower temperature, polymorphs may be affected by the exsolution of impurities. This latter hypothesis receives strong support from subsequent observations that attack begins at exsolution lamellae and grain boundaries, and that these may contain phases, such as C_3A, that are much more reactive than the C_2S (Chan *et al.*, 1988). Without these lamellae, β-C_2S might be much less reactive than it is.

There is wide agreement that substitution of alkali metal ions retards the early reaction of the aluminate phase, which is thus less for the orthorhombic than for the cubic polymorphs (Boikova *et al.*, 1977; Regourd *et al.*, 1980; Spierings and Stein, 1976). The effect has been attributed to structural differences, but the early reaction of pure C_3A is also retarded by adding NaOH to the solution; the OH^- ion concentration in the solution may be the determining factor (Spierings and Stein, 1976). The reaction of C_3A is also retarded by iron substitution and by close admixture with ferrite; formation of a surface layer of reaction products may be a determining factor, at least in later stages of reaction, and the retarding effect of such a layer may be greater if it contains Fe^{3+} (Boikova *et al.*, 1981).

4.6.3 Ferrite reactivity; effect of complexing in solution

The reactivities of ferrite phases in the pure $Ca_2(Al_xFe_{1-x})_2O_5$ series have been widely found to increase with Al/Fe ratio; material of composition similar to that in clinkers is more reactive than either C_4AF or C_6A_2F (e.g. Boikova *et al.*, 1977, 1981; de Keyser and Tenoutasse, 1969). Anomalies in the relation between rate and Al/Fe ratio found by some workers can possibly be attributed to zoning or other variations in crystallinity. As with the aluminate phase, the decrease in reactivity with Fe content is possibly explainable by differences in the properties of a retarding layer of product. The relatively high reactivity of the clinker material may be related to its semiconducting properties (Miyazawa and Tomita, 1969) and to disorder arising from the complex ionic substitutions (Boikova *et al.*, 1981).

The rate of reaction of the ferrite phase is much increased by the presence of tri-isopropanolamine in the solution. This is not strongly adsorbed on the cement grains but complexes the iron in solution, thereby inhibiting the formation of a layer of reaction product and favouring the passage of iron into solution (Gartner and Myers, 1993). This illustrates the important fact that the rate of reaction is not solely controlled by properties of the solid phase but depends also on what is present in the solution.

REFERENCES

Albinati A and Willis TM (2006) The Rietveld method. In *International Tables for Crystallography, Section 8.6* **1**: 710–712.

Aldridge LP (1982) Accuracy and precision of phase analysis in Portland cement by Bogue, microscopic and X-ray diffraction methods. *Cement and Concrete Research* **12(3)**: 381–398.

Allen T (1974) *Particle Size Measurement*, 2nd edn. Chapman and Hall, London, UK.

Allen T (2003) *Powder Sampling and Particle Size Determination*. Elsevier Science, Amsterdam, the Netherlands.

Anderegg FO and Hubbell DS (1929) The rate of hydration of cement clinker. *Proceedings of ASTM* **29**: 554–564.

Anderegg FO and Hubbell DS (1930) The rate of hydration of cement clinker$_1$. Part II. Portland cement at nine and twelve months. Part III. Three minerals found in Portland cement. *Proceedings of ASTM* **30**: 572–578.

Aranda MAG, de la Torre ÁG and León-Reina L (2012) Rietveld quantitative phase analysis of OPC clinkers, cements and hydration products. *Reviews in Mineralogy and Geochemistry* **74(1)**: 169–209.

Barret P, Menetrier D and Bertrandie D (1983) Mechanism of C_3S dissolution and problem of the congruency in the very initial period and later on. *Cement and Concrete Research* **13(5)**: 728–738.

Bartl H and Scheller T (1970) Zur Struktur des $12CaO{\cdot}7Al_2O_3$. *Neues Jahrbuch für Mineralogie, Monatshefte*: 547–552. (In German.)

Bauschinger -, Belk W, Bömches F, Dyce Cay W, Colson C, Dyke T, Gostling W, Heintzel C, Higgin G, Mann IJ, Meÿer CH, Michaëlis -, Porter W, Reid W, Studt -, Scott HYD, Grant J (1880) Correspondence. Portland cement. *Minutes of Proceedings of the Institution of Civil Engineers* **62**: 208–249.

Bensted J (1976) Uses of Raman spectroscopy in cement chemistry. *Journal of the American Ceramic Society* **59(3–4)**: 140–143.

Bensted J (1980) Some applications of thermogravimetry to Portland cement. *Il Cemento* **77**: 169–182.

Bensted J and Varma SP (1974) Some applications of infrared and Raman spectroscopy in cement chemistry part 2. *Cement Technology* **5**: 378–382.

Bentz DP, Stutzman PE, Haecker CJ and Remond S (1999) SEM/X-ray imaging of cement-based materials. In *Proceedings of 7th Euroseminar on Microscopy Applied to Building Materials* (Pietersen HS, Larbi JA and Janssen HHA (eds)). Delft University of Technology, Delft, the Netherlands, pp. 457–466.

Berger RL, Frohnsdorff GJC, Harris PH and Johnson PD (1966) Application of X-ray diffraction to routine mineralogical analysis of Portland cement. *Highway Research Board Special Report* **90**: 234–253.

Black L (2009) Raman spectroscopy of cementitious materials. *Spectroscopic Properties of Inorganic and Organometallic Compounds* **40**: 72–127.

Blaine RL (1943) A simplified air permeability fineness apparatus. *ASTM Bulletin* **123**: 51–55.

Blount BJ (1906) Soc. Chem. Ind. **25** 1083. Reported by Desch CH (1911) *The Chemistry and Testing of Cement*. Edward Arnold, London, UK.

Boikova AI (1986) Chemical composition of raw materials as the main factor responsible for the composition, structure and properties of clinker phases. *Proceedings of 8th International Congress on the Chemistry of Cement* **1**: 19–34.

Boikova AI, Domansky AI, Paramonova VA, Stavitskaja GP and Nikushchenko VM (1977) The influence of Na_2O on the structure and properties of $3CaO{\cdot}Al_2O_3$. *Cement and Concrete Research* **7(5)**: 483–492.

Boikova AI, Degen MG, Paramonova VA and Sud'ina VV (1978) Defect nature and hydration activity of solid solutions of tricalcium silicate with zinc oxide. *Tsement* **3(5)**. (In Russian.)

Boikova AI, Grishchenko LV and Domansky AI (1981) Hydration of C_3A and solid solutions of various composition. *Proceedings of 7th International Congress on the Chemistry of Cement* **4**: 460–464.

Butt JM, Kolbasov VM and Melniekij GA (1974) *Zement Kalk Gips* **27**: 27.
Campbell DH (1986) Microscopical examination and interpretation of Portland cement and clinker. *Construction Technology Laboratories*, Skokie, IL, USA.
Chan CJ, Kriven WM and Young JF (1988) Analytical electron microscopic studies of doped dicalcium silicates. *Journal of the American Ceramic Society* **71(9)**: 713–719.
Colson C (1875) Experiments on the Portland cement used in the Portsmouth dockyard extension works. *Minutes of Proceedings of the Institution of Civil Engineers* **41**: 125–157.
Commission chimique du CETIC (1978) Determination de la composition mineralogique du clinker par analyse microscopique et dissolution des phases. *Ciments, Bétons, Plâtres, Chaux* **713**: 205–211.
Conjeaud M and Boyer H (1980) Some possibilities of Raman microprobe in cement chemistry. *Cement and Concrete Research* **10(1)**: 61–70.
Courtial M, de Noirfontaine MN, Dunstetter F, Gasecki G and Signes-Frehel M (2003) Polymorphism of tricalcium silicate in Portland cement: a fast visual identification of structure and superstructure. *Powder Diffraction* **18(1)**: 7–15.
Danilov VV (1976) Mechanism of hydration in cement pastes. *Proceedings of 6th International Congress on the Chemistry of Cement* **2**: 73–76. Part 1. (In Russian.)
de Keyser WL and Tenoutasse N (1969) The hydration of the ferrite phase of cements. *Proceedings of 5th International Symposium on the Chemistry of Cement* **2**: 379–386.
de la Torre ÁG, de Vera RN, Cuberos AJM and Aranda MAG (2008) Crystal structure of low magnesium-content alite: application to Rietveld quantitative phase analysis. *Cement and Concrete Research* **38(11)**: 1261–1269.
de Noirfontaine MN, Dunstetter F, Courtial M, Gasecki G and Signes-Frehel M (2006) Polymorphism of tricalcium silicate, the major compound of Portland cement clinker 2. Modelling alite for Rietveld analysis, an industrial challenge. *Cement and Concrete Research* **36(1)**: 54–64.
de Noirfontaine MN, Courtial M, Dunstetter F, Gasecki G and Signes-Frehel M (2012) Tricalcium silicate Ca_3SiO_5 superstructure analysis: a route towards the structure of the M_1 polymorph. *Zeitschrift für Kristallographie* **227(2)**: 102–112.
Desch CH (1911) *The Chemistry and Testing of Cement*. Edward Arnold, London, UK.
Diamond S and Bonen O (1992) A chemical image analysis method for Portland cements. *Materials Research Society Symposium Proceedings* **245**: 291–301.
Fierens P and Verhaegen JP (1972) Structure and reactivity of chromium-doped tricalcium silicate. *Journal of the American Ceramic Society* **55(6)**: 309–312.
Fierens P, Tirlocq J and Verhaegen JP (1973) Luminescence et hydration du silicate tricalcique. *Cement and Concrete Research* **3(5)**: 549–560. (In French.)
Franke B (1941) Bestimmung von Calciumoxyd und Calciumhydroxyd neben wasserfreiem und wasserhaltigem Calciumsilikat. *Zeitschrift für anorganische und allgemeine Chemie* **247(1–2)**: 180–184. (In German.)
Gartner E and Myers D (1993) Influence of tertiary alkanolamines on Portland cement hydration. *Journal of the American Ceramic Society* **76(6)**: 1521–1530.
Gille F, Dreizler I, Grade K, Kramer H and Woermann E (1965) *Mikroskopie des Zementklinkers-Bilderatlas, with English translation of text*. Beton Verlag, Dusseldorf, Germany.
Goldstein JI, Newbury DE, Echlin P, Joy DC, Fiori C and Lifshin E (1981) *Scanning Electron Microscopy and X-Ray Microanalysis: A Text for Biologists, Materials Scientists, and Geologists*. Plenum Press, New York, NY, USA.
Golovastikov NI, Matveeva RG and Belov NV (1975) Crystal structure of the tricalcium silicate $3CaO{\cdot}SiO_2{=}C_3S$. *Soviet Physics. Crystallography* **20**: 441–445 (Translated from Kristallografiya 20,: 721).

Goodwin DW and Lindop AJ (1970) The crystal structure of $CaO.2Al_2O_3$. *Acta Crystallographica B* **26(9)**: 1230–1235.

Grant J (1866) Experiments on the strength of cement, chiefly in reference to the Portland cement used in the Southern Main Drainage Works. *Minutes of Proceedings of the Institution of Civil Engineers* **25**: 66–111.

Grant J (1880) Portland cement; its nature, tests and uses. *Minutes of Proceedings of the Institution of Civil Engineers* **62**: 98–179.

Gutteridge WA (1979) On the dissolution of the interstitial phases in Portland cement. *Cement and Concrete Research* **9(3)**: 319–324.

Gutteridge WA (1984) Quantitative X-ray powder diffraction in the study of some cementive materials. *Proceedings of the British Ceramic Society* **35**: 11–23.

Harrisson AM, Taylor HFW and Winter NB (1985) Electron-optical analyses of the phases in a Portland cement clinker, with some observations on the calculation of quantitative phase composition. *Cement and Concrete Research* **15(5)**: 775–780.

Hill RJ and Howard CJ (1987) Quantitative phase analysis from neutron powder diffraction data using the Rietveld method. *Journal of Applied Crystallography* **20(6)**: 467–474.

Hjorth L and Lauren KG (1971) Belite in Portland cement. *Cement and Concrete Research* **1(1)**, 27–40.

Hoerkner W and Mueller Buschbaum H (1976) Zur Kristallstruktur von Ca Al_2 O_4. *Journal of Inorganic and Nuclear Chemistry* **38**, 983–984. (In German.)

Hofmänner F (1975) *Microstructure of Portland Cement Clinker*. Rheintaler Druckerei, Heerbrugg, Switzerland.

Hubbard CR, Evans EH and Smith DK (1976) The reference intensity ratio, I/I_c, for computer simulated powder patterns. *Journal of Applied Crystallography* **9(2)**, 169–174.

Il'inets AM, Malinovskii YA and Nevskii NN (1985) Crystal structure of the rhombohedral modification of tricalcium silicate, Ca_3SiO_5. *Doklady Akademii Nauk SSSR* **281**, 332–336.

Javellana MP and Jawed I (1982) Extraction of free lime in Portland cement and clinker by ethylene glycol. *Cement and Concrete Research* **12(3)**, 399–403.

Jeffery JW (1952) The crystal structure of tricalcium silicate. *Acta Crystallographica* **5(1)**, 26–35.

Jost KH and Ziemer B (1984) Relations between the crystal structures of calcium silicates and their reactivity against water. *Cement and Concrete Research* **14(2)**, 177–184.

Klug HP and Alexander LE (1974) *X-Ray Diffraction Procedures for Polycrystalline and Amorphous Materials*, 2nd edn. Wiley, New York, NY, USA.

Knudsen T (1976) Quantitative analysis of the compound composition of cement and cement clinker by X-ray diffraction. *American Ceramic Society Bulletin* **55**: 1052–1054.

Kristmann M (1977) Portland cement clinker: Mineralogical and chemical investigations: Part I microscopy, X-ray fluorescence and X-ray diffraction. *Cement and Concrete Research* **7(6)**: 649–658.

Kuhlmann K, Ellerbrock HG and Sprung S (1985) Particle size distribution and properties of cement. Part I: Strength of Portland cement. *Zement Kalk Gips* **38**: 169 (partial English translation: p. 136).

Le Saoût G, Kocaba V and Scrivener KL (2011) Application of the Rietveld method to the analysis of anhydrous cement. *Cement and Concrete Research* **41(2)**: 133–148.

Lea FM and Desch CH (1956) *The Chemistry of Cement and Concrete*, 2nd edn. Edward Arnold, London, UK.

Lea FM and Nurse RW (1939) The specific surface of fine powders. *Journal of the Society of Chemical Industry* **58(35)**: 277Y–283T.

Lebiedzik J, Lebiedzik J and Gouda GR (1981) Clinker analysis by automated SEM. *Proceedings of 3rd International Congress Cement Microscopy*: 154–170.

León-Reina L, de la Torre AG, Porras-Vásquez JM *et al.* (2009) Round robin on Rietveld quantitative phase analysis of Portland cements. *Journal of Applied Crystallography* **42**: 1–11.

Lerch W and Bogue RH (1926) Determination of uncombined lime in Portland cement. *Industrial and Engineering Chemistry* **18(7)**: 739–743.

Lerch WC and Ford CL (1948) Long-time study of cement performance in concrete – Chapter 3. Chemical and physical tests of the cements. *American Concrete Society Journal, Proceedings* **44**: 743–795.

Locher FW, Sprung S and Korf P (1973) The effect of the particle size distribution on the strength of Portland cement. *Zement Kalk Gips* **26**: 349.

Long GR (1982a) Clinker quality characterization by reflected light techniques. *Proceedings of 4th International Conference on Cement Microscopy*: 92–109.

Long GR (1982b) The effect of burning environment on the microscopic characteristics of Portland cement clinker. *Proceedings of 4th International Conference on Cement Microscopy*: 128–140.

Long GR (1983) Microstructure and chemistry of unhydrated cements. *Philosophical Transactions of the Royal Society London A* **310**: 43–51.

Märten A, Knöfel D and Strunge J (1994) Quantitative structure analysis via image analysis. A suitable tool for quality control of Portland cement clinker? *World Cement* **25**: 49–52.

Méric JP (1980) Influence of grinding and storage of conditions of clinker. *Proceedings of 7th International Congress on the Chemistry of Cement* **1**: I-4/1-I-4/16.

Midgley H (1954) Discussion comment after Malquori GV and Cirilli V (1954) The ferrite phase. *Proceedings of the 3rd International Symposium on the Chemistry of Cement* 120–136; 140–143.

Miyazawa K and Tomita K (1969) On the color change of Portland cement. *Proceedings of 5th International Symposium on the Chemistry of Cement* **1**: 252–261.

Mondal P and Jeffery JW (1975) The crystal structure of tricalcium aluminate, $Ca_3Al_2O_6$. *Acta Crystallographica* **B31**: 689–697.

Mumme WG (1995) Crystal-structure of tricalcium silicate from a Portland-cement clinker and its application to quantitative XRD analysis. *Neues Jahrbuch für Mineralogie, Monatshefte* **4**: 145–160.

Mumme WG, Hill RJ, Bushnell-Wye G and Segnit ER (1995) Rietveld crystal structure refinements, crystal chemistry and calculated powder diffraction data for the polymorphs of dicalcium silicate and related phases. *Neues Jahrbuch für Mineralogie – Abhandlungen* **169**: 35–68.

Norris AC (1981) *Computational Chemistry*. Wiley, Chichester, UK.

Odler I, Abdul-Maula S, Nudling P and Richter T (1981) Mineralogical and oxidic composition of industrial Portland cement clinkers. *Zement Kalk Gips* **34**: 445 (partial English translation, p. 240).

Ono Y (1981) Microscopical observation of clinker, for the estimation of burning condition, grindability and hydraulic activity. *Proceedings of 3rd International Conference on Cement Microscopy*: 198–210.

Osbaeck B and Jøns ES (1980) The influence of the content and distribution of alkalies on the hydration properties of Portland cement. *Proceedings of 7th International Congress on the Chemistry of Cement* **2**:135–140.

Pearson JC and Sligh WH (1915) An air analyzer for determining the fineness of cement. *Technologic Papers of the Bureau of Standards* No. 48, Washington Government Printing Office, Washington, DC, USA.

Popovics S (1979) *Concrete-Making Materials*. McGraw Hill, London, UK.

Potgieter-Vermaak SS, Potgieter JH and Van Grieken R (2006) The application of Raman spectrometry to investigate and characterize cement, part I: a review. *Cement and Concrete Research* **36(4)**: 656–662.

Poulsen SL, Kocaba V, Le Saoût G, Jakobsen HJ, Scrivener KL and Skibsted J (2009) Improved quantification of alite and belite in anhydrous Portland cements by ^{29}Si MAS NMR: Effects of paramagnetic ions. *Solid State Nuclear Magnetic Resonance*, **36(1)**: 32–44.

Redgrave GR (1895) *Calcerous Cements: their Nature and Uses, with Some Observations upon Cement Testing*. Charles Griffin, London, UK.

Regourd M, Hornain H and Mortureux B (1980) Hydration of C_3A in synthetic mixtures and in Portland cements. *Proceedings of 7th International Congress on the Chemistry of Cement* **4**: 477–482. (In French.)

Reuter W (1972) The ionization function and its application to the electron probe analysis of thin films. *Proceedings of 6th International Conference on X-Ray Optics and Microanalysis* (Shinoda G, Kohra K and Ichinokawa T (eds.)), University of Tokyo Press, Tokyo, Japan, p. 121.

Richardson IG (1991) *Microstructure and Composition of Hardened Cement Pastes*. DPhil thesis, University of Oxford, Oxford, UK.

Richardson IG (2002) Electron microscopy of cement. In *Structure and Performance of Cements*, 2nd edn (Barnes P and Bensted J (eds.)). E & FN Spon, London, UK, pp. 500–556.

Ritzmann H (1968) On the relation between the particle size distribution and the strength of Portland cement. *Zement Kalk Gips* **21**: 390.

Roode-Gutzmer QI and Ballim Y (2001) Phase composition and quantitative X-Ray powder diffraction analysis of Portland cement and clinker. In *Materials Science of Concrete VI* (Mindess S and Skalny J (eds.)). American Ceramic Society, Westerville, OH, USA.

Rosin P and Rammler E (1933) The laws governing the fineness of powdered coal. *Journal of the Institute of Fuel* **7**, 29–36.

Saburi S, Kusachi I, Henmi C, Kawahara A, Henmi K and Kawada I (1976) Refinement of the structure of rankinite. *Mineralogical Journal (Japan)* **8(4)**: 240–246.

Sakurai T, Sato T and Yoshinaga A (1969) The effect of minor components on the early hydraulic activity of the major phases of Portland cement clinker. *Proceedings of 5th International Symposium on the Chemistry of Cement* **1**: 300–321.

Scott HYD and Redgrave GR (1880) The manufacture and testing of Portland cement. *Minutes of Proceedings of the Institution of Civil Engineers* **62**: 67–86.

Scrivener KL (1987) The microstructure of anhydrous cement and its effect on hydration. *Materials Research Society Symposium Proceedings* **85**: 39–46.

Scrivener KL (1989) The microstructure of concrete. *Materials Science of Concrete* **1**: 127–161.

Scrivener KL and Pratt PL (1984) Backscattered electron images of polished cement sections in the SEM. *Proceedings of 6th International Conference on Cement Microscopy*: 145–155.

Skalny JP and Young JF (1980) Mechanisms of Portland cement hydration. *Proceedings of 7th International Congress on the Chemistry of Cement* **1**: II-1/3-II-1/45.

Smallwood TB and Wall CD (1981) The quantitative determination of hydrated calcium sulphates in cement by use of self-generating atmosphere thermogravimetry. *Talanta* **28(4)**: 265–267.

Snellings R (2015) X-ray powder diffraction applied to cement. In *A Practical Guide to Microstructural Analysis of Cementitious Materials* (Scrivener K, Snellings R and Lothenbach B (eds.)). CRC Press, Boca Raton, FL, USA.

Spalding FP (1900) *Hydraulic Cement. Its Properties, Testing and Use*. Wiley, London, UK.

Spierings GACM and Stein HN (1976) The influence of Na_2O on the hydration of C_3A. II. Suspension hydration. *Cement and Concrete Research* **6(4)**: 265, 487–496.

Sprung S, Kuhlmann K and Ellerbrock H-G (1985) Particle size distribution and properties of cement. Part II: Water demand of Portland cement. *Zement Kalk Gips* **38**: 528 (partial English translation, p. 275).

St John DA and McGavin PN (1981) The use of ONO's technique to predict the strength of Portland cement. *Proceedings of 3rd International Conference on Cement Microscopy*: 193–197.

St John DA, Poole AW and Sims I (1998) *Concrete Petrography: A Handbook of Investigative Techniques*. Arnold, London, UK.

Struble LJ (1983) Quantitative X-ray diffraction analysis of cement and clinker: a bibliography. *Cement, Concrete and Aggregates* **5(1)**: 62–70.

Struble LJ (1985) The effect of water on maleic acid and salicylic acid extractions. *Cement and Concrete Research* **15(4)**: 631–636.

Stutzman PE (1991a) Cement clinker characterization by scanning electron microscopy. *Cement, Concrete and Aggregates* **13(2)**: 109–115.

Stutzman P (1991b) Quantitative phase abundance analysis of three cement clinker reference materials by scanning electron microscopy. *Ceramic Transactions: Advances in Cementitious Materials* **16**: 199–210.

Stutzman PE (1992) X-ray powder diffraction analysis of three Portland cement reference material clinkers. *Proceedings of 14th International Conference on Cement Microscopy*: 290–308.

Stutzman P (2004a) Scanning electron microscopy imaging of hydraulic cement microstructure. *Cement and Concrete Composites* **26(8)**: 957–966.

Stutzman P (2004b) Development of an ASTM standard test method on X-ray powder diffraction analysis of hydraulic cements. *Advances in X-Ray Analysis* **47**: 206–211.

Stutzman P (2005) Powder diffraction analysis of hydraulic cements: ASTM Rietveld round robin results on precision. *Advances in X-Ray Analysis* **48**: 33–38.

Stutzman PE, Bullard JW and Feng P (2015) *Quantitative Imaging of Clinker and Cement Microstructure*. NIST Technical Note 1877.

Stutzman PE, Feng P and Bullard JW (2016) Phase analysis of Portland cement by combined quantitative X-ray powder diffraction and scanning electron microscopy. *Journal of Research of the National Institute of Standards and Technology* **121**: 47–107.

Sumner MS, Hepher NM and Moir GK (1986) The influence of a narrow cement particle size distribution on cement paste and concrete water demand. *Proceedings of 8th International Congress on the Chemistry of Cement* **2**: 310–315; and private communication (HFWT).

Tabikh A and Weht RJ (1971) An X-ray diffraction analysis of Portland cement. *Cement and Concrete Research* **1(3)**: 317–328.

Takashima S (1958) Selective dissolution of alite with organic acids. *Semento Gijutsu Nenpo* **12**: 49–55.

Tamas FD, Sarkar AK and Roy DM (1976) Effect of variables upon silylation products of hydrated cements. In *Hydraulic Cement Pastes: Their Structure and Properties*. Cement and Concrete Association, Slough, UK, pp. 55–72.

Taylor HFW (1989) Modification of the Bogue calculation. *Advances in Cement Research* **2(6)**: 73–77.

Taylor HFW (1990) Reactions and reactivities of compounds in hydraulic cements. *Solid State Ionics* **43**: 31–35.

Taylor HFW (1991) Use of calculated x-ray powder diffraction patterns in the quantitative phase analysis of Portland cement or clinker. *Ceramic Transactions: Advances in Cementitious Materials* **16**: 185–198.

Taylor JC and Aldridge LP (1993) Full-profile Rietveld quantitative XRD analysis of Portland cement: Standard XRD profiles for the major phase tricalcium silicate (C_3S: $3CaO.SiO_2$). *Powder Diffraction* **8(3)**: 138–144.

Taylor JC, Aldridge LP, Matulis CE and Hinczak I (2002) X-ray powder diffraction analysis of cements. In *Structure and Performance of Cements*, 2nd edn (Bensted J and Barnes P (eds.)). Spon, Abingdon, UK.

Thorvaldson T (1970) In *The Chemistry of Cement and Concrete* (Lea FM), 3rd edn. Arnold, London, UK.

Trojer FJ (1968) The crystal structure of parawollastonite. *Zeitschrift für Kristallographie* **127(1–4)**: 291–308.

Utsunomiya A, Tanaka K, Morikawa H, Marumo F and Kojima H (1988) Structure refinement of $CaO{\cdot}6(Al_2O_3)$. *Journal of Solid State Chemistry* **75**: 197–200.

Vincent MG and Jeffery JW (1978) The crystal structure of pentacalcium trialuminate, $5CaO.3Al_2O_3$. *Acta Crystallographica B* **34(5)**: 1422–1428.

Wagner LA (1933) A rapid method for the determination of the specific surface of Portland cement. *ASTM Bulletin* **33**: 553–570.

Walenta G and Füllmann T (2004) Advances in quantitative XRD analysis for clinker, cements, and cementitious additions. *Advances in X-Ray Analysis* **47**: 287–296.

Weigand WA (1994) Progress toward a standard procedure for point-counting the phases in cement clinker with reflected light microscopy. *Petrography of Cementitious Materials, ASTM STP 1215*. (DeHayes SM and Stark D (eds)). American Society for Testing and Materials International, West Conshohocken, PA, USA.

Yamaguchi G and Takagi S (1969) The analysis of Portland cement clinker. *Proceedings of 5th International Symposium on the Chemistry of Cement* **1**: 181–218.

Zur Strassen H (1959) The problem of non-selective hydration of cement minerals. *Zement und Beton* **16**: 32–34. (In German.)

Ian G Richardson and HFW Taylor
ISBN 978-1-83549-777-7
https://doi.org/10.1108/978-1-83549-776-020253005

Chapter 5
Calcium silicate hydrates: natural minerals

5.1. Introduction

Over 4 bn tonnes of Portland cement are currently manufactured annually worldwide (CEMBUREAU, 2018; US Geological Survey, 2019) from which an estimated 2·1–2·3 m^3 or 4·8–5·5 tonnes per capita of cement-based materials are produced (Scrivener *et al.*, 2016). The principal binding phase in all Portland cement-based materials is a calcium silicate hydrate (C–S–H) phase. This C–S–H is virtually X-ray amorphous, compositionally and structurally very variable, and generally finely intermixed with other phases, all of which make it difficult to study. Researchers have consequently looked for compositional and structural similarity with natural crystalline calcium silicate hydrates – most commonly 14 Å tobermorite ($C_5S_6H_9$) and jennite ($C_9S_6H_{11}$) – and have attempted to synthesise a single-phase C–S–H in the laboratory that is similar to the phase that forms in concrete. The results of studies of the C–S–H that forms in hardened pastes of tricalcium silicate and β-dicalcium silicate are discussed in detail in Chapter 7. The natural minerals are the subject of this chapter and the laboratory-synthesised material is the subject of Chapter 6. Since diffraction techniques are of only limited utility in studies of the C–S–H that forms in real cements (see §7.3.5), there is increasing interest in using atomistic modelling techniques, which offer the prospect of significant advances in understanding its nanostructure. The results of such studies should of course be consistent with the results of the experimental studies that are described in the aforementioned chapters, as well as with crystal chemical principles. While atomistic modelling studies have usually started with the structure of 14 Å tobermorite (see §6.6.2), inspection of the structures of all of the crystalline calcium silicate hydrate phases is necessary to establish the crystal-chemical principles; the results of such an inspection (Richardson, 2014) are also reiterated in this chapter.

5.2. Natural minerals

More than 30 crystalline calcium silicate hydrates are known, in most cases both as natural minerals and as laboratory products; many of them are relevant to the subject of this volume. The composition and crystal structure of 42 crystalline calcium silicate hydrates and related phases are given in Table 5.1, which includes for each phase chemical composition, lattice parameters and calculated density (it is a modified version of Table 1 in Richardson (2008)). In nature, calcium silicate hydrates are most often found in metamorphic contact zones between igneous rocks and limestone (Burnham, 1959). They also occur in cavities in basalt or dolerite. In an apparently unique region in Israel and Jordan, they appear to have been formed by processes remarkably like those taking place in the manufacture and use of Portland cement (Kolodny and Gross, 1974). None of the calcium silicate hydrate minerals occur in very many localities and all are only minor or trace constituents of the rocks in which they are present. Amorphous and quasi-crystalline calcium silicate hydrates, known collectively as plombièrite, occur naturally and are essentially identical with forms of C–S–H. Some calcium silicate binders show even closer relationships to natural phases. Autoclaved aerated concrete contains a high proportion of a crystalline 11 Å tobermorite, while pastes of some other types of autoclaved materials and of alkali-activated slags contain quasi-crystalline forms of C–S–H that are relatively close to tobermorite.

5.3. The tobermorite phases

Tobermorites have layer structures and may be classified by the basal spacing of the individual layers, which is represented by a strong XRD (X-ray diffraction) powder peak. There are three principal types, with layer thicknesses of approximately 14 Å, 11 Å and 9 Å and compositions approximating to $C_5S_6H_9$, $C_5S_6H_5$ and C_5S_6H, respectively. Tobermorites with intermediate basal spacings also occur. The 11 Å forms are often called, simply, tobermorite. Because it is formed at the lowest temperatures and has the highest water content, 14 Å tobermorite is the form that is probably the most relevant to the C–S–H of normally cured cement pastes, but many more data are available for the 11 Å forms. A new nomenclature scheme for minerals in the tobermorite group was approved in 2014 by the IMA Commission on New Minerals, Nomenclature and Classification (Williams *et al.*, 2014; details reported by Biagioni *et al.*, 2015). That nomenclature is included in Table 5.1 for information, but it is not used in this book because of the need to consider synthetic phases as well as minerals, and also because of the need to discuss plombièrite in the context of its original meaning as gelatinous quasicrystalline or amorphous calcium silicate hydrates (§5.6.1), rather than as 14 Å tobermorite (Biagioni *et al.* (2015); Table 5.1).

5.3.1 11 Å tobermorites

5.3.1.1 Normal and anomalous varieties

All 11 Å tobermorites lose most of their water by 300 °C. For some, known as normal tobermorites, the loss is accompanied by a decrease in layer thickness to around 9·3 Å. For others, known as anomalous tobermorites, the layer thickness does not decrease (Gard and Taylor, 1957; Mitsuda and Taylor, 1978). Various types of intermediate behaviour have been observed in both natural and synthetic specimens, which have been described as 'mixed' (El-Hemaly *et al.*, 1977; Mitsuda and Taylor, 1978). The commonest behaviour of this type is possibly the appearance on heating of a broad XRD peak at about 10·5 Å.

There are no major differences between the XRD powder patterns of normal and anomalous 11 Å tobermorites, although small differences in cell parameters have been reported (Hara and Inoue, 1980a). In single crystal diffraction patterns of both, certain groups of reflections are systematically weak and often streaked, indicating disorder. The first selected area electron diffraction (SED) study of an anomalous tobermorite (Gard and Taylor, 1957) showed an unusual pattern of these weak reflections, but later work showed that this was a feature of the particular specimen and not of anomalous tobermorites in general.

The terms normal and anomalous arose because the first tobermorites to be studied in detail, both natural and synthetic, were normal. In nature, at least, anomalous tobermorites are much the commoner of the two. An explanation for normal and anomalous behaviour is given in §5.3.3.6.

Table 5.1 Composition and crystal structure data for calcium silicate hydrates, calcium aluminosilicate hydrates and related phases. In most cases the listed reference reports the full crystal structure solution. The crystal system (Sy) is indicated by: C = cubic; M = monoclinic; O = orthorhombic; Tg = trigonal; Tc = triclinic. SG = full international space group. D_x is the calculated density.

Table 5.1(a) Wollastonite group

Name	Formula	Reference	Sy	SG	Z	*M* formula: g mol^{-1}	*V*: Å^3	D_x: g cm^{-3}	*a*: Å	*b*	*c*	α	β	γ
Foshagite	$Ca_4(Si_3O_9)(OH)_2$	Gard and Taylor (1960)	Tc	$P\bar{1}$	2	422·59	513·0	2·736	10·32	7·36	7·04	90	106·4	90
Hillebrandite	$Ca_2(SiO_3)(OH)_2$	Dai and Post (1995)	O	$Cmc2_1$	6	190·25	702·1	2·700	3·6389	16·311	11·829	90	90	90
Nekoite	$Ca_3Si_6O_{15}·7H_2O$	Alberti and Galli (1980)	Tc	$P1$	1	654·85	490·3	2·217	7·588	9·793	7·339	111·77	103·5	86·53
Okenite	$Ca_{10}Si_{18}O_{46}·18H_2O$	Merlino (1983)	Tc	$P\bar{1}$	1	1966·60	1418·5	2·302	9·69	7·28	22·02	92·7	100·1	110·9
Pectolite	$Ca_2NaHSi_3O_9$	Takéuchi and Kudoh (1977)	Tc	$P\bar{1}$	2	332·41	382·2	2·888	7·98	7·023	7·018	90·54	95·14	102·55
Wollastonite-Tc	$Ca_3Si_3O_9$	Ohashi (1984)	Tc	$P\bar{1}$	2	348·49	396·96	2·916	7·9258	7·3202	7·0653	90·055	95·217	103·43
Xonotlite[a]	$Ca_6Si_6O_{17}(OH)_2$	Hejny and Armbruster (2001)	M	$P1\,2/a\,1$	2	713·99	879·3	2·700	17·032	7·363	7·012	90	90·36	90

[a] *Ma2bc* polytype (an MDO_2 polytype).

Table 5.1(b) Tobermorite group

Name	Mineral, IMA 2014[c]	Formula	References	Sy	SG	Z	*M* formula: g mol^{-1}	*V*: Å^3	D_x: g cm^{-3}	*a*: Å	*b*	*c*	α	β	γ
Oyelite		$Ca_{10}B_2Si_8O_{29}·12·5H_2O$	Heller and Taylor (1956)	O	n.d.	2	1336·26	1668·77	2·659	11·25	7·25	20·46	90	90	90
14 Å tobermorite[a]	plombièrite-4*O*	$Ca_5Si_6O_{16}(OH)_2·7H_2O$	Bonaccorsi *et al.* (2005)	O	*F* 2 *d d*	8	785·03	4578·56	2·278	11·2	7·3	56·0	90	90	90
14 Å tobermorite[b]	plombièrite-2*M*	$Ca_5Si_6O_{16}(OH)_2·7H_2O$	Bonaccorsi *et al.* (2005)	M	*B* 1 1 *b*	2	785·03	1170·4	2·228	6·735	7·425	27·987	90	90	123·25
Anomalous 11 Å tobermorite[a]	kenotobermorite-4*O*	$Ca_4Si_6O_{15}(OH)_2·5H_2O$	Merlino *et al.* (2001)	O	*F* 2 *d d*	8	692·92	3741·65	2·460	11·265	7·386	44·970	90	90	90
Anomalous 11 Å tobermorite[b]	kenotobermorite-2*M*	$Ca_4Si_6O_{15}(OH)_2·5H_2O$	Merlino *et al.* (1999; 2000)	M	*B* 1 1 *m*	2	692·92	935·4	2·460	6·735	7·385	22·487	90	90	123·25
Normal 11 Å tobermorite[b]	tobermorite-2*M*	$Ca_{4·5}Si_6O_{16}(OH)·5H_2O$	Merlino *et al.* (2001)	M	*B* 1 1 *m*	2	711·94	941·68	2·511	6·732	7·369	22·680	90	90	123·18
Clinotobermorite[a]	clinotobermorite-2*M*	$Ca_5Si_6O_{17}·5H_2O$	Merlino *et al.* (2000)	M	*C 1 c 1*	4	730·98	1859·64	2·611	11·276	7·343	22·642	90	97·28	90
Clinotobermorite[b]	clinotobermorite-1*A*	$Ca_5Si_6O_{17}·5H_2O$	Merlino *et al.* (1999; 2000)	Tc	*C* 1	2	730·98	929·8	2·611	11·274	7·344	11·468	99·18	97·19	90·03
9 Å tobermorite[a]	(riversideite)	$Ca_5Si_6O_{16}(OH)_2$	Bonaccorsi and Merlino (2005)	O	*F d* 2 *d*	8	658·92	3055·25	2·865	11·16	7·32	37·40	90	90	90
9 Å tobermorite[b]	(riversideite)	$Ca_5Si_6O_{16}(OH)_2$	Bonaccorsi and Merlino (2005)	M	*P 1 1 2_1/a*	2	658·92	764·78	2·861	6·7	7·32	18·70	90	90	123·5
Clinotobermorite 9 Å[a]		$Ca_5Si_6O_{16}(OH)_2$	Bonaccorsi and Merlino (2005)	M	*C 1 2/c 1*	4	658·92	1528·03	2·864	11·161	7·303	18·771	90	92·91	90
Clinotobermorite 9 Å[b]		$Ca_5Si_6O_{16}(OH)_2$	Merlino *et al.* (1999; 2000)	Tc	*C* $\bar{1}$	2	658·92	763·9	2·865	11·156	7·303	9·566	101·08	92·83	89·98

[a] MDO_1 polytype.
[b] MDO_2 polytype.
[c] The nomenclature scheme for minerals in the tobermorite group that was approved in June 2014 by the IMA Commission on New Minerals, Nomenclature and Classification (Williams *et al.*, 2014). Riversideite was redefined as a questionable species and plombièrite was redefined as 14 Å tobermorite (Biagioni *et al.*, 2015), but plombièrite is used in this book with its original meaning as gelatinous quasicrystalline or amorphous calcium silicate hydrates (§5.6.1).

Table 5.1(c) Jennite group

Name	Formula	References	Sy	SG	Z	*M* formula: g mol^{-1}	*V*: Å^3	D_x: g cm^{-3}	*a*: Å	*b*	*c*	α	β	γ
Jennite	$Ca_9Si_6O_{18}(OH)_6{\cdot}8H_2O$	Bonaccorsi *et al.* (2004)	Tc	$P\bar{1}$	1	1063·44	759·5	2·325	10·576	7·265	10·931	101·30	96·98	109·65
Metajennite	$Ca_9Si_6O_{18}(OH)_6{\cdot}2H_2O$	Taylor (2000) reported in Richardson (2008)	M	*I* 1 2/*m* 1	1	955·30	631·1	2·513	9·944	3·638	17·72	90	100·09	90

Table 5.1(d) Gyrolite group

Name	Formula	References	Sy	SG	Z	M formula: g mol^{-1}	V: Å^3	D_x: g cm^{-3}	a: Å	b	c	α	β	γ
Fedorite	$(Na,K)2(Ca,Na)7(Si,Al)_{16}O_{38}(F,OH)_2\cdot 3\cdot 5H_2O$[a]	Mitchell and Burns (2001)	Tc	$P\bar{1}$	1	1448·66	961·24	2·503	9·645	9·6498	12·617	102·43	96·247	119·89
Gyrolite	$NaCa_{16}Si_{23}AlO_{60}(OH)_8\cdot 14H_2O$	Merlino (1988)	Tc	$P\bar{1}$	1	2685·47	1824·1	2·445	9·74	9·74	22·40	95·7	91·5	120
K-phase	$Ca_7Si_{16}O_{38}(OH)_2$	Gard *et al.* (1981)	Tc	$P\bar{1}$	1	1371·91	944·39	2·412	9·70	9·70	12·25	101·9	96·5	120
Reyerite	$Na_2Ca_{14}Si_{22}Al_2O_{58}(OH)_8\cdot 6H_2O$	Merlino (1988)	Tg	$P\bar{3}$	1	2451·07	1574·6	2·585	9·765	9·765	19·067	90	90	120
Truscottite	$Ca_{14}Si_{24}O_{58}(OH)_8\cdot 2H_2O$	Gard *et al.* (1975)	Tg	$P\bar{3}$	1	2335·20	1545·5	2·509	9·735	9·735	18·83	90	90	120
Z-phase	$Ca_9Si_{16}O_{40}(OH)_2\cdot 14H_2O$	Garbev (2003)[b]	Tc	$P\bar{1}$	1	1736·27	1238·3	2·328	9·70	9·70	15·24	90·94	93·19	120

[a] Actual composition: $(Na_{1\cdot 29}K_{0\cdot 79})(Ca_{4\cdot 48}Na_{2\cdot 52})Si_{16}O_{38}F_2\cdot 3\cdot 47H_2O$.
[b] Garbev's model structure.

Table 5.1(e) γ-C_2S group

Name	Formula	References	Sy	SG	Z	*M* formula: g mol^{-1}	*V*: Å^3	D_x: g cm^{-3}	*a*: Å	*b*	c	α	β	γ
Calcium chondrodite[a]	$Ca_5(SiO_4)_2(OH)_2$	Kuznetsova *et al.* (1980)	M	*P* 1 1 $2_1/b$	2	418·58	492·1	2·825	8·9207	11·4481	5·0759	90	90	108·32
Chegemite	$Ca_7(SiO_4)_3(OH)_2$	Galuskin *et al.* (2009)	O	*P* $2_1/b$ $2_1/n$ $2_1/m$	4	590·81	1360·91	2·884	5·0696	11·3955	23·5571	90	90	90
Kilchoanite	$Ca_6(SiO_4)(Si_3O_{10})$	Taylor (1971)	O	*I* 2 c *m*	4	576·80	1275·9	3·003	11·42	5·09	21·95	90	90	90

[a]Synthetic phase similar to the mineral reinhardbraunsite, $Ca_5[SiO_4]_2(OH,F)_2$.

Table 5.1(f) Other calcium silicate phases

Name	Formula	References	Sy	SG	Z	M formula: $g\,mol^{-1}$	V: Å^3	D_x: $g\,cm^{-3}$	a: Å	b	c	α	β	γ
Afwillite	$Ca_3(SiO_3OH)_2{\cdot}2H_2O$	Malik and Jeffrey (1976)	M	*C* 1 *c* 1	4	342·46	859·6	2·646	16·278	5·6321	13·236	90	134·90	90
Bultfonteinite	$Ca_4(SiO_3OH)_2F_2{\cdot}2H_2O$	McIver (1963); Biagioni *et al.* (2010)	Tc	$P\bar{1}$	2	420·52	499·89	2·794	10·992	8·185	5·671	93·95	91·32	89·65
α-C_2SH	$Ca_2(HSiO_4)(OH)$	Marsh (1994)	O	*P* $2_1/b$ $2_1/c$ $2_1/a$	8	190·26	928·8	2·721	9·487	9·179	10·666	90	90	90
Cuspidine[a]	$Ca_4(F_{1{\cdot}5}(OH)_{0{\cdot}5})Si_2O_7$	Saburi *et al.* (1977)	M	*P* 1 $2_1/c$ 1	4	365·49	814·2	2·982*	7·518	10·521	10·906	90	70·7	90
Dellaite	$Ca_6(Si_2O_7)(SiO_4)(OH)_2$	Safronov *et al.* (1981)	Tc	$P\bar{1}$	2	534·74	598·8	2·966	6·825	6·931	12·907	90·68	97·57	98·18
Jaffeite	$Ca_6[Si_2O_7](OH)_6$	Yamnova *et al.* (1993)	Tg	*P* 3	2	510·69	654·0	2·593	10·035	10·035	7·499	90	90	120
Killalaite	$Ca_{6{\cdot}4}(H_{0{\cdot}6}Si_2O_7)_2(OH)_2$	Taylor (1977)	M	*P* 1 $2_1/m$ 1	2	629·24	710·2	2·943	6·807	15·459	6·811	90	97·76	90
Poldervaartite[b]	$Ca(Ca_{0{\cdot}67}Mn_{0{\cdot}33})(HSiO_4)(OH)$	Dai *et al.* (1993)	O	*P* $2_1/b$ $2_1/c$ $2_1/a$	8	195·21	904·8	2·866	9·398	9·139	10·535	90	90	90
Rosenhahnite	$Ca_3Si_3O_8(OH)_2$	Wan *et al.* (1977)	Tc	$P\bar{1}$	2	366·51	420·2	2·897	6·955	9·484	6·812	108·64	94·84	95·89
Suolunite	$CaSiO_{2{\cdot}5}(OH){\cdot}0{\cdot}5H_2O$	Ma *et al.* (1999)	O	*F d* 2 *d*	16	134·18	1317·1	2·707	19·776	5·99	11·119	90	90	90
Tilleyite	$Ca_5Si_2O_7(CO_3)_2$	Grice (2005)	M	*P* 1 2/*a* 1	4	488·59	1131·8	2·867	15·108	10·241	7·579	90	105·17	90

[a]Cuspidine: cell parameters given in the paper transformed by the following matrix (00-1/010/100).
[b]Isostructural with α-C_2SH.

Table 5.1(g) Other high temperature cement phases

Name	Formula	References	Sy	SG	Z	M formula: g mol^{-1}	V: Å^3	D_x: g cm^{-3}	a: Å	b	c	α	β	γ
Bicchulite	$Ca_2(Al_2SiO_6)(OH)_2$	Dann *et al.* (1996)	C	$I\bar{4}3m$	4	292·22	688·9	2·818	8·8318	8·8318	8·8318	90	90	90
Fukalite[a]	$Ca_4(Si_2O_6)(CO_3)(OH)_2$	Merlino *et al.* (2009)	M	$P\,1\,2_1/c\,1$	8	406·50	1929·5	2·799	7·573	23·364	11·544	90	109·15	90
Katoite Hydrogarnet[b]	$Ca_{1\cdot46}AlSi_{0\cdot55}O_6H_{3\cdot78}$	Ferro *et al.* (2003)	C	$I\,4_1/a\,\bar{3}\,2/d$	16	200·73	1847·3	2·887	12·27	12·27	12·27	90	90	90
Rustumite	$Ca_{10}(Si_2O_7)_2(SiO_4)Cl_2(OH)_2$	Howie and Ilyukhin (1977)	M	$C\,1\,2/c\,1$	4	934·14	2124·1	2·921	7·620	18·550	15·510	90	104·33	90
Scawtite[c]	$Ca_7(Si_6O_{18})(CO_3){\cdot}2H_2O$	Zhang *et al.* (1992); Grice (2005)	M	$C\,1\,m\,1$	2	833·11	1003·2	2·758	11·0394	15·1935	6·6344	90	115·645	90
Strätlingite	$Ca_2Al(AlSi)_{2\cdot22}O_2(OH)_{12}\cdot 2{\cdot}25\,H_2O$	Rinaldi *et al.* (1990)	Tg	$R\,\bar{3}\,2/m$	3	444·88	1079·6	2·053	5·745	5·745	37·77	90	90	120

[a] MDO_1 polytype.
[b] Composition is variable.
[c] Scawtite: cell parameter *a* given by Zhang *et al.* (1992) as 10·0394 is a mistake and has been corrected to 11·0394 Å.

5.3.1.2 Conditions of formation

11 Å tobermorites may be synthesised from mixtures of calcium hydroxide and quartz in aqueous suspensions under saturated steam pressures at 105–180 °C (Kalousek, 1955). Using CaO and a clinoptilolite rock containing also minor proportions of quartz and felspar, they have been obtained at temperatures as low as 90 °C (Mitsuda, 1970). The reaction first gives C–S–H, and then in succession normal, mixed and anomalous 11 Å tobermorite. Factors that tend to stop the reaction at the stage of normal tobermorite include, beside short time, high Ca/Si ratio (which for tobermorite is ≈1·0), presence of Al^{3+} without alkali, and low temperature (El-Hemaly *et al.*, 1977; Hara and Inoue, 1980a). Relatively long time, low Ca/Si ratio (≈0·8), higher temperatures and the presence of Al^{3+} together with alkali favour the formation of anomalous tobermorite. Alkali without Al_2O_3 is variously reported to impede crystallisation (El-Hemaly *et al.*, 1977) and to favour formation of anomalous tobermorite (Hara and Inoue, 1980a). These results indicate that normal tobermorite is a precursor of anomalous tobermorite and suggest that it has a higher Ca/Si ratio than anomalous tobermorite. Normal 11 Å tobermorite can also be obtained by heating 14 Å tobermorite at 55–120 °C (Farmer *et al.*, 1966; McConnell, 1954).

Tobermorite forms much less readily if amorphous forms of silica are used, and the reaction then tends to stop with formation of C–S–H (Kalousek, 1955). Addition of Al_2O_3 has the reverse effect (Kalousek, 1957). At water pressures of 80–170 MPa, tobermorite is stable at temperatures up to ≈285 °C (Buckner *et al.*, 1960). At saturated steam pressures, it is unstable relative to xonotlite ($Ca_6Si_6O_{17}(OH)_2$) above ≈ 140 °C, but the transition is very slow below 180–190 °C (El-Hemaly *et al.*, 1977; Taylor, 1962).

5.3.1.3 Crystal data

11 Å tobermorite exhibits several types of structural variation and disorder. Some research teams (Bonaccorsi *et al.*, 2005; Bonaccorsi and Merlino, 2005; Merlino *et al.*, 1999, 2000) discussed these in terms of the OD (order-disorder) theory of Dornberger-Schiff (1956). The variations arise in part because adjacent layers can be stacked together in different ways, which gives rise to the possibility of an infinite variety of ordered and disordered sequences. Different arrangements are also possible within individual layers (Hamid, 1981). Two broad groups or families are known to exist.

The most common family is characterised by a body-centred orthorhombic sub- or pseudocell, with approximate parameters $a_s = 5{\cdot}65$ Å, $b_s = 3{\cdot}66$ Å, $c_s = 22{\cdot}6$ Å ($V = 467$ Å^3) (McConnell, 1954) and space group $I\,m\,m\,2$ (No. 44, setting **abc**) (Hamid, 1981; Merlino *et al.*, 1999). The members of this family are referred to in this book as *ortho*tobermorites. The subcell corresponds to a structure in which disorder is complete. XRD powder patterns show few peaks, if any, that cannot be indexed on this subcell. Single crystal XRD and SED patterns show additional reflections, which are systematically weak and usually streaked parallel to **c***. These indicate a doubled *b*-dimension and varying values of the other cell parameters. For one specimen that displayed anomalous thermal behaviour, all the weak reflections had nonintegral indices, suggesting a modulated structure (Gard and Taylor, 1957). Urabe *et al.* (1983) reported lattice images of the (100) and (001) planes obtained by high-resolution transmission electron microscopy (HRTEM).

Members of the other family are called *clino*tobermorites (Henmi and Kusachi, 1992) because the subcell is monoclinic, with parameters $a_s = 5{\cdot}593$ Å, $b_s = 3{\cdot}645$ Å, $c_s = 22{\cdot}456$ Å, $\beta_s = 96{\cdot}97°$ ($V = 454$ Å^3) and space group $I\,2/m$ (No. 12, unique axis *b*, cell choice 3) (Hoffmann and Armbruster, 1997). As with the orthotobermorites, an infinite number of ordered or disordered modifications is theoretically possible, with a doubled true value of *b* and varying values of the other true cell parameters (Merlino *et al.*, 1999, 2000). All may be expected to show the same pattern of systematically strong reflections and thus to give essentially identical X-ray powder patterns. The pattern of systematically strong reflections, and thus also the powder pattern (Henmi and Kusachi, 1992), differ significantly from those of orthotobermorites. This is quickly evident on comparison of Figures 5.1(a) and 5.1(b), which show the calculated XRD patterns for 11 Å ortho- and clinotobermorite, respectively. Henmi and Kusachi (1992), who discovered clinotobermorite, likened its relationship to orthotobermorite to that existing between clino- and orthopyroxenes. Both the clinotobermorites described so far were found in nature and were thermally normal (Henmi and Kusachi, 1992; Merlino *et al.*, 1999, 2000). So far, no synthesis has been reported. Henmi and Kusachi (1992) thought it likely that clinotobermorite formed at lower temperatures than orthotobermorite, which would seem to support the conclusion in §6.6 that clinotobermorite is a better candidate than orthotobermorite as a basis for models for the structure of C–S–H(I).

If the idealised formula $C_5S_6H_5$ is assumed, the subcells of both ortho- and clinotobermorite contain one formula unit (i.e. Z = 1) and the X-ray densities are 2·60 and 2·67 g cm^{-3}, respectively. In all cases, the structural layers are parallel to (001), and doubling or higher multiples of *c* relative to the layer thickness arises from the manner in which they are stacked. Information about the subcells of the phases in the tobermorite group is summarised in Table 5.2, which is based on Table 4

Figure 5.1 Calculated X-ray powder diffraction pattern for the MDO_2 polytypes of (a) anomalous 11 Å orthotobermorite and (b) 11 Å clinotobermorite. The patterns were calculated using CrystalDiffract® (CrystalMaker, n.d.); Cu Kα radiation, wavelength 1.5406 Å; structures from (a) Merlino *et al.* (1999), and (b) Merlino *et al.* (2000)

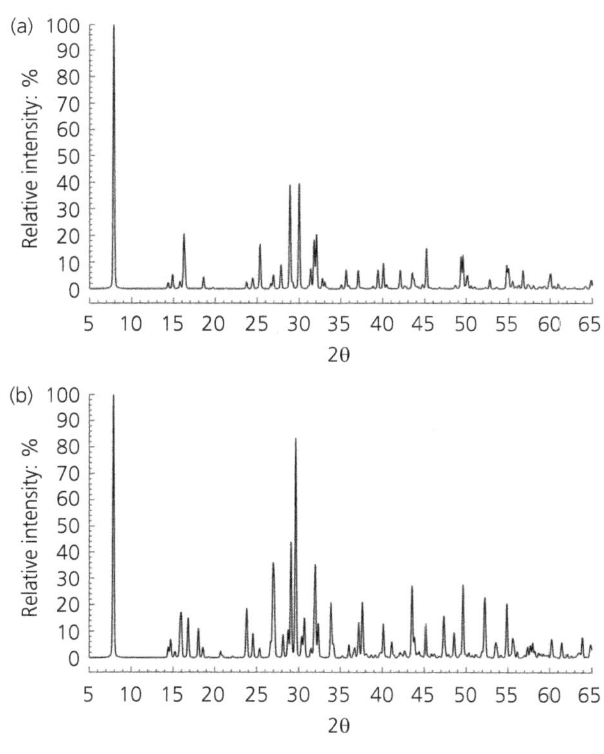

Table 5.2 Composition and subcell parameters (in Å and °) of the phases in the tobermorite group (a slightly modified version of Table 4 of Bonaccorsi and Merlino, 2005)

Composition	Phase	Subcell	SG	a_S	b_S	c_S	β_S
$Ca_5Si_6O_{16}(OH)_2$	9Å clinotobermorite	monoclinic	*A 2/m*	5·58	3·65	18·78	92·8
$Ca_5Si_6O_{16}(OH)_2$	9Å tobermorite	orthorhombic	*P n m m*	5·58	3·65	18·78	
$Ca_5Si_6O_{17}.5H_2O$	clinotobermorite	monoclinic	*A 2/m*	5·638	3·671	22·642	97·3
$Ca_5Si_6O_{17}.5H_2O$	11Å tobermorite	orthorhombic	*I 2 m m*	5·632	3·692	22·487	
$Ca_5Si_6O_{16}(OH)_2.7H_2O$	14Å tobermorite	orthorhombic	*I 2 m m*	5·63	3·71	27·99	

of Bonaccorsi and Merlino (2005). It should be noted that the subcell parameters and contents are different from those that are discussed above and so the calculated densities are also different.

Some natural and apparently most synthetic 11 Å tobermorites form platey crystals with (001) cleavage and a tendency to elongation parallel to **b**. Most natural tobermorites, however, form crystals showing both (001) and (100) cleavages and are more strongly elongated parallel to **b**. For both natural and synthetic specimens, the crystals are usually of micrometre dimensions, but those of some natural tobermorites are larger. The optical properties for one normal 11 Å tobermorite are: biaxial +, low 2V; $\alpha = 1{\cdot}570$, $\beta = 1{\cdot}571$, $\gamma = 1{\cdot}575$; $X = \gamma$, $Y = \beta$, $Z = \alpha$, where *X*, *Y* and *Z* are the axial directions of the orthorhombic pseudo cell (McConnell, 1954). The refractive indices of a clinotobermorite were slightly higher (Henmi and Kusachi, 1992).

5.3.1.4 Chemical compositions and cell contents

Kalousek (1957) showed that 11 Å tobermorite can accommodate Al and suggested that it substitutes for Si in the structure. X-ray spectroscopy (Diamond *et al.*, 1966) and ^{27}Al MAS NMR (magic-angle spinning nuclear magnetic resonance) (Komarneni *et al.*, 1985a) studies show that the Al is tetrahedrally coordinated, thereby supporting these conclusions. Al substitution slightly increases the layer thickness, but analyses of natural specimens show no clear correlation between layer thickness and Al content (Mitsuda and Taylor, 1978). Probably all the naturally occurring anomalous tobermorites that have been analysed contain significant proportions of Al and alkalis, but most naturally occurring normal tobermorites appear to be substantially unsubstituted. The limit of substitution in both natural and synthetic preparations is around 7% Al_2O_3, referred to the ignited mass, which corresponds to about one sixth of the Si atoms (Kalousek, 1957).

For synthetic preparations, Hara and Inoue (1980a) reported basal spacings of 11·30–11·55 Å for those showing normal and 11·23–11·25 Å for those showing anomalous behaviour. There were also much smaller differences in *a* and *b*, which were inversely correlated with each other. The normal specimens were obtained using high Ca/Si ratios, addition of Al_2O_3 (without alkali) or short processing times, and the anomalous ones by using low Ca/Si ratios, addition of NaOH (without Al) or longer processing times. Preparations showing mixed thermal behaviour tended to have intermediate values of the subcell parameters.

The observed densities of both normal and anomalous orthotobermorites are 2·4–2·5 g cm^{-3}, showing that considerable deficiencies of atomic contents relative to the idealised composition $C_5S_6H_5$ may occur. From a study of six natural 11 Å tobermorites, Zadov *et al.* (1995) concluded that the general formula was $Ca_{4+x}Si_6O_{17}H_{2-2x}{\cdot}5H_2O$, where $0 < x < 1$ (i.e. $C_4S_6H_6$–$C_5S_6H_5$). For naturally occurring anomalous tobermorites, the subcell contents are typically about $Ca_{4{\cdot}2}Na_{0{\cdot}1}Si_{5{\cdot}5}Al_{0{\cdot}5}O_{21{\cdot}5}H_{11}$ (Mitsuda and Taylor, 1978). This suggests that for anomalous tobermorites, the number of Ca atoms per formula unit is little above 4 ($x \lesssim 0{\cdot}3$) and that there is substantial substitution of Si by Al and minor substitution of Ca by Na. The Ca/(Si + Al) ratios are thus well below 5:6. However, although the formation of anomalous tobermorite is favoured by the presence of Al and alkalis, these are not essential, as it can be synthesised without them (El-Hemaly *et al.*, 1977).

There are fewer data for natural normal 11 Å tobermorites. For normal orthotobermorites, subcell contents of $\approx Ca_{4{\cdot}7}Si_{5{\cdot}4}O_{21}H_{11}$ (McConnell, 1954) and $\approx Ca_{4{\cdot}5}Si_6O_{17}H{\cdot}5H_2O$ (Bonaccorsi and Merlino, 2000) have been reported for specimens from Northern Ireland and the Ural Mountains (Russia), respectively. For normal clinotobermorites, the data indicate subcell contents of $\approx Ca_{5{\cdot}1}Si_{5{\cdot}7}O_{22}H_{11}$ (Henmi and Kusachi, 1992) and $\approx Ca_5Si_6O_{22}H_{10}$ (Hoffmann and Armbruster, 1997). In all cases, the number of Ca atoms per formula unit is substantially greater than 4·0, but there can be a deficiency in Si, giving rise to Ca/Si ratios of 5:6 or more. Most normal tobermorites show only minor degrees of ionic substitution. These results are consistent with the observations that the formation of normal tobermorite is favoured by high Ca/Si ratio and absence of both Al and alkalis. Synthetic normal 11 Å tobermorites appear to be formed free from other phases, with Ca/Si ratios up to ≈1·0 (El-Hemaly *et al.*, 1977; Kalousek, 1955).

5.3.1.5 Crystal structures of 11 Å orthotobermorites

The first X-ray structure investigation was by Megaw and Kelsey (1956), who determined the approximate crystal structure of a normal 11 Å orthotobermorite from Ballycraigy (N. Ireland). Because of the disorder, this determination, and a subsequent refinement by Hamid (1981) for another specimen, were carried out using only the systematically strong reflections, and are thus related to a disordered substructure. The structure was found to be based on Ca–O layers, of which the O atoms on both sides are shared with single chains having the empirical formula SiO_3^{2-} and kinked to produce the dreierkette conformation, which is characterised by a repeat of three tetrahedra. Water molecules and additional Ca atoms were assumed to occur in the interlayers between the composite Ca–Si–O layers. The dreierkette conformation of the silicate chains is imposed by the coordination requirements of the Ca atoms (and thus O–O distances) and occurs in a number of other crystalline calcium silicates, including pectolite (Buerger, 1956; Buerger and Prewitt, 1961; Müller, 1976; Prewitt, 1967; Prewitt and Buerger, 1963; Takeuchi and Kudoh, 1977), wollastonite (Buerger and Prewitt, 1961; Ohashi, 1984; Ohashi and Finger, 1978; Prewitt and Buerger, 1963; Trojer, 1968), bustamite (Ohashi and Finger, 1978), hillebrandite (Dai and Post, 1995), foshagite (Gard and Taylor, 1960) and jennite (Bonaccorsi *et al.*, 2004). Mamedov and Belov (1958) proposed the formula $Ca_{10}(Si_{12}O_{31})(OH)_6{\cdot}8H_2O$ and a structure in which the dreierketten were condensed into sheets. The formula and structure were based only on crystal chemical reasoning and differed radically from those established by the X-ray structure determinations and spectroscopic evidence. Later, it was suggested that the presence of interlayer Si–O–Si linkages in anomalous tobermorites might account for the absence of lattice shrinkage on dehydration and for the morphology and tendency to show (100) cleavage (Mitsuda and Taylor, 1978; Taylor, 1962).

Merlino *et al.* (1999) reported a more detailed investigation in which the systematically weak reflections were also used. The specimen was from Wessels Mine, South Africa, and the determination related to a polytype with $a = 6{\cdot}735$ Å, $b = 7{\cdot}385$ Å, $c = 22{\cdot}87$ Å, $\gamma = 123{\cdot}25°$, space group $B\,1\,1\,m$ (No. 8, unique axis c, cell choice 2). The thermal behaviour was anomalous. Figure 5.2(a) shows the unit cell of this ordered structure (the MDO_2 polytype) projected along its **b** axis.

Figure 5.2(a) The structure of the MDO_2 polytype of anomalous 11 Å orthotobermorite projected along its **b** axis showing the positions of the atoms in the unit cell

(a)

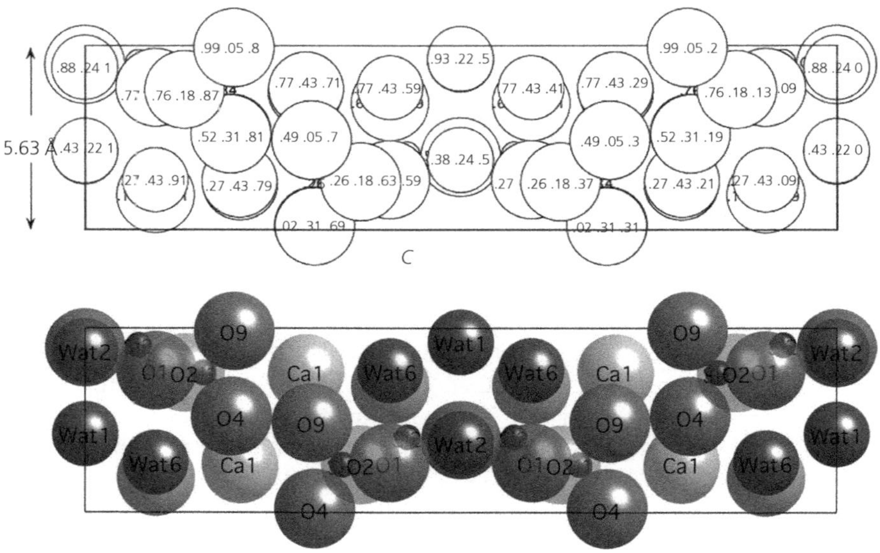

The structure is from Merlino *et al.* (1999). Coordinates (*x/a y/b z/c*) and atom labels are given in the top and bottom parts, respectively. Unit edge lengths are $c = 22{\cdot}487$ Å and $a \sin \gamma = 5{\cdot}632$ Å. Atoms O1 and O2 are translucent on the lower figure so as to reveal the positions of the Si atoms.

The arrangement of the Si–O tetrahedra and Ca–O polyhedra are illustrated in Figures 5.2(b) and 5.2(c), which show the structure viewed along [010] and [210], respectively. The Ca atoms in the core are coordinated by seven oxygen atoms, of which six are always shared with the Si–O chains, which form ribs covering each surface. The coordination of the seventh oxygen alternates along each row parallel to **b**. In every second case it is a water molecule, and in the intervening ones it is part of a silicate chain. Within the silicate chains, the tetrahedra closest to the Ca–O sheet are called paired tetrahedra (and so abbreviated as 'PT'), and those further away are called bridging tetrahedra (BT). The chains of adjacent layers share oxygen atoms, thus forming double chains of empirical formula $Si_6O_{17}H_2$, which run parallel to **b** (Figures 5.2(b) and 5.2(c)) and which display $2mm$ symmetry, Figures 5.3(a) and 5.3(b). The composition of the part of the structure thus far described is $Ca_4(Si_6O_{17}H_2){\cdot}2H_2O$.

The Ca–O cores of the layers can be referred to the subcell with $a = 5{\cdot}6$ Å, $b = 3{\cdot}6$ Å and may be regarded as layers of extremely distorted $Ca(OH)_2$ due to replacement of their OH groups by the silicate chains, the O–O distances of which have to be respected. Calcium hydroxide, referred to orthogonal axes, has $a = 6{\cdot}2$ $(3{\cdot}6\sqrt{3})$ Å, $b = 3{\cdot}6$ Å. The Ca atoms, which lie in a single plane in $Ca(OH)_2$, are split between two planes in tobermorite, and the oxygen positions are also markedly different. The structure refined by Merlino *et al.* (1999) differs from those found by Megaw and Kelsey (1956) and by Hamid (1981) in showing Si–O–Si linkages between the layers, thus converting two single chains of formula Si_3O_9 into a double chain of formula Si_6O_{17}. The difference corresponds to a shift of $b/2$ in the silicate chains (compare Figures 5.3(b) and 5.3(d) with 5.3(e)); it is questionable whether studies in which only the systematically strong reflections were used

Figure 5.2(b) and Figure 5.2(c) Polyhedral representations of the structure of the MDO_2 polytype of anomalous 11 Å orthotobermorite viewed (b) along [010] (edge lengths are c = 22·487 Å and $a \sin \gamma$ = 5·632 Å) and (c) along [210]

(b)

(c)

could provide reliable information on this feature of the structure. In addition, both of the refined structures showed the bridging tetrahedra to be tilted around their edges parallel to **b** in a way differing markedly from that found by Megaw and Kelsey (1956).

The structure contains channels, parallel to the silicate chains, which may contain water molecules and additional Ca atoms. The rows of channels in the (001) plane correspond to the interlayer regions of the single-chain structure found in the earlier determinations. Although the term 'interlayer' is not strictly applicable for the double-chain structure, it is useful to retain it. Spectroscopic and chemical evidence shows that some types of tobermorite and almost all types of tobermorite-like C–S–H do not contain interlayer Si–O–Si linkages. The term 'interlayer' is well established in these cases, and the use of a different term in that of double-chain tobermorites would be highly inconvenient.

Figure 5.3 Silicate double chains in 11 Å orthotobermorite: (a) and (b) displaying 2 *mm* symmetry and in 11 Å clinotobermorite; (c) and (d) displaying 2/*m* symmetry; (e) a shift of *b*/2 in the silicate chains results in single chains

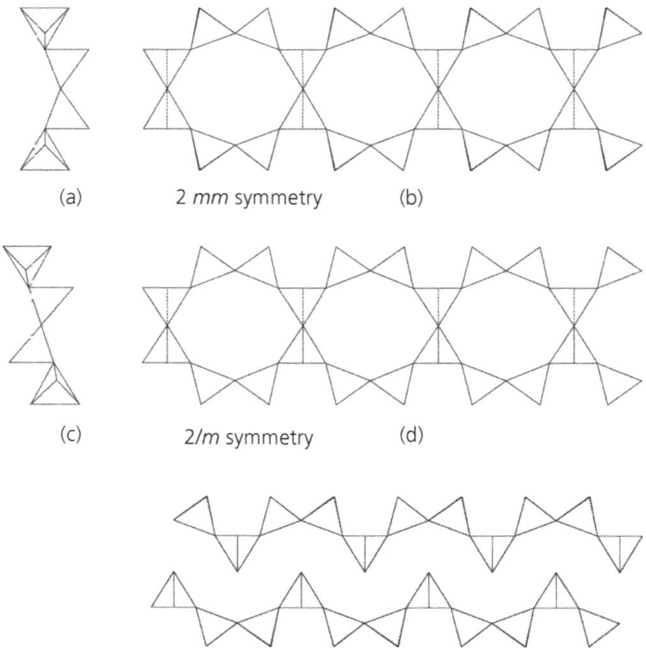

In the anomalous 11 Å orthotobermorite specimen examined by Merlino *et al.* (1999), the interlayers contained three water molecules per subcell, which corresponds to the formula $Ca_4(Si_6O_{17}H_2)\cdot 5H_2O$ or $C_4S_6H_6$, Table 5.2. The possible presence of a small content of interlayer Ca^{2+} was not excluded, the excess positive charge then being balanced by partial substitutions of Al^{3+} for Si^{4+}, or of O^{2-} for OH^- or both.

Bonaccorsi and Merlino (2000) determined the structure of a normal 11 Å orthotobermorite from the Ural Mountains (Russia). This had a double-chain structure similar to that of the Wessels specimen, from which it differed essentially in having a substantial content of interlayer Ca (0.5 atoms per formula unit).

It follows from the above that a generalised structural-chemical formula for double-chain tobermorite can be represented by Formula 5.1. The variable *i* is equal to $1 - x$ in the general formulae for double-chain tobermorites given by Zadov *et al.* (1995) and Biagioni *et al.* (2015).

$$Ca_4\left[Si_6O_{17}\right]H_{2i}Ca_{1-i}\cdot mH_2O \qquad (5.1)$$

The equivalent formula for single-chain tobermorite is

$$Ca_4\left[Si_6O_{18}\right]H_{2i}Ca_{2-i}\cdot mH_2O \qquad (5.2)$$

In these formulae, the contents of the square brackets represent the silicate part of the structure. There are four main layer Ca atoms for every six tetrahedral sites. The two formulae can be combined to represent a mixture of single- and double-chain phases

$$Ca_4\left[Si_6O_{18-d}\right]H_{2i}Ca_{2-i-d}\cdot mH_2O \tag{5.3}$$

where d represents the fraction of double-chain structure – that is, $0 \leq d \leq 1$, where if $d = 0$ the structure is entirely single chain, and if $d = 1$ it is entirely double chain.

If the idealised formula $C_5S_6H_5$ is assumed, the subcells of both 11 Å orthotobermorite and 11 Å clinotobermorite correspond to $i = 0$ in Formula 5.1, whereas the subcell contents given in Table 5.2 for the orthotobermorite correspond to $i = 1$.

5.3.1.5.1 MODEL CRYSTAL STRUCTURES FOR 11 Å ORTHOTOBERMORITES

It is evident from Table 5.1(b) that Merlino and his co-workers conducted a very thorough experimental study of the crystal structures of the phases in the tobermorite group. While broad similarities between the different structures are obvious on inspection of their schematic structure diagrams (and are discussed in detail by the authors), (i) the closeness of the relationship for structures that have the same subcell (i.e. orthorhombic or monoclinic) is not particularly clear from the tables of the atomic coordinates for the refined structures and (ii) the authors have not published a straightforward method that can be used by other workers to quickly produce models that possess different layer spacings, which would be useful for those interested in tobermorite-based models for the structure of the C–S–H phase in hardened cements. Richardson (2014) attempted to address both of these points with the provision of a number of model structures, and that work is reiterated and expanded on in this chapter. Some of the models are purely theoretical, in that as yet nothing similar has been observed in nature. A method to change the layer spacing while preserving the relative positions of the atoms in the central core of the structure is presented. The lattice parameters for the model structures are in some cases slightly different from those of the refined structures in order to help demonstrate the structural similarities; they would no doubt change slightly if a model was used as the starting point for a structure refinement.

A model structure for 11 Å orthotobermorite that has double silicate chains – that is, that are cross linked – is given in Table 5.3. This model covers both the normal and anomalous 11 Å tobermorites because the contents of the interlayer region can be adjusted. It is based on the structures that are given in Table 2 of Merlino *et al.* (2001) for the MDO_2 polytypes. The Ca2 site is vacant in the case of anomalous tobermorite and O6 becomes OH6. The powder XRD patterns calculated from the model structures and those of Merlino *et al.* (2001) are closely similar if the lattice parameters are adjusted to be the same. Merlino *et al.* (1999) proposed extreme formulations for the composition of 11 Å orthotobermorites (cross linked) of $Ca_4(Si_6O_{15}(OH)_2)(H_2O)_5$ and $Ca_5(Si_6O_{17})(H_2O)_5$. These were based on Zadov *et al.*'s (1995) experimental observations of natural tobermorites and correspond to $i = 1$ and 0, respectively, in Formula 5.1. Selected bond distances are given in Table 5.4.

A model structure for 11 Å orthotobermorite that has staggered silicate chains is given in Table 5.5 (Richardson, 2014). This model is very similar to the double-chain structure that is given in Table 5.3 but with a shift of $b/2$ in the silicate chains. The atom positions are the same except that the y/b values for Ca1 and O5 are reduced by 0·5 and the space group is $B\ 1\ 1\ b$ rather than $B\ 1\ 1\ m$. The model is essentially equivalent to Hamid's (1981) structure for 11 Å orthotobermorite, which has single chains (for the reason given earlier), although the interlayer Ca atom (Ca2 in the model) is in a different position (note that Hamid's (1981) structure is in a different space group, *P* 1 1 2 (No. 3, unique axis *c*)). The bond distances are the same as for the double-chain structure given in Table 5.4.

The structure given here corresponds to $i = 1$ in Formula 5.2. Hamid (1981) gave formulae that correspond to $i = 0$, 1, 1·5 and 2, although with fewer water molecules than in the present model. The variation in the value of *i* relates to variation in the amount of interlayer Ca, described by Merlino *et al.* (2001) as 'zeolitic' Ca, and this is easily adjusted in the model structure by simply adjusting the occupancy of the Ca2 site and the fraction of O5 and O6 atoms that carry an H atom, as indicated in Table 5.6. It is relevant to the material on models for the structure of C–S–H that is presented in §6.6 to note here that the formulae that are given by Hamid (1981) and in Table 5.6 can also be described in terms of Richardson and Groves' (1992) model for C–S–H, specifically to the T∞ structural units that have values for the degree of protonation, *w*/*n*, of 0, 1, 1·5 and 2; it is evident from this that $\frac{w}{n} = i$ and that these model structures can be used to represent the T∞ structural units in Richardson and Groves' (1992) model (see §7.5.1).

Table 5.3 Model structure for 11 Å orthotobermorite that has cross-linked silicate chains (continued on next page)

Systematic name:	calcium hexasilicate hydroxide hydrate
Structural formulae:	$Ca_4(Si_6O_{15}(OH)_2)(H_2O)_5$ (anomalous) i.e. $i = 1$, $m = 5$ in Formula 5.1
	$Ca_{4\cdot5}(Si_6O_{16}(OH))(H_2O)_5$ (normal) i.e. $i = 0\cdot5$, $m = 5$ in Formula 5.1
Chemical formula (sum):	H_{12} Ca_4 O_{22} Si_6 (anomalous); H_{11} $Ca_{4\cdot5}$ O_{22} Si_6 (normal)
a	6·748 Å
b	7·3 Å
c	22·68 Å
α	90°
β	90°
γ	122·75°
Z	2
Space group:	No. 8, *B* 1 1 *m*
Unique axis:	*c*
Cell choice:	2
Symmetry equivalent positions:	(0, 0, 0)+ (½, 0, ½)+
4*b*:	(1) + *x*, + *y*, + *z* (2) + *x*, + *y*, −*z*
2*a*:	+*x*, + *y*, 0

Atom	Label	Position	Occ.	*x*/*a*	*y*/*b*	*z*/*c*
Ca	Ca1	4*b*	1	0·75	0·9286	0·2941
Ca	Ca2	4*b*	0·25 or 0[a]	0·55	0·075	0·045

Table 5.3 Continued

Atom	Label	Position	Occ.	*x/a*	*y/b*	*z/c*
Ca	Ca3	4*b*	1	0·25	0·4286	0·206
O	O1	4*b*	1	0·75	0·5	0·0957
O	O2	4*b*	1	0·75	0·173	0·1354
O	O3	4*b*	1	0·9833	0·5556	0·1966
O	O4	4*b*	1	0·515	0·315	0·1966
O	O5	2*a*	1	0·9	0·75	0
O	O6 or OH6[a]	4*b*	1	0·17	0·887	0·0957
O	O7	4*b*	1	0·75	0·85	0·0957
O	O8	4*b*	1	0·515	0·815	0·1966
O	O9	4*b*	1	0·9833	0·05	0·1966
O	Wat1	2*a*	1[b]	0·4	0·25	0
O	Wat2	2*a*	1	0·9	0·25	0
O	Wat3	2*a*	1[b]	0·4	0·75	0
O	Wat6	4*b*	1	0·25	0·4286	0·0957
Si	Si1	4*b*	1	0·75	0·386	0·1577
Si	Si2	4*b*	1	0·9	0·75	0·071
Si	Si3	4*b*	1	0·75	0·96	0·1577

[a]The Ca2 site is vacant in Merlino *et al.*'s (2001) structure for anomalous 11 Å orthotobermorite and O6 is an OH^- ion (and so in that case should be labelled OH6); in normal 11 Å orthotobermorite occupancy of O6 is half OH^- and half O^{2-}.

[b]The occupancy for these sites is 0·5 in Merlino *et al.*'s (2001) structure for normal tobermorite because $z/c \neq 0$ in their structure and so the multiplicity is 4 rather than 2 (i.e. the Wyckoff position is 4*b* rather than 2*a*).

Table 5.4 Selected bond distances (Å) for the structure given in Table 5.3. The interlayer Ca atom is indicated by 'I'

Ca1	Ca2 (I)	Ca3		Si1	Si2	Si3
2·323	2·233	2·353		1·634	1·610	1·603
2·352	2·258	2·354		1·635	1·632	1·612
2·396	2·258	2·397		1·635	1·634	1·619
2·407	2·344	2·439		1·637	1·634	1·635
2·540	2·444	2·502				
2·578	2·870	2·577				
2·583	2·871	2·581				
2·454	2·468	2·457	Average for polyhedron	1·635	1·628	1·617
	2·460		Overall average		1·627	

Table 5.5 Model structure for 11 Å orthotobermorite that has staggered silicate chains (Richardson, 2014; CIF datablock T∞_11so)

Systematic name:	calcium hexasilicate dihydroxide trihydrate
Structural formula:	$Ca_5(Si_6O_{16}(OH)_2)(H_2O)_3$ i.e. $i = 1$, $m = 3$ in Formula 5.2
Chemical formula (sum):	H_8 Ca_5 O_{21} Si_6
a	6·748 Å
b	7·3 Å
c	22·68 Å
α	90°
β	90°
γ	122·75°
Z	2
Space group:	No. 9, *B* 1 1 *b*
Unique axis:	*c*
Cell choice:	2
Symmetry equivalent positions:	(0, 0, 0)+ (½ , 0, ½)+
4*a*:	(1) + *x*, + *y*, + *z* (2) + *x*, ½+*y*, −*z*

Atom	Label	Position	Occ.	*x/a*	*y/b*	*z/c*
Ca	Ca1	4*a*	1	0·75	0·4286	0·2941
Ca	Ca2	4*a*	0·5	0·55	0·075	0·045
Ca	Ca3	4*a*	1	0·25	0·4286	0·206
O	O1	4*a*	1	0·75	0·5	0·0957
O	O2	4*a*	1	0·75	0·173	0·1354
O	O3	4*a*	1	0·9833	0·5556	0·1966
O	O4	4*a*	1	0·515	0·315	0·1966
O	OH5	4*a*	0·5, O 0·5	0·9	0·25	0
O	OH6	4*a*	0·5, O 0·5	0·17	0·887	0·0957
O	O7	4*a*	1	0·75	0·85	0·0957
O	O8	4*a*	1	0·515	0·815	0·1966
O	O9	4*a*	1	0·9833	0·05	0·1966
O	Wat3	4*a*	0·5	0·4	0·75	0
O	Wat6	4*a*	1	0·25	0·4286	0·0957
Si	Si1	4*a*	1	0·75	0·386	0·1577
Si	Si2	4*a*	1	0·9	0·75	0·071
Si	Si3	4*a*	1	0·75	0·96	0·1577

Table 5.6 The alternative formulae for single-chain 11 Å orthotobermorite that are given by Hamid (1981)

i	Formula	Density: $g\,cm^{-3}$	Ca2/Wat3 Occ.	O5: OH5 O6: OH6 Occ.
0	$Ca_6(Si_6O_{18})(H_2O)_4$	2·718	1·0	1 : 0
1	$Ca_5(Si_6O_{16}(OH)_2)(H_2O)_3$	2·520	0·5	0·5 : 0·5
1·5	$Ca_{4\cdot5}(Si_6O_{15}(OH)_3)(H_2O)_{2\cdot5}$	2·421	0·25	0·25 : 0·75
2	$Ca_4(Si_6O_{14}(OH)_4)(H_2O)_2$	2·322	0	0 : 1

i is a variable in Formulae 5.2 and 5.3 and it is equal to the degree of protonation, *w/n*, in Richardson and Groves' (1991) model for the C–S–H that is present in hardened cements.
O5/OH5 in the model represent the same atoms as O8/OH8 and O14/OH14 in Hamid's (1981) structure and O6/OH6 represent O9/OH9 and O18/OH18.
$H_2O/Si = 0{\cdot}67$ for all four formulae.

5.3.1.6 Crystal structure of a normal 11 Å clinotobermorite

Hoffmann and Armbruster (1997) determined the substructure of a thermally normal clinotobermorite; Merlino *et al.* (1999, 2000) made a more detailed determination, which yielded the real structures of two polytypes. The composition was $\approx Ca_5Si_6O_{23}H_{12}$. Figure 5.4(a) shows the unit cell of this ordered structure (the MDO_2 polytype) projected along its **b** axis. The arrangement of the Si–O tetrahedra and Ca–O polyhedra are illustrated in Figures 5.4(b) and 5.4(c), which show the structure viewed along the **b** and **a** axes, respectively. This polytype has $a = 11{\cdot}274$ Å, $b = 7{\cdot}344$ Å, $c = 11{\cdot}468$ Å, $\alpha = 99{\cdot}18°$, $\beta = 97{\cdot}19°$, $\gamma = 90{\cdot}03°$, space group $C\bar{1}$ (No. 2, setting 2). Individual layers are closely similar to those of the orthotobermorite described above, but the conformation of the double dreierketten differs in a way that alters the subcell from orthorhombic to monoclinic. The double chains in 11 Å clinotobermorite display $2/m$ symmetry compared with $2mm$ in 11 Å orthotobermorite, which is illustrated in Figure 5.3. None of the O atoms of the silicate chains carries an H atom. As with the orthotobermorite, there are channels parallel to the chain direction. These were found to contain one calcium atom and three water molecules per formula unit, consistent with the formula $Ca_5Si_6O_{17}{\cdot}5H_2O$.

5.3.1.6.1 MODEL CRYSTAL STRUCTURES FOR 11 Å CLINOTOBERMORITES

A model structure for 11 Å clinotobermorite that has cross-linked silicate chains is given in Table 5.7 (Richardson, 2014). It is a slightly idealised version of the monoclinic MDO_1 polytype that is given in Table 4 of Merlino *et al.* (2000). Merlino *et al.*'s (2000) structure is in space group Cc (No. 9, unique axis *b*, cell choice 1); the idealisation of the structure produced an increase in symmetry to space group $C\,2/c$ and therefore to a reduction in the number of atoms in the asymmetric unit. Merlino *et al.* (2000) actually commenced their refinement in space group $C\,2/c$ but rejected it on the basis that there were some short Ca–O distances between the interlayer Ca atom and the water molecules located in the interlayer. Selected bond distances are given for the model structure in Table 5.8. The minimum Ca–O distance for the interlayer Ca atom in the model structure is quite short (2·203 Å) but it is longer than the minimum distance that has been observed by experiment in calcium silicate hydrates in general (§5.7) and it is similar to the minimum in Merlino *et al.*'s (2000) structure (2·236 Å). The maximum distance in the model is shorter than in Merlino *et al.*'s (2000) structure (2·862 Å compared with 2·957 Å), which is more consistent with the data in §5.7. The powder XRD patterns calculated from the model structure and Merlino *et al.*'s (2000)

Figure 5.4(a) Structure of the MDO_2 polytype of normal 11 Å clinotobermorite projected along its **b** axis showing the positions of the atoms in the unit cell

(a)

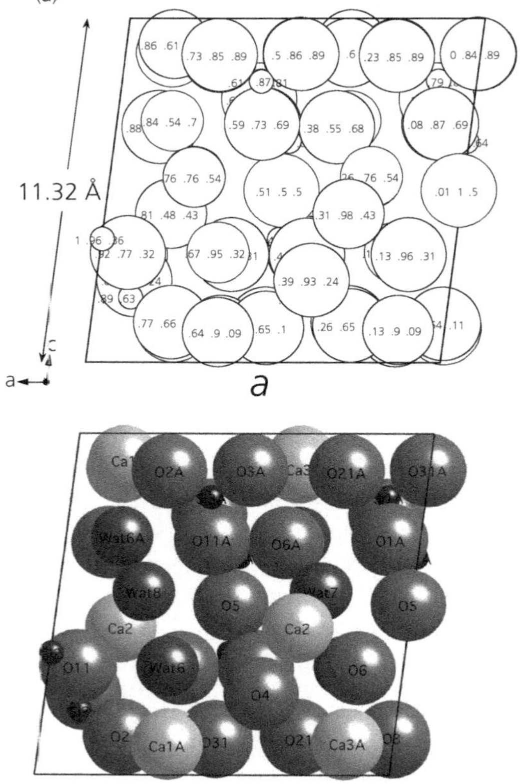

The structure is from Merlino *et al.* (1999, 2000). Coordinates (*x/a y/b z/c*) and atom labels are given in the top and bottom parts, respectively. Edge lengths are $c = 11{\cdot}274$ Å and $c \sin \alpha = 11{\cdot}321$ Å.

Figure 5.4(b) and Figure 5.4(c) Polyhedral representations of the structure of the MDO_2 polytype of normal 11 Å clinotobermorite viewed (b) along its **b** axis and (c) along its **a** axis (continued on next page)

(b)

Figure 5.4(b) and Figure 5.4(c) Continued

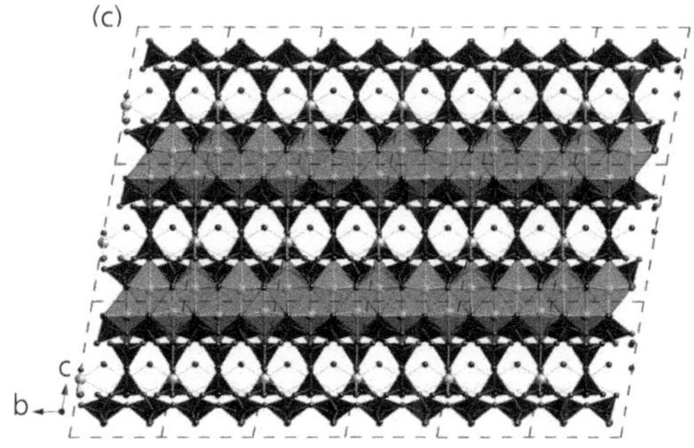

Unit edge lengths are $b = 7{\cdot}344$ Å and $c \sin \beta = 11{\cdot}378$ Å.

Table 5.7 Model structure for 11 Å clinotobermorite that has cross-linked silicate chains (Richardson, 2014; CIF datablock T∞_11ac) (continued on next page)

Systematic name:	pentacalcium hexasilicate trihemihydrate
Structural formula:	$Ca_5(Si_6O_{17})(H_2O)_{3\cdot5}$ i.e. $i = 0$, $m = 3{\cdot}5$ in Formula 5.1
Chemical formula (sum):	$H_7\ Ca_5\ O_{20\cdot5}\ Si_6$
a	11·35 Å
b	7·3 Å
c	22·68 Å
α	90°
β	97·28°
γ	90°
Z	4
Space group:	No. 15, *C* 1 2/*c* 1
Unique axis:	*b*
Cell choice:	1
Symmetry equivalent positions:	(0, 0, 0)+ (½, ½, 0)+
8*f*:	(1) + *x*, + *y*, + *z* (2) −*x*, + *y*, ½−*z* (3) −*x*, −*y*, −*z* (4) + *x*, −*y*, ½+*z*
4*e*:	0, + *y*, ¼ 0, −*y*, ¾

Atom	Label	Position	Occ.	*x*/*a*	*y*/*b*	*z*/*c*
Ca	Ca1	8*f*	1	0·8667	0·375	0·4556
Ca	Ca2	8*f*	1	0·8667	0·875	0·4556
Ca	Ca3	8*f*	0·5	0·817	0·375	0·2
O	O1	8*f*	1	0·9167	0·9529	0·6579

Table 5.7 Continued

Atom	Label	Position	Occ.	*x/a*	*y/b*	*z/c*
O	O2	8*f*	1	0·9167	0·3077	0·6579
O	O3	8*f*	1	0·7692	0·375	0·5556
O	O4	8*f*	1	0·7692	0·875	0·5556
O	O5	8*f*	1	0	0·375	0·5556
O	O6	8*f*	1	0	0·875	0·5556
O	O7	8*f*	1	0·8919	0·625	0·615
O	O8	4*e*	1	0	0·125	0·75
O	Wat9	4*e*	0·5	0	0·375	0·25
O	O10	8*f*	1	0·13	0·125	0·6579
O	Wat11	8*f*	1	0·1579	0·625	0·6579
O	Wat12	8*f*	0·5	0·75	0·125	0·25
Si	Si1	8*f*	1	0·8929	0·837	0·5949
Si	Si2	8*f*	1	0	0·125	0·679
Si	Si3	8*f*	1	0·8958	0·4167	0·5949

Table 5.8 Selected bond distances (Å) for the structure given in Table 5.7. The interlayer Ca atom is indicated by 'I'

Ca1	Ca2	Ca3 (I)		Si1	Si2	Si3
2·382	2·382	2·203		1·589	1·608	1·590
2·382	2·382	2·214		1·613	1·608	1·590
2·404	2·404	2·237		1·621	1·610	1·622
2·404	2·404	2·327		1·653	1·670	1·625
2·553	2·562	2·327				
2·562	2·578	2·798				
2·646	2·646	2·863				
2·476	2·480	2·424	Average for polyhedron	1·619	1·624	1·609
	2·460		Overall average		1·617	

Table 5.9 Model structure for 11 Å clinotobermorite that has staggered silicate chains (Richardson, 2014; CIF datablock T∞_11sc) (continued on next page)

Systematic name:	pentacalcium hexasilicate dihydroxide trihydrate
Structural formula:	$Ca_5(Si_6O_{16}(OH)_2)(H_2O)_3$ i.e. $i = 1$, $m = 3$ in Formula 5.2
Chemical formula (sum):	$H_8\ Ca_5\ O_{21}\ Si_6$
a	11·35 Å

Table 5.9 Continued

b	7·3 Å
c	22·68 Å
α	90°
β	97·28°
γ	90°
Z	4
Space group:	No. 14, $P\ 1\ 2_1/c\ 1$
Unique axis:	b
Cell choice:	1
Symmetry equivalent positions:	(0, 0, 0)+
4e:	(1) $+x, +y, +z$ (2) $-x, ½+y, ½-z$ (3) $-x, -y, -z$ (4) $+x, ½-y, ½+z$

Atom	Label	Position	Occ.	x/a	y/b	z/c
Ca	Ca1	4e	1	0·8667	0·375	0·4556
Ca	Ca2	4e	1	0·8667	0·875	0·4556
Ca	Ca3	4e	0·5	0·817	0·875	0·2
Ca	Ca1A	4e	1	0·3667	0·875	0·4556
Ca	Ca2A	4e	1	0·3667	0·375	0·4556
Ca	Ca3A	4e	0·5	0·317	0·375	0·2
O	O1	4e	1	0·9167	0·9529	0·6579
O	O2	4e	1	0·9167	0·3077	0·6579
O	O3	4e	1	0·7692	0·375	0·5556
O	O4	4e	1	0·7692	0·875	0·5556
O	O5	4e	1	0	0·375	0·5556
O	O6	4e	1	0	0·875	0·5556
O	O7	4e	1	0·8919	0·625	0·615
O	O(H)8	4e	1	0	0·125	0·75
O	O(H)10	4e	1	0·13	0·125	0·6579
O	Wat11	4e	1	0·1579	0·625	0·6579
O	Wat12	4e	0·5	0·75	0·125	0·25
O	O(H)10A	4e	1	0·63	0·625	0·6579
O	Wat11A	4e	1	0·6579	0·125	0·6579
O	Wat12A	4e	0·5	0·75	0·625	0·25
O	O1A	4e	1	0·4167	0·4529	0·6579
O	O2A	4e	1	0·4167	0·8077	0·6579
O	O3A	4e	1	0·2692	0·875	0·5556

Table 5.9 Continued

Atom	Label	Position	Occ.	*x/a*	*y/b*	*z/c*
O	O4A	4*e*	1	0·2692	0·375	0·5556
O	O5A	4*e*	1	0·5	0·875	0·5556
O	O6A	4*e*	1	0·5	0·375	0·5556
O	O7A	4*e*	1	0·3919	0·125	0·615
O	O(H)8A	4*e*	1	0·5	0·625	0·75
Si	Si1	4*e*	1	0·8929	0·837	0·5949
Si	Si2	4*e*	1	0	0·125	0·679
Si	Si3	4*e*	1	0·8958	0·4167	0·5949
Si	Si1A	4*e*	1	0·3929	0·337	0·5949
Si	Si2A	4*e*	1	0·5	0·625	0·679
Si	Si3A	4*e*	1	0·3958	0·9167	0·5949

Table 5.10 Alternative formulae for single-chain 11 Å clinotobermorite.

i	Formula	Density g cm^{-3}	Ca3/3A Wat12/12A Occ.	O8/8A : OH8/8A O10/10A : OH10/10A Occ.
0	$Ca_6(Si_6O_{18})(H_2O)_4$	2·740	1·0	1 : 0
1	$Ca_5(Si_6O_{16}(OH)_2)(H_2O)_3$	2·541	0·5	0·5 : 0·5
1·5	$Ca_{4\cdot5}(Si_6O_{15}(OH)_3)(H_2O)_{2\cdot5}$	2·441	0·25	0·25 : 0·75
2	$Ca_4(Si_6O_{14}(OH)_4)(H_2O)_2$	2·341	0	0 : 1

i is a variable in Formula 5.2 and is equal to the degree of protonation, *w/n*, in Richardson and Groves' (1991) model for the C–S–H that is present in hardened cements.
$H_2O/Si = 0{\cdot}67$ for all four formulae.

are closely similar if the lattice parameters are adjusted to be the same; the pattern calculated for Merlino *et al.*'s (2000) MDO_2 polytype structure is shown in Figure 5.1(b). The foregoing indicates that the model is suitable for use in the development of models for C–S–H. The structural formula is $Ca_5Si_6O_{17}(H_2O)_{3\cdot5}$, which corresponds to $i = 0$ in Formula 5.1.

11 Å clinotobermorite that has staggered silicate chains has not been observed in nature, but by analogy with the orthotobermorites it seems entirely possible that it could occur; it could also be relevant to models for the structure of C–S–H (see §6.6). A model structure is given in Table 5.9 (Richardson, 2014). It is very similar to the double-chain structure that is given in Table 5.7 but with a shift of *b*/2 in the silicate chains. The displacement of the chains forces a reduction in symmetry to space group $P\,2_1/c$ and a consequent increase in the number of atoms in the asymmetric unit. The atom positions are essentially the same except that the *y*/*b* value for Ca3 is increased by ½ and Wat9 is deleted; O10/10A become OH10/10A. These changes result in a structural formula of $Ca_5(Si_6O_{16}(OH)_2)(H_2O)_4$, which corresponds to $i = 1$ in Formula 5.2 but it is of course possible to

vary the contents of the interlayer in the same way as for the corresponding orthotobermorite structure: the result of this in terms of the values of *i*; the occupancy of sites Ca3/3A, O8/8A (OH8/8A) and O10/10A (OH10/10A) is given in Table 5.10.

5.3.1.7 Silicate anions in 11 Å tobermorites: chemical methods and nuclear magnetic resonance

The first studies on silicate anion type other than by crystal structure determination employed chemical methods, of which the most widely used was based on the rate of reaction with molybdate in an acid solution. Subsequent work has mainly employed ^{29}Si MAS NMR, which gives less direct information but, being a purely physical method, does not suffer from significant uncertainties caused by the possibility of changes in the anion during the investigation. It provides quantitative information on the connectivities of the silicate tetrahedra (Q), from which hypotheses regarding anion type may be tested. The connectivities are referred to using the terms Q^0 … Q^4, where Q^0 denotes isolated tetrahedra, Q^1 end–group tetrahedra (and thus including silicon present in Si_2O_7 groups), Q^2 middle groups and so on. Thus, an infinite dreierkette with no substitution of the Si would show only Q^2 tetrahedra, an infinite double dreierkette will show a Q^2/Q^3 ratio of 2 and, with decreasing chain lengths, the end groups of the chains will give rise to increasing proportions of Q^1 tetrahedra.

The Q^2/Q^3 ratio is modified if Al is substituted for some Si. If, as is probable, Al occurs only in bridging tetrahedra, and never in adjacent tetrahedra, $Q^2/Q^3 = 2/(1 - 3f)$ for an infinite double dreierkette, where f is the fraction of the total number of tetrahedral sites occupied by Al and Q^2 and Q^3 each include Si atoms in tetrahedra sharing O atoms with any combination of Si and Al. Values of Q^2/Q^3 above $2/(1–3f)$ suggest the partial or complete absence of interlayer Si–O–Si or Si–O–Al linkages. The maximum possible value of f is $\frac{1}{6}$; at this degree of Al substitution, Q^2/Q^3 is 4 for a fully connected, double-chain structure.

Using the molybdate method and also a technique employing paper chromatography, Winkler and Wieker (1979) concluded that the anions in a number of 11 Å tobermorites prepared at 120 °C or 140 °C were largely or entirely double chains. In contrast, the anions in one obtained by heating a 14 Å tobermorite at 80 °C were found to be single chains (Wieker, 1968). This latter conclusion was confirmed using ^{29}Si NMR, which showed only Q^2 tetrahedra in a sample heated at 100 °C (Wieker *et al.*, 1982). 11 Å tobermorites made from CaO and quartz at 140 °C were also studied and gave Q^2/Q^3 ratios which increased with the Ca/Si ratio, from 0·8 at Ca/Si = 0·7 to 5·0 at Ca/Si = 1·0 (Wieker *et al.*, 1982). These variations were interpreted as arising from single chains joined with varying proportions of interlayer Si–O–Si linkages but could also have arisen from varying proportions of single- and double-chain structures. Komarneni *et al.* (1985b) studied a number of preparations all having Ca/(Si + Al) = 0·83 by ^{29}Si and ^{27}Al NMR. Most probably contained both Al^{3+} and alkali, either added deliberately or present in zeolites used as starting materials. Those made at 175–180 °C, which were thermally anomalous, showed substantial contents of Q^3 tetrahedra, whereas one obtained from a 120-day treatment at 80 °C, which was thermally normal, showed only Q^2 tetrahedra. The results also showed that Al substituted for Si in Q^3 sites and in some cases also in Q^2 sites. In C–S–H, Al substitution is restricted to bridging sites (Andersen *et al.*, 2003, 2004; Richardson, 1999; Richardson *et al.*, 1993) and this is likely to be the case also with tobermorite.

Mitsuda *et al.* (1989) examined a number of preparations made at 180 °C from various starting materials. Analytical electron microscopy showed that the Ca/(Si + Al) ratios were in most cases 0·8–0·9. The results showed that the silicate anions in the C–S–H formed as a precursor to tobermorite were predominantly single chains and that those in the tobermorites formed at 180 °C, whether or not Al-substituted, were predominantly double chains. The thermal behaviour of these tobermorites was normal or, less frequently, mixed. Bell *et al.* (1990) found an Al-free preparation made at 160 °C to have a Q^2/Q^3 ratio of $\frac{67}{20}$, indicating a degree of polymerisation intermediate between single and double chains. They also found some Q^1, which was intensified when ^1H cross-polarisation was used, suggesting the presence of breaks in the chains associated with water or hydroxyl groups – that is, vacant bridging sites. Sato and Grutzeck (1992) obtained broadly similar results for a sample probably made at 180 °C. Cong and Kirkpatrick (1996) confirmed that the silicate anions in a thermally normal 11 Å tobermorite obtained by heating the 14 Å form at 105 °C were predominantly single chains, whereas those in a thermally anomalous sample synthesised at 175 °C were predominantly double chains. They also confirmed the presence of some Q^1 sites and showed that Q^1 and Q^3 sites were more effectively cross-polarised by ^1H than were Q^2 sites. There were indications that the protons were more mobile in the sample obtained by heating 14 Å tobermorite, perhaps because the paucity of cross-linkages made the structure less rigid.

These results may be summarised as follows.

- 11 Å tobermorite formed by heating the 14 Å form at temperatures up to at least 105 °C contains predominantly single chains.
- In those formed by direct synthesis, the proportion of Q^3 increases with increasing temperature, with decreasing Ca/Si ratio and probably with time. At 80 °C, the anions are single chains, even after prolonged treatment (120 days).
- Tobermorites containing single chains are thermally normal. Those containing double chains may be normal, mixed or anomalous. As noted earlier, the presence of Al^{3+} favours normal and that of Al^{3+} with alkali favours anomalous behaviour.

5.3.1.8 Infrared absorption spectra

Farmer *et al.* (1966) reported the mid IR spectrum (600–4000 cm^{-1}) of a normal 11 Å tobermorite obtained by heating the 14 Å form. The silicate anions were single chains (Wieker, 1968). The main Si–O stretching band had a maximum at 979 cm^{-1} with subsidiary peaks at 1046 cm^{-1} and 932 cm^{-1}. A peak at 675 cm^{-1} could be attributed to Si–O–Si bending. These data are consistent with the presence of Si–O chains. Mid IR spectra for other 11 Å tobermorites reported in the literature (Henmi and Kusachi, 1992; Henning, 1974; Kalousek and Roy, 1957; Mitsuda, 1970; Mitsuda and Taylor, 1978; Yu *et al.*, 1999), including that of a clinotobermorite (Henmi and Kusachi, 1992), are generally similar but many include bands at 890 and ≈1180 cm^{-1}. There does not appear to be any clear correlation between these or other features of the spectrum and whether the thermal behaviour is normal or anomalous (Mitsuda and Taylor, 1978). The 1180 cm^{-1} band, at least, could be a feature arising from the presence of double dreierketten. Xonotlite, in which the anions are of this type, gives a spectrum in which the main Si–O stretching band has a maximum at 975 cm^{-1} and a strong subsidiary band at 1200 cm^{-1} (Henning, 1974). The far infrared spectrum of a 11 Å tobermorite includes bands at 450–480 cm^{-1} and 200–350 cm^{-1}, arising from O–Si–O bending and from the Ca–O parts of the structure, respectively (Yu *et al.*, 1999).

The mid infrared spectra of 11 Å ortho- and clinotobermorites include an OH stretching band at ≈3500 cm^{-1} and an H–O–H bending band at ≈1670 cm^{-1}. The near infrared spectrum includes a combination band at 4500 cm^{-1}, diagnostic for Si–OH (Yu *et al.*, 1999). The band at ≈3500 cm^{-1} persists after all the molecular water has been driven off (Farmer *et al.*, 1966) and thus includes one or more components from hydroxyl groups. Its profile varies somewhat between specimens. In some cases (e.g. Farmer *et al.*, 1966), there is a sharp peak at ≈3450 cm^{-1} and a low-frequency tail extending to $\approx 2500\,cm^{-1}$, but the peak can be less sharp and more symmetrical (Yu *et al.*, 1999). The variations could arise from differences in the content of Si–OH and possibly also from the presence of OH groups attached only to Ca. The frequency of the sharp peak is only slightly less than those of $3480–3644\,cm^{-1}$ for non-hydrogen-bonded OH groups attached only to Ca in other calcium silicate hydrates (Henning, 1974). The crystal structure determinations (Merlino *et al.*, 1999) show that, in addition to molecular water, double-chain 11 Å ortho- and clinotobermorites contain varying proportions of OH groups attached to both Ca and Si. Calculation from the data of Merlino *et al.* (1999) shows that the nearest oxygen neighbour of this OH group is a water molecule at 2·93 Å. The hydrogen bonds formed by this group are therefore extremely weak, and the group could partly or wholly account for the sharp peak at $\approx 3450\,cm^{-1}$. In the absence of crystal structure data, it is uncertain whether this applies also to single-chain tobermorites. The frequency of the sharp peak is equally compatible with the presence of OH attached only to Ca, which might arise from structural defects or sites on crystallite edges. The low frequency tail shows the presence of more strongly hydrogen bonded OH and, at least for double-chain tobermorites, is attributable to molecular water.

The spectra of some 11 Å tobermorites include weak bands at $\approx 1430\,cm^{-1}$, attributable to small contents of CO_3^{2-} or BO_3^{3-} (Farmer *et al.*, 1966; Henmi and Kusachi, 1992). There were indications that the ions were substituents in the structure and that the CO_3^{2-} was not present in admixed calcite (Farmer *et al.*, 1966).

5.3.1.9 Raman spectra, ^{17}oxygen NMR and Ca X-ray absorption spectra

Raman spectroscopy resembles infrared spectroscopy in providing information about vibrational modes, and thereby on local structures, but the bands excited often differ. Kirkpatrick *et al.* (1997a) determined the Raman spectrum of an anomalous 11 Å tobermorite, which had been synthesised at 175 °C and had been shown by ^{29}Si NMR to contain a high proportion of Q^3 sites and a small proportion of Q^1 sites. The data were interpreted empirically using the substantial body of reference data for comparable substances. Significant features included bands at $800–1080\,cm^{-1}$ (Si–O stretching), $610–700\,cm^{-1}$ (Si–O–Si bending), $430–540\,cm^{-1}$ (O–Si–O bending) and $300–350\,cm^{-1}$ (Ca–O polyhedra). Within the Si–O stretching group, individual peaks at $850\,cm^{-1}$ (very weak) and $1010\,cm^{-1}$ (strong) were assigned to Q^1 and Q^2 tetrahedra, respectively. A moderately strong peak at $1040\,cm^{-1}$ could be partly due to Q^3 tetrahedra but was overlapped by a carbonate peak. The Si–O–Si bending peaks included a very strong peak at $680\,cm^{-1}$, assigned mainly to Q^2 tetrahedra, and a weaker one at $620\,cm^{-1}$, which could be due to Q^3 tetrahedra. Further strong peaks at $<250\,cm^{-1}$ were assigned to complex vibrations, as yet not understood. The conclusions are fully compatible with those from ^{29}Si NMR.

^{17}O MAS NMR of 11 Å tobermorites obtained either by direct synthesis or by heating the 14 Å form were similar to each other (Cong and Kirkpatrick, 1996). Tentative modelling of the spectra indicated that a high proportion of the intensity could be attributed to O atoms bonded to Ca atoms and only one Si, which agrees with the results of the crystal structure determinations. The more detailed assignments differed in some ways from those found in the structure determinations and the origin of the discrepancy requires further examination.

Kirkpatrick *et al.* (1997b) reported calcium X-ray absorption spectra (extended X-ray absorption fine structure (EXAFS) and X-ray absorption near-edge structure (XANES)) for an 11 Å tobermorite. They showed that the Ca environments were similar to those in C–S–H preparations and significantly different from those in calcite or xonotlite. Data from the crystal structure (Hamid, 1981) allowed the calculation of parameters that were used in interpreting the spectra of the C–S–H preparations and of a 14 Å tobermorite.

5.3.1.10 Ionic substitutions

The substitution of Si by Al in tobermorite has already been discussed in §5.3.1.4. The only other significant substitution commonly observed is of alkali cations for interlayer Ca. Chemical analyses of naturally occurring mixed or anomalous tobermorites show up to 0·13 atoms of Na and 0·06 atoms of K per subcell (Mitsuda and Taylor, 1978). Alkali cations can be charge balanced by decrease in interlayer Ca, substitution of tetrahedral Al for Si, substitution of SiO^- for SiOH, or any combination of these.

Komarneni *et al.* (1982) showed that various mixtures of Al-substituted tobermorites with other phases underwent cation exchange with K^+, especially with Cs^+, which was selectively sorbed. Observations by Hara *et al.* (1997) confirm that alkali cations can be taken up and suggest the possibility that interlayer anions might also contribute to the balance. Other postulated substitutions appear more doubtful. Komarneni *et al.* (1986a) concluded that reaction of tobermorite with Co or Ni solutions could result in almost complete replacement of Ca^{2+} by Co^{2+} or Ni^{2+}, but the products of the near complete substitution were amorphous and, although pseudomorphs were formed, the crystal structure was clearly destroyed. The XRD patterns of the products showed broad bands peaking at about 24° 2θ (Cu Kα radiation), suggesting that highly siliceous residues had been formed. The persistence of the tobermorite pattern at lesser degrees of substitution could, as the authors noted, arise from incomplete reaction. Smaller cations, such as Co^{2+} or Ni^{2+}, are unlikely to substitute for the 7-coordinated Ca in the cores of the layers but could conceivably occupy sites in the interlayers. Substitution of Ca^{2+} by Mg^{2+} in tobermorite has from time to time been postulated but is unlikely to occur to more than a minor extent for the same reason. Mitsuda (1973) described a naturally occurring mixture of a low-Mg tobermorite and a poorly crystalline magnesium silicate hydrate and reached the same conclusion.

5.3.2 14 Å Tobermorite

5.3.2.1 Natural occurrence, synthesis and crystal data

Naturally occurring 14 Å tobermorite is rare and usually occurs in small quantities (Farmer *et al.*, 1966; McConnell, 1954; Mitsuda *et al.*, 1972; Murdoch, 1961; Stephens and Bray, 1973; Taylor, 1953). It is sometimes intergrown with 11 Å tobermorite (Taylor, 1953) or with other forms of tobermorite and an apatite mineral in oriented, submicroscopic intergrowths called crestmoreite. McConnell (1954) reported that 14 Å tobermorite could be obtained by moistening the 11 Å form with water, but other workers have failed to confirm this. It can readily be synthesised in the laboratory from $Ca(OH)_2$ and silica in aqueous suspensions at 60–90 °C (Cong and Kirkpatrick, 1996; Hara and Inoue, 1979, 1980b; Hara *et al.*, 1978; Kalousek and Roy, 1957). One of the natural specimens contained 0·8% of B_2O_3 and 0·7% of CO_2, both of which appeared to be substituents in the crystal structure (Farmer *et al.*, 1966; Mitsuda and Taylor, 1978; Murdoch, 1961).

Bonaccorsi *et al.* (2005) determined the crystal structure of the 14 Å tobermorite from Crestmore. The individual layers are of the single-chain type shown in Figure 5.3(e) but are more widely separated than in an 11 Å tobermorite. Additional interlayer water molecules can thus be incorporated, giving the approximate composition $Ca_5Si_6O_{18}H_2 \cdot 8H_2O$ or $C_5S_6H_9$. The TG curve for this specimen (Farmer *et al.*, 1966) showed $8 \cdot 2H_2O$ per subcell; some water could have been lost by storage in a dry atmosphere. Chemical analysis of a synthetic preparation indicated Ca/Si ratios of 0·74–0·82 and H_2O/SiO_2 ratios of 1·35–1·42 (Hara and Inoue, 1980b).

As with 11 Å orthotobermorites, there is a strongly marked body-centred orthorhombic subcell with $a = 5 \cdot 63$ Å, $b = 3 \cdot 71$ Å, $c = 27 \cdot 99$ Å (Bonaccorsi *et al.*, 2004) (Table 5.2). Two main polytypes, both with a doubled value of *b*, were found (Bonaccorsi *et al.*, 2005). Apart from one weak reflection probably indexable on a larger true cell, the XRD powder pattern (ICDD card 29–331) can be fully indexed on the subcell given above, and the general distribution of intensities is similar to that shown by 11 Å tobermorites.

The crystals of the Crestmore material are very small laths with (001) cleavage and elongation parallel to **b**. They have positive elongation, low birefringence, mean refractive index 1·542 in sodium light and density 2·20 g cm^{-3} (Farmer *et al.*, 1966). The composition and subcell contents do not correspond exactly to the idealised formula $C_5S_6H_9$. The subcell contents calculated from the density, composition and cell parameters were approximately $Ca_5Si_{5\cdot5}O_{24}H_{16}$, whereas those corresponding to the idealised formula are $Ca_5Si_6O_{26}H_{18}$. Analyses of other natural specimens (McConnell, 1954; Mitsuda *et al.*, 1972) in which 14 Å tobermorite is mixed with 11 Å tobermorite suggest Ca/Si ratios near to 5:6.

Analyses of natural 14 Å tobermorites show negligible contents of Al and alkalis (Mitsuda and Taylor, 1978; Mitsuda *et al.*, 1972) and the presence of Al^{3+} in synthetic samples appears to favour the formation of less highly hydrated forms of tobermorite (Hara and Inoue, 1980b; Hara *et al.*, 1978).

5.3.2.2 Transmission electron microscopy, diffraction and microanalysis

Hara and Inoue (1980b) reported TEM studies of 14 Å tobermorites synthesised from CaO and silica glass at a Ca/Si ratio of 0·7. The crystals were laths or plates elongated parallel to **b**, closely similar in morphology to those of the Crestmore material. They were mixed with poorly crystalline foils, apparently of C–S–H. The SED pattern of a crystal lying on (001) was similar to those obtained from 11 Å tobermorites in this orientation (Gard and Taylor, 1957; Mitsuda and Taylor, 1978). It would not be possible to distinguish between an 11 Å and a 14 Å tobermorite from the pattern obtained in this orientation, and the data could have related either to the 14 Å tobermorite or to an 11 Å form produced by dehydration in the high vacuum. X-ray microanalyses indicated that the Ca/Si ratio of the tobermorite was 0·78–0·81 and that of the admixed C–S–H varied between 0·65 and 1·0. A different synthetic 14 Å tobermorite studied by TEM contained laths and submicrometre particles (Viehland *et al.*, 1997). X-ray microanalyses indicated local Ca/Si ratios varying between 0·75 and 0·90. The (100) repeat varied between 4·95 Å and 5·45 Å, and the (010) repeat varied between 3·65 and 3·80 Å. A reproduction of part of the SED pattern is consistent with these values, and, except for a relatively intense 100 reflection, with a body-centred subcell. Weak 100 reflections have also been observed in patterns from the Ballycraigy and Crestmore tobermorites (Taylor, unpublished).

Maeshima *et al.* (2003) examined a sample of natural 14 Å tobermorite from Crestmore, California, by analytical TEM, which gave a mean Ca/Si ratio of 0·85 (standard deviation = 0·03, $n = 28$); the analyses ranged between 0·80 and 0·91. In comparison, another sample from Crestmore – which was intermixed with small amounts of oyelite and scawtite (on a host rock containing grossular, calcite, wollastonite, Mg silicate hydrate and scawtite) – gave a significantly lower mean Ca/Si of 0·80 (standard deviation = 0·06, $n = 19$) with a broader range of composition – between 0·70 and 0·96 – which suggests significant variation in protonation of the silicate chains (which ^{29}Si NMR showed were essentially infinitely long (no Q^1)) (Richardson, 2008). The layer structure in that sample was shown clearly in a high-resolution transmission electron micrograph, which is reproduced in Figure 5.5. The micrograph was taken as the crystal lost interlayer water under the electron beam, which resulted in a reduction in the layer spacing from 14·2 Å (e.g. A) to 12·4 Å (e.g. C) to 11·3 Å (e.g. D). Further exposure to the beam resulted in decomposition of the tobermorite structure (Richardson, 2008). It was noted that this sequence of dehydration steps

Figure 5.5 High resolution transmission electron micrograph of 14 Å tobermorite from Crestmore, California. The image was obtained using a Philips CM200 field emission TEM equipped with light element EDX detector. Reproduced from Richardson (2008)

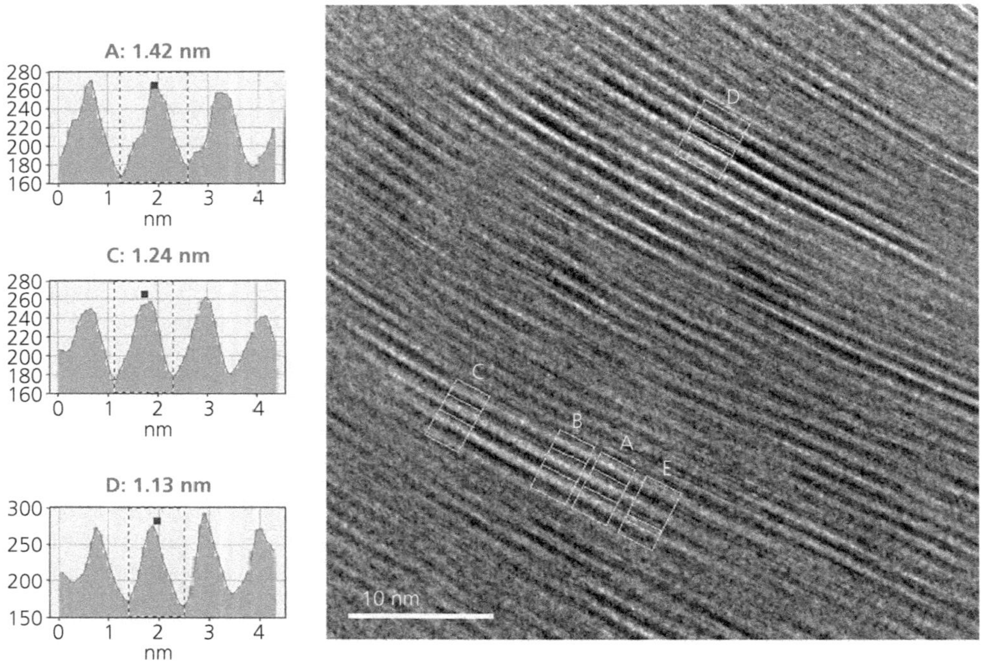

is direct confirmation of the thermal analysis and XRD results reported by Yu and Kirkpatrick (1999), which showed that on heating, water was lost in steps that corresponded to decreases in layer spacing from 14 to 12, 11 and 9·6 Å; the same steps have been observed subsequently by Biagioni *et al.* (2013) for a natural 14 Å sample but a 12 Å stage was not noted by Bonaccorsi and Merlino (2005). Yu and Kirkpatrick's (1999) data are discussed in greater detail in §5.3.3.5. It is of interest to note that the intermediate spacing of 12·4 Å in Figure 5.5 is the same as the value typical of the basal spacing for semicrystalline C–S–H that is often observed in alkali-activated hardened cements (Richardson *et al.*, 1994) and in many synthetic C–S–H preparations – that is, phases that give XRD patterns that match C–S–H(I) (§6.2.2). An HRTEM image of a crystal of a 14 Å tobermorite synthesised hydrothermally (with a target Ca/Si = 0·9) showed only a 10 Å interlayer distance because the crystal had dehydrated during specimen preparation (Gmira *et al.*, 2002).

5.3.2.3 Model crystal structures for 14 Å orthotobermorites

The structure of 14 Å orthotobermorite has been solved by Bonaccorsi *et al.* (2005). In contrast with their structures for the 11 Å orthotobermorites, the 14 Å phase has staggered silicate chains where the layers have *C2m(b)* symmetry, as illustrated in Figure 5.6 – that is, there is a *b*/2 shift in the chains when compared with their structures for the 11 Å orthotobermorites. The model structure that is given in Table 5.11 was derived from the equivalent model structure for the 11 Å orthotobermorite that is given in Table 5.5. While it is very similar to Bonaccorsi *et al.*'s (2005) structure – because

Figure 5.6 The two possible layers of 14 Å tobermorite and their symmetry elements (based on Figure 2 in Bonaccorsi *et al.*, 2005)

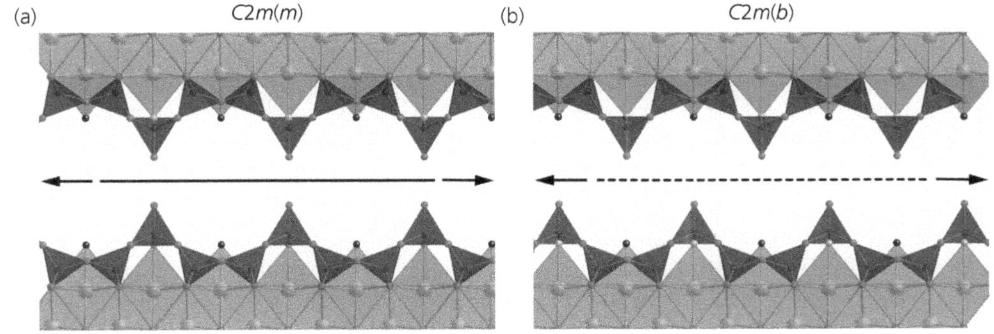

that structure was derived from their earlier work on the 11 Å phase – it is nevertheless somewhat idealised. The structure of the central Ca–O slab was preserved when increasing the layer spacing from 11 to 14 Å by recalculation of the *z/c* coordinate of the atoms in the asymmetric unit so that the relative *z* positions of the atoms in the slab were retained when the *c* parameter was increased. This was achieved by using the simple transformation that is given in Equation 5.4.

$$z_{14} = \frac{z_{11} \cdot c_{11} + \left(\frac{c_{14} - c_{11}}{4} \right)}{c_{14}} \tag{5.4}$$

where z_{14} and c_{14} are the *z/c* coordinate and *c* parameter of the 14 Å phase (c_{14} is set here as 28·0 Å), respectively, and z_{11} and c_{11} are the values for the 11 Å phase that are given in Table 5.5.

In addition, it was necessary to move Ca2 to $\left(\frac{9}{10}, 0, 0\right)$; to change *z/c* for O5 to 0·0531; and to add two additional water molecules in order to complete the coordination of Ca2: these were placed in the model at $\left(\frac{9}{10}, \frac{1}{4}, 0{\cdot}0531\right)$ (Wat1) and at $\left(\frac{9}{10}, \frac{1}{2}, 0\right)$ (Wat4), which gives reasonable Ca–O bond distances and polyhedral volume; the values are given in Table 5.12 and can be seen to be reasonable when compared with the data for other calcium silicate hydrates that are discussed in §5.7. The occupancy factors for Wat1 and Wat4 are, respectively, 1 and (1 – $Occ._{Ca2}$). If $Occ._{Ca2} = 0{\cdot}5$, these changes result in the structural formula $Ca_5(Si_6O_{16}(OH)_2)(H_2O)_7$.

As with the model for the 11 Å phase, this structure for a 14 Å phase corresponds to $i = 1$ in Formula 5.2 but it is of course again possible to vary the contents of the interlayer in the same way as for the orthotobermorite structure: the result of this in terms of the values of *i* and the occupancy of sites Ca2, O5/OH5 and Wat2/3/4 is given in Table 5.13, as is the calculated density, which reduces as *i* is increased. As with the 11 Å structures, these models can again be used to represent the T∞ structural units in Richardson and Groves' (1992) model for C–S–H that is present in hardened cements. As noted, the variable *i* in Formulae 5.2 and 5.3 is equal to the degree of protonation, *w/n*, in Richardson and Groves' (1992) model (§7.5.1). The appropriate layer spacing will depend on the context and can be changed to an intermediate value, which is demonstrated in §6.6. $i < 1$ is not possible because in that case the Wat2 and Wat3 sites would both need to be occupied but they are too close for that to occur. A lower value could be accommodated by replacing those two molecules by a single water molecule at $\left(\frac{2}{5},\frac{3}{4},0\right)$ that would coordinate to the adjacent Ca ions with the long Ca–O distance of 2·84 Å; however, that would result in a Ca–O polyhedral volume that is much too big for six-fold coordination and so it is unlikely. There is also a problem for $i = 2$. In that case there are no Ca atoms in the interlayer region and so the layer spacing is most likely to shrink, the extent of the shrinkage being dependent on the number of water molecules that remain in the interlayer; removal of all the water molecules presumably resulting in essentially a totally protonated 9 Å tobermorite-like structure. The H_2O/Si is 1·333 for both $i = 1$ and $i = 1{\cdot}5$.

14 Å orthotobermorite that has bridging tetrahedra that are adjacent to one another (i.e. where the layers have *C* 2 *m* (*m*) symmetry, Figure 5.6) has not as yet been observed in nature, although Bonaccorsi *et al.* (2005) did note that the possible occurrence of domains with this layer symmetry in their crystal might have contributed to the relatively high value of the *R* index achievable during their refinement. Certainly, by analogy with the 11 Å orthotobermorites, it seems entirely possible that it could occur and it could in particular be relevant to models for the structure of C–S–H (see §6.6). A model structure is given in Table 5.14. The chains are of course not cross-linked as in the 11 Å structure because they are too far apart and are instead separated by a larger interlayer region of Ca ions and water molecules. This model is very similar to the single-chain structure that is given in Table 5.11 but with a shift of *b*/2 in the silicate chains. The displacement of the chains forces a change in space group to *B* 1 1 *m*, which results in a reduction in the multiplicity of the Ca2, Wat2 and Wat4 sites (i.e. the Wyckoff position changes from 4*b* to 2*a*) and thus necessitates the addition of atoms Ca2A, Wat2A and Wat4A at the positions of Ca2, Wat2 and Wat4 plus $\left(0,\frac{1}{2},0\right)$. The value of *y*/*b* for Ca1 is also increased by 0·5.

Table 5.11 Model structure for 14 Å orthotobermorite that has staggered silicate chains (i.e. where the layers have *C*2*m*(*b*) symmetry) (continued on next page)

Systematic name:	pentacalcium hexasilicate dihydroxide heptahydrate
Structural formula:	$Ca_5(Si_6O_{16}(OH)_2)(H_2O)_7$ i.e. $i = 1$, $m = 7$ in Formula 5.2
Chemical formula (sum):	$H_{16}\ Ca_5\ O_{25}\ Si_6$
a	6·748 Å
b	7·3 Å
c	28·8 Å
α	90°
β	90°
γ	122·75°
Z	2
Space group:	No. 9, *B* 1 1 *b*
Unique axis:	*c*
Cell choice:	2
Symmetry equivalent positions:	(0, 0, 0)+ (½ , 0, ½)+
4*a*:	(1) + *x*, + *y*, + *z* (2) + *x*, ½+*y*, −*z*

Table 5.11 Continued

Atom	Label	Position	Occ.	*x/a*	*y/b*	*z/c*
Ca	Ca1	*4a*	1	0·75	0·4286	0·2847
Ca	Ca2	*4a*	0·5	0·9	0	0
Ca	Ca3	*4a*	1	0·25	0·4286	0·2154
O	O1	*4a*	1	0·75	0·5	0·1285
O	O2	*4a*	1	0·75	0·173	0·1598
O	O3	*4a*	1	0·9833	0·5556	0·2079
O	O4	*4a*	1	0·515	0·315	0·2079
O	O5	*4a*	1	0·9	0·75	0·0531
O	O7	*4a*	1	0·75	0·85	0·1285
O	O8	*4a*	1	0·515	0·815	0·2079
O	O9	*4a*	1	0·9833	0·05	0·2079
O	OH6	*4a*	1	0·17	0·887	0·1285
O	Wat1	*4a*	1	0·9	0·25	0·0531
O	Wat2	*4a*	0·5	0·325	0·71	0
O	Wat3	*4a*	0·5	0·475	0·29	0
O	Wat4	*4a*	0·5	0·9	0·5	0
O	Wat6	*4a*	1	0·25	0·4286	0·1285
Si	Si1	*4a*	1	0·75	0·386	0·1773
Si	Si2	*4a*	1	0·9	0·75	0·109
Si	Si3	*4a*	1	0·75	0·96	0·1773

Table 5.12 Selected bond distances (Å) for the structure given in Table 5.11. The interlayer Ca atom is indicated by 'I'

Ca1	Ca2 (I)	Ca3		Si1	Si2	Si3
2·323	2·381	2·353		1·634	1·610	1·603
2·353	2·381	2·354		1·633	1·632	1·611
2·396	2·381	2·397		1·635	1·635	1·619
2·407	2·381	2·439		1·637	1·635	1·635
2·541	2·412	2·503				
2·579	2·412	2·576				
2·583		2·581				
2·455	2·391	2·458	Average for polyhedron	1·635	1·628	1·617
	2·437		Overall average		1·627	

Table 5.13 Alternative formulae for single-chain 14 Å orthotobermorite

i	Formula	Density: g cm^{-3}	Ca2 Wat2/3 Occ.	Wat4	O5 : OH5 Occ.
1	$Ca_5(Si_6O_{16}(OH)_2)(H_2O)_7$	2·185	0·5	0·5	1 : 0
1·5	$Ca_{4\cdot5}(Si_6O_{15}(OH)_3)(H_2O)_{6\cdot5}$	2·107	0·25	0·75	0·5 : 0·5

Table 5.14 Model structure for 14 Å orthotobermorite that has bridging tetrahedra that are adjacent to one another (i.e. where the layers have *C* 2 *m* (*m*) symmetry) (continued on next page)

Systematic name:	pentacalcium hexasilicate dihydroxide heptahydrate
Structural formula:	$Ca_5(Si_6O_{16}(OH)_2)(H_2O)_7$ i.e. $i = 1$, $m = 7$ in Formula 5.2
Chemical formula (sum):	H_{16} Ca_5 O_{25} Si_6
a	6·748 Å
b	7·3 Å
c	28·0 Å
α	90°
β	90°
γ	122·75°
Z	2
Space group:	No. 8, *B* 1 1 *m*
Unique axis :	*c*
Cell choice:	2
Symmetry equivalent positions:	(0, 0, 0)+ (½, 0, ½)+
4*b*:	(1) + *x*, + *y*, + *z* (2) + *x*, + *y*, –*z*
2*a*:	+*x*, + *y*, 0

Atom	Label	Position	Occ.	*x/a*	*y/b*	*z/c*
Ca	Ca1	4*b*	1	0·75	0·9286	0·2857
Ca	Ca2	2*a*	0·5	0·9	0	0
Ca	Ca2A	2*a*	0·5	0·9	0·5	0
Ca	Ca3	4*b*	1	0·25	0·4286	0·2144
O	O1	4*b*	1	0·75	0·5	0·125
O	O2	4*b*	1	0·75	0·173	0·1572
O	O3	4*b*	1	0·9833	0·5556	0·2067
O	O4	4*b*	1	0·515	0·315	0·2067
O	O5	4*b*	1	0·9	0·25	0·9524
O	O7	4*b*	1	0·75	0·85	0·125

Table 5.14 Continued

O	O8	4*b*	1	0·515	0·815	0·2067
O	O9	4*b*	1	0·9833	0·05	0·2067
O	OH6	4*b*	1	0·17	0·887	0·125
O	Wat1	4*b*	1	0·9	0·75	0·9524
O	Wat2	2*a*	1	0·4	0·75	0
O	Wat2A	2*a*	1	0·4	0·25	0
O	Wat4	2*a*	0·5	0·9	0·5	0
O	Wat4A	2*a*	0·5	0·9	0	0
O	Wat6	4*b*	1	0·25	0·4286	0·125
Si	Si1	4*b*	1	0·75	0·386	0·1752
Si	Si2	4*b*	1	0·9	0·75	0·105
Si	Si3	4*b*	1	0·75	0·96	0·1752

5.3.2.4 Model crystal structures for theoretical 14 Å clinotobermorites

14 Å clinotobermorite has not been observed in nature (regardless of the relative position of adjacent silicate chains), but by analogy with the orthotobermorites it seems possible that it could occur, and again it could also be relevant to models for the structure of C–S–H (see §6.6). Table 5.15 gives the model structure for hypothetical 14 Å clinotobermorite that has bridging tetrahedra that are adjacent to one another (Richardson, 2014); they are of course not cross-linked because they are too far apart and so are separated by an interlayer region of Ca ions and water molecules. It was derived from the structure of the 11 Å phase that is given in Table 5.7 and therefore indirectly from Merlino *et al.*'s (2000) structure (in their Table 4). As with the models for the 14 Å orthotobermorites, the structure of the central Ca–O slab was preserved by recalculation of the *z*/*c* coordinate of the atoms in the asymmetric unit so that the relative *z* positions of the atoms in the slab were retained when the *c* parameter was increased, which was achieved by again using the simple transformation that is given in Equation 5.4, except that the denominator within the brackets is 2 rather than 4 (i.e. it is 4 for the orthotobermorite structure but 2 for clinotobermorite). In addition, it was necessary to move the interlayer calcium atom (Ca3) and the value of *y*/*b* for Wat9 had to be increased by ¼. As with the 14 Å orthotobermorites, Ca3 is placed on the reflection plane at the centre of the interlayer region (which in this case is at $z/c = ¼$) at $\left(0, \frac{1}{8}, \frac{1}{4}\right)$. Since the multiplicity of this site is half that of Ca3 in the 11 Å structure (i.e. a Wyckoff position of 4*e* instead of 8*f*) an additional atom, Ca3A, is needed at $\left(0, \frac{5}{8}, \frac{1}{4}\right)$. Ca3/3A are coordinated to six oxygen atoms, with an average distance of 2·39 Å, which is close to the ideal value for six-fold coordinated Ca in calcium silicate hydrates (§5.7) (4 × 2·381 Å, two with O8 and two with Wat9; 2 × 2·406 Å, both with Wat12); the polyhedral volume (18·33 $Å^3$) is also consistent with the data for calcium silicate hydrates that are discussed in §5.7. Wat13 occupies the same site as Ca3 and so has an occupancy factor of $1 - Occ._{Ca3}$. If $Occ._{Ca3} = 0{\cdot}5$, these changes result in the structural formula $Ca_5(Si_6O_{16}(OH)_2)(H_2O)_7$, which corresponds to $i = 1$ in Formula 5.1; the hydroxyl ion is O10 (and so is OH10), which is the equivalent atom to O6 in the orthotobermorite.

A model structure for a theoretical 14 Å clinotobermorite that has staggered silicate chains is given in Table 5.16 (Richardson, 2014). It is very similar to the structure that is given in Table 5.15 but with a shift of $b/2$ in the silicate chains. As with the 11 Å phase, the displacement of the chains forces a reduction in symmetry to space group $P\,2_1/c$. The structural formula is $Ca_5(Si_6O_{16}(OH)_2)(H_2O)_7$, which corresponds to $i = 1$ in Formula 5.2 but it is of course possible to vary the contents of the interlayer: the result of this in terms of the values of i, the density and the occupancy of sites Ca3/3A and O8/8A (OH8/8A) is given in Table 5.17. Selected bond distances for the structure are given in Table 5.18. The average Ca–O distance for the interlayer Ca (i.e. Ca3) is 2·39 Å, which is close to the value that would be expected for Ca in six-fold coordination; indeed, the maximum, average and minimum Ca–O and Si–O distances are all consistent with the values that are calculated from known structures of related phases (§5.7).

As for the orthotobermorite, $i < 1$ is not possible because in that case adjacent Wat12 and Wat14 sites would have to be occupied, which are too close (Richardson, 2014). It would be possible to accommodate a lower value by replacing those molecules by a single water molecule at $\left(\frac{3}{4},\frac{1}{8}/\frac{5}{8},\frac{1}{4}\right)$ that would coordinate to adjacent Ca ions with the long Ca–O distance of 2·84 Å. However, it is unlikely because it would result in a Ca–O polyhedral volume of 20·98 $Å^3$, which is too big for six-fold coordination. There is again a problem for $i = 2$: since there are no Ca atoms in the interlayer region, the layer spacing is likely to shrink, with the extent of the shrinkage dependent on the number of water molecules that remain in the interlayer; removal of all the water molecules would presumably result in essentially a totally protonated 9 Å clinotobermorite-like structure. Two model structures that have no interlayer Ca atoms are given in §6.6.3.3, which have layer spacings of 14·15 Å and 13·30 Å; atomistic modelling studies can be expected to provide insight into the most likely position for interlayer water molecules. The H_2O/Si is $\frac{4}{3}$ for both $i = 1$ and $i = 1{\cdot}5$.

These 14 Å or 11 Å model structures can be used to represent the T∞ structural units in Richardson and Groves' (1992) model for C–S–H; the appropriate layer spacing will depend on the context and can of course be changed to an intermediate value, which is discussed in §6.6.

5.3.2.5 Spectroscopic investigations

Observations of synthetic 14 Å tobermorites by ^{29}Si MAS NMR (Cong and Kirkpatrick, 1996; Wieker *et al.*, 1982) show dominant Q^2 peaks with little or no detectable Q^1 or Q^3, showing that the anions are long, single chains. A study of the Crestmore material by the molybdate method (Wieker, 1968) reached a similar conclusion, while ^{29}Si NMR (Komarneni *et al.*, 1987; Maeshima *et al.*, 2003) showed mainly Q^2 with very minor Q^1, although Maeshima *et al.* (2003) were uncertain as to whether a small Q^1 peak was due to the 14 Å tobermorite or oyelite. The presence of Q^1 is compatible with the observation that the atomic cell contents are deficient in Si, but the amount does not appear sufficient to account for the deficiency indicated by the atomic cell contents.

From the absence of any detectable ^{29}Si NMR peak at –90·1 ppm, Komarneni *et al.* (1987) concluded that the Crestmore material did not contain Si–OH. In contrast, Wieker (1968) concluded from broad-line ^{1}H NMR data that such groups were present. For synthetic preparations, a ^{29}Si MAS NMR study (Wieker *et al.*, 1982) indicated that a little Si–OH was present. Cong and Kirkpatrick (1996) reached the same conclusion from ^{29}Si and ^{17}O NMR. ^{1}H–^{29}Si cross polarisation spectra suggested relatively strong motion of protons in the 10^4–10^5 Hz range. A ^{1}H MAS NMR spectrum showed a strong signal from molecular water, a weaker one from Ca–OH and a weak shoulder from Si–OH (Yu and Kirkpatrick, 1999). The hydroxyl OH6 in Bonaccorsi *et al.*'s (2005) structure for 14 Å tobermorite is bonded to both Ca1 and Si2 atoms, which would account for both of these groups (Si2–OH6 = 1·66 Å; Ca1–OH6 = 2·74 Å). The ^{27}Al NMR spectrum of the Crestmore material showed peaks attributable to both tetrahedral and octahedral Al, showing that NMR is more sensitive than chemical analysis for detecting this element (Komarneni *et al.*, 1987).

Table 5.15 Model structure for a theoretical 14 Å clinotobermorite that has bridging tetrahedra that are adjacent to one another (they are not cross-linked because they are separated by an interlayer region of Ca ions and water molecules) (Richardson, 2014; CIF datablock T∞_14ac) (continued on next page)

Systematic name:	pentacalcium hexasilicate dihydroxide heptahydrate
Structural formula:	$Ca_5(Si_6O_{16}(OH)_2)(H_2O)_7$ i.e. $i = 1$, $m = 7$ in Formula 5.2
Chemical formula (sum):	$H_{16}\ Ca_5\ O_{25}\ Si_6$
a	11·35 Å
b	7·3 Å
c	28·8 Å
α	90°
β	95·5°
γ	90°
Z	4
Space group:	No. 15, *C* 1 2/*c* 1
Unique axis,:	*b*
Cell choice:	1
Symmetry equivalent positions:	(0, 0, 0)+ (½, ½, 0)+
8*f*:	(1) + *x*, + *y*, + *z* (2) −*x*, + *y*, ½−*z* (3) −*x*, −*y*, −*z* (4) + *x*, −*y*, ½+*z*
4*e*:	0, + *y*, ¼ 0, −*y*, ¾

Atom	Label	Position	Occ.	*x/a*	*y/b*	*z/c*
Ca	Ca1	8*f*	1	0·8667	0·375	0·465
Ca	Ca2	8*f*	1	0·8667	0·875	0·465
Ca	Ca3A	4*e*	0·5	0	0·625	0·25
Ca	Ca3	4*e*	0·5	0	0·125	0·25
O	O1	8*f*	1	0·9167	0·9529	0·6243
O	O2	8*f*	1	0·9167	0·3077	0·6243
O	O3	8*f*	1	0·7692	0·375	0·5438
O	O4	8*f*	1	0·7692	0·875	0·5438
O	O5	8*f*	1	0	0·375	0·5438
O	O6	8*f*	1	0	0·875	0·5438
O	O7	8*f*	1	0·8919	0·625	0·5906
O	O8	8*f*	1	0	0·125	0·6969
O	Wat9	8*f*	1	0	0·625	0·6969
O	OH10	8*f*	1	0·13	0·125	0·6243
O	Wat11	8*f*	1	0·1579	0·625	0·6243

Table 5.15 Continued

Atom	Label	Position	Occ.	*x/a*	*y/b*	*z/c*
O	Wat12	8*f*	1	0·75	0·125	0·25
O	Wat13	4*e*	0·5	0	0·625	0·25
O	Wat13A	4*e*	0·5	0	0·125	0·25
Si	Si1	8*f*	1	0·8929	0·837	0·5747
Si	Si2	8*f*	1	0	0·125	0·641
Si	Si3	8*f*	1	0·8958	0·4167	0·5747

Table 5.16 Model structure for a theoretical 14 Å clinotobermorite that has staggered silicate chains (Richardson, 2014; CIF datablock T∞_14sc) (continued on next page)

Systematic name:	pentacalcium hexasilicate dihydroxide heptahydrate
Structural formula:	$Ca_5(Si_6O_{16}(OH)_2)(H_2O)_7$ i.e. $i = 1$, $m = 7$ in Formula 5.2
Chemical formula (sum):	H_{16} Ca_5 O_{25} Si_6
a	11·35 Å
b	7·3 Å
c	28·8 Å
α	90°
β	95·5°
γ	90°
Z	4
Space group:	No. 14, $P\ 1\ 2_1/c\ 1$
Unique axis,	*b*
Cell choice	1
Symmetry equivalent positions:	(0, 0, 0)+
4*e*:	(1) + *x*, + *y*, + *z* (2) –*x*, ½+*y*, ½–*z* (3) –*x*, –*y*, –*z* (4) + *x*, ½–*y*, ½+*z*

Atom	Label	Position	Occ.	*x/a*	*y/b*	*z/c*
Ca	Ca1	4*e*	1	0·8667	0·375	0·465
Ca	Ca2	4*e*	1	0·8667	0·875	0·465
Ca	Ca3	4*e*	0·5	0	0·625	0·25
Ca	Ca1A	4*e*	1	0·3667	0·875	0·465
Ca	Ca2A	4*e*	1	0·3667	0·375	0·465
Ca	Ca3A	4*e*	0·5	0·5	0·125	0·25
O	O1	4*e*	1	0·9167	0·9529	0·6243

Table 5.16 Continued

Atom	Label	Position	Occ.	*x/a*	*y/b*	*z/c*
O	O2	4*e*	1	0·9167	0·3077	0·6243
O	O3	4*e*	1	0·7692	0·375	0·5438
O	O4	4*e*	1	0·7692	0·875	0·5438
O	O5	4*e*	1	0	0·375	0·5438
O	O6	4*e*	1	0	0·875	0·5438
O	O7	4*e*	1	0·8919	0·625	0·5906
O	O8	4*e*	1	0	0·125	0·6969
O	Wat9	4*e*	1	0	0·625	0·6969
O	OH10	4*e*	1	0·13	0·125	0·6243
O	Wat11	4*e*	1	0·1579	0·625	0·6243
O	Wat12	4*e*	0·5	0·712	0·125	0·25
O	Wat13	4*e*	0·5	0	0·625	0·25
O	Wat14	4*e*	0·5	0·212	0·125	0·25
O	Wat9A	4*e*	1	0·5	0·125	0·6969
O	OH10A	4*e*	1	0·63	0·625	0·6243
O	Wat11A	4*e*	1	0·6579	0·125	0·6243
O	Wat12A	4*e*	0·5	0·712	0·625	0·25
O	Wat13A	4*e*	0·5	0·5	0·125	0·25
O	Wat14A	4*e*	0·5	0·212	0·625	0·25
O	O1A	4*e*	1	0·4167	0·4529	0·6243
O	O2A	4*e*	1	0·4167	0·8077	0·6243
O	O3A	4*e*	1	0·2692	0·875	0·5438
O	O4A	4*e*	1	0·2692	0·375	0·5438
O	O5A	4*e*	1	0·5	0·875	0·5438
O	O6A	4*e*	1	0·5	0·375	0·5438
O	O7A	4*e*	1	0·3919	0·125	0·5906
O	O8A	4*e*	1	0·5	0·625	0·6969
Si	Si1	4*e*	1	0·8929	0·837	0·5747
Si	Si2	4*e*	1	0	0·125	0·641
Si	Si3	4*e*	1	0·8958	0·4167	0·5747
Si	Si1A	4*e*	1	0·3929	0·337	0·5747
Si	Si2A	4*e*	1	0·5	0·625	0·641
Si	Si3A	4*e*	1	0·3958	0·9167	0·5747

Table 5.17 Alternative formulae for single-chain 14 Å clinotobermorite

i	Formula	Density: g cm^{-3}	Ca3/3A Wat12/12A Wat14/14A Occ.	Wat13/13A Occ.	O8/8A : OH8/8A Occ.
1	$Ca_5(Si_6O_{16}(OH)_2)(H_2O)_7$	2·195	0·5	0·5	1 : 0
1·5	$Ca_{4.5}(Si_6O_{15}(OH)_3)(H_2O)_{6.5}$	2·117	0·25	0·75	0·5 : 0·5

Table 5.18 Selected bond distances (Å) for the structure given in Table 5.16. The interlayer Ca atom is indicated by 'I'

Ca1/1A	Ca2/2A	Ca3/3A (I)		Si1/1A	Si2/2A	Si3/3A
2·387	2·387	2·381		1·599	1·595	1·576
2·387	2·387	2·381		1·612	1·610	1·590
2·400	2·400	2·381		1·614	1·618	1·632
2·400	2·400	2·381		1·660	1·678	1·644
2·560	2·576	2·406				
2·604	2·604	2·406				
2·618	2·618					
2·481	2·483	2·389	Average for polyhedron	1·621	1·625	1·611
	2·454		Overall average		1·619	

The IR absorption spectra of natural (Farmer *et al.*, 1966; Mitsuda and Taylor, 1978; Stephens and Bray, 1973) and synthetic (Yu *et al.*, 1999) 14 Å tobermorites are generally similar to those of single-chain 11 Å tobermorites, although that of the Crestmore material (Farmer *et al.*, 1966) shows distinct differences in the Si–O band at 800–1100 cm^{-1}. The OH stretching bands of the natural specimens comprise a sharp peak at ≈3400 cm^{-1} with a low-frequency tail, but that of a synthetic specimen (Yu *et al.*, 1999) was less sharp and more symmetrical. These IR data and the available Raman (Kirkpatrick *et al.*, 1997a) and Ca X-ray absorption (Kirkpatrick *et al.*, 1997b) spectra of synthetic specimens seem compatible with the results of the X-ray structure determination (Bonaccorsi *et al.*, 2005). The ^{17}O NMR spectrum (Cong and Kirkpatrick, 1996) of a synthetic specimen also showed an additional peak in the region assigned to O atoms attached to Ca and Si, which disappeared on dehydration to the 11 Å hydrate.

5.3.3 Thermal behaviour of tobermorites and structures of related products

5.3.3.1 Natural 14 Å tobermorites

Farmer *et al.* (1966) reported a dehydration study of a natural 14 Å tobermorite from Crestmore, California, using static mass loss data, XRD and mid IR spectra. A sharp loss of water occurred at 55 ± 5 °C, giving normal 11 Å tobermorite, with a water content of ≈5 molecules per subcell. A further gradual loss above 55 °C led to the formation at ≈200 °C of a 9·35 Å tobermorite, with a water content of ≈1 molecule per subcell. The XRD pattern of the 11 Å hydrate was substantially unchanged up to 210 °C, when it disappeared, showing that the 11 Å hydrate can lose much of its water without undergoing substantial change in the structure of the calcium silicate layers. Little further water loss occurred until ≈ 450 °C, above which there was a gradual loss of the remaining water and of a small content of CO_2, and a gradual increase in the basal spacing to 9·7 Å. These changes were substantially complete by ≈700 °C. An intermediate phase showing some similarities to wollastonite was formed at 730–770 °C; this changed to wollastonite at 850–900 °C. Both the 11 Å and the 9·3 Å tobermorites were shown by the molybdate method to contain single chains but, in a nearly anhydrous sample formed at 600 °C, the anions were found to be of a sheet type (Wieker, 1968).

The DTA (differential thermal analysis) curve showed endotherms at approximately 130 °C and 220 °C and a very weak exotherm at ≈ 800 °C (Mitsuda and Taylor, 1978). The endotherms are attributable to the losses of molecular water associated with the formation of the 11 Å and 9·3 Å hydrates. The TG curve (10 °C min^{-1}) (Mitsuda and Taylor, 1978) is similar to the static mass loss curve, with the temperatures somewhat raised. As with dynamic methods in general, the temperatures are affected by heating rates and other experimental conditions.

In the IR spectrum of the 9·3 Å tobermorite formed by heating at 520 °C, a sharp band at 3490 cm^{-1} remained, but that at ≈1670 cm^{-1} was absent, showing that hydroxyl groups remained, but that the molecular water had been lost (Farmer *et al.*, 1966). There were also marked changes in the band at 800–1200 cm^{-1}, indicating that significant changes had occurred in the silicate ions. The spectrum of the 9·7 Å material formed at 710 °C contained a weak peak from hydroxyl, and the silicate peaks showed further changes. The spectrum resembled one reported earlier for a synthetic specimen heated at 650 °C (Kalousek and Roy, 1957). These observations suggest renewed polymerisation of the silicate ions. The spectrum of the intermediate phase formed at 800 °C was near to that of wollastonite, and material heated at 940 °C gave a spectrum characteristic of that phase.

The available data on the thermal behaviour of the other known natural 14 Å tobermorites are essentially similar to those for the Crestmore specimen (McConnell, 1954; Mitsuda and Taylor, 1978; Mitsuda *et al.*, 1972). Probably the only significant difference lies in the DTA curves for those from Fuka (Japan) (Mitsuda *et al.*, 1972) and Ballycraigy (Northern Ireland) (McConnell, 1954), which show weak shoulders at ≈200 °C on the DTA endotherm at ≈220 °C. A synthetic specimen showed more distinct triple low-temperature endotherms (Yu and Kirkpatrick, 1999). The middle one was shown to arise from formation of a 12 Å tobermorite as an intermediate stage in that of the 11 Å material.

5.3.3.2 Structure of 9.3 Å tobermorite

Megaw and Kelsey (1956) suggested that the silicate chains on both sides of each layer in 9·3 Å orthotobermorite bedded down into the grooves between those of the adjacent layers. This situation would contrast with that existing in 11 Å tobermorites, in which the chains of adjacent layers butt against each other, with or without sharing of oxygen atoms. Early work indicated that the subcell of the 9·3 Å tobermorite formed from 11 Å orthotobermorite is *A*-centred, which would introduce the necessary shift in the **a**-direction (Taylor, 1959). A full determination of the crystal structure confirmed the mode of packing of adjacent layers, but showed the subcell to be only pseudo *A*-centred, its space group being really *P n m m* (No. 59, setting $\bar{\mathbf{c}}\mathbf{ba}$) (Bonaccorsi *et al.*, 2002; Bonaccorsi and Merlino, 2005). The crystal structure of the 9·3 Å tobermorite formed by heating a clinotobermorite has also been determined (Merlino *et al.*, 1999, 2000); the structure is illustrated in Figure 5.7. Its relation to the 9·3 Å orthotobermorite is similar to that existing between the 11 Å clino- and orthotobermorites, the two forms thus giving different XRD powder patterns.

^{29}Si NMR of a 9 Å tobermorite obtained by heating a synthetic 14 Å tobermorite at 300 °C showed only Q^2 sites, indicating that no significant formation of Si–O–Si cross-linkages had occurred (Wieker *et al.*, 1982). In contrast, in a 9 Å sample formed from an 11 Å tobermorite of apparently mixed type, Q^2 and Q^3 sites were predominant. It was suggested that the single and double chains present in the 11 Å starting material were wholly or partly retained in the 9 Å products, but this is not easily reconciled with the layer thickness of the 9·3 Å hydrate. The lack of Q^1 indicates an absence of vacant tetrahedral sites, which is in contrast to the 11 Å phase.

5.3.3.3 The conversion to wollastonite

A single crystal XRD study of the thermal decomposition of the natural 11 Å tobermorite from Ballycraigy showed that the formation of wollastonite was topotactic – that is, there was a definite relationship between the crystallographic orientations of the starting material and the product (Taylor, 1959). A mechanism was suggested, but this needs to be reassessed in view of the subsequent observation that sheet silicate ions are formed at 600 °C.

Figure 5.7(a) Structure of the MDO_2 polytype of 9 Å clinotobermorite projected along its **b** axis showing the positions of the atoms in the unit cell

(a)

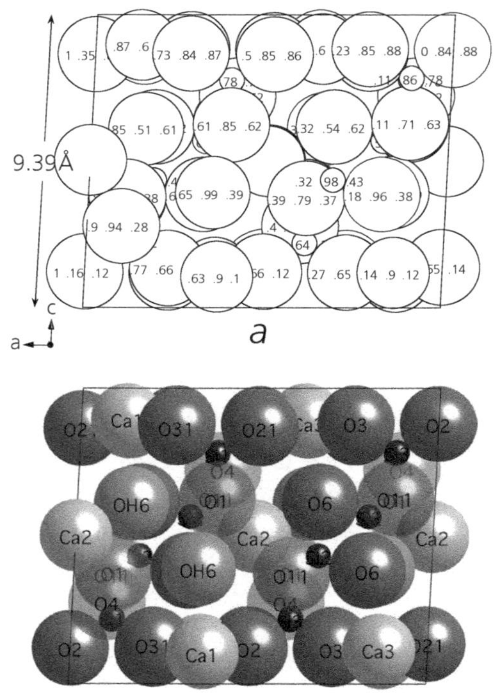

The structure is from Merlino *et al.* (1999, 2000). Coordinates (*x/a y/b z/c*) and atom labels are given in the top and bottom parts, respectively. Edge lengths are $c = 11{\cdot}156$ Å and $c \sin \alpha = 9{\cdot}388$ Å. Atoms O1, O11 and O4 are translucent on the lower figure so as to reveal the positions of the Si atoms.

Figure 5.7(b) and 5.7(c) Polyhedral representations of the structure of the MDO_2 polytype of 9 Å clinotobermorite viewed (b) along the **b** axis (unit edge lengths are $a = 11{\cdot}156$ Å and $c \sin \alpha = 9{\cdot}388$ Å) and along its **a** axis (unit edge lengths are $b = 7{\cdot}303$ Å and $c \sin \beta = 9{\cdot}554$ Å)

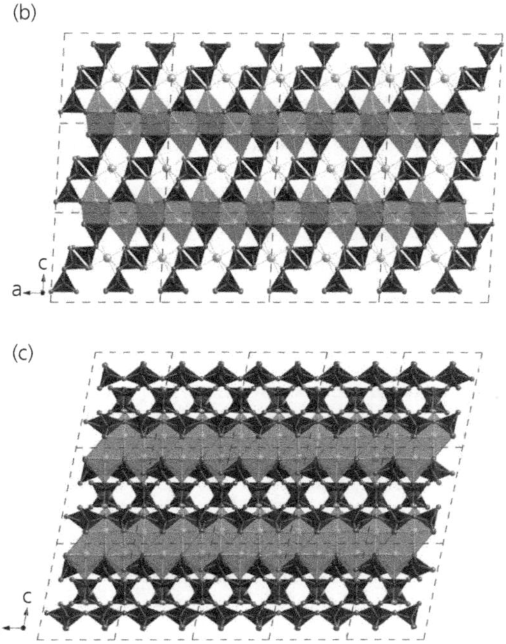

DTA curves of all tobermorites show one or more low-temperature endotherms at 220 °C or lower associated with loss of molecular water, and an exotherm at 800–850 °C which varies greatly in sharpness and intensity (Henmi and Kusachi, 1992; McConnell, 1954; Mitsuda and Taylor, 1978; Yu and Kirkpatrick, 1999). The exotherm is associated with crystallisation of wollastonite;, its intensity and breadth presumably reflect differences in the mechanism of this process. The occurrence of a strong, sharp exotherm seems to be associated with the presence of more than $\approx$1% of Al_2O_3 (Mitsuda and Taylor, 1978). For Al-free tobermorites, especially the Crestmore and Ballycraigy specimens, the IR and XRD evidence shows that the transition to wollastonite is gradual, and that the final stage of that process is topotactic. This may account for the absence of any strong exotherm. No detailed studies of the transition have been reported for tobermorites containing Al, but this element probably cannot be incorporated in the wollastonite structure, and a different and more abrupt mechanism might result. Tobermorite-like forms of C–S–H pass through an amorphous stage before wollastonite is formed, and in this case also, crystallisation occurs abruptly and strong exotherms are observed (see §6.5.1 and §6.7).

The difference between the Ca/Si ratios of tobermorites and wollastonite implies that, if the latter is stoichiometric, a phase higher in SiO_2 will also be formed. Yu and Kirkpatrick (1999) observed a weak Q^4 signal in the ^{29}Si NMR spectrum of a 14 Å tobermorite heated at 1000 °C, suggesting that amorphous silica was formed.

5.3.3.4 10 Å Tobermorites; oyelite

10 Å tobermorites are rare and compositionally complex. Natural and synthetic samples are known. In nature, they have been found as intergrowths, occasionally as an almost pure phase in Crestmore material (Murdoch, 1961; Heller and Taylor, 1956), and as a minor constituent intergrown with the 11 Å tobermorite from Ballycraigy (McConnell, 1954; Taylor, 1959). The Crestmore phase has been identified with oyelite, a tobermorite-like mineral of composition $CaO(B_2O_3)_{0{\cdot}1}(SiO_2)_{0{\cdot}8}(H_2O)_{1{\cdot}25}$ found at Fuka, Japan (Kusachi *et al.*, 1984) and, like it, contains a substantial proportion of B_2O_3. The Fuka specimen had a density of 2·620 g cm^{-3} and refractive indices $\alpha = 1{\cdot}602$, $\beta = 1{\cdot}606$, $\gamma = 1{\cdot}613$. Its XRD powder pattern has been indexed on an orthorhombic cell with $a = 11{\cdot}25$ Å, $b = 7{\cdot}25$ Å, $c = 20{\cdot}46$ Å, but most reflections can be referred to a primitive subcell with a, b and c all halved. This subcell has the approximate contents $Ca_5Si_4BO_{14{\cdot}5}{\cdot}6H_2O$.

Hara and Inoue (1980b) described the direct synthesis of a 10 Å tobermorite similar to the Crestmore mineral. They obtained it from a volcanic glass and CaO at 80 °C. The glass contained 74% of SiO_2 and 13% of Al_2O_3, with significant proportions of Na_2O, K_2O, CaO and Fe_2O_3. The product gave a tobermorite-type XRD pattern with a basal spacing of 10 Å. TEM microanalyses of a preparation obtained by a 30-day treatment indicated the empirical composition $Ca_{1{\cdot}15}SiAl_{0{\cdot}20}O_{3{\cdot}45}{\cdot}xH_2O$. TEM showed the crystals to be laths with (100) cleavage; 10 Å lattice planes were resolved in a high-resolution micrograph of a crystal lying on (100). The SED pattern in this projection showed a primitive sublattice with $a \approx 10$ Å, $b \approx 3{\cdot}7$ Å. In the true cell, b was doubled and reflections with k odd were continuous streaks parallel to **c***, indicating disorder. The XRD powder pattern could be indexed on a primitive orthorhombic subcell with $a = 5{\cdot}62$ Å, $b = 3{\cdot}66$ Å, $c = 10{\cdot}25$ Å. With prolonged reaction time, the 10 Å tobermorite was replaced by an anomalous 11·8 Å tobermorite. On being heated in air at 300 °C, the 10 Å tobermorite became amorphous.

Some 11 Å tobermorites of 'mixed' type yield 10 Å tobermorites on heating, possibly through formation of randomly interstratified structures (Mitsuda and Taylor, 1978). The 11 Å tobermorite from Wessels mine studied by Bonaccorsi and Merlino (2000) behaves in this way; the lattice parameters did not change at temperatures up to 400 °C, but a pseudo-orthorhombic phase with subcell parameters $a = 5{\cdot}48$ Å, $b = 3{\cdot}66$ Å, c 20·31 Å is formed at ≈ 500 °C.

Biagioni *et al.* (2012a) investigated the thermal behaviour of an anomalous 11 Å tobermorite from N'Chwaning II mine (Kuruman, Kalahari manganese field, Northern Cape Province, South Africa). They showed by in situ XRD that the 11 Å tobermorite transformed at ≈ 420 °C into a phase that has a layer spacing of 10·1 Å. Ex situ powder and single-crystal XRD studies showed that an intermediate, clinotobermorite-like phase formed during the transformation from the 11 Å to the 10 Å tobermorite phase. The diffraction pattern of the 10 Å phase had a pseudo-orthorhombic subcell with $a = 11{\cdot}2$ Å, $b = 3{\cdot}67$ Å, $c = 20{\cdot}1$ Å, the a parameter being doubled with respect to that of the other tobermorite-group minerals (Biagioni *et al.*, 2012b). They concluded that its structure consists of layers of six- and seven-fold coordinated Ca cations that are linked together by double silicate chains, which have a conformation that is similar to the chains that are present in xonotlite. Comparison of structural, XRD and microRaman data for oyelite and the 10 Å tobermorite indicated that oyelite is not the natural counterpart of the 10 Å tobermorite.

The 9·7 Å phase formed from some normal tobermorites at 700 °C probably differs from any of the 10 Å phases described above, as it forms at high temperatures and has a low water content and high degree of silicate ion polymerisation. It thus appears that basal spacings of ≈ 10 Å can arise from a number of different causes. 10 Å tobermorites are also clearly different from jennite, which has a similar layer thickness and is discussed in §5.4.

5.3.3.5 12 Å Tobermorites

Tobermorites with basal spacings of ≈12 Å have not been found as pure, natural minerals, apart from a related but clearly distinct mineral known as tacharanite, which is described in §5.5. However, some specimens from Crestmore (Heller and Taylor, 1956; Murdoch, 1961) and from Ballycraigy (McConnell, 1954; Taylor, 1959) yield weak peaks in this region with stronger spacings of 14 Å or 11 Å or both. These could represent 12 Å tobermorite intergrown with the other hydrates.

Synthetic samples have been described. Hara and Inoue (1979, 1980b), in a study of the thermal dehydration of synthetic 14 Å and 11 Å tobermorites prepared using CaO and silica glass at 80 °C, observed various kinds of anomalous or mixed thermal behaviour. X-ray microanalyses showed the materials to have Ca/Si ratios of 0·78–0·89. In a few cases, products obtained by heating 14 Å tobermorites at 90 °C or 200 °C gave single basal spacings of 12·3–12·9 Å. Further heating at temperatures up to 700 °C caused the basal spacings to decrease gradually to ≈11 Å. At 700–800 °C wollastonite was formed. They also obtained products containing 14 Å and 12·0–12·6 Å tobermorites by direct reaction using CaO and silica glass at 90–100 °C.

Komarneni *et al.* (1986b) synthesised a 12·6 Å tobermorite by reacting CaO and amorphous silica (Ca/Si = 1·0) in 4 M NaOH solutions at 80–98 °C. TEM and SEM showed that the product contained elongated platey crystals and crumpled foils a few nm thick. These were probably C–S–H. The XRD pattern of wet material showed a highly crystalline tobermorite-type phase, with a basal spacing of 12·6 Å. The latter shrank to ≈ 10·6 Å on drying in air and rehydrated to the 12·6 Å phase in NaOH solution. On heating at 100 °C, the basal spacing decreased irreversibly to 9·3 Å. ^{29}Si NMR showed only Q^2 sites, and at 300 °C the material became amorphous. The IR spectrum of the 10·6 Å phase was similar to those of 11 or 14 Å tobermorites, and the ^{29}Si NMR spectrum showed only a broad Q^2 peak; no Q^3 was detected. ^{27}Al NMR showed both tetrahedral and octahedral Al, introduced as an impurity in the starting materials. The SED pattern, which probably relates to the 10·6 Å dehydration product, appears to be consistent with a primitive tobermorite-type lattice or sublattice with $a \approx 5{\cdot}6$ Å, $b = 3{\cdot}7$ Å. Reflections with $(h + k)$ odd appear to be systematically weak.

Yu and Kirkpatrick (1999) described a 12 Å tobermorite formed by heating a synthetic 14 Å tobermorite for 1 hour at 110 °C. This reduction in layer spacing has been observed directly in the TEM, as discussed in §5.3.2.2. ^{29}Si NMR showed only Q^2 sites. This spectrum contrasts with that of the unheated material, which also showed a little Q^1. On further heating for 1 h periods, the basal spacing changed to 11·2 Å at 250 °C and to 9·8 Å at 350–600 °C. The DSC (differential scanning calorimetry) curve showed well-defined low-temperature endotherms at 80 °C and 190 °C and a shoulder at ≈250 °C, which were attributed to the successive changes to the 11·8 Å, 11·2 Å and 9·8 Å states. The Q^2 peak broadened progressively at higher temperatures, indicating increasing disorder. At 250 °C a little Q^3 appeared, indicating some cross-linking of the silicate chains. It was not clear whether this persisted at higher temperatures because of the broadening of the Q^2 peak. ^{1}H NMR spectra indicated a gradual loss of molecular water and, above 350 °C, loss of Si–OH and Ca–OH.

Together, these observations for 12 Å tobermorites show that they have marked variability in properties and probably also in composition. As for 10 Å tobermorites, the same spacing may arise in different ways.

5.3.3.6 Determinants and mechanisms of normal and anomalous thermal behaviour

It has been widely assumed that normal thermal behaviour is due to the absence of cross-layer Si–O–Si linkages and anomalous behaviour to their presence (El-Hemaly *et al.*, 1977; Taylor, 1962). However, Merlino *et al.* (1999, 2000) demonstrated the occurrence of double-chain structures in thermally normal 11 Å clino- and orthotobermorites. They attributed normal behaviour to high contents of interlayer calcium ions, which would have been left in an unsatisfactory coordination state if the cross-linked 11 Å structure were retained. As a result, decondensation occurred, giving a single-chain 9·3 Å tobermorite. They considered that if the interlayer cation content was smaller, the double-layer structure could withstand loss of the interlayer water molecules, and that this gave rise to anomalous thermal behaviour. This hypothesis is consistent with the observation that high Ca/Si ratios favour normal behaviour. The effects of alkali and of Al_2O_3 are less readily explained and require further consideration.

Cation exchange also appears to affect thermal behaviour. Hara *et al.* (1997) found that a normal 11 Å tobermorite made by autoclaving a lime–quartz slurry with Ca/Si = 0·8 for 12 h at 180 °C was rendered anomalous by immersing the sample in solutions of KOH or KCl at 20 °C. The quartz used was reported to contain small amounts of Al_2O_3 and alkali. Even brief immersion was effective, and NaOH or NaCl solutions had a similar, although weaker, effect. If the anomalous tobermorites thus obtained were immersed in water, they showed partial reversion to normal behaviour. Hara *et al.* (1997) attributed the anomalous behaviour to the presence of intercalated alkali cations. However, anomalous behaviour is not always associated with the presence of alkali cations since anomalous tobermorites can be prepared from alkali-free materials (El-Hemaly *et al.*, 1977).

Based on the crystal structure determinations, Merlino *et al.* (2000) considered the hypothesis that normal thermal behaviour could arise from a single-chain structure to be untenable. However, chemical and NMR data for normal 11 Å tobermorites formed by heating the 14 Å form, or directly synthesised at 80 °C, show them to contain single chains. It is unfortunate that adequately detailed crystal structure and NMR determinations have not yet been made on the same specimens. The results available at present suggest that the factors governing normal and anomalous behaviour are complex, and that no one explanation accounts for all the experimental data. There appear to be two distinct determinants of thermal behaviour.

- 11 Å tobermorites formed by heating 14 Å tobermorites, or by direct synthesis at sufficiently low temperatures, are thermally normal because they contain single chains. Tobermorite-type C–S–H preparations, which contain single chains of varying lengths, are thermally normal; single-chain tobermorites may represent an intermediate stage in the formation of double-chain tobermorite from C–S–H.
- In tobermorites formed at higher temperatures, at least at longer times, most or all the chains are double. The great majority of natural specimens appear to be in this category. These samples can be either normal or anomalous. The determinant of thermal behaviour in these cases is, in general, that proposed by Merlino *et al.* (1999): low contents of interlayer Ca^{2+} produce anomalous thermal behaviour, while high contents produce normal behaviour. If the interlayer cation content is sufficiently high, however, and especially if it includes large amounts of K^+, shrinkage to the 9·3 Å hydrate may be impossible. This results in the reversibly anomalous behaviour observed by Hara *et al.* (1997).

Tobermorites in the first category probably all have the markedly platey morphology, with (001) as the only significant cleavage. For those in the second category, synthetic preparations tend also to have this morphology, but natural specimens seem usually to form the more elongated crystals with both (100) and (001) cleavages. The morphology is thus not a reliable property for distinguishing either normal from anomalous character or single- from double-chain structures.

5.3.3.7 Some unresolved problems

An outstanding problem is how widely single-chain structures occur in natural and directly synthesised 11 Å tobermorites. The evidence for single-chain structures is strong for C–S–H preparations, 9·3 Å tobermorites, 14 Å tobermorites and 11 Å tobermorites obtained from the latter by heating, but there seems to be only one clear case of positive evidence for such a structure in a directly synthesised 11 Å tobermorite, which was a product synthesised at 80 °C (Komarneni *et al.*, 1986b). It would also be instructive to know whether highly crystalline natural 11 Å tobermorites associated with the 14 Å form, such as that from Ballycraigy (McConnell, 1954), contain single or double chains.

The primitive, orthorhombic subcells reported for oyelite (Kusachi *et al.*, 1984) and possibly 12 Å (Komarneni *et al.*, 1986) tobermorite are as yet unexplained. The 100 reflections (subcell indexing) shown by some 11 Å and 14 Å tobermorites (Viehland *et al.*, 1997; Taylor, unpublished data) and by quasi-crystalline plombièrite (McConnell, 1971) also suggest tendencies to form a primitive sublattice. This sublattice appears to require systematic removal of some silicate tetrahedra from one side of each Ca–O layer. Likely possibilities in this respect are entire chains or bridging tetrahedra; model structures that have the latter are given in Chapter 6. The composition of oyelite and analogies with C–S–H suggest absence of bridging tetrahedra, but some of the data for the Crestmore material suggest absence of longer segments or whole chains. McConnell (1971) explained the presence of a 100 reflection in the diffraction pattern of plombièrite by suggesting that some of the silicate chains on one side of each Ca–O layer were missing. This hypothesis is discussed later.

Other unresolved problems include the observation that a specimen of 11 Å tobermorite could be rehydrated (McConnell, 1954); the differences between the Si–O stretching bands in the IR spectra of the Crestmore 14 Å and 11 Å tobermorites (Farmer *et al.*, 1966); the variable nature of 12 Å tobermorites (Hara and Inoue, 1980b; Komarneni *et al.*, 1986b; Yu and Kirkpatrick, 1999); the differing ^{17}O NMR spectra of synthetic 11 Å and 14 Å tobermorites (Cong and Kirkpatrick, 1996); the differing SED patterns reported for synthetic 14 Å tobermorites (Hara and Inoue, 1980b; Viehland *et al.*, 1997) and the varying thermal behaviour of 14 Å tobermorites (Farmer *et al.*, 1966; McConnell, 1954; Mitsuda and Taylor, 1978; Mitsuda *et al.*, 1972; Yu and Kirkpatrick, 1999). Some of the variations in 14 Å tobermorites could be explained by the presence of differing proportions of 12 Å tobermorite, or differing tendencies to form this hydrate on dehydration, but the reason for such differences is not clear.

5.3.4 Structural-chemical formulae for tobermorite and C–A–S–H phases

The compositional and spectroscopic data that are discussed in previous sections mean that the structural-chemical Formulae 5.1–5.3 need to be modified to reflect the observed substitution of Al^{3+} ions for Si^{4+} (and any associated mechanism of charge compensation), as well as the possible presence of vacant tetrahedral sites, which must occur if Q^1 sites are present in single-phase samples. A generalised structural-chemical formula for double-chain tobermorites or, indeed, for cross-linked C–(A–)S–H phases (which are discussed in Chapter 6) is (Richardson, 2014)

$$Ca_4\left[Si_{1-f-v}Al_f\square_v O_{\frac{17}{6}-2v}\right]_6 H_{2i}Ca_{1-i}\left(Ca, Na_2, K_2\right)_{3f}\cdot mH_2O \qquad (5.5)$$

where the contents of the square brackets represent the aluminosilicate part of the structure;□ represents a vacant tetrahedral site and Al can substitute for Si; v is the fraction of tetrahedral sites that are vacant; f is the fraction of sites that are occupied by Al. There are four main-layer Ca atoms for every six tetrahedral sites.

The double silicate chains that are present in 11 Å tobermorite – that is, '*dreierdoppelketten*' in Liebau's (1985) classification – are illustrated schematically in Figure 5.8(a); it should be noted that the chains are infinite and so the chain-end triangles on the diagram represent Q^2 Si, not Q^1. The effect of introducing Al or vacant tetrahedral sites is illustrated in Figures 5.8(b) to 5.8(f). Figure 5.8(b) shows the same chains as in (a) but with 1 in 12 tetrahedral sites occupied by Al $\left(f = \frac{1}{12}\right)$; Al–O tetrahedra are represented by shaded triangles. It is assumed that Al can substitute for Si only at bridging sites and that bridging sites that are adjacent cannot both be occupied by Al. The assumption that Al cannot substitute for Si at paired sites is well supported by a number of experimental studies on single-chain structures (Andersen *et al.*, 2003, 2004; Richardson *et al.*, 1993; Sun *et al.*, 2006). Figure 5.8(c) shows the same chains as in (b) but with 1 in 48 tetrahedral sites vacant $\left(v = \frac{1}{48}\right)$. It is assumed that only bridging tetrahedral sites can be vacant. If the fraction of vacant sites that are adjacent to a bridging site that is occupied by Si is p, and the fraction of vacant sites that are adjacent to a bridging site that is occupied by Al is q, then since in the case of Figure 5.8(c) the sole vacant site is adjacent to a site that is occupied by Si, $p = 1$ and $q = 0$. Figure 5.8(d) shows the same chains as in (c) but in this case the vacant site is adjacent to a site that is occupied by Al and so $p = 0$ and $q = 1$. Vacant sites could of course also occur that are opposite other vacant sites, and the fraction of such sites would equal $1 - p - q$. This is illustrated in Figure 5.8(e), which shows the same chains as in (b) but with 1 in 24 tetrahedral sites vacant

Figure 5.8(a) Schematic diagram that illustrates the '*dreierdoppelketten*' silicate chains that are present in 11 Å tobermorite

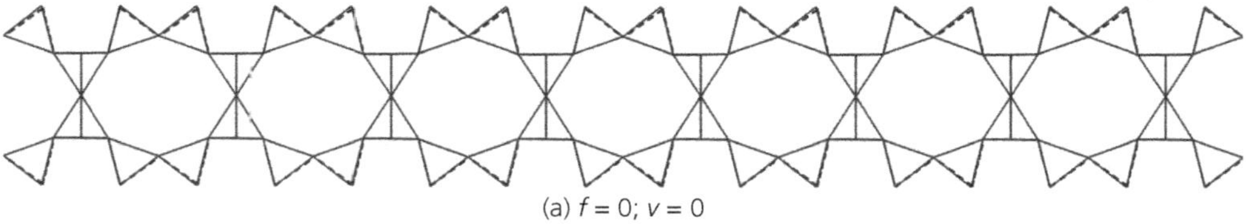

(a) $f = 0$; $v = 0$

Note that the chains are infinite and so the chain-end triangles on the diagram represent Q^2 Si, not Q^1.

Figure 5.8(b) As Figure 5.8(a) but with 1 in 12 tetrahedral sites occupied by Al $\left(f = \frac{1}{12}\right)$

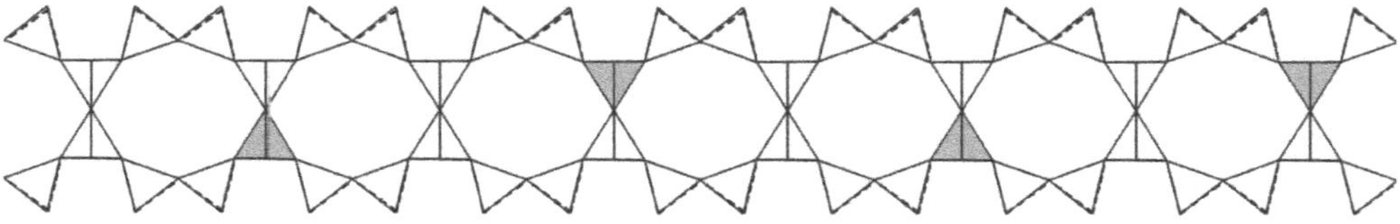

(b) $f = \frac{1}{12}$; $v = 0$

It is assumed that Al can only substitute for Si at bridging sites. Al–O tetrahedra are represented by shaded triangles.

$\left(\text{that is } v = \frac{1}{24}\right)$ and, since the two vacant sites are adjacent to one another, $p = 0$ and $q = 0$. Figure 5.8(f) shows the same chains as in (e) but in this case one vacant site is adjacent to a site that is occupied by Si and the other to a site that is occupied by Al and so $p = ½$ and $q = ½$. If a situation existed where vacant sites occurred opposite other vacant sites as well as opposite occupied sites, then obviously $(p + q) < 1$.

Figure 5.8(c) As Figure 5.8(b) but with 1 in 48 tetrahedral sites vacant $\left(v = \frac{1}{48}\right)$

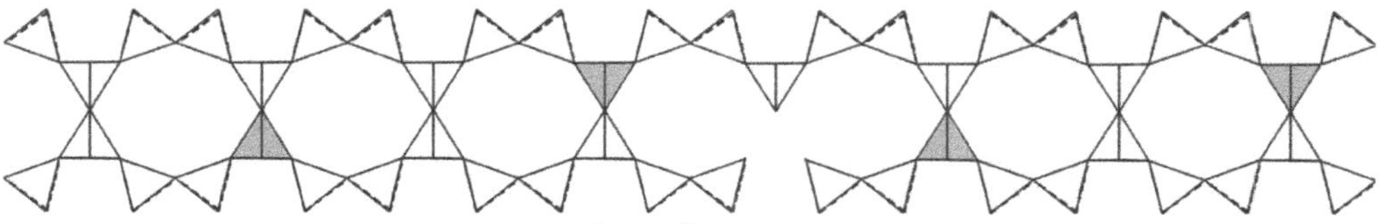

(c) $f = \frac{1}{12}$; $v = \frac{1}{48}$; $p = 1$; $q = 0$

It is assumed that only bridging sites can be vacant; in this case the vacant site is adjacent to a site that is occupied by Si ($p = 1$; $q = 0$).

Figure 5.8(d) As Figure 5.8(c) but in this case the vacant site is adjacent to a site that is occupied by Al ($p = 0$; $q = 1$)

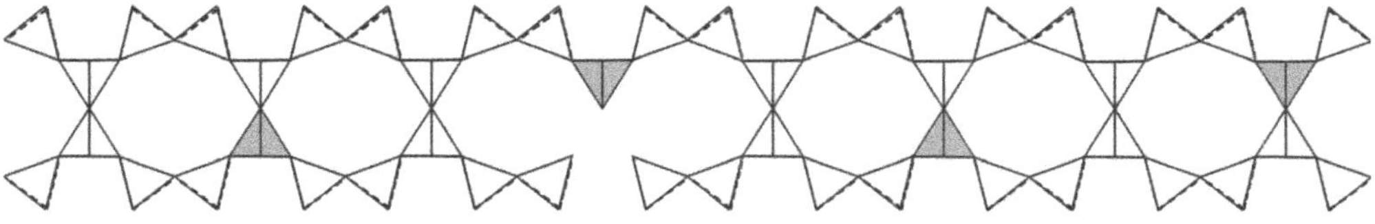

(d) $f = \frac{1}{12}$; $v = \frac{1}{48}$; $p = 0$; $q = 1$

Figure 5.8(e) As Figure 5.8(b) but with 1 in 24 tetrahedral sites vacant $\left(v = \frac{1}{24}\right)$; in this case the two vacant sites are adjacent to one another ($p = 0$; $q = 0$)

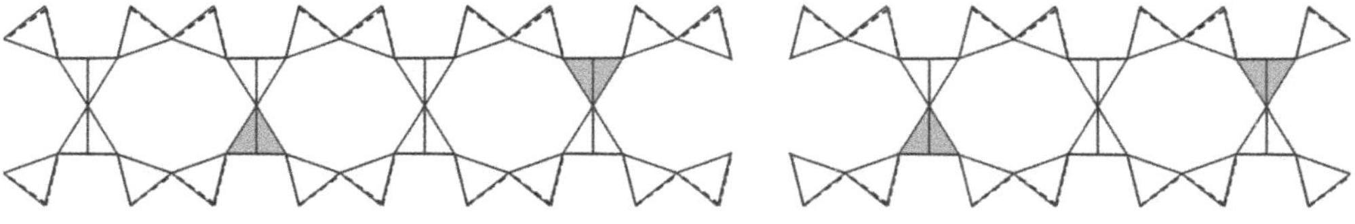

(e) $f = \frac{1}{12}$; $v = \frac{1}{24}$; $p = 0$; $q = 0$

Figure 5.8(f) As Figure 5.8(e) but in this case one vacant site is adjacent to a site that is occupied by Si and the other to a site that is occupied by Al $\left(p=\frac{1}{2}; q=\frac{1}{2}\right)$

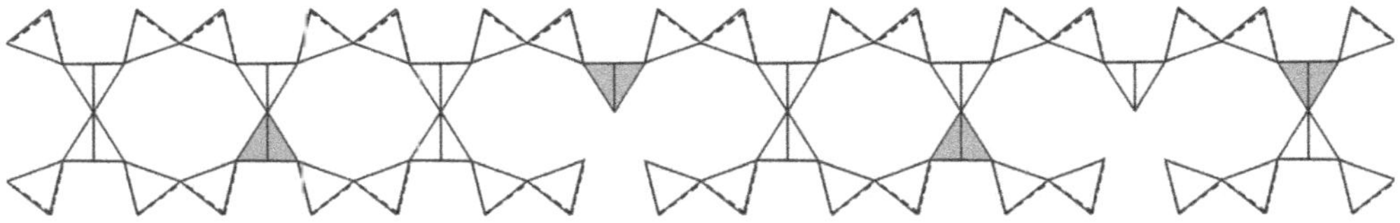

(f) $f=\frac{1}{12}$; $v=\frac{1}{24}$; $p=\frac{1}{2}$; $q=\frac{1}{2}$

As an example, Yamazaki and Toraya (2001) studied the positions of zeolitic calcium atoms and water molecules in a hydrothermally formed aluminium-substituted 11 Å tobermorite using synchrotron radiation powder diffraction, thermal analysis and analytical transmission electron microscopy. The preparation had double chains with no vacant tetrahedral sites and a composition of $Ca_{4{\cdot}9}Si_{5{\cdot}5}Al_{0{\cdot}5}O_{16{\cdot}3}(OH)_{0{\cdot}7}{\cdot}5H_2O$. f is therefore equal to $(0{\cdot}5/6)=\frac{1}{12}$, $i = 0{\cdot}35$ and $m = 5$, and so the formula can be rewritten in terms of Formula 5.5 as $Ca_4\left[Si_{0{\cdot}917}Al_{0{\cdot}083}O_{2{\cdot}833}\right]_6 H_{0{\cdot}7}Ca_{0{\cdot}65}(Ca)_{0{\cdot}25}\cdot 5H_2O$. Yamazaki and Toraya's (2001) structure therefore has alumino-silicate chains that are represented by Figure 5.8(b).

The formula for single-chain tobermorite or C–(A–)S–H phases that is equivalent to Formula 5.5 for double chains is (Richardson, 2014)

$$Ca_4\left[Si_{1-f-v}Al_f\square_v O_{3-2v}\right]_6 H_{2i}Ca_{2-i}(Ca,Na_2,K_2)_{3f}\cdot mH_2O \qquad (5.6)$$

This formula is applicable to the C–(A–)S–H(I) phases that are discussed in Chapter 6. Figure 5.9(a) illustrates the single silicate chains that are present in some tobermorite phases – that is, '*dreiereinfachketten*' in Liebau's (1985) classification. It should again be noted that the chains are infinite and so the chain-end triangles on the diagram represent Q^2 Si, not Q^1. Figure 5.9(b) shows the same chains as in (a) but with 1 in 12 tetrahedral sites occupied by Al $\left(f=\frac{1}{12}\right)$. It is again assumed that Al can only substitute for Si at bridging sites; Al–O tetrahedra are represented by shaded triangles. Figure 5.9(c) shows the same chains as in (b) but with 1 in 48 tetrahedral sites vacant $\left(v=\frac{1}{48}\right)$; it is assumed that only bridging sites can be vacant. Formula 5.6 is very similar to Equation 4 in Taylor (1993) for purely tobermorite-like structures (that is $q = 1$ in Taylor's formulation; his other variables are related to those in Formula 5.6 by $A = 6f$, $\Delta = 6v$ and $p = i$; it is important to note that the variables p and q in Taylor's formulation do not have the same meaning as used in this chapter).

Formulae 5.5 and 5.6 can be combined to represent a mixture of single- and double-chain phases (Richardson, 2014)

$$Ca_4\left[Si_{1-f-v}Al_f\square_v O_{\frac{18-d}{6}-2v}\right]_6 H_{2i}Ca_{2-i-d}(Ca,Na_2,K_2)_{3f}\cdot mH_2O \qquad (5.7)$$

Figure 5.9(a) Schematic diagram illustrating the '*dreiereinfachketten*' silicate chains that are present in single-chain 11 Å tobermorites

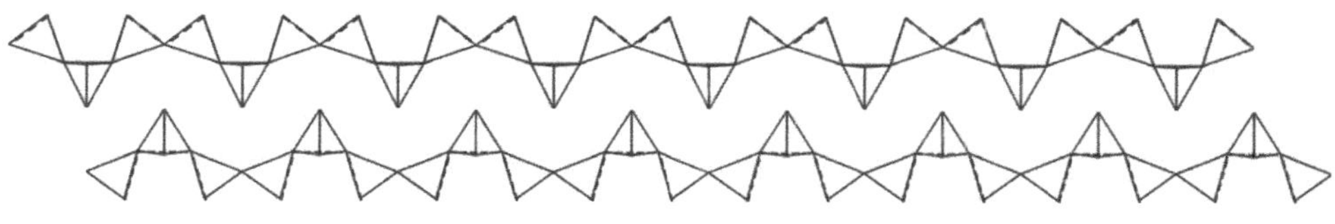

(a) $f = 0$; $v = 0$

Note that the chains are infinite and so the chain-end triangles on the diagram represent Q^2 Si not Q^1.

Figure 5.9(b) As Figure 5.9(a) but with 1 in 12 tetrahedral sites occupied by Al $\left(f = \frac{1}{12}\right)$

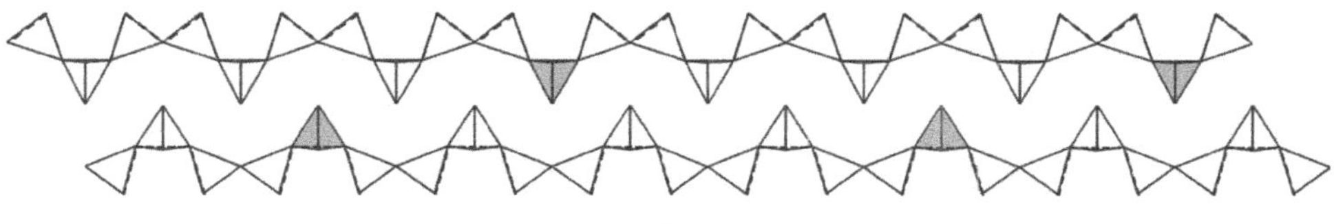

(b) $f = \frac{1}{12}$; $v = 0$

It is assumed that Al can only substitute for Si at bridging sites. Al–O tetrahedra are represented by shaded triangles.

Figure 5.9(c) As Figure 5.9(b) but with 1 in 48 tetrahedral sites vacant $\left(v = \frac{1}{48}\right)$

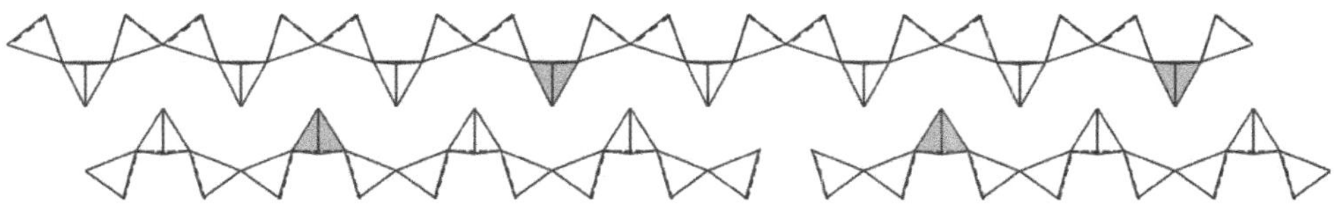

(c) $f = \frac{1}{12}$; $v = \frac{1}{48}$

It is assumed that only bridging sites can be vacant.

where d represents the fraction of double-chain structure – that is, $0 \leq d \leq 1$, where if $d = 0$ the structure is entirely single chain and if $d = 1$ it is entirely double chain.

These equations and those given in the next section are applicable to C–(A–)S–H phases in general and so are used in Chapter 6.

5.3.5 Useful equations for the interpretation of single-pulse ^{29}Si MAS NMR data

In principle, values for the variables in Formulae 5.5 to 5.7 can be determined by experiment, although in practice problems can arise where samples contain more than one phase. ^{29}Si MAS NMR can be used to provide quantitative information on the fractions of silicon that are present in different tetrahedral environments because an increase in polymerisation results in characteristic up-field ^{29}Si chemical shifts. In alumino-silicates these characteristic chemical shifts are further influenced by the replacement of Si by Al. There are 15 possible $Q^n(m\text{Al})$ structural units where the silicate tetrahedron, Q, is connected by way of oxygen bridges to mAl and $n-m$ other Si atoms, with $n = 0$ to 4 and $m = 0$ to n (Engelhardt and Michel, 1987). The Q tetrahedra that are present in single-chain tobermorites or C–A–S–H phases are illustrated in Figure 5.10. Double-chain tobermorites also have Q^3(0Al) and Q^3(1Al) sites.

Expressions are given in Table 5.19 for the fractions of the tetrahedral sites in tobermorite – or indeed in C–(A–)S–H phases in general – that are occupied by Si or Al or that are vacant. As noted above, it is assumed that Al only substitutes for Si at bridging sites, that Al cannot occur at adjacent bridging sites and that nonbridging sites cannot be vacant.

Figure 5.10 Schematic diagram illustrating the nature of the linear aluminosilicate chains that are present in C–S–H

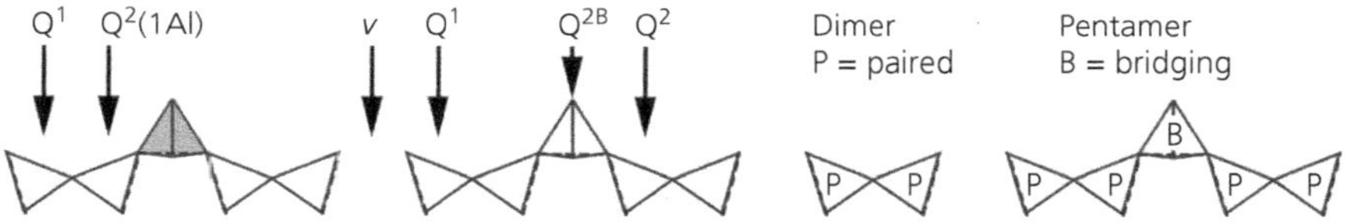

The unfilled triangles represent Si–O tetrahedra and the shaded triangle represents an Al–O tetrahedron. $Q^2 = Q^{2P}$; Q^2 is sometimes written as Q^2(0Al) to indicate the lack of Al at one of the two adjacent tetrahedral sites. v = vacant tetrahedral site.

Table 5.19 Fractions of the tetrahedral sites that are present in tobermorites and C–(A–)S–H phases that are occupied by Si or Al or that are vacant (from Richardson, 2014) (continued on next page)

Sites occupied by:	Value* for double-chain tobermorite	Value for single-chain tobermorite and C–(A–)S–H
Al	f	f
vacancy	v	v
Q^1 Si	$2v$	$2v$
Q^2(0Al) Si	$\frac{2}{3}-2f-v(2-p)$	$1-3(f+v)$
Q^2(1Al) Si	$2f$	$2f$
Q^3(0Al) Si	$\frac{1}{3}-2f-v(1+p-q)$	-

Table 5.19 Continued

Sites occupied by:	Value* for double-chain tobermorite	Value for single-chain tobermorite and C–(A–)S–H
Q^3(1Al) Si	$f-qv$	-
Total	1	1
Paired Q^2(0Al) Si	$\frac{2}{3}-2f-2v$	$\frac{2}{3}-2f-2v$
Bridging Q^2(0Al) Si	pv	$\frac{1}{3}-f-v$
Paired Q^2(0Al) Si adjacent to Q^3	$\frac{2}{3}-2f-2v(1+p)$	-
Paired Q^2(0Al) Si adjacent to Q^{2B}	$2pv$	$\frac{2}{3}-2f-2v$
Bridging Q^2(0Al) Si	pv	$\frac{1}{3}-f-v$

It is assumed that Al only substitutes for Si at bridging sites and that nonbridging sites cannot be vacant. The theoretical maximum values of f, v and $(f+v)$ are $\frac{1}{6}, \frac{1}{3}$ (that is dimer) and $\frac{1}{3}$, respectively, for tobermorite that has condensed double chains and $\frac{1}{3}, \frac{1}{3}$ (that is dimer) and $\frac{1}{3}$ for single-chain tobermorite. The actual minimum and maximum values of f are dependent on the fraction (v) and distribution of vacant sites, as illustrated in Figure 5.11.
*p = fraction of vacant sites that are adjacent to a bridging site that is occupied by Si; q = fraction of vacant sites that are adjacent to a bridging site that is occupied by Al; it follows that the fraction of vacant sites that are opposite other vacant sites = $1-p-q$.

The fraction of vacant tetrahedral sites that are present in double- or single-chain tobermorite or C–(A–)S–H can be calculated from ^{29}Si NMR data using the following expression (Richardson, 2014)

$$v=\frac{\frac{1}{2}Q^1}{\frac{3}{2}Q^1+Q^2(0Al)+\frac{3}{2}Q^2(1Al)+Q^3(0Al)+Q^3(1Al)} \tag{5.8}$$

In this equation – and those that follow – $Q^n(m\text{Al})$ represents the relative intensity of a peak determined from the deconvolution of a ^{29}Si single-pulse MAS NMR spectrum (assuming of course that the spectrum was collected using quantitative conditions). $Q^3(m\text{Al})$ are obviously zero when the structure is single chain.

The mean aluminosilicate chain length (MCL) for all tobermorites – both single- and double-chain – can be calculated from NMR data using the following equation (Richardson, 2014)

$$\text{MCL}=\frac{2}{\left(\frac{Q^1}{Q^1+Q^2(0Al)+\frac{3}{2}Q^2(1Al)+Q^3(0Al)+Q^3(1Al)}\right)} \tag{5.9}$$

For double-chain structure, MCL corresponds to the average number of linked tetrahedra in sections of the chains along **b** that are separated by vacant sites, and thus includes paired Q^1 and Q^2 tetrahedra and bridging Q^2 and Q^3 tetrahedra, the Q^3 being cross-linked with other bridging tetrahedra (Richardson, 2014). Again, the values for $Q^3(m\text{Al})$ are obviously zero for single-chain tobermorite and in that case the expression reduces to that given by Richardson *et al.* (1994) for Al-substituted C–S–H. The version given by Coleman (2006) for double-chain tobermorite is not quite correct because it requires that all vacant sites must be either adjacent to sites which are occupied by Si (i.e. bridging $Q^2(0\text{Al})$) or that are also vacant; tetrahedra which are occupied by Al that are adjacent to vacant sites are not counted in Coleman's (2006) version. The version given by Reinik *et al.* (2007) (their Equation (1)) is also not quite correct, giving values that are slightly too long ($Q^3(1\text{Al})$ is multiplied by 1·5 in their version, which is incorrect because all of the Al–O tetrahedra are already accounted for by the multiplication by 1·5 of the $Q^2(1\text{Al})$).

For both single- and double-chain tobermorite, the MCL and the fraction of vacant tetrahedral sites, v, are related simply by (Richardson, 2014)

$$\text{MCL} = \frac{\text{Fraction of tetrahedral sites that are occupied}}{\text{Fraction of tetrahedral sites that are vacant}} = \frac{1-v}{v} \tag{5.10}$$

Since one third of all tetrahedral sites are bridging sites, if all were vacant, $v = \frac{1}{3}$ and $\text{MCL} = \frac{1-\frac{1}{3}}{\frac{1}{3}} = 2$ (i.e. the chains would be entirely dimeric); if one half were vacant, then $v = \frac{1}{2} \times \frac{1}{3} = \frac{1}{6}$ and $\text{MCL} = \frac{1-\frac{1}{6}}{\frac{1}{6}} = 5$ (i.e. the chains would on average be entirely pentameric) and so on. While short chains such as these are not relevant to the crystalline tobermorite phases, they are relevant to the C–S–H(I) phases that are discussed in Chapter 6 and also to tobermorite-based models for the C–S–H that forms in hardened cements (Chapters 6 and 7).

Equation 5.10 can be rearranged to show how v can be calculated easily using quantitative data from single-pulse ^{29}Si MAS NMR

$$v = \frac{1}{\text{MCL}+1} \tag{5.11}$$

The mean length of just the silicate part of the chains is

$$\text{MCL}_{\text{Si}} = \frac{\text{Fraction of tetrahedral sites that are occupied by Si}}{\text{Fraction of tetrahedral sites that are vacant or occupied by Al}} = \frac{1-(v+f)}{v+f} \tag{5.12}$$

which for single-chain tobermorite can be calculated from NMR data using

$$\text{MCL}_{\text{Si}} = \frac{2}{\left(\frac{Q^1+Q^2(1\text{Al})}{Q^1+Q^2(0\text{Al})+Q^2(1\text{Al})}\right)} \tag{5.13}$$

This is the same as the equation given by Andersen *et al.* (2004) for Al-substituted C–S–H (their Equation (2)).

The Al/Si ratio for double-chain tobermorite can be calculated from ^{29}Si NMR data using the following expression (Richardson, 2014)

$$\mathrm{Al/Si} = \frac{\frac{1}{2}\mathrm{Q}^2(1\mathrm{Al})}{\mathrm{Q}^1+\mathrm{Q}^2(0\mathrm{Al})+\mathrm{Q}^2(1\mathrm{Al})+\mathrm{Q}^3(0\mathrm{Al})+\mathrm{Q}^3(1\mathrm{Al})} \tag{5.14}$$

The values for $\mathrm{Q}^3(m\mathrm{Al})$ are obviously zero for single-chain tobermorite and so in that case the expression reduces to that given by Richardson *et al.* (1994) for Al-substituted C–S–H. The $\mathrm{Q}^3(1\mathrm{Al})$ is not included in the numerator because all the Al is accounted for by the $\mathrm{Q}^2(1\mathrm{Al})$.

For both single- and double-chain tobermorite, the Al/Si ratio is related to the fractions of tetrahedral sites that are occupied by Al and that are vacant by

$$\mathrm{Al/Si} = \frac{f}{1-f-v} \tag{5.15}$$

The fraction of tetrahedral sites that are occupied by Al can be calculated from ^{29}Si NMR data using (Richardson, 2014)

$$f = \frac{\frac{1}{2}\mathrm{Q}^2(1\mathrm{Al})}{\frac{3}{2}\mathrm{Q}^1+\mathrm{Q}^2(0\mathrm{Al})+\frac{3}{2}\mathrm{Q}^2(1\mathrm{Al})+\mathrm{Q}^3(0\mathrm{Al})+\mathrm{Q}^3(1\mathrm{Al})} \tag{5.16}$$

The fraction of occupied *bridging* sites in single-chain tobermorite that are occupied by Si is

$$\left(\frac{\mathrm{Si}}{\mathrm{Al+Si}}\right)_{\mathrm{BT}} = \frac{\text{Fraction of tetrahedral sites that are bridging and occupied by Si}}{\text{Fraction of tetrahedral sites that are bridging and occupied by Si or Al}}$$

$$\left(\frac{\mathrm{Si}}{\mathrm{Al+Si}}\right)_{\mathrm{BT}} = \frac{\frac{1}{3}-f-v}{\frac{1}{3}-v} \tag{5.17}$$

The fraction of occupied *bridging* sites in single-chain tobermorite that are occupied by Al rather than Si is

$$\left(\frac{\mathrm{Al}}{\mathrm{Al+Si}}\right)_{\mathrm{BT}} = \frac{\text{Fraction of tetrahedral sites that are bridging and occupied by Al}}{\text{Fraction of tetrahedral sites that are bridging and occupied by Si or Al}}$$

$$\left(\frac{\mathrm{Al}}{\mathrm{Al+Si}}\right)_{\mathrm{BT}} = \frac{f}{\frac{1}{3}-v} \tag{5.18}$$

which can be calculated from ^{29}Si NMR data using the expression given by Richardson, (2014)

$$\left(\frac{\text{Al}}{\text{Al+Si}}\right)_{\text{BT}} = \frac{\text{Q}^2(1\text{Al})}{\frac{2}{3}\text{Q}^2(0\text{Al})+\text{Q}^2(1\text{Al})} \tag{5.19}$$

This is very similar to the variable a in Richardson and Groves' (1993) model for Al-substituted C–S–H, which refers to all tetrahedral sites – that is, $a = f/(1 - v)$.

The site occupancy factor (SOF) for the bridging site is (Richardson, 2014)

$$\text{SOF}_{\text{BT}} = 3\times\left[\begin{pmatrix}\text{Fraction of tetrahedral sites}\\ \text{that are bridging}\end{pmatrix} - \begin{pmatrix}\text{Fraction of tetrahedral sites}\\ \text{that are vacant}\end{pmatrix}\right]$$

$$\text{SOF}_{\text{BT}} = 3\times\left(\frac{1}{3} - v\right) = 1 - 3v \tag{5.20}$$

If Al ions are present at the bridging site as well as Si, then site occupancy factors for Si and Al are

$$\text{SOF}_{\text{BT}}^{\text{Si}} = 1 - 3f - 3v \tag{5.21}$$

$$\text{SOF}_{\text{BT}}^{\text{Al}} = 3f \tag{5.22}$$

Values for SOF_{BT} can be calculated from NMR data by first calculating v and f using Equations 5.8 and 5.16 and then substituting these into Equations 5.20, 5.21 and 5.22. Such values should be used to inform refinements of XRD data but are obviously most useful if it has been established that the NMR data are for a single phase. The practical use of SOF_{BT} (Equation 5.20) is demonstrated in Chapter 6 for C–S–H(I) phases.

The ratio of Q^2/Q^3 sites for double-chain tobermorite is

$$\frac{\text{Q}^2}{\text{Q}^3}\left(i.e.\ \frac{\text{Q}^2(0\text{Al})+\text{Q}^2(1\text{Al})}{\text{Q}^3(0\text{Al})+\text{Q}^3(1\text{Al})}\right) = \frac{\frac{2}{3}-v(2-p)}{\frac{1}{3}-f-v(1+p)} \tag{5.23}$$

When there are no vacant sites (i.e. $v = 0$), this equation reduces to that given in §5.3.1.7, namely

$$\frac{\text{Q}^2}{\text{Q}^3} = \frac{\frac{2}{3}}{\frac{1}{3}-f} = \frac{2}{1-3f} \tag{5.24}$$

Reinik *et al.*'s (2007) expression for the Q^2/Q^3 ratio (their Equation (2)) refers to the ratio of sites that are occupied by Si or Al, rather than just by Si, which therefore gives the following

$$\frac{\text{Q}^2}{\text{Q}^3}\left(i.e.\ \frac{\text{Q}^2(0\text{Al})+\frac{3}{2}\text{Q}^2(1\text{Al})-\text{Q}^3(1\text{Al})}{\text{Q}^3(0\text{Al})+2\text{Q}^3(1\text{Al})}\right) = \frac{\frac{2}{3}-v(2-p-q)}{\frac{1}{3}-v(1+p+q)} \tag{5.25}$$

These equations can be used predictively or to test whether the measured chemical composition and the details of the aluminosilicate structure of a sample are compatible with one another. This is illustrated for double-chain tobermorite in Figure 5.11, which shows how the ratio of Q^2/Q^3 sites and the fractions of $Q^n(m\text{Al})$ sites vary with the fraction of tetrahedral sites that are occupied by Al (i.e. f) and with the fraction and nature of the tetrahedral sites that are vacant (i.e. with v, p and q).

Figure 5.11 Plots for double-chain tobermorite-like structures showing how the predicted relative intensity of $Q^n(m\text{Al})$ peaks and the Q^2/Q^3 ratio vary with the fraction of tetrahedral sites that are occupied by Al for different fractions and arrangements of vacant sites (continued on next page)

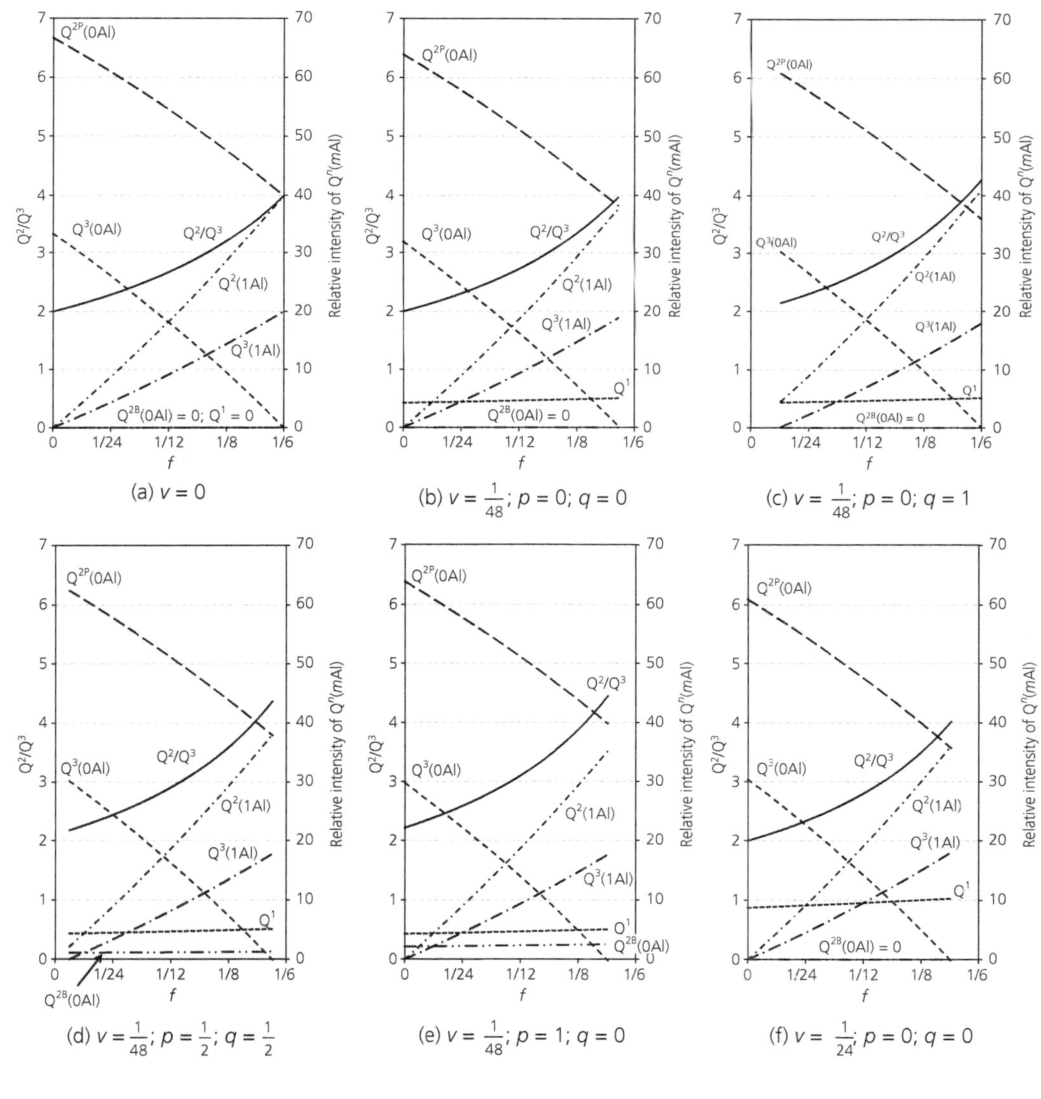

Figure 5.11 Continued

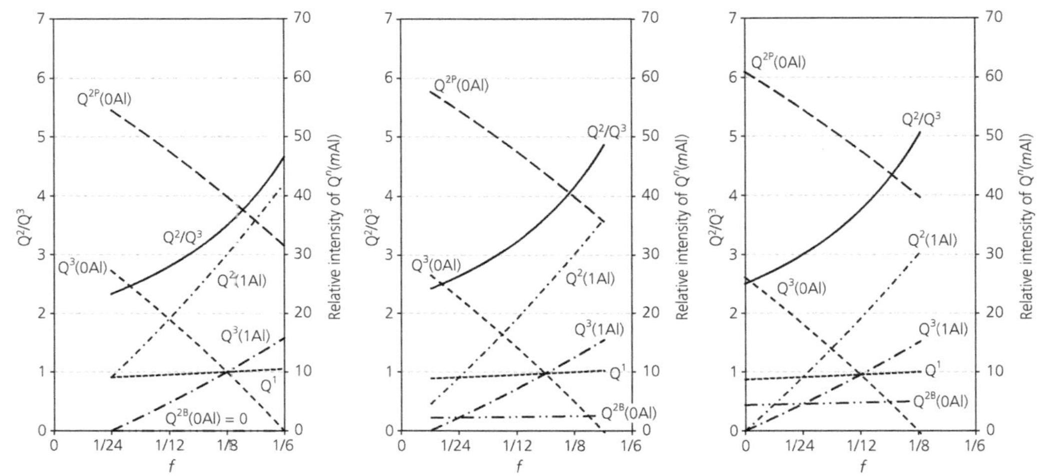

(g) $v = \frac{1}{24}$; $p = 0$; $q = 1$ (h) $v = \frac{1}{24}$; $p = \frac{1}{2}$; $q = \frac{1}{2}$ (i) $v = \frac{1}{24}$; $p = 1$; $q = 0$

v = fraction of tetrahedral sites that are vacant; f = fraction of tetrahedral sites that are occupied by Al; p = fraction of vacant sites that are adjacent to a bridging site that is occupied by Si; q = fraction of vacant sites that are adjacent to a bridging site that is occupied by Al; it follows that the fraction of vacant sites that are opposite other vacant sites = $1 - p - q$. f can be zero if the fraction of vacant sites that are adjacent to a bridging site that is occupied by Al is zero i.e. $q = 0$. The figures include curves for Q^1, Q^{2P}(0Al), Q^{2B}(0Al), Q^2(1Al), Q^3(0Al) and Q^3(1Al). The separate parts of the figure are discussed in the text.

The figures include curves for Q^1, Q^{2P}(0Al), Q^{2B}(0Al), Q^2(1Al), Q^3(0Al) and Q^3(1Al) sites. It would be straightforward to further divide paired Q^2 between Q^2(0Al) Si that are adjacent to Q^3 ('Q^{2v}'in the terminology of Klur *et al.* (1998) and Brunet *et al.* (2004)) and those that are adjacent to Q^{2B}, using the expression given in Table 5.19. Q^{2B}(0Al) sites are only present if there are vacant sites that are opposite to bridging sites that are occupied by Si – that is, when $p > 0$. Q^3(1Al) is obviously zero when $f = 0$, but it is also zero at minimum f because in that case all the bridging sites that are occupied by Al are adjacent to vacant sites. Q^3(0Al) is always zero at maximum f because in that case there are no two adjacent bridging sites that are both occupied by Si.

Figure 5.11(a) shows how the relative proportions of $Q^n(m\text{Al})$ sites and the Q^2/Q^3 ratio vary with f when there are no vacant tetrahedral sites – that is, $v = 0$. The (alumino)silicate chains that have $f = 0$ and $f = \frac{1}{12}$ are illustrated in Figures 5.8(a) and 5.8(b), respectively. Figures 5.11(b) to 5.11(i) illustrate the effect on the curves of introducing some vacant tetrahedral sites: for Figures 5.11(b) to 5.11(e) 1 in 48 of the sites are vacant $\left(\text{i.e. } v = \frac{1}{48}\right)$, and for Figures 5.11(f) to

5.11(i) 1 in 24 are vacant $\left(\text{i.e. } v = \frac{1}{24}\right)$. Four figures are used for each value of v to show the different ways that vacant sites can be distributed. Figure 5.11(b) illustrates the case when all the vacant sites are opposite other vacant sites – that is, $p = 0$ and $q = 0$. Since $q = 0$, f can be zero; and since the maximum value of $(f + v)$ is $\frac{1}{6}$, the maximum value of f, $f_{\max}$, is $\frac{1}{6} - \left(\frac{1}{48} \div 2\right) = \frac{15}{96}$. The chains that have $f = \frac{1}{12}$ are represented by an equal mixture of the chains that are illustrated in Figures 5.8(b) and 5.8(e) – that is, the chains are the same as those that are shown in Figure 5.8(e) ($p = 0$ and $q = 0$) except that $v = \frac{1}{48}$ instead of $v = \frac{1}{24}$. In Figure 5.11(c) all the vacant sites are opposite bridging sites that are occupied by Al ($p = 0$ and $q = 1$) and so f has a minimum value of $\frac{1}{48}$; $f_{\max}$ is $\frac{1}{6}$. The chains that have $f = \frac{1}{12}$ are illustrated in Figure 5.8(d). Half of the vacant sites in Figure 5.11(d) are adjacent to bridging sites that are occupied by Al and half by Si and so $p = \frac{1}{2}$ and $q = \frac{1}{2}$. As a consequence, the minimum value of f, $f_{min} = \frac{1}{48} \div 2 = \frac{1}{96}$ and $f_{max} = \frac{1}{6} - \left(\frac{1}{48} \div 2\right) = \frac{15}{96}$. All the vacant sites in Figure 5.11(e) are opposite sites that are occupied by Si ($p = 1$ and $q = 0$) and so $f_{\max}$ is limited to $\frac{1}{6} - \frac{1}{48} = \frac{7}{48}$, that is, $\frac{1}{48}$ is subtracted because there are $\frac{1}{48}$ bridging Si adjacent to the $\frac{1}{48}$ vacant sites. The structure of the chains that have $f = \frac{1}{12}$ is illustrated in Figure 5.8(c).

The effect on the curves of increasing the fraction of vacant tetrahedral sites to 1 in 24 is illustrated in Figures 5.11(f) to 5.11(i), which are equivalent to Figures 5.11(b) to 5.11(e), respectively. The most obvious effect of increasing the fraction of vacant sites is the increase in the fraction of Q^1 sites, which occurs because every vacant bridging site results in the two Q^{2P} sites that are adjacent to it becoming Q^1 sites. Q^{2B} is of course again present when $p > 0$, and in a larger amount because of the increase in v. Figure 5.11(f) includes the composition for the structure that is illustrated in Figure 5.8(e) (i.e. for $f = \frac{1}{12}$; $v = \frac{1}{24}$; $p = 0$; $q = 0$). Since $q = 0$ for Figure 5.11(f), $f_{\min}$ can be zero and $f_{max} = \frac{1}{6} - \left(\frac{1}{24} \div 2\right) = \frac{7}{48}$. Figure 5.11(h) includes the composition for the structure that is illustrated in Figure 5.8(f) (i.e. for $f = \frac{1}{12}$; $v = \frac{1}{24}$; $p = \frac{1}{2}$; $q = \frac{1}{2}$). This is the same situation as in Figure 5.11(d) except that the fraction of vacant sites is $\frac{1}{24}$. The minimum value of f is therefore $f_{min} = \frac{1}{24} \div 2 = \frac{1}{48}$ and the maximum value of f, $f_{max} = \frac{1}{6} - \left(\frac{1}{24} \div 2\right) = \frac{7}{48}$.

As an example, Figure 5.12 shows the Q^2/Q^3 ratios that are predicted from compositional data for nine synthesised samples of tobermorite plotted against the value calculated from the deconvolution of ^{29}Si MAS NMR spectra; the data are from Sakiyama *et al.* (2000). The compositional and structural data that are given by Sakiyama *et al.* (2000) are not compatible with one another for a pure double-chain tobermorite and so the interpretation of the data needs to be revisited. The values that are calculated from ^{29}Si NMR data are in most cases higher than those that are predicted from the compositional data, which could be due to the sample consisting of a mixture of double- and single-chain structures or perhaps because the NMR data were not collected using quantitative conditions. In either case, it is evident that the structural-chemical treatment given in this section can be used to test the consistency of data from different techniques. As another example, the

Figure 5.12 The Q^2/Q^3 ratio predicted from compositional data for nine synthesised samples of tobermorite plotted against the value calculated from the deconvolution of ^{29}Si MAS NMR spectra; data are from Sakiyama *et al.* (2000)

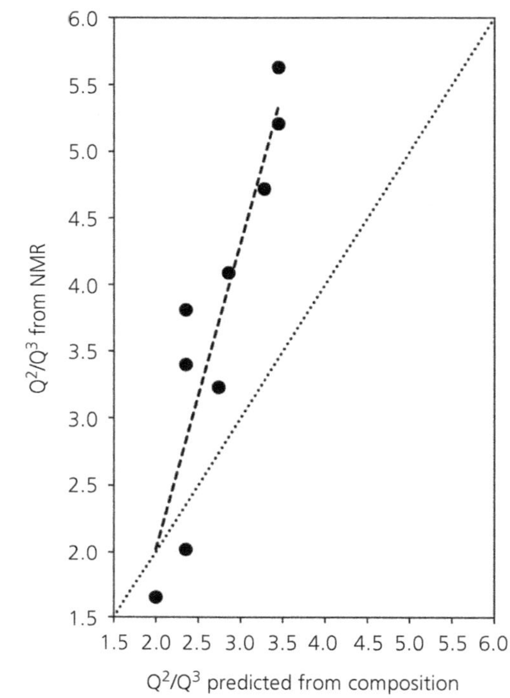

values at $f = \frac{1}{12}$ on Figure 5.11(a) are those that are predicted for Yamazaki and Toraya's (2001) preparation that was discussed earlier and so ^{29}Si NMR could have been used to test the validity of their structure determination.

Figure 5.13 shows plots that are equivalent to those in Figure 5.11 but that are for single-chain structures rather than for double-chain – that is, they illustrate how the fractions of $Q^n(m\text{Al})$ sites vary with v and f and thus how the equations that are given in this section can be used predictively. As an example, the full circles on Figure 5.13(g) represent the data for a C–S–H(I) that formed in a KOH-activated slag cement paste, which was evidently almost saturated with respect to the incorporation of Al (Richardson *et al.*, 1994).

Coleman (2006) gave an expression for the ratio of bridging tetrahedra that are cross-linked to the total number of tetrahedra in double-chain tobermorite, which she called the '*cross-linking quotient*' (CLQ). The published version has a typographical error (i.e. the absence of $Q^3(0\text{Al})$ from the denominator) and also has the same problem as her version of the equation for MCL (i.e. tetrahedra that are occupied by Al that are adjacent to vacant sites are not counted); a corrected expression for CLQ is

$$\mathrm{CLQ} = \frac{\mathrm{Q}^3(0\mathrm{Al})+2\mathrm{Q}^3(1\mathrm{Al})}{\mathrm{Q}^1+\mathrm{Q}^2(0\mathrm{Al})+\frac{3}{2}\mathrm{Q}^2(1\mathrm{Al})+\mathrm{Q}^3(0\mathrm{Al})+\mathrm{Q}^3(1\mathrm{Al})} \tag{5.26}$$

Figure 5.13 Plots for single-chain tobermorite-like structures showing how the predicted relative intensity of $Q^n(m\mathrm{Al})$ peaks vary with the fraction of tetrahedral sites that are occupied by Al (i.e. f) for different fractions and arrangements of vacant sites. The figures include curves for Q1, $Q^{2P}(0\mathrm{Al})$, $Q^{2B}(0\mathrm{Al})$ and $Q^2(1\mathrm{Al})$ (continued on next page)

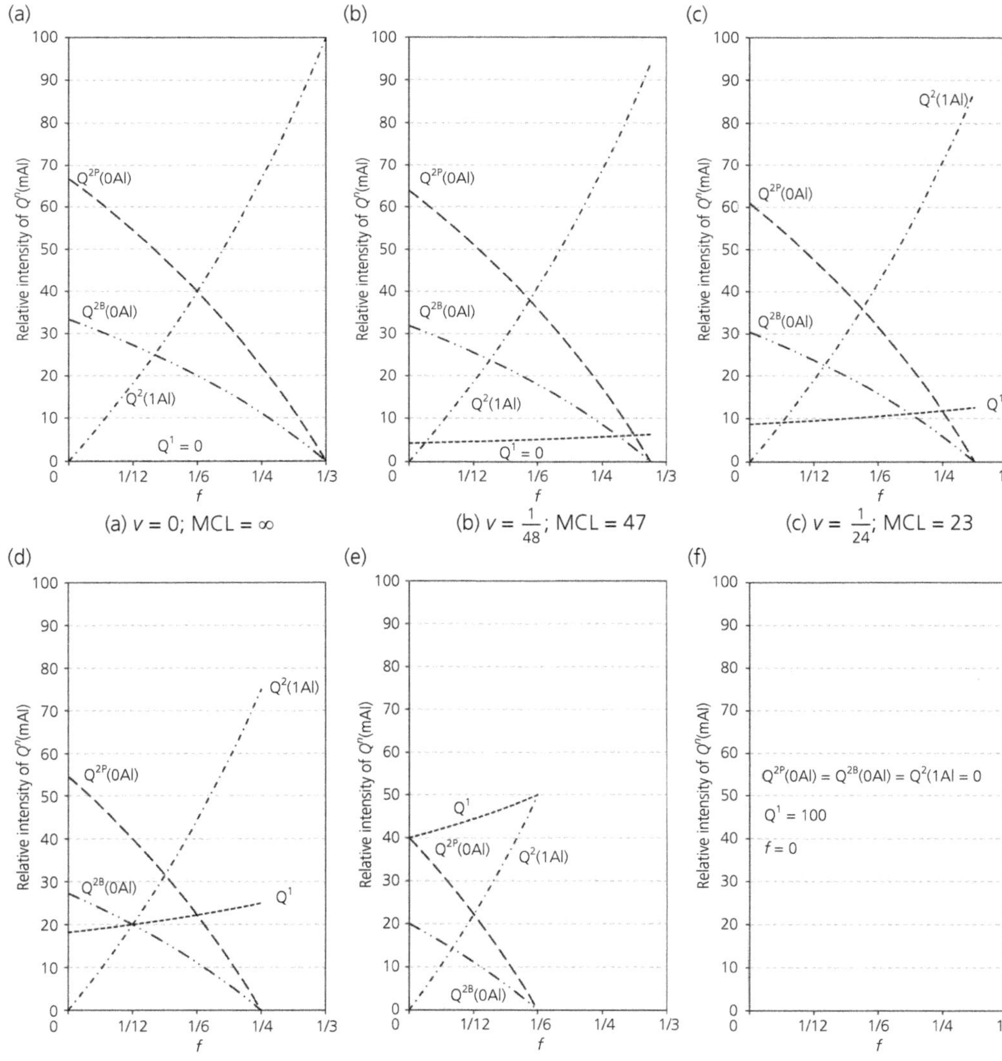

Figure 5.13 Continued

(d) $v = \frac{1}{12}$; MCL = 11 (e) $v = \frac{1}{6}$; MCL = 5 (f) $v = \frac{1}{3}$; MCL = 2

(g)

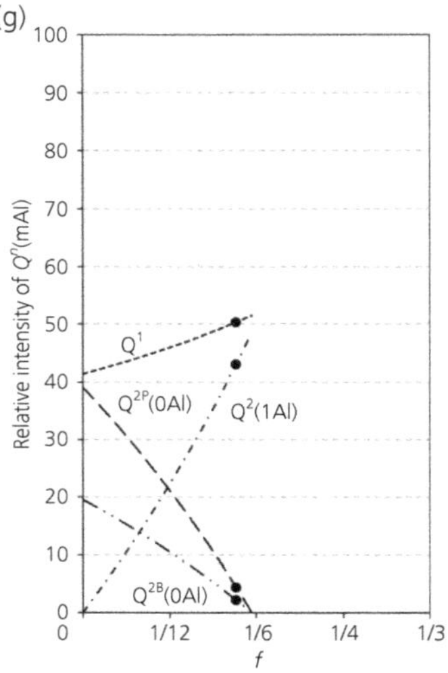

CLQ is not relevant to single-chain tobermorite because in that case there are no Q^3 tetrahedra. CLQ can be expressed alternatively in terms of vacant tetrahedral sites

$$\mathrm{CLQ} = \frac{\begin{pmatrix}\text{Fraction of tetrahedral}\\ \text{sites that are bridging}\end{pmatrix} - \begin{pmatrix}\text{Fraction of tetrahedral sites}\\ \text{that are vacant + fraction}\\ \text{of tetrahedral sites that are}\\ \text{occupied and opposite vacant sites}\end{pmatrix}}{\text{Fraction of tetrahedral sites that are occupied}} \tag{5.27}$$

$$\mathrm{CLQ} = \frac{\frac{1}{3} - v(1+p+q)}{1-v} \tag{5.28}$$

It is also useful to relate the fraction of cross-linked tetrahedra to the total number of tetrahedral sites, rather than to just those that are occupied, in which case the equation is simply modified to

$$\mathrm{CLQ}^* = \frac{1}{3} - v(1+p+q) \tag{5.29}$$

where the asterisk is used to indicate the modification.

CLQ* can also be calculated from NMR data by using

$$\mathrm{CLQ}^* = \frac{\mathrm{Q}^3(0\mathrm{Al}) + 2\mathrm{Q}^3(1\mathrm{Al})}{\frac{3}{2}\mathrm{Q}^1 + \mathrm{Q}^2(0\mathrm{Al}) + \frac{3}{2}\mathrm{Q}^2(1\mathrm{Al}) + \mathrm{Q}^3(0\mathrm{Al}) + \mathrm{Q}^3(1\mathrm{Al})} \tag{5.30}$$

The equation for CLQ is changed only slightly, with Q^1 in the denominator multiplied by 1·5 to account for the adjacent vacant sites (because two Q^1 sites are separated by one vacant site): the sum in this denominator thus accounts for all tetrahedral sites, whether vacant or occupied by Si or Al. CLQ and MCL are related by:

$$\mathrm{CLQ} = \frac{\frac{\mathrm{MCL}}{3} - \left(\frac{2}{3} + p + q\right)}{\mathrm{MCL}} \tag{5.31}$$

5.4. Jennite and metajennite

5.4.1 Occurrence, synthesis, composition and crystal data

Jennite ($C_9S_6H_{11}$) is a rare natural mineral and is a key model compound for high-lime C–S–H (see Chapter 6). The type specimen from Crestmore, California (Carpenter *et al.*, 1966; Gard *et al.*, 1977) is closely accompanied by 14 Å tobermorite. Jennite has been synthesised (e.g. Cong and Kirkpatrick, 1996; Hara and Inoue, 1980c; Hara *et al.*, 1986; Hong and Glasser, 2004; Pourbeik *et al.*, 2013) in suspensions or slurries from CaO or $Ca(OH)_2$ and various forms of silica or β-C_2S.

The composition given in the original description of the type specimen (Carpenter *et al.*, 1966) was seriously incorrect. Later analyses suggested the formula $C_9S_6H_{11}$ (Gard *et al.*, 1977) and the most satisfactory analysis of a natural specimen, from Fuka, Japan, gave a closely similar composition (Kusachi *et al.*, 1989). However, several investigators have found that apparently pure and highly crystalline jennite can be obtained from starting materials with Ca/Si ratios of 1·40–1·45 but not from those with a ratio of 1·50 (Cong and Kirkpatrick, 1996; Hara and Inoue, 1980c; Hong and Glasser, 2004). This partly explains why the preparation has proved difficult to reproduce. Hara *et al.* (1986) found the optimum temperature to be 80 °C, and that highly reactive silica, such as rice husk ash or silica fume, is to be preferred. Hong and Glasser (2004) found that afwillite ($C_3S_2H_3$) was formed at a ratio of 1·5 and obtained evidence that jennite coexists hydrothermally with afwillite and also that jennite reacts with $Ca(OH)_2$ to form afwillite. The synthetic evidence thus suggests a Ca/Si ratio lower than 1·5, but this is inconsistent with the data for the natural specimens. Possibly, the Ca/Si ratio of jennite is variable and the transition to afwillite occurs under the conditions that have been used for synthesis unless it is somewhat below 1·5.

Single crystal X-ray and electron diffraction investigations (Carpenter *et al.*, 1966; Gard *et al.*, 1977) show that jennite is triclinic with a = 10·593 Å, b = 7·284 Å, c = 10·839 Å, α = 99·67°, β = 97·65°, γ = 110·11° (V = 757·5 $Å^3$). As for the tobermorite phases, there is a strongly marked subcell, which is A-centred monoclinic with a = 9·947 Å, b = 3·642 Å, c = 21·37 Å, β = 101·90° (V = 757·5 $Å^3$). The crystals are blade-shaped, with length b and cleavage (001), and are optically negative with 2V = –74°, α = 1·552, β = 1·564, γ = 1·571. The cell parameters and optical data for the Fuka specimen (Kusachi *et al.*, 1989) are very similar.

On static heating, jennite undergoes a sharp mass loss of 7–8% at 70–90 °C, which was shown by XRD to be associated with lattice shrinkage normal to (001) to give a product called metajennite (Gard *et al.*, 1977). This has an A-centred subcell, with a = 9·945 Å, b = 3·639 Å, c = 18·67 Å, β = 111·44° (V = 628·9 $Å^3$) (McConnell, 1971). As with jennite, the true cell is triclinic.

Based on the Ca/Si ratio of 1·5, H_2O content, density, thermal behaviour and IR spectra, the structural formulae $Ca_9(Si_6O_{18}H_2)(OH)_8{\cdot}6H_2O$ and $Ca_9(Si_6O_{18}H_2)(OH)_8{\cdot}2H_2O$ were suggested for jennite and metajennite, respectively (Gard *et al.*, 1977). The calculated densities are 2·33 g cm^{-3} and 2·52 g cm^{-3}, respectively. The value for jennite agrees well with those of 2·32 and 2·33 g cm^{-3}, observed for the Crestmore and Fuka specimens, respectively (Carpenter *et al.*, 1966; Kusachi *et al.*, 1989).

5.4.2 Electron microscopy and diffraction

The original TEM study of the Crestmore material (Carpenter *et al.*, 1966) concluded that dehydration to metajennite occurs in high vacuum. Later work (Gard *et al.*, 1977) using a goniometric stage did not support this, and confirmed the unit cell given by the original XRD examination (Carpenter *et al.*, 1966). Most crystals of both jennite and metajennite were euhedral and gave sharp reflections with odd values of true *k*. A few crystals were subeuhedral and gave patterns in which the *k*-odd reflections were streaks parallel to **a*** of the subcell. High-resolution electron microscopy of a synthetic specimen (Hara *et al.*, 1986) resolved the (200), (020) and (420) planes.

TEM investigation of a synthetic sample (Viehland *et al.*, 1997) with the beam normal to (001) gave results similar to those for the natural mineral in which the *k*-odd reflections were streaked. Maxima on the streaks indicated a modulated structure. Dark-field images obtained with the objective aperture over a (100) streak showed superperiodicity along (100), varying between 20 and 40 Å in different parts of the specimen. The average value was 34 Å. The lattice parameters and the Ca/Si ratios obtained by TEM microanalyses showed local variations: 9·7–10·1 Å for *a*, 3·55–3·65 Å for *b* and 1·25–1·50 for the Ca/Si ratio. It was concluded that local fluctuations in the Ca/Si ratio led to disorder in the layering along (100), which was attributed to differing modes of stacking of the silicate chains, although it is possible that the lower values are simply a result of damage under the electron beam. By increasing the beam intensity, progressively greater beam damage could be induced. This led first to loss of all but the strongest *hk*0 reflections (or *hk*· streaks?) together with the appearance of a diffuse ring at ≈ 3·0 Å. At higher intensities, only the diffuse ring remained. Its spacing and random orientation suggests residual short-range order mainly located in Ca–O parts of the structure.

5.4.3 Thermal dehydration

A static mass loss curve (Carpenter *et al.*, 1966) showed the change to metajennite at 70–90 °C, noted above. Further loss was gradual, with some steepening at 300–450 °C, and was almost complete by 600 °C. Metajennite remained detectable up to 300 °C, but the products at 400–750 °C were poorly crystalline. At 900 °C, wollastonite and β-C_2S (larnite) were formed. The specimen from Fuka behaved similarly (Kusachi *et al.*, 1989). A TG and XRD study of a synthetic preparation (Yu and Kirkpatrick, 1999) gave similar results, with the expected increase in temperature of the conversion to metajennite, which occurred at 80–150 °C. DSC showed an endotherm corresponding to this loss, a weak, broad one at 400 °C corresponding to the steepening in the TG curve, and a small one at 700 °C from loss of CO_2. A strong, sharp exotherm at 880 °C marked the crystallisation of wollastonite and β-C_2S.

5.4.4 Vibrational spectroscopy

The mid IR spectrum of the Crestmore material (Carpenter *et al.*, 1966) included bands peaking at 3565 cm^{-1} and 3490 cm^{-1}, attributable to weakly hydrogen bonded OH groups, and thus presumably coordinated to Ca. A broad band peaking at 3170 cm^{-1} could be due to SiOH, molecular water or both. The band at 1630 cm^{-1} shows it is at least partly due to molecular water. Other bands were attributable to the silicate part of the structure or to a small content of carbonate. These included the major Si–O stretching band peaking at 905–1070 cm^{-1} and bands at 710 and 740 cm^{-1}, assigned to Si–O–Si bending. This suggests some tetrahedral condensation. The spectra of synthetic samples (Hara *et al.*, 1986; Yu *et al.*, 1999) are closely similar. The jennite spectra have broad similarities to those of 11 Å and 14 Å tobermorites, but also some distinct differences. This is also true of Raman (Kirkpatrick *et al.*, 1997a) and far IR (Yu *et al.*, 1999) spectra.

The infrared spectra of metajennites formed from the natural mineral at 180 °C (Carpenter *et al.*, 1966) and from a synthetic preparation heated at 90 °C (Hara *et al.*, 1986) are generally similar to those of unheated samples. However, the synthetic specimen heated at 180 °C showed substantial changes in the Si–O stretching band at 900–1100 cm^{-1}, demonstrating that important changes had taken place in the silicate parts of the structure. The poorly crystalline material formed by heating at 500 °C gave broad bands peaking at 400, 950–970 and 2420–3500 cm^{-1} (Carpenter *et al.*, 1966; Hara *et al.*, 1986), showing that some OH remained. There were also weak bands from carbonate.

5.4.5 Nuclear magnetic resonance

The first study using ^{29}Si NMR (Komarneni *et al.*, 1987) showed the presence of Q^2 sites together with a substantial proportion of Q^1 sites, but no Q^3 sites. It was also interpreted as providing little or no evidence of Si–OH. However, the synthetic specimen used had been prepared from CaO and fumed silica in a 1:1 molar ratio and cannot have been pure. No XRD or other tests for the absence of other phases were reported. In another study, ^{29}Si, ^{1}H–^{29}Si and ^{17}O NMR were used (Cong and Kirkpatrick, 1996). The specimen was also synthetic and had a Ca/Si ratio of 1·40. The results showed the presence of Q^2 Si sites, with only minor Q^1, which was attributed to the end groups of dreierketten with missing bridging tetrahedra.

The ^{17}O spectrum provided strong evidence of Ca–OH, and both it and the ^{1}H–^{29}Si spectrum (Cong and Kirkpatrick, 1996) led to the conclusion that Si–OH was also present. The ^{1}H MAS NMR spectrum (Yu and Kirkpatrick, 1999) provided especially clear evidence for these sites. Signals at 4·8, 1 and 8 ppm, due respectively to molecular water, Ca–OH and Si–OH, were all present.

^{29}Si NMR spectra (Cong and Kirkpatrick, 1996) showed that the Q^2 peak broadens with increasing temperature up to 250 °C. From 350 °C to 600 °C its width increases drastically, and its mean value becomes less negative, indicating progressive depolymerisation of the silicate chains (Yu and Kirkpatrick, 1999). By 600 °C, it is spread over the known range of Q^0, Q^1 and Q^2 silicate tetrahedra, indicative of a highly disordered material. At 800 °C, Q^1 and Q^2 peaks were observed, consistent with a mixture of wollastonite and β-C_2S. The ^{1}H MAS NMR spectra showed that the Ca–OH signal was drastically weakened in a sample heated at 350 °C, suggesting an intimate connection between loss of Ca–OH and formation of the disordered structure observed by XRD. The signal from Si–OH was negligible above 450 °C, and that from molecular water decreased with increasing temperature. Yu and Kirkpatrick (1999) attributed the lower thermal stability of jennite compared with tobermorite to the presence of Ca–OH, which, when lost, produced oxide ions which reacted with the silicate chains, which were thus depolymerised. The formation of wollastonite and β-C_2S necessarily occurred by nucleation and growth.

5.4.6 Crystal structures of jaffeite, jennite and metajennite

The crystal structure of jennite remained unsolved for many years because of the small size and poor quality of the crystals available from either natural or synthetic sources. The literature contains models having a tobermorite structure modified only by substitution of hydroxyl groups for silicate chains. These are clearly incorrect, because the different lengths of the *a*-axes of the subcells show that the Ca–O parts of the structures must be markedly different. The subcells of the tobermorite phases have $a \approx 5{\cdot}6$ Å, $b \approx 3{\cdot}6$ Å, and contain, for each single layer parallel to (001), two atoms of Ca, not counting any in the interlayer. The corresponding volume in jennite has $a/2 \approx 5{\cdot}0$ Å, $b \approx 3{\cdot}6$ Å. The shorter *a*-repeat implies that if the layer has a Ca–O core, this must be strongly kinked in lines parallel to **b**, and thus radically different from that in the tobermorites.

Taylor (1969) suggested a tentative crystal structure for jennite, based largely on X-ray fibre rotation data. It was based on corrugated Ca–O layers formed from linked octahedra. Many of the oxygen atoms were shared with dreierketten, others being parts of hydroxyl groups or water molecules. It assumed an incorrect chemical composition and contained other errors, but similar patterns of linked Ca–O octahedra were later found to occur in jaffeite (tricalcium silicate hydrate; $Ca_6(Si_2O_7)(OH)_6$) (Kazak *et al.*, 1974; Sarp and Peacor, 1989). The jaffeite structure is based on a three-dimensional framework of linked Ca–O octahedra in which some of the O atoms are shared with Si_2O_7 groups and others form parts of OH groups. The Ca–O framework can alternatively be described as consisting of three intersecting sets of layers arranged around a threefold axis; these layers are identical with those in the suggested jennite structure. The structure of jaffeite is illustrated in Figure 5.14.

Consideration of the jaffeite structure and correction of the jennite composition prompted further work by Taylor in which the suggested structure was modified to give a trial substructure for metajennite (reported by Richardson, 2008). Each individual Ca–O octahedron in the corrugated layer lies with two bonds parallel to the plane of the figure, two slanting upwards and two slanting downwards; the octahedra share edges in some cases and corners in others. The Si atoms of the bridging tetrahedra are present only in alternate subcells in the **b**-direction, but the O atoms are present in every subcell at coordinates that are treated as identical in the substructure. The placing of the interlayer Ca was tentative. The data were inadequate to locate the H atoms, but consideration of local charge balance suggested that, of the O atoms not bonded to Si, those bonded to either two or

Figure 5.14 (a) Structure of jaffeite projected along its **c** axis showing the positions of the atoms in the unit cell. The structure is from Sarp and Peacor (1989). Coordinates (*x/a y/b z/c*) and atom labels are given in the left and right figures, respectively. (b) Structure of jaffeite viewed along the **c** axis; $a = 10{\cdot}035$ Å, $b = 10\ 035$ Å. (c) Structure of jaffeite viewed along the **b** axis. Unit edge length in the **a** direction $= a \sin \gamma = 8{\cdot}691$ Å; $c = 7{\cdot}499$ Å (continued on next page)

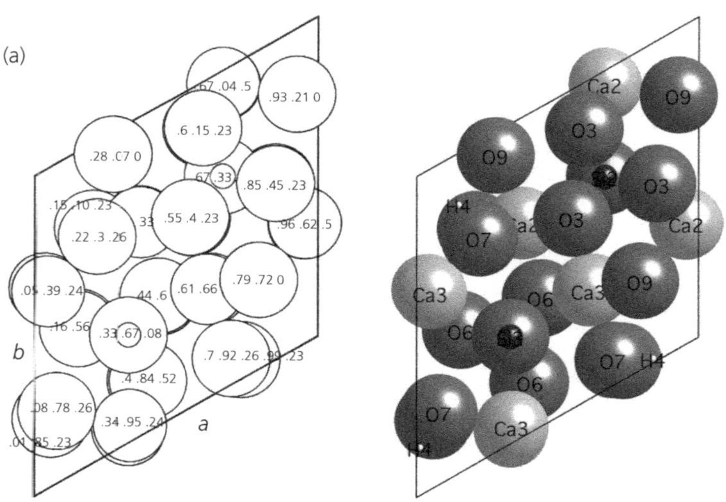

Figure 5.14 Continued

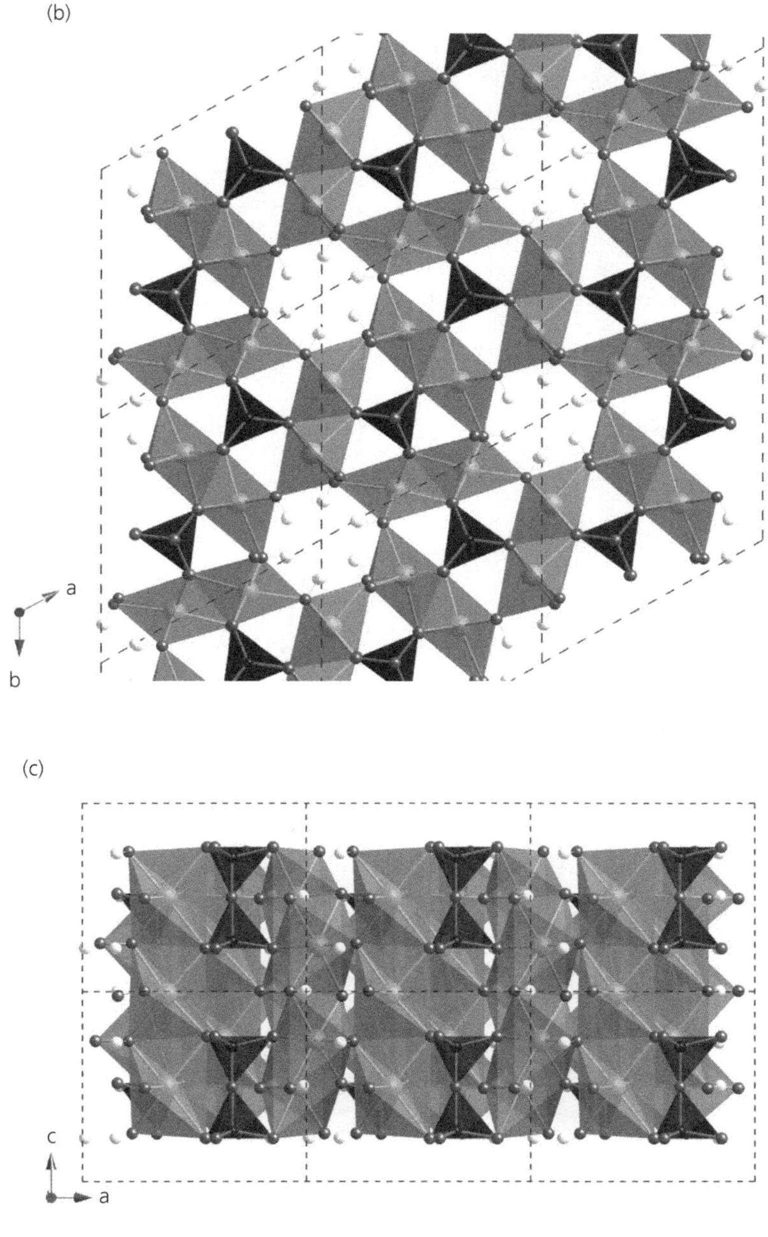

three Ca atoms of the corrugated layer are in OH groups and the others in H_2O molecules. Two O atoms per six Si, probably of water molecules in the interlayer region, could not be located. These assignments would make the ionic constitution $Ca_9(Si_6O_{18})(OH)_6{\cdot}4H_2O$, but in synthetic material Si–OH could compensate for a deficiency in interlayer Ca. The atomic positions were adjusted to give reasonable interatomic distances, and an XRD powder pattern calculated from the structure agreed moderately well in spacings and relative intensities with that observed. The structure is consistent with the thermal behaviour and with the IR and NMR spectra (Cong and Kirkpatrick, 1996).

The crystal structure of jennite is related to that of metajennite by the addition of interlayer water molecules and was finally solved by Bonaccorsi *et al.* (2004) using an extremely small single crystal from Fuka, Japan, and synchrotron radiation. The solution was obtained by X-ray crystallographic direct methods, and the results showed that the structural features suggested by Taylor (1969) are substantially correct; features of the structure are shown in Figure 5.15.

Figure 5.15(a) The structure of jennite projected along its **b** axis showing the positions of the atoms in the unit cell. The structure is from Bonaccorsi *et al.* (2004)

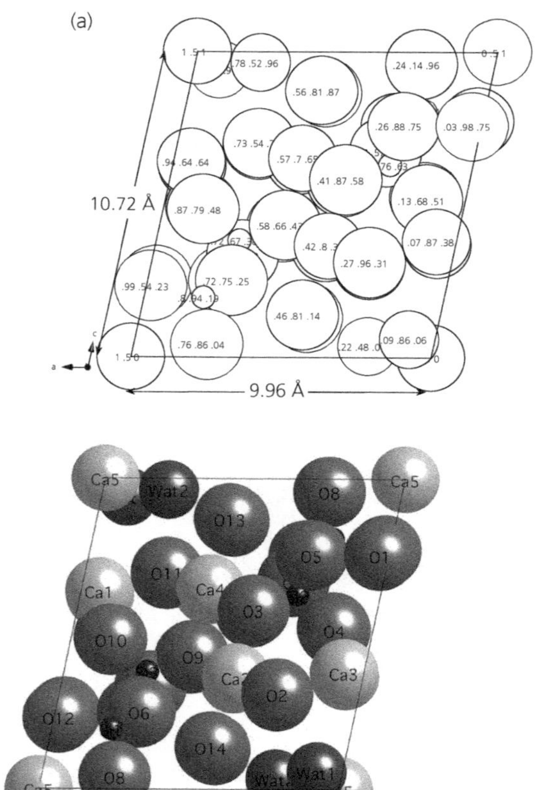

Coordinates (*x/a y/b z/c*) and atom labels are given in the top and bottom figures respectively. Edge lengths are $a \sin\gamma = 9{\cdot}960$ Å and $c \sin\alpha = 10{\cdot}719$ Å.

Figure 5.15(b) Structure of jennite viewed along the **b** axis

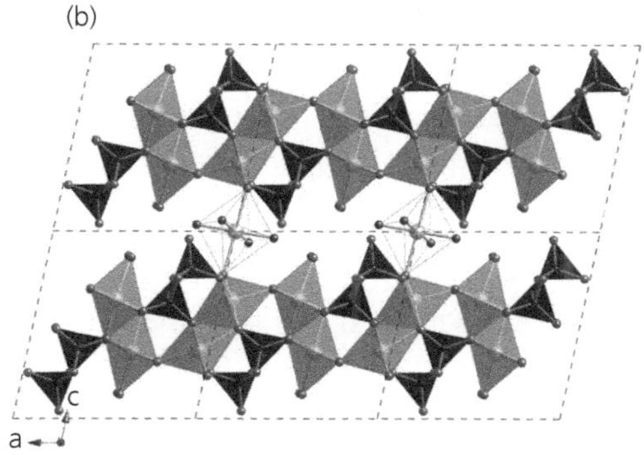

The faces of the Ca5-O polyhedra are not shaded.

Figure 5.15(c) Jennite structure viewed along the **a** axis

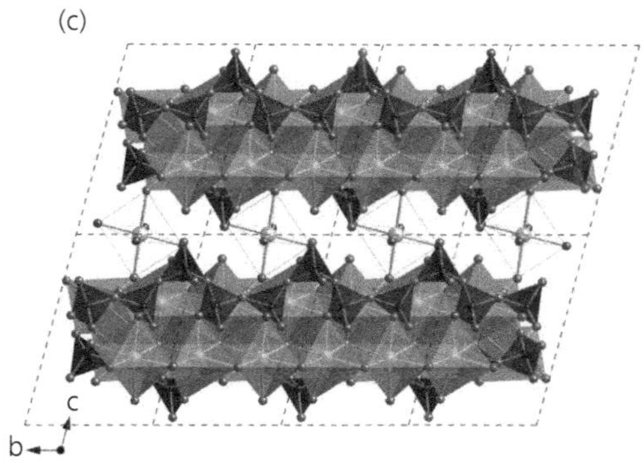

Edge lengths of units are $b \sin \gamma = 6{\cdot}842$ Å and $c \sin \beta = 10{\cdot}850$ Å.

5.5. Tacharanite and fukalite

Both these minerals have features suggesting similarity to tobermorite and both show structural features similar to ones that have been considered to occur in various forms of C–S–H. No laboratory preparation of either has been reported.

Tacharanite (Cliff *et al.*, 1975; Sweet, 1961) has a composition approximating to $Ca_{12}Al_2Si_{18}O_{69}H_{36}$; X-ray and electron diffraction patterns indicate a unit cell and subcell related to those of the tobermorite phases. The mid IR spectrum is similar to that of 11 Å tobermorite. Only small changes in the XRD pattern occur on heating at temperatures up to at least 580 °C. A volume comparable to the pseudo cell of orthorhombic tobermorites has $a = 5{\cdot}69$ Å, $b = 3{\cdot}65$ Å, $c = 27{\cdot}9$ Å, $\beta = 114{\cdot}1°$ ($d_{001} = 25{\cdot}5$ Å) and approximate contents $Ca_4Al_{0{\cdot}67}Si_6O_{23}H_{12}$. The layer thickness of 12·7 Å suggests a possible relation to the 12 Å tobermorites described earlier. The content of (Si + Al) is too great to be accommodated entirely in dreierketten. The data suggest the possibility that tacharanite is a tobermorite modified by the presence of interlayer aluminate, but this remains an untested speculation. It is nevertheless of interest in view of findings that Al can occupy interlayer sites in C–S–H (Lognot *et al.*, 1998; Sun *et al.*, 2006).

Fukalite has the ideal composition $Ca_4Si_2O_6(OH)_2(CO_3)$. It was first found at Fuka in Japan (Okayama Prefecture) occurring as an alteration product of spurrite ($Ca_5(SiO_4)_2CO_3$) and associated with calcite and/or xonotlite (Henmi *et al.* 1977). Henmi *et al.* (1977) reported a *B*-centred orthorhombic subcell with $a = 5{\cdot}48$ Å, $b = 3{\cdot}78$ nm, $c = 23{\cdot}42$ Å, Z = 2 (Henmi *et al.*, 1977), with possible space groups $B\ m\ 2_1\ b$ (No. 36, setting $\mathbf{a\bar{c}b}$), $B\ 2m\ b$ (No. 40, setting **cab**) or $B\ m\ m\ b$ (No. 63, setting $\mathbf{a\bar{c}b}$). The measured density is $2{\cdot}770\,g\,cm^{-3}$. It loses water and CO_2 at 600 °C. Rastsvetaeva *et al.* (2005) reported a structure for a sample from the Gumeshevsk deposit in the Urals, Russia, that has an orthorhombic cell, with $a = 3{\cdot}786$ Å, $b = 10{\cdot}916$ Å, $c = 23{\cdot}379$ Å, Z = 4 and space group $P\ 2_1\ 2_1\ 2_1$ (No. 19). Rastsvetaeva *et al.*'s (2005) work provided the basis for a full solution by Merlino *et al.* (2009), who refined synchrotron radiation diffraction data from a single crystal and applied order–disorder (OD) theory. The average structure is orthorhombic, space group *A m m a* (No. 63, setting **cab**), with $a = 3{\cdot}786$ Å, $b = 5{\cdot}458$ Å, $c = 23{\cdot}379$ Å, Z = 2, which is consistent with the results of Henmi *et al.* (1977) who assumed a reference frame with interchanged **a** and **b** vectors – that is, *A m m a* is space group No. 63, cell setting **cab**, whereas *B m m b* is No. 63, cell setting $\mathbf{a\bar{c}b}$. Merlino *et al.* (2009) established that there are six possible maximum degree of order (MDO) polytypes; the refined monoclinic cell parameters for one of the six (the MDO_1 polytype) are: $a = 7{\cdot}573$ Å, $b = 23{\cdot}364$ Å, $c = 11{\cdot}544$ Å, $\beta = 109{\cdot}15°$, Z = 8, space group $P\ 2_1/c$ (No. 14, unique axis *b*, cell choice 1). The structure consists of: (a) a Ca–O polyhedral framework, formed by tobermorite-type polyhedral layers that alternate along **b** with tilleyite-type zigzag polyhedral layers; (b) infinite single silicate chains which have a conformation that repeats after every fifth tetrahedron (i.e. '*vierereinfachketten*' in Liebau's (1985) classification) and that run along **a** and are linked to the Ca–O polyhedral layers on opposite sides; and (c) rows of CO_3 groups that are parallel to (100) and stacked along **a**.

5.6. Plombièrite

5.6.1 Occurrence and general features

In 1858, Daubrée (1858) discovered and analysed a gelatinous calcium silicate hydrate, which he named plombièrite. Similar materials have been found in nature at several localities, but the only specimens to have been studied in any detail have been those from several localities in Northern Ireland. McConnell (1954) used the name 'plombièrite' for both crystalline and quasicrystalline forms of minerals similar in composition to Daubrée's material, but subsequent usage has restricted the name to quasicrystalline or amorphous calcium silicate hydrates, irrespective of composition, with the notable exception of Bonaccorsi *et al.* (2005) who associated the name with 14 Å tobermorite. Unfortunately this led to a proposal that 'plombièrite' should be used for 14 Å tobermorite as part of a nomenclature scheme for minerals in the tobermorite group, and this scheme was approved by the IMA Commission on New Minerals, Nomenclature and Classification (Williams *et al.*, 2014; Biagioni *et al.*, 2015). While this definition is included in Table 5.1 for information, it is not appropriate and 'plombièrite' is used in this book with its original meaning – that is, as gelatinous quasicrystalline or amorphous calcium silicate hydrates of variable composition.

The plombièrite specimens described by McConnell (1954, 1955) gave XRD patterns closely similar to those of synthetic tobermorite-like C–S–H preparations, but with no detectable basal reflection and few reflections with other than *hk*0 indices. The compositions were approximately $C_5S_6H_{11-12}$, with minor contents of Al_2O_3 and other components. The plombièrites are transparent to translucent and brittle, with densities $2{\cdot}018$–$2{\cdot}186\,g\,cm^{-3}$. The mean refractive indices of

1·49–1·53 are highly correlated with the densities, in approximate accordance with the Gladstone–Dale equation. Nawaz (1977) reported additional optical and other data.

5.6.2 Mode of formation in nature

The plombièrites studied by McConnell (1954, 1955) typically occurred in nodules some 100 mm in diameter, with cores of larnite (β-C_2S) or bredigite (C_7MS_4). The nodules were surrounded by chalk in the near vicinity of dolerite, and the silica in them had originated either in the dolerite or in nearby nodules of flint. The quasicrystalline plombièrite had formed topotactically from the larnite or bredigite, with its chain direction (**b**) sub parallel to the 6·8 Å axis of the latter, and was randomly oriented around this direction. From the compositions and densities of the plombièrite and C_2S, McConnell (1954; 1955) noted that substantial fractions of the Ca and Si in the latter must be expelled during the reaction.

Long and McConnell (1959) used X-ray microabsorption spectroscopy to determine calcium contents in small areas of plombièrite in a thin section of a larnite-spinel rock. The rock contained cavities of millimetre dimensions, filled mainly with afwillite and portlandite (CH). The plombièrite formed a zone some 150 μm thick, which lined the inner wall of the cavity. From its Ca content and those of other phases present, its Ca/Si ratio was estimated at ≈1·5. The plombièrite formed in this situation had been effectively isolated from circulating ground waters and was considered to be the initial hydration product of the larnite. Except in the peripheral zone of the cavity, it had altered further to afwillite and portlandite, which was considered to be the thermodynamically stable assemblage. In other situations, the plombièrite had not been isolated from circulating ground waters, and reaction with them had lowered its Ca/Si ratio to the observed values of about 5:6. In some cases, this low-lime material had recrystallised to form 14 Å tobermorite (McConnell, 1971). The latter formed lath-like crystals which were appreciably bent and intimately intermeshed, and the preferred orientation was completely lost. Clearly, total reorganisation had occurred. This recrystallisation probably explains the occurrence of a 14 Å peak in the XRD patterns of some specimens of plombièrite.

Because the high-lime plombièrite was only found in very small amounts, more detailed examinations have so far been confined to the low-lime material.

5.6.3 X-ray and electron diffraction

A more detailed XRD study of low-lime plombièrite formed as a pseudomorph after bredigite (McConnell, 1971) confirmed that the plombièrite reflections showed considerable angular spread. This was considerably less on SED patterns. The individual diffracting regions are thus much more nearly aligned on the scale of ≈ 1 μm studied by electron diffraction than on that of ≈ 50 μm studied by XRD.

The XRD fibre rotation and SED patterns included a very strong reflection indexable as 100 referred to the orthotobermorite subcell with $a \approx 5{\cdot}6$ Å. This is inconsistent with the body-centred lattice of the crystalline 11 Å and 14 Å phases and suggests a primitive lattice. McConnell (1971) suggested that the silicate chains attached to one side of the Ca–O sheet were absent in a substantial proportion of the Ca–Si–O structural units. The Ca/Si ratios of the low-lime plombièrites are much too low for this situation to exist throughout the structure; it was suggested that this could have been the case in the high-lime precursor and that the resulting structure was partly preserved when the Ca/Si ratio decreased. As noted earlier in this chapter, a 100 reflection has also been observed in some crystalline tobermorites.

Apart from their considerable angular spread, the *hk*0 reflections were sharp, with no marked broadening in other directions. At the same time, the SED patterns showed regions of diffuse intensity which appeared to occur on rows parallel to **c*** passing through the *hk*0 reflections. These data were tentatively interpreted as indicating the presence of separate layers which were randomly spaced along **c** but which were stacked in the correct vertical register over distances of at least 10 nm. It was recognised that the actual position was probably more complicated.

The XRD patterns showed no discrete basal reflection, but there was an intense low-angle streak normal to the fibre axis (**b***). Its profile showed that there was no favoured periodicity in the 10–30 Å range. Low-angle scattering XRD gave results compatible with a model in which the scattering volumes are lamellae with one extremely short dimension and longer dimensions of the order of 600 Å or more. The results showed that, whatever the exact interpretation, there must be fluctuations in electron density on a scale of $\approx$ 800 Å.

5.6.4 Water sorption and thermal analysis

Water sorption and desorption isotherms of low-Ca material formed from larnite showed characteristics typical of cement paste or silica gel (McConnell, 1955). A marked loss of water at 40% RH (for 16 °C) indicated the loss of a large fraction of the water from sites of equivalent energy. At the end of this desorption step, the H_2O/SiO_2 ratio was $\approx 1{\cdot}5$, which is comparable to the ratio for 14 Å tobermorite. The water lost at higher humidities is thus presumably that held in pores larger than interlayer spaces.

Further work (McConnell, 1971) showed that water loss in a high vacuum at 100 °C was very slow. It reached 17·9% after several days and extrapolation suggested that it would eventually reach 20%, both values being referred to the initial mass. BET (Brunauer–Emmett–Teller) analysis of sorption isotherms at various temperatures indicated a specific surface area of $320\,m^2\,g^{-1}$. Assuming the pores to be lamellar and using the known density of $2{\cdot}042\,g\,cm^{-3}$, this indicated that the pores were $\approx$ 12 Å wide and separated by an average distance of 30 Å, which agrees broadly with the conclusions from low-angle X-ray scattering. The sorption data also provided data on the isosteric heat of sorption, which was found to be $47\,kJ\,mol^{-1}$. This value indicated that the water was moderately strongly bound.

A TG curve (Cong *et al.*, unpublished) showed no distinct steps and was similar to those of synthetic C–S–H preparations. The DSC curve showed a very strong endotherm near 100 °C, attributable to loss of water molecules, and two small endotherms at 500–600 °C, attributed to decomposition. It was not extended beyond 620 °C.

5.6.5 Nuclear magnetic resonance

A ^{29}Si NMR spectrum (Cong *et al.*, unpublished) differed significantly from those of 14 Å tobermorite or synthetic C–S–H preparations in indicating a higher degree of polymerisation and broader peaks. There was a strong Q^2 peak, a weaker Q^3 peak and a possible small Q^1 peak hidden under that of Q^2, indicating the presence of very long, cross-linked silicate chains. From the Q^2/Q^3 ratio, it was estimated that about two-thirds of the bridging tetrahedra were linked to form Q^3 sites. The sample gave good ^1H–^{29}Si cross-polarisation (CP) signals, but only the Q^2 peak existed for all contact times and was well resolved. A charge balance calculation indicated that there were at least 0·26 Si–OH groups per tetrahedron. This value is smaller than for comparable synthetic C–S–H samples because the higher degree of polymerisation lowers the content of Si–OH needed for charge balance.

5.6.6 Relations to cement paste

The plombièrite specimens so far studied resemble the C–S–H of cement paste in some respects but not in others. At the nanometre level, the XRD and SAED (selected area electron diffraction) data show that they are less well ordered than many synthetic C–S–H preparations but much more so than the C–S–H in hardened pastes of tricalcium silicate or Portland cement. The Ca/Si ratios of the low-lime plombièrites, to which almost all the data relate, are similar to those of the C–S–H formed by precipitation from aqueous solution (see Chapter 6) or formed in cement pastes containing high proportions of silica fume or other highly reactive pozzolanas. They are much lower than those of the C–S–H of unblended Portland cement pastes and their average anion sizes are much larger. There is no direct evidence for the nanostructure of the high-lime precursor plombièrite, which could be similar to that of the C–S–H of unblended pastes. McConnell's (1971) suggestion that the high-lime precursor contains long silicate chains on only one side of the Ca–O sheets of a tobermorite-like structure differs from the results of subsequent studies of silicate anion type in cement or calcium silicate pastes by ^{29}Si NMR and chemical methods. These studies show initial formation of silicate monomer, followed by significantly more dimer, and afterwards a gradual increase in mean chain length. In a 26-year-old C_3S paste, 42 % of the Si was in Q^1 sites and 58 % in Q^2 sites (Rodger *et al.*, 1987), which corresponds to a mean chain length of 4·8 tetrahedra. These data are discussed in detail in Chapter 7.

The observations on plombièrite possibly provide important information about the long-term stability of the phase composition of cement pastes. If correct, they suggest that, in the absence of water, a C–S–H with quite high Ca content can persist in the presence of the thermodynamically stable products over periods of many millions of years, and that a low-lime C–S–H is highly persistent even under conditions where some leaching has occurred. It should be recognised, however, that these data are few and concern indirect compositional evidence, which needs to be confirmed by direct observation.

The plombièrites resemble the C–S–H of cement pastes in their gelatinous nature. Their densities and H_2O/SiO_2 ratios are broadly similar to those of the C–S–H equilibrated at 11 % RH, for which the composition is $\approx C_{1\cdot7}SH_{2\cdot1}$ and the density is $\approx 2{\cdot}180\,g\,cm^{-3}$.

The mechanism by which the plombièrite is formed is of much interest. Funk (1962) found that, under some conditions, synthetic β-C_2S is converted to pseudomorphs of C–S–H, and suggested that the process is topotactic. From XRD powder evidence, he concluded that the structural change was gradual. The possible existence of some kind of intermediate product structurally related to the starting material is of interest in view of observations that the calcium silicate hydrate initially formed on hydration of cement contains monomeric silicate ions, as discussed in §7.4.2.

The formation of the inner product from alite or belite in a cement paste proceeds through the inward movement of an interface, which implies a topochemical mechanism, although not necessarily a topotactic one. As noted in §7.7.4, comparison of the compositions and densities of alite or belite and saturated C–S–H shows that the net effect of the process is loss of Ca^{2+} and Si^{4+} and gain of H^+, with little or no net gain or loss of O^{2-} (Taylor, 1985). This relationship is similar to that proposed by McConnell (1955) for the formation of plombièrite.

5.7. General aspects of the crystal chemistry of calcium silicate hydrates and related phases

X-ray diffraction techniques are of only limited utility in studies of the C–S–H that forms in real cements because it is virtually X-ray amorphous. It is noted in §7.3.5 that poorly-defined diffraction patterns are to be expected because C–S–H occurs in cement pastes as aggregations of very small particles (see §7.3.2). As a consequence, there is increasing interest in using atomistic modelling techniques which offer the prospect of significant advances in understanding its nanostructure. The results of such studies should of course be consistent with the results of the experimental studies on the C–S–H that forms in pastes (e.g. those that are described in Chapter 7) but they must also adhere to crystal chemical principles. While atomistic modelling studies have usually started with the structure of 14 Å tobermorite (see §6.6.2), inspection of the structures of all the crystalline calcium silicate hydrate phases is necessary to establish the crystal-chemical principles. Such a treatment is reported by Richardson (2014), and the following is an augmented reiteration of that work.

The information provided in Table 5.1 is extended in Table 5.20 with additional crystal-chemical information. There are a total of 132 unique Ca atoms in the structures of the 35 crystalline calcium silicate hydrates and related phases that are listed in Table 5.20. Richardson (2014) showed that out of 859 Ca–O distances there is only one that is less than 2·2 Å (as shown in Figure 5.16(a)), and only five that are greater than 2·9 Å; 2·9 Å and 2·2 Å were thus considered to be reasonable maximum and minimum values in structural studies of calcium silicate hydrates. Inspection of the structures of the 35 crystalline calcium silicate hydrates using these values reveals that the 'natural' coordination for calcium cations in these phases is either six- or seven-fold, which is illustrated in Figure 5.16(b). A small number of Ca atoms are coordinated to eight oxygen atoms, but none to fewer than six, which is an important observation for atomistic modelling studies. For example, more than half of the Ca atoms in the model of Pellenq *et al.* (2009) (which has 99 unique Ca atoms) are coordinated to fewer than six oxygen atoms (their model has Ca in five-, four- and even three-fold coordination), which is implausible; their model is discussed in §6.6.2.

The average Si–O distances in the silicate tetrahedra that are present in the phases detailed in Table 5.20 are shown in Figure 5.17(a) plotted against average Si–O–Si connectivity. There are

Figure 5.16 (a) Normalised frequency histogram for the Ca–O distances that are present in the structures of the crystalline calcium silicate hydrates and related phases that are listed in Table 5.20. The 35 crystal structures that are reported in Table 5.20 involve a total of 132 unique Ca atoms with 859 Ca–O distances. (b) Histogram showing the percentage of the calcium atoms in these phases that are coordinated to different numbers of oxygen atoms. Reproduced from Figure 6 of Richardson (2014)

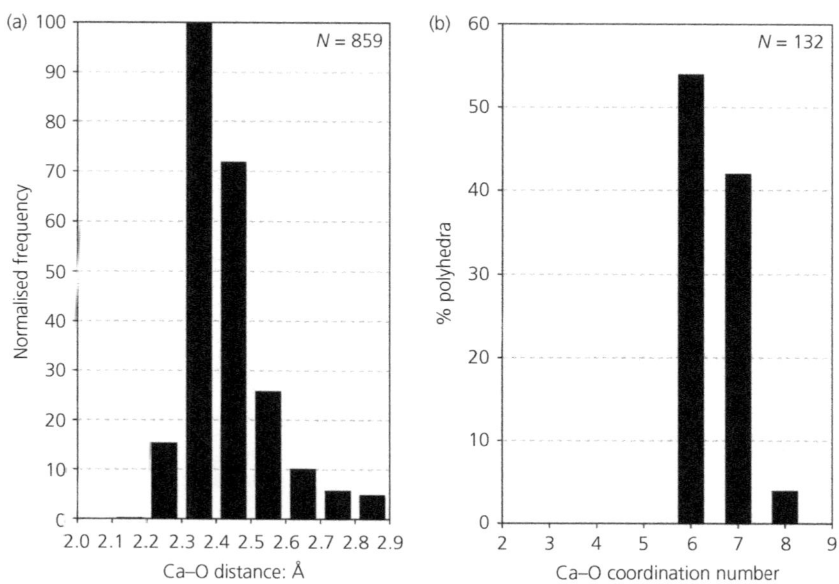

Figure 5.17 (a) Average Si–O distances in the silicate tetrahedra that are present in the phases detailed in Table 5.20 plotted against average Si–O–Si connectivity; (b) average Ca–O distance plotted against average Ca–O coordination number; (c) Ca–O polyhedral volume plotted against average Ca–O distance; (d) Ca–O polyhedral volume plotted against average Ca–O coordination number; (e) calculated density plotted against H_2O/Ca ratio for all the phases in Table 5.20 that have dreierketten, with the exception of xonotlite (continued on next page)

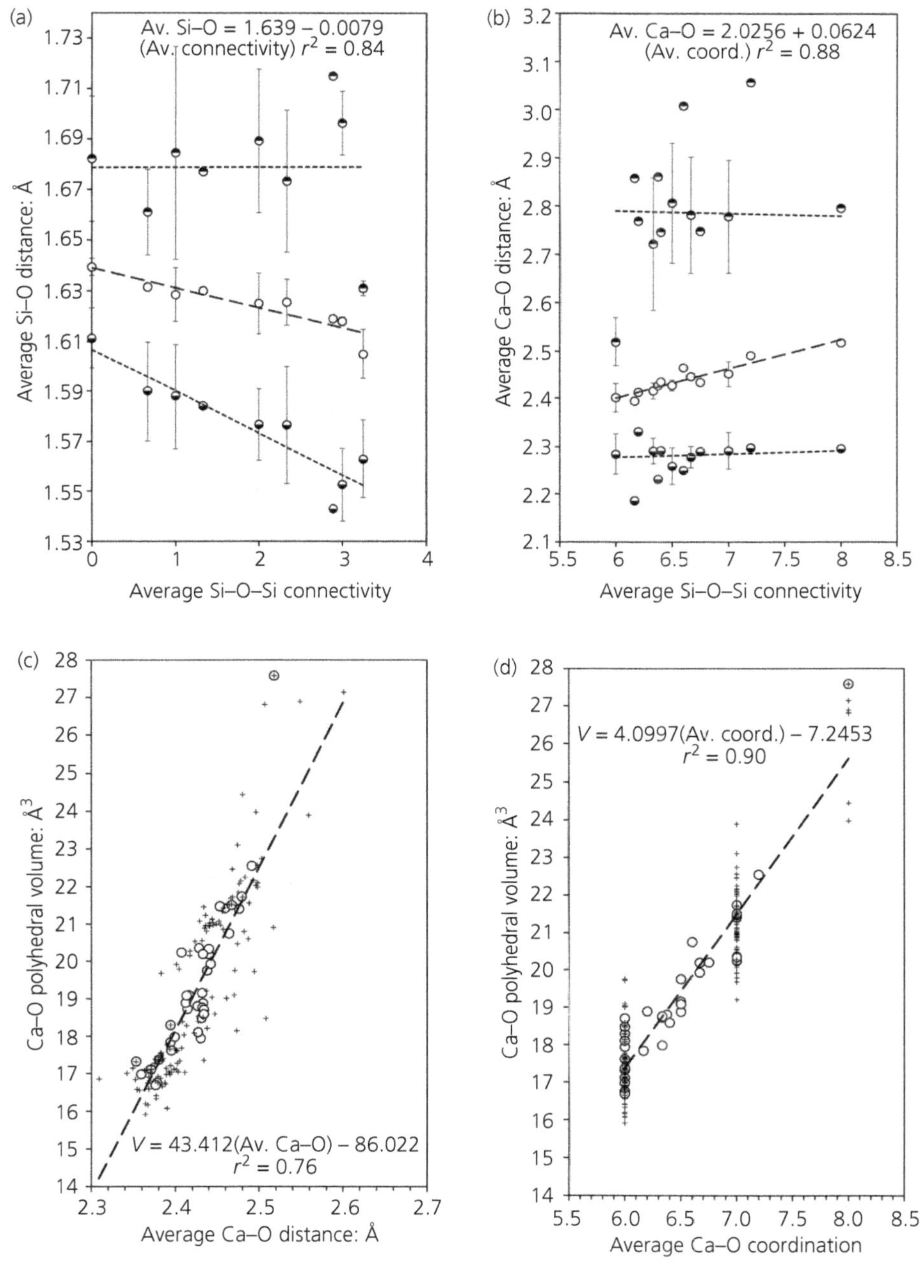

Figure 5.17 Continued

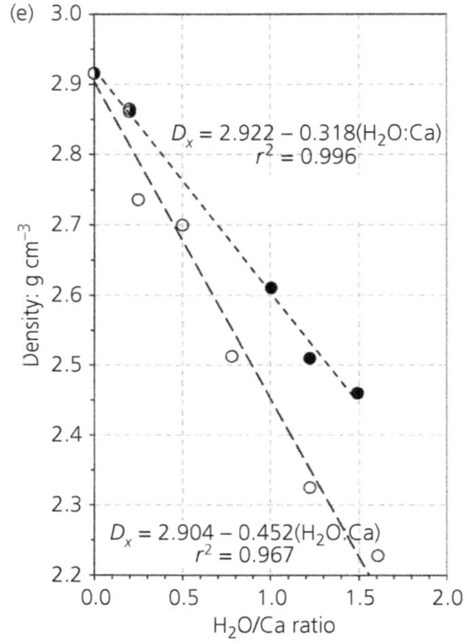

The circled outlying data point that is at the top of (c) is for suolunite, which is the only natural calcium silicate hydrate that is stoichiometrically CSH. (a), (b) and (d) are reproduced from Figure 7 of Richardson (2014); (c) and (e) are new.

actually only nine unique values of average connectivity in those phases, which is the reason why some of the points have error bars – that is, the points that have error bars represent more than one phase. The three sets of points for each value of connectivity show the average of the average distance present in the phases (the middle set), the average of the maximum distance (upper set) and the average of the minimum distance (lower set). The long-dash line for the middle set shows that the average Si–O distance in a silicate tetrahedron present in a calcium silicate hydrate (or related phases) decreases linearly with increasing connectivity of the tetrahedron. The regression equation (Equation 5.32; $r^2 = 0{\cdot}84$) can be used as a reasonable guide in structural and atomistic modelling studies that are concerned with such phases.

$$\text{Average Si–O} = 1{\cdot}639 - 0{\cdot}0079(\text{Average connectivity}) \quad (\text{Å}) \tag{5.32}$$

The data indicate that the silicate tetrahedra become increasingly distorted as the connectivity of the tetrahedra increases.

Figure 5.17(b) shows that there is a good linear correlation between the average Ca–O distance and the average Ca–O coordination, Equation 5.33 ($r^2 = 0{\cdot}88$), which should also provide a useful guide for atomistic modelling studies, together with the maximum and minimum values that are listed in Table 5.20.

$$\text{Average Ca–O} = 2\cdot0256 - 0\cdot0624(\text{Average coordination})\ (\text{Å}) \tag{5.33}$$

Good linear correlations also exist between the volume of the Ca–O polyhedra and the average Ca–O distance (Figure 5.17(c)), and between the volume and the average Ca–O coordination number (Figure 5.17(d)); the regression equation for the latter (Equation 5.34; $r^2 = 0\cdot90$) should be particularly useful.

$$V_{\text{Ca-O}} = 4\cdot0997(\text{Average coordination}) - 7\cdot2453\ (\text{Å}^3) \tag{5.34}$$

A number of the phases that are discussed in this chapter (and listed in Tables 5.1 and 5.20) have silicate chains that have the dreierkette conformation, which is particularly relevant to models for the nanostructure of C–S–H. The measured density of those phases decreases with an increase in H_2O/Ca ratio. This is illustrated in Figure 5.17(e), which shows the calculated density plotted against H_2O/Ca ratio for all the relevant phases in Table 5.20, with the exception of xonotlite. The filled circles represent the phases that have double chains and the open and half-filled circles represent those that have single chains. The lines represent the linear regression fits to the two sets of data, with the data points with half-filled symbols used in both fits; the correlation is good in both cases. The data with half-filled symbols are for wollastonite (no water) and the 9 Å forms of both ortho- and clinotobermorite, which have very little water and single silicate chains that are close to one another. The two regression equations intercept the density axis at a similar value, which in both cases is close to the density for wollastonite. As a consequence, an approximate value for the density of a calcium silicate hydrate that has dreierkette silicate can be calculated using Equation 5.35, where the density of wollastonite, $\rho_{\text{woll}} = 2\cdot92\ \text{g cm}^{-3}$, and $s = 0\cdot46$ for single chains and $0\cdot32$ for double chains.

$$\rho_{\text{CSH}} = \rho_{\text{woll}} - s(H_2O/Ca) \tag{5.35}$$

Alternatively, in cases where the H_2O/Ca ratio and the density are known, in principle, for those phases that have water contents >3 wt.%, the calculation of s can provide an indication as to whether the silicate chains are condensed or not – that is, whether the phase has double or single chains, respectively. In that case, Equation 5.35 is more meaningful if rearranged as Equation 5.36, where single chains are likely if $s > 0\cdot38$ and double chains if $s < 0\cdot38$.

$$s = \frac{\rho_{\text{woll}} - \rho_{\text{CSH}}}{(H_2O/Ca)} \tag{5.36}$$

In practice, this can be complicated by the fact that observed densities can be affected by deviations in atomic contents relative to the idealised composition (e.g. caused by the presence of vacant sites) and that single- and double-chain phases are often intimately mixed in volumes that are smaller than is practicable for some tests. The relationship between the density and H_2O/Ca ratio is different for the single- and double-chain phases because there is a greater prevalence in the single-chain structures of oxygen atoms of water molecules that are involved in completing the coordination requirements of Ca atoms.

Table 5.20 Selected compositional and crystal chemical data for calcium silicate hydrates, calcium aluminosilicate hydrates and related phases for which there are refined crystal structures; the data complement those in Table 5.1. References are given in Table 5.1. D_x is the calculated density. For each phase, the sum of the numbers in the columns for silicate connectivity is the number of Si sites in the asymmetric unit and the sum of the numbers in the columns for Ca–O coordination is the number of Ca sites; for example, there are three Si and six Ca sites in the asymmetic unit of dellaite.

Table 5.20(a) Wollastonite group

Name	Formula	Ca/Si	H_2O/Si	%H_2O (mass)	Connectivity of silicate tetrahedra					Si–O distance: Å			Ca–O coordination			Ca–O distance: Å		
					Q^0	Q^1	Q^2	Q^3	Q^4	Min.	Average	Max.	6	7	8	Min.	Av.	Max.
Foshagite	$Ca_4(Si_3O_9)(OH)_2$	1·33	0·33	4·3			3			1·572	1·599	1·669	4			2·209	2·430	2·561
Hillebrandite	$Ca_2(SiO_3)(OH)_2$	2·00	1·00	9·5			2			1·598	1·639	1·732	2	1		2·288	2·414	2·646
Nekoite	$Ca_3Si_6O_{15}·7H_2O$	0·50	1·17	19·3				6		1·537	1·617	1·710	3			2·243	2·380	2·508
Okenite	$Ca_{10}Si_{18}O_{46}·18H_2O$	0·56	1·00	16·5			1	8		1·543	1·619	1·715	5	1		2·186	2·394	2·858
Pectolite	$NaCa_2Si_3O_8(OH)$	1·00[a]	0·17	2·7			3			1·596	1·630	1·675	2			2·299	2·359	2·435
Wollastonite-Tc	$Ca_3Si_3O_9$	1·00	0	0			3			1·578	1·624	1·664	2	1		2·266	2·399	2·639
Xonotlite[b]	$Ca_6Si_6O_{17}(OH)_2$	1·00	0·17	2·5			2	1		1·565	1·620	1·649	2	2		2·308	2·433	2·770

[a] (Na + Ca)/Si.
[b] *Ma2bc* polytype (an MDO_2 polytype).

Table 5.20(b) Tobermorite group (all are MDO_2 polytypes)

Name	Formula	*i*	Ca/Si	H_2O/Si	%H_2O (mass)	Connectivity of silicate tetrahedra					Si–O distance: Å			Ca–O coordination			Ca–O distance: Å		
						Q^0	Q^1	Q^2	Q^3	Q^4	Min.	Average	Max.	6	7	8	Min.	Av.	Max.
14 Å tobermorite	$Ca_5Si_6O_{16}(OH)_2{\cdot}7H_2O$	1	0·83	1·33	18·4			3			1·565	1·630	1·735	1	2		2·252	2·440	2·742
Anomalous 11 Å tobermorite	$Ca_4Si_6O_{15}(OH)_2{\cdot}5H_2O$	1	0·67	1·00	15·6			2	1		1·597	1·621	1·655		2		2·357	2·459	2·646
Normal 11 Å tobermorite	$Ca_{4{\cdot}5}Si_6O_{16}(OH){\cdot}5H_2O$	0·5	0·75	0·92	13·9			2	1		1·595	1·639	1·711		3		2·277	2·476	2·872
Clinotobermorite	$Ca_5Si_6O_{17}{\cdot}5H_2O$	0	0·83	0·83	12·3			4	2		1·549	1·622	1·678		4	1	2·298	2·491	3·057
Clinotobermorite 9 Å	$Ca_5Si_6O_{16}(OH)_2$		0·83	0·17	2·7			3			1·554	1·627	1·677	1	2		2·292	2·453	2·686

Table 5.20(c) Jennite

Name	Formula	Ca/Si	H_2O/Si	%H_2O (mass)	Connectivity of silicate tetrahedra					Si–O distance: Å			Ca–O coordination			Ca–O distance: Å		
					Q^0	Q^1	Q^2	Q^3	Q^4	Min.	Average	Max.	6	7	8	Min.	Av.	Max.
Jennite	$Ca_9Si_6O_{18}(OH)_6{\cdot}8H_2O$	1·50	1·83	18·6			3			1·569	1·637	1·709	4	1		2·331	2·412	2·769

Table 5.20(d) Gyrolite group

Name	Formula	Ca/Si	H_2O/Si	%H_2O (mass)	Connectivity of silicate tetrahedra					Si–O distance: Å			Ca–O coordination			Ca–O distance: Å		
					Q^0	Q^1	Q^2	Q^3	Q^4				6	7	8	Min.	Av.	Max.
Fedorite	$(Na,K)_2(Ca,Na)_7(Si,Al)_{16}O_{38}(F,OH)_2\cdot3\cdot5H_2O$[b]	0·56[a]	0·22[b,c]	4·3				6	2	1·574	1·612	1·633	4			2·316	2·431	2·584
Gyrolite	$NaCa_{16}Si_{23}AlO_{60}(OH)_8\cdot14H_2O$	0·71[a]	0·75[c]	12·1				12		1·566	1·617	1·685	5	3		2·231	2·426	2·861
K-phase	$Ca_7Si_{16}O_{38}(OH)_2$	0·44	0·06	1·3				6	2	1·552	1·598	1·629	4			2·349	2·427	2·483
Reyerite	$Na_2Ca_{14}Si_{22}Al_2O_{58}(OH)_8\cdot6H_2O$	0·67[a]	0·42[c]	7·4				4[d]	1[e]	1·569	1·615	1·636	2	1		2·319	2·432	2·880
Z-phase	$Ca_9Si_{16}O_{40}(OH)_2\cdot14H_2O$	0·56	0·94	15·6				8		1·555	1·620	1·694	5			2·264	2·434	2·555

[a] (Na + K+Ca)/(Al + Si).
[b] Actual composition: $(Na_{1\cdot29}K_{0\cdot79})(Ca_{4\cdot48}Na_{2\cdot52})Si_{16}O_{38}F_2\cdot3\cdot47H_2O$.
[c] $H_2O/(Si + Al)$.
[d] Q^3(1Al);
[e] Q^4(1Al).

Table 5.20(e) γ-C_2S group

Name	Formula	Ca/Si	H_2O/Si	%H_2O (mass)	Connectivity of silicate tetrahedra					Si–O distance: Å			Ca–O coordination			Ca–O distance: Å		
					Q^0	Q^1	Q^2	Q^3	Q^4	Min.	Average	Max.	6	7	8	Min.	Av.	Max.
Calcium chondrodite*	$Ca_5(SiO_4)_2(OH)_2$	2·50	0·5	4·3	1					1·620	1·640	1·653	3			2·313	2·378	2·502
Chegemite	$Ca_7(SiO_4)_3(OH)_2$	2·33	0·33	3·1	2					1·623	1·640	1·650	4			2·291	2·376	2·476
Kilchoanite	$Ca_6(SiO_4)(Si_3O_{10})$	1·50	0	0	1	1	1			1·576	1·647	1·717	3		1	2·241	2·416	2·613

*Synthetic phase similar to the mineral reinhardbraunsite, $Ca_5(SiO_4)_2(OH,F)_2$.

Table 5.20(f) Other calcium silicate phases

Name	Formula	Ca/Si	H_2O/Si	%H_2O (mass)	Connectivity of silicate tetrahedra					Si–O distance: Å			Ca–O/F coordination			Ca–O/F distance: Å		
					Q^0	Q^1	Q^2	Q^3	Q^4	Min.	Average	Max.	6	7	8	Min.	Av.	Max.
Afwillite	$Ca_3(SiO_3OH)_2{\cdot}2H_2O$	1·50	1·50	15·8	2					1·607	1·635	1·688		3		2·325	2·467	2·870
Bultfonteinite	$Ca_4(SiO_3OH)_2F_2{\cdot}2H_2O$	2·00	1·50	12·9	2					1·592	1·644	1·695		4		2·245	2·407	2·604
α-C_2SH	$Ca_2(SiO_3OH)(OH)$	2·00	1·00	9·5	1					1·606	1·641	1·710	1	1		2·266	2·438	2·936
Cuspidine	$Ca_4(F_{1{\cdot}5}(OH)_{0{\cdot}5})Si_2O_7$	2·00	0·13	1·2		2				1·600	1·619	1·658	1	2	1	2·275	2·428	2·843
Dellaite	$Ca_6(Si_2O_7)(SiO_4)(OH)_2$	2·00	0·33	3·4	1	2				1·604	1·631	1·649	3	3		2·276	2·432	2·833
Jaffeite	$Ca_6[Si_2O_7](OH)_6$	3·00	1·50	10·6		4				1·552	1·635	1·756	4			2·275	2·395	2·567
Killalaite	$Ca_{6{\cdot}4}(H_{0{\cdot}6}Si_2O_7)_2(OH)_2$	1·60	0·40	4·6		2				1·602	1·623	1·655	3	2		2·292	2·434	2·746
Poldervaartite[a]	$Ca(Ca_{0{\cdot}67}Mn_{0{\cdot}33})$-$(SiO_3OH)(OH)$	2·00[b]	1·00	9·2	1					1·620	1·636	1·697	1	1		2·204	2·413	2·881
Rosenhahnite	$Ca_3Si_3O_8(OH)_2$	1·00	0·33	4·9		2	1			1·584	1·630	1·677	1	2		2·290	2·442	2·917
Suolunite	$CaSiO_{2{\cdot}5}(OH){\cdot}0{\cdot}5H_2O$	1·00	1·00	13·4		1				1·590	1·622	1·662			1	2·296	2·518	2·796
Tilleyite	$Ca_5Si_2O_7(CO_3)_2$	2·50	0	0		1				1·608	1·625	1·659	1	3	1	2·265	2·440	2·733

[a]Isostructural with α-C_2SH.
[b](Ca + Mn)/Si.

Table 5.20(g) Other high temperature cement phases

Name	Formula	Ca/Si	H_2O/Si	%H_2O (mass)	Connectivity of silicate tetrahedra					Si–O distance: Å			Ca–O/Cl coordination			Ca–O/Cl distance: Å		
					Q^0	Q^1	Q^2	Q^3	Q^4	Min.	Average	Max.	6	7	8	Min.	Av.	Max.
Fukalite[b]	$Ca_4(Si_2O_6)(CO_3)(OH)_2$	2·00	0·50	4·4			4			1·576	1·617	1·679	2	6		2·290	2·433	2·748
Rustumite	$Ca_{10}(Si_2O_7)_2(SiO_4)Cl_2(OH)_2$	2·00	0·33	1·9	1	2				1·576	1·632	1·673	2	3		2·250	2·480	3·008[a]
Scawtite	$Ca_7(Si_6O_{18})(CO_3)\cdot 2H_2O$	1·17	0·33	4·3			3			1·582	1·622	1·663	1	2	1	2·296	2·479	2·882

[a] Ca–Cl.
[b] MDO_1 polytype.

REFERENCES

Alberti A and Galli E (1980) The structure of nekoite, $Ca_3Si_6O_{15}\cdot 7H_2O$, a new type of sheet silicate. *American Mineralogist* **65**: 1270–1276.

Andersen MD, Jakobsen HJ and Skibsted J (2003) Incorporation of aluminum in the calcium silicate hydrate (C–S–H) of hydrated Portland cements: a high-field 27Al and 29Si MAS NMR investigation. *Inorganic Chemistry* **42(7)**: 2280–2287.

Andersen AD, Jakobsen HJ and Skibsted J (2004) Characterization of white Portland cement hydration and the C–S–H structure in the presence of sodium aluminate by ^{27}Al and ^{29}Si MAS NMR spectroscopy. *Cement and Concrete Research* **34(5)**: 857–868.

Bell GMM, Bensted J, Glasser FP, Lachowski EE, Roberts DR and Taylor MJ (1990) Study of calcium silicate hydrates by solid state high resolution ^{29}Si nuclear magnetic resonance. *Advances in Cement Research* **3(9)**: 23–27.

Biagioni C, Bonaccorsi E and Merlino S (2010) Crystal structure of bultfonteinite, $Ca_4[SiO_3(OH)]_2F_2\cdot 2H_2O$, from N'Chwaning II mine. *Atti della Società Toscana di Scienze Naturali, Memorie, Serie A* **115**: 9–15.

Biagioni C, Bonaccorsi E, Lezzerini M, Merlini M and Merlino S (2012a) Thermal behaviour of tobermorite from N'Chwaning II mine (Kalahari Manganese Field, Republic of South Africa). I. Thermo-gravimetric and X-ray diffraction studies. *European Journal of Mineralogy* **24(6)**: 981–989.

Biagioni C, Bonaccorsi E, Merlino S, Bersani D and Forte C (2012b) Thermal behaviour of tobermorite from N'Chwaning II mine (Kalahari Manganese Field, Republic of South Africa) II. Crystallographic and spectroscopic study of tobermorite 10 Å. *European Journal of Mineralogy* **24(6)**: 991–1004.

Biagioni C, Bonaccorsi E, Merlino S and Bersani D (2013) New data on the thermal behavior of 14 Å tobermorite. *Cement and Concrete Research* **49**: 48–54.

Biagioni C, Merlino S and Bonaccorsi E (2015) The tobermorite supergroup: a new nomenclature. *Mineralogical Magazine* **79(2)**: 485–495.

Bonaccorsi E and Merlino S (2005) Modular microporous minerals: cancrinite-davyne group and C–S–H phases. *Reviews in Mineralogy and Geochemistry* **57(1)**: 241–290.

Bonaccorsi E, Merlino S and Taylor HFW (2002) Crystal structures of jennite and tobermorites: new data for understanding the nanostructure of Portland cement. *Congress of the Italian Association of Crystallography*.

Bonaccorsi E, Merlino S and Taylor HFW (2004) The crystal structure of jennite, $Ca_9Si_6O_{18}(OH)_6\cdot 8H_2O$. *Cement and Concrete Research* **34(9)**: 1481–1488.

Bonaccorsi E, Merlino S and Kampf AR (2005) The crystal structure of tobermorite 14Å (plombierite), a C–S–H phase. *Journal of the American Ceramic Society* **88(3)**: 505–512.

Brunet F, Bertani P, Charpentier T, Nonat A and Virlet J (2004) Application of ^{29}Si homonuclear and ^{1}H-^{29}Si heteronuclear NMR correlation to structural studies of calcium silicate hydrates. *The Journal of Physical Chemistry B* **108(40)**: 15494–15502.

Buckner DA, Roy DM and Roy R (1960) Studies in the system CaO–Al_2O_3–SiO_2–H_2O. II. The System $CaSiO_3$–H_2O. *American Journal of Science* **258**: 132–147.

Buerger MJ (1956) The determination of the crystal structure of pectolite, $Ca_2NaHSi_3O_9$. *Zeitschrift für Kristallographie* **108(3–4)**: 248–262.

Buerger MJ and Prewitt CT (1961) The crystal structure of wollastonite and pectolite. *Proceedings of the National Academy of Sciences* **47**: 1884–1888.

Burnham CW (1959) Contact metamorphism of magnesian limestones at Crestmore, California. *Geological Society of America Bulletin* **70(7)**: 879–920.

Carpenter AB, Chalmers RA, Gard JA, Speakman K and Taylor HFW (1966) Jennite, a new mineral. *American Mineralogist* **51**: 56–74.

CEMBUREAU (2018) *Activity Report 2017*. CEMBUREAU, The European Cement Association, Brussels, Belgium. https://cembureau.eu/media/1716/activity-report-2017.pdf.

Cliff G, Gard JA, Lorimer GW and Taylor HFW (1975) Tacharanite. *Mineralogical Magazine* **40(310)**: 113–126.

Coleman NJ (2006) Interactions of Cd(II) with waste-derived 11 Å tobermorites. *Separation and Purification Technology* **48(1)**: 62–70.

Cong X and Kirkpatrick RJ (1996) ^{29}Si and ^{17}O NMR investigation of the structure of some crystalline calcium silicate hydrates. *Advanced Cement Based Materials* **3(3–4)**: 133–143.

CrystalMaker (n.d.) Crystal-structure images generated using CrystalMaker®; XRD patterns calculated using CrystalDiffract®. CrystalMaker Software, Oxford, UK. See www.crystalmaker.com

Dai YS and Post JE (1995) Crystal structure of hillebrandite: a natural analogue of calcium silicate hydrate (CSH) phases in Portland cement. *American Mineralogist* **80(7–8)**: 841–844.

Dai YS, Harlow GE and McGhie AR (1993) Poldervaartite, $Ca(Ca_{0.5}Mn_{0.5})(SiO_3(OH))(OH)$, a new acid nesosilicate from the Kalahari manganese field, South Africa: Crystal structure and description. *American Mineralogist* **78**: 1082–1087.

Dann SE, Mead PJ and Weller MT (1996) Lowenstein's rule extended to an aluminium rich framework. The structure of bicchulite, $Ca_8(Al_2SiO_6)_4(OH)_8$, by MASNMR and neutron diffraction. *Inorganic Chemistry* **35(6)**: 1427–1428.

Daubrée GA (1858) Sur la relation des sources thermales de Plombières avec les filons métallifères et la formation contempraine des zéolites. *Annales des Mines* **s5 13**: 227–256.

Diamond S, White JL and Dolch WL (1966) Effects of isomorphous substitution in hydrothermally-synthesized tobermorite. *American Mineralogist* **51**: 388–401.

Dornberger-Schiff K (1956) On order-disorder structures (OD-structures). *Acta Crystallographica* **9(7)**: 593–601.

El-Hemaly SAS, Mitsuda T and Taylor HFW (1977) Synthesis of normal and anomalous tobermorites. *Cement and Concrete Research* **7(4)**: 429–438.

Engelhardt G and Michel D (1987) *High Resolution Solid-State NMR of Silicates and Zeolites*. Wiley, New York, NY, USA.

Farmer VC, Jeevaratnam J, Speakman K and Taylor HFW (1966) Thermal decomposition of 14 Å tobermorite from Crestmore. In *Symposium on Structure of Portland Cement Paste and Concrete*, Highway Research Board, Washington, DC, USA, *Special Report 90* pp. 291–299.

Ferro O, Galli E, Papp G, Quartieri S, Szakáll S and Vezzalini G (2003) A new occurrence of katoite and re-examination of the hydrogrossular group. *European Journal of Mineralogy* **15(2)**: 419–426.

Funk H (1962) Two different ways of hydration in the reaction of β-C_2S with water at 25 to 120°C. *Proceedings of 4th International Symposium on the Chemistry of Cement* **1**: 291–295.

Galuskin EV, Gazeev VM, Lazic B *et al.* (2009) Chegemite $Ca_7(SiO_4)_3(OH)_2$ – a new humite-group calcium mineral from the Northern Caucasus, Kabardino-Balkaria, Russia. *European Journal of Mineralogy* **21(5)**: 1045–1059.

Gard JA and Taylor HFW (1957) A further investigation of tobermorite from Loch Eynort, Scotland. *Mineralogical Magazine* **31(236)**: 361–370.

Gard JA and Taylor HFW (1960) The crystal structure of foshagite. *Acta Crystallographica* **13(10)**: 785–793.

Gard JA, Mitsuda T and Taylor HFW (1975) Some observations on Assarsson's Z-phase and its structural relations to gyrolite, truscottite, and reyerite. *Mineralogical Magazine* **40**: 325–333.

Gard JA, Taylor HFW, Cliff G and Lorimer GW (1977) A reexamination of jennite. *American Mineralogist* **62**: 365–368.

Gard JA, Luke K and Taylor HFW (1981) Crystal structure of K-phase, $Ca_7Si_{16}O_{40}H_2$. *Soviet physics. Crystallography* **26**: 691–695.

Gmira A, Pellenq R, Rannou JM *et al.* (2002) A structural study of dehydration/rehydration of tobermorite, a model cement compound. *Studies in Surface Science and Catalysis* **144**: 601–608.

Grice JD (2005) The structure of spurrite, tilleyite and scawtite, and relationships to other silicate-carbonate minerals. *The Canadian Mineralogist* **43(5)**: 1489–1500.

Hamid SA (1981) The crystal structure of the 11 Å natural tobermorite $Ca_{2.25}[Si_3O_{7.5}(OH)_{1.5}]\cdot 1H_2O$. *Zeitschrift für Kristallographie* **154**: 189–198.

Hara N and Inoue N (1979) Hydrothermal reaction of silica glass with lime at 70°~100°C. *Yogyo-Kyokai-Shi* **87**: 135–141. (In Japanese with English abstract.)

Hara N and Inoue N (1980a) Thermal behaviour of 11 Å tobermorite and its lattice parameters. *Cement and Concrete Research* **10(1)**: 53–60.

Hara N and Inoue N (1980b) Formation of 10 Å and 14 Å tobermorite from pozzolanic glassy silica. *Proceedings of the 7th International Congress on the Chemistry of Cement* **3**: 13–18.

Hara N and Inoue N (1980c) Formation of jennite from fumed silica. *Cement and Concrete Research* **10(5)**: 677–682.

Hara N, Chan CF and Mitsuda T (1978) Formation of 14 Å tobermorite. *Cement and Concrete Research* **8(1)**: 113–115.

Hara N, Inoue N and Noma H (1986) Formation and characterization of jennite. *Proceedings of the 8th International Congress on the Chemistry of Cement* **3**: 160–166.

Hara N, Inoue N and Noma H (1997) Thermal behaviour of tobermorite intercalated with alkali ions. *Proceedings of the 10th International Congress on the Chemistry of Cement* **2**, Paper **2(ii)**065, 5 pp.

Hejny C and Armbruster T (2001) Polytypism in xonotlite $Ca_6Si_6O_{17}(OH)_2$. *Zeitschrift für Kristallographie* **216**: 396–408. (Note: polytype derived from structure in Kudoh Y and Takéuchi Y (1979) Polytypism of xonotlite: (I) Structure of an $A\bar{1}$ polytype. *Mineralogical Journal* **9(6)**: 349–373.)

Heller L and Taylor HFW (1956) *Crystallographic Data for the Calcium Silicates.* HMSO, London, UK.

Henmi C and Kusachi I (1992) Clinotobermorite, $Ca_5Si_6(O,OH)_{18}\cdot 5H_2O$, a new mineral from Fuka, Okayama Prefecture, Japan. *Mineralogical Magazine* **56(384)**: 353–358.

Henmi C, Kusachi I, Kawahara A and Henmi K (1977) Fukalite, a new calcium carbonate silicate hydrate mineral. *Mineralogical Journal* **8(7)**: 374–381.

Henning O (1974) Cements: the hydrated silicates and aluminates. In *The Infrared Spectra of Minerals* (Farmer VC (ed.)). Mineralogical Society, London, UK. pp. 445–463.

Hoffmann C and Armbruster T (1997) Clinotobermorite, $Ca_5[Si_3O_8(OH)]_2\cdot 4H_2O$ – $Ca_5[Si_6O_{17}]\cdot 5H_2O$, a natural C–S–H(I) type cement mineral: determination of the substructure. *Zeitschrift für Kristallographie* **212**: 864–873.

Hong SY and Glasser FP (2004) Phase relations in the $CaO\text{-}SiO_2\text{-}H_2O$ system to 200°C at saturated steam pressure. *Cement and Concrete Research* **34(9)**: 1529–1534.

Howie RA and Ilyukhin VV (1977) Crystal structure of rustumite. *Nature* **269**: 231–231.

Kalousek GL (1955) Tobermorite and related phases in the system $CaO–SiO_2–H_2O$. *Journal of the American Concrete Institute (Proceedings)* **51**: 989–1011.

Kalousek GL (1957) Crystal chemistry of hydrous calcium silicates: I, substitution of aluminum in lattice of tobermorite. *Journal of the American Ceramic Society* **40(3)**: 74–80.

Kalousek GL and Roy R (1957) Crystal chemistry of hydrous calcium silicates: II, characterization of interlayer water. *Journal of the American Ceramic Society* **40(7)**: 236–239.

Kazak VF, Blinov VA, Ilyukhin VV and Belov NV (1974) Crystal structure of hydrated tricalcium silicate – TSH. *Doklady Akademii Nauk SSSR* **219**: 340–343. (In Russian.)

Kirkpatrick RJ, Yarger JL, McMillan PF, Ping Y and Cong X (1997a) Raman spectroscopy of C–S–H, tobermorite and jennite. *Advanced Cement Based Materials* **5(3–4)**: 93–99.

Kirkpatrick RJ, Brown GE, Xu N and Cong X (1997b) Ca X-ray absorption spectroscopy of C–S–H and some model compounds. *Advances in Cement Research* **9(33)**: 31–36.

Klur I, Pollet B, Virlet J and Nonat A (1998) C–S–H structure evolution with calcium content by multinuclear NMR. In *Nuclear Magnetic Resonance Spectroscopy of Cement-Based Materials* (Colombet P, Grimmer AR, Zanni H and Sozzani P (eds)). Springer, Berlin, Germany, pp. 119–141.

Kolodny Y and Gross S (1974) Thermal metamorphism by combustion of organic matter: isotopic and petrological evidence. *The Journal of Geology* **82(4)**: 489–506.

Komarneni S, Roy DM and Roy R (1982) Al-substituted tobermorite: shows cation exchange. *Cement and Concrete Research* **12(6)**: 773–780.

Komarneni S, Roy R, Roy DM, Fyfe CA and Kennedy GJ (1985a) Al-substituted tobermorite – the coordination of aluminum as revealed by solid-state ^{27}Al magic angle spinning (MAS) NMR. *Cement and Concrete Research* **15(4)**: 723–728.

Komarneni S, Roy R, Roy DM *et al.* (1985b) ^{27}Al and ^{29}Si magic angle spinning nuclear magnetic resonance spectroscopy of Al-substituted tobermorites. *Journal of Materials Science* **20(11)**: 4209–4214.

Komarneni S, Breval E, Roy DM and Roy R (1986a) Synthesis and characterization of a 12.6 Å calcium silicate hydrate. *Cement and Concrete Research* **16(4)**: 580–586.

Komarneni S, Roy R and Roy DM (1986b) Pseudomorphism in xonotlite and tobermorite with Co^{2+} and Ni^{2+} exchange for Ca^{2+} at 25°C. *Cement and Concrete Research* **16(1)**: 47–58.

Komarneni S, Roy DM, Fyfe CA and Kennedy GJ (1987) Naturally occurring 1.4 nm tobermorite and synthetic jennite: characterization by ^{27}Al and ^{29}Si MASNMR spectroscopy and cation exchange properties. *Cement and Concrete Research* **17(6)**: 891–895.

Kusachi I, Henmi C and Henmi K (1984) An oyelite-bearing vein at Fuka, the town of Bitchu, Okayama Prefecture. *The Journal of the Japanese Association of Mineralogists, Petrologists and Economic Geologists* **79(7)**: 267–275.

Kusachi I, Henmi C and Henmi K (1989) Afwillite and jennite from Fuka, Okayama Prefecture, Japan. *Mineralogical Journal* **14(7)**: 279–292.

Kuznetsova TP, Nevskii NN, Ilyukhin VV and Belov NV (1980) Refinement of the crystal structure of calcium chondrodite, $Ca_5[SiO_4]_2(OH)_2 = Ca(OH)_2{\cdot}2Ca_2SiO_4$. *Soviet Physics. Crystallography* **25**: 91–92 (translation of *Kristallografiya* **25**: 159–160 (1980)).

Liebau F (1985) *Structural Chemistry of Silicates – Structure, Bonding, and Classification.* Springer, Berlin, Germany.

Lognot I, Klur I and Nonat A (1998) NMR and infrared spectroscopies of C–S–H and Al-substituted C–S–H synthesised in alkaline solution. In Colombet P, Grimmer AR, Zanni H and Sozzani P (eds), *Nuclear Magnetic Resonance Spectroscopy of Cement-Based Materials*. Springer, Berlin, Germany, pp. 189–196.

Long JVP and McConnell JDC (1959) A mineralogical application of X-ray absorption microspectroscopy: the hydration of larnite. *Mineralogical Magazine* **32(245)**: 117–127.

Ma Z, Shi N, Mou G and Liau L (1999) Crystal structure refinement of suolunite and its significance to the cement techniques. *Chinese Science Bulletin* **44(23)**: 2125–2130.

Maeshima T, Noma H, Sakiyama M and Mitsuda T (2003) Natural 1·1 and 1·4 nm tobermorites from Fuka, Okayama, Japan: Chemical analysis, cell dimensions, ^{29}Si NMR and thermal behavior. *Cement and Concrete Research* **33**: 1515–1523.

Malik KMA and Jeffery JW (1976) A re-investigation of the structure of afwillite. *Acta Crystallographica* **32(2)**: 475–480.

Mamedov KS and Belov NV (1958) On the crystal structure of tobermorite (tobermorites). *Doklady Akad. Nauk SSSR* **123**: 163–165. (In Russian.)

Marsh RE (1994) A revised structure for α-dicalcium silicate hydrate. *Acta Crystallographica C* **50(7)**: 996–997.

McConnell JDC (1954) The hydrated calcium silicates riversideite, tobermorite, and plombierite. *Mineralogical Magazine* **30(224)**: 293–305.

McConnell JDC (1955) The hydration of larnite (β-Ca_2SiO_4) and bredigite (α_1-Ca_2SiO_4) and the properties of the resulting gelatinous mineral plombierite. *Mineralogical Magazine* **30(229)**: 672–680.

McConnell JDC (1971) The microstructure and properties of the bonding medium in set cement: a restudy of the natural hydrogel. *Transactions and Journal of the British Ceramic Society* **70**: 171–179.

McIver EJ (1963) The structure of bultfonteinite, $Ca_4Si_2O_{10}F_2H_6$. *Acta Crystallographica* **16(6)**: 551–558.

Megaw HD and Kelsey CH (1956) Crystal structure of tobermorite. *Nature* **177**: 390–391.

Merlino S (1983) Okenite, $Ca_{10}Si_{18}O_{46}\cdot 18H_2O$: the first example of a chain and sheet silicate. *American Mineralogist* **68**: 614–622.

Merlino S (1988) Gyrolite: its crystal structure and crystal chemistry. *Mineralogical Magazine* **52**: 377–387.

Merlino S, Bonaccorsi E and Armbruster T (1999) Tobermorites: their real structure and order-disorder (OD) character. *American Mineralogist* **84(10)**: 1613–1621.

Merlino S, Bonaccorsi E and Armbruster T (2000) The real structures of clinotobermorite and tobermorite 9Å: OD character, polytypes, and structural relationships. *European Journal of Mineralogy* **12(2)**: 411–429.

Merlino S, Bonaccorsi E and Armbruster T (2001) The real structure of tobermorite 11 Å: normal and anomalous forms, OD character and polytypic modifications. *European Journal of Mineralogy* **13(3)**: 577–590.

Merlino S, Bonaccorsi E, Grabezhev AI, Zadov AE, Pertsev NN and Chukanov NV (2009) Fukalite: an example of an OD structure with two-dimensional disorder. *American Mineralogist* **94(2–3)**: 323–333.

Mitchell RH and Burns PC (2001) The structure of fedorite: a reappraisal. *The Canadian Mineralogist* **39(3)**: 769–777.

Mitsuda T (1970) Synthesis of tobermorite from zeolite. *Mineralogical Journal* **6(3)**: 143–158.

Mitsuda T (1973) Paragenesis of 11 Å tobermorite and poorly crystalline hydrated magnesium silicate. *Cement and Concrete Research* **3(1)**: 71–80.

Mitsuda T and Taylor HFW (1978) Normal and anomalous tobermorites. *Mineralogical Magazine* **42**: 229–235.

Mitsuda T, Kusachi I and Henmi K (1972) Mixtures of 14 Å and 11 Å tobermorite from Fuka, Japan. *Cement Association Japan Review* **26**: 47–68.

Mitsuda T, Toraya H, Okada Y and Shimoda M (1989) Synthesis of tobermorite: NMR spectroscopy and analytical electron microscopy. *Ceramic Transactions* **5**: 206–213.

Müller WF (1976) On stacking disorder and polytypism in pectolite and serandite. *Zeitschrift für Kristallographie* **144**: 401–408.

Murdoch J (1961) Crestmore, past and present. *American Mineralogist* **46**: 245–257.

Nawaz R (1977) A second occurrence of killalaite. *Mineralogical Magazine* **41**: 546–548.

Ohashi Y (1984) Polysynthetically-twinned structures of enstatite and wollastonite. *Physics and Chemistry of Minerals* **10(5)**: 217–229.

Ohashi Y and Finger LW (1978) The role of octahedral cations in pyroxenoid crystal chemistry. I. Bustamite, wollastonite, and the pectolite-schizolite-serandite series. *American Mineralogist* **63**: 274–288.

Pellenq RJM, Kushima A, Shahsavari R *et al.* (2009) A realistic molecular model of cement hydrates. *Proceedings of the National Academy of Sciences* **106(38)**: 16102–16107.

Pourbeik P, Beaudoin JJ, Alizadeh R and Raki L (2013) Mechanical property–porosity relationships of layered calcium silicate hydrate phases. *Materials and Structures* **46**: 1489–1495.

Prewitt CT (1967) Refinement of the structure of pectolite, $Ca_2NaHSi_3O_9$. *Zeitschrift für Kristallographie* **125**: 298–316.

Prewitt CT and Buerger MJ (1963) Comparison of the crystal structures of wollastonite and pectolite. *American Mineralogical Society / 3rd General Meeting of the International Mineralogical Association, Special Paper 1:* 293–302.

Rastsvetaeva RK, Bolotina NB, Zadov AE and Chukanov NV (2005) Crystal structure of fukalite dimorph $Ca_4(Si_2O_6)(CO_3)(OH)_2$ from the Gumeshevsk deposit, the Urals. *Doklady Earth Sciences* **405**: 1347–1351.

Reinik J, Heinmaa I, Mikkola JP and Kirso U (2007) Hydrothermal alkaline treatment of oil shale ash for synthesis of tobermorites. *Fuel* **86**: 669–676.

Richardson IG (1999) The nature of C–S–H in hardened cements. *Cement and Concrete Research* **29(8)**: 1131–1147.

Richardson IG (2008) The calcium silicate hydrates. *Cement and Concrete Research* **38(2)**: 137–158.

Richardson IG (2014) Model structures for C–(A)–S–H(I). *Acta Crystallographica B* **70**: 903–923. https://doi.org/10.1107/S2052520614021982. Licensed under CC BY 2.0: https://creativecommons.org/licenses/by/2.0/

Richardson IG and Groves GW (1992) Models for the composition and structure of calcium silicate hydrate (C–S–H) gel in hardened tricalcium silicate pastes. *Cement and Concrete Research* **22(6)**: 1001–1010.

Richardson IG and Groves GW (1993) The incorporation of minor and trace elements into calcium silicate hydrate (C–S–H) gel in hardened cement pastes. *Cement and Concrete Research* **23(1)**: 131–138.

Richardson IG, Brough AR, Brydson R, Groves GW and Dobson CM (1993) Location of aluminum in substituted calcium silicate hydrate (C–S–H) gels as determined by ^{29}Si and ^{27}Al NMR and EELS. *Journal of the American Ceramic Society* **76(9)**: 2285–2288.

Richardson IG, Brough AR, Groves GW and Dobson CM (1994) The characterization of hardened alkali-activated blast-furnace slag pastes and the nature of the calcium silicate hydrate (C–S–H) phase. *Cement and Concrete Research* **24(5)**: 813–829.

Rinaldi R, Sacerdoti M and Passaglia E (1990) Strätlingite: crystal structure, chemistry, and a reexamination of its polytype vertumnite. *European Journal of Mineralogy* **2(6)**: 841–850.

Rodger SA, Groves GW, Clayden NJ and Dobson CM (1987) A study of tricalcium silicate hydration from very early to very late ages. *Proceedings of the Symposium of the Materials Research Society* **85**: 13–20.

Saburi S, Kawahara A, Henmi C, Kusachi I and Kihara K (1977) The refinement of the crystal structure of cuspidine. *Mineralogical Journal (Japan)* **8(5)**: 286–298.

Safronov AN, Nevsky NN, Ilyukhin VV and Belov NV (1981) The refinement of the crystal structure of the cement phase gamma-C_6S_3H. *Doklady Akademii Nauk SSSR* **256**: 1387–1389.

Sakiyama M, Maeshima T and Mitsuda T (2000) Synthesis and crystal chemistry of Al-substituted 11 Å tobermorite. *Journal of the Society of Inorganic Materials of Japan* **7**: 413–419.

Sarp J and Peacor DR (1989) Jaffeite, a new hydrated calcium silicate from the Kombat mine, Namibia. *American Mineralogist* **74**: 1203–1206.

Sato H and Grutzeck M (1992) Effect of starting materials on the synthesis of tobermorite. *Proceedings of the Symposium of the Materials Research Society* **245**: 235–240.

Scrivener KL, John VM and Gartner EM (2016) *Eco-Efficient Cements: Potential, Economically Viable Solutions for a low-CO_2, Cement-Based Materials Industry*. United Nations Environment Programme, Paris, France.

Stephens JD and Bray E (1973) Occurrence and infrared analysis of unusual zeolitic minerals from bingham. *Utah Mineralogical Record* **4**: 67–72.

Sun GK, Young JF and Kirkpatrick RJ (2006) The role of Al in C–S–H: NMR, XRD, and compositional results for precipitated samples. *Cement and Concrete Research* **36(1)**: 18–29.

Sweet JM, Bothwell DI and Williams DL (1961) Tacharanite and other hydrated calcium silicates from Portree, Isle of Skye. *Mineralogical Magazine* **32(253)**: 745–753.

Takéuchi Y and Kudoh Y (1977) Hydrogen bonding and cation ordering in Magnet Cove pectolite. *Zeitschrift für Kristallographie* **146**: 281–292.

Taylor HFW (1953) Crestmoreite and riversideite. *Mineralogical Magazine* **30(222)**: 155–165.

Taylor HFW (1959) The dehydration of tobermorite. *Proceedings of 6th National Conference on Clays and Clay Minerals, Berkeley, 1957*. Pergamon, London, UK, pp. 101–109.

Taylor HFW (1962) Hydrothermal reactions in the system $CaO–SiO_2–H_2O$ and the steam curing of cement and cement-silica products. *Proceedings of the 4th International Symposium on the Chemistry of Cement* **1**: 167–190 (specifically, p. 173).

Taylor HFW (1969) The calcium silicate hydrates. *Proceedings of the 5th International Symposium on the Chemistry of Cement* **2**: 1–26.

Taylor HFW (1971) The crystal structure of kilchoanite, $Ca_6(SiO_4)$ (Si_3O_{10}), with some comments on related phases. *Mineralogical Magazine* **38**: 26–31.

Taylor HFW (1977) The crystal structure of killalaite. *Mineralogical Magazine* **41**: 363–369.

Taylor HFW (1985) The reactions of cement compounds with water. In *Materials Science Monographs 28A: Reactivity of Solids, part A*: 39–45.

Taylor HFW (1993) A discussion of the papers 'Models for the composition and structure of calcium silicate hydrate (C–S–H) gel in hardened tricalcium silicate pastes' and 'The incorporation of minor and trace elements into calcium silicate hydrate (C–S–H) gel in hardened cement pastes' by Richardson IG and Groves GW. *Cement and Concrete Research* **23(4)**: 995–998.

Trojer FJ (1968) The crystal structure of parawollastonite. *Zeitschrift für Kristallographie* **127(1–4)**: 291–308.

Urabe K, Udagawa S and Ikawa H (1983) High resolution electron microscopy of tobermorite. *Cement Association of Japan Review* **37**: 29–35.

US Geological Survey (2019) *Mineral Commodity Summaries 2019*. US Geological Survey, Reston VA, USA, https://doi.org/10.3133/70202434

Viehland D, Yuan LJ, Xu Z, Cong XD and Kirkpatrick RJ (1997) Structural studies of jennite and 1.4 nm tobermorite: disordered layering along the [100] of jennite. *Journal of the American Ceramic Society* **80(12)**: 3021–3028.

Wan C, Ghose S and Gibbs GV (1977) Rosenhahnite, $Ca_3Si_3O_8(OH)_2$: crystal structure and the stereochemical configuration of the hydroxylated trisilicate group, $[Si_3O_8(OH)]$. *American Mineralogist* **62(2)**: 503–512.

Wieker W (1968) Silikatanionenstruktur des 14 Å Tobermorits von Crestmore und seiner Entwässerungsprodukte. *Zeitschrift für anorganische und allgmeine Chemie* **360**: 307–316. (In German.)

Wieker W, Grimmer AR, Winkler A, Mägi M, Tarmak M and Lippmaa E (1982) Solid-state high-resolution ^{29}Si NMR spectroscopy of synthetic 14 Å, 11 Å and 9 Å tobermorites. *Cement and Concrete Research* **12(3)**: 333–339.

Williams PA, Hatert F, Pasero M and Mills SJ (2014) IMA commission on new minerals, nomenclature and classification (CNMNC) newsletter 21. New minerals and nomenclature modifications approved in 2014. *Mineralogical Magazine* **78(4)**: 797–804.

Winkler A and Wieker W (1979) Über Synthese, Aufbau und thermisches Verhalten von 11 Å Tobermorit. *Zeitschrift für anorganische und allgemeine Chemie* **451(1)**: 45–56. (In German.)

Yamazaki S and Toraya H (2001) Determination of positions of zeolitic calcium atoms and water molecules in hydrothermally formed aluminium-substituted tobermorite-1.1 nm using synchrotron radiation powder diffraction data. *Journal of the American Ceramic Society* **84(11)**: 2685–2690.

Yamnova NA, Sarp K, Egorov-Tismenko YK, Pushcharovskii DY and Dasgupta G (1993) Crystal structure of jaffeite. *Crystallography Reports (c/c of Kristallografiia)* **38**: 464–466.

Yu P and Kirkpatrick RJ (1999) Thermal dehydration of tobermorite and jennite. *Concrete Science and Engineering* **1**: 185–191.

Yu P, Kirkpatrick RJ, Poe B, McMillan PF and Cong X (1999) Structure of calcium silicate hydrate (C–S–H): near-, mid-, and far-infrared spectroscopy. *Journal of the American Ceramic Society* **82**: 742–748.

Zadov AE, Chukanov NV, Organova NI *et al.* (1995) The new findings and study of the tobermorite group minerals. *Proceedings of the Russian Mineralogical Society* **124**: 36–54. (In Russian.)

Zhang L, Fu P, Yang H, Yu K and Zhou Z (1992) Crystal structure of scawtite. *Chinese Science Bulletin* **37**: 930–934.

Ian G Richardson and HFW Taylor
ISBN 978-1-83549-777-7
https://doi.org/10.1108/978-1-83549-776-020253006

Chapter 6
Calcium (alumino)silicate hydrates: synthetic preparations

6.1. Introduction

The C–(A–)S–H that forms as the main binding phase in hardened cements is not a thermodynamically stable phase but it is kinetically persistent under the normal service conditions for concrete, which is the reason why some thermodynamically stable phases (e.g. the crystalline calcium silicate hydrates tobermorite and jennite) are suppressed in thermodynamic equilibrium calculations for cementitious systems (Damidot *et al.*, 2011; Lothenbach, 2010) – that is, the kinetics of any transformation are so slow that the stable phases do not occur over relevant timescales. This chapter deals with the compositions and structures of synthetic C–(A–)S–H preparations, defined as artificially prepared materials other than those formed in cement or calcium silicate pastes. They provide important model materials for understanding the C–(A–)S–H of cement pastes, which is generally intermixed on such a fine scale with other phases that characterisation is difficult. Synthetic C–(A–)S–H preparations are structurally and compositionally complex and vary widely in crystallinity. At one extreme, they are highly disordered, yielding poor diffraction patterns similar to those of the C–(A–)S–H in typical cement pastes. At the other, they are quite highly ordered, with well-defined particle shapes in transmission or scanning electron microscopy (TEM, SEM) and giving diffraction patterns approaching those of related crystalline compounds. Because these compounds are structurally and compositionally diverse, it is convenient to treat separately those materials with Ca/Si ratios less than 1·5, greater than 1·5 and formed at elevated temperatures.

6.2. Room temperature preparations with Ca/Si ratios of 0·7–1·5

6.2.1 Conditions of formation and metastable equilibria in the $CaO–SiO_2–H_2O$ system at ordinary temperatures

C–S–H preparations of this type are known as C–S–H(I), calcium silicate hydrate(I), tobermorite-type C–S–H or defect-tobermorite C–S–H. They are typically formed by bringing together calcium and silicate ions in the appropriate proportions in dilute aqueous suspensions at temperatures below ≈60 °C. They also form at higher temperatures as intermediate products that are replaced by crystalline phases on further processing. Many synthesis routes start with CH or anhydrous C_3S or β-C_2S and reactive forms of silica such as silica gel or silica fume. Ethyl orthosilicate has also been used, in particular in methods that were designed to maintain an essentially constant lime concentration in solution, but the results from those studies are treated separately in §6.3. In another method, solutions of a soluble calcium salt (typically nitrate) and an alkali silicate are mixed together, followed by washing to remove the soluble product.

Equilibria in the $CaO–SiO_2–H_2O$ system at ordinary temperatures have been studied widely for the light they may cast on cement hydration and on leaching of concrete by ground waters. Steinour (1947; 1954) reviewed early work, and further reviews have appeared regularly – for example,

Jennings (1986), Chen *et al.* (2004) and Walker *et al.* (2016). Most of the work has concerned synthetic preparations of C–S–H(I). If C–S–H(I) is placed in water or CH solutions, its Ca/Si ratio decreases or increases, depending on the concentration of the solution. Except at low CaO concentrations, the SiO_2 concentrations in solution are very low, so that transfer between solid and solution is almost entirely of CaO. The data may be plotted on a triangular CaO–SiO_2–H_2O diagram but, because of the low concentrations, it is more useful to employ separate plots of the Ca/Si ratio of the solid against the analytical concentration of CaO in solution (C_{CaO}) (Figure 6.1) or C_{SiO_2} against C_{CaO} (Figures 6.2(a) and 6.2(b)). The similarities between the six typical data sets plotted in Figure 6.1 that relate the Ca/Si mole ratios of C–S–H preparations and C_{CaO} show that a reasonably clear metastable equilibrium can be obtained. The references are given on the figure, and the temperature that was used in each study is also indicated. The concentration of a solution saturated with CH varies with temperature, and the appropriate value for each data set is indicated using a vertical dotted line, its position calculated using Equation 7.1 (Chapter 7). It is evident that the abrupt rise in each curve that occurs at high lime concentration coincides with the dotted line and the rise can therefore be attributed to the formation of CH as the solutions became saturated (Steinour, 1947). The horizontal full line is drawn at a Ca/Si ratio of 1·5: in all but one case (that of Greenberg and Chang, 1965); it is clear that the abrupt rise in the curves occurs at Ca/Si ratio

Figure 6.1 Metastable equilibrium curves relating the Ca/Si mole ratios of C–S–H preparations and the analytical concentration of CaO in solution, selected as six typical studies

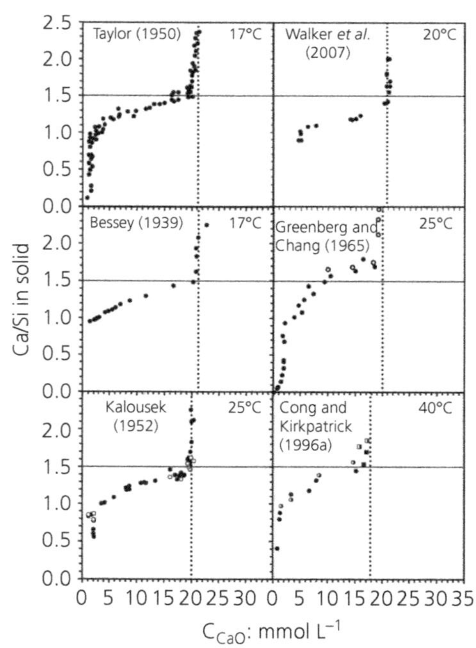

The data are from the references that are indicated on the figure. The temperature that was used in each study is also indicated, and the concentration of a solution saturated with CH at that particular temperature is indicated using the vertical dotted line, which was calculated using Equation 7.1 (Chapter 7).

≤1·5, which is thus indicative of the maximum Ca/Si ratio of C–S–H(I) that it is possible to prepare using these methods.

Figure 6.2(a) is a plot of the analytical concentrations of SiO_2 (C_{SiO_2}) and CaO (C_{CaO}) in solution for C–S–H preparations; it is part of the water-rich region of the CaO–SiO_2–H_2O phase diagram. The data are from studies that did not involve the use of anhydrous calcium silicates. The eye-guide curves A and B are due to Jennings (1986) (who also considered preparations that involved C_3S), curve C is due to Chen *et al.* (2004) and the two thick-dashed lines are from Kulik (2011). The full vertical line represents the solubility of CH at 25 °C calculated using Equation 7.1 (Chapter 7), which is from the regression analysis of data for solutions that were made using large crystals, Figure 7.2 (Chapter 7). The dotted vertical line represents the higher apparent solubility that occurs with finely crystalline material, Figure 7.2. The points that fall along and between these two vertical lines correspond to the steep step at high C_{CaO} on the curves of Figure 6.1, and thus are likely to include CH, whether crystalline or amorphous. There are no data points to the right of the dotted

Figure 6.2(a) Plot of the analytical concentrations of SiO_2 (C_{SiO_2}) and CaO (C_{CaO}) in solution for C–S–H preparations from studies that did not involve the use of anhydrous calcium silicates

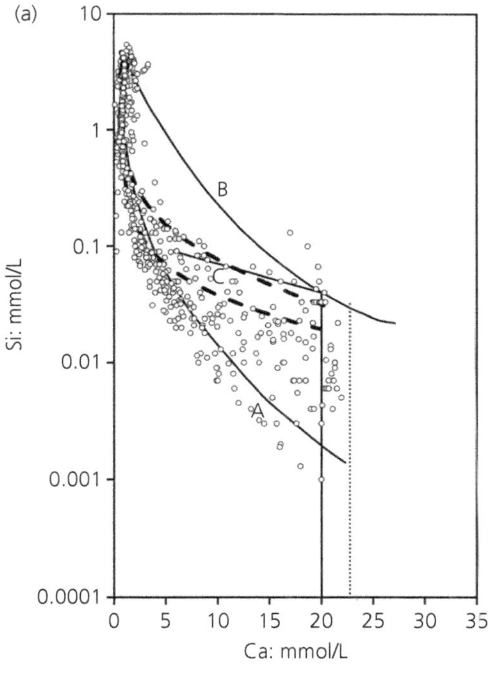

The data are from: Atkins *et al.* (1992); Atkinson *et al.* (1989); Barbarulo (2002); Chambers *et al.* (2005); Chen *et al.* (2004); Flint and Wells (1934); Fujii and Kondo (1981, 1983); Glasser *et al.* (1999); Greenberg and Chang (1965); Grutzeck *et al.* (1989); Haas (2012); Haas and Nonat (2015); Halstead and Lawrence (1960); Harris *et al.* (2002); Henocq (2005); Kalousek (1954); Plusquellec (2014); Pointeau (2000); Roller and Ervin (1940); Swanton *et al.* (2005, 2016); Sugiyama (2008); Suzuki *et al.* (1985); Taylor (1950); Tits *et al.* (2006); Walker *et al.* (2007, 2016).

vertical line, which indicates that the solutions do not become supersaturated with respect to CH when using these methods of preparation – that is, CH precipitates on saturation of the solution; the data on Figure 6.1 indicate that the Ca/Si ratio of the C–S–H is almost always <1·5. XRD powder evidence shows that the central part of the curves of Figure 6.1 (i.e. from $C_{CaO} \approx 2$ to $\approx$ 18–21 mmol L^{-1}) and curve A of Figure 6.2(a) represent the solubilities of C–S–H(I) preparations and that there are invariant points for hydrous silica, C–S–H(I) with Ca/Si ≈ 1·0 and solution with $C_{CaO} \approx 2\,mmol\,L^{-1}$ and $C_{SiO_2} \approx 2\,mmol\,L^{-1}$, and for CH, C–S–H(I) with Ca/Si ≈ 1·3–1·5 and solution with $C_{CaO} \approx$ 18–21 mmol L^{-1} and $C_{SiO_2} \approx 1\,\mu mol\,L^{-1}$ (Jennings, 1986; Steinour, 1954; Taylor, 1950). Curve A is clearly the solubility curve of C–S–H(I).

Figure 6.2(b) shows the same plot as (a) but only includes data from studies that involved the use of tricalcium silicate. Since some of the data points occur to the right of the dotted line, it is evident

Figure 6.2(b) As Figure 6.2(a) but for studies that involved the use of tricalcium silicate

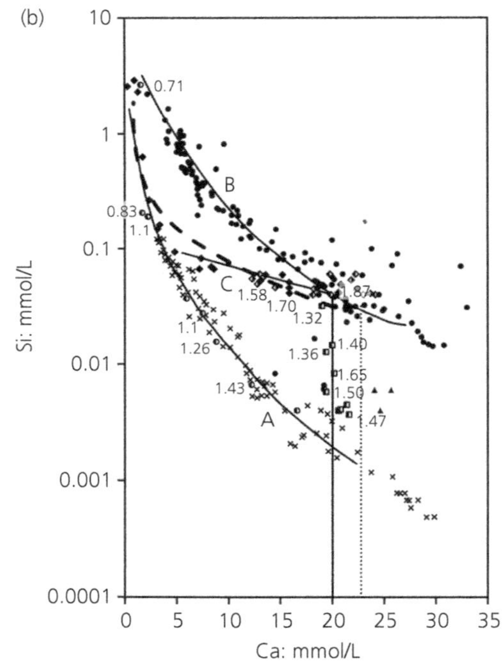

The solutions represented by the data points that fall to the right of the dotted line are supersaturated with respect to CH. For values of [Ca] > 3·5 mmol/L, all of the × data points that fall along curve A are from Brown *et al.* (1984) except one, which is from Greenberg and Chang (1965). The half-filled circles and squares are from Courault (2000) and Nonat *et al.* (2001) (squares if CH present); the filled circles that are close to the square-symbol points are from Greenberg and Chang (1965). The filled-symbol data points that fall along curve B are from: Barret and Bertrandie (1986); Bellmann *et al.* (2015); de Jong *et al.* (1968); Fierens and Verhaegen (1976); Flint and Wells (1934); Fujii and Kondo (1968, 1975); Jennings *et al.* (1986); Naber *et al.* (2019, 2020); Nicoleau and Nonat (2016); Stein and Stevels (1964); Tadros *et al.* (1976). The data that fall close to curve C and merge with curve B are from Chen *et al.* (2004; filled diamonds) Rodger (1984; right-half-filled diamonds) and Thomas and Double (1981; left-half-filled diamonds).

Figure 6.2(c) Metastable equilibrium curve relating the Ca/Si mole ratio of C–S–H preparations and the analytical concentration of CaO in solution for a study that involved the use of tricalcium silicate

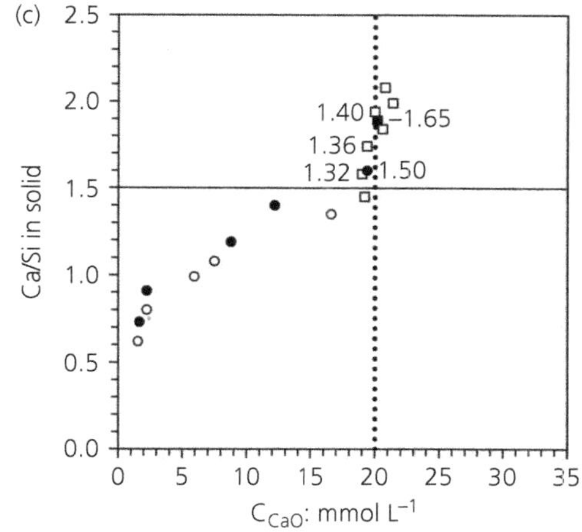

The data are from Courault (2000) and Nonat *et al.* (2001). Note that the values for the Ca/Si of the reactants were recalculated from the molar quantities given in Tables 1 and 2 of Nonat *et al.* (2001), which do not match the Ca/Si values given in the same tables. The numbers next to some of the points on (b) and (c) show the values of the Ca/Si of the C–S–H that was determined by TEM-EDX, which are lower than the bulk values, thus confirming that those particular preparations consisted of C–S–H intermixed with CH.

that in these methods solutions can become supersaturated with respect to CH. The points largely fall on one or other of two curves, Jennings' (1986) curves A and B, with relatively few in between. For values of $C_{CaO} > 3{\cdot}5\,\mathrm{mmol\,L^{-1}}$, almost all of the points that fall along curve A are from Brown *et al.* (1984; $\times$), except one from Greenberg and Chang (1965) and a few that are from Courault (2000) and Nonat *et al.* (2001); the latter are represented using half-filled symbols, circles unless CH was present, in which case they are plotted as squares. The numbers next to their data points on Figures 6.2(b) and 6.2(c) are the Ca/Si ratio of the C–S–H measured directly by TEM-EDX; it is evident on inspection that their data are consistent with C–S–H(I) at $C_{CaO} < 20\,\mathrm{mmol\,L^{-1}}$ and intermixture of C–S–H(I) with CH at $C_{CaO} \approx 20\,\mathrm{mmol\,L^{-1}}$. Bullard *et al.* (2018) noted that Brown *et al.* (1984) used colorimetry without treating the solution with a dilute acid and speculated that, after the C_3S was removed by filtering, additional precipitation could have occurred in the solution to drive it toward curve A. All of the data that contribute to curve B were obtained by reaction of C_3S with water for periods not exceeding a few days and, probably in all cases, unreacted C_3S was present (Jennings, 1986).

Curve B is most unlikely to represent the metastable solubility of anhydrous C_3S; thermodynamic calculations (Stein, 1972) indicate that such a curve would be very much higher, as would be expected from the high reactivity. Jennings (1986, 1988) and Gartner and Jennings (1987) considered it to be the metastable solubility curve of a product formed as a layer on the C_3S surface. Thermodynamic calculations based on this assumption indicated Ca/Si ratios of 1·1 at low and 1·65 at high CaO concentrations. Scrivener *et al.* (2023) demonstrated that the use of different silica speciation models that result in a range of Ca/Si ratio for the initial product (1·27–1·72) can all yield good fits to curve B and, as a consequence, they concluded that '*pore solution composition alone is not sufficient to predict the chemical composition of the initially precipitating hydrates*'. Direct measurement of the Ca/Si is of course preferable to prediction, and all the works considered by Scrivener *et al.* (2023) have the problem that none of the Ca/Si ratios are compatible with the XPS (X-ray photoelectron spectroscopy) evidence, which suggests that the product has a very high Ca/Si of about 2·0–2·5 at ages from a few minutes up to 1 to 4 hours (see §7.7.2). Barret and Bertrandie (1986; 1988) considered that curve B was not a solubility curve, but represented a quasi-steady state in which the concentrations were such as to equalise the rate of dissolution of a hydroxylated surface layer on the C_3S and the rate of precipitation of C–S–H. Jennings (1986) speculated that curve B may be associated with a mixture of tobermorite-like and jennite-like forms or C–S–H (i.e. C–S–H(I) and C–S–H(II) (as described in §6.2.2)) but that such a conclusion required further work. While he later (Jennings, 1988) stepped back from that speculation – associating it instead with the purely monomeric silicate that had been observed by NMR to form during the initial hydration of C_3S (Rodger *et al.*, 1988) – he was subsequently involved with work that associated jennite-like structure with a region close to curve B (Chen *et al.*, 2004). Chen *et al.* (2004) combined solubility results with a reassessment of existing data in the CaO–SiO_2–H_2O system and proposed a family of solubility curves that they designated as curves C, C′ and C″; curve C is included on Figures 6.2(a) and 6.2(b). They proposed that departures in solubility from curve A onto curves C, C′ and C″ are associated with the presence of Ca–OH bonds in C–S–H: the highest mean silicate chain lengths and Ca–OH contents at a given Ca/Si ratio above 1·2 would occur along curve C, with progressively lower values of both on curves C′, C″ and A. On the assumption that high Ca–OH contents are accommodated by the presence of jennite-like structure, they proposed that the family of solubility curves represented a spectrum of metastable phases that range from purely tobermorite-like structure on curve A (C–S–H(I)) to a largely jennite-like structure on curve C (C–S–H(II)). At CH saturation, the structure of C–S–H might therefore resemble C–S–H(I), C–S–H(II) or a mixture of the two. While the data collated on Figures 6.2(a) and 6.2(b) do not support the proposition of a well-defined family of curves, Chen *et al.*'s (2004) conclusions regarding structural variability nevertheless appear to be compelling – that is, that the scatter that occurs with increasing C_{CaO} arises because a given Ca/Si ratio can be associated with more than one structural arrangement.

Many thermodynamic models of C–S–H solubility have been developed to try to account for experimental data such as those that are collated on Figures 6.1 and 6.2, with perhaps the most successful being due to Kulik (2011) who revised the ideal solid solution model of C–S–H to make it consistent with the Richardson–Groves structural model of C–S–H (Richardson and Groves, 1992a; Chapter 7). The two thick-dashed lines on Figure 6.2(a) represent Kulik's (2011) CSHQ (top) and CSH3T models, the former incorporating elements of both tobermorite- and jennite-like structure. It is evident from inspection of Figure 6.2(b) that the curve for Kulik's (2011) CSHQ model is qualitatively similar to Chen *et al.*'s (2004) curve C, and that both are close to curve B at CH saturation: the C–S–H would include many Ca–OH bonds and would have high Ca/Si ratio. It is relevant to note that both Jennings (1986) and Chen *et al.* (2004) considered that there are no data regarding the solubility of jennite, C–S–H(II) or jennite-like C–S–H, but this is probably not correct. It is shown in §6.3 that Ishida *et al.* (1992a, 1992b, 1992c, 1992d) characterised a product that appears to be C–S–H(II), and their solubility data for that product fall close to the region just described. It thus seems likely that Ca–OH bonds can be present in high Ca/Si C–S–H as part of a jennite-like structure (i.e. as Si–O–Ca–OH), with the highest content where curves B and C converge at CH saturation.

6.2.2 The average structure of C–S–H(I): X-ray powder diffraction patterns

The XRD powder pattern for C–S–H(I), first interpreted in 1950 (Taylor, 1950), indicates a layer structure with, typically, reasonable two-dimensional order but with apparently poorly ordered layer stacking. The individual layers have a centred orthogonal lattice, with $a \approx 5{\cdot}6$ Å and $b \approx 3{\cdot}65$ Å. The basal spacing, which represents the layer thickness and is defined as 001 or *c*, varies from 9 to 14 Å. The peaks are often referred to cells with doubled values of *a*, *b* or *c*. As an example, Table 6.1 lists the diffraction peaks that were observed for a relatively highly ordered preparation of composition $C_{0{\cdot}94}SH_{2{\cdot}3}$ that was obtained by double decomposition of calcium nitrate and sodium silicate (Heller and Taylor, 1956). Most peaks can be assigned *hk*0 indices but are perhaps more properly regarded as *hk* band heads.

A key conclusion of further early work (Claringbull and Hey, 1952; Taylor, 1953) was that C–S–H preparations of this type are structurally related to the phase later known as 14 Å tobermorite. All later work has confirmed this central conclusion. Details of the tobermorite phases are given in §5.3.

Synthetic C–S–H preparations in this compositional range vary in crystallinity. Some give patterns similar to those of low-lime natural plombièrites (a tobermorite-like phase; §5.6), with a full complement of *hk*0 peaks but no detectable basal peak. The degree of crystallinity tends to decrease with increasing Ca/Si ratio and to be lower for samples made by mixing solutions of alkali silicate and a calcium salt than for those made from lime and hydrous silica or from anhydrous calcium silicates. Diagnostic features which distinguish them from high lime forms of C–S–H include the clear separation of the peaks at approximately 3·07 Å and 2·80 Å, the much greater relative intensity of the former and the presence of a peak at 1·67 Å.

An early summary of X-ray powder patterns for C–S–H(I) phases is given by Heller and Taylor (1956). They noted the following points, which have been modified slightly with information from other early work (Bernal *et al.*, 1952; Grudemo, 1955; Kalousek and Prebus, 1958; Taylor, 1950, 1964, 1969; Taylor and Howison, 1956).

(1) Preparations tend to have a single, broadened basal reflection (which was indexed as 002) that has a maximum at varying positions between 9 and 14 Å; this was interpreted as being due to mixtures of hydrates that have different layer spacings that are randomly interstratified in sheets normal to **c**.

Table 6.1 Typical observed X-ray powder diffraction data for a tobermorite-type C–S–H (i.e. C–S–H(I)) (Cu Kα): spacings, intensities and indices (Heller and Taylor, 1956)

d: Å	I_{rel}	*hkl*	*d*: Å	I_{rel}	*hkl*	*d*: Å	I_{rel}	*hkl*
12·5	vs	002	2·1	w/d	?	1·23	vw	?
5·3	vvw	201	1·83	s	040	1·17	vw	260
3·07	vvs	220	1·67	mw	620	1·11	w	840
2·80	s	400	1·53	vw	440	1·07	vw	10·2·0
2·4	w/d	?	1·40	w	800			

The data are for a preparation of composition $(CaO)_{0{\cdot}94}{\cdot}SiO_2{\cdot}(H_2O)_{2{\cdot}3}$ that was obtained by double decomposition of calcium nitrate and sodium silicate. The method of drying is not given but it must have been very gentle. All peaks are broad.
s = strong, m = moderate, w = weak, v = very, d = diffuse.

(2) The basal spacing depends on both the Ca/Si and H_2O/Si ratios.
(3) Preparations that have been dried gently show a rough correlation between the basal spacing and the Ca/Si ratio: it decreases from about 13–14 Å at Ca/Si = 0·8 to about 10 Å at Ca/Si = 1·5.
(4) There are few reflections other than 002 and *hk*0s.
(5) There is often a peak at ≈5·3 Å (indexed as 201) that varies markedly in intensity from very, very weak to moderate.
(6) Diffuse bands at ≈2·1 and ≈2·4 Å become resolved into separate lines in preparations that are better crystallised.
(7) Additional reflections can occur, in particular between 3·2 and 3·6 Å.
(8) Relative changes in intensity and general weakening of the pattern occurs in less well-crystallised material.
(9) In extreme cases, peaks are only visible at ≈3·03–3·07 Å (220 reflection) and ≈1·80–1·83 Å (040), which can be of comparable intensity. This was interpreted as being due to random stratification together with very small crystal size in both the **a** and **c** dimensions.

These points are still applicable and have been slowly built on by later workers.

A number of studies have reported a decrease in *b* with increasing Ca/Si ratio (Cong, 1994; Cong and Kirkpatrick, 1996a; Garbev *et al.*, 2008a; Renaudin *et al.*, 2009a). These workers all observed a value of ≈3·67 Å at a Ca/Si ≈ 0·7 but different reductions on increasing Ca/Si, with the smallest decrease observed by Cong and Kirkpatrick (1996a) (≈ 0·03 Å) and the largest by Garbev *et al.* (2008a) (≈ 0·08 Å). These studies also show a consistent trend of increasing *a* with increasing Ca/Si, from ≈ 5·60 Å at Ca/Si ≈ 0·7 to ≈ 5·65 Å at Ca/Si ≈ 1·4.

Many investigators have confirmed Grudemo's (1955) observation that the basal spacing decreases with increasing Ca/Si ratio, although the specific results of different investigations differ considerably, which is evident in a compilation by Grangeon *et al.* (2013b); their Figure 3 shows much scatter. Grangeon *et al.* (2013b) had apparently missed an earlier compilation by Matsuyama and Young (2000) (their Figure 8(a)) who attributed some of the scatter to differences in the degree of sample drying, an explanation that had been noted much earlier by Taylor (1964). A similar compilation of literature data is given in Figure 6.3, which – following Matsuyama and Young (2000) – includes two approximately parallel lines; in the present case the lines are the result of linear regression analysis of subsets of the data. The data that were used for the top regression analysis line (i.e. the top long-dashed line) are indicated on the figure using white crosses and those for the bottom (long-dashed) line using black crosses. On inspection, it is evident that the two lines are separated by about 2 Å over the full range of Ca/Si ratio (i.e. 0·67 to 1·5), which – following Matsuyama and Young (2000) – is interpreted here as being caused by drying – that is, the layer spacing shrinks by about 2 Å because of the removal of water molecules from the interlayer region, the extent of the shrinkage being largely unaffected by the Ca/Si ratio. This interpretation is supported by other data, discussed in §6.7. There are many data points that fall between the two regression analysis lines and also some data that fall to the left of the bottom line or to the right of the top line. Some of the variation could be due to random or systematic errors in some of the studies, either in the determination of the Ca/Si ratio or in the XRD data, such as specimen displacement and zero shift. Certainly some of the studies do not report a method of calibration for the XRD experiments, including Matsuyama and Young (2000), whose data form the bulk of the points that are clustered around the bottom regression line, whereas the data from Garbev *et al.* (2008a) – who do report careful calibration – fall to the right of this line. It is relevant to this point to note that it is difficult to reconcile the position of the bottom regression analysis line with crystal-chemical arguments (see §6.6.3), in

Figure 6.3 Plot of the layer spacing against molar Ca/Si ratio for C–S–H(I) preparations

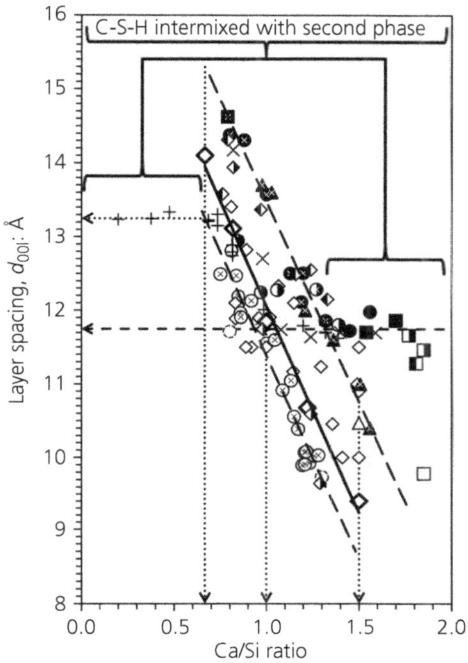

Data are from: Matsuyama and Young (2000) (unfilled circles); Matsuyama and Young (1999) (unfilled dotted-line circles); Cong and Kirkpatrick (1996a) (filled symbols for 'SCFUM' and 'CSHFS' preparations except 'CSHSF2' (unfilled triangle) and CSHFS0 (unfilled square); half-filled (right) symbols for 'SEWCS' preparations); Garbev *et al.* (2008) (+), compositions from Black *et al.* (2008); Renaudin *et al.* (2009a) (×); Taylor (1953) and Taylor and Howison (1956) (small unfilled diamonds); Grudemo (1955) (filled triangles); Tajuelo Rodriguez *et al.* (2015) (half-filled diamonds). Preparations that contain crystalline CH are indicated with squares except for the filled square at Ca/Si = 0·79, which includes 16 % Q^3 Si. Regression analysis line data are indicated using white crosses (top line) and black crosses (bottom). The large bold unfilled diamonds (and the associated full line) represent the model structures of Richardson (2014) and §6.6.3; from top to bottom: T∞, T11, T5, T3 and T2. Modified from Figure 1 of Richardson (2014).

particular with the fact that that the regression equation gives a basal spacing of 8·6 Å at Ca/Si = 1·5, which seems unlikely to be possible (i.e. it is about 0·7 Å too small). In addition to the possibility of experimental errors, the most important explanations are likely to be as follows (Richardson, 2014).

(1) The data points with Ca/Si $< \frac{2}{3}$ are due to intermixture of C–S–H that has Ca/Si $\geq \frac{2}{3}$ with unreacted silica. The layer spacing of the C–S–H is indicated by the horizontal dotted-line arrow.
(2) The data points that fall to the right of the upper long-dashed line that have layer spacings between ≈11·5–12·0 Å are due to intermixture of C–S–H that has Ca/Si ≤1·32 with CH, evidently either amorphous or crystalline (the latter represented using square symbols).

Intermixture of C–S–H(I) with an amorhous Ca-rich phase has been demonstrated by Richardson *et al.* (2010) using TEM-EDX. The layer spacing of the C–S–H is indicated by the horizontal dashed-line arrow.

(3) The data points that fall between the two regression analysis lines are due to either (i) an intermediate level of drying or (ii) to intermixture of a C–S–H that has a Ca/Si that falls on the lower regression line with CH, again either amorphous or crystalline.

Explanations (1) and (2) amount to essentially the same situation as occurs in synthetic layered double hydroxide phases (Richardson, 2013). Explanation 3(ii) appears to be particularly relevant to the data of Garbev *et al.* (2008a) and Renaudin *et al.* (2009a) and, as a consequence, any discussion in those articles that relates to samples that have a Ca/Si ratio greater than around 1·00–1·25 should be considered in this context – that is, the C–S–H in the samples has a Ca/Si ratio of about 1·00–1·25 and it is intermixed with (varying amounts of) a Ca-rich phase.

6.2.3 Transmission electron microscopy: morphology and diffraction patterns

Pioneer TEM studies by Grudemo (1955, 1962) and confirmed by others (e.g. Copeland and Schulz, 1962; Gard *et al.*, 1959; Henderson and Bailey, 1986; Kalousek and Prebus, 1958) showed that samples of tobermorite-type C–S–H with Ca/Si ratios of 0·8–1·4 have a characteristic morphology that has been described as crinkly or crumpled foils, some 24–67 Å thick (Taylor, 1964). These samples give weak selected area powder diffraction patterns showing the 110, 200 and 020 and sometimes other *hk*0 reflections. A TEM micrograph that illustrates the typical crumpled-foil morphology of a synthetic C–S–H is shown in Figure 6.4(a). The Ca/Si ratio was determined by TEM-EDX to be $0{\cdot}80 \pm 0{\cdot}04$ ($n = 40$). A selected area electron diffraction pattern of the C–S–H is shown in Figure 6.4(b), which is consistent with microcrystalline C–S–H(I). At higher Ca/Si ratios approaching 1·5, the foils tend to be shredded into smaller pieces and the morphology becomes increasingly fibrous.

Much care is needed to avoid alteration of C–S–H in the TEM. For example, Melzer and Eberhard (1989) reported the existence of several crystalline forms of C–S–H on the basis of SAED (selected area electron diffraction) single crystal patterns, but these were shown to have been due to CaO (Groves, 1990), which had evidently formed through heating and possibly also radiation damage.

6.2.4 Silicate anion structure of C–S–H(I): early studies

The wide variability of Ca/Si ratio has long excited interest, with some of the key questions being the origin of this variation, the structure of the silicate anionic units and their connectivity to the Ca-containing portion of the structure. The early literature, reviewed by Steinour (1947), showed differences of opinion as to whether the compositional variation arose from solid solution or adsorption of CH onto a low-lime phase. The arguments were not conclusive. Determinations of densities by hydrogen displacement and by calculation from mean refractive indices (Taylor and Howison, 1956) showed that changes in Ca/Si ratio have a relatively slight effect on the mass of the atomic cell contents. By this time (mid-1950s), the structural relation to the tobermorite minerals was known, and it was suggested that increase in Ca/Si ratio above its minimum value was due to a combination of absences of tetrahedra in the silicate chains and replacement of hydrogen atoms by Ca. Specifically, the missing tetrahedra were considered to be the so-called 'bridging' tetrahedra (Figure 5.10). This conclusion is supported by observations that, in the C–S–H formed in calcium silicate or cement pastes, the chain lengths follow the sequence 2, 5, 8...($3n$–1) (§7.4.1; Hirljac

Figure 6.4 (a) TEM micrograph illustrating the crumpled-foil morphology of a synthetic C–S–H; the Ca/Si ratio was determined by TEM-EDX to be 0·80 ± 0·04 ($n = 40$). (b) Selected area electron diffraction pattern of the C–S–H. The *d*-spacings for the three strongest reflections correspond to 3·07 Å (vs), 2·78 Å (s) and 1·83 Å (s), which are consistent with the 110, 200 and 020 reflections for C-S-H(I) (see Table 6.1). The figures are reproduced from Richardson *et al.* (2010)

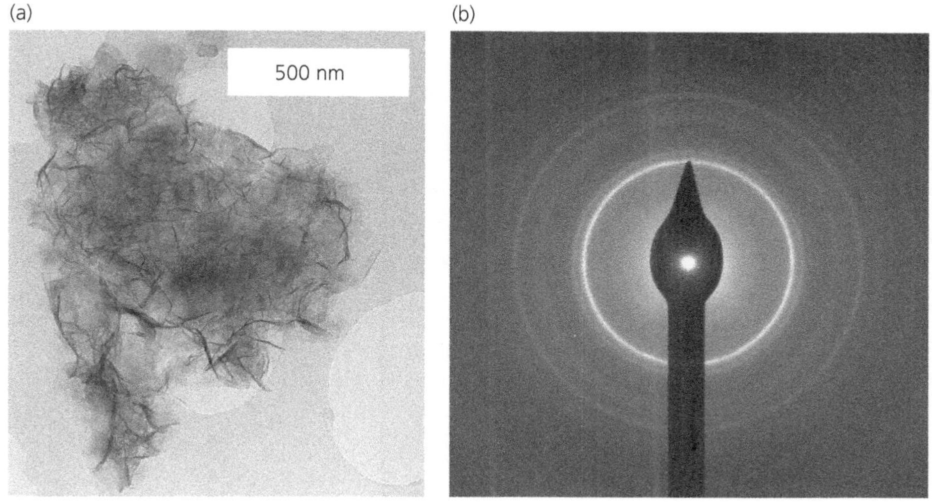

et al., 1983; Richardson, 1999). These sequences imply that each silicate unit is either a dimer of 'paired' tetrahedra or a chain terminated at each end by two paired tetrahedra (Figure 5.10).

Funk (1967) reported estimates of the silicate anion sizes, obtained from rates of reaction of solutions in methanolic HCl with molybdate. For a preparation with Ca/Si = 1·01, he estimated that 80% of the Si was present in anions containing six to ten tetrahedra, with 10% in smaller units and 10% in larger units. At Ca/Si = 1·1 the ions were less highly condensed.

6.2.5 Silicate anion structure of C–S–H(I): ^{29}Si MAS NMR

^{29}Si MAS NMR (magic-angle spinning nuclear magnetic resonance spectroscopy) gives data on the proportions of Si atoms that are present in SiO_4 tetrahedra of different connectivities, designated Q^0–Q^4; some useful equations for the interpretation of single-pulse ^{29}Si MAS NMR data are given in §5.3.5. The present section provides a collation and interpretation of NMR data for C–S–H(I) preparations. Many studies (e.g. Cong and Kirkpatrick, 1996a; Damidot *et al.*, 1995; Grutzeck *et al.*, 1989; Klur *et al.*, 1998; Lequeux *et al.*, 1999; Nonat and Lecoq, 1998; Rassem *et al.*, 1992a, 1992b) have shown that the silicate tetrahedra in preparations with Ca/Si ratios less than ≈1·0 are predominantly Q^2 and that at higher Ca/Si ratios substantial proportions of Q^1 are also present. Some preparations with Ca/Si ratios of 0·6–0·8 also contain Q^3 (Grutzeck *et al.*, 1989). Figure 6.5 includes results from a number of studies for which quantitative relative peak intensities and compositional data were reported. Data for preparations containing Q^3 tetrahedra or unreacted silica are excluded and those that contain crystalline CH are represented using square symbols.

Figure 6.5 Plots of (a) MCL, (b) net charge per Si atom, (c) $1/(1\text{-}v)$, (e) SOF_{BT} against the bulk Ca/Si ratio of C–S–H(I) preparations; (d) is a plot of the layer spacing against SOF_{BT} (continued on next page)

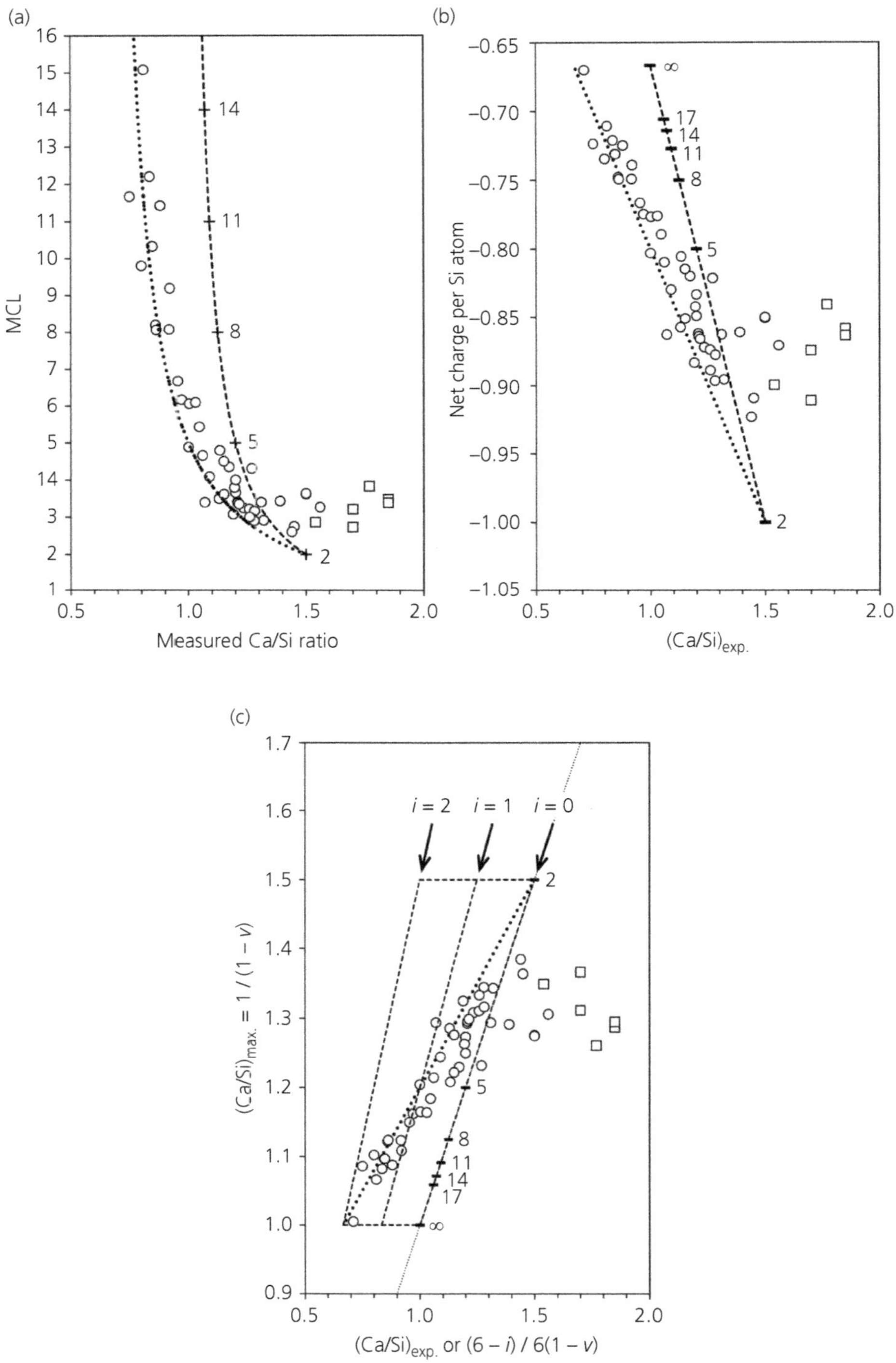

Figure 6.5 Continued

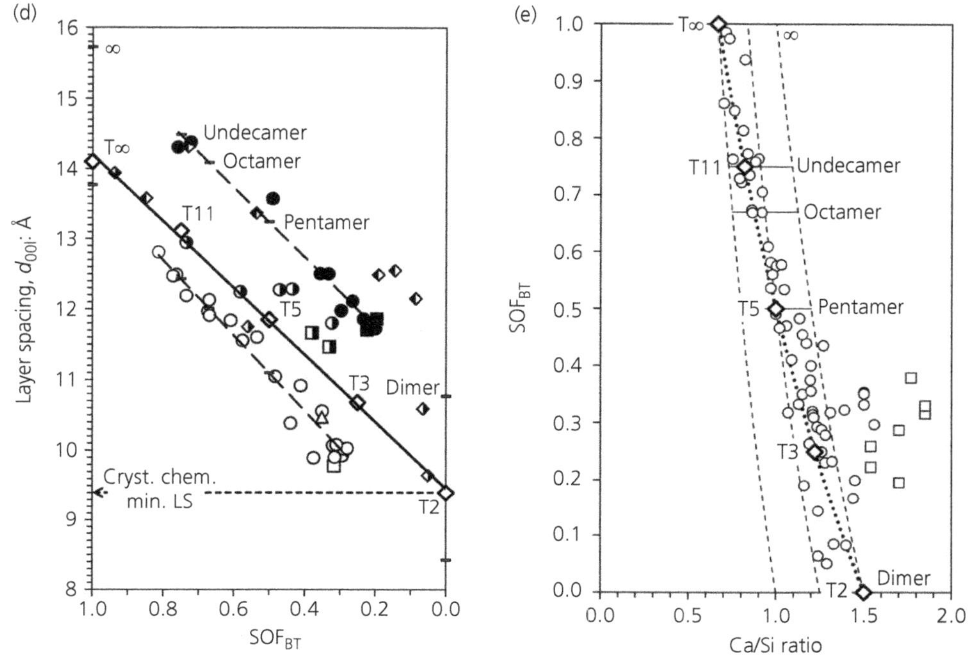

The data on (a) – (c) and (e) are from Chen *et al.* (2004), Cong and Kirkpatrick (1996), Damidot *et al.* (1995), Grutzeck *et al.* (1989), Matsuyama and Young (2000) and Nonat and Lecoq (1998). The symbols on (d) have the same meaning as in Figure 6.3. Square symbols indicate preparations that contain crystalline calcium hydroxide. The dashed lines in (a) and (b) and the furthest-right dashed lines in (c) and (e) correspond to the negative charge of the silicate anions being balanced entirely by Ca^{2+} ions. The dotted line in all parts of the figure represents Equations 6.7 and 6.8. Figures 6.5 (a), (b), (c) and (e) are reproduced from Richardson (2014) while (b) and (d) are new. The large bold unfilled diamonds on (d) and (e) represent the model structures of Richardson (2014) and §6.6.3.

In the following discussion – which is an augmented reiteration of that given in Richardson (2014) – it is assumed that the structure is based on that of 14 Å tobermorite, modified by some or all of the following effects

- omission of some or all of the bridging tetrahedra
- variation in the content of interlayer Ca
- variation in the content of OH attached to Ca or Si.

Variation in the content of molecular water is not relevant to the calculations. It is further assumed that the ordered regions are sufficiently large that edge effects can be neglected, that it is a single phase – that is, no other phases, such as CH or silica, are present – and that no Si is present on sites other than Q^1 or Q^2. Several quantities may then be calculated from the $\frac{Q^1}{(Q^1+Q^2)}$ ratio, which will be called q (after Richardson and Groves, 1992a).

(1) The mean chain length (MCL) is given by $\frac{2}{q}$ (Richardson and Groves, 1992a). This and related quantities are number means.

(2) The mean number of tetrahedral Si atom sites per chain is equal to MCL + 1 because each chain is associated with 1 vacant bridging site. It is therefore equal to the reciprocal of the fraction of vacant tetrahedral sites (Equation 5.11) – that is, the number of tetrahedral sites per vacant site; so

$$\frac{1}{v} = \mathrm{MCL} + 1 = \frac{2}{q} + 1. \tag{6.1}$$

(3) The net charge per three tetrahedral sites in a single-chain tobermorite-based structure is always –2, irrespective of whether there are any vacant bridging sites. This is because when a bridging site is vacant the two negative charges that are associated with the bridging tetrahedron are simply replaced at the two resulting chain-end Q^1 tetrahedra. The net charge per Si atom, z_{Sil}, does of course change with the presence of vacant sites: the –2 charge is simply divided by 3 × the fraction of the tetrahedral sites that are occupied, and so

$$z_{\mathrm{Sil}} = \frac{-2}{3(1-v)} \tag{6.2}$$

This can be calculated from ^{29}Si NMR data by substituting for v, which gives,

$$z_{\mathrm{Sil}} = \frac{-(q+2)}{3} \tag{6.3}$$

(4) From Formula 5.6 the Ca/Si ratio for Al-free C–S–H is

$$\frac{\mathrm{Ca}}{\mathrm{Si}} = \frac{6-i}{6(1-v)} \tag{6.4}$$

The maximum Ca/Si ratio is obtained when the net charge per Si atom is balanced entirely by Ca^{2+} ions – that is, when $i = 0$, and so

$$\left(\frac{\mathrm{Ca}}{\mathrm{Si}}\right)_{\max} = \frac{1}{1-v} \tag{6.5}$$

This means that when a variation in composition in a single-chain tobermorite-based structure occurs by the omission of some or all of the bridging tetrahedra and by variation in the content of interlayer Ca ions, the maximum Ca/Si ratio that is possible is numerically equal to the number of tetrahedral sites per occupied site. It can therefore be calculated easily from ^{29}Si NMR data using

$$\left(\frac{\mathrm{Ca}}{\mathrm{Si}}\right)_{\max} = \frac{q+2}{2} \tag{6.6}$$

Some of these relationships are equivalent to ones used by Chen *et al.* (2004), Cong and Kirkpatrick (1996a, 1996b), Nonat and Lecoq (1998), Richardson (2004) and Richardson and Groves (1992a). Figures 6.5(a)– 6.5(c) show values of MCL, z_{Sil} and $\frac{1}{1-v}$ calculated from data from several sources (the references are given in the figure captions). The positions for silicate chains of finite length are indicated – that is, of length $3n - 1$, where n is integer. The mean chain length decreases sharply from ≈11 at a Ca/Si ratio of 0·8 to ≈5 at Ca/Si of 1·0 and then more slowly to ≈3 at Ca/Si of 1·2, where it remains on further increase in Ca/Si. The maximum Ca/Si ratio that is possible without the presence of Ca–OH linkages in the C–S–H structure, $(\mathrm{Ca/Si})_{\max}$ (i.e. $\frac{1}{1-v}$), is represented by short-dash lines on Figure 6.5(a) (i.e. the plot of MCL against Ca/Si) and Figure 6.5(b) (z_{Sil} against Ca/Si). It corresponds to the situation where the negative charge of the silicate anions is balanced entirely by Ca^{2+} ions; Si–OH linkages must be present if a point is to the left of these lines and Ca–OH linkages must be present if a point is to the right of them. In principle, the quantity of either could be augmented if an equivalent additional quantity of the other was also present. The imbalance of charge therefore requires the presence of Si–OH at low Ca/Si ratios and of Ca–OH at high ratios, neither being needed at ≈1·3. The simplest hypothesis is that only Si–OH is present at Ca/Si <1·3 and only Ca–OH at Ca/Si >1·3. This is demonstrated on Figure 6.5(c), which in addition to a line for $(Ca/Si)_{max}$ (i.e. $i = 0$ in Formula 5.6) also shows lines for $i = 1$ and 2. These two lines correspond respectively to the situation where the negative charge is balanced equally by protons and Ca^{2+} ions and totally by protons. The area bounded on Figure 6.5(c) by the short-dashed lines thus encompasses all of the compositions that are possible for single-chain tobermorite-based structures when the structural variation is the omission of some or all of the bridging tetrahedra and variation in the content of interlayer Ca ions. This is just an alternative way of representing the information in Figure 3 of Richardson and Groves (1992a) (which has $\frac{1}{MCL}$ plotted against Ca/Si instead of $\frac{1}{1-v}$, which is $\frac{MCL+1}{MCL}$; the line for $i = 1$ is also therefore equivalent to the line for the tobermorite-based structure in Figure 2 of Taylor (1986). It is evident that the data for these synthetic C–S–H preparations broadly follow the diagonal dotted line that is plotted on Figure 6.5(c) from $i = 2$ at low Ca/Si ratio to $i = 0$ at high ratio. This line represents the equation

$$\frac{Ca}{Si} = \frac{\text{Fraction of paired tetrahredral sites+fraction of vacant tetrahedral sites}}{\text{Fraction of occupied tetrahedral sites}} = \frac{\frac{2}{3}+v}{1-v} \tag{6.7}$$

The fraction of tetrahedral sites in tobermorite-based structures that are 'paired' is numerically the same as the number of main-layer Ca atoms per three tetrahedral sites (that is $\frac{2}{3}$; as shown in Formulae 5.2 and 5.6). The dotted line therefore represents the situation where one Ca^{2+} ion is added to the interlayer region for each bridging site that is vacant, and there are consequently no interlayer calcium ions when the chains are of infinite length. This situation is also represented by dotted lines on Figures 6.5(a) and 6.5(b). On inspection, it can be seen that if expressed in terms of six tetrahedral sites, Equation 6.7 is the same as Equation 6.4 with $i = 2 - 6v$ – that is, $i = 2$ for infinite chains ($v = 0$) and $i = 0$ for dimeric chains $\left(v = \frac{1}{3}\right)$.

Matsuyama and Young (2000) compared their data with those of Cong and Kirkpatrick (1996a) and confirmed that the layer spacing decreases with Ca/Si; they calculated that the decrease in layer spacing with increasing Ca/Si is largely due to missing bridging tetrahedra. Their data and those of Cong and Kirkpatrick (1996a) are replotted in Figure 6.5(d) but as SOF_{BT} rather than as the ratio $\frac{Q^1}{\left(Q^1+Q^2\right)}$. As with Figure 6.3, the data are largely clustered around two parallel lines, supporting the view of early workers (§6.2.2) that the observed layer spacing is dependent on the water content as well as the Ca/Si ratio.

The variation in the silicate anion structure that is represented by Equation 6.7 is perhaps more easily envisaged if the equation is recast in terms of the occupancy of tetrahedral bridging sites, SOF_{BT}, which is defined in Equation 5.20. Substitution for v in Equation 6.7 results in Equation 6.8, which can be used to calculate the Ca/Si ratio from single-pulse ^{29}Si NMR data. The data in Figure 6.5(a) are replotted (with some additional data) in Figure 6.5(e) in terms of SOF_{BT}, together with a line that represents Equation 6.8 (the dotted line) which, given the various errors that are possible in the experimental data, is evidently a good explanation for most of the data, which as a consequence provides a useful constraint for model structures (as discussed in §6.6).

$$\frac{Ca}{Si} = \frac{\left(3-SOF_{BT}\right)}{\left(2+SOF_{BT}\right)} = \frac{\frac{3}{2}Q^1+\frac{2}{3}Q^2}{Q^1+Q^2} \tag{6.8}$$

^1H–^{29}Si cross polarisation spectra show that protons are associated with both Q^1 and Q^2 sites (Cong and Kirkpatrick, 1995). For specimens with Ca/Si >0·9 the difference in relaxation times for the two types of site increases with Ca/Si. This observation can be interpreted to indicate that at low Ca/Si ratios, the chain length distribution is unimodal, with predominantly chains with MCL = 11 at Ca/Si = 0·8, while at higher Ca/Si ratios, the distribution becomes increasingly bimodal, with relatively abundant dimers and chains of ≈11 tetrahedra and relatively few chains of intermediate lengths. The results also indicated that the ^1H in Ca–OH linkages contribute to the cross-polarisation of Q^1 sites at the higher Ca/Si ratios. These conclusions support the conclusion from the density evidence (Taylor and Howison, 1956) that increase in Ca/Si ratio is associated with both a decrease in Si content and an increase in Ca content.

Figure 6.6(a) shows data from a ^{29}Si double quantum homonuclear CP/MAS correlation NMR experiment on a ^{29}Si-enriched C–S–H(I) sample that had a target Ca/Si atom ratio of 0·9 (bottom). The sample was made from a mixture of CaO and ^{29}Si-enriched SiO_2 in stirred suspension with w/s (water/solid) = 50, stored at 40 °C for 4 weeks, filtered, washed in acetone and then dried under low vacuum for 3 minutes. The figure is modified from Figure 4(b) of Brunet *et al.* (2004). A

Figure 6.6(a) Data from a ^{29}Si double quantum homonuclear CP/MAS correlation NMR experiment on a ^{29}Si-enriched C–S–H(I) sample that had a target Ca/Si atom ratio of 0·9 (bottom). The figure is modified from Figure 4(b) of Brunet *et al.* (2004). A schematic diagram illustrating the interpretation of the data is given at the top

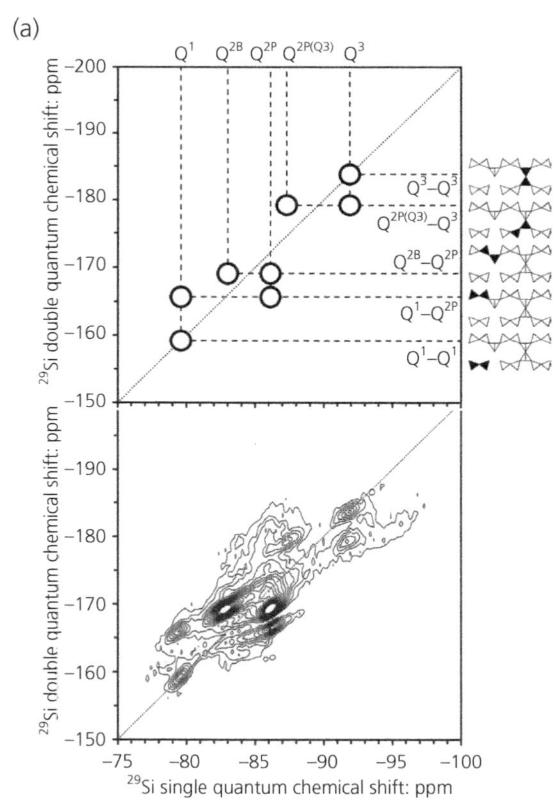

Figure 6.6(b) Data from a 2D heteronuclear ^{29}Si–^{1}H chemical shift correlation (HETCOR) NMR experiment on a ^{29}Si-enriched C–S–H(I) sample that had a target Ca/Si atom ratio of 1·5. The figure is modified from Figure 5(c) of Brunet *et al.* (2004). $Q^2 = Q^{2P}$

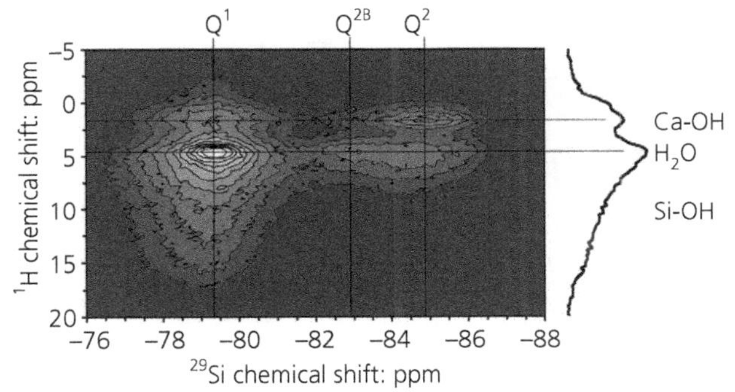

schematic diagram that illustrates the interpretation of the data is given at the top; the correlating tetrahedra are indicated by the filled triangles. There is the potential for confusion in the literature regarding terminology: Klur *et al.* (1998), Brunet *et al.* (2004) and other, generally French, workers use 'Q^{2P}' to refer to Q^2 'bridging' tetrahedra (i.e. 'P' for 'pontants') whereas English-speaking workers generally use Q^{2P} to refer to 'paired' tetrahedra. In this chapter, 'P' is used for paired and 'B' for bridging. $Q^{2P(Q3)}$ is used where the paired tetrahedron is adjacent to a Q^3 tetrahedron; Klur *et al.* (1998) and Brunet *et al.* (2004) used Q^{2v} for such tetrahedra but its meaning is less obvious and '*v*' is used in this chapter to refer to vacant sites. With reference to the schematic diagram, the peaks in the figure are interpreted as follows (from top to bottom), with the caution that the data are not quantitative.

- There is some low intensity associated with Q^3–Q^3 and $Q^{2P(Q3)}$ –Q^3 and so this particular sample must have included silicate structure where bridging tetrahedra were adjacent to one another – that is, some double *dreierkette* chains.
- The most intense peaks are due to Q^{2B}–Q^{2P}. This means that the preparation included long linear chains – that is, single *dreierketten* – and perhaps also, as illustrated in the schematic diagram, occupied bridging sites that are opposite vacant bridging sites.
- There is strong Q^1 –Q^{2P} correlation. This means that there must have been some vacant tetrahedral bridging sites – that is, linear chains of finite length.
- There is strong Q^1 –Q^1 correlation. This is very significant because it means that there must have been some dimeric structure, even at the low Ca/Si ratio of 0·9, which is consistent with the conclusions of Cong and Kirkpatrick (1996a) discussed above from their ^{1}H–^{29}Si cross-polarisation experiments. Any model(s) for the structure of C–S–H(I) must therefore account for the presence of dimeric structure over most of the compositional range.

6.2.6 ^{17}O,^{43}Ca and ^{1}H MAS NMR

^{17}O MAS NMR spectra (Cong and Kirkpatrick, 1996b) show a relatively sharp band containing one or two narrow peaks attributable to nonbridging sites (SiO^-) and a broader band attributable to bridging sites (Si–O–Si), Si–OH, Ca–OH and H_2O. The proportions of bridging tetrahedral sites and Si–OH linkages decrease with increasing Ca/Si ratio, while that of Ca–OH linkages increases. These conclusions agree with those from the ^{29}Si NMR data. Both Si–OH and Ca–OH were found to be present at all Ca/Si ratios, although Si–OH usually accounted for <10% of the total intensity at Ca/Si >1·3 and Ca–OH for a similar proportion at Ca/Si <1·3. Quantitative simulation of the ^{17}O spectra is difficult due to peak overlap.

Klur *et al.* (1998) reported the ^{43}Ca NMR spectrum of a preparation with Ca/Si = 0·7. A strong peak was assigned to the atoms in the main layer and a very weak one to those in the interlayer, confirming that little interlayer Ca is present at this composition. They also reported a detailed deconvolution of the peaks in the ^{29}Si NMR spectrum. Klur *et al.* (1998) also obtained ^{1}H CRAMPS NMR spectra, which contain peaks attributed to Ca–OH. These are very weak at Ca/Si ratios ≤1·2 but become progressively stronger as the Ca/Si ratio increases towards 1·5. A ^{1}H-^{29}Si 2-D correlation spectrum for a preparation with Ca/Si 1·5 was interpreted as indicating the presence of both Si–OH and Ca–OH.

Brunet *et al.* (2004) also conducted 2D heteronuclear ^{29}Si–^{1}H chemical shift correlation (HETCOR) NMR experiments on ^{29}Si-enriched C–S–H(I) preparations. For a preparation with Ca/Si = 0·9, the main resonance in the ^{1}H projection is at a chemical shift of 4·7 ppm, which they attributed to the protons of water molecules and which is correlated with Q^{2P}, Q^{2B} and Q^{1} in the ^{29}Si projection. There is also a broad shoulder, extending from 7 to 12 ppm, which Brunet *et al.* (2004) considered to be due to protons in Si–OH groups, although they also associated Si–OH protons with the intense peak at 4·7 ppm, attributed mainly to the protons of water molecules. A small signal at 1·9 ppm that they attributed to Ca–OH protons is correlated with Q^{2}. Brunet *et al.* (2004) considered that these Ca–OH groups are present in the interlayer of a tobermorite-like structure but it is relevant to note that 14 Å tobermorite has a main-layer Ca–O polyhedron that carries an H atom (OH6 in Bonaccorsi *et al.*'s (2005) structure).

The highest Ca/Si ratio targeted in Brunet *et al.*'s (2004) study is 1·5, although it is possible that the real Ca/Si would be lower than 1·5; while they did not report XRD results and did not comment on whether any phases other than C–S–H were present, it seems likely from other studies that used similar synthesis techniques (e.g. Klur *et al.*, 1998) that CH would have been present in the sample, which has been the case in many studies, as is demonstrated in Figure 6.3. Brunet *et al.* (2004) identified three ^{1}H resonances at chemical shifts of 1·8 ppm, 4·6 ppm and between 7 and 17 ppm, which they assigned to protons in Ca–OH, H_2O and Si–OH, respectively. They stated that the H_2O protons correlate with Q^{1}, Q^{2B} and Q^{2} Si sites, and that those present in both Ca–OH and Si–OH correlate with all silicon sites, by which they presumably meant the six sites given for that sample in their Table 1, namely two Q^{1} sites (at –77·4 and –79·6 ppm), Q^{2B} (at –82·2 ppm), Q^{2i} (at –83·8 ppm, associated with interlayer Ca) and two Q^{2P} sites (at –85 and –86 ppm). The figure from their HETCOR NMR experiment for this sample is reproduced in Figure 6.6(b), which has been modified to facilitate interpretation. It is clear from inspection of Figure 6.6(b) that the H_2O protons do indeed correlate with Q^{1}, Q^{2B} and Q^{2P} Si sites. However, it is difficult to understand how they could consider Si–OH protons to be correlated with all Si sites unless the proton of a Q^{2} Si–OH has a ^{1}H chemical shift that is very similar to that for an H_2O molecule because there is no intensity evident on the figure that is associated with chemical shifts of between 7 and 17 ppm and –80 and –86 ppm in the ^{1}H and ^{29}Si projections, respectively: Si–OH protons are certainly correlated with Q^{1}. The Ca–OH protons appear to be correlated mainly with Q^{1} and Q^{2P}. The contours in the ^{1}H projection are spaced essentially symmetrically around the vertical lines for the Q^{1} and Q^{2} (i.e. paired, Q^{2P}) sites except for some very low intensity (i.e. at the level of the first contour) in the region for Q^{2B}. Correlation with Q^{2B} therefore seems less strong than for Q^{1} and Q^{2P} and the lack of asymmetry around Q^{2} does not support the presence of Q^{2i} – that is, a Q^{2} peak at –83·8 ppm would have resulted in an asymmetrical peak; Brunet *et al.* (2004) associated a peak at that chemical shift with interlayer Ca atoms. So, while Brunet *et al.*'s (2004) data do provide evidence for Ca–OH

Figure 6.7 A view down **c** of a part of the structure of 14 Å tobermorite (structure from Bonaccorsi *et al.*, 2005) showing that the oxygen atoms of the Wat6 water molecules are closer than 4 Å to the paired Si atoms (Si1 and Si2; the distances are between sites); the protons of the water molecules are thus likely to be within 5 Å of the Si atoms

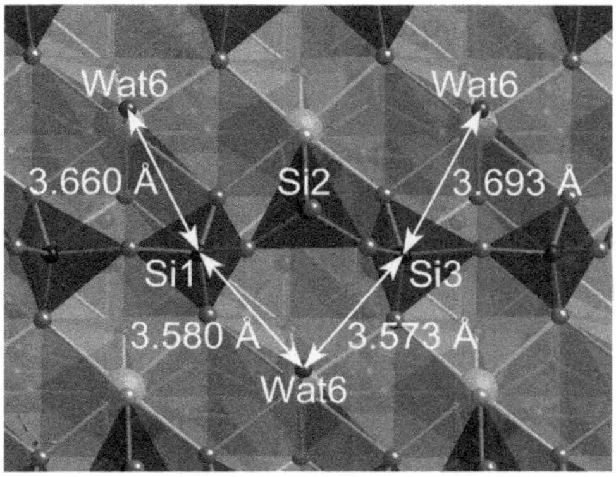

groups that are close to Si that is in the C–S–H, they do not provide conclusive support for their and Klur *et al.*'s (1998) proposed interlayer Ca–OH groups. This point is discussed further in §6.9.

Klur *et al.*'s (1998) and Brunet *et al.*'s (2004) observation that Q^1 Si correlates strongly with protons of water molecules is consistent with tobermorite- or jennite-like structure and so provides no evidence for either. As an example, regarding the former, the relevant part of the structure of 14 Å tobermorite (Bonaccorsi *et al.*, 2005) is illustrated in Figure 6.7. The removal of the Si2 bridging tetrahedron would necessitate charge compensation of the terminal oxygen atoms of the Si1 and Si3 tetrahedra: this could be achieved either by protons (i.e. Si–OH groups) or by Ca^{2+} ions. The coordination requirements of the Ca^{2+} ion would need to be accommodated and this would most likely involve Wat6; it is evident that protons of water molecules would be within 5 Å of Si1 and Si3, which would be Q^1 Si (i.e. protons of Wat6), regardless of whether charge compensation was achieved by protons (Si–OH) or Ca^{2+} ions. Inspection of the structure of jennite (Bonaccorsi *et al.*, 2004) results in the same conclusion – that is, protons of water molecules would be within 5 Å of Q^1 Si.

6.2.7 Infrared absorption spectroscopy

Mid IR spectra for tobermorite-type C–S–H were first reported by van Bemst (1955). Yu *et al.* (1999) reported spectra for the preparations that Cong and Kirkpatrick (1996a) had examined by NMR. The mid-IR region includes bands attributable to Si–O stretching vibrations of Q^2 tetrahedra which are generally similar to those given by 14 Å tobermorite. With increase in Ca/Si ratio between 0·8 and 1·5, several trends were attributed to the decreasing polymerisation of the silicate chains. These include a shift in the maximum of the main Si–O stretching band for Q^2 tetrahedra from $\approx 980\,cm^{-1}$ to $\approx 960\,cm^{-1}$, increasing strength of a band at $811\,cm^{-1}$ attributable to stretching of Q^1 tetrahedra, and weakening and broadening of the Si–O–Si bending band at $670\,cm^{-1}$. Bands near $500\,cm^{-1}$ became better defined due to internal vibrations of SiO_4 tetrahedra. In the spectra of the low-lime samples, the O–H stretching band is very broad and has a maximum at $\approx 3500\,cm^{-1}$, as for 14 Å tobermorite. At higher Ca/Si ratios there are two broad maxima at $\approx$ 3600 and $\approx 3300\,cm^{-1}$. The $3600\,cm^{-1}$ component was attributed to less strongly H-bonded water molecules and Ca–OH groups.

In the near IR spectra, bands at 4567 and 3740 cm^{-1}, considered diagnostic for Si–OH, weaken progressively with increasing Ca/Si ratio, indicating that such groups are almost completely absent at Ca/Si ratios above ≈1·25. Far IR spectra contain bands at 450 and 480 cm^{-1}, attributable to SiO_4 deformations and at 200–350 cm^{-1}, assigned to lattice vibration modes of Ca polyhedra. Increased intensity in the 300 cm^{-1} region for a sample with Ca/Si 1·45 indicate increased abundance of regions with a portlandite-like local structure. The authors considered the sample to be single phase because of the absence of crystalline CH, but inspection of Figure 6.3 suggests that it includes a small amount of a second phase that is Ca rich (i.e. the layer spacing of filled-circle data point at Ca/Si = 1·45 is consistent with phase intermixture).

The broad conclusion from the IR evidence was that for Ca/Si ratios up to ≈1·3, the dominant mechanism for accommodating increasing Ca/Si ratio is loss of bridging tetrahedra and Si–OH. At ratios above ≈1·3, further depolymerisation is less important and increases in Ca/Si ratio are mainly due to incorporation of Ca–OH, which from Figure 6.3 is in some cases likely to be due to intermixture of C–S–H(I) with CH, whether crystalline or amorphous, or perhaps with a jennite-based phase.

6.2.8 Raman and Ca X-ray absorption spectra

For preparations with Ca/Si <1·0, Raman spectra of the 200–1200 cm^{-1} region (Kirkpatrick *et al.*, 1997a) are very similar to those of 14 Å tobermorite, described in Chapter 5. With increasing Ca/Si ratio above this value, absorption at ≈970 cm^{-1} increases and is attributable to Si–O stretching of Q^1 tetrahedra. The main Si–O–Si bending band is at 664–669 cm^{-1} in all cases. This value lies between those of the tobermorites (675–680 cm^{-1}) and jennite (665 cm^{-1}). These data are consistent with decreasing polymerisation of the chains as Ca/Si increases, and with the possible existence of jennite-like environments in the high lime preparations. The absence of any significant shift in the Si–O–Si bending band with Ca/Si ratio suggests that any such regions are not extensive.

Ca X-ray absorption spectra were reported for tobermorite-type C–S–H preparations made from CH and silica, β-C_2S or β-C_2S and silica (Kirkpatrick *et al.*, 1997b) and from CH and silica or $CaCl_2$ and sodium silicate solutions (Lequeux *et al.*, 1999). All were consistent with a structure derived from that of 14 Å tobermorite and having a number of defects that increases with Ca/Si ratio but differed from those of other crystalline calcium silicate hydrates, including jennite. The Ca coordination does not vary with Ca/Si ratio between 0·7 and 1·4 (Lequeux *et al.*, 1999). The spectra differ from that of a fully hydrated β-C_2S preparation, largely because of the presence in the latter of CH, suggesting that increasing Ca/Si ratio is not associated with the presence of volumes high in CH (Kirkpatrick *et al.*, 1997b).

6.2.9 Conclusions

The studies discussed in §6.2 used methods to synthesise C–S–H which involve solutions that are undersaturated with respect to calcium; methods such as the double decomposition of a calcium salt and an alkali silicate in aqueous solution, or direct silica–lime reaction. The results concerning preparations with molar Ca/Si ratios <1·3–1·5 are useful in helping to better understand the C–S–H that is formed in real cements that contain supplementary cementitious materials.

Attempts to use these methods to obtain products with Ca/Si ratios greater than ≈1·5 yield C–S–H of approximately that Ca/Si ratio (or lower) and CH. C–S–H preparations are sometimes discussed in the literature as having Ca/Si >1·5, when inspection of the data reveals that this is clearly not true. Renaudin *et al.*'s (2009a) study of synthetic C–S–H preparations using Rietveld refinement of X-ray powder patterns is a good example of this: it is stated in the conclusion to the paper that they observed '... *structural continuity from Ca/Si = 0·8 (i.e. C–S–H(I) type) up to Ca/Si = 1·7 (i.e. the C–S–H(II) type)*', yet inspection of their Table 1 shows clearly that the actual maximum Ca/Si was only 1·4. Preparations that have a bulk Ca/Si >1·5 can contain an amorphous Ca-rich phase as well as C–S–H (Richardson *et al.*, 2010) and so it might not always be evident that a two-phase mixture is being studied. This is a serious issue that is best addressed by direct measurement of the Ca/Si ratio of the C–S–H by microanalysis in the TEM; the same issue arises with synthetic preparations of layered double hydroxides (Richardson, 2013). Since the compositions of most of the preparations reported in this chapter were not checked by TEM, it must be recognised that some of the results might not relate to a single phase; indeed, some preparations are shown to be most likely not single phase. It is important that the actual compositions of synthesised C–S–H are reported prominently in publications, rather than the target compositions, and that the distinction is clear and unambiguous; for example, target compositions could be reported in the experimental section of an article, with actual compositions given in the abstract and conclusions.

6.3. Room temperature preparations with Ca/Si ratios greater than 1·5

It is shown in §6.2 that the maximum molar Ca/Si ratio of C–S–H that can be synthesised is 1·3–1·5 when using methods such as the double decomposition of a calcium salt and an alkali silicate in aqueous solution or by direct silica–lime reaction. It has foil-like morphology and a structure that is derived from that of the mineral 14 Å tobermorite. When Ca/Si ratios >1·5 are targeted using those methods, CH precipitates when the solution becomes saturated (the CH may be crystalline or amorphous) and as a consequence the target value is not achieved. In hydrating pastes of C_3S, the C–S–H that forms has fibrillar morphology, an average Ca/Si ratio of about 1·7–1·8 and the pore solution is supersaturated with CH (Chapter 7), but the C–S–H is mixed intimately with crystals of CH and any unreacted C_3S. It follows that in order to obtain single-phase C–S–H that has a Ca/Si ratio >1·5, a synthetic route must be conceived where the solution is supersaturated with CH. The present section is concerned with the small number of studies that have successfully prepared alkali-free phase-pure C–S–H that has an average Ca/Si ratio greater than 1·3–1·5.

6.3.1 Preparations of low crystallinity

A straightforward method of synthesising C–S–H using a solution that is supersaturated with CH was developed by Hedin (1945) and, while his method was included in Steinour's (1947) comprehensive review (Steinour considered it to be well-conceived), it appears to have been subsequently little used, apparently only by Grudemo (1955) and by Vinnakota *et al.* (2025). Hedin (1945) demonstrated that a high degree of supersaturation can be obtained from a saturated carbonate-free CH solution by careful evaporation by distillation in vacuum. His results are replotted in Figure 6.8(a). The distillation temperature was 30 °C. The long-dashed horizontal line represents the concentration of a saturated solution at 30 °C, calculated using Equation 7.1 for a solution prepared using large crystals, and the horizontal dotted line shows the value when using fine crystals (Figure 7.2). Hedin's (1945) starting solution had a concentration of about 21 mmol L^{-1} and it is evident that it must have been prepared using fine crystals. Distillation was stopped after 3·5 hours. The short-dashed curve shows the theoretical increase in concentration if there is no CH precipitation. The actual experimental curve deviated from the theoretical curve at a concentration of about 30 mmol L^{-1}, indicating that precipitation of CH occurred only after a high level of supersaturation had been reached. Hedin (1945) noted that at a concentration of 23 mmol L^{-1} crystallisation is so slow that such a supersaturated solution can be kept for several hours, even in the presence of crystals of CH, and that if no CH crystals are present, a solution is completely stable up to a concentration of about 26·7–28·5 mmol L^{-1}. Hedin (1945) synthesised C–S–H by the slow addition of drops of alcoholic tetra-ethyl silicate solution to lime solutions maintained at essentially

Figure 6.8(a) The changes on vacuum distillation of a supersaturated lime solution. The data points were extracted from diagram 3 on page 25 of Hedin (1945)

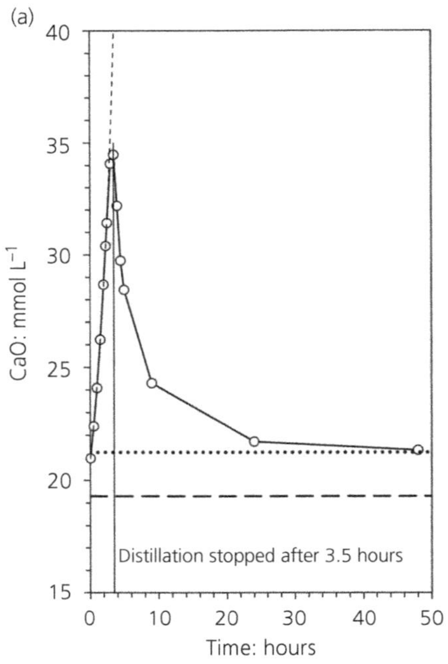

constant concentration by distillation in vacuum at 30 °C. His data are plotted on the lower part of Figure 6.8(b) using unfilled circles. The vertical dotted line is drawn at the concentration for a solution at 30 °C that is saturated with CH (calculated using Equation 7.1), and it is clear that Hedin's (1945) data all fall to its right, consistent with the solutions being supersaturated. Hedin's (1945) experiments were reproduced by Grudemo (1955) but he used 35 °C and covered a wider range of concentration that included both undersaturated and supersaturated solutions. His data are plotted on Figure 6.8(b) as filled circles, which can be seen to overlap with Hedin's (1945) data. The distillation-constant lime method does not seem to have been used again until a PhD project commenced at the University of Leeds in 2014 (Vinnakota, 2019), which is perhaps surprising given the number of studies published that report attempts to synthesise high Ca/Si C–S–H. Some data from the Leeds study are represented on Figure 6.8(b) using the $\times$ symbol, and the points can again be seen to overlap with Hedin's (1945) data. The general trend for the three sets of data taken together is shown using a short-dashed eye-guide line. It is clear that the rise in the trend occurs at a much higher concentration than the dotted line, at about 27 $mmol\,L^{-1}$ and a Ca/Si ratio of about 1·75. This is in complete contrast to Figure 6.1, where abrupt rises in the curves coincide with the dotted lines, indicating that CH precipitated when the solutions became saturated.

Grudemo (1955) reported X-ray diffraction data for preparations with Ca/Si ratios between 0·98–1·56. The strongest lattice spacings were for a layer spacing peak and for peaks at $\approx$ 3·06 Å, $\approx$ 2·81 Å, $\approx$ 1·83 Å, $\approx$ 1·67 Å, and $\approx$ 1·40 Å (averages for the six samples in his Table 14); the layer spacings for his preparations are shown on Figure 6.3 using filled triangles, which fall along the upper trend line, the layer spacing decreasing with increasing Ca/Si. The XRD data are

Figure 6.8(b) Metastable equilibrium curves relating the Ca/Si mole ratios of C–S–H preparations and the analytical concentration of CaO in solution

(b)

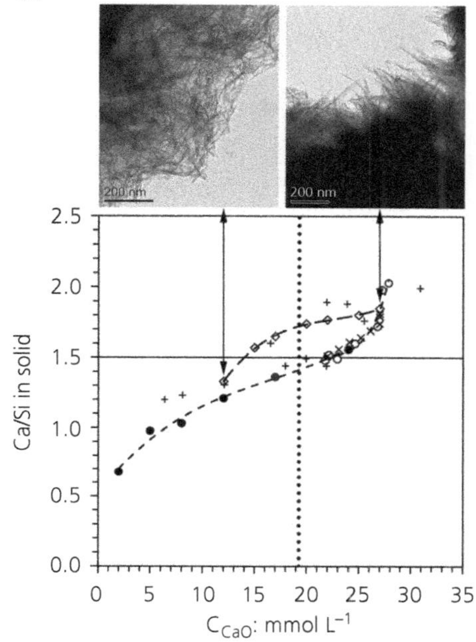

The data are from Hedin (1945; unfilled circles), Grudemo (1955; filled circles), Nonat and Lecoq (1998; +) Tajuelo Rodriguez *et al.* (2015; open diamonds) and Vinnakota *et al.* (2025; ×). The Ca/Si ratios of the C–S–H in the studies of Tajuelo Rodriguez *et al.* (2015) and Vinnakota *et al.* (2025; ×) were determined unambiguously by using TEM-EDX. The TEM micrographs at the top show C–S–H produced by hydration of C_3S at controlled lime concentration: (left) a micrograph that shows foil-like C–S–H for C_{CaO} = 12 mmol/L and (right) a micrograph that shows fibrillar C–S–H for C_{CaO} = 27 mmol/L. The micrographs are modified from Figure 3 of Tajuelo Rodriguez *et al.* (2015).

consistent with C–S–H(I), Table 6.1. Grudemo (1955) examined his preparations using bright-field imaging and selected area electron diffraction in the TEM (SAED); the SAED results are consistent with those from XRD. He reported that the preparations with Ca/Si ratios between 0·98–1·36 had wrinkled-foil morphology, a sample with Ca/Si = 1·51 that had been prepared in a saturated lime solution consisted of mixed morphologies, including fibres, and a sample with Ca/Si = 1·56 that had been prepared in a supersaturated lime solution consisted of fibres or needles. Vinnakota *et al.* (2025) confirmed the change in morphology from foils to fibres with increasing Ca/Si ratio, which was determined directly by TEM-EDX. The morphology is evidently dependent on the absolute concentration of Ca in solution: foil-like and fibrillar morphologies are associated with alkali-free solutions that are undersaturated and supersaturated, respectively; mixed morphologies can be observed for values around saturation.

Vinnakota *et al.*'s (2025) samples were also examined by ^{29}Si MAS NMR and the results are represented (using open-triangle symbols) in terms of SOF_{BT} on Figure 6.9(a), which is the same as

Figure 6.9(a) Plot of SOF_{BT} against the molar Ca/Si ratio of C–S–H preparations that were synthesised using a variety of techniques

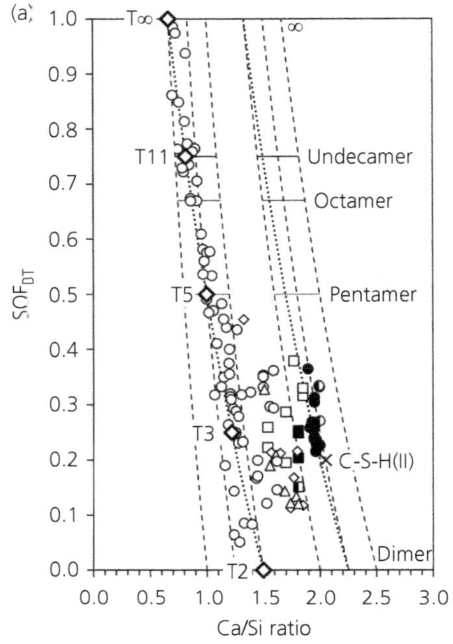

The data that are plotted as unfilled circles are from: Brunet *et al.*, 2004; Chen *et al.*, 2004; Cong and Kirkpatrick, 1996; Damidot *et al.*, 1995; García-Lodeiro *et al.*, 2012 (uncarbonated samples); Grutzeck *et al.*, 1989; Matsuyama and Young, 1999, 2000; Nonat and Lecoq, 1998, Tajuelo Rodriguez *et al.*, 2015 (mechanochemical and CaO-SiO_2 solution samples). Filled-symbol data are from: Ishida *et al.*, 1992b; Ishida *et al.*, 1992d (static reaction); Sasaki *et al.*, 1993. Other data are from: Tajuelo Rodriguez *et al.*, 2015 (unfilled diamonds; 'deceleration period' samples); Vinnakota *et al.*, 2025 (unfilled triangles; constant lime method samples); Ishida *et al.*, 1992d (half-filled (left) symbols; ball-mill hydrated samples). Square symbols indicate preparations that contain crystalline calcium hydroxide. The point for C–S–H(II) is from Gard and Taylor (1976). The large bold unfilled diamonds represent the C–S–H(I) model structures of Richardson (2014) and §6.6.

Figure 6.5(e) but with the Ca/Si axis extended to 3. The data points are far to the right of the region that covers C–S–H(I) structure (i.e. the region of the three dashed lines at low Ca/Si) and so – as explained in §6.2.5 – the C–S–H must contain Ca–OH groups. Figure 6.9(a) is discussed further in §6.3.3.

Another method of synthesising high Ca/Si C–S–H using a solution that is supersaturated with CH involves the use of C_3S (e.g. Nonat and Lecoq, 1998; Tajuelo Rodriguez *et al.*, 2015). In this method, C_3S is hydrated in a lime solution that has a fixed concentration, with the concentration maintained by the addition of deionised water and the simultaneous removal of an equal volume of the solution, so that the overall volume does not change. Nonat and Lecoq (1998) synthesised C–S–H with bulk Ca/Si ratios between 1·2–2·0 using solutions with C_{CaO} between 6·5–30 mmol/L: their data are shown on Figure 6.8(b) using the + symbol. Experiments that required up to 20 mmol/L were conducted using filtered solutions prepared using undersaturated lime solutions, while experiments that required supersaturated concentrations were conducted using filtered solutions obtained from hydrating C_3S. Nonat and Lecoq (1998) stated that their XRD patterns are consistent with C–S–H(I) structure; their patterns are discussed further in §6.6.4. Tajuelo Rodriguez *et al.* (2015) used the same method as Nonat and Lecoq (1998) and characterised their samples using many techniques, including by ^{29}Si MAS NMR and by bright-field imaging and EDX analysis in the TEM. Their results are shown on Figure 6.8(b) as open diamonds and the trend in their data is illustrated using the long-dashed line as an eye-guide. As with the distillation method, Tajuelo Rodriguez *et al.*'s (2015) TEM micrographs show a change in morphology from foil-like to fibrillar as the solutions change from undersaturated to supersaturated, which is illustrated on the figure using two micrographs (modified from Figure 3 of Tajuelo Rodriguez *et al.*, 2015). Nonat and Lecoq's (1998) data (+) are quite scattered on Figure 6.8(b), but most of their points fall close to one or the other of the two trend lines.

Ishida *et al.* (1992a, 1992b, 1992c, 1992d) produced high Ca/Si C–S–H by the complete hydration of a highly reactive β-C_2S that was obtained from the dehydration of hillebrandite (composition and crystal structure data for hillebrandite are given in Table 5.1(a) and some crystal chemical data in Table 5.20(a)) (Ishida *et al.*,1992a, 1992b, 1992c, 1992d; Sasaki *et al.*, 1993). The product normally obtained from aqueous suspensions is typically very poorly crystalline, with a Ca/Si ratio of ≈ 1·7. The diffractometer traces reported by Ishida *et al.* (1992d) for their 'static hydration' (at 25 °C) samples show a relatively sharp peak at ≈ 3·05 Å, with a pronounced high angle tail including a subsidiary peak at ≈ 2·8 Å and a very weak peak at ≈ 1·82 Å; the patterns are essentially the same as those found for fully reacted β-C_2S or C_3S pastes (e.g. see Figure 7.1). Peaks from CH are also present. Hydration was completed by 8 days, giving C–S–H with a mean Ca/Si ratio of 1·81 together with CH. TEM microanalyses showed that the mean Ca/Si ratio of the C–S–H was constant from 2 days, when unreacted β-C_2S was still present, to at least 64 days. ^{29}Si NMR spectra showed that after all the β-C_2S appeared to have reacted, only Q^1 and Q^2 peaks were clearly detectable, although there was a slight indication that some Q^0 might still be present at 8 and 16 days. The Q^1/Q^2 ratios imply mean chain lengths of ≈ 2·7 at 2–8 days, increasing to ≈3·1 at 18–28 days. Analyses of the solutions showed that the CaO concentration passed through a maximum of 27 mmol L^{-1} at 2 days and decreased to ≈12 mmol L^{-1} at 16 days. These results are shown using open diamonds on Figure 6.9(b), which is a plot of C_{SiO_2} against C_{CaO} that also includes Jennings' (1986) curve B (the upper curve) and Chen *et al.*'s (2004) curve C. The full vertical line represents the solubility of CH at 25 °C calculated using Equation 7.1. The points that correspond to complete hydration are close to curve B; the Ca/Si ratios determined by TEM-EDX are indicated.

On the assumption that the silicate anions are based on *dreierketten*, Ishida *et al.*'s (1992a, 1992b, 1992c, 1992d) NMR data for high Ca/Si preparations can be recalculated in terms of the occupancy of bridging sites and compared with the data for C–S–H(I) preparations that are presented on Figure 6.5(c). Such a comparison is given in Figure 6.9(a), which is the same as Figure 6.5(e) but with the Ca/Si axis extended to 3. Ishida *et al.* (1992a, 1992b, 1992c, 1992d) published their work on highly reactive β-C_2S in a series of papers and results from a number of them are included on

Figure 6.9(b) Plot of the analytical concentrations of SiO_2 (C_{SiO_2}) and CaO (C_{CaO}) in solution for C–S–H preparations that were synthesised using various methods involving β-C_2S

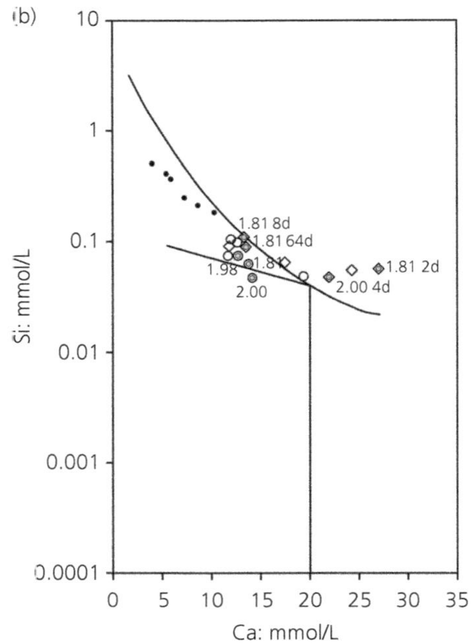

The top curve is Jennings' (1986) curve B and the lower one is Chen *et al.*'s (2004) curve C. The full vertical line represents the solubility of CH at 25 °C calculated using Equation 7.1. The open symbols are for results from Ishida *et al.*'s (1992d) study of the hydration of a highly reactive β-C_2S that was obtained from the dehydration of hillebrandite. The diamonds represent their 'static hydration' samples and the circles represent their 'ball mill hydration' samples. The Ca/Si ratios of the C–S–H determined unambiguously by TEM-EDX are indicted where available, together with the age of the 'static hydration' samples. The small filled circles are from Trettin *et al.* (1991) who studied pastes of β-C_2S (w/c = 0·5; 20 °C) hydrated for between 0·5 and 30 minutes, moving from left to right with increasing time. The liquid phases were separated from C_2S-water mixtures by very fast filtration.

Figure 6.9(a). The filled symbols include the static hydration samples just discussed from Ishida *et al.* (1992d; w/c = 10; 25 °C) together with their data for pastes hydrated at 25 °C (w/c = 0·25 and 0·50; Ishida *et al.*, 1992b; Okada *et al.*, 1994a; Sasaki *et al.*, 1993). The results are notable for the fact that the high average Ca/Si ratios of the C–S–H in the samples were confirmed by direct analysis in the TEM. The data are not consistent with a structure that is based purely on tobermorite because the points are very far to the right of the lines for C–S–H(I) and so Ca–OH groups must be present; this is discussed further in §6.3.3 and §6.9.

TEM study of apparently similar material formed in a paste from the highly-reactive β-C_2S (Okada *et al.*, 1994a; w/c = 0·5; 25 °C) showed fibres that gave a weak diffraction pattern that disappeared after a few seconds' exposure to the beam. These patterns contained relatively sharp peaks at 1·82 Å parallel to the fibre direction and indexable as 020 and a diffuse ring at $\approx$ 2·95 Å, with maxima on the zero and first layer lines with spacings of 2·99 Å and 2·91 Å, respectively. Grudemo (1962) described a similar pattern, also from a fibre in a C_2S paste. These patterns differ significantly from those of tobermorite-type C–S–H, for which the corresponding peaks on the zero and first layers have spacings of 2·82 Å and 3·05 Å, respectively. Their interpretation is uncertain, but they could represent poorly crystalline forms of the material described in the next section.

6.3.2 Preparations of higher crystallinity

More highly ordered C–S–H preparations with a Ca/Si ratio significantly greater than 1·50 were first obtained by prolonged reaction of C_3S (Taylor, 1950) or β-C_2S (Brunauer and Greenberg, 1960) in aqueous suspensions. In the former case, repeated replacement of solution by water almost completely eliminated the CH. These preparations gave identical XRD powder patterns (replotted in Figure 6.10(a)) which, although somewhat similar to those of tobermorite-type C–S–H, are clearly distinct from them. The profile of the complex band at 25–35° 2θ (for Cu Kα radiation), the absence of a peak at 1·67 Å and the presence of one at 1·565 Å are especially distinctive.

TEM showed a characteristic morphology, described as cigar-shaped bundles of fibres, typically a few micrometres long (Brunauer and Greenberg, 1960; Copeland and Schulz, 1962; Grudemo, 1962). Figure 6.10(c) shows a typical TEM bright-field image of the 'cigar-shaped' fibre-bundle morphology of high-Ca C–S–H. The main micrograph is modified from Brunauer (1962), Brunauer and Greenberg (1960) and Copeland and Schulz (1962); it is of C–S–H that was formed by the reaction of β-C_2S hydrated for 778 days at 25 °C in a rotating bottle with w/c = 9, which was their sample D-69. Brunauer and Greenberg (1960) gave the composition of the C–S–H in a D-dried sample of D-69 as $C_{1\cdot73}SH_{1\cdot37}$. Brunauer (1962) gave a portion of D-69 to Gard and Taylor who presented additional results over a decade later (Gard and Taylor, 1976), including new TEM results that confirmed the fibre-bundle morphology, which is notably different from the foil-like morphology that is characteristic of C–S–H(I), a typical example of which is shown in Figure 6.4(a). The fibre-bundle morphology was actually observed in the very first TEM studies of cement and clay minerals that were published between 1939 and 1942 (as discussed by Richardson, 2002), which included the examination of dispersed specimens of hydrated tricalcium silicate; the micrograph that is inset on Figure 6.10(c) is modified from Figure 3 of Radczewski *et al.* (1939). Gard and Taylor (1976) also reported the results of direct compositional analysis in the TEM, which gave a mean Ca/Si ratio of 2·06 $\pm$ 0·09 ($n = 6$) which, taken together with the very small amount of CH that was detected, indicates that with aging the Ca/Si ratio of the C–S–H increases at the expense of CH. Gard and Taylor (1976) showed that the bundles of fibres in D-69 gave a distinctive preferred orientation SAED pattern, different from that of tobermorite-type C–S–H. The XRD and SAED patterns were indexed assuming an A-centred monoclinic subcell with a = 9·93 Å, b = 3·64 Å, c = 20·36 Å, β = 106·1°. These parameters and the broad distribution of intensity in reciprocal space were similar to those of jennite, and there was an indication of imperfect ordering in the c-direction. The composition of this preparation, after equilibration over saturated $CaCl_2$ solution at $\approx$15 °C, was approximately $C_2SH_{3\cdot2}$. Determinations by the molybdate and trimethylsilylation methods showed dimer and more highly condensed silicate ions, with possibly 68 % of the Si present as dimer.

The mid-IR spectrum (Gard and Taylor, 1976) resembles that of jennite, but the presence of a relatively strong band at 820 cm^{-1} that can be attributed to stretching of Q^1 tetrahedra is a notable

Figure 6.10(a) X-ray powder diffraction pattern of a high-Ca C–S–H preparation of relatively high crystallinity with approximate composition $C_2SH_{3.2}$; the pattern was adapted from ICDD card 29–373 (from a microdensitometer trace of Guinier film) and the thin vertical bars correspond to the *d*-spacings and intensities that are given on the card, which are referenced to C–S–H(II) (Gard and Taylor, 1976). The figure includes peaks for CH, which are indicated using a calculated pattern at the bottom of the figure

Figure 6.10(b) X-ray powder diffraction pattern of a high-Ca C–S–H preparation formed from highly-reactive β-C_2S (made by heating hillebrandite) that had been hydrated in a ball mill at 30–35 °C for 64 days (Ishida *et al.*, 1992d). The thin vertical bars have the same meaning as in Figure 6.10(a). The figure includes peaks for calcite, which are indicated using a calculated pattern at the bottom of the figure

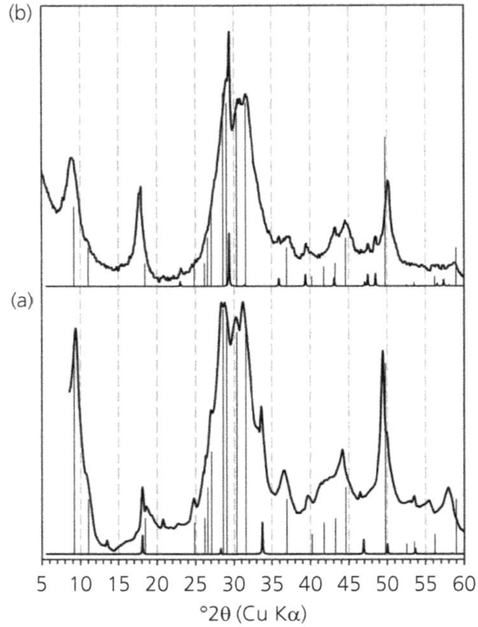

difference. Thermal analysis (DTA and thermogravimetry) showed a substantial and partly reversible water loss by 200 °C. The XRD pattern does not change significantly below 230 °C, but at higher temperatures the crystallinity decreases up to 650 °C, above which temperature β-C_2S gradually crystallises. The DTA curve shows endotherms at 145 °C (strong), 260 °C (weak) and 477 °C (weak, and due to CH) along with a broad, weak exotherm at 620 °C attributable to crystallisation of β-C_2S. These data are consistent with the conclusion that these preparations were of a phase, structurally derived from jennite by the omission of a high proportion of the bridging tetrahedra, and having something less than full three-dimensional order. The material is more highly ordered than tobermorite-type C–S–H but otherwise its relationship to jennite resembles that of the latter to 14 Å tobermorite. Gard and Taylor (1976) called the high-Ca C–S–H present in D-69 'C–S–H(II)', which is essentially identical with Taylor's 'calcium silicate hydrate (II)' (Taylor, 1950).

Figure 6.10(c) TEM bright field image showing the typical 'cigar-shaped' fibre-bundle morphology of high-Ca C–S–H. The main micrograph is modified from Brunauer (1962), Brunauer and Greenberg (1960) and Copeland and Schulz (1962); it is of C–S–H that was formed by the reaction of C_2S hydrated for 778 days at 25 °C in a rotating bottle with w/c = 9, which was their sample D-69. Brunauer and Greenberg (1960) gave the composition of the C–S–H in a D-dried sample of D-69 as $C_{1\cdot73}SH_{1\cdot37}$. Scale bars = 1 μm. The inset micrograph is modified from Figure 3 of Radczewski *et al.* (1939)

(c)

Attempts to repeat these preparations have yielded only poorly crystalline material of the type described in §6.3.1, but Ishida *et al.*, (1992d) obtained similar material by a different method and their results provide some clues as to the conditions needed for its formation. Highly reactive β-C_2S, made by heating hillebrandite, was hydrated in a ball mill at 30–35 °C. The β-C_2S reacted completely within 2 days, giving a product for which the XRD powder pattern shows tobermorite-type C–S–H and CH. Within a further 2 days of ball mill reaction, the CH was no longer detectable by XRD or thermal methods and after 64 days the XRD pattern was as shown in Figure 6.10(b), which clearly resembles that shown in Figure 6.10(a) for C–S–H(II). TEM microanalyses of the samples made by Ishida *et al.*, (1992d) showed that the mean Ca/Si ratio of the 2-day product was 1·81, with a standard deviation of 0·157 for analyses of different particles, while that of the 64-day product was 1·98 and the standard deviation had decreased to 0·058. In both cases, the number of analyses was between 30 and 40 (the exact numbers are not given). ^{29}Si NMR showed the presence of Q^1 and Q^2 tetrahedra in both cases; the mean chain lengths were 2·5, 3·2 and 3·5 at 2, 4 and 16–28 days, respectively. The Ca/Si ratio and NMR data for these samples are included on Figure 6.9(a) and can be seen to follow the same trend as the data for the preparations with low crystallinity; the Ca/Si-NMR data are discussed further in §6.3.3. Ishida *et al.* (1992d) showed that the C_{CaO} rose to the very high value of 38 mmol L^{-1} at 12 hours before decreasing to an apparently steady value of ≈12 mmol L^{-1} at 12–64 days. These latter results are shown using open circles on Figure 6.9(b): the points fall close to the region where curves B and C converge, suggesting strongly that C–S–H(II) is associated with this part of the C_{SiO_2}–C_{CaO} diagram.

These results suggest that the formation of relatively highly crystalline C–S–H with Ca/Si ratios of or approaching 2·0 is a difficult process that requires either prolonged treatment or favourable conditions that are not yet fully understood. The initial formation of material with a wide distribution of local Ca/Si ratios and a high content of dimeric silicate ions, followed by a narrowing of the compositional distribution and increasing silicate polymerisation is similar to the sequence of changes that occurs in calcium silicate or Portland cement pastes, but the ultimate product is much more highly crystalline.

6.3.3 Structural-chemical formulae for high-Ca C–S–H phases

It was noted in §6.3.2 that C–S–H(II) appears to be a structurally imperfect form of jennite, related to it in the same way as C–S–H(I) is to 14 Å tobermorite (Gard and Taylor, 1976). Its Ca/Si ratio is higher than that of jennite (as discussed in §6.3.2), which resembles the situation for C–S–H(I) and 14 Å tobermorite. Gard *et al.* (1977) suggested that jennite has the structural formula $Ca_9(Si_6O_{18}H_2)(OH)_8{\cdot}6H_2O$, which was based on an ideal Ca/Si ratio of 1·5, the measured H_2O content and density and on the thermal behaviour and IR spectra. They tentatively explained the structure as being similar to jennite '*but in which a substantial proportion of the silicate chains are replaced by pyrosilicate groups through omission of tetrahedra*'. The suggested omission of bridging tetrahedra is the same mechanism as suggested earlier by Taylor and Howison (1956) for C–S–H(I), and so the above formula for jennite can be modified in the same way as done in Chapter 5 and Richardson (2014) for tobermorite, giving

$$Ca_8\left[Si_{1-v}\square_v O_{3-2v}\right]_6 (OH)_8\, H_{2i}Ca_{2-i}\cdot mH_2O \tag{6.9}$$

where v is again the fraction of vacant tetrahedral sites and $0 \leq i \leq 2$ (i.e. as for the tobermorite and C–A–S–H phases).

The Ca/Si ratio can therefore be calculated using

$$\frac{Ca}{Si} = \frac{10-i}{6(1-v)} = \frac{10-i}{2(2+SOF_{BT})} \tag{6.10}$$

The maximum Ca/Si ratio is obtained when $i = 0$, and so

$$\left(\frac{Ca}{Si}\right)_{max} = \frac{\frac{5}{3}}{1-v} \tag{6.11}$$

There is a slight contradiction in Gard and Taylor's (1976) paper: while stating that the omission of bridging tetrahedra occurs '*probably in a largely random way*', they also conclude from the results of two different experimental techniques that the silicate anions were either polysilicate chains (i.e. $\left(SiO_3^{2-}\right)_n$) or pyrosilicate groups (i.e. $Si_2O_7^{6-}$), with an '*absence of small, finite groups other than pyrosilicate*', which is not consistent with random omission of bridging tetrahedra. Quantification of their evidence on the silicate anions led them to propose the following formula, which includes elements of both poly- and pyrosilicate structure

$$Ca_9\left[Si_3O_9H\right]_{2x}\left[Si_2O_7H\right]_{2(1-x)}(OH)_8\cdot mH_2O \tag{6.12}$$

Gard and Taylor (1976) stated that the value of x was about 0·2 for their particular specimen, which therefore gives the formula

$$Ca_9\left[Si_3O_9H\right]_{0\cdot 4}\left[Si_2O_7H\right]_{1\cdot 6}(OH)_8\cdot mH_2O \tag{6.13}$$

As noted by Gard and Taylor (1976), this indicates that about 73% of Si in the sample was present as pyrosilicate. Recasting this in terms of Formula 6.9 gives

$$Ca_8\left[Si_{0\cdot 733}\square_{0\cdot 267}O_{2\cdot 467}\right]_6(OH)_8\,H_2Ca\cdot mH_2O \tag{6.14}$$

It is evident that $i = 1$, $v = 0{\cdot}267$ and so MCL = 2·75, $SOF_{BT} = 0{\cdot}2$ and Ca/Si = 2·05. These data are represented by a cross (×) on Figure 6.9(a), labelled 'C–S–H(II)'.

There are six dashed lines on Figure 6.9(a): the three that are below Ca/Si = 1·5 represent the relationship between SOF_{BT} and Ca/Si for tobermorite-based structure, from left to right for $i = 2$, 1 and 0 – that is, Equation 6.4, as discussed in §6.2.5; the three that are above Ca/Si = 1·5 represent the same relationship for jennite-based structure, again from left to right for $i = 2$, 1 and 0 – that is, Equation 6.10. These six lines are equivalent to those in Richardson and Groves (1992a) for tobermorite- (T) and jennite/jaffeite- (J) based structure (§7.5). As noted already, the cross shows the position for Gard and Taylor's (1976) sample of C–S–H(II) and the filled symbols represent the room-temperature data from the work of Ishida *et al.* (1992a, 1992b, 1992c, 1992d) on the hydration of highly reactive β-C_2S (that they made by heating hillebrandite, $Ca_2(SiO_3)(OH)_2$, which had been synthesised hydrothermally from CaO and quartz), including high w/c (=10) static hydration samples from Ishida *et al.* (1992d) and their data for pastes with w/c = 0·25 and 0·50 reported in Ishida *et al.* (1992b), Sasaki *et al.* (1993) and Okada *et al.* (1994a). Their work is notable because the Ca/Si ratios of the C–S–H that were calculated from thermal analysis data were confirmed by direct measurement in the TEM. As a consequence, the high reported values of between 1·8 and 2·0 must correspond to the C–S–H itself, rather than to intermixture of C–S–H that has a lower Ca/Si ratio with a second phase that is Ca-rich, which was shown to be a serious issue for the methods of synthesis discussed in §6.2. The half-filled symbols also represent data from the work of Ishida *et al.* (1992a, 1992b, 1992c, 1992d) but are for samples of higher crystallinity that were produced by hydrating the hillebrandite-derived β-C_2S in a ball mill (w/c = 10; rotated at 110 rpm). The ball-mill experiments were again conducted at room temperature, although the sample temperature was in the range 30–35 °C due to rotation of drum. In this case the β-C_2S was reacted completely by 2 days to yield a C–S–H with Ca/Si ≈ 1·8 together with CH, after which the CH disappeared and the Ca/Si ratio of the C–S–H increased to ≈ 2. Gard and Taylor's (1976) sample of C–S–H(II) was actually synthesised many years earlier by Brunauer and Greenberg (1960) by hydrating β-C_2S in a rotating bottle (w/c = 9) for over 2 years. Inspection of the two papers shows that in the period between the two studies, aging of the sample resulted in the disappearance of most of the CH and an increase in the Ca/Si ratio of the C–S–H, which is therefore the same as Ishida *et al.* (1992d) observed with their samples.

Inspection of Figure 6.9(a) shows that the data for the preparations that are free of CH all fall close to the 'J' line for $i = 1$ – that is, equivalent to the jennite-based line in Figure 2 of Taylor (1986) or to Richardson and Groves' (1992a) jennite-based structural units that have $w/n = 1$ (see §7.5). It should be noted that the formula from Bonaccorsi *et al.*'s (2004) refined structure of jennite does not conform to Formula 6.9: it is $Ca_9Si_6O_{18}(OH)_6\cdot 8H_2O$; based on this formula,

C–S–H(II) would be represented by Formula 6.15, which is shown on Figure 6.9(a) by the right-hand dotted line.

$$Ca_8\left[Si_{1-v}\square_v O_{3-2v}\right]_6 (OH)_6\, H_{2i}Ca_{1-i}\cdot mH_2O \tag{6.15}$$

where $1 \leq i \leq 2$.

Since data from Ishida *et al.*'s (1992c) studies on hillebrandite-derived β-C_2S pastes that were hydrated at higher temperatures (up to 80 °C; Ishida *et al.* 1992c) fall between the dashed 'J' lines on Figure 6.9(a) for $i = 1$ and $i = 0$ (i.e. the two dashed lines at the highest Ca/Si ratio), Formula 6.9 is to be preferred to Formula 6.15 when considering C–S–H(II) and related phases that are less crystalline.

6.4. Preparations containing monomeric or high proportions of dimeric silicate ion

6.4.1 Reactions in C_3S or cement pastes

Using ^1H–^{29}Si MAS NMR, Rodger *et al.* (1988) found the silicate ions in the initial product in C_3S pastes to be monomeric. It has been suggested that the monomeric ions result from protonation of the C_3S surface (Damidot *et al.*, 1995; Rassem *et al.*, 1992b) but, while this process may well occur, the amounts present, estimated at 2% of the total Si at the end of the induction period (Rodger *et al.*, 1988), are much too great to be wholly explained in this way (see §7.4.2 and §7.5.3). It is well established that, in C_3S or Portland cement pastes, dimeric ions quickly predominate and that the average silicate ion size subsequently increases on a time scale of months or years (§7.4.1 and §7.4.2).

6.4.2 C–S–H obtained by precipitation reactions

Some early reports suggested that the silicate ions in synthetic C–S–H prepared in aqueous suspensions at ordinary temperatures are monomeric (e.g. Funk and Thilo, 1955; Greenberg and Chang, 1965; Roller and Ervin, 1940), but no substantial evidence was given and the conditions under which the preparations were formed suggest that they were probably tobermorite-like. However, there is evidence that in the initial products formed by mixing solutions of an alkali silicate and a calcium salt, the silicate ions are monomeric, and that the monomer is quickly replaced by dimer, which is subsequently partly replaced by larger groups.

Stade and Wieker (1980) concluded from determinations using the molybdate method that the silicate ions in the first-formed solids in reactions of sodium silicate and calcium chloride solutions at ≈0 °C were those present in the silicate solution, which were monomeric. This conclusion is consistent with evidence that the anions in silicate solutions are predominantly monomeric at concentrations up to ≈2 mmol L^{-1} at pH 9 and up to ≈300 mmol L^{-1} at pH 13·5 (Barby *et al.*, 1977). On storage in the mother liquor at 0 °C, dimeric ions were formed and the solids containing them could be isolated by drying at –10 °C. Consistent with this result, Michaux *et al.* (1983) found using trimethylsilylation that the silicate ions in preparations made by mixing very dilute solutions supplying CaO and SiO_2 were almost entirely dimeric.

A ^{17}O and ^{29}Si MAS NMR study of preparations obtained by mixing potassium silicate and calcium chloride solutions (Cong and Kirkpatrick, 1993) showed that for four specimens containing only Q^1 and Q^2 tetrahedra, the mean chain lengths decreased with increasing Ca/Si ratio from 2·8 at Ca/Si 0·79 to 2·2 at Ca/Si 1·50. Three of these specimens contained crystalline CH, implying that the Ca/Si ratios of the C–S–H were lower than the bulk values given above. The mean chain lengths are considerably less than those shown in Figure 6.5(a) for samples obtained by reaction of CH or anhydrous calcium silicates with amorphous silica and water. Two other preparations, with Ca/Si ratios of 0·75–0·79, contained high proportions of Q^3 and in one case no Q^1.

6.4.3 Vibratory milling of lime–silica slurries

Sasaki *et al.* (1996) hydrated a mixture of CH and amorphous silica in a vibratory mill at temperatures below ≈40 °C. XRD and ^{29}Si MAS NMR showed that the starting materials had completely reacted in 14 h. XRD of the product showed only a single, very diffuse band peaking at 2·8–3·0 Å, while NMR showed a single, broad peak at –74 ppm, which was attributed to a mixture of Q^0 and Q^1. TEM showed irregular agglomerates and local Ca/Si ratios of 1·7–2·5.

6.4.4 Conclusions

In possibly all reactions giving C–S–H, the initial products contain monomeric silicate ions. These products are quickly replaced by ones that contain dimeric silicate ions. The subsequent formation of fragments or longer chains of *dreierkette* type is sufficiently slow that C–S–H preparations obtained by mixing solutions of an alkali silicate and a calcium salt can have smaller mean chain lengths for a given Ca/Si ratio than those formed by other methods. They also tend to be less crystalline. There is no evidence regarding the structures of the products containing monomer. It is most unlikely that monomer could be the sole or major silicate component of a tobermorite- or jennite-like structure. Tobermorite and jennite both contain *dreierketten* silicate, the conformation of which is dictated by the O–O distances of the Ca–O layers; monomeric silicate ions are thus only likely to occur at the edges of particles (adjacent to a vacant tetrahedral bridging site). For products with Ca/Si ratios up to ≈1·5, the formation of dimeric ions or longer chains is associated with that of a tobermorite-like structure. The morphology of C–S–H is determined largely by the concentration of Ca^{2+} ions in solution: in alkali-free methods, foil-like and fibrillar morphologies are associated with solutions that are undersaturated and supersaturated, respectively; mixed morphologies can be observed for values around saturation. Preparations that involve alkali almost inevitably result in foil-like morphology because the value of C_{CaO} is too low to obtain C–S–H that has fibrillar morphology (see §6.10.2).

6.5. Preparations made at elevated temperatures

6.5.1 Formation, XRD and other data

Reactions in the $CaO–SiO_2–H_2O$ system give C–S–H as the initial product under a wide range of conditions at temperatures up to at least 200 °C (Heller and Taylor, 1951; Kalousek and Prebus, 1958; Mitsuda *et al.*, 1986; Stade, 1980). At temperatures above ≈ 60 °C, the C–S–H is replaced at widely varying rates by one or more of the many crystalline hydrated calcium silicate phases. C–S–H formed at elevated temperatures is essentially similar to that formed near room temperature. Specimens with Ca/Si ratios up to ≈ 1·3, if sufficiently highly ordered, give XRD powder patterns similar to those of tobermorite-type C–S–H formed at lower temperatures – that is, C–S–H(I) (Heller and Taylor, 1951; Kalousek and Prebus, 1958; Mitsuda *et al.*, 1986; Renaudin *et al.*, 2009a; Stade, 1980). They also have similar particle morphologies, DTA curves and IR absorption spectra (Mitsuda *et al.*, 1986). DTA of a specimen with Ca/Si = 1·5 showed exotherms at both 830 °C and 890 °C, which were attributed to crystallisation of β-C_2S and wollastonite, respectively (Kalousek and Prebus, 1958).

Figure 6.11(a) shows the variation in the peak temperature of the exotherm that is due to the crystallisation of β-wollastonite that occurs when C–S–H(I) preparations are heated. The sources of the data are given in the figure caption. There is some scatter in the data, which is unsurprising because the exact temperature at which a particular thermal event is observed by thermal analysis can vary between studies due to sample and instrumental factors – for example, particle size and heating rate. This makes comparisons of exact peak temperatures measured in different studies problematic, particularly when experimental details are limited and, as a consequence, analysis beyond general trends is best restricted to individual data sets that are most complete. In this particular case, the most extensive data are due to Mitsuda *et al.* (1986) who reported results for C–S–H(I) preparations with $0{\cdot}5 \leq$ Ca/Si $\leq 1{\cdot}5$ synthesised from CaO and silicic acid at 120 °C (open circles), 150 °C (half-filled circles) or 180 °C (filled circles); the results are very similar for the three temperatures. The dashed lines show the results of linear regression analyses of Mitsuda *et al.*'s (1986) data: the regression line on the left of the figure is for the data points with Ca/Si $\leq 0{\cdot}9$ and the line on the right is for data with Ca/Si $> 0{\cdot}9$; data outliers that were excluded from the regression analyses are indicated using square symbols. It is evident that the peak temperature decreases linearly from 835 °C for Ca/Si = 0·5 to 820 °C for Ca/Si = 0·96, with 820 °C being the minimum peak temperature of the exotherm. The minimum peak temperature thus occurs when

Figure 6.11(a) Variation in the peak temperature of the exotherm that is due to the crystallisation of β-wollastonite on heating C–S–H(I) preparations

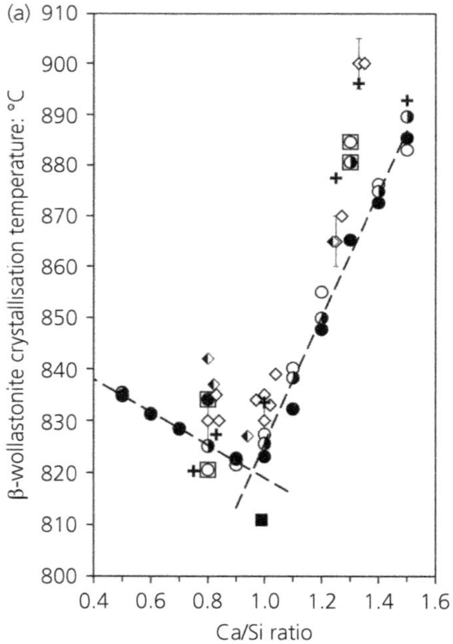

The samples were prepared hydrothermally except for those that are represented by the filled square, the + symbol or the unfilled diamonds that have error bars, which were prepared at room temperature or thereabouts. The data are from Tajuelo Rodriguez *et al.* (2017; +), Mitsuda *et al.* (1986; circles), Kalousek (1954, 1955; unfilled diamonds), Klimesch and Ray (2002; half-filled diamonds) and Taylor (1953; filled square). The values for the Ca/Si ratio are for the preparation as a whole, not just the C–S–H, unless the preparation is single phase.

the Ca/Si ratio of the C–S–H is approximately the same as that of wollastonite. Further increase in Ca/Si ratio results in a linear increase in the peak temperature to a maximum of ≈ 890 °C for Ca/Si = 1·5. The regression equations are given in Equations 6.16 and 6.17.

$$T_{\text{exo}} = 703{\cdot}1 + 122{\cdot}5(\text{Ca/Si})\, r^2 = 0{\cdot}99 \tag{6.16}$$

$$T_{\text{exo}} = 850{\cdot}6 - 31{\cdot}5(\text{Ca/Si})\, r^2 = 0{\cdot}99 \tag{6.17}$$

Stade (1980) described preparations obtained at 80 °C and 150 °C, later called C–S–H(di, poly), and reported silicate anion distributions obtained using the molybdate and trimethylsilylation methods. The starting materials were either prepared from sodium silicate and calcium chloride solutions at 0 °C and then stored at 80 °C under the mother liquid (Stade and Wieker, 1980) or by hydrothermal reaction of lime and silica at 150 °C. The Ca/Si ratios of the preparations that were free of crystalline CH were 1·0–1·9 for the 80 °C samples and 1·0–1·5 for those made at 150 °C. In the 80 °C preparations the Si was approximately equally divided between dimeric and larger anions at most ratios. An 80 °C preparation with Ca/Si 1·0 and 150 °C preparations with Ca/Si 1·1 and 1·5 were somewhat lower in dimer. The silicate ions in a 150 °C preparation with Ca/Si 1·0 were more highly condensed and included some double chains.

Figure 6.11(b) Plot of the fraction of bridging sites that are occupied, SOF_{BT}, against bulk molar Ca/Si ratio for C–S–H(I) preparations that were synthesised hydrothermally

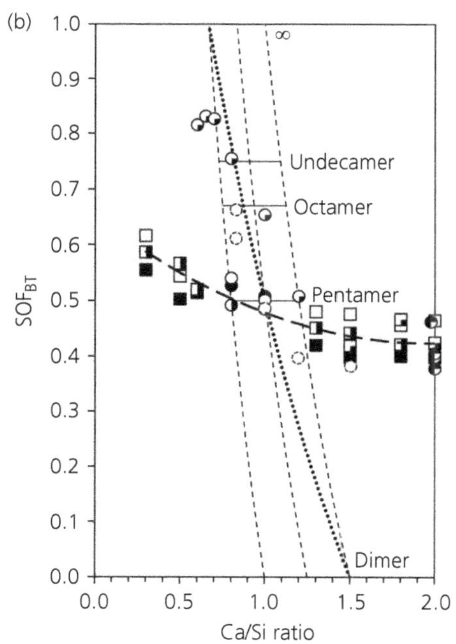

The data are from: Okada *et al.*, 1994b (unfilled symbols for preparations made at 180 °C; half-filled symbols for 150 °C; and filled symbols for 120 °C); Noma *et al.*, 1998 (quarter-filled symbols); Haga *et al.*, 2002 (dotted-line unfilled symbols; samples reported as prepared hydrothermally but no experimental details given); the 80 °C data from Ishida *et al.*, 1992c (three quarter-filled circles).

6.5.2 Nuclear magnetic resonance

^{29}Si MAS NMR studies of the preparations described in the previous section mostly showed the presence only of Q^1 and Q^2 tetrahedra (Mitsuda *et al.*, 1989; Noma *et al.*, 1998; Okada *et al.*, 1994b; Stade *et al.*, 1985). Okada *et al.* (1994b) made their C–S–H preparations by hydrothermal reaction of CaO and silicic acid in stirred suspensions with a water/solid ratio of 20 at three temperatures (120, 150 and 180 °C) for 2 hours; longer reaction times at 150 and 180 °C resulted in crystalline calcium silicate hydrates, as described by Mitsuda *et al.* (1986). The samples were dried under vacuum at 80 °C. ^{29}Si NMR data show that the C–S–H present in their samples is quite similar in most of them, regardless of the Ca/Si ratio; this is illustrated in Figure 6.11(b), which is a plot of the fraction of bridging sites that are occupied, SOF_{BT}, against Ca/Si ratio. The silicate anions in the C–S–H made at 120 °C are on average pentameric for Ca/Si ratios of 0·8 and 1·0 and on average tetrameric for Ca/Si $\geq$1·3, with the higher Ca/Si ratios being due to intermixture with CH. Ca/Si ratios below 0·8 are due to intermixture with unreacted silica. Increasing the temperature results in slightly longer chains for preparations with Ca/Si $\geq$1·3. Comparison of Figure 6.11(b) with Figure 6.5(e) demonstrates clearly the difference in the product produced hydrothermally by Okada *et al.* (1994b) compared with preparations made at lower temperatures. The most striking difference is that it does not appear to be possible to make C–S–H(I) hydrothermally that has silicate anions with an average length that is shorter than about 4. Okada *et al.*'s (1994b) preparations all contain C–S–H(I) with MCL $\approx$ 4–5, regardless of the temperature of synthesis: the variation in Ca/Si ratio is consistent with changes in the value of *i* in Formula 5.6 (in this case with no Al). Preparations with Ca/Si ratios outside the range permitted by varying *i* (i.e. outside the area on Figure 6.11(b) that is bounded by the dashed lines) contain second phases. In contrast, Noma *et al.* (1998) did achieve longer chain lengths in the hydrothermal preparations at lower Ca/Si ratio (their data are represented by the quarter-filled symbols on Figure 6.11(b)). The reason for the difference in the two hydrothermal studies is not clear because Noma *et al.* (1998) used a very similar preparation method to Okada *et al.* (1994b), the only differences being a reaction temperature of 130 °C (rather than 120, 150 and 180 °C), a water/solid ratio of 40 (rather than 20) and their samples were dried at 60 °C (instead of at 80 °C).

Stade *et al.* (1985) concluded from ^{1}H–^{29}Si cross-polarisation spectra that, in 'C–S–H(di, poly)' preparations with Ca/Si ratios of 1·0–1·5 (Stade, 1980), both Q^1 and Q^2 tetrahedra carried OH groups. They found the ratio of Q^1–OH to Q^2–OH to be independent of Ca/Si ratio in this range, but to decrease with increasing temperature of formation. They concluded that the materials had in all cases tobermorite-type structures with missing bridging tetrahedra and varying contents of SiOH and of interlayer Ca^{2+} and OH^-. Okada *et al.* (1994b) similarly concluded from cross-polarisation and GHD (gate proton decoupling) spectra that substantial fractions of both Q^1 and Q^2 tetrahedra carried H atoms, especially at Ca/Si <1·0, which is in contrast with an increase in the value of *i* in Formula 5.6.

6.5.3 Calcium extraction

Stade (1980) conducted calcium extraction experiments on C–S–H preparations with a wide range of target Ca/Si ratios. The C–S–H phases were prepared from sodium silicate and calcium chloride solutions at 0 °C but stored at 80 °C under the mother liquid (Stade and Wieker, 1980). The preparations that were free of crystalline CH had bulk Ca/Si ratios of 1·0–1·9. Calcium was extracted from the preparations using an ethyl acetoacetate-isobutyl alcohol mixture at ambient temperature or at 80 °C. Figure 6.12 shows a plot of the reduction in the Ca/Si ratio that occurred during the extraction procedures against the initial Ca/Si of the preparations. Data for extraction at ambient temperature are represented by circles and those at 80 °C by squares. The results of linear regression analysis of the two sets of data are shown on the figure. In both cases the lines intersect the x-axis close to the expected point – that is, at the Ca/Si ratio of single-chain tobermorite that has minimum Ca/Si ratio (0·67) and therefore no interlayer Ca. It is evident on inspection of the figure that the extraction procedure conducted at room temperature was very gentle but that it was nevertheless sufficient to reduce the Ca/Si ratio of high-Ca preparations to about 1·4–1·5 (those data are indicated by the bold circles on Figure 6.12 that fall close to the upper trend line, which is for the 80 °C data). This Ca was of course also removed by extraction at 80 °C, which in addition removed a much greater amount of interlayer Ca than when the extraction was conducted at room temperature; as an example, the difference between the two procedures is shown on the figure for a preparation with Ca/Si ratio of 1·6. Extraction at 80 °C would reduce the Ca/Si ratio to 1·35 whereas extraction at room temperature would reduce it to only 1·5. Consideration of the NMR results discussed in §6.5.2 indicates that extraction at 80 °C probably removed about 40% of the interlayer Ca. The relatively modest severity of even the 80 °C procedure can be gathered by comparing the data with some from Harris *et al.* (2002) that involved leaching C–S–H preparations (synthesised by lime–silica reaction at laboratory temperature) in de-mineralised water; the 80 °C

Figure 6.12 Results of calcium extraction experiments conducted on C–S–H preparations with a wide range of target Ca/Si ratios (Stade, 1980)

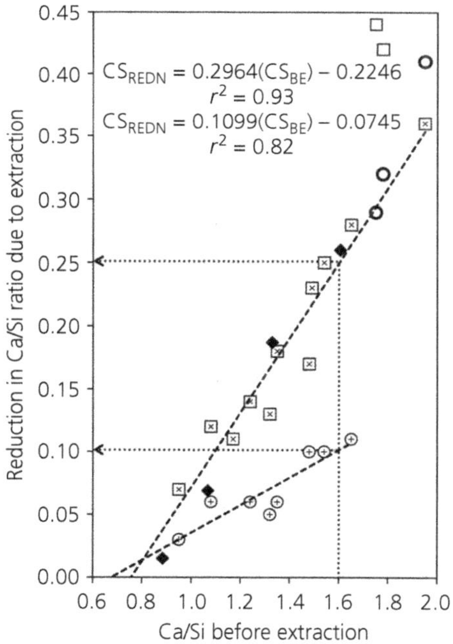

The samples and method are described in the text. The data that were included in the regression analyses are indicated by an '×' in the square symbols or a '+' in the circles. The diamonds are data from Harris *et al.* (2002) (corresponding to the second leaching-stage data on their Figure 5(a)).

data correspond quite closely with a very early stage in Harris *et al.*'s (2002) experiments (their second leaching stage), which is represented on the figure by the filled diamonds. The results discussed here support other evidence (§6.2) that the Ca/Si ratios of preparations that are greater than 1·3–1·5 are due to intermixture of the C–S–H with a second, Ca-rich phase, which can be X-ray amorphous (unless synthesised using the methods described in §6.3).

6.6. Tobermorite-based structural models for C–S–H(I)

All the available data are consistent with the view that, apart from some very early products of reaction, C–S–H preparations that have Ca/Si ratios up to ≈1·5 have structures that are derived from 14 Å tobermorite, but with defects of the types described in §6.2.5 and in Chapter 5. The resulting structural-chemical formula (Formula 5.6, but in this case with no Al i.e. $f = 0$) is compatible with many of the models that have been proposed for the C–S–H that forms in cement pastes (for this range of composition), as discussed by Richardson (2004, 2008) and which is demonstrated in Table 6.2, which includes the values of *v*. For the four shortest mean silicate chain lengths together with the values of the equivalent variables in the models of Richardson and Groves (1992a), Taylor (1993) and Cong and Kirkpatrick (1996a,b) for single-chain tobermorite-based C–(A–) S–H phases, it is evident that the three models are very similar. It should be remembered of course that in the general case, the anion sizes represented in these models are means – that is, in MCL = ($3n$–1), n will generally be a positive noninteger >1, while for individual silicate anions it is a positive integer ≥1. At Ca/si = $\frac{2}{3}$ the chains must be effectively infinite single *dreierketten* and the $\frac{\text{Si–OH}}{\text{Si}}$ ratio is the maximum possible. Ca/Si ratios $\frac{2}{3}$ are incompatible with a tobermorite-type structure. If Ca–OH groups are assumed to be associated with a second phase, then Ca contents in excess of four per six tetrahedral sites must be assigned to interlayer sites. There are two such sites per six tetrahedral sites in 14 Å tobermorite but it is unlikely that each can be more than 50%

Table 6.2 Values of the variables in models for single-chain tobermorite-based C–(A–) S–H phases for the four shortest mean silicate chain lengths (MCL). Reproduced from Table 2 in Richardson (2014)

	MCL	2	5	8	11
Richardson (2014)	v	$\frac{1}{3}$	$\frac{1}{6}$	$\frac{1}{9}$	$\frac{1}{12}$
Richardson and Groves (1992)	n	1	2	3	4
Taylor (1993)	Δ	2	1	$\frac{2}{3}$	$\frac{1}{2}$
Cong and Kirkpatrick (1996)	m	1	$\frac{1}{2}$	$\frac{1}{3}$	$\frac{1}{4}$
	$\frac{1}{1-v}$	1·5	1·2	1·125	1·091
	SOF_{BT}	0	$\frac{1}{2}$	$\frac{2}{3}$	$\frac{3}{4}$

The variable v in Equation 5.6 is the fraction of tetrahedral sites in a silicate chain that are vacant. n, Δ and m are the equivalent variables in the models of Richardson and Groves (1992), Taylor (1993) and Cong and Kirkpatrick (1996), respectively; it is evident that $v = \frac{1}{3n} = \frac{\Delta}{6} = \frac{m}{3}$. Values are also included for the site occupancy factor for the bridging site (Equation 5.20) and the reciprocal of the fraction of occupied tetrahedral sites, $\frac{1}{1-v}$, which corresponds to the maximum Ca/Si ratio that is possible without the presence of Ca–OH groups (Equation 6.5; see §6.2.5) – that is, the Ca/Si ratio in Richardson and Groves' (1992) model for T-based structural units when $w/n = 0$.

occupied (as discussed in §5.3.2.3) and so this will need to be considered in any structural model that involves infinite chains.

6.6.1 Approaches that involve the refinement of powder XRD data

The XRD results of the early studies on C–S–H(I) preparations that are summarised in §6.2.2 have been advanced by a number of workers who have used one or more of Bonaccorsi *et al.*'s (2005) and Merlino *et al.*'s (1999, 2001) tobermorite structures as models for whole-pattern or Rietveld-type refinements (e.g. Garbev *et al.*, 2008a, 2008b; Renaudin *et al.*, 2009a) or for the application of calculations that have been developed specifically for defect structures (Grangeon, *et al.*, 2013a, 2013b). Garbev *et al.* (2008a) studied calcium silicate hydrate preparations with bulk Ca/Si ratios of between 0·2 and 1·5 that they made mechanochemically. Their study involved the refinement of the unit cell parameters using a whole powder pattern decomposition (WPPD) method. The *c* parameter was essentially constant at ≈ 26·5 Å for Ca/Si ratios $< \frac{2}{3}$ and again constant at ≈ 23·5 Å for Ca/Si ratios >1·25, with a sharp drop observed between those two values at intermediate compositions (they are represented by the '+' symbol on Figure 6.1). They concluded that there are two ordered end members that have Ca/Si $= \frac{2}{3}$ and Ca/Si = 1.25, the former consisting of tobermorite slabs linked by H-bonds with no interlayer Ca atoms, with the composition $Ca_4H_4[Si_3O_9]_2 \cdot 4H_2O$, and the latter having structural features of a tobermorite-based dimer with composition $Ca_4H_2[Si_2O_7]_2 \cdot Ca \cdot mH_2O$. They considered that preparations with intermediate Ca/Si ratio '*... may be well described by the defect-tobermorite model*'. These conclusions are consistent with the view of earlier workers (summarised in §6.2.2) and with models for C–S–H that are discussed in §6.6.3. Garbev *et al.* (2008a) utilised two structural models in their fit for a sample with Ca/Si $= \frac{2}{3}$: one with an orthorhombic subcell (space group *I* 2 *m m*) that they used to fit the sharper peaks, and a second with a monoclinic cell (space group *B* 1 1 *b*) that was used to fit broader components. They did not attempt to refine the cell contents.

Renaudin *et al.* (2009a) also studied C–S–H(I) preparations with a range of Ca/Si ratios which were made by reacting silica and CaO at 20 °C with a w/s ratio of 50 and stored under N_2 for 3 weeks (with stirring). Their samples were prepared with target Ca/Si ratios of between 0·8 and 1·7, but the maximum value obtained was 1·4 – that is, after accounting for CH. The variation in the *c* parameter with Ca/Si ratio was similar to that reported by Garbev *et al.* (2008a) (Renaudin *et al.*'s (2009a) data are represented by the '×' symbol on Figure 6.3) and the results of their refinements of the XRD patterns were also consistent with a defect-tobermorite model, although they found it necessary to reduce the occupancy of the paired Si site to below 1, which would result in a deviation from the $3n - 1$ sequence of silicate chains. They did not refine the atomic positions '*... because of the numerous 32 independent atomic sites and the largely unresolved diffraction lines*'.

Grangeon *et al.* (2013b) considered that there are at least three main phenomena that could explain the decrease in layer spacing that is observed to occur with decreasing Ca/Si ratio (as illustrated in Figure 6.3).

(1) The first, which they attributed to Garbev *et al.* (2008a, 2008b), is that the site occupancy of tetrahedral bridging sites decreases with increasing Ca/Si and that this results in a decrease in the layer-to-layer distance because of the reduced contents of the interlayer region. They drew attention to Figure 8 in Garbev *et al.* (2008b), which is a schematic illustration of three structural arrangements: (a) 14 Å tobermorite; (b) single-chain tobermorite with no interlayer

Ca; (c) tobermorite with no bridging tetrahedra – that is, dimeric silicate anions. While this figure does not in fact demonstrate Grangeon *et al.*'s (2013b) point very well – in that the schematic structures that are illustrated in the figure actually show the layer spacing first increasing from 13·3 Å to 14·0 Å as the Ca/Si increases from $\frac{2}{3}$ (i.e. tobermorite with infinite chains but no interlayer Ca) to $\frac{5}{6}$ (14 Å tobermorite) and then decreasing to 11·7 Å at Ca/Si = 1·25 (i.e. tobermorite with no bridging tetrahedra and so totally dimeric) – diagrams (b) (Ca/Si = $\frac{2}{3}$) and (c) (Ca/Si = 1·25) do nevertheless illustrate plausible end-member structures, with the important proviso that the layer spacing given for the dimer (11·7 Å) is most likely to be far too big, with an implausibly long average Ca–O distance for the interlayer Ca atom, which also appears to be in eight-fold coordination (which is unusual, but has been observed previously; Figure 5.16(b)); the same objection is applicable to Grangeon *et al.* (2013b) who removed all of the bridging tetrahedra from an 11 Å orthotobermorite but made no other adjustments to the cell and its contents other than to increase occupancy at the site for interlayer Ca to its maximum value. Taylor and Howison (1956) considered that a proportion of the bridging tetrahedra must be retained to maintain the tobermorite structure and that the layer would collapse to a different but related structure if all of them were removed. It is unlikely that this situation would result in the linear decrease in the layer spacing that is observed to occur with increasing Ca/Si ratio (Figure 6.3); an abrupt reduction at the Ca/Si ratio of the dimeric structure would be more likely, and so an alternative explanation is necessary.

(2) Secondly, Grangeon *et al.* (2013b) considered that a reduction in the size of the average crystal in the direction parallel with **c** leads to a shift of the 001 reflection towards a lower 2θ angle because the average number of layers that are stacked within the crystals is reduced (Grangeon *et al.*, 2013a). They state that this effect '*... mainly originates from the fact that in the low-angle region of XRD patterns, the Lorentz polarisation factor is strongly rising towards the low-angle part and thus this artificially displaces the diffraction maxima*', referencing Reynolds (1986). Grangeon *et al.* (2013a) used this as an explanation for much of the scatter in their compilation of layer spacing-Ca/Si data – that is, the quite large differences in layer spacing that have been observed for preparations that have essentially the same Ca/Si ratio is due in large part to differences in the number of stacked layers. They did state that the method of synthesis and sample preparation might also be a factor, and inspection of the more extensive compilation made by Richardson (2014) and in Figure 6.3 would seem to indicate that the observed variation is adequately explained by differences in the extent of drying or by intermixture of C–S–H(I) with second phases, although the crystals are without doubt often very small indeed (i.e. on the order of just a few layers thick).

(3) Grangeon *et al.*'s (2013b) third hypothesis to explain the decrease in the layer spacing with increase in Ca/Si ratio is that layers of low Ca/Si ratio are interstratified with others of higher Ca/Si ratio, which was originally proposed by Taylor and Howison (1956) and Heller and Taylor (1956). Grangeon *et al.* (2013b) performed calculations to illustrate this effect using a theoretical 14 Å tobermorite with interlayer calcium omitted as the low Ca/Si end-member (i.e. as illustrated by Garbev *et al.* (2008b) in their Figure 8(b)), and a theoretical 11 Å orthotobermorite (MDO_2 polytype) with all of the bridging Si tetrahedra omitted as the high Ca/Si end-member. Again, Grangeon *et al.* (2013b) did not refine atomic coordinates.

The studies discussed above have clearly provided valuable additional insight into the structure of C–S–H(I) phases. However, a significant drawback of the methodologies is that none of them attempted to refine the atomic positions, essentially because of the lack of information in the diffraction data. The refinements were therefore restricted to the lattice parameters, site occupancy, crystal size, preferred orientation and, in the case of Grangeon *et al.* (2013b), to examining the implications of the interstratification of layers that have different structure. While the omission of atom-coordinate refinement in these studies is entirely understandable, it can easily lead to crystal-chemical problems if the lattice parameters are changed by more than only a small amount. As an example of this issue, Renaudin *et al.* (2009a) – who did not refine atomic coordinates – obtained an average Si–O distance of 1·77 Å for a preparation that had Ca/Si = 0·8, which they considered to be an acceptable value for the purpose of their discussion. However, inspection of Figure 5.17(a) reveals that it is in fact implausibly long. This is a particularly serious issue for C–S–H(I) phases when the C–S–H(I) phase that is of interest has a layer spacing that is significantly bigger or smaller than the starting model, which is usually either an 11 Å or 14 Å tobermorite. An increase or decrease in the c parameter results in expansion or compression of the central Ca–O core of the structure and so it is important that this issue is addressed if crystal-chemically sensible distances and coordinations are to be retained in the model structure.

6.6.2 Approaches that involve atomistic modelling

Faucon *et al.* (1996, 1997) described molecular dynamics (MD) simulations of C–S–H structures. In the absence of interlayer Ca at a Ca/Si ratio of $\frac{2}{3}$, the calculations indicated infinite *dreierketten* with possible interlayer linkages through Q^3 tetrahedra. Introduction of interlayer Ca caused breaks in the chains, which were considered to occur between a paired and a bridging tetrahedron. No tetrahedral sites were vacated, but the bridging tetrahedron was displaced and the elements of water were added, which is not consistent with the $3n - 1$ sequence of silicate anions that is observed by experiment (see §7.4.1). There have been many subsequent atomistic modelling studies, many of which have been discussed in review articles (e.g. Duque-Redondo *et al.*, 2022). Pellenq *et al.* (2009) proposed a molecular model for C–S–H with Ca/Si of 1·65 that was developed using atomistic simulations; they state that the model '*includes both glass-like short-range order and crystalline features of the mineral tobermorite*'. The unit cell contains 60 Si atoms. Inspection of the equilibrium structure reveals that 40 are in chain-end tetrahedra (Q^1), 13 in middle-chain (Q^2) and 7 in monomers (Q^0). The tetrahedra are arranged in eight rows of linear chains; 50 % of the Si atoms are in dimer, 5 % in trimer, 33 % pentamer and 12 % monomer. Pellenq *et al.* (2009) state that the distribution of Q^n units in their model is '*guided by the NMR results*' in reference 10 of the article, but there are no peaks for hydrated monomer (Q^{0H}) on any of the ^{29}Si NMR spectra in that reference. In fact, only about 2 % of Q^{0H} is observed in hardened cements using NMR (e.g. Brough *et al.*, 1994; Rodger *et al.*, 1988; see §7.4.2) and so a realistic model should not include 12 % hydrated monomer. The presence of trimer is also not consistent with the experimentally established sequence of silicate chain lengths – that is, 2, 5, 8, … ($3n - 1$), where n is integer; see §7.4.1. There are no Ca–OH groups in the model, which should be present at such high Ca/Si ratio (e.g. Thomas *et al.*, 2003), and a large proportion of the Ca–O distances are either shorter than the minimum distance or longer than the maximum distance that are present in known crystal structures of calcium silicate hydrates and related phases, as established by Richardson (2014) and reiterated in §5.7. Calculation of the Ca–O coordination numbers based on these distances shows that nearly two thirds (64 %) of the Ca atoms in the model (which has 99 unique Ca atoms) are coordinated to fewer than six oxygen atoms. Four other versions of the same model have between 36 % and 59 % of the Ca atoms coordinated to fewer than six oxygen atoms; selected details of all five versions are given in Table 6.3.

Table 6.3 Selected details for five versions (crystal-structure files) of the C–S–H model described in Pellenq *et al.* (2009)

Version	a: Å	b: Å	c: Å	α	β	γ	V: Å^3	D_x: g cm^{-3}	C/S	H/S	Ca–O coordination					
											3	4	5	6	7	8
MIT1[a]	13·309	29·521	23·691	92·020	88·520	123·580	7749·65	2·34	1·65	1·63	0	9	54	26	9	1
MIT2	13·341	29·591	23·982	91·685	88·697	123·341	7905·36	2·32	1·65	1·73	2	14	42	32	9	0
MIT3[b]	13·336	29·613	23·803	91·628	88·717	123·337	7850·41	2·33	1·65	1·73	2	13	41	33	10	0
MIT4	13·179	29·231	23·393	92·653	88·010	123·367	7517·69	2·44	1·65	1·73	1	7	43	32	14	2
MIT5[c]	13·023	28·916	22·757	93·181	88·035	123·229	7157·17	2·56	1·65	1·73	0	3	33	34	28	1

MIT1 is the version that is published in supplementary table S7 of Pellenq *et al.* (2009) but with the *a* parameter halved; the formula that is given in table S7 also corresponds to a doubled cell and needs to be halved to be consistent with the number of atoms in the table. The other four files were supplied to Richardson by Pellenq (2009/2010). The unit cell contents for the MIT1 version give the composition as $Ca_{99}Si_{60}O_{317}H_{196}$; the other four versions have an additional 18 atoms (as six water molecules) giving $Ca_{99}Si_{60}O_{323}H_{208}$. All H atoms are associated with water molecules – that is, there are no Si–OH or Ca–OH groups. The formula given in the text of Pellenq *et al.* (2009) is $(CaO)_{1{\cdot}65}(SiO_2)(H_2O)_{1{\cdot}75}$, which on the basis of 60 Si atoms is $Ca_{99}Si_{60}O_{324}H_{210}$ – that is, it has seven more water molecules than the version given in the supplementary table and one more than the other four files. The 'Ca–O coordination' columns show the number of Ca atoms for each version that are coordinated to 3, 4, 5, 6, 7 or 8 O atoms based on Ca–O distances ≤2·90 Å.

[a]Version published in Pellenq *et al.* (2009).

[b]Supplied in response to a request for the definitive version of the model.

[c]The highest density version (MIT5); Pellenq *et al.* (2009) stated that it was not the most stable (i.e. lowest energy) solution and on full NPT simulation (i.e. where the volume can relax) the density reduced to that of MIT4 (Pellenq, 2009/2010).

Unsurprisingly, the highest density version of Pellenq *et al.*'s (2009) model has the lowest percentage of Ca atoms that are coordinated to fewer than six oxygen atoms (36%), but it was not the most stable solution. There is a large peak for the interlayer spacing on the calculated XRD pattern at 11·9 Å, which is not consistent with values observed by experiment for synthetic C–S–H with Ca/Si of 1·65; a much smaller value would be expected, as shown in Figure 6.3 (Richardson, 2014). Hou *et al.* (2015) conducted an MD study of C–S–H that was concerned with Ca/Si ratios of 1·00 and 1·75. They refer to '*distorted calcium tetrahedrons*', which – like Pellenq *et al.* (2009) – they concluded must be due to C–S–H possessing glass-like structure. Kumar *et al.* (2017) conducted MD simulations of tobermorite-based structures that had Ca/Si ratios of 1·25, 1·50, 1·75 and 2·00, all of which also included a significant proportion of the Ca atoms coordinated to fewer than six oxygen atoms: about 30–40% coordinated to five and a few % coordinated to four. While the atomistic modelling studies of Pellenq *et al.* (2009), Hou *et al.* (2015) and Kumar *et al.* (2017) all resulted in a significant proportion of the Ca atoms being coordinated to fewer than six oxygen atoms, Ca atoms have never been observed by experiment in calcium silicate hydrates and related phases that are coordinated to fewer than six oxygen atoms (as shown in Figure 5.16(b)); indeed, Gagné and Hawthorne's (2016) extensive study of bond-length distributions for alkali and alkaline-earth ions bonded to oxygen shows that the minimum and maximum coordination numbers for Ca are six and 12. The minimum value of six is well-illustrated by consideration of the crystal structures of carbonates of divalent cations, where a switch occurs with calcium from a rhombohedral unit cell (i.e. the calcite-type structure) with $^{[6]}M^{2+}$ to orthorhombic (i.e. aragonite-type structure) with $^{[9]}M^{2+}$. The evidence is thus compelling that the Ca^{2+} ion is simply too big to be coordinated to fewer than six oxygen atoms. The explanation of Pellenq *et al.* (2009) that Ca^{2+} ions coordinated to fewer than six oxygens is due to glass-like structure is further undermined by the odd distribution of those atoms around the Ca in many of the polyhedra in their equilibrium structure. It seems more likely that there was a problem with their MD simulations, which Kumar *et al.* (2017) also seem to have experienced since their simulation of the structure of 14 Å tobermorite resulted in about 30% $^{[5]}$Ca and 70% $^{[6]}$Ca rather than 20% $^{[6]}$Ca and 80% $^{[7]}$Ca, leading them to conclude that '*we expect the results ... to systematically underestimate a proper coordination number by nearly one*' (this quotation is from a Supporting Information file for Kumar *et al.*, 2017). While Kumar *et al.*'s (2017) model has been improved in subsequent papers from the same group (Kunhi Mohamed *et al.*, 2018, 2020), crystal-chemical issues persist. For example, problems with the Ca–O polyhedra in a later version of the model (Kunhi Mohamed *et al.*, 2020) are readily demonstrated using Bauer's (1974) polyhedral distortion indices ('DI'), three of which can be calculated using Equations 6.18–6.20 for a polyhedron with a Ca centroid atom, *n* vertex O atoms and *m* O–O edges.

$$\mathrm{DI}(\mathrm{CaO}) = \frac{\sum_{i=1,n}\left|\mathrm{CaO}_i - \overline{\mathrm{CaO}}\right|}{n \times \overline{\mathrm{CaO}}} \tag{6.18}$$

$$\mathrm{DI}(\mathrm{OCaO}) = \frac{\sum_{i=1,m}\left|\mathrm{OCaO}_i - \overline{\mathrm{OCaO}}\right|}{m \times \overline{\mathrm{OCaO}}} \tag{6.19}$$

$$\mathrm{DI}(\mathrm{OO}) = \frac{\sum_{i=1,m}\left|\mathrm{OO}_i - \overline{\mathrm{OO}}\right|}{m \times \overline{\mathrm{OO}}} \tag{6.20}$$

Values calculated using Equation 6.18 represent distortion of the Ca–O polyhedra in terms of the variation in the bond distances from the centroid Ca to vertex O atoms (CaO_i) relative to the mean bond distance, $\overline{\mathrm{CaO}}$ (CrystalMaker, 2022). Those from Equation 6.19 represent it in terms of the variation in the bond angles ($OCaO_i$) between shared-edge O atoms and the centroid Ca atom relative to the mean bond angle, $\overline{OCaO}$ (CrystalMaker, 2022). Values from Equation 6.20 represent distortion in terms of the variation in polyhedral edge lengths (OO_i) relative to the mean edge length, $\overline{OO}$ (CrystalMaker, 2022). Figure 6.13 includes plots of (a) DI(OO) against DI(CaO) and (b) DI(OO) against DI(OCaO) that compare the Ca–O polyhedra in the models of Pellenq *et al.* (2009; – symbol) and Kunhi Mohamed *et al.* (2020; × symbol) with those in the crystal structures of calcium hydroxide (Busing and Levy, 1957; open triangle), wollastonite (Ohashi, 1984; open diamonds; see Tables 5.1(a) and 5.20(a)), 14 Å tobermorite (Bonaccorsi *et al.*, 2005; open squares; see Tables 5.1(b) and 5.20(b)), and jennite (Bonaccorsi *et al.*, 2004; filled circle; see Tables 5.1(c) and 5.20(c)). On inspection, it is evident that the distributions of the values for the models of Pellenq *et al.* (2009) and Kunhi Mohamed *et al.* (2020) are similar, and that in both cases the DI values are mostly much larger than those for the other structures. This indicates that many of the Ca–O polyhedra in the atomistic models are very distorted when compared with those that are present in known structures of crystalline calcium silicates and calcium silicate hydrates that have *dreierkette* silicate, as well as in calcium hydroxide. For comparison, the open circles on Figure 6.13 represent a clinotobermorite-based

Figure 6.13(a) O–O and Ca–O polyhedral distortion indices (DI) for two atomistic-simulation models compared with those for four known crystal structures

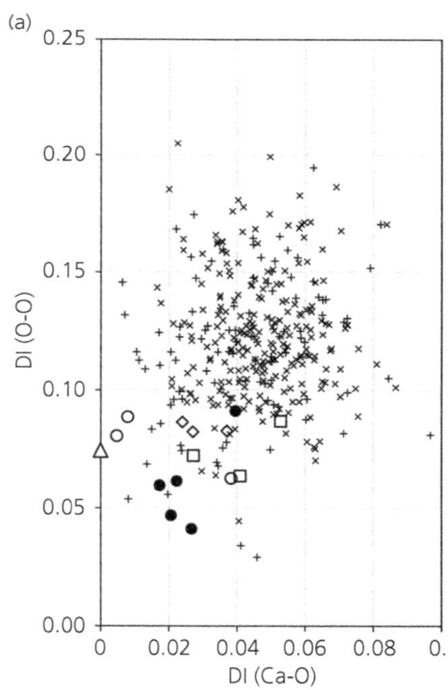

The structures are from: Bonaccorsi *et al.*, 2004 (filled circles; jennite); Bonaccorsi *et al.*, 2005 (open squares, 14Å tobermorite); Busing and Levy, 1957 (open triangle; calcium hydroxide), Kunhi Mohamed *et al.*, 2020 (× symbol; CASH model); Ohashi, 1984 (open diamonds; wollastonite-Tc); and Pellenq *et al.*, 2009 (+ symbol; CSH model). DI values calculated using CrystalMaker® (CrystalMaker Software Ltd, 2022).

Figure 6.13(b) O–O and O–Ca–O polyhedral distortion indices (DI) for two atomistic-simulation models compared with those for four known crystal structures

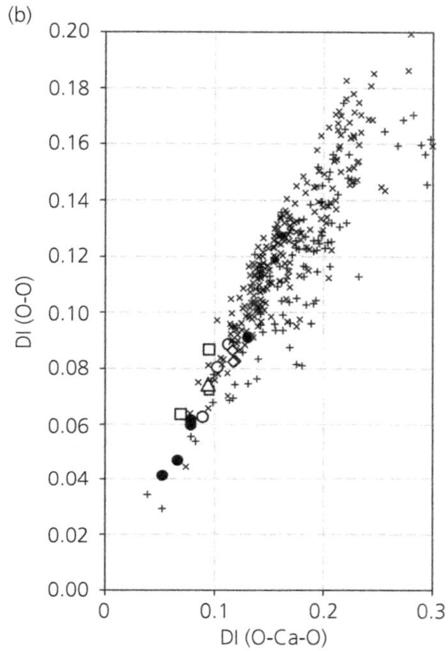

References for the structures are given in the caption to Figure 6.13(a). DI values calculated using CrystalMaker® (CrystalMaker Software Ltd, 2022).

model for dimeric C–S–H(I) from Richardson (2014; CIF datablock T2_sc_LS2; open circles; Ca/Si = 1·50) and that is described in more detail in §6.6.3.2. In that case, the values of the polyhedral distortion indices are similar to those for the known structures. It is possible that the crystal-chemical difficulties described above for the models of Pellenq *et al.* (2009) and Kunhi Mohamed *et al.* (2020) are not due to any fundamental issues with the atomistic modelling methods used by the workers, but rather that it is simply not possible geometrically to produce a high Ca/Si ratio model for C–S–H that is derived entirely from tobermorite while at the same time maintaining essential structural features and crystal-chemical plausibility – that is, it is likely to be inevitable that attempts to accommodate the required number of Ca^{2+} ions in the structure will result in undercoordination of some Ca^{2+} ions and/or bond distances that are either too long or too short and/or implausibly distorted polyhedra. It appears most likely that high Ca/Si ratio models for C–S–H must involve some other way of raising the Ca/Si ratio above 1·50 – that is, by incorporating elements of structure based on calcium hydroxide or jennite, as discussed in §6.3.3 and §7.5.

It is important that structures that result from atomistic modelling studies are published in such a way that the crystal-chemical details can be scrutinised readily. The atomic positions are sometimes provided using Cartesian coordinates (dimension of length) whereas the details of the structure may be more easily inspected using fractional (dimensionless) coordinates. It is as a consequence sometimes necessary to convert between the two. The transformation matrices given in Equations 6.21–6.23 can be used to convert between Cartesian coordinates (orthonormal axes) and fractional coordinates (triclinic axes) if (*a*) the fractional coordinate system has the same origin as the Cartesian coordinate system, (*b*) the **a**-axis is collinear with the *x*-axis and (*c*) the **b**-axis lies in the *xy*-plane (Dunitz, 1979).

Fractional coordinates are converted to Cartesian using

$$\begin{bmatrix} x_{\text{cart}} \\ y_{\text{cart}} \\ z_{\text{cart}} \end{bmatrix} = \begin{bmatrix} a & b\cos(\gamma) & c\cos(\beta) \\ 0 & b\sin(\gamma) & \frac{c(\cos(\alpha)-\cos(\beta)\cos(\gamma))}{\sin(\gamma)} \\ 0 & 0 & \frac{cv}{\sin(\gamma)} \end{bmatrix} \begin{bmatrix} x_{\text{frac}} \\ y_{\text{frac}} \\ z_{\text{frac}} \end{bmatrix} \tag{6.21}$$

$$\text{where } v = \sqrt{1-\cos^2(\alpha)-\cos^2(\beta)-\cos^2(\gamma)+2\cos(\alpha)\cos(\beta)\cos(\gamma)} \tag{6.22}$$

Cartesian coordinates are converted to fractional using

$$\begin{bmatrix} x_{\text{frac}} \\ y_{\text{frac}} \\ z_{\text{frac}} \end{bmatrix} = \begin{bmatrix} \frac{1}{a} & \frac{-\cos(\gamma)}{a\sin(\gamma)} & \frac{\cos(\gamma)\cos(\alpha)-\cos(\beta)}{av\sin(\gamma)} \\ 0 & \frac{1}{b\sin(\gamma)} & \frac{\cos(\gamma)\cos(\beta)-\cos(\alpha)}{bv\sin(\gamma)} \\ 0 & 0 & \frac{\sin(\gamma)}{cv} \end{bmatrix} \begin{bmatrix} x_{\text{cart}} \\ y_{\text{cart}} \\ z_{\text{cart}} \end{bmatrix} \tag{6.23}$$

6.6.3 Structural models for C–S–H(I) derived using crystal-chemical and geometrical reasoning

It is noted in §6.6.1 that refinements of XRD data that involve even quite small changes in lattice parameters can lead to crystal-chemically implausible distances and coordinations if atomic coordinates are not included in the refinement, which is generally not possible because of the lack of information in the diffraction data. While atomic coordinates are of course changed in atomistic simulations, it is apparent from §6.6.2 they too often result in implausible distances and coordinations. Richardson (2014) showed how this issue could be addressed by the use of straightforward crystal-chemical and geometrical reasoning. Models were developed that are consistent with the experimental observations that have been discussed in this chapter, in particular:

- the linear decrease in layer spacing with increasing Ca/Si ratio (Figure 6.3)
- the decrease in layer spacing with decreasing SOF_{BT} (Figure 6.5(d))
- the observed trend for the decrease in SOF_{BT} with increasing Ca/Si ratio (i.e. Figure 6.5(e), Equation 6.8)
- the presence of dimeric silicate anions at Ca/Si ratios as low as 0·9 (e.g. Figure 6.6(a))
- the observed variation in H_2O/Si with increasing Ca/Si (discussed in §6.7)
- the main features on XRD patterns for C–S–H(I) preparations.

The rest of §6.6.3 and §6.6.4, §6.6.5 and §6.7 consist of an augmented reiteration of Richardson's (2014) treatment. A crystallographic information file (CIF) that contains data blocks for the model structures described in Richardson (2014) and in this section is available from the International

Union of Crystallography's (IUCr) electronic archives, reference HW5035. The CIF file contains a separate data block for each model structure and the names of the data blocks are given in the captions to the relevant figures and tables in this chapter.

The problem with changes in the *c* parameter resulting in expansion or compression of the central Ca–O core of the structure is easily resolved by recalculation of the *z/c* coordinate of the atoms in the asymmetric unit so that the relative *z* positions of the atoms in the central Ca–O core are retained. This is achieved by using a simple transformation that depends slightly on whether the layer spacing of the C–S–H(I) phase is smaller or larger than that of the starting model structure, as well as on the number of layers that are in the unit cell. For example, Equation 6.24 should be used when the model is an orthotobermorite structure that has a larger *c* parameter than the C–S–H(I) phase (e.g. 14 Å orthotobermorite), while Equation 6.25 should be used when the model orthotobermorite structure has a smaller *c* parameter than the C–S–H(I) phase (e.g. 11 Å orthotobermorite). z_{Tob} and c_{Tob} are the *z/c* coordinate and *c* parameter of the tobermorite phase and z_{CSH} and c_{CSH} are for the C–S–H(I) phase.

$$z_{\mathrm{CSH}} = \frac{z_{\mathrm{Tob}} \cdot c_{\mathrm{Tob}} - \left(\frac{c_{\mathrm{Tob}} - c_{\mathrm{CSH}}}{4} \right)}{c_{\mathrm{CSH}}} \tag{6.24}$$

$$z_{\mathrm{CSH}} = \frac{z_{\mathrm{Tob}} \cdot c_{\mathrm{Tob}} + \left(\frac{c_{\mathrm{CSH}} - c_{\mathrm{Tob}}}{4} \right)}{c_{\mathrm{CSH}}} \tag{6.25}$$

These transformations are appropriate for most of the atoms in the asymmetric unit but there are special cases, depending on the structure that is used as the starting model. For example, a *z/c* coordinate equal to zero may need to remain as zero and negative *z/c* coordinates should be multiplied by –1 before and after the transformation. Equations 6.24 and 6.25 should be used when the starting model structure is an orthotobermorite; the '4' in the equations should be replaced by '2' if the starting model is a clinotobermorite.

6.6.3.1 Dimeric C–S–H(I) structures that are based on orthotobermorite

Richardson (2014) noted that the end-members of Equation 6.8 (i.e. the dotted line on Figure 6.5(e)) are a structure with infinite silicate chains and no interlayer Ca at Ca/Si $= \frac{2}{3}$, and an entirely dimeric structure at Ca/Si $= \frac{3}{2}$, which were referred to respectively as T∞ and T2 (after Richardson and Groves, 1992a). Models were consequently developed for these two end-member structures and, following the earlier studies discussed above, the initial development concerned orthotobermorite structures.

As noted earlier, it was Taylor and Howison (1956) who first suggested that an entirely dimeric structure could be achieved by removing all of the bridging tetrahedra from silicate chains that have *dreierkette* conformation and that this would likely result in the collapse of the layer to give a different but related structure. Figure 6.14 shows two alternative hypothetical structures for dimer that are derived from an orthotobermorite structure that had staggered silicate chains. Figures 6.14(a) – 6.14(d) are for a version where the interlayer Ca atom is placed close to the vacant tetrahedral bridging site and Figures 6.14(e) – 6.14(h) are for a version where the interlayer Ca atom is offset from that position. The first version – which will be referred to as 'T2_so_LS1' ('so' meaning that the T2 structure is derived from an orthotobermorite that has staggered chains; LS1 = layer spacing 1) – represents the situation where the same Ca atom charge compensates the negative charges

Figure 6.14 Two alternative hypothetical structures for C–S–H(I) dimer that are derived from an orthotobermorite structure that had staggered silicate chains: (a) – (d) are for a version where the interlayer Ca atom is placed close to the vacant tetrahedral bridging site and (e) – (h) are for a version where the interlayer Ca atom is offset from that position; (a), (c), (d), (g) and (h) are from Figure 8 of Richardson (2014); (b) and (f) are new. The CIF file data blocks are T2_so_LS1 and T2_so_LS2 (continued on next page)

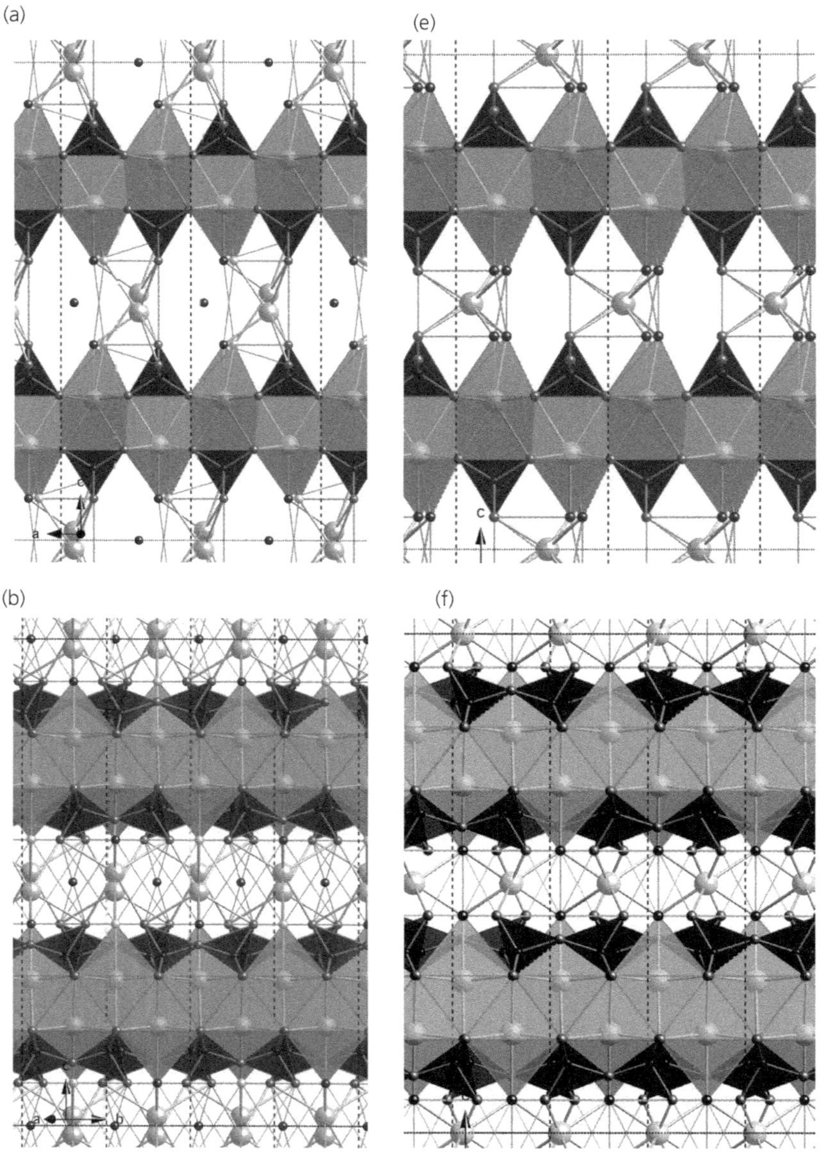

Figure 6.14 Continued

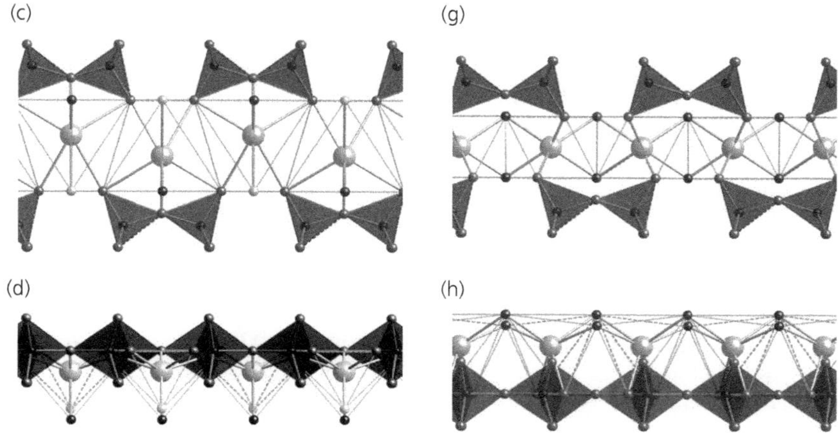

of the chain-end O atoms that result from the omission of a single bridging tetrahedron, whereas the second version (T2_so_LS2) involves two Ca atoms. T2_so_LS1 is a slight modification of the structure that is presented in Figure 28 of Richardson (2004). It is derived from the structure of the model for 11 Å orthotobermorite that is given in Table 5.5. The space group is $B\ 1\ 1b$ (No. 9, unique axis c, cell choice 2) with lattice parameters $a = 6{\cdot}748$ Å, $b = 7{\cdot}3$ Å, $c = 21{\cdot}5$ Å, $\gamma = 122{\cdot}75°$. The z/c coordinate for all of the atoms was recalculated using Equation 6.24 so that the relative positions of the atoms in the main layer were preserved on reduction of the c parameter. The Si atom of the bridging tetrahedron (Si2) was of course deleted and the interlayer Ca atom was moved to the position of the apical O atom of the bridging tetrahedron (O5, which was obviously deleted), with a very slight adjustment of the z/c coordinate necessary (0 to 0·02) in order to preserve reasonable Ca–O distances. The T2_so_LS2 structure is very similar to T2_so_LS1 but with the following adjustments: c was reduced further to 19·0 Å, and so the z/c coordinate for all of the atoms was again recalculated using Equation 6.24; the interlayer Ca atom was moved (0·5 added to both the x/a and y/b coordinates and the z/c coordinate returned to 0); O6 was replaced by a water molecule and Wat3 deleted.

The two structures are shown in Figures 6.14(a) and 6.14(e) viewed along the **b** axis (the unit cell is indicated by black dotted lines), and in Figures 6.14(b) and 6.14(f) viewed perpendicular to the **b** axis. The faces of the interlayer Ca–O polyhedra are not shaded. Views for the two versions that have just the silicate chains and the interlayer Ca atoms are shown perpendicular to the **b** axis in Figures 6.14(c) and 6.14(g) and along the **c** axis in (d) and (h).

It is evident from inspection of Figure 6.14(c) that the interlayer Ca atom for T2_so_LS1 is in seven-fold coordination. The regression analysis equation on Figure 5.17(d) indicates that the ideal average Ca–O polyhedral volume for seven-fold coordinated Ca in calcium silicate hydrates is $21{\cdot}45$ Å^3 and so the c parameter was adjusted to achieve a value that is close to this; in fact it is $21{\cdot}31$ Å^3 for $c = 21{\cdot}5$ Å. While this Ca–O polyhedral volume is close to the ideal value, the average Ca–O distance for this polyhedron is far too long at 2·72 Å, which compares very badly with the ideal distance for this coordination of 2·46 Å, Figure 5.17(b).

The situation is very similar for the T2_so_LS2 model structure. In that case the Ca atom is in six-fold coordination and so the ideal Ca–O distance and polyhedral volume are 2·40 Å and 17·35 $Å^3$, respectively, Figure 5.17. A *c* parameter of 19·0 Å – as represented by the illustrations in Figures 6.14(e) to 6.14(h) – results in an average Ca–O distance that is slightly too long (2·45 Å) but a polyhedral volume that is much too small (15·63 $Å^3$). The latter can of course be improved easily by increasing the *c* parameter (and recalculating the *z*/*c* coordinates accordingly) but that results in a further increase in the average Ca–O distance. For example, an appropriate polyhedral volume (17·18 $Å^3$) results from an increase in the *c* parameter to 19·5 Å but gives an average Ca–O distance that is much too long (2·52 Å).

The problem in both versions is that while the interlayer Ca atoms are coordinated to an appropriate number of oxygen atoms, the Ca atoms are in both cases in distorted trigonal prism coordination – which is evident from Figures 6.14(d) and 6.14(h) – rather than in hexagonal coordination, which is commonly observed for calcium silicate hydrates. The consequence of this is that it does not seem to be possible to generate a dimeric structure from an orthotobermorite structure that has staggered silicate chains that is crystal-chemically consistent with known calcium silicate hydrates.

Richardson (2014) noted that dimeric structures can be generated easily from an orthotobermorite that has silicate chains where the bridging tetrahedra are adjacent to one another instead of staggered – that is, using the structure that is given in Table 5.3 – but they suffer from the same Ca-coordination issue. As an example, a dimeric structure can be derived easily from the infinite-chain structure that is given in Table 5.3 by reducing the *c* parameter to 21·39 Å (and recalculating the *z*/*c* coordinates accordingly), deleting the Si atom of the bridging tetrahedron (Si2) and moving the interlayer Ca atom to the position of the apical O atom of the bridging tetrahedron (O5, which was obviously deleted). This results in the ideal value for the average Ca–O for the interlayer Ca atom, which is in six-fold coordination (i.e. 2·40 Å, Figure 5.17) but the volume of the polyhedron is much too small (11·23 $Å^3$ compared with the ideal value of 17·35 $Å^3$). The reason for this is that the interlayer Ca atom is in triangular prism coordination rather than in hexagonal coordination.

Richardson (2014) concluded from his trial structures that it is not possible to generate a dimeric model that is crystal-chemically consistent with known calcium silicate hydrates by using an orthotobermorite starting structure that has bridging tetrahedra that are adjacent to one another or that are staggered by *b*/2.

6.6.3.2 Dimeric C–S–H(I) structures that are based on clinotobermorite (T2)

The reason for the Ca-coordination problem that is encountered with the orthotobermorite models is immediately evident when the O atoms that are coordinated to the interlayer Ca atom are identified; the issue derives from the fact that the silicate chains that are clasped to opposite Ca–O main layers are directly adjacent to one another (i.e. across the interlayer), as is illustrated in Figures 6.14(a) and 6.14(e). Since this is not the case in clinotobermorite, the problem is as a consequence resolved easily by using a clinotobermorite as the starting structure instead of an orthotobermorite. As with the orthotobermorites, starting models can be chosen that have bridging tetrahedra that are either adjacent to one another or that are staggered by *b*/2. The structure of a hypothetical dimer that is derived from a clinotobermorite structure that has bridging tetrahedral sites that are adjacent to one another is shown in Figure 6.15(a) viewed along the **a** axis, and in Figure 6.15(b) viewed along the **b** axis. The faces of the Ca3-O polyhedra are not shaded (i.e. the interlayer Ca). The structure is monoclinic, space group *C* 1 2/*c* 1 (No. 15, unique axis *b*, cell choice 1); full details are given in Table 6.4 at the end of this chapter on p. 335, including lattice parameters and atom positions. It is evident from Figure 6.15(b) that the chains that are on opposite main layers are not

Figure 6.15 The structure of a hypothetical C–S–H(I) dimer derived from a clinotobermorite structure that had bridging tetrahedral sites that were adjacent to one another: (a) viewed along the **a** axis and (b) viewed along the **b** axis. The faces of the Ca8–O polyhedra are not shaded (i.e. the interlayer Ca). Edge lengths are *b* and *c* sin $\beta = 21{\cdot}27$ Å. Reproduced from Figure 9 in Richardson (2014). The CIF file data block is T2_ac

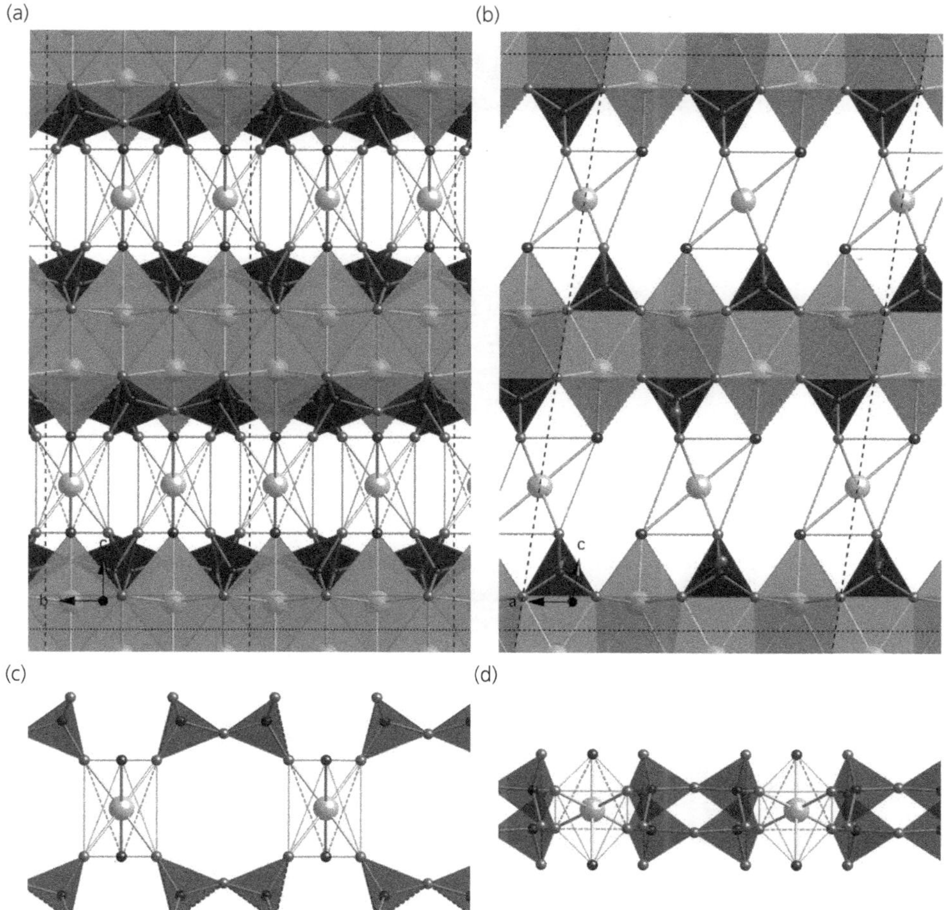

directly adjacent to one another. The change in the position of the silicate chains has resulted in the interlayer Ca atom being in six-fold hexagonal coordination, which is clear in Figure 6.15(d). The *c* parameter for this structure (i.e. 21·5 Å) was set at a compromise value for the Ca–O distance and polyhedral volume: the average Ca–O distance for the interlayer Ca atom (Ca3) is slightly longer than the ideal value for six-fold coordinated Ca (2·43 Å rather than 2·40 Å) and the Ca3–O polyhedral volume is slightly too small (16·57 Å^3 compared with 17·35 Å^3); nevertheless, inspection of Figure 5.17(d) shows that both are plausible values. This model structure has the formula $Ca_4H_2[Si_2O_7]_2{\cdot}Ca{\cdot}4H_2O$ and so Ca/Si = 1·25 and H_2O/Si = 1·25; the calculated density is 2·302 g cm^{-3}. Richardson (2014) noted that this structure is similar to Garbev *et al.*'s (2008b) dimer (compare Figure 6.15(c) to the schematic illustration in their Figure 8(c)) but without the water molecules that put their interlayer Ca in eight-fold coordination; he also observed that their layer spacing of 11·7 Å, when compared with the value for this structure (d_{002} = 10·63 Å i.e. c sin β), must result in an implausibly long average Ca–O distance. It is relevant to note that a layer spacing of 10·63 Å is consistent with the top-line dimer on Figure 6.3.

As with the orthotobermorite models (i.e. Figure 6.14), model dimer structures can be derived from a clinotobermorite structure that has staggered silicate chains. Richardson (2014) found that there are again two alternative positions for the interlayer Ca atom that seem plausible. Figures 6.16(a) – 6.16(d) illustrate the version where the interlayer Ca atom is placed close to the vacant tetrahedral bridging site – that is, the equivalent of T2_so_LS1 for the orthotobermorite, while Figures 6.17(a) – 6.17(d) are for a version where the interlayer Ca atom is offset from that position – that is, the equivalent of T2_so_LS2 for the orthotobermorite. Again, the first version (data block 'T2_sc_LS1'; 'sc' for clinotobermorite and staggered silicate chains) represents the situation where the same Ca atom charge compensates for the negative charges of the chain-end O atoms that result from the omission of a single bridging tetrahedron, whereas the second version (T2_sc_LS2) involves two Ca atoms.

The T2_sc_LS1 structure is monoclinic, space group $P\,1\,2_1/c\,1$ (No. 14, unique axis b, cell choice 1); full details are given in Table 6.5 at the end of this chapter on p. 336, including lattice parameters and atom positions. The structure is illustrated in Figure 6.16: (a) viewed along the **a** axis (the unit cell is indicated by black dotted lines and the edge lengths are b and c sin β = 20·09 Å) and (b) viewed along the **b** axis (edge lengths are a and c). The faces of the Ca3–O polyhedra are not shaded (i.e. the interlayer Ca). The relationship of the interlayer Ca with the silicate tetrahedra is illustrated in (c) and (d). The formula is in this case $Ca_4[Si_2O_7]_2{\cdot}Ca_2{\cdot}4H_2O$ and so Ca/Si = 1·5 and H_2O/Si = 1·00, with no contribution from hydroxyl groups. Richardson (2014) stated that this model represents a more plausible version of the structure that is illustrated in Figure 28 of Richardson (2004) for tobermorite-based dimeric C–S–H that had the minimum degree of protonation of the silicate chains (w/n = 0 in Richardson and Groves' (1992a) model). The layer spacing is 10·04 Å (i.e. (c/2) sin β) and the calculated density is 2·590 g cm^{-3}. The interlayer Ca atom (Ca3/3A) is coordinated to seven oxygen atoms, the average Ca3/3A–O = 2·455 Å and the Ca3/3A–O polyhedral volume = 21·58 Å^3; both are consistent with the values from the regression analysis of the data for crystalline calcium silicate hydrates (Figure 5.17), which are, respectively, 2·462 Å and 21·45 Å^3.

Richardson's (2014) second staggered-chain clinotobermorite-derived dimer (data block T2_sc_LS2) is very similar to his first (T2_sc_LS1), but it has a different arrangement in the interlayer; full details are given in Table 6.6 at the end of this chapter on p. 337, including lattice parameters and atom positions. The structure is illustrated in Figure 6.17: (a) viewed along the **a** axis (the unit cell is indicated by black dotted lines and the edge lengths are b and c sin β = 18·796 Å) and (b) viewed along the **b** axis (edge lengths are a and c). The faces of the Ca3–O polyhedra are not shaded (i.e. the interlayer Ca). The formula is again $Ca_4[Si_2O_7]_2{\cdot}Ca_2{\cdot}4H_2O$; Ca/Si = 1·5 and H_2O/Si = 1·00, with no contribution from hydroxyl groups. The c parameter is 19·0 Å, and so the layer spacing is 9·40 Å (i.e. (c/2)sin β). The calculated density is 2·767 g cm^{-3}. The interlayer Ca atom (Ca3) is coordinated to six oxygen atoms, the average Ca3–O = 2·40 Å and the Ca3/3A–O polyhedral volume = 17·44 Å^3; both are close to the ideal values for Ca in six-fold coordination in crystalline calcium silicate hydrates (Figure 5.17; ideal values are 2·40 Å and 17·35 Å^3). Richardson (2014) noted that these facts, together with the O–O distances in the structure, suggest strongly that the interlayer spacing for this structure must at 9·4 Å be close to the minimum that is crystal-chemically possible, and that this value is essentially the same value as in the 9 Å tobermorites (§5.3.3.2).

Figure 6.16 Structure of a hypothetical C–S–H(I) dimer derived from a clinotobermorite structure that had staggered silicate chains: (a) viewed along the **a** axis (the unit cell is indicated by black dotted lines and the edge lengths are *b* and *c* sin $\beta = 20{\cdot}09$ Å); (b) viewed along the **b** axis (edge lengths are *a* and *c*) (the faces of the Ca3–O polyhedra are not shaded i.e. the interlayer Ca); (c) and (d) relationship of the interlayer Ca with the silicate tetrahedra. Reproduced from Figure 10 in Richardson (2014). The CIF file data block is T2_sc_LS1

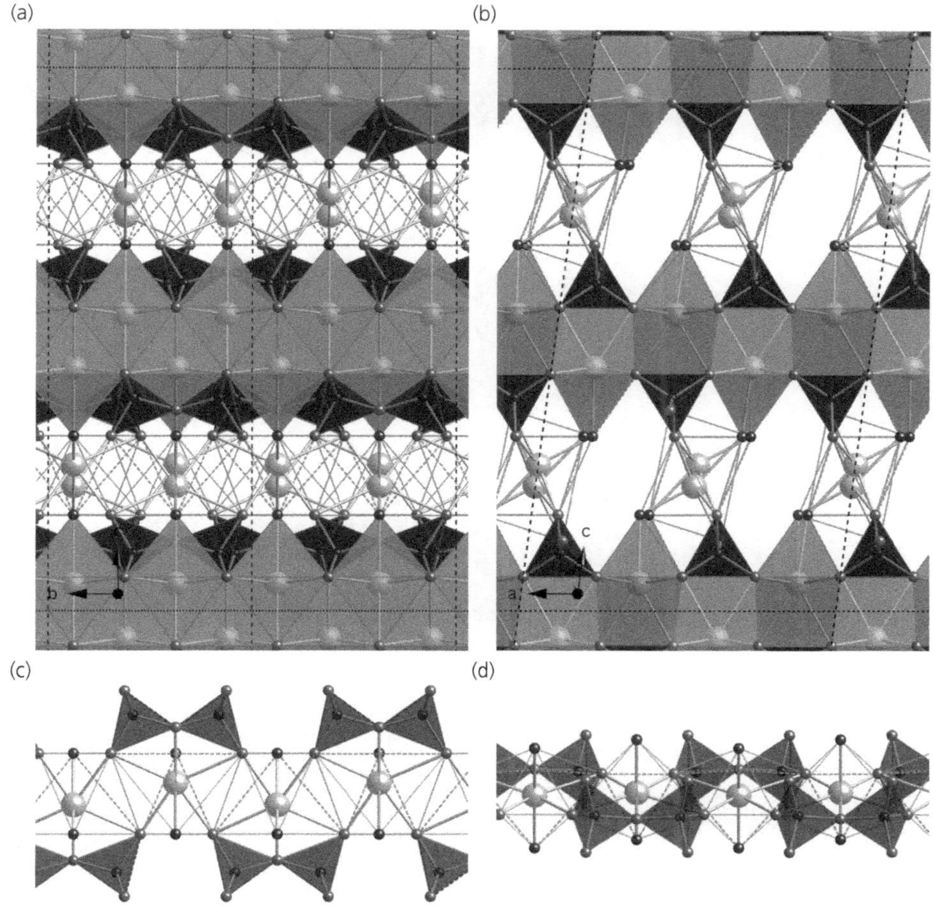

The interlayer region of the T2_sc_LS2 structure consists of ribbons of Ca–O octahedra that run parallel to the **b** axis. Richardson (2014) observed that these octahedra are very similar to those that are present in calcium hydroxide, but distorted slightly due to the need to respect the O–O distances of the silicate dimers that are clasped to either side. These ribbons of Ca–O octahedra and the associated silicate dimers are illustrated in Figures 6.17 (c) and (d): viewed along **a** in (c), with **c** up the page; and perpendicular to the layer in (d).

The close similarity of these octahedra with those present in CH is illustrated in Figure 6.18, which has been drawn to illustrate a possible topotactic relationship between the main layer of the CH structure and the interlayer of this hypothetical dimeric C–S–H(I). This offers a possible

Figure 6.17 Structure of a hypothetical C–S–H(I) dimer derived from a clinotobermorite structure that had staggered silicate chains: (a) viewed along the **a** axis (the unit cell is indicated by black dotted lines and the edge lengths are *b* and *c* sin $\beta = 18{\cdot}796$ Å); (b) viewed along the **b** axis (edge lengths are *a* and *c*). The faces of the Ca3–O polyhedra are not shaded (i.e. the interlayer Ca). Reproduced from Figure 11 in Richardson (2014). The CIF file data block is T2_sc_LS2

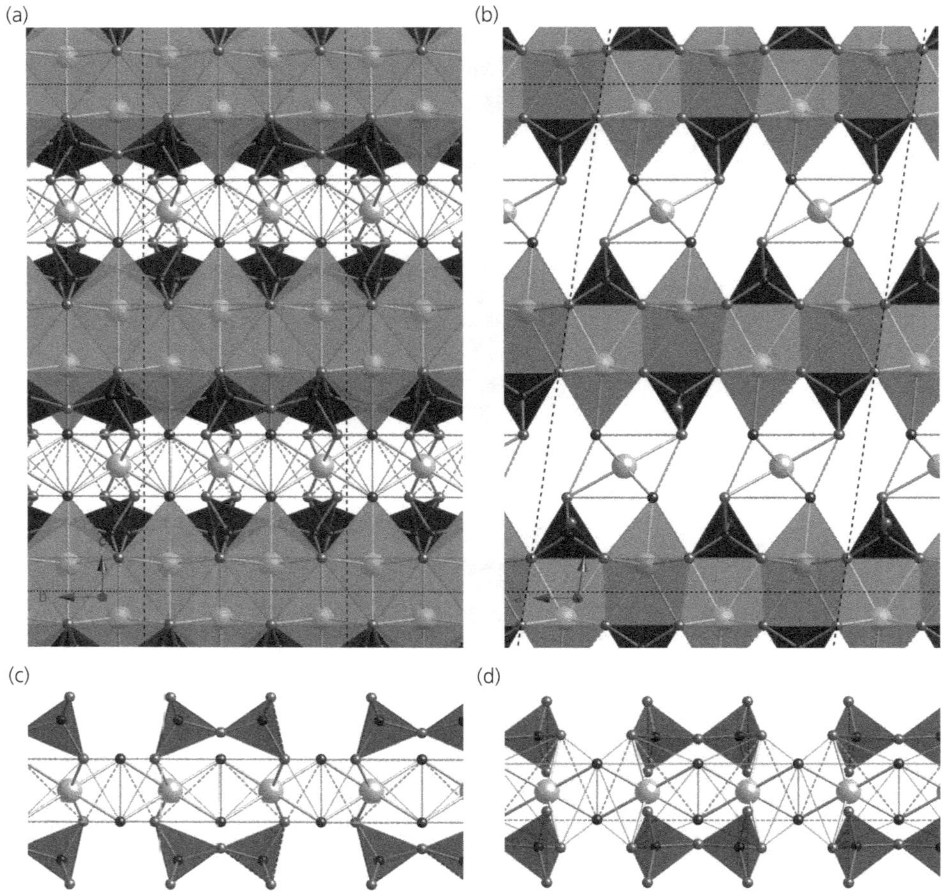

explanation for the intergrowth of C–S–H(I) and CH that has been observed by TEM to occur in alkali-activated cements (Richardson, 2004; Richardson and Groves, 1997). Topotactic growth and interstratification can be taken to represent the way that elements of CH-like structure are incorporated into C–S–H in the T/CH interpretation of Richardson and Groves' (1992a) model for C–S–H (§7.5.1).

Full occupancy of the interlayer Ca site (Ca3) in the T2_sc_LS2 model corresponds to $i = 0$ in Formula 5.6 (and so $w/n = 0$ in Richardson and Groves' (1992a) model for C–S–H). Reducing occupancy of Ca3 to 0·5 requires one half of the O1/O1A/O2/O2A sites to carry an H atom, and the details change to: Ca/Si = 1·25; H_2O/Si = 1·25; $D_x = 2{\cdot}605\,\mathrm{g\,cm^{-3}}$. The composition then corresponds to $i = 1$ in Formula 5.6 (and to $w/n = 1$ in Richardson and Groves' (1992a) model).

Figure 6.18 Schematic diagram illustrating the possible topotactic relationship between the interlayer of a hypothetical clinotobermorite-derived C–S–H(I) and the main layer of CH. Reproduced from Figure 12 in Richardson (2014). The CIF file data block is T2_sc_LS2

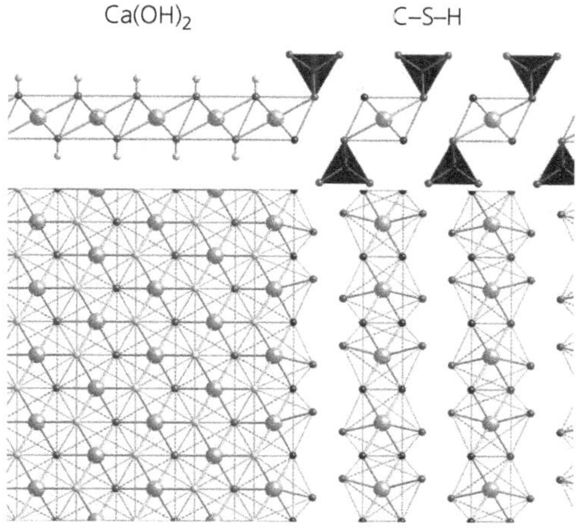

Occupancy of the Wat8/Wat9 sites is unaffected by a change in the occupancy of the interlayer Ca site because these water molecules are at the apices of main-layer Ca–O polyhedra. The H_2O/Si against Ca/Si data that are discussed in §6.7 indicate that the occupancy of the Ca3 site is most likely to be 0·75 – that is, $i = 0{\cdot}5$ in Formula 5.6, in which case a quarter of the O1/O1A/O2/O2A sites carry a H atom and the details change to: Ca/Si = 1·375; H_2O/Si = 1·125; $D_x = 2{\cdot}686\,\mathrm{g\,cm^{-3}}$.

The T2_sc_LS2 model structure with full occupancy of the interlayer Ca site is represented by a large unfilled, bold-outlined triangle on Figure 6.3 at Ca/Si = 1·5. It is evident that it is in a reasonable position with regard to the experimental data (indeed, a slight increase in the value of i from 0 to 0·5 would move the point to the lower regression line). The same point is also represented on Figure 6.5(e), where it is at the Ca-rich end of the dotted line that represents Equation 6.8. Richardson (2014) chose the T2_sc_LS2 model to represent the dimeric tobermorite-based structure because of its consistency with the experimental data on these figures and because of the possible explanation that it offers for the observed intergrowth of layers of C–S–H(I) and CH; it is therefore the structure that is referred to simply as 'T2'.

6.6.3.3 Infinite-chain clinotobermorite that has no interlayer Ca (T∞)

Richardson (2014) followed his structure for the dimeric end-member of Equation 6.8 (i.e. the dotted line on Figure 6.5(e)) with the development of a model for the other end-member – that is, a structure with infinite silicate chains and no interlayer Ca (Ca/Si $= \frac{2}{3}$), referred to as T∞ (after Richardson and Groves, 1992a). It is also based on clinotobermorite because the best model for dimer was T2_sc_LS2 (Table 6.6), and he used the same arrangement of staggered silicate chains. A structure is given in Table 5.16 for a hypothetical staggered-chain 14 Å clinotobermorite; the steps taken in the development of the model are explained in §5.3.2.4. The Ca/Si ratio of this structure is reduced easily to $\frac{2}{3}$ by setting the occupancy of the interlayer Ca site to zero but, since the value of the layer spacing is in part determined by the coordination requirements of the interlayer Ca atom, it is likely to reduce if that site is vacated completely, with the size of the reduction likely to be dependent on the number of water molecules that are retained in the interlayer, which is dependent on the extent of drying and so will be variable. Two example structures are given in Tables 6.7 and 6.8 at the end of this chapter on pp. 339–340. The removal of all molecular water results in a H_2O/Si ratio of just $\frac{1}{3}$ (i.e. Si–OH/Si = $\frac{2}{3}$), which is the value established by experiment for the most highly dried samples of C–S–H(I) (see §6.7). Atomistic modelling studies can be expected to provide insight into the most likely position for interlayer water molecules.

The clinotobermorite-based models described for dimeric and infinite-chain silicate structures have SOF_{BT} values of 0 (T2) and 1 (T∞), respectively. A straightforward explanation for the trend on the plot of SOF_{BT} against Ca/Si ratio (i.e. the dotted line on Figure 6.5(e) that represents Equation 6.8) is that a mix of these two structure types could account for C–S–H preparations that have intermediate fractions of vacant bridging sites (Richardson, 2014): a 1:1 mix of dimeric (T2) and infinite-chain (T∞) structure would give the same fraction of vacant bridging sites as pentamer (i.e. T5, $SOF_{BT} = 0{\cdot}5$); a 1:2 mix would give the same fraction as octamer (T8, $SOF_{BT} = \frac{2}{3}$); a 1:3 mix the same as undecamer (T11, $SOF_{BT} = 0{\cdot}75$) and so on. Combining layers of T2 and T∞ in different ratios to form T5, T8 and so on provides a plausible and compelling explanation for the presence of dimeric structure down to Ca/Si ratios as low as 0·9, as observed using ^{29}Si double quantum homonuclear CP/MAS correlation NMR experiments (Figure 6.6(a), data from Brunet *et al.*, 2004). The observation of Q^1–Q^{2P} correlation as well as Q^1–Q^1 would be accommodated by the presence of some vacant bridging sites in the T∞ layers. Richardson (2014) gave examples that demonstrate how the T2 and T∞ layers can be combined in practice to produce structures that have on *average* pentameric, undecameric and trimeric structure; his treatment is reiterated and expanded on in the following sections. It is important to stress that the names refer to average rather than absolute chain lengths. The inclusion of a 'trimer' structure should not be taken to imply the presence of linear chains that consist of three silicate tetrahedra, which would not be consistent with the $3n - 1$ sequence that results from the assumption that paired tetrahedral sites cannot be vacant.

6.6.3.4 'Pentameric' C–S–H(I) structures that are based on clinotobermorite (T5)

Richardson's (2014) structure for a hypothetical 'pentamer' that is derived from a clinotobermorite structure that has staggered silicate chains is shown in Figure 6.19: the structure is viewed (a) along the **a** axis (the unit cell is indicated by black dotted lines and the edge lengths are b and c sin $\beta = 23{\cdot}790$ Å) and (b) along the **b** axis (edge lengths are a and c). The faces of the interlayer Ca–O polyhedra are not shaded. Single silicate chains and associated interlayer Ca atoms are viewed (c) along the **a** axis. As an illustration that the chains do not have to be staggered, Figure 6.19(d) shows the same as in (c) but for an alternative hypothetical 'pentamer' that – while still derived from a clinotobermorite structure – is derived from a structure where the bridging tetrahedra of the silicate chains are adjacent to one another; the edge lengths are $b = 7{\cdot}3$ Å and c sin $\beta = 24{\cdot}947$ Å. The dimeric part of the structure is the same as given earlier as one of the alternatives for clinotobermorite-based dimer (in Table 6.4), which was shown to be crystal-chemically plausible in terms of Ca–O distances and polyhedral volume.

The details of the staggered-chain clinotobermorite 'pentamer' are given in Table 6.9 at the end of this chapter on p. 342, including lattice parameters and atom positions. While it is evident that this 'pentamer' model structure – that will be referred to as 'T5' – still has the monoclinic cell of the T2 and T∞ models from which it is derived, the resulting structure is actually triclinic (space group C 1) because the rotational symmetry has been lost. The Ca/Si ratio is 1·0, which is consistent with Equation 6.8 (i.e. substitution of $SOF_{BT} = 0{\cdot}5$ into Equation 6.8 results in Ca/Si = 1·0). Ca3 is

Figure 6.19 Structure of a hypothetical C–S–H(I) 'pentamer' (T5) derived from a clinotobermorite structure that had staggered silicate chains: (a) viewed along the **a** axis (the unit cell is indicated by black dotted lines and the edge lengths are *b* and *c* sin $\beta = 23{\cdot}790$ Å); (b) viewed along the **b** axis (edge lengths are *a* and *c*). The faces of the interlayer Ca–O polyhedra are not shaded (i.e. Ca3 for the T2 part of the structure and Ca2 for the T∞ part). The CIF file data block is T5_14sc. Single silicate chains and associated interlayer Ca atoms are shown in (c) viewed along the **a** axis; (d) the same as in (c) except for an alternative hypothetical 'pentamer' that is still derived from a clinotobermorite structure but where the bridging tetrahedra of the silicate chains are adjacent to one another; the edge lengths are $b = 7{\cdot}3$ Å and *c* sin $\beta = 24{\cdot}947$ Å). (a) – (c) are reproduced from Figure 13(a) – (c) of Richardson (2014); (d) is new

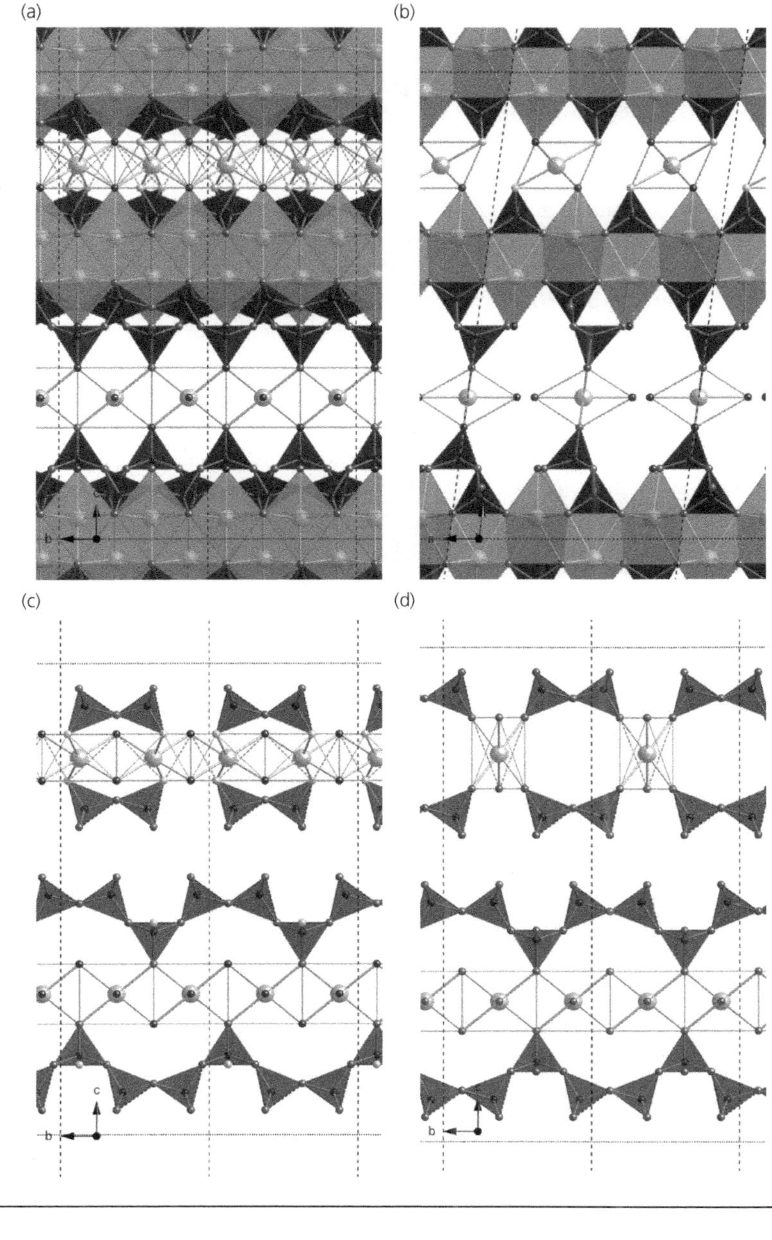

the interlayer Ca that is in the T2 part of the structure and Ca2 is in the T∞ part that is between the chains that are nominally of infinite length (nominally because in some preparations there will be tetrahedral bridging sites that are vacant). The site occupancy factor for the interlayer Ca sites in both parts of the structure $= 1 - SOF_{BT} = 0{\cdot}5$. For the T2 part, the fraction of O atoms that are at the end of the silicate dimers that carry a H atom $= 1 - SOF_{BT} = 0{\cdot}5$ (i.e. the OH1 atoms), while for the T∞ part the fraction of O atoms that are at the apex of the bridging tetrahedra that carry a H atom $= 1 - (2 \times SOF_{BT}) = 0$ (i.e. atoms O3_4 and O3_9). In terms of Formula 5.6, the formula is $Ca_4[Si_5\square_1O_{16}]H_2Ca{\cdot}(H_2O)_{4{\cdot}5}$ – that is, $i = 2 \times SOF_{BT} = 1$. The H_2O/Si ratio is 1·10 and the calculated density is 2·291 g cm^{-3}. Most of the interatomic distances are as reported in Table 5.18 for the model 14 Å clinotobermorite that has staggered chains, and the average Ca–O distances and polyhedral volumes for the interlayer Ca atoms are consistent with the values that are expected for six-fold coordinated Ca in calcium silicate hydrates – that is, with the regression analysis equations on Figure 5.17 (average Ca3–O = 2·40 Å; Ca3–O polyhedral volume = 17·48 Å^3; average Ca2–O = 2·39 Å; Ca2–O polyhedral volume = 17·82 Å^3). The layer spacing, $d_{002} = 11{\cdot}895$ Å (i.e. $(c/2)\sin\beta$), is represented by a large unfilled, bold-outlined triangle on Figure 6.3 at Ca/Si = 1·0. As with the points for the T2 and T∞ models, it is again evident that the T5 point is in a reasonable position with regard to the experimental data. The same point is also represented on Figure 6.5(e), where it is at the central position on the dotted line that represents Equation 6.8.

Richardson (2014) provided a second example of a T5 pentamer model, which he derived using 11 Å rather than 14 Å clinotobermorite to give a smaller-spacing structure that may be relevant to preparations that are more highly dried (i.e. where the structure has collapsed during drying). The lattice parameters and the positions of the atoms in the asymmetric unit of this structure are given in Table 6.10 at the end of this chapter on p. 344. The space group is again *C* 1. The structure is illustrated in Figure 6.20: (a) viewed along the **a** axis (the unit cell is indicated by black dotted lines and the edge lengths are *b* and *c* sin $\beta = 20{\cdot}672$ Å); (b) viewed along the **b** axis (edge lengths are *a* and *c*). The faces of the interlayer Ca–O polyhedra are not shaded. Single silicate chains and associated interlayer Ca atoms are shown in (c) viewed along the **a** axis. The Ca ions that are in the interlayer spaces that are associated with the dimeric (Ca3) or infinite-chain structures (Ca2) are, respectively, in six- and seven-fold coordination. The average Ca–O distance and the polyhedral volume for the six-fold coordinated ions (i.e. Ca3) are both very close to the values that are calculated from the regression analysis equations of Figure 5.17: the values calculated from the structure are 2·40 Å and 17·40 Å^3 compared with 2·40 Å and 17·35 Å^3 from the regression analysis equations. The average Ca–O distance and the polyhedral volume for the seven-fold coordinated ions (i.e. Ca2) are not quite as close but are perfectly acceptable: the values calculated from the structure are 2·43 Å and 20·32 Å^3 compared with 2·46 Å and 21·45 Å^3. In terms of Formula 5.6, the formula is $Ca_4[Si_5\square_1O_{16}]H_2Ca(H_2O)_{3{\cdot}5}$ (i.e. $i = 1$). It is very similar to the formula for the structure that is derived from 14 Å clinotobermorite but with one less water molecule per six tetrahedral sites. In going from the T5 structure in which the T∞ layer is derived from 14 Å clinotobermorite to that derived from the 11 Å structure, the H_2O/Si ratio decreases from 1·1 to 0·9, the layer spacing decreases from 11·90 Å to 10·34 Å and the calculated density increases from 2·291 to 2·567 g cm^{-3}.

6.6.3.5 'Trimeric' C–S–H(I) structure that is based on clinotobermorite (T3)

Combining the T2 and T∞ layers in the ratio of 3:1 produces a structure that has only a quarter of the tetrahedral bridging sites occupied (i.e. $SOF_{BT} = 0{\cdot}25$; $v = 0{\cdot}25$) and so the average length of the silicate chains is three (which can be calculated by substituting $v = 0{\cdot}25$ into Equation 5.10). It is of course important to remember that this is an average rather than absolute chain length. The lattice

Figure 6.20 Structure of a hypothetical C–S–H(I) 'pentamer' derived from a clinotobermorite structure that had staggered silicate chains, where the T∞ part of the structure is collapsed compared with that illustrated in Figure 6.19 i.e. derived from 11 Å clinotobermorite rather than 14 Å: (a) viewed along the **a** axis (the unit cell is indicated by black dotted lines and the edge lengths are *b* and *c* sin $\beta = 20{\cdot}672$ Å); (b) viewed along the **b** axis (edge lengths are *a* and *c*). The faces of the interlayer Ca–O polyhedra are not shaded (i.e. Ca3 for the T2 part of the structure and Ca2 for the T∞ part). Single silicate chains and associated interlayer Ca atoms are shown in (c) viewed along the **a** axis. (b) and (c) are reproduced from Figure 13(d) and (e) of Richardson (2014); (a) is new. The CIF file data block is T5_11sc

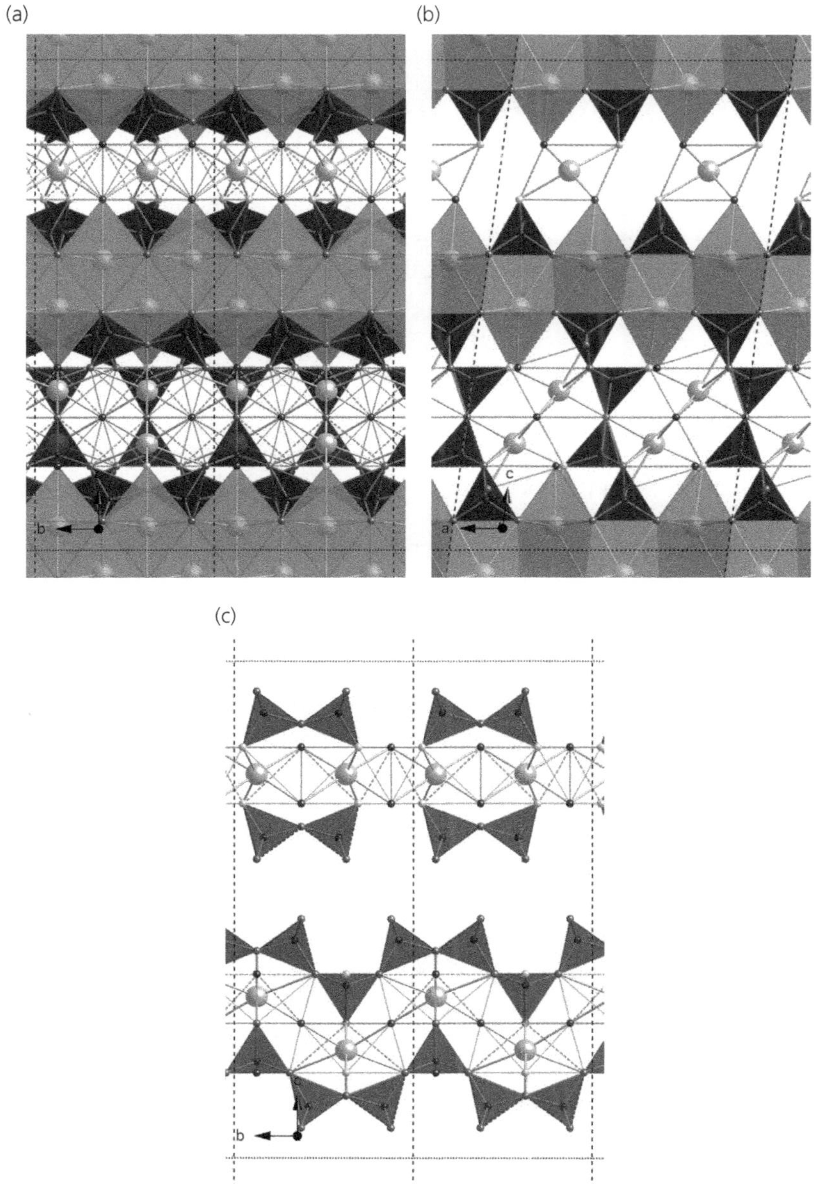

parameters and the positions of the atoms in the asymmetric unit of a hypothetical 'trimer' that is derived from a 14 Å clinotobermorite that had staggered silicate chains structure are given in Table 6.11 at the end of this chapter on p. 347 and the structure is illustrated in Figure 6.21: (a) viewed along the **a** axis (the unit cell is indicated by black dotted lines and the edge lengths are b and c sin $\beta = 42{\cdot}702$ Å); (b) viewed along the **b** axis (edge lengths are a and c). The faces of the interlayer Ca–O polyhedra are again not shaded – that is, Ca4 for the T2 part of the structure and Ca3 for the T∞ part; the average Ca–O distances and polyhedral volumes are again sensible (average Ca4–O = 2·41 Å; Ca4–O polyhedral volume = 17·45 $Å^3$; Ca3–O = 2·39 Å; Ca3–O polyhedral volume = 17·85 $Å^3$).

The Ca/Si ratio is 1·22, calculated by substituting $SOF_{BT} = 0{\cdot}25$ into Equation 6.8, and with a site occupancy factor of 0·75 for the interlayer Ca sites in both the T2 (Ca4) and T∞ (Ca3) parts of the structure (i.e. $1 - SOF_{BT} = 0{\cdot}75$). The value of i is $2 \times SOF_{BT} = 0{\cdot}5$. However, it was established in §5.3.2.4 that a value of i that is less than 1 is unlikely for the T∞ part of the structure because that would require the occupancy of adjacent sites for the O atom of water molecules that are too close to one another (replacing those molecules by a single water molecule at an intermediate position results in a Ca–O polyhedral volume that is much too big for six-fold coordination and so is unlikely). The solution to this problem is to fix the site occupancy for the Ca3 sites to 0·5 and to

Figure 6.21 The structure of a hypothetical C–S–H(I) 'trimer' derived from a clinotobermorite structure that had staggered silicate chains: (a) viewed along the **a** axis (the unit cell is indicated by black dotted lines and the edge lengths are b and c sin $\beta = 42{\cdot}702$ Å); (b) viewed along the **b** axis (edge lengths are a and c). The faces of the interlayer Ca–O polyhedra are not shaded (i.e. Ca4 for the T2 part of the structure and Ca3 for the T∞ part). Reproduced from Figure 14 of Richardson (2014). The CIF file data block is T3_14sc

(a) (b)

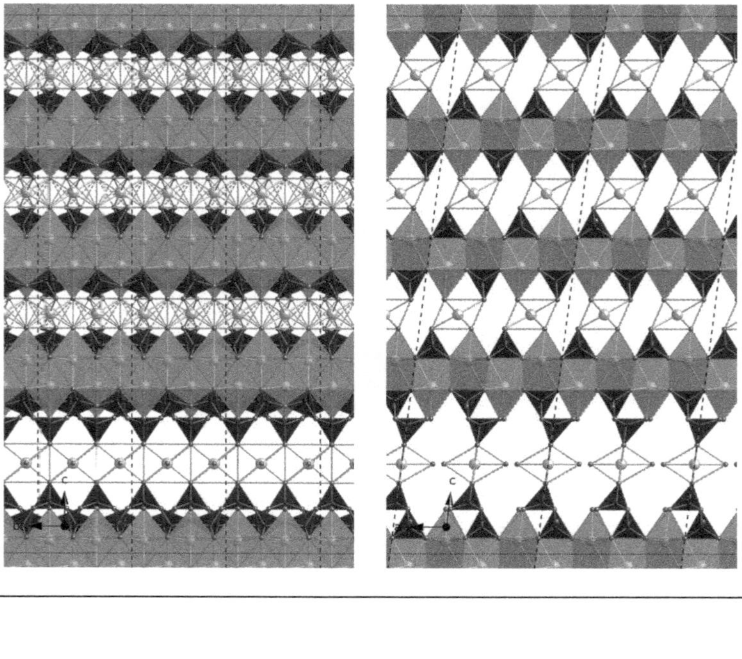

adjust the occupancy of Ca4 to $\frac{5}{6}$ to compensate. This means that the fraction of O atoms that are at the end of the silicate dimers that carry a H atom is $\frac{1}{6}$ and that the T∞ part of the structure is the same as in the 'pentamer' model.

The enhanced occupancy of the Ca sites in the T2 interlayer hints at a possible explanation for the fact that few of the NMR data for C–S–H(I) preparations that are collated on Figure 6.5(e) fall much beyond the point for T3, with the data for Ca/Si ratios greater than about 1·3 involving intermixture of C–S–H(I) with a second phase that is Ca rich. It seems plausible that full occupancy of the Ca sites in the T2 interlayer could lead to the formation of layers that have CH-like structure, as illustrated in Figure 6.18, which would occur at a Ca/Si ratio below 1·5 because of the limitation on the occupancy of the Ca sites in the T∞ interlayer.

In terms of Formula 5.6, the formula for the T3 model is $Ca_4[Si_{4\cdot5}\square_{1\cdot5}O_{15}]H_1Ca_{1\cdot5}(H_2O)_{4\cdot25}$ (i.e. $i = 2 \times SOF_{BT} = 0{\cdot}5$) and so the H_2O/Si ratio is 1·06. The calculated density is 2·494 g cm^{-3}. There are two layer-spacing peaks for this structure that fall between 5 and 10°2θ with Cu Kα radiation and that have significant intensity: $d_{003} = 14{\cdot}234$ Å (i.e. $(c/3)\sin\beta$; $2\theta_{Cu} = 6{\cdot}204°$) and $d_{004} = 10{\cdot}676$ Å (i.e. $(c/4)\sin\beta$; $2\theta_{Cu} = 8{\cdot}276°$). The layer spacing that is at the larger 2θ angle ($d_{004} = 10{\cdot}676$ Å) is represented by a large unfilled, bold-outlined triangle on Figures 6.3 and 6.5(e) at Ca/Si = 1·22. As with the points for the T2, T5 and T∞ models, it is again evident that the points for T3 are in a reasonable position with regard to the experimental data. The point for T3 ($i = 0{\cdot}5$) is represented on Figure 6.5(e) halfway between T5 ($i = 1$) and T2 ($i = 0$) on the dotted line that represents Equation 6.8.

6.6.3.6 'Undecameric' C–S–H(I) structure that is based on clinotobermorite (T11)

The T3 model was developed by combining T2 and T∞ structures in the ratio of 3:1 and, as a consequence, T3 has only a quarter of the tetrahedral bridging sites occupied (i.e. $SOF_{BT} = 0{\cdot}25$; $v = 0{\cdot}25$). If the T2 and T∞ structures are instead combined in the ratio of 1:3, then the resulting model has three quarters of the tetrahedral bridging sites occupied ($SOF_{BT} = 0{\cdot}75$, $v = \frac{1}{12}$) and the average length of the silicate chains is 11 (which can be calculated by substituting $v = \frac{1}{12}$ into Equation 5.10). The lattice parameters and the positions of the atoms in the asymmetric unit of such a hypothetical 'undecamer' are given in Table 6.12 at the end of this chapter on p. 351 and the structure is illustrated in Figure 6.22, viewed along the **b** axis (edge lengths are *a* and *c*). The T2 and T∞ parts of the structure are indicated and the faces of the interlayer Ca–O polyhedra are not shaded (i.e. Ca3 for the T2 part of the structure and Ca2 for the T∞ part). The Ca/Si ratio is $\frac{9}{11}$, calculated by substituting $SOF_{BT} = 0{\cdot}75$ into Equation 6.8, and with a site occupancy factor of 0·25 for the interlayer Ca sites in both the T2 (Ca3) and T∞ (Ca2) parts of the structure (i.e. $1 - SOF_{BT} = 0{\cdot}25$). Most of the interatomic distances are as reported in Table 5.18 for the model 14 Å clinotobermorite that has staggered chains. The average Ca–O distances and polyhedral volumes for the interlayer Ca atoms are again consistent with the values that are expected for six-fold coordinated Ca in calcium silicate hydrates – that is, with the regression analysis equations on Figure 5.17 (average Ca3–O = 2·40 Å; Ca3–O polyhedral volume = 17·44 Å^3; average Ca2–O = 2·39 Å; Ca2–O polyhedral volume = 17·78 Å^3).

In terms of Formula 5.6, the formula for the T11 model is $Ca_4[Si_{5\cdot5}\square_{0\cdot5}O_{17}]H_3Ca_{0\cdot5}(H_2O)_{3\cdot625}$ (i.e. $i = 2 \times SOF_{BT} = 1{\cdot}5$) and so the H_2O/Si ratio is $\frac{41}{44}$. The calculated density is 2·063 g cm^{-3}. There are two layer-spacing peaks for this structure that fall between 5 and 10° 2θ with Cu Kα

Figure 6.22 Structure of a hypothetical C–S–H(I) 'undecamer' derived from a clinotobermorite structure that had staggered silicate chains, viewed along the **b** axis (edge lengths are *a* and *c*). The faces of the interlayer Ca–O polyhecra are not shaded (i.e. Ca3 for the T2 part of the structure and Ca2 for the T∞ part). Reproduced from Figure 15 of Richardson (2014). The CIF file data block is T11_14sc

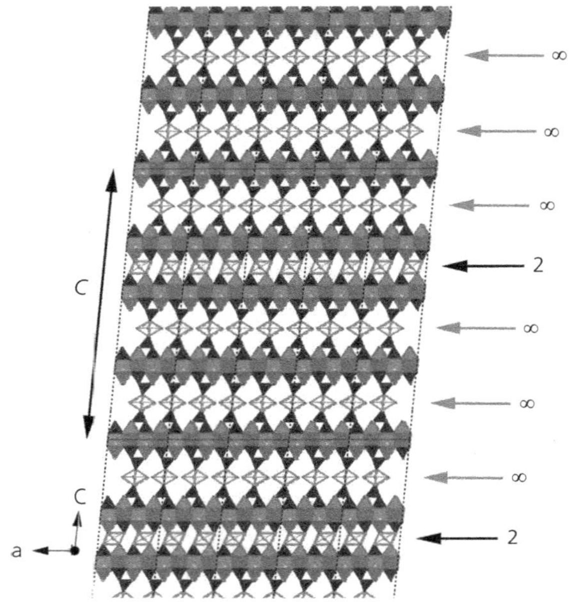

radiation and that have significant intensity: d_{003} = 17·486 Å (i.e. (*c*/3)sin *β*; $2\theta_{Cu}$ = 5·050°) and d_{004} = 13·114 Å (i.e. (*c*/4)sin *β*; $2\theta_{Cu}$ = 6·735°). The layer spacing that is at the larger 2θ angle (d_{004} = 13·114 Å) is represented by a large unfilled, bold-outlined triangle on Figures 6.3 and 6.5(e) at Ca/Si = $\frac{9}{11}$. As with the points for the T2, T3, T5 and T∞ models, it is again evident that the points for T11 are in a reasonable position with regard to the experimental data. The point for T11 (*i* = 0·5) is represented on Figure 6.5(e) halfway between T∞ (*i* = 2) and T5 (*i* = 1) on the dotted line that represents Equation 6.8.

6.6.4 Calculated powder XRD patterns for the model structures that are based on clinotobermorite

The main layer-spacing peak for each of the models is represented by a large unfilled bold-outlined diamond on Figure 6.3; it is evident that they are consistent with experimental XRD data for C–S–H(I) phases, in particular with the linear decrease that is observed to occur with increasing Ca/Si ratio. It is of course important that the model structures are also consistent with the general features of the powder XRD patterns that are discussed in §6.2.2. Figure 6.23(a) shows the calculated powder XRD patterns for the model structures from 5 to 70° 2θ and Figure 6.23(b) shows an expanded section between 2 to 10° 2θ (monochromatic Cu Kα radiation, λ = 1·5406 Å). The patterns were calculated using CrystalDiffract® assuming isotropic crystals of size 10 nm, although it is important to note that the crystals in real samples are likely to be anisotropic, which would

Figure 6.23(a) Calculated powder XRD patterns for the model C–S–H(I) structures

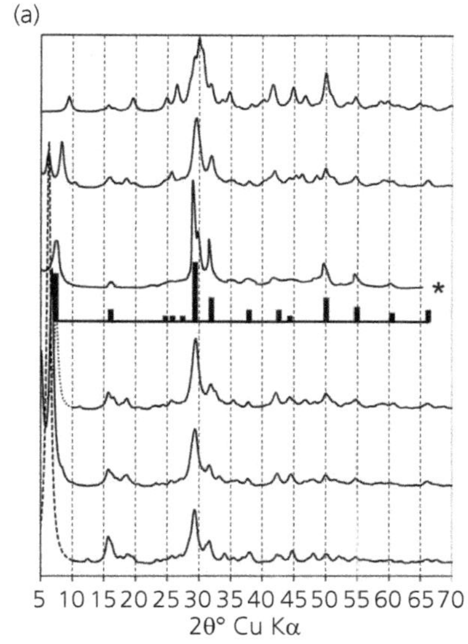

Monochromatic Cu Kα radiation, $\lambda = 1{\cdot}5406\,\text{Å}$. The patterns were calculated using CrystalDiffract® assuming isotropic crystals of size 10 nm. The inset bar-pattern represents typical experimental data and incorporates the points that are listed in §6.2.2. An experimental pattern is also included that was extracted from Figure 3 of Cong and Kirkpatrick (1996a; sample SCFUMd, indicated by *). Reproduced from Figure 16 of Richardson (2014).

Figure 6.23(b) Low-angle section of the patterns for the T3 (top), T5 (middle) and T11 (bottom) models (continued on next page)

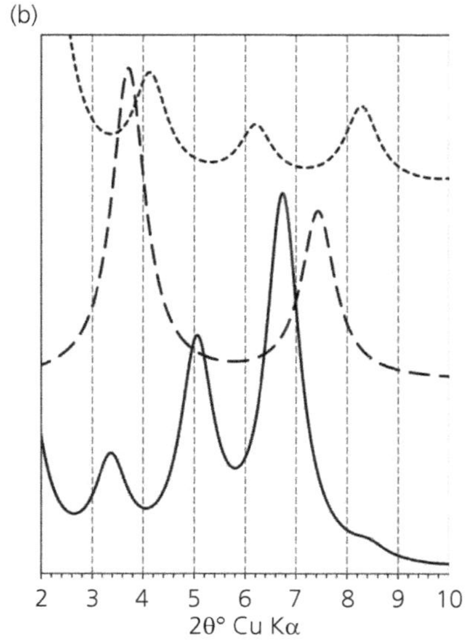

Figure 6.23(b) Continued

There are three layer-spacing peaks for the T3 structure that fall between 2 and 10° 2θ with Cu Kα radiation and that have significant intensity: d_{002} = 21·351 Å (i.e. (*c*/2)sin *β*; $2\theta_{Cu}$ = 4·135°); d_{003} = 14·234 Å (i.e. (*c*/3)sin *β*; $2\theta_{Cu}$ = 6·204°) and d_{004} = 10·676 Å (i.e. (*c*/4)sin *β*; $2\theta_{Cu}$ = 8·276°). There are two layer-spacing peaks for the T5 structure: d_{001} = 23·790 Å (i.e. *c* sin *β*; $2\theta_{Cu}$ = 3·711°); d_{002} = 11·895 Å (i.e. (*c*/2)sin *β*; $2\theta_{Cu}$ = 7·426°). There are three layer-spacing peaks for the T11 structure: d_{002} = 26·229 Å (i.e. (*c*/2)sin *β*; $2\theta_{Cu}$ = 3·366°); d_{003} = 17·486 Å (i.e. (*c*/3)sin *β*; $2\theta_{Cu}$ = 5·050°) and d_{004} = 13·114 Å (i.e. (*c*/4)sin *β*; $2\theta_{Cu}$ = 6·735°).

Figure 6.23(c) The same low-angle section for four experimental patterns that were published in Figure 3.2 of Nonat and Lecoq (1998)

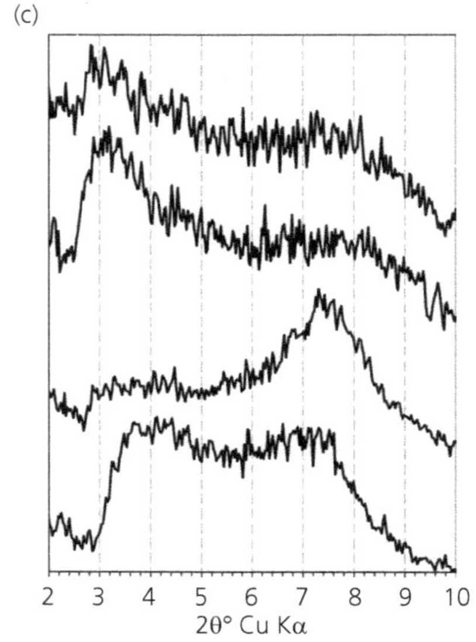

The patterns could not be extracted from the figure in Nonat and Lecoq (1998) and so Nonat gave the raw data to Richardson in 2019. Nonat and Lecoq's (1998) method of synthesis is outlined in §6.3.1. The patterns from top to bottom are for the C–S–H prepared in lime solutions that had fixed concentrations of 24, 22, 12 and 7 mmol L^{-1}. Nonat and Lecoq's XRD patterns were acquired using an INEL CPS 120 wide-angle diffractometer that had a 120° curved detector and a Cu Kα X-ray source (monochromatic radiation, Kα1 = 1·54051 Å), over a 2θ range of 0·026217° to 74·974359°, with a step size of 0·029242° and so 2564 points.

affect the relative intensities and widths of the peaks (Renaudin *et al.*, 2009a). Experimental data for C–S–H(I) are inset on Figure 6.23(a), including a pattern that was extracted from Figure 3 of Cong and Kirkpatrick (1996a) and bars that take into account the points that are listed in §6.2.2; the bar that represents the basal reflection is set at 12 Å to facilitate comparison with the pattern calculated for the T5 model. It is evident that the main features of the experimental data are present in the calculated patterns.

Powder XRD patterns that are presented in the literature are most often acquired starting from 4 or 5° 2θ (Cu Kα radiation) and generally display only one basal peak. However, Figure 3.2 in Nonat and Lecoq (1998) includes four patterns that have intensity from 1° 2θ, which – although not particularly good quality – do clearly have more than one basal peak, as shown in Figure 6.23(c). Nonat and Lecoq (1998) did not discuss these multiple basal peaks, stating simply that all of their patterns were '*characteristic of C–S–H(I)*'. While the Ca/Si ratios given by Nonat and Lecoq (1998) for their samples are higher than those for the model structures presented here and in Richardson (2014), it is nevertheless evident that the general nature of the low-angle region is reproduced quite well by the models (details of the peaks in Figure 6.23(b) are given in the figure caption).

More accurate simulations of powder XRD patterns – or indeed the use of the models reported here and in Richardson (2014) for Rietveld refinements – would need to take account of anisotropy in the dimensions of small crystals, and the experimental value for SOF_{BT} and the Ca/Si ratio of the C–S–H should be determined unambiguously by direct measurement (e.g. Nonat and Lecoq's (1998) compositions are for the bulk samples and so it is possible that they do not accurately reflect the Ca/Si ratio of the C–S–H). An experimental value for SOF_{BT} is needed for a simulation because it determines to a large extent the ratio of the T2 to T∞ modules that are necessary in the structural model, which has a big effect on the low-angle peaks, although there are also likely to be vacant bridging sites present in the T∞ layers. Other types of defects can be expected that will affect the XRD pattern – for example, stacking faults.

6.6.5 Density of the clinotobermorite-based model structures

The calculated density (D_x) for the model structures is plotted against Ca/Si ratio, SOF_{BT} and H_2O/Si ratio in Figures 6.24(a) to 6.24(c) respectively. It is evident from Figure 6.24(a) that the density increases linearly with Ca/Si ratio. The linear regression equation is D_x $(g\,cm^{-3}) =$ $1{\cdot}0298(Ca/Si) + 1{\cdot}233$ $(r^2 = 0{\cdot}998)$ but the dashed line on the figure shows that the data are described adequately and much more simply by D_x $(g\,cm^{-3}) = Ca/Si + 1{\cdot}25$. The density also decreases linearly with an increase in the fraction of bridging tetrahedral sites that are occupied (Figure 6.24(b)), which is to be expected given that the models consist of different ratios of T2/T∞ structure. The maximum in the H_2O/Si ratio (Figure 6.24(c)) occurs with the T5 structure because in that case there is maximum interlayer Ca in the half of the structure that nominally contains infinite silicate chains and those particular ions require additional water molecules to satisfy their coordination requirements. The specific gravity data (filled circles) on Figures 6.24(a) and 6.24(c) indicate that the calculated density values cover the range required for C–S–H(I) preparations. The data are from Taylor and Howison (1956) and are for samples prepared at room temperature from CH and SiO_2 sol, which were dried to constant mass at 108 °C and $p(H_2O)$ 6 mm. The agreement is probably as good as can be expected given the uncertainties in the reported compositions.

The crossed-square symbol in Figures 6.24(a) and 6.24(c) represents a bottle-hydrated C_3S sample that was almost fully reacted (96·3 %) (Brunauer, 1962; Brunauer *et al*., 1958). The w/c ratio was 9 and the sample was hydrated for 162 days at 23·5 ± 0·5 °C and continuously rotated at 30 rpm. (sample reference D-35). It was D-dried and the density was determined at 27·2 °C by measuring the displacement of a saturated calcium hydroxide solution in a Weld type pycnometer. The XRD pattern for Brunauer *et al.*'s (1958) preparation did not show a basal peak but they calculated a layer spacing of 9·3 ± 0·1 Å based on the measured density and composition and assuming an orthorhombic pseudo-cell with $a = 5{\cdot}59$ Å and $b = 3{\cdot}64$ Å. Inspection of the figures shows that the point for Brunauer's (1962) sample is close to that for the T2 ($i = 0$) model.

Figure 6.24 Calculated density of the clinotobermorite-based model structures for C–S–H(I) plotted against (a) Ca/Si ratio, (b) H_2O/Si ratio and (c) SOF_{BT}

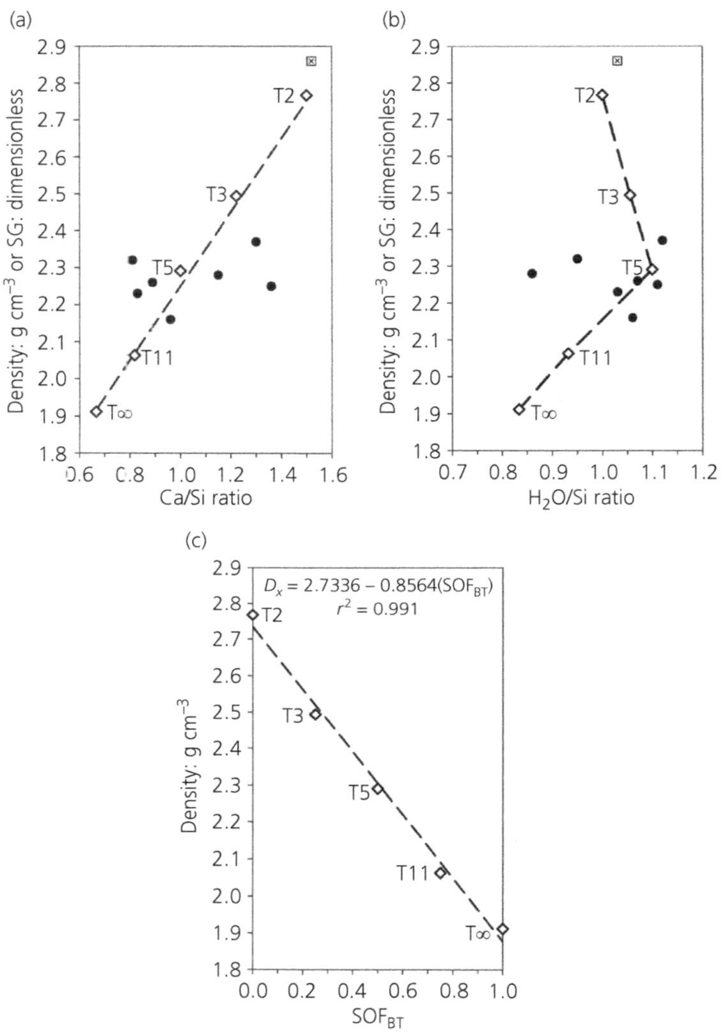

6.7. Thermal behaviour and water content of C–S–H(I)

Early work, reviewed by Steinour (1947), indicated that for C–S–H with Ca/Si ≈1·0 the molar H_2O/SiO_2 ratio is 2·0–2·5 for material equilibrated at 50% RH at ordinary temperature, falling to ≈1·0 at 105 °C. Taylor (1953) determined isotherms and isobars which were consistent with these values. The isobars showed a continuous loss of water with increasing temperature, with only a slight indication of a step at H_2O/SiO_2 ≈0·5 and 250 °C. The curves varied little with Ca/Si ratio for Ca/Si ratios of 0·99–1·37. In contrast, later work showed that H_2O/SiO_2 ratios at a given temperature increase with Ca/Si ratio (Brunauer and Greenberg, 1960; Cong and Kirkpatrick, 1996a; El-Hemaly *et al.*, 1978; Gard *et al.*, 1959). Data from a number of sources are compiled on Figure 6.25(a), which is a plot of H_2O/Si against Ca/Si for C–S–H(I) preparations that were made using a

Figure 6.25(a) Plot of H_2O/Si against Ca/Si for C–S–H(I) preparations

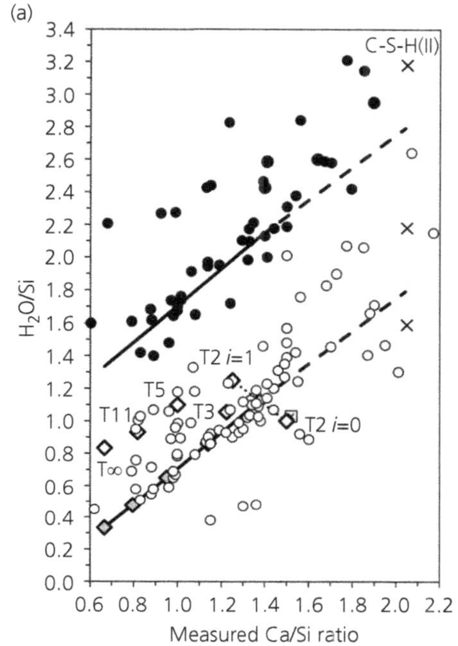

The filled circles represent samples that were lightly dried (e.g. in flowing N_2 at room temperature) and the unfilled circles represent samples that were more harshly dried – that is, at around 110 °C (or by using a method that is approximately equivalent) or more severely than that (e.g. at 236 °C and $p(H_2O) = 6\,mm$). In some cases, data points have been placed in a category that does not relate to the extent of drying that was reported in the relevant reference because it seems likely that they were dried more or less aggressively than as reported. Data are from: Brunauer and Greenberg, 1960; Cong and Kirkpatrick, 1996a; Copeland *et al.*, 1967; El-Hemaly *et al.*, 1978; Fujii and Kondo, 1981; Gard *et al.*, 1959; Hannawayya (1980); Hedin (1945); Renaudin *et al.*, 2009a; Taylor, 1950, 1953; Taylor and Howison, 1956. Samples with Ca/Si >1·5 often contained crystalline CH. The lower thick line represents Equation 6.29 and the upper black line is the same with one additional water molecule per Si atom. The bold unfilled diamonds represent the model structures that are described in §6.6.3 and in Richardson (2014); the five grey diamonds represent the least hydrated versions of them (from left to right, the T∞, T11, T5, T3 and T2 structures). The long-dashed lines correspond to intermixture of T2(i = 0·5) with CH. The cross symbols are for C–S–H(II) (Gard and Taylor, 1976) with different degrees of drying. The crossed-square symbol is for an almost fully reacted bottle-hydrated C_3S (Brunauer, 1962; Brunauer *et al.*, 1958). Figure 6.25(a) is modified from Figure 2 in Richardson (2014).

number of methods. The sources of the data are given in the figure caption. The filled circles represent samples that were lightly dried (e.g. in flowing N_2 at room temperature) and the unfilled circles represent samples that were more harshly dried – that is, at around 110 °C (or by using a method that is approximately equivalent) or more severely than that (e.g. at 236 °C and $p(H_2O)$ 6 mm). In some cases, data points have been placed in a category that does not relate to the extent of drying that was reported in the relevant reference because it seems likely that they were dried more or less aggressively than as reported. While there is quite a lot of scatter in the data, it is evident that the

Figure 6.25(b) Plot of the layer spacing against the H_2O/Si for C-S-H(I) preparations

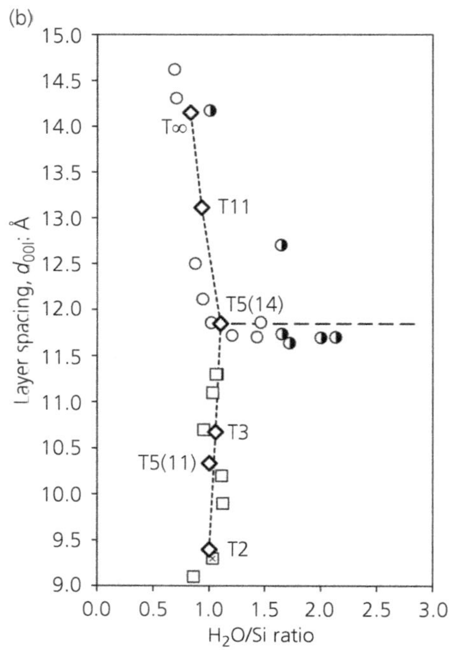

Data are from: Brunauer *et al.*, 1958 (crossed square); Cong and Kirkpatrick, 1996a (unfilled circles: SCFUM series dried at 110°C); Renaudin *et al.*, 2009a (half-filled circle); Taylor and Howison, 1956 (unfilled squares: group B samples). The bold unfilled diamonds represent the model structures that are that are described in §5.6 and in Richardson (2014). T5(14) is the model structure where the T∞ part is from 14 Å clinotobermorite and T5(11) is where the T∞ part is from 11 Å clinotobermorite; both with staggered chains. Figure 6.25 (b) is reproduced from Figure 17 in Richardson (2014).

H_2O/SiO_2 ratio does increase with increasing Ca/Si and that the value depends on the severity of the drying.

The H_2O/SiO_2 ratio of the most highly dried samples (i.e. those in the lower part of the figure) can be seen to increase essentially linearly with an increase in the Ca/Si ratio. The lowest value is consistent with a totally collapsed tobermorite-like phase – that is, essentially a fully protonated 9 Å single-chain tobermorite. This phase would have four Ca atoms per six tetrahedral sites, no interlayer Ca, and so four hydroxyl groups per six tetrahedral sites, and no interlayer water molecules. Since there would be no vacant tetrahedral sites, this gives a formula of $Ca_4Si_6O_{14}(OH)_4$ (i.e. $i = 2$ and $m = 0$ in Formula 5.2) and so $H_2O/SiO_2 = \frac{1}{3}$ and $Ca/Si = \frac{2}{3}$. The observed increase in H_2O/SiO_2 ratio with increasing Ca/Si ratio must involve the presence of vacant tetrahedral sites, a reduced number of hydroxyl groups and the introduction of interlayer Ca. On inspection of the data for the most highly dried samples it transpires that any dimeric structure present at higher Ca/Si ratio must have slightly less than the maximum amount of interlayer Ca that is theoretically possible; in fact, for purely dimeric structure, $i = 0{\cdot}5$ rather than $i = 0$. The structural-chemical formula for the most highly dried C–S–H(I) (i.e. without destroying the tobermorite-like structure) is therefore as given in §6.2, with $i = 2–4·5v$ (i.e. $i = 2$ for infinite chains ($v = 0$) and $i = 0·5$ for dimeric chains ($v = \frac{1}{3}$)).

$$Ca_4 \left[Si_{1-v} \square_v O_{3-2v}\right]_6 H_{2i} Ca_{2-i} \cdot 12v H_2O \tag{6.26}$$

The Ca/Si ratio can be calculated by substituting for i in Equation 6.4

$$\frac{Ca}{Si} = \frac{4+\frac{9}{2}v}{6(1-v)} \tag{6.27}$$

which when expressed per occupied tetrahedral site reduces to

$$\frac{Ca}{Si} = \frac{\frac{2}{3}+\frac{3}{4}v}{1-v} \tag{6.28}$$

and so can be expressed in words as

$$\frac{Ca}{Si} = \frac{\text{Fraction of paired tetrahredral sites}+\frac{3}{4}(\text{fraction of vacant tetrahedral sites})}{\text{Fraction of occupied tetrahedral sites}} \tag{6.29}$$

Equation 6.28 can alternatively be written as

$$\frac{Ca}{Si} = \frac{11-(3\times SOF_{BT})}{8+(4\times SOF_{BT})} = \frac{\frac{11}{8}Q^1+\frac{2}{3}Q^2}{Q^1+Q^2} \tag{6.30}$$

The H_2O/Si ratio can be calculated using

$$\frac{H_2O}{Si} = \frac{2+\frac{15}{2}v}{6(1-v)} \tag{6.31}$$

which when expressed per occupied tetrahedral site reduces to

$$\frac{H_2O}{Si} = \frac{\frac{1}{3}+\frac{5}{4}v}{1-v} \tag{6.32}$$

and so can be expressed in words as

$$\frac{H_2O}{Si} = \frac{\text{Fraction of bridging tetrahredral sites}+\frac{5}{4}(\text{fraction of vacant tetrahedral sites})}{\text{Fraction of occupied tetrahedral sites}} \tag{6.33}$$

Equation 6.32 can alternatively be written as

$$\frac{H_2O}{Si} = \frac{9-(5\times SOF_{BT})}{8+(4\times SOF_{BT})} = \frac{\frac{9}{8}Q^1+\frac{1}{3}Q^2}{Q^1+Q^2} \tag{6.34}$$

Combining Equations 6.28 and 6.32 gives

$$\frac{H_2O}{Si} = \frac{19}{17} \cdot \frac{Ca}{Si} - \frac{7}{17} \tag{6.35}$$

Equation 6.35 is represented by the lower thick line on Figure 6.25(a). It is evident that this equation satisfactorily represents the data for the more highly dried preparations, apart from three outlying data points. For example, the position of the line is almost identical to the straight line on Figure 1 of Fujii and Kondo (1981), which they approximately represented by the formula $n\text{CaO}\cdot\text{SiO}_2\cdot(n + 0{\cdot}8)\text{H}_2\text{O}$. The five grey diamonds represent the positions for the least hydrated versions of – from left to right – the T∞, T11, T5, T3 and T2 model structures. The upper black line represents the same as the lower line but with one additional water molecule per Si atom.

Figure 6.25(b) is a plot of the layer spacing against the H_2O/Si ratio for C–S–H(I) preparations. The sources of the data are given in the figure caption and the bold unfilled diamonds again represent the model structures that are described in this chapter and in Richardson (2014). T5(14) is the model 'pentamer' structure where the T∞ part is derived from a hypothetical 14 Å clinotobermorite (i.e. Figure 6.19; Table 6.9) and T5(11) (i.e. Figure 6.20; Table 6.10) is where the T∞ part is from 11 Å clinotobermorite, both with staggered chains. The points for the 14 Å clinotobermorite-derived structures are joined by the short-dash line. It is evident that the models are consistent with the observed variation in the layer spacing and H_2O/Si ratio. The figure also supports the view given earlier that many C–S–H(I) preparations are in fact mixtures of C–S–H with a second phase; in this case the long-dash line indicates intermixture with T5(14).

XRD powder diffraction shows that the basal spacing decreases to 10–11 Å at 120 °C and 9 Å at 240–450 °C (Taylor, 1953). Lawrence (1965) found that at higher temperatures, the basal spacing disappears for specimens with Ca/Si = 0·5–0·9, but that the remaining reflections are substantially preserved up to 700 °C, although with some decrease in *a* and *b*. At 770 °C, wollastonite is formed, initially in a poorly crystalline state. He found that formation of β-C_2S as an intermediate product, reported earlier (Taylor, 1953), occurred only if the sample was partly carbonated. Specimens with Ca/Si 1·0–1·5 became almost amorphous at 550°–770 °C. Those with Ca/Si = 1·0–1·2 gave more highly crystalline wollastonite together with β-C_2S.

DTA (differential thermal analysis) curves show a broad endotherm at ≈100 °C and a strong, sharp exotherm at ≈800 °C, attributable to loss of molecular water and crystallisation of wollastonite, respectively (Taylor, 1953). The temperature of the exotherm increases with Ca/Si, values of 830 °C, 865 °C and 900 °C being obtained at Ca/Si = 0·8–1·0, 1·25 and 1·33, respectively (Kalousek, 1954). These data are compared with others on Figure 6.11(a) and in §6.5.1. As with DTA in general, the exact temperatures depend on the heating rate and other experimental factors. The endotherm is absent if the sample has been previously dried to a sufficient extent.

A significant problem that occurs when attempting to assess studies of C–S–H(I) is that – as with layered double hydroxides (Richardson, 2013) – the product of a synthesis procedure is sometimes not single-phase C–S–H, but is instead a mixture of phases. For example, Richardson *et al.* (2010) showed by analytical TEM that a preparation that was supposed to be single-phase C–S–H was in fact a mixture of C–S–H and an amorphous Ca-rich phase – that is, the C–S–H had a lower Ca/Si ratio than the target composition.

It was demonstrated in §6.2.2 that the layer spacing of C–S–H(I) is dependent both on the Ca/Si ratio and the extent of drying, with the Ca/Si ratio being directly related to the fraction of 'bridging' sites that are vacant. The layer spacing, d_{002}, decreases linearly with increasing Ca/Si ratio, while drying the preparation reduces d_{002} by 2 Å at all values of Ca/Si ratio. It seems that this observation of the effect of drying on the layer spacing can be used as a test for the likely composition of the C–S–H(I). This is demonstrated in Figure 6.26, which shows a d_{002}-Ca/Si plot with data from four studies that measured the layer spacing both before and after drying. The data are from Alizadeh *et al.* (2007), Beaudoin *et al.* (2009), Gutteridge and Parrott (1976) and Komarneni *et al.* (1986). The reported Ca/Si ratios range from 0·965 (Gutteridge and Parrott, 1976) to 1·6 (Beaudoin *et al.*, 2009) and the data for each study are shown as two diamonds joined by a full line; the top diamond indicates the layer spacing before drying and the lower one after a modest drying step (the layer spacing can often be reduced further by more severe drying). The two coarse dashed lines are the trend lines from Figure 6.3, which represent the layer spacing before and after drying, with drying having the effect of reducing d_{002} by 2 Å. On inspection, it is evident that the 2 Å difference between these two lines is essentially the same as between the values for the four pairs of data, except that each pair is offset from the lines; the extent of the offset is indicated by dotted-line arrows and varies considerably between preparations. The most straightforward interpretation of these data is that the reported Ca/Si ratios are not the correct values for the C–S–H(I) phases and that the actual values can be estimated from the extent of the offset on the figure. The numbers that

Figure 6.26 Plot of d_{002}-Ca/Si showing data from four studies that measured the layer spacing both before and after drying. Each pair is shown with diamond symbols that are connected with a full line, the layer spacing before drying is indicated by the top diamond and the lower one after a modest drying step. The value for the Ca/Si ratio that is reported and the value estimated using the diagram are, respectively, 0·965 and 0·92 (Gutteridge and Parrott, 1976), 1·0 and 1·15 (Komarneni *et al.*, 1986), 1·2 and 1·32 (Alizadeh *et al.*, 2007) and 1·6 and 1·17 (Beaudoin *et al.*, 2009)

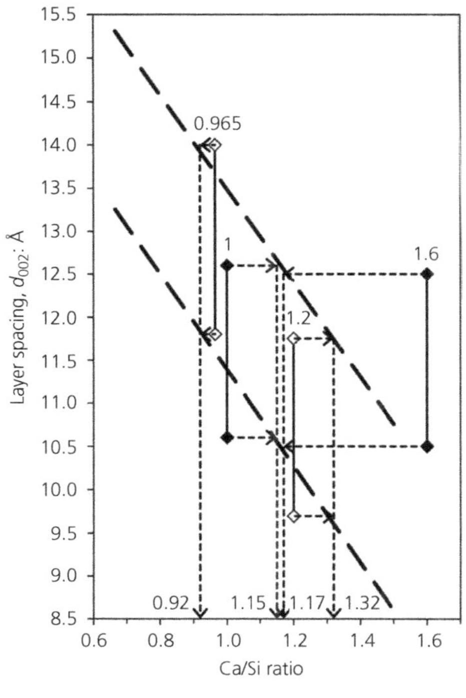

are close to the x-axis therefore indicate the real Ca/Si ratios; this would mean, for example, that Gutteridge and Parrott's (1976) preparation had a composition that was close to the reported value, but that Beaudoin *et al.*'s (2009) was very different (1·17 rather than 1·6).

6.8. Preparations containing monomer or high proportions of dimer

As noted in §6.4.4, it is difficult to see how forms of C–S–H containing only monomeric silicate ions could have a structure related to that of tobermorite or jennite, although they might contain Ca–O sheets with a different conformation. There is also a problem with combinations of low Ca/Si ratio and low mean chain length, which occur in some C–S–H preparations made by mixing solutions of an alkali silicate and a calcium salt (§6.4.2) and which do not contain enough Ca to occupy all the sites in the central Ca layer.

Grutzeck (1999) suggested that the C–S–H formed at early ages in calcium silicate pastes has a structure based on Si_2O_7 groups and octahedrally coordinated Ca ions, having some structural relation to such sorosilicate structures as cuspidine ($Ca_4(Si_2O_7)(OH,F)$), but no substantial evidence for this was given, and a statement that sorosilicates normally contain rather large channels that could host Ca^{2+}, OH^- or other ions or water molecules is inconsistent with the known crystal structures. The hypothesis is nevertheless of interest as jennite has a structure with some features in common with sorosilicate phases (§5.4).

6.9. Ca–OH groups in C–S–H preparations with Ca/Si ratios >1·3–1·5

The equations in §6.2.5 and experimental evidence in §6.2.6 and §6.3 indicate that Ca–OH groups become increasingly significant at Ca/Si ratios >1·3 and it is important to know how these groups are incorporated into the structure. For preparations that are confirmed by analytical TEM to be single-phase of high Ca/Si ratio – such as Brunauer and Greenberg's (1960) sample D-69 (i.e. C–S–H(II); Brunauer and Greenberg, 1960; Gard and Taylor, 1976) and the products obtained by Ishida *et al.* (1992a, 1992b, 1992c, 1992d) from the hydration of highly reactive β-C_2S (§6.3) – the position of the data points on Figure 6.9(a) suggests that they may be based on jennite-like structure. If this is the case, then the most straightforward explanation for the data that fall between the 3rd and 4th dashed lines on Figure 6.9(a) is that the preparations consist of intermixture of tobermorite- and jennite/jaffeite-like structures. This will be referred to as Explanation 1 because there are two other possibilities that have been proposed: (2) that the C–S–H consists of tobermorite-like structure, intermixed with layers of CH; or (3) that it is possible to have tobermorite-based structure that incorporates –OH groups that are bonded only to Ca. Establishing which explanation is correct – or combination of explanations – is central to understanding the structure of the C–S–H that forms in hardened pastes of tricalcium silicate (Chapter 7) or Portland cement.

Brunet *et al.* (2004) considered that their data for a C–S–H(I) with bulk Ca/Si = 1·5 supports the presence of Ca^{2+} and OH^- ions in the interlayer region of a tobermorite-like structure – that is, Explanation 3 – although the reinterpretation of their data given in §6.2.6 is that the protons of Ca–OH groups did not correlate with Si atoms that are close to interlayer Ca. Nevertheless, Explanation 3 has previously received much support, for example from: Klur *et al.* (1998); Kurczyk and Schwiete (1960, 1962); Nonat (2004); Nonat and Lecoq (1998); Stade (1980); Stade and Müller (1987); Stade and Wieker (1980); Stade *et al.* (1983, 1984, 1985). With reference to Stade's model (see Table 7.4), Taylor (1990) found it difficult from a crystal chemical standpoint to see how anions as well as cations could be present in the interlayer region of a tobermorite-like structure.

However, he considered that the differences were '... *largely semantic, since the probable corrugation of the layers in a jennite-type structure could bring the additional Ca^{2+} and OH^- ions into what could alternatively be described as part of an interlayer region*'. This 'corrugation' of the Ca–O layer in jennite has of course since been confirmed by Bonaccorsi *et al.*'s (2004) determination of its crystal structure. A view of the corrugated layer – which is often referred to as a tilleyite ribbon – is shown in Figure 6.27(a). The *dreierketten* are seen end on and the positions of the paired (P)

Figure 6.27(a) Part of the jennite structure

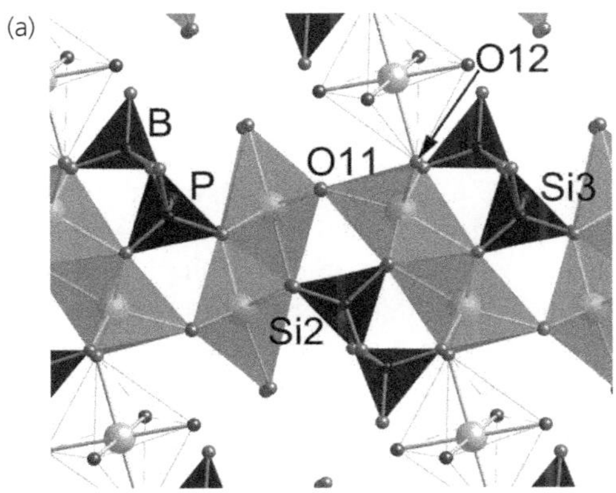

The data are from Bonaccorsi *et al.*, 2004. Si2 and Si3 are paired tetrahedra; O11 and O12 are hydroxyl groups. The distance between Si2 and O11 is 3·757 Å and between Si3 and O12 is 3·679 Å.

Figure 6.27(b) Part of the jaffeite structure

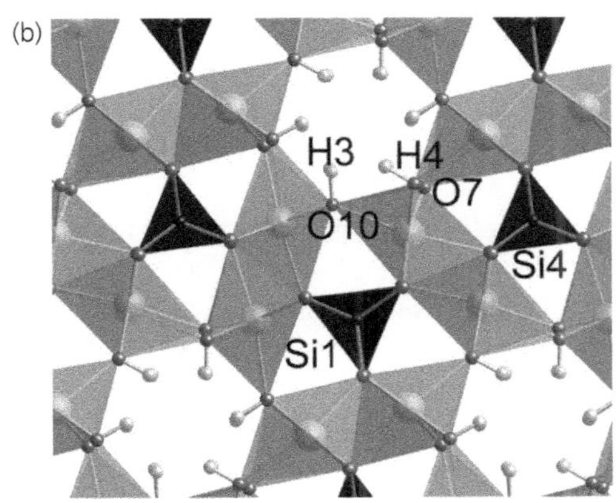

The data are from Yamnova *et al.*, 1993. The distances between Si1 and O10 and H3 are 3·800 and 4·658 Å, respectively; the distances between Si4 and O7 and H4 are 3·588 and 4·768 Å, respectively. Si1 and Si4 are equivalent to Si2 and Si3 in the jennite structure shown in Figure 6.27(a); O10 and O7 are equivalent to O11 and O12.

and bridging (B) tetrahedra are indicated; the Si2 and Si3 atoms are therefore paired Q^2 Si in jennite but would be Q^1 Si in a J-based dimer (i.e. in a structure derived from jennite by omission of the bridging tetrahedra). In jennite, O11 and O12 are OH^- ions; the distance between O11 and Si2 is 3·757 Å and between Si3 and O12 is 3·679 Å. While the positions of the H atoms in jennite were not located by Bonaccorsi *et al.* (2004), comparison with the equivalent part of the jaffeite structure (Figure 6.27(b)) shows that the protons of the O11 and O12 hydroxyl groups in jennite would be less than 5 Å from Si2 and Si3, respectively (the original model for the jennite structure was derived by Taylor from the jaffeite structure (Richardson, 2004)). Since these distances are less than 5 Å, the correlation between protons of Ca–OH groups and Q^1 Si shown by Klur *et al.* (1998) and Brunet *et al.* (2004) is consistent with the presence of jennite-like structure.

So the questions arise as to whether (*a*) crystal-chemically plausible sites can be found in a tobermorite-based structure for OH groups bonded only to interlayer Ca and (*b*) elements of CH-like structure can be incorporated into a tobermorite-based structure. The evidence from the model structures that are presented in §6.6 suggests that the answer to (*a*) is no, but that the answer to (*b*) is yes. The view that jennite-type regions can occur in C–S–H preparations is based largely on two considerations. First, the ionic constitutions increasingly approach that of C–S–H(II) as the Ca/Si ratio increases. Second, the XRD and electron diffraction evidence for the relatively highly ordered forms of high-Ca C–S–H suggest a jennite-based rather than a tobermorite-based structure. It seems reasonable to conclude that Ca–OH groups in C–S–H do not occur in the interlayer of tobermorite-based structure, but that they can occur as CH- or jennite/jaffeite-based structure, depending on the method of synthesis.

6.10. Structural substitution and ion exchange

6.10.1 General

Structural substitutions play an important role in cement hydration and concrete durability. They are also likely to significantly affect anion and cation sorption and exchange capacities, which are important for applications of cement-based materials as containment barriers and waste forms. Studies on crystalline tobermorite have complemented those on synthetic C–S–H and are reviewed in Chapter 5. Among the possible structural substitutions, that of Al^{3+} is the most extensively studied. Synthetic preparations generally have foil-like morphology and X-ray diffraction patterns that are consistent with C–S–H(I), §6.2.2; the structural-chemical formulae and equations that are given in §5.3.4 and §5.3.5 are applicable. Incorporation of Fe^{3+}, alkalis and SO_4^{2-} is less well understood. The sorption behaviour for Zn^{2+}, Pb^{2+} and Cl^- has also been investigated.

6.10.2 Incorporation of alkali metal cations

Regardless of the target Ca/Si ratio and experimental method, the inclusion of only modest quantities of alkali hydroxide during the synthesis of C–S–H preparations usually results in a tobermorite-like phase that has the morphology of crumpled foils. It was shown in §6.3 that the morphology of C–S–H is determined largely by the concentration of Ca^{2+} ions in solution and, since the solubility of CH decreases very rapidly with the addition of sodium or potassium hydroxide, foil-like morphology is to be expected, and is indeed observed (e.g. Kumar *et al.*, 2017; Macphee *et al.*, 1989). This dramatic reduction in CH solubility is illustrated in Figure 6.28, which includes data from Duchesne and Reardon (1995), Lawrence (1966) and Yuan *et al.* (2010). It is evident that the solubility is one order of magnitude lower at only about $0{\cdot}2\,mol\,L^{-1}$ of alkali hydroxide and two orders of magnitude lower at about $4\,mol\,L^{-1}$. This means that methods of synthesis that involve alkali (with the associated reduction in the concentration of Ca^{2+} ions in solution) – such as that employed by Kumar *et al.* (2017) and Harris *et al.* (2022) – will not produce a product that has the

Figure 6.28 Calcium hydroxide solubility in KOH and NaOH solutions at 25 °C

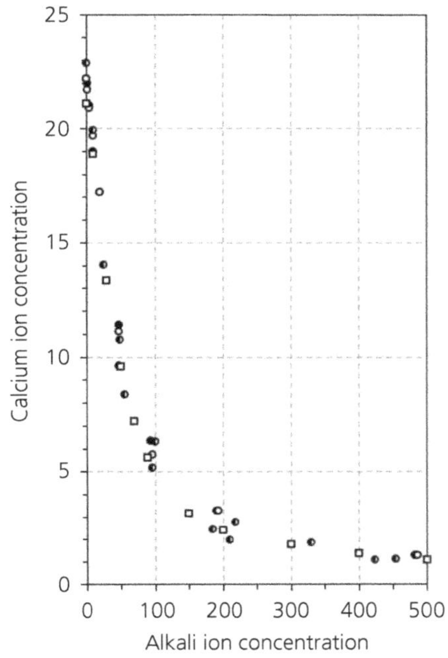

Data are from Duchesne and Reardon (1995; open circles for NaOH and filled circles for KOH), Lawrence (1966; KOH + NaOH, half-filled circles; extracted from Figure 6) and Yuan *et al.* (2010; NaOH, open squares). The units are mmol L^{-1} for Lawrence (1966) but mmol kg^{-1} for Duchesne and Reardon (1995) and Yuan *et al.* (2010). In all three cases, the value for the solubility with no alkali is about 22 mmol L^{-1}, which is the value that is obtained at 25 °C when a solution is made using fine rather than coarse crystals; see §7.1.3 and Figure 7.2.

fibrillar morphology (or the nanostructure) that is similar to that observed for the Op C–S–H in hardened C_3S pastes because the value of C_{CaO} is too low.

Studies on tobermorite, reviewed in §5.3, provide indications of possible interactions of alkali metal cations with C–S–H. Analyses of most natural tobermorites, and the effects of cation exchange on the thermal behaviour of synthetic preparations, can only reasonably be explained by supposing that small contents of Na^+ and K^+ can be accommodated in the interlayers. The incorporation of Rb^+ or Cs^+ in tobermorite was interpreted in the same way.

Experimental and thermodynamic studies of phase equilibria in the Na_2O–CaO–SiO_2–H_2O system at ordinary temperatures (Brown, 1990; Kalousek, 1944; Macphee *et al.*, 1989) indicated that C–S–H can take up Na^+ and K^+ and provide data on solubility relations, but included no detailed investigation of the solid phases. Stade (1989) studied the reactions of tobermorite-type C–S–H preparations made at 22°, 80° and 150 °C with KOH, NaOH, LiOH and $(CH_3)_4NOH$ solutions. The amount of the cation taken up was calculated from the change in concentration of the solution.

It increased with the initial concentration, decreasing Ca/Si ratio and decreasing cation radius, although with no detectable difference between K^+ and Na^+ and with much scatter for samples prepared at 22 °C. The highest values of cation/Si ratio observed were ≈0·3. The greater uptake at low Ca/Si ratios was attributed to the presence of SiOH groups, the protons of which were easily replaced even at low concentrations of the hydroxide solution. At higher concentrations, further uptake depended on exchange with interlayer Ca^{2+}. An increase in alkali binding with decrease in Ca/Si ratio has been observed by other workers – for example, Hong and Glasser (1999) and L'Hôpital *et al.* (2016).

Lognot *et al.* (1998) studied synthetic tobermorite-type C–S–H and found a maximum atomic Na/Si ratio of 0·45 and a maximum atomic Na/Ca ratio of 0·49 for Al-free samples. In agreement with Stade (1989), they proposed that, with increasing levels of Na^+ substitution, the latter progressively replaces silanol hydrogens, occupies vacant interlayer sites and replaces interlayer Ca. Sun *et al.* (2006) found lower levels of Na substitution (maximum Na/Si ratio 0·13) for Al-free C–S–H and also observed that the extent of Na substitution generally decreases with increasing (Ca + Na)/Si ratio.

A ^{23}Na NMR study of moist and freeze-dried Al-free C–S–H samples by Viallis *et al.* (1999) indicates that Na^+ directly associated with the C–S–H becomes mobile under moist conditions, suggesting that it is surface sorbed and in the diffuse ion swarm. The dried samples contained crystalline NaCl, which dissolved in the solution under moist conditions. The authors suggest that Na^+ ions occur in outer sphere surface complexes, retaining their hydration shells even under freeze-dried conditions.

There remains considerable uncertainty about the structural roles of alkalis in C–S–H and related phases: how these roles vary with composition, structure and mode of synthesis, and the dynamic behaviour both on particle surfaces and in the interlayer space.

6.10.3 Incorporation of aluminium

Aluminium is the most important substituent in the C–S–H that occurs in real cements and, as a consequence, it is commonly referred to as C–A–S–H. Preparations of C–A–S–H generally have foil-like morphology and X-ray diffraction patterns have the characteristics of C–S–H(I), §6.2.2 (e.g. L'Hôpital *et al.*, 2015, 2016). The structural-chemical formulae and equations that are given in §5.3.4 and §5.3.5 are applicable.

Early studies on natural and synthetic samples of 11 Å tobermorite, described in §5.3.1, showed that Al could substitute for up to a limit of about one in six of the Si atoms. The first study of the incorporation of Al (and other guest atoms) in C–S–H was by Copeland *et al.* (1967). They processed mixtures of C_3S paste with C_4AH_{13} or C_3A in a ball mill, and could detect no products by XRD other than C–S–H and CH. They also observed removal of Al from C–S–H formed by alite hydration by reaction with gypsum to form ettringite. From the relations between the Al/Si, Ca/Si and H_2O/Si ratios, they concluded that the main replacement was of Si by (Al + H), but that there was also some replacement of $3Ca^{2+}$ by $2Al^{3+}$. As with 11 Å tobermorite, up to about one Si atom in six could be replaced by Al, the incorporation of which modified the morphology as observed by TEM.

In the 1980s, it became possible to probe Al coordination by way of ^{27}Al NMR. Although the relatively low-field spectrometers of the time offered poor spectral resolution, these data provided important insights. Studies on natural and synthetic tobermorites by Komarneni *et al.* (1985), discussed more fully in §5.3.1.7, showed that Al could substitute for Si in both Q^2 and Q^3 sites. They made the important observation that the substitution leads to additional peaks in the ^{29}Si NMR spectrum, arising from Si atoms on both Q^2 and Q^3 sites in which one next nearest neighbour (NNN) is tetrahedral Al. These peaks, called Q^2(1Al) and Q^3(1Al), are shifted to values ≈3 ppm less negative than the comparable Q^2(0Al) and Q^3(0Al) peaks due to Si with no NNN Al.

From an early study using the molybdate method and thermal dehydration data, Stade and Wieker (1982) concluded that Al could enter the structure of 'C–S–H(di, poly)' (§6.5.1) and that it was mainly present, not as a substituent for Si, but in the interlayer. Later, Stade and Müller (1987) and Stade *et al.* (1984) used ^{27}Al NMR to investigate synthetic C–S–H preparations made by several routes. They found that Al occurs at both tetrahedral ($Al^{[4]}$) and octahedral ($Al^{[6]}$) sites, and that the fraction of $Al^{[4]}$ (i.e. the $Al^{[4]}/Al^{[4,6]}$ ratio) decreases linearly with increasing Ca/Si ratio, as illustrated in Figure 6.29(a). For C–S–H(di, poly), they observed no $Al^{[4]}$ at Ca/Si = 1·5. In pastes made from C_3S and C_3A at a water/solid ratio of 0·5, however, they observed both $Al^{[4]}$ and $Al^{[6]}$;

Figure 6.29 (a) Fraction of $Al^{[4]}$ (i.e. the $Al^{[4]}/Al^{[4,5,6]}$ ratio) in C–A–S–H preparation plotted against Ca/Si ratio. Data are from: Faucon *et al.* (1999b; half-filled circles); Pardal *et al.* (2009; squares); Renaudin *et al.* (2009a, 2009b; filled circles); Stade and Müller (1987; open circles). (b) Comparison of ^{29}Si MAS NMR experimental (bottom) and simulated spectra of a C–A–S–H preparation that had Ca/(Si + Al) = 0·95 and Al/Si = 0·11. The experimental spectrum was extracted from Figure 10(d) of Pardal *et al.* (2012). The simulated spectrum (top) consists of peaks for Q^1, Q^2(1Al), Q^2(0Al) and Q^3(1Al) at chemical shifts of –79·2, –81·7, –85·2 and –91·0 ppm, respectively. All peaks have the same shape and width at half height

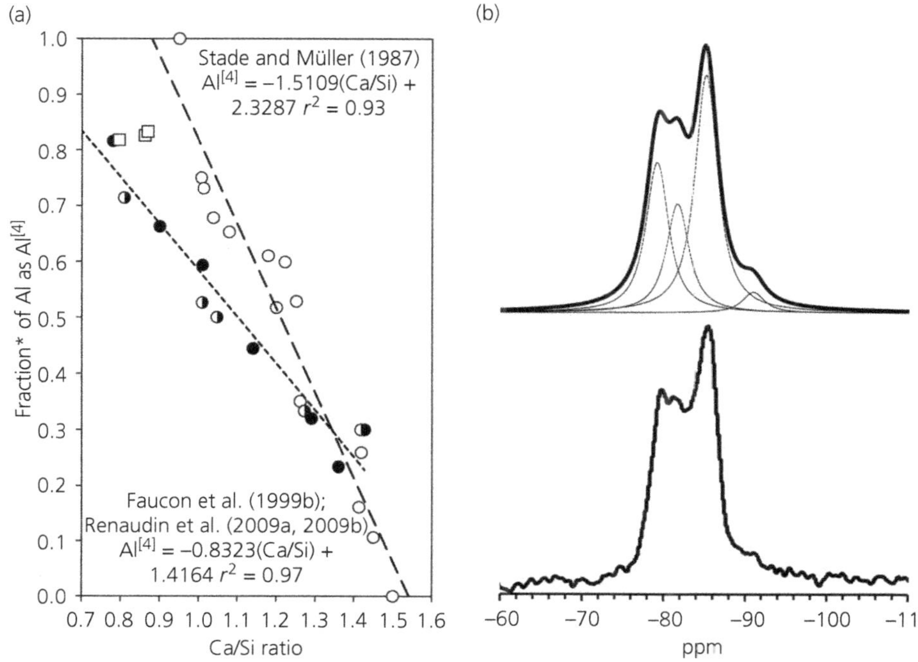

this was attributed to the wide spread of local Ca/Si ratio (Stade and Müller, 1987). The data for the tetrahedrally coordinated Al indicated that the Al atoms generally formed two or more Al–O–Si bonds. This is consistent with the presence of Al in Q^2 or Q^3, but not Q^1 sites.

Kwan *et al.* (1996) prepared aluminous C–S–H samples with a wide range of compositions. They reported pH-composition relations, XRD patterns and ^{29}Si and ^{27}Al MAS NMR spectra. The results showed that Al enters the C–S–H without altering the general features of the structure. The Al was present at tetrahedral sites, at which it substituted for Si, and possibly also at octahedral sites, at which, if present, it most probably substituted for interlayer Ca. At low Ca/Si ratios, the Si predominantly occupied Q^2 sites, indicating the presence of long chains, whereas at higher Ca/Si ratios it occupied Q^1 and Q^2 sites, indicating that both chains and dimers were present.

Faucon *et al.* (1998a) studied the incorporation of Al in synthetic C–S–H using multiple quantum (MQ) ^{27}Al MAS NMR. The C–A–S–H had calculated Ca/Si ratios of 0·88–1·00, $Al^{[4]}$/Si ratios of 0·08–0·35, Ca/(Si + $Al^{[4]}$) ratios of 0·66–0·82 and Na/(Si + $Al^{[4]}$) ratios of 0·19–0·35; the C–A–S–H was present in solids that had bulk Ca/(Si + Al) = 0·59–0·75, Al/Si = 0·33 or 0·5 and Na/Si = 0·20–0·47. The Al/Si of the solids was higher than for the C–A–S–H because all of the samples contained gibbsite and one also contained a trace of C_4AH_{13}. Evidence of Al substitution was found only for preparations that contained Na. Two peaks attributable to $Al^{[4]}$ were observed. One, with an isotropic chemical shift of 72·5–76·3 ppm and a quadrupole frequency of 0·45–0·48 MHz, was attributed to Al replacing Si in bridging tetrahedra (Q^{2L} in the authors' nomenclature). The other, for which the corresponding values were 63·7 ppm and 0·35 MHz, was attributed to Al replacing Si in paired tetrahedra. A peak at 11·2 ppm and 0·35 MHz was attributed to $Al^{[6]}$. This could have arisen from gibbsite or other impurity phases, but the possibility that it included a contribution from Al in the C–A–S–H – either in the Ca sites of the main layer (regarded as octahedral) or in the interlayer – or in an AFm phase not detected by XRD, was not excluded; the latter would be consistent with Richardson and Groves' (1993a) model for C–A–S–H. Charge balance of the $Al^{[4]}$ was attributed to interlayer Na^+ or replacement of SiO^- by AlOH, depending on the presence or absence of Na and the pH. A maximum $Al^{[4]}$/(Si + $Al^{[4]}$) ratio of 0·26 ($Al^{[4]}$/Si = 0·35) was observed and was attributed to the need to distribute the excess negative charge of an AlO_4 tetrahedron across the structure. High $Al^{[4]}$/(Si + $Al^{[4]}$) ratios favoured substitution at bridging tetrahedral sites.

In closely similar work Faucon *et al.* (1999a) dealt with C–A–S–H preparations that had maximum contents of Ca that were slightly higher, with Ca/Si ratios of 0·71–1·24, Ca/(Si + Al) of 0·61–1·19 and Al/Si ratios of 0·04–0·20; the samples were prepared using mixes that had bulk Ca/Si ratios of 1·0–1·7 and Al/Si ratios of 0·2–0·4. ^{27}Al MAS NMR showed peaks attributed to C_2AH_8 and to $Al^{[4]}$ in the C–S–H. For a preparation with Ca/(Si + $Al^{[4]}$) 0·84 (after correction for the presence of C_2AH_8) the MQ spectrum showed two $Al^{[4]}$ peaks, which were explained as in Faucon *et al.* (1998a). No clear evidence was obtained for or against the presence of $Al^{[6]}$ in the C–S–H. The maximum $Al^{[4]}$/Si ratio in the C–A–S–H was ≈0·17 for Ca/Si ratios up to ≈1·0 and was much lower at Ca/Si ≈1·2.

Faucon *et al.* (1999b) successfully increased the maximum Ca/(Si + $Al^{[4]}$) ratio of their preparations, reporting ^{29}Si and ^{27}Al NMR results for C–A–S–H with calculated ratios of Ca/Si = 0·80–1·52, $Al^{[4]}$/Si = 0·03–0·21 and Ca/(Si + $Al^{[4]}$) = 0·74–1·47; the samples were prepared using mixes that had bulk Ca/Si ratios of 0·64–1·71 and Al/Si ratios of 0·1 or 0·3. Most preparations also contained C_2AH_8 and $NaNO_3$ and some contained SiO_2 gel. The maximum $Al^{[4]}$/(Si + $Al^{[4]}$)

ratio found in the C–A–S–H was 0·17. Al signals with isotropic chemical shifts of approximately 4·5, 10·3, 38·8, 58–67 and 74·5 ppm were attributed to $Al^{[6]}$, Al in C_2AH_8, $Al^{[5]}$, $Al^{[4]}$ in paired tetrahedra and $Al^{[4]}$ in bridging tetrahedra, respectively. In preparations with the highest Ca/(Si + $Al^{[4]}$) (e.g. 1·31, 1·47), little Al entered the C–S–H; this was attributed to the presence of a high proportion of the Si in dimeric (Si_2O_7) groups, which could not accommodate Al. The small amount of Al that was present was partly in bridging tetrahedra and partly as $Al^{[5]}$ and $Al^{[6]}$, which were considered to occur in interlayers and to substitute for Ca in main layers, respectively. XRD of some freeze-dried samples showed an increase in basal spacing from the normal 11 Å to 12·5 Å. At intermediate ratios (e.g. 0·93, 1·11), much of the Al substituted for Si in bridging tetrahedra, but limited amounts of $Al^{[5]}$ and $Al^{[6]}$ were also present. At the lowest Ca/(Si + $Al^{[4]}$) ratios (e.g. 0·58, 0·74), the Al was considered to substitute for Si in paired tetrahedra. Inspection of their data shows that the fraction of Al that is $Al^{[4]}$ (i.e. the $Al^{[4]}/Al^{[4,5,6]}$ ratio) decreases linearly with increasing Ca/Si ratio, as illustrated in Figure 6.29(a). The linear trend is slightly different from the one from Stade and Müller's (1987) data, which were obtained from lower resolution spectra. C_2AH_8, when present, tended to disappear after prolonged storage at 100% RH. The Al was presumed to enter the C–S–H, but no details of its distribution were reported.

Pardal *et al.* (2009, 2012) studied C–A–S–H with Ca/(Si + Al) ≈0·7 or ≈0·95 and with varying Al/Si ratio that had been prepared by equilibrating C–S–H synthesised from CaO, SiO_2 and water (at water/solid = 50) in the hydration solution of tricalcium aluminate. ^{27}Al NMR again indicated two peaks attributable to $Al^{[4]}$, with isotropic chemical shifts of 71·0–72·0 ppm and 60·3–69·2 ppm. However, in contrast to Faucon *et al.* (1998a, 1999b), the former was attributed to Al substituted in paired tetrahedra and the latter to Al in bridging tetrahedra, which led to the conclusion that there was greater substitution in paired tetrahedra at the higher of the two Ca/Si ratios, which is the opposite to Faucon *et al.*'s (1998a, 1999b) view and is also not supported by molecular dynamics calculations (e.g. Churakov and Labbez, 2017). Pardal *et al.* (2009, 2012) interpreted their ^{29}Si NMR spectra in such a way as to support their alternative assignment for the $Al^{[4]}$ peaks. However, since there is no peak at the chemical shift that would be expected for Q^1(1Al) and the intensity suggested by the authors for Q^{2B}(1Al) is at the level of the noise, their ^{29}Si NMR spectra can be convincingly interpreted without the need to have Al in paired tetrahedra. This is demonstrated in Figure 6.29(b), which compares their spectrum for a sample with Ca/(Si + Al) = 0·95 and Al/Si = 0·11 (bottom spectrum; extracted from Figure 10(d) of Pardal *et al.*, 2012) with a spectrum created using a simple simulation (top); it is evident that the two are very similar. The simulated spectrum just consists of peaks for Q^1, Q^2(1Al), Q^2(0Al) and Q^3(1Al) at chemical shifts of –79·2, –81·7, –85·2 and –91·0 ppm, respectively. Aside from the chemical shift, the only difference between the four peaks is the intensity – that is, they have the same shape and width at half height. The intensities for the four peaks are (in the same order) 29, 21, 46 and 4%, which gives an Al/Si ratio – calculated using Equation 5.14 – of 0·11, which agrees with the experimental value. The ratio of $Al^{[4]}/Al^{tot}$ for Pardal's (2009) samples falls between the data of Stade and Müller (1987) and Faucon *et al.* (1999b)/Renaudin *et al.* (2009a, 2009b), Figure 6.29(a).

Workers have in general struggled to incorporate as much Al in synthetic preparations of C–A–S–H as occurs in real cements. This is because second phases tend to precipitate when higher levels are attempted – for example, katoite, AH_3 or strätlingite or other AFm phases (e.g. Kwan *et al.*, 1996; L'Hôpital *et al.*, 2015; Lognot *et al.*, 1998; Lothenbach and Nonat, 2015; Pardal *et al.*, 2009; Roosz *et al.* 2018). This generally lower level of incorporation is demonstrated in Figure 6.30, which is a plot of $Al^{[4]}$/Ca against Si/Ca ratio for C–A–S–H(I) preparations; the references for the data are given in the figure caption. The dashed lines represent the C–A–S–H that forms

Figure 6.30 Plot of $Al^{[4]}$/Ca against Si/Ca ratio for C–A–S–H(I) preparations

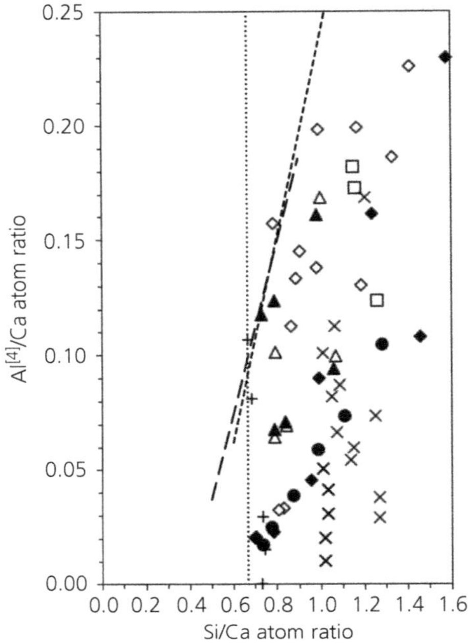

Data are from Copeland *et al.* (1967; +); Faucon *et al.* (1999a; unfilled diamond); Faucon *et al.* (1999b, filled diamond); L'Hôpital *et al.* (2015; ×); Pardal *et al.* (2012; unfilled square); Renaudin *et al.* (2009b; filled circle); Sun *et al.* (2006; unfilled and filled triangles represent data at 1 week and 4 weeks, respectively). The dashed lines represent the C–A–S–H that forms in real cements: they are the result of linear regression analysis of TEM-EDX data for neat Portland cement pastes and blends of Portland cement with blast-furnace slag, pulverised fuel ash and metakaolin; the data are from Girão (2007), Girão *et al.* (2007a), Girão *et al.* (2007b), Girão *et al.* (2010), Love *et al.* (2007), Richardson and Groves (1992b, 1993b, 1997) and Taylor *et al.* (2010). Where possible, $Al^{[5]}$ and $Al^{[6]}$ are excluded. $Al^{[5]}$ and $Al^{[6]}$ are excluded from the data for the synthetic preparations because the compositions for the C–A–S–H in real cements determined by TEM-EDX analysis must correspond only to $Al^{[4]}$ since the Al/Si ratios agree with those calculated from the ^{29}Si NMR data; i.e. a calculation that assumes Al substituting for Si at bridging sites (Equation 5.14; Richardson *et al.*, 1993, 1994, 2016). In the case of the data of Sun *et al.* (2006), the percentage of Al atoms in their Table 1 was multiplied by a value for the fraction of Al in four-fold coordination that was extracted from their Figure 4. Modified from Figure 18 in Richardson (2014).

in real cements: they are the result of linear regression analysis of TEM-EDX data for neat Portland cement pastes and blends of Portland cement with blast-furnace slag, pulverised fuel ash and metakaolin. The vertical dotted line simply indicates the minimum value on the Si/Ca axis that is possible for tobermorite-based structure. On inspection of Figure 6.30 it is evident that the points for the synthetic preparations mostly fall to the Si-rich side of the lines representing the C–A–S–H that forms in real cements, with a few data points falling essentially on the lines. The data for the latter points can be taken to best reflect the material that forms in real cements. The fact that the data for the synthetic preparations fall to the right of the lines suggests that the amount of Al that is incorporated into the C–A–S–H in real cements represents a practical maximum. Sun *et al.* (2006) and Faucon *et al.* (1999b) considered that some $Al^{[5]}$ and $Al^{[6]}$ would be present either in the interlayer region of the C–A–S–H or on the surface of C–A–S–H particles; Pardal *et al.* (2012) suggested that they are associated with the edges of C–A–S–H particles. Either case would mean that $Al^{[5]}$ and $Al^{[6]}$ would be expected to contribute to the TEM-EDX analyses as well as $Al^{[4]}$ because the analysis volume used in that technique is quite large compared with the size of the particles. Inclusion of $Al^{[5]}$ and $Al^{[6]}$ in the data for the synthetic preparations on Figure 6.30 obviously affects the position of the data points on the figure, but they nevertheless remain to the right of the lines – that is, on the Si-rich side (data not shown).

Andersen *et al.* (2006) prepared C–A–S–H samples with bulk Ca/Si ratios of 0·66, 0·83, 1·00, 1·25, 1·50 and 1·75, all with the Al/Si ratio fixed at 0·05, which they characterised with samples of hardened Portland cement using ^{27}Al MAS NMR. They assigned a peak with an isotropic chemical shift of 5·0 ppm and a quadrupole product parameter, P_Q, of 1·2 MHz to a new phase that they called the '*third aluminate hydrate*' (TAH) ('third' because it was in addition to AFt and AFm phases). They proposed that TAH is '*either an amorphous/disordered aluminate hydroxide or a calcium aluminate hydrate, produced as a separate phase or as a nanostructured surface precipitate on the C-S-H phase*'. In contrast to Andersen *et al.* (2006), Kunhi Mohamed *et al.* (2020) assigned the peak at 5 ppm on ^{27}Al NMR spectra to bridging-site $Al^{[6]}$ in the C–A–S–H phase, based on the results of a DNP-enhanced 2D $\{^{29}Si\}^{27}Al$ refocused dipolar INEPT MAS spectrum (their Figure 4(a)) and DFT-calculated ^{29}Si chemical shifts for different structural units (their Figure 4(b)). They state that the 5 ppm peak is involved in '*weaker correlation*' to '*silicates with chemical shifts around –77 ppm*', which is evident on the figure, although the ^{29}Si intensity is low at that chemical shift. There is a 2·1 ppm difference between –77 ppm and their calculated mean chemical shift for the $Q^{2P}(Al^{[6]})$ unit, which is given as –79·1 ppm, which is therefore situated firmly under the Q^1 peak. This discrepancy does not appear to have been discussed. Yang *et al.* (2021) ultimately supported Kunhi Mohamed *et al.*'s (2020) assignment of a peak at –77 ppm to $Q^{2P}(Al^{[6]})$ by using results from quantitative ^{27}Al MAS NMR to inform their fitting of 1D ^{29}Si MAS NMR spectra. However, as well as taking intensity from the Q^1 peak, the incorporation of a peak at –77 ppm in their fitting procedure also produced negative dips in the resulting residuals at around –77 ppm. This is indicative of a problem with the procedure, and probably a false peak in those cases where the dips are visually almost as large as the fitted peak, as on their Figure 9. As a consequence of these observations, the evidence presented as supporting this assignment does not appear to be compelling, and it should be noted that Yang *et al.* (2021) suggested a plausible alternative explanation for Kunhi Mohamed *et al.*'s (2020) Figure 4(a), namely that a −77 ppm peak may be due to end-chain Q^1 sites that are close to nonbonded $AlO_x(OH)_{6-x}$ units. Regardless of this scepticism about Kunhi Mohamed *et al.*'s (2020) assignment of a peak at –77 ppm, it is evident that such a peak cannot be observed on 1D ^{29}Si NMR spectra because of the significant overlap with Q^1 (as pointed out by Yang *et al.*, 2021), while the absence of significant amounts of $Al^{[6]}$ at bridging sites of the C–A–S–H that is present in hardened cement pastes is indicated by the general agreement between the Al/Si ratios determined by TEM-EDX with those calculated from ^{29}Si NMR when assuming the incorporation of only $Al^{[4]}$ at bridging sites (Zhu and Richardson, 2023).

6.10.4 Incorporation of aluminium and alkali

Preparations that incorporate aluminium and alkali generally give X-ray diffraction patterns that are consistent with C–S–H(I), Table 6.1 (e.g. L'Hôpital *et al.*, 2015, 2016), and have foil-like morphology. Lognot *et al.* (1998) studied several samples of Na- and Al-substituted synthetic C–S–H using compositional analysis, IR spectroscopy and ^{27}Al and ^{29}Si MAS NMR. The samples were made from CH, reactive SiO_2, gibbsite (AH_3) and water or NaOH solutions; most contained unreacted AH_3. They concluded that the sites occupied by the Al depended on the NaOH concentration. For samples made in water or 0·5 mol kg^{-1} NaOH, the Al occupied octahedral sites in the interlayer. With an NaOH concentration of 1 mol kg^{-1}, it was in tetrahedral coordination and replaced Si in the bridging tetrahedra. At the highest concentration studied (5 mol kg^{-1}), it was concluded that it also substituted for Si in paired tetrahedra. A sample from an initial mix with Ca/(Si + Al) = 0·66 and Al/Si = 0·33 had relative intensities for peaks with ^{29}Si chemical shifts of –78, –80·8 and –84·2 ppm of 29, 67 and 5%, respectively. Assignment of these peaks to Q^1, Q^2(1Al) and Q^2(0Al) gives a calculated Al/Si ratio of 0·33 (using Equation 5.14) which, given the errors that are associated, the NMR data is consistent with the value that is calculated from the initial mix and the amount of Al that is in tetrahedral coordination (i.e. $Al/Si^{C\text{–}S\text{–}H} = Al/Si^{mix} \times$ fraction of $Al^{[4]} = 0{\cdot}33 \times 0{\cdot}9 = 0{\cdot}30$). Lognot *et al.* (1998) actually assigned some of the intensity at –78 ppm to Q^2(2Al), but that would not be consistent with their statement that for that sample Al only substituted at bridging sites. Since the C–A–S–H had tobermorite-like structure (i.e. an XRD pattern characteristic of C–S–H(I)), the ^{29}Si NMR data can be used to calculate a structural-chemical formula: Equations 5.8 and 5.16 are used to calculate values for the fractions of vacant tetrahedral sites (v) and sites that are occupied by Al (f). Substitution of those values into Formula 5.6 gives $Ca_4\left[Si_{0{\cdot}678}Al_{0{\cdot}225}\square_{0{\cdot}097}O_{2{\cdot}806}\right]_6 H_4\left(Na_2\right)_{0{\cdot}675}\cdot mH_2O$, or rearranged slightly, $Ca_4\left[Si_{0{\cdot}678}Al_{0{\cdot}225}\square_{0{\cdot}097}O_{2{\cdot}139}\left(OH\right)_{0{\cdot}667}\right]_6 Na_{1{\cdot}350}\cdot mH_2O$. This formula has a Ca/(Si + Al) ratio of 0·74, which is higher than the value for the mix (i.e. 0·66); it is not possible for C–A–S–H to have a Ca/(Si + Al) ratio as low as 0·66 when Q^1 sites are present because, in the absence of Q^3 sites, 0·66 corresponds to infinite single chains.

Sun *et al.* (2006) investigated how the structural role of Al in synthetic C–A–S–H varies with composition and reaction time. The samples were made from calcium and aluminium nitrates, sodium silicate and sufficient NaOH to maintain the pH at >12, and had Ca/(Si + Al) ratios of 0·86–1·2. XRD showed patterns of tobermorite-type C–S–H. The basal spacings increased consistently with Al content by up to 4 Å, whereas the *hk*0 peaks were unaffected. The authors concluded that the $Al^{[5]}$ and $Al^{[6]}$ occur in the interlayer and not in the central Ca–O layer.

^{27}Al MAS NMR at a higher magnetic field strength than in previous investigations (17·5 T) gave improved resolution. It showed three different $Al^{[4]}$ sites with maxima or shoulders near 58, 66 and 74 ppm and also maxima at 35 and 4 ppm, attributable to $Al^{[5]}$ and $Al^{[6]}$, respectively. The decrease in relative abundance of $Al^{[4]}$ with increasing Ca/(Si + Al) observed by others (Faucon *et al.*, 1999b; Renaudin *et al.*, 2009a, 2009b; Stade and Müller, 1987; Stade *et al.*, 1984) was confirmed, but the reduction was not very pronounced: even at Ca/(Si + Al) = 1·1 to 1·2, 65–92% of the Al as determined from peak intensities was $Al^{[4]}$, depending on the reaction time. The equivalent values from the studies of Faucon *et al.* (1999b) and Renaudin *et al.* (2009a, 2009b) are about 50% and 40%, respectively, at Ca/(Si + Al) = 1·1.

Based on the X-ray diffraction data, the variation of the ^{27}Al and ^{29}Si NMR spectra with composition and reaction time, the known variation of ^{27}Al NMR chemical shifts with polymerisation and NNN composition, and comparison with the data of Komarneni *et al.* (1985) for Al-substituted tobermorite (Komarneni *et al.*, 1985), Sun *et al.* (2006) assigned the observed ^{27}Al peak at 58 ppm to Al at crosslinking Q^3 sites, that at 66 ppm to Al at bridging Q^2 sites with at least some of its charge balanced by interlayer Na^+, and that at 74 ppm to $Al^{[4]}$ at bridging tetrahedral sites charge balanced by interlayer $Al^{[5]}$ and $Al^{[6]}$ through $Al^{[4]}$–O–$Al^{[5]}$ or $Al^{[4]}$–O–$Al^{[6]}$ linkages. The fitted relative abundance of $Al^{[4]}$ increased with increasing magnetic field strength due to second-order quadrupolar effects. This agrees with the relatively large quadrupole frequency observed for some

$Al^{[4]}$ by Faucon *et al.* (1998a). The authors concluded that quantitative use of peak intensities in ^{27}Al NMR spectra remains problematic.

Because of the high magnetic field used, the reported peak maxima in the MAS spectra are only 1–2 ppm less positive than the isotropic chemical shifts. Thus, the peaks near 66 and 74 ppm may be comparable to those with isotropic chemical shifts of 58–67 ppm and 72–76 ppm, respectively, observed by Faucon *et al.* (1998a, 1999a, 1999b). The structural model of Sun *et al.* (2006) does not require the presence of Al in the paired tetrahedra. This agrees with the conclusions of Faucon *et al.* (1999b) for C–A–S–H of all but very low Ca/(Si + $Al^{[4]}$) ratios and with studies on C–A–S–H in cement pastes (Andersen *et al.*, 2004; Richardson, 1999; Richardson *et al.*, 1993, 1994; Skibsted and Andersen, 2013). The peak for Al at Q^3 sites is most prominent at low Ca/(Si + Al) ratios, as would be expected from the work on tobermorite; the prominence of the peak for $Al^{[4]}$ charge balanced by interlayer $Al^{[5]}$ and $Al^{[6]}$ increases with increasing observed $Al^{[5]}$ and $Al^{[6]}$ peak intensity. The total (Ca + Na)/(Si + Al) ratio remains nearly constant with increasing Al content and polymerisation, suggesting that interlayer $Al^{[5]}$ or $Al^{[6]}$ is accompanied by Ca for charge balance. Na contents generally decreased with increasing Al/(Si + Al) and (Ca + Na)/(Si + Al) ratios, suggesting that in this case the substitution of $Na^+ + Al^{3+}$ for Si^{4+} was not the dominant charge balancing mechanism for incorporation of $Al^{[4]}$. This is in contrast with the situation for C–S–H in hardened pastes of Portland cement where it has been shown that the substitution of Si^{4+} by Al^{3+} ions at bridging sites is balanced by alkali ions in the interlayer region (Skibsted and Andersen, 2013). Yang *et al.* (2021) found a correlation between the Na content and the $Al^{[4]}$/Si ratio for synthetic C–A–S–H that had high Ca/Si ratio, but there was no relationship for preparations with low Ca/Si.

6.10.5 Aluminium in C–S–H: summary

The currently available data suggest the following conclusions, which are consistent with the results of studies on cement pastes. Under many conditions of synthesis, most of the Al occurs as $Al^{[4]}$ substituting for Si at the bridging tetrahedral sites. Substitution for Si at paired (nonbridging) sites occurs, if at all, only at very low Ca/(Si + $Al^{[4]}$) ratios. Because of Al substitution, polymerisation of the aluminosilicate chains tends to be greater than in Al-free samples. Al may also occur as $Al^{[5]}$ and $Al^{[6]}$ in the interlayer space, possibly charge balancing $Al^{[4]}$. Al occupancy of main layer Ca sites is most unlikely, as these sites are quite large and are seven-coordinated. Alkalis can charge-balance the substitution of Al^{3+} for Si^{4+} at bridging sites of synthetic C–A–S–H preparations that have quite high target Ca/Si ratios, but further work is needed to establish their role over a wide range of confirmed Ca/Si ratios. Evidence for the incorporation of $Al^{[6]}$ at bridging sites is questionable.

6.10.6 Incorporation of iron(III) and other cations

The structural role of iron (essentially all Fe^{3+} in Portland cements) is less well understood than that of Al. Based on studies similar to those used to investigate Al substitution, Copeland *et al.* (1967) found that Fe^{3+} is incorporated into C–S–H and that at the maximum Fe content, the Si/Fe ratio is about 6 (essentially the same as the maximum Si/Al ratio). On the basis of analysed Ca/Si and Fe/Si ratios, they proposed a coupled substitution of $2Fe^{3+}$ for $Ca^{2+} + Si^{4+}$, with equal amounts of Fe in the tetrahedral and interlayer sites. Most analyses of natural tobermorites show <0·5% of Fe_2O_3 (Mitsuda and Taylor, 1978); attempts to make a synthetic tobermorite containing Fe gave inconclusive results (Diamond *et al.*, 1966). Stade *et al.* (1984b) prepared 'C–S–H(di, poly)' at 80 °C in the presence of Fe^{3+}. They concluded that Fe could be incorporated up to a limiting Fe_2O_3/SiO_2 ratio of at least 0·09. The incorporation did not affect the silicate anion structure; studies using electron spin resonance and Mössbauer spectroscopy showed that over 99% of the Fe^{3+} was octahedrally coordinated. Faucon *et al.* (1998b) reported briefly on a study of Fe incorporation in tobermorite-type C–S–H using XRD, ^{29}Si NMR, Mössbauer spectroscopy and TEM. They concluded that Fe could replace Ca in the main layers and interlayer sites up to an Fe/Si limit of 0·33.

Pytel and Malolepszy (1997) claimed that hydrothermally prepared C–S–H can accommodate up to 6% of MgO. A study by sorption isotherms and ^{29}Si NMR (Moulin *et al.*, 1999) suggests that both Pb^{2+} and Zn^{2+} are incorporated into the C–S–H structure at the 10^{-4} mol g^{-1} range by way of formation of Si-O-M linkages. It was suggested that zinc could occupy tetrahedral sites at chain ends.

As with Al^{3+}, neither Mg^{2+} nor Fe^{3+} is likely to substitute for Ca^{2+} in the main layers at more than a very minor level, as these ions are too small to occupy the seven-coordinate sites, but they could possibly enter the interlayers.

6.10.7 Anion substitution and exchange: SO_4^{2-} and Cl^-

Copeland *et al.* (1967) investigated the reactions of SO_4^{2-} with synthetic C–S–H by methods similar to those used with Al^{3+} and Fe^{3+}. They reported a maximum Si/S ratio in the C–S–H of about 6, comparable to the values for Al and Fe, and proposed that S^{6+} substitutes for Si^{4+} at tetrahedral sites. This is, however, improbable, as Si–O–S linkages are readily hydrolysable and, to our knowledge, do not occur in the presence of water.

Many studies show that the C–S–H of cement pastes contains sulfate, SEM or electron probe microanalysis typically indicating S/Ca ratios of $\approx$0·02 for material cured at ordinary temperature, or up to $\approx$ 0·10 directly after curing at elevated temperatures. The sulfate is very loosely bound and is probably sorbed, along with an equivalent quantity of Ca^{2+} (Fu *et al.*, 1994; Odler, 1981). A study of the interaction of sulfate with tobermorite-type C–S–H (Divet and Randriambololona, 1998) showed that the uptake followed a Langmuir isotherm and that it was favoured by increases in temperature and pH.

Chloride has been studied in conjunction with C–S–H because of the potential role of C–S–H in binding chloride in cement pastes and thus retarding its rate of transport through concrete. A study of the interaction of C–S–H with Ca-saturated alkali chloride solutions by exchange isotherms and ^{35}Cl NMR (Yu, 1990) suggests that the Cl^- exchange capacity of C–S–H increases with Ca/Si ratio and decreases with increasing Al content. The Cl^- occurs as outer sphere complexes that are readily removed by washing with deionised water. The results also indicate that Cl sorption may be substantially enhanced by cosorption with cations. The latter conclusion agrees well with those of Viallis *et al.* (1999) on Na^+ and Cs^+.

6.11. Relation of synthetic C–S–H to the C–S–H in cement pastes

Tobermorite-type C–S–H is an excellent model for the material present in products high in slag or pozzolanas and in some kinds of autoclaved materials. The alkali-free high-lime varieties of C–S–H offer the closest available model for the C–S–H that is present in tricalcium silicate pastes. The wide, bimodal distribution of Ca/Si ratio in early age pastes of neat Portland cement (Richardson, 1999) suggests that local structures of both types may be present. A form of C–S–H with purely monomeric silicate ions is present as an initial product and is the sole product until the end of the induction period. Many of the materials described in this chapter are much more highly ordered than those formed in pure Portland cement pastes and caution is needed in extrapolating between them.

Table 6.4 Model structure for a hypothetical C–S–H(I) dimer derived from a clinotobermorite that had bridging tetrahedral sites that were adjacent to one another. The CIF file data block is T2_ac

Systematic name:	pentacalcium tetrasilicate dihydroxide tetrahydrate
Structural formula:	$Ca_5(Si_4O_{12}(OH)_2)(H_2O)_4$
Chemical formula (sum):	H_{10} Ca_5 O_{18} Si_4
a	11·35 Å
b	7·3 Å
c	21·5 Å
α	90°
β	98·4°
γ	90°
Z	4
Space group:	No. 15, *C* 1 2/*c* 1
Unique axis:	*b*
Cell choice:	1
Symmetry equivalent positions:	(0, 0, 0)+ (½, ½, 0)+
8*f*:	(1) + *x*, + *y*, + *z* (2) –*x*, + *y*, ½–*z* (3) –*x*, –*y*, –*z* (4) + *x*, –*y*, ½+*z*
4*e*:	0, + *y*, ¼ 0, –*y*, ¾

Atom	Label	Position	Occ.	*x*/*a*	*y*/*b*	*z*/*c*
Ca	Ca1	8*f*	1	0·8667	0·375	0·4532
Ca	Ca2	8*f*	1	0·8667	0·875	0·4532
Ca	Ca3	4*e*	1	0	0·125	0·75
O	O1	8*f*	0·5, 0·5 OH	0·9167	0·9529	0·6666
O	O2	8*f*	0·5, 0·5 OH	0·9167	0·3077	0·6666
O	O3	8*f*	1	0·7692	0·375	0·5587
O	O4	8*f*	1	0·7692	0·875	0·5587
O	O5	8*f*	1	0	0·375	0·5587
O	O6	8*f*	1	0	0·875	0·5587
O	O7	8*f*	1	0·8919	0·625	0·6213
O	Wat10	8*f*	1	0·155	0·125	0·6666
O	Wat11	8*f*	1	0·1579	0·625	0·6666
Si	Si1	8*f*	1	0·8929	0·837	0·6
Si	Si3	8*f*	1	0·8958	0·4167	0·6

Ca/Si = 1·25; D_x = 2·302 g cm^{-3}; av. Ca3–O = 2·430 Å; Ca3–O polyhedral volume = 16·57 Å^3; d_{002} = 10·63 Å (i.e. c sin β).

Table 6.5 Model structure for a hypothetical C–S–H(I) dimer derived from a clinotobermorite structure that had staggered silicate chains. The CIF file data block is T2_sc_LS1 (continued on next page)

Systematic name:	hexacalcium tetrasilicate tetrahydrate
Structural formula:	$Ca_6(Si_4O_{14})(H_2O)_4$
Chemical formula (sum):	H_8 Ca_6 O_{18} Si_4
a	11·35 Å
b	7·3 Å
c	20·25 Å
α	90°
β	97·28°
γ	90°
Z	4
Space group:	No. 14, P 1 2_1/*c* 1
Unique axis:	*b*
Cell choice:	1
Symmetry equivalent positions:	(0, 0, 0)+
4*e*:	(1) + *x*, + *y*, + *z* (2) −*x*, ½+*y*, ½−*z* (3) −*x*, −*y*, −*z* (4) + *x*, ½−*y*, ½+*z*

Atom	Label	Position	Occ.	*x*/*a*	*y*/*b*	*z*/*c*
Ca	Ca1	4*e*	1	0·8667	0·375	0·4503
Ca	Ca2	4*e*	1	0·8667	0·875	0·4503
Ca	Ca3	4*e*	1	0	0·125	0·77
Ca	Ca1A	4*e*	1	0·3667	0·875	0·4503
Ca	Ca2A	4*e*	1	0·3667	0·375	0·4503
Ca	Ca3A	4*e*	1	0·5	0·625	0·77
O	O1	4*e*	1	0·9167	0·9529	0·6768
O	O2	4*e*	1	0·9167	0·3077	0·6768
O	O3	4*e*	1	0·7692	0·375	0·5623
O	O4	4*e*	1	0·7692	0·875	0·5623
O	O5	4*e*	1	0	0·375	0·5623
O	O6	4*e*	1	0	0·875	0·5623
O	O7	4*e*	1	0·8919	0·625	0·6288
O	Wat10	4*e*	1	0·13	0·125	0·6768
O	Wat11	4*e*	1	0·1579	0·625	0·6768
O	Wat10A	4*e*	1	0·63	0·625	0·6768
O	Wat11A	4*e*	1	0·6579	0·125	0·6768

Table 6.5 Continued

Atom	Label	Position	Occ.	*x/a*	*y/b*	*z/c*
O	O1A	4e	1	0·4167	0·4529	0·6768
O	O2A	4e	1	0·4167	0·8077	0·6768
O	O3A	4e	1	0·2692	0·875	0·5623
O	O4A	4e	1	0·2692	0·375	0·5623
O	O5A	4e	1	0·5	0·875	0·5623
O	O6A	4e	1	0·5	0·375	0·5623
O	O7A	4e	1	0·3919	0·125	0·6288
Si	Si1	4e	1	0·8929	0·837	0·6063
Si	Si3	4e	1	0·8958	0·4167	0·6063
Si	Si1A	4e	1	0·3929	0·337	0·6063
Si	Si3A	4e	1	0·3958	0·9167	0·6063

Ca/Si = 1·5; H_2O:Si = 1·00; D_x = 2·590 g cm^{-3}; av. Ca3/3A–O = 2·455 Å and Ca3/3A–O polyhedral volume = 21·58 $Å^3$; d_{002} = 10·04 Å (i.e. (*c*/2)sin *β*).

Table 6.6 Model structure for a hypothetical C–S–H(I) dimer derived from a clinotobermorite structure that had staggered silicate chains. The CIF file data block is T2_sc_LS2 (continued on next page)

Systematic name:	hexacalcium tetrasilicate tetrahydrate
Structural formula:	$Ca_6(Si_4O_{14})(H_2O)_4$
Chemical formula (sum):	H_8 Ca_6 O_{18} Si_4
a	11·35 Å
b	7·3 Å
c	19·0 Å
α	90°
β	98·4°
γ	90°
Z	4
Space group:	No. 14, P 1 2_1/*c* 1
Unique axis:	*b*
Cell choice:	1
Symmetry equivalent positions:	(0, 0, 0)+
4e:	(1) + *x*, + *y*, + *z* (2) –*x*, ½+*y*, ½–*z* (3) –*x*, –*y*, –*z* (4) + *x*, ½–*y*, ½+*z*

Table 6.6 Continued

Atom	Label	Position	Occ.	*x/a*	*y/b*	*z/c*
Ca	Ca1	4*e*	1	0·8667	0·375	0·447
Ca	Ca2	4*e*	1	0·8667	0·875	0·447
Ca	Ca3	4*e*	1	0·75	0·125	0·25
Ca	Ca1A	4*e*	1	0·3667	0·875	0·447
Ca	Ca2A	4*e*	1	0·3667	0·375	0·447
Ca	Ca3A	4*e*	1	0·75	0·625	0·25
O	O1	4*e*	1	0·9167	0·9529	0·6885
O	O2	4*e*	1	0·9167	0·3077	0·6885
O	O3	4*e*	1	0·7692	0·375	0·5664
O	O4	4*e*	1	0·7692	0·875	0·5664
O	O5	4*e*	1	0	0·375	0·5664
O	O6	4*e*	1	0	0·875	0·5664
O	O7	4*e*	1	0·8919	0·625	0·6373
O	Wat8	4*e*	1	0·145	0·125	0·6885
O	Wat9	4*e*	1	0·145	0·625	0·6885
O	Wat8A	4*e*	1	0·645	0·625	0·6885
O	Wat9A	4*e*	1	0·645	0·125	0·6885
O	O1A	4*e*	1	0·4167	0·4529	0·6885
O	O2A	4*e*	1	0·4167	0·8077	0·6885
O	O3A	4*e*	1	0·2692	0·875	0·5664
O	O4A	4*e*	1	0·2692	0·375	0·5664
O	O5A	4*e*	1	0·5	0·875	0·5664
O	O6A	4*e*	1	0·5	0·375	0·5664
O	O7A	4*e*	1	0·3919	0·125	0·6373
Si	Si1	4*e*	1	0·8929	0·837	0·6133
Si	Si2	4*e*	1	0·8958	0·4167	0·6133
Si	Si1A	4*e*	1	0·3929	0·337	0·6133
Si	Si2A	4*e*	1	0·3958	0·9167	0·6133

Ca/Si = 1·5; H_2O:Si = 1·00; D_x = 2·767 g cm^{-3}; av. Ca3–O = 2·398 Å; Ca3–O polyhedral volume = 17·44 Å^3; d_{002} = 9·40 Å (i.e. $(c/2)\sin\beta$).

Table 6.7 Model structure for a theoretical 14 Å clinotobermorite that has staggered silicate chains and NO interlayer Ca atoms. The CIF file data block is T∞_sc_noCa_LS1 (continued on next page)

Systematic name:	tetracalcium hexasilicate tetrahydroxide tetrahydrate
Structural formula:	$Ca_4(Si_6O_{14}(OH)_4)(H_2O)_4$
Chemical formula (sum):	H_{12} Ca_4 O_{22} Si_6
a	11·35 Å
b	7·3 Å
c	26·715 Å
α	90°
β	95·5°
γ	90°
Z	4
Space group:	No. 14, P 1 $2_1/c$ 1
Unique axis:	*b*
Cell choice:	1
Symmetry equivalent positions:	(0, 0, 0)+
4e:	(1) + *x*, + *y*, + *z* (2) −*x*, ½+*y*, ½−*z* (3) −*x*, −*y*, −*z* (4) + *x*, ½−*y*, ½+*z*

Atom	Label	Position	Occ.	*x/a*	*y/b*	*z/c*
Ca	Ca1	4*e*	1	0·8667	0·375	0·4623
Ca	Ca2	4*e*	1	0·8667	0·875	0·4623
Ca	Ca1A	4*e*	1	0·3667	0·875	0·4623
Ca	Ca2A	4*e*	1	0·3667	0·375	0·4623
O	O1	4*e*	1	0·9167	0·9529	0·6341
O	O2	4*e*	1	0·9167	0·3077	0·6341
O	O3	4*e*	1	0·7692	0·375	0·5472
O	O4	4*e*	1	0·7692	0·875	0·5472
O	O5	4*e*	1	0	0·375	0·5472
O	O6	4*e*	1	0	0·875	0·5472
O	O7	4*e*	1	0·8919	0·625	0·5976
O	OH8	4*e*	1	0	0·125	0·7122
O	Wat9	4*e*	1	0	0·625	0·69
O	OH10	4*e*	1	0·13	0·125	0·6341
O	Wat11	4*e*	1	0·1579	0·625	0·6341
O	Wat9A	4*e*	1	0·5	0·125	0·69
O	OH10A	4*e*	1	0·63	0·625	0·6341
O	Wat11A	4*e*	1	0·6579	0·125	0·6341

Table 6.7 Continued

Atom	Label	Position	Occ.	*x/a*	*y/b*	*z/c*
O	O1A	4*e*	1	0·4167	0·4529	0·6341
O	O2A	4*e*	1	0·4167	0·8077	0·6341
O	O3A	4*e*	1	0·2692	0·875	0·5472
O	O4A	4*e*	1	0·2692	0·375	0·5472
O	O5A	4*e*	1	0·5	0·875	0·5472
O	O6A	4*e*	1	0·5	0·375	0·5472
O	O7A	4*e*	1	0·3919	0·125	0·5976
O	OH8A	4*e*	1	0·5	0·625	0·7122
Si	Si1	4*e*	1	0·8929	0·837	0·5806
Si	Si2	4*e*	1	0	0·125	0·652
Si	Si3	4*e*	1	0·8958	0·4167	0·5806
Si	Si1A	4*e*	1	0·3929	0·337	0·5806
Si	Si2A	4*e*	1	0·5	0·625	0·652
Si	Si3A	4*e*	1	0·3958	0·9167	0·5806

Ca/Si = 0·67; H_2O/Si = 1·00; D_x = 2·089 g cm^{-3}; LS = 13·30 Å.

Table 6.8 Model structure for a theoretical 14 Å clinotobermorite that has staggered silicate chains and NO interlayer Ca atoms. The CIF file data block is T∞_sc_noCa_LS2 (continued on next page)

Systematic name:	tetracalcium hexasilicate tetrahydroxide trihydrate
Structural formula:	$Ca_4(Si_6O_{14}(OH)_4)(H_2O)_3$
Chemical formula (sum):	H_{10} Ca_4 O_{21} Si_6
a	11·35 Å
b	7·3 Å
c	28·43 Å
α	90°
β	95·5°
γ	90°
Z	4
Space group:	No. 14, P 1 2_1/*c* 1
Unique axis:	*b*
Cell choice:	1
Symmetry equivalent positions:	(0, 0, 0)+
4e:	(1) + *x*, + *y*, + *z* (2) −*x*, ½+*y*, ½−*z* (3) −*x*, −*y*, −*z* (4) + *x*, ½−*y*, ½+*z*

Table 6.8 Continued

Atom	Label	Position	Occ.	*x/a*	*y/b*	*z/c*
Ca	Ca1	4*e*	1	0.8667	0·375	0·4646
Ca	Ca2	4*e*	1	0·8667	0·875	0·4646
Ca	Ca1A	4*e*	1	0·3667	0·875	0·4646
Ca	Ca2A	4*e*	1	0·3667	0·375	0·4646
O	O1	4*e*	1	0·9167	0·9529	0·626
O	O2	4*e*	1	0·9167	0·3077	0·626
O	O3	4*e*	1	0·7692	0·375	0·5444
O	O4	4*e*	1	0·7692	0·875	0·5444
O	O5	4*e*	1	0	0·375	0·5444
O	O6	4*e*	1	0	0·875	0·5444
O	O7	4*e*	1	0·8919	0·625	0·5917
O	OH8	4*e*	1	0	0·125	0·6994
O	Wat9	4*e*	0·5	0	0·5	0·71
O	OH10	4*e*	1	0·13	0·125	0·626
O	Wat11	4*e*	1	0·1579	0·625	0·626
O	Wat9A	4*e*	0·5	0·5	0	0·71
O	OH10A	4*e*	1	0·63	0·625	0·626
O	Wat11A	4*e*	1	0·6579	0·125	0·626
O	O1A	4*e*	1	0·4167	0·4529	0·626
O	O2A	4*e*	1	0·4167	0·8077	0·626
O	O3A	4*e*	1	0·2692	0·875	0·5444
O	O4A	4*e*	1	0·2692	0·375	0·5444
O	O5A	4*e*	1	0·5	0·875	0·5444
O	O6A	4*e*	1	0·5	0·375	0·5444
O	O7A	4*e*	1	0·3919	0·125	0·5917
O	OH8A	4*e*	1	0·5	0·625	0·6994
Si	Si1	4*e*	1	0·8929	0·837	0·5757
Si	Si2	4*e*	1	0	0·125	0·6428
Si	Si3	4*e*	1	0·8958	0·4167	0·5757
Si	Si1A	4*e*	1	0·3929	0·337	0·5757
Si	Si2A	4*e*	1	0·5	0·625	0·6428
Si	Si3A	4*e*	1	0·3958	0·9167	0·5757

Ca/Si = 0·67; H_2O/Si = 0·83; D_x = 1·912 g cm^{-3}; LS = 14·15 Å.

Table 6.9 Model structure for a hypothetical C–S–H(I) 'pentamer' derived from a 14 Å clinotobermorite structure that had staggered silicate chains. The CIF file data block is T5_14sc (continued on next page)

Systematic name:	pentacalcium pentasilicate dihydroxide tetrahemihydrate
Structural formula:	$Ca_5(Si_5O_{14}(OH)_2)(H_2O)_{4\cdot5}$
Chemical formula (sum):	H_{11} Ca_5 $O_{20\cdot5}$ Si_5
a	11·35 Å
b	7·3 Å
c	23·9 Å
α	90°
β	95·5°
γ	90°
Z	4
Space group:	No. 1 (setting 2), *C* 1
Symmetry equivalent positions:	(0, 0, 0)+
2*a*:	(1) + *x*, + *y*, + *z* (2) ½+*x*, ½+*y*, *z*

Atom	Label	Position	Occ.	*x/a*	*y/b*	*z/c*
Ca	Ca1_1	2*a*	1	0·3667	0·125	0·9578
Ca	Ca1_2	2*a*	1	0·3667	0·375	0·5603
Ca	Ca1_3	2*a*	1	0·3667	0·625	0·9578
Ca	Ca1_4	2*a*	1	0·3667	0·875	0·5603
Ca	Ca1_5	2*a*	1	0·6333	0·125	0·6447
Ca	Ca1_6	2*a*	1	0·6333	0·375	0·0422
Ca	Ca1_7	2*a*	1	0·6333	0·625	0·6447
Ca	Ca1_8	2*a*	1	0·6333	0·875	0·0422
Ca	Ca2_1	2*a*	0·5	0·5	0·125	0·3013
Ca	Ca2_3	2*a*	0·5	0·5	0·625	0·3013
Ca	Ca3_1	2*a*	0·5	0·25	0·375	0·8
Ca	Ca3_2	2*a*	0·5	0·25	0·875	0·8
O	O1_1	2*a*	1	0·4167	0·0471	0·1498
O	O1_3	2*a*	1	0·4167	0·6923	0·1498
O	O1_5	2*a*	1	0·5833	0·1923	0·4527
O	O1_7	2*a*	1	0·5833	0·5471	0·4527
O	O2_1	2*a*	1	0·7308	0·125	0·5497
O	O2_2	2*a*	1	0·7308	0·375	0·9472
O	O2_3	2*a*	1	0·7308	0·625	0·5497

Table 6.9 Continued

Atom	Label	Position	Occ.	*x/a*	*y/b*	*z/c*
O	O2_4	2*a*	1	0·7308	0·875	0·9472
O	O2_5	2*a*	1	0·2692	0·125	0·0528
O	O2_6	2*a*	1	0·2692	0·375	0·6553
O	O2_7	2*a*	1	0·2692	0·625	0·0528
O	O2_8	2*a*	1	0·2692	0·875	0·6553
O	O3_1	2*a*	1	0·5	0·125	0·0528
O	O3_2	2*a*	1	0·5	0·125	0·5497
O	O3_4	2*a*	1	0·5	0·375	0·3652
O	O3_5	2*a*	1	0·5	0·375	0·6553
O	O3_6	2*a*	1	0·5	0·375	0·9472
O	O3_7	2*a*	1	0·5	0·625	0·0528
O	O3_8	2*a*	1	0·5	0·625	0·5497
O	O3_9	2*a*	1	0·5	0·875	0·2373
O	O3_11	2*a*	1	0·5	0·875	0·6553
O	O3_12	2*a*	1	0·5	0·875	0·9472
O	O4_1	2*a*	1	0·3919	0·125	0·7117
O	O4_2	2*a*	1	0·3919	0·375	0·1092
O	O4_3	2*a*	1	0·6081	0·625	0·8908
O	O4_4	2*a*	1	0·6081	0·875	0·4933
O	OH5_1	2*a*	1	0·37	0·375	0·4527
O	OH5_2	2*a*	1	0·63	0·875	0·1498
O	OH1_2	2*a*	0·5, 0·5O	0·4167	0·4529	0·7523
O	OH1_4	2*a*	0·5, 0·5O	0·4167	0·8077	0·7523
O	OH1_6	2*a*	0·5, 0·5O	0·5833	0·3077	0·8502
O	OH1_8	2*a*	0·5, 0·5O	0·5833	0·9529	0·8502
O	W3_3	2*a*	0·5	0·5	0·375	0·2373
O	W3_10	2*a*	0·5	0·5	0·875	0·3652
O	W6_1	2*a*	1	0·3421	0·875	0·4527
O	W6_2	2*a*	1	0·6579	0·375	0·1498
O	W7_1	2*a*	1	0·355	0·125	0·8502
O	W7_2	2*a*	1	0·355	0·625	0·8502
O	W7_3	2*a*	1	0·645	0·125	0·7523
O	W7_4	2*a*	1	0·645	0·625	0·7523
O	W8_1	2*a*	0·5	0·212	0·625	0·3013
O	W8_2	2*a*	0·5	0·288	0·125	0·3013

Table 6.9 Continued

Atom	Label	Position	Occ.	*x/a*	*y/b*	*z/c*
O	W8_3	2*a*	0·5	0·712	0·625	0·3013
O	W8_4	2*a*	0·5	0·788	0·125	0·3013
Si	Si1_1	2*a*	1	0·3929	0·163	0·09
Si	Si1_2	2*a*	1	0·3929	0·337	0·6925
Si	Si1_3	2*a*	1	0·6071	0·663	0·5125
Si	Si1_4	2*a*	1	0·6071	0·837	0·91
Si	Si2_1	2*a*	1	0·3958	0·5833	0·0907
Si	Si2_2	2*a*	1	0·3958	0·9167	0·6932
Si	Si2_3	2*a*	1	0·6042	0·0833	0·5118
Si	Si2_4	2*a*	1	0·6042	0·4167	0·9093
Si	Si3_1	2*a*	1	0·5	0·375	0·4326
Si	Si3_2	2*a*	1	0·5	0·875	0·1699

Ca/Si = 1·0; H_2O:Si = 1·10; D_x = 2·291 g cm^{-3}; av. Ca3–O = 2·405 Å; Ca3–O polyhedral volume = 17·477 $Å^3$; av. Ca2–O = 2·389 Å; Ca2–O polyhedral volume = 17·815 $Å^3$; d_{002} = 11·895 Å (i.e. (c/2)sin β).
Ca3 is the interlayer Ca between the dimeric silicate chains and Ca2 is between the chains that are nominally of infinite length (nominally because there are quite possibly tetrahedral bridging sites that are vacant).

Table 6.10 Model structure for a hypothetical C–S–H(I) 'pentamer' derived from an 11 Å clinotobermorite structure that had staggered silicate chains. The CIF file data block is T5_11sc (continued on next page)

Systematic name:	pentacalcium pentasilicate dihydroxide trihemihydrate
Structural formula:	$Ca_5(Si_5O_{14}(OH)_2)(H_2O)_{3.5}$
Chemical formula (sum):	H_9 Ca_5 $O_{19.5}$ Si_5
a	11·35 Å
b	7·3 Å
c	20·84 Å
α	90°
β	97·28°
γ	90°
Z	4
Space group:	No. 1 (setting 2), *C* 1
Symmetry equivalent positions:	(0, 0, 0)+
2*a*:	(1) + *x*, + *y*, + *z* (2) ½+*x*, ½+*y*, *z*

Table 6.10 Continued

Atom	Label	Position	Occ.	*x/a*	*y/b*	*z/c*
Ca	Ca1_1	2*a*	1	0·3667	0·125	0·9517
Ca	Ca1_2	2*a*	1	0·3667	0·375	0·4958
Ca	Ca1_3	2*a*	1	0·3667	0·625	0·9517
Ca	Ca1_4	2*a*	1	0·3667	0·875	0·4958
Ca	Ca1_5	2*a*	1	0·6333	0·125	0·5925
Ca	Ca1_6	2*a*	1	0·6333	0·375	0·0483
Ca	Ca1_7	2*a*	1	0·6333	0·625	0·5925
Ca	Ca1_8	2*a*	1	0·6333	0·875	0·0483
Ca	Ca2_1	2*a*	0·5	0·317	0·375	0·2177
Ca	Ca2_2	2*a*	0·5	0·683	0·875	0·3265
Ca	Ca3_1	2*a*	0·5	0·25	0·375	0·775
Ca	Ca3_2	2*a*	0·5	0·25	0·875	0·775
O	O1_1	2*a*	1	0·4167	0·0471	0·1718
O	O1_3	2*a*	1	0·4167	0·6923	0·1718
O	O1_5	2*a*	1	0·5833	0·1923	0·3723
O	O1_7	2*a*	1	0·5833	0·5471	0·3723
O	O2_1	2*a*	1	0·7308	0·125	0·4836
O	O2_2	2*a*	1	0·7308	0·375	0·9395
O	O2_3	2*a*	1	0·7308	0·625	0·4836
O	O2_4	2*a*	1	0·7308	0·875	0·9395
O	O2_5	2*a*	1	0·2692	0·125	0·0605
O	O2_6	2*a*	1	0·2692	0·375	0·6047
O	O2_7	2*a*	1	0·2692	0·625	0·0605
O	O2_8	2*a*	1	0·2692	0·875	0·6047
O	O3_1	2*a*	1	0·5	0·125	0·0605
O	O3_2	2*a*	1	0·5	0·125	0·4836
O	O3_5	2*a*	1	0·5	0·375	0·6047
O	O3_6	2*a*	1	0·5	0·375	0·9395
O	O3_7	2*a*	1	0·5	0·625	0·0605
O	O3_8	2*a*	1	0·5	0·625	0·4836
O	O3_11	2*a*	1	0·5	0·875	0·6047
O	O3_12	2*a*	1	0·5	0·875	0·9395
O	O4_1	2*a*	1	0·3919	0·125	0·6693

Table 6.10 Continued

Atom	Label	Position	Occ.	*x*/*a*	*y*/*b*	*z*/*c*
O	O4_2	2*a*	1	0·3919	0·375	0·1252
O	O4_3	2*a*	1	0·6081	0·625	0·8748
O	O4_4	2*a*	1	0·6081	0·875	0·419
O	OH5_1	2*a*	1	0·37	0·375	0·3723
O	OH5_2	2*a*	1	0·63	0·875	0·1718
O	OH1_2	2*a*	0·5, 0·5O	0·4167	0·4529	0·716
O	OH1_4	2*a*	0·5, 0·5O	0·4167	0·8077	0·716
O	OH1_6	2*a*	0·5, 0·5O	0·5833	0·3077	0·8282
O	OH1_8	2*a*	0·5, 0·5O	0·5833	0·9529	0·8282
O	O8_1	2*a*	1	0	0·375	0·2721
O	O8_2	2*a*	1	0	0·875	0·2721
O	W6_1	2*a*	1	0·3421	0·875	0·3723
O	W6_2	2*a*	1	0·6579	0·375	0·1718
O	W7_1	2*a*	1	0·355	0·125	0·8282
O	W7_2	2*a*	1	0·355	0·625	0·8282
O	W7_3	2*a*	1	0·645	0·125	0·716
O	W7_4	2*a*	1	0·645	0·625	0·716
O	W9_1	2*a*	0·5	0·25	0·625	0·2721
O	W9_2	2*a*	0·5	0·75	0·625	0·2721
Si	Si1_1	2*a*	1	0·3929	0·163	0·1033
Si	Si1_2	2*a*	1	0·3929	0·337	0·6474
Si	Si1_3	2*a*	1	0·6071	0·663	0·4409
Si	Si1_4	2*a*	1	0·6071	0·837	0·8967
Si	Si2_1	2*a*	1	0·3958	0·5833	0·1033
Si	Si2_2	2*a*	1	0·3958	0·9167	0·6474
Si	Si2_3	2*a*	1	0·6042	0·0833	0·4409
Si	Si2_4	2*a*	1	0·6042	0·4167	0·8967
Si	Si3_1	2*a*	1	0·5	0·375	0·3493
Si	Si3_2	2*a*	1	0·5	0·875	0·1948

Ca/Si = 1·0; H_2O:Si = 0·90; D_x = 2·567 g cm^{-3}; av. Ca3–O = 2·400 Å; Ca3–O polyhedral volume = 17·40 $Å^3$; av. Ca2–O = 2·425 Å; Ca2–O polyhedral volume = 20·32 $Å^3$; d_{002} = 10·34 Å (i.e. (*c*/2)sin *β*).

Ca3 is the interlayer Ca between the dimeric silicate chains and Ca2 is between the chains that are nominally of infinite length (nominally because there are quite possibly tetrahedral bridging sites that are vacant).

Table 6.11 Model structure for a hypothetical C–S–H(I) 'trimer' C-S-H ($SOF_{BT} = 0·25$; MCL = 3) derived from a 14 Å clinotobermorite structure that had staggered silicate chains. The CIF file data block is T3_14sc (continued on next page)

Systematic name:	undecacalcium nonasilicate dihydroxide octahemihydrate
Structural formula:	$Ca_{11}(Si_9O_{28}(OH)_2)(H_2O)_{8·5}$
Chemical formula (sum):	$H_{19}\ Ca_{11}\ O_{38·5}\ Si_9$
a	11·35 Å
b	7·3 Å
c	42·9 Å
α	90°
β	95·5°
γ	90°
Z	4
Space group:	No. 1 (setting 2), *C* 1
Symmetry equivalent positions:	(0, 0, 0)+
2*a*:	(1) + *x*, + *y*, + *z* (2) ½+*x*, ½+*y*, *z*

Atom	Label	Position	Occ.	x/a	y/b	z/c
Ca	Ca1_1	2*a*	1	0·6333	0·375	0·5806
Ca	Ca1_2	2*a*	1	0·6333	0·375	0·0235
Ca	Ca1_3	2*a*	1	0·6333	0·125	0·3591
Ca	Ca1_4	2*a*	1	0·6333	0·125	0·8021
Ca	Ca1_5	2*a*	1	0·3667	0·625	0·5336
Ca	Ca1_6	2*a*	1	0·3667	0·625	0·9765
Ca	Ca1_7	2*a*	1	0·3667	0·875	0·755
Ca	Ca1_8	2*a*	1	0·3667	0·875	0·3121
Ca	Ca2_1	2*a*	1	0·3667	0·125	0·9765
Ca	Ca2_2	2*a*	1	0·3667	0·125	0·5336
Ca	Ca2_3	2*a*	1	0·3667	0·375	0·755
Ca	Ca2_4	2*a*	1	0·3667	0·375	0·3121
Ca	Ca2_5	2*a*	1	0·6333	0·875	0·5806
Ca	Ca2_6	2*a*	1	0·6333	0·875	0·0235
Ca	Ca2_7	2*a*	1	0·6333	0·625	0·3591
Ca	Ca2_8	2*a*	1	0·6333	0·625	0·8021
Ca	Ca3_1	2*a*	0·5	0·5	0·125	0·1678
Ca	Ca3_2	2*a*	0·5	0·5	0·625	0·1678
Ca	Ca4_1	2*a*	0·8333	0·25	0·375	0·8886

Table 6.11 Continued

Atom	Label	Position	Occ.	x/a	y/b	z/c
Ca	Ca4_2	*2a*	0·8333	0·25	0·375	0·4457
Ca	Ca4_3	*2a*	0·8333	0·25	0·875	0·4457
Ca	Ca4_4	*2a*	0·8333	0·25	0·875	0·8886
Ca	Ca4_5	*2a*	0·8333	0·25	0·125	0·6679
Ca	Ca4_6	*2a*	0·8333	0·25	0·625	0·6679
O	O1_1	*2a*	1	0·5833	0·5471	0·2521
O	O1_2	*2a*	1	0·4167	0·0471	0·0835
O	O2_1	*2a*	1	0·5833	0·1923	0·2521
O	O2_2	*2a*	1	0·4167	0·6923	0·0835
O	O3_1	*2a*	1	0·7308	0·375	0·5277
O	O3_2	*2a*	1	0·7308	0·375	0·9706
O	O3_3	*2a*	1	0·7308	0·125	0·7492
O	O3_4	*2a*	1	0·7308	0·125	0·3063
O	O3_5	*2a*	1	0·2692	0·625	0·5865
O	O3_6	*2a*	1	0·2692	0·625	0·0294
O	O3_7	*2a*	1	0·2692	0·875	0·3651
O	O3_8	*2a*	1	0·2692	0·875	0·808
O	O4_1	*2a*	1	0·2692	0·125	0·5865
O	O4_2	*2a*	1	0·2692	0·125	0·0294
O	O4_3	*2a*	1	0·2692	0·375	0·3651
O	O4_4	*2a*	1	0·2692	0·375	0·808
O	O4_5	*2a*	1	0·7308	0·875	0·9706
O	O4_6	*2a*	1	0·7308	0·875	0·5277
O	O4_7	*2a*	1	0·7308	0·625	0·3063
O	O4_8	*2a*	1	0·7308	0·625	0·7492
O	O5_1	*2a*	1	0·5	0·875	0·808
O	O5_2	*2a*	1	0·5	0·875	0·3651
O	O5_3	*2a*	1	0·5	0·375	0·9706
O	O5_4	*2a*	1	0·5	0·375	0·5277
O	O5_5	*2a*	1	0·5	0·125	0·3063
O	O5_6	*2a*	1	0·5	0·125	0·7492
O	O5_7	*2a*	1	0·5	0·625	0·0294
O	O5_8	*2a*	1	0·5	0·625	0·5865
O	O6_1	*2a*	1	0·5	0·375	0·3651
O	O6_2	*2a*	1	0·5	0·375	0·808

Table 6.11 Continued

Atom	Label	Position	Occ.	x/a	y/b	z/c
O	O6_3	2*a*	1	0·5	0·875	0·5277
O	O6_4	2*a*	1	0·5	0·875	0·9706
O	O6_5	2*a*	1	0·5	0·625	0·7492
O	O6_6	2*a*	1	0·5	0·625	0·3063
O	O6_7	2*a*	1	0·5	0·125	0·5865
O	O6_8	2*a*	1	0·5	0·125	0·0294
O	O7_1	2*a*	1	0·3919	0·125	0·3966
O	O7_2	2*a*	1	0·3919	0·125	0·8393
O	O7_3	2*a*	1	0·6081	0·625	0·4963
O	O7_4	2*a*	1	0·6081	0·625	0·9392
O	O7_5	2*a*	1	0·6081	0·875	0·7177
O	O7_6	2*a*	1	0·6081	0·875	0·2748
O	O7_7	2*a*	1	0·3919	0·375	0·0608
O	O7_8	2*a*	1	0·3919	0·375	0·6179
O	O8_1	2*a*	1	0·5	0·875	0·1321
O	O8_2	2*a*	1	0·5	0·375	0·2035
OH	OH1_1	2*a*	OH 0·1667 O 0·8333	0·4167	0·4529	0·4192
OH	OH1_2	2*a*	OH 0·1667 O 0·8333	0·4167	0·8077	0·4192
OH	OH1_3	2*a*	OH 0·1667 O 0·8333	0·5833	0·3077	0·4737
OH	OH1_4	2*a*	OH 0·1667 O 0·8333	0·5833	0·9529	0·4737
OH	OH2_1	2*a*	OH 0·1667 O 0·8333	0·4167	0·0471	0·6406
OH	OH2_2	2*a*	OH 0·1667 O 0·8333	0·4167	0·6923	0·6406
OH	OH2_3	2*a*	OH 0·1667 O 0·8333	0·5833	0·1923	0·6951
OH	OH2_4	2*a*	OH 0·1667 O 0·8333	0·5833	0·5471	0·6951
OH	OH3_1	2*a*	OH 0·1667 O 0·8333	0·4167	0·4529	0·8621
OH	OH3_2	2*a*	OH 0·1667 O 0·8333	0·4167	0·8077	0·8621
OH	OH3_3	2*a*	OH 0·1667 O 0·8333	0·5833	0·3077	0·9165

Table 6.11 Continued

Atom	Label	Position	Occ.	x/a	y/b	z/c
OH	OH3_4	2*a*	OH 0·1667 O 0·8333	0·5833	0·9529	0·9165
OH	OH4_1	2*a*	1	0·37	0·375	0·2521
OH	OH4_2	2*a*	1	0·63	0·875	0·0835
Wa	Wa1_1	2*a*	1	0·645	0·625	0·4192
Wa	Wa1_2	2*a*	1	0·645	0·625	0·8621
Wa	Wa1_3	2*a*	1	0·355	0·125	0·4737
Wa	Wa1_4	2*a*	1	0·355	0·125	0·9165
Wa	Wa2_1	2*a*	1	0·645	0·125	0·8621
Wa	Wa2_2	2*a*	1	0·645	0·125	0·4192
Wa	Wa2_3	2*a*	1	0·355	0·625	0·4737
Wa	Wa2_4	2*a*	1	0·355	0·625	0·9165
Wa	Wa2_5	2*a*	1	0·355	0·875	0·6951
Wa	Wa2_6	2*a*	1	0·645	0·375	0·6406
Wa	Wa2_7	2*a*	1	0·355	0·375	0·6951
Wa	Wa2_8	2*a*	1	0·645	0·875	0·6406
Wa	Wa3_1	2*a*	0·5	0·712	0·625	0·1678
Wa	Wa3_2	2*a*	0·5	0·288	0·125	0·1678
Wa	Wa3_3	2*a*	0·5	0·212	0·625	0·1678
Wa	Wa3_4	2*a*	0·5	0·788	0·125	0·1678
Wa	Wa4_1	2*a*	0·5	0·5	0·875	0·2035
Wa	Wa4_2	2*a*	0·5	0·5	0·375	0·1321
Wa	Wa5_1	2*a*	1	0·3421	0·875	0·2521
Wa	Wa5_2	2*a*	1	0·6579	0·375	0·0835
Si	Si1_1	2*a*	1	0·3929	0·163	0·0501
Si	Si1_2	2*a*	1	0·3929	0·163	0·6072
Si	Si1_3	2*a*	1	0·3929	0·337	0·3857
Si	Si1_4	2*a*	1	0·3929	0·337	0·8286
Si	Si1_5	2*a*	1	0·6071	0·837	0·507
Si	Si1_6	2*a*	1	0·6071	0·837	0·9499
Si	Si1_7	2*a*	1	0·6071	0·663	0·2855
Si	Si1_8	2*a*	1	0·6071	0·663	0·7285
Si	Si2_1	2*a*	1	0·5	0·375	0·241
Si	Si2_2	2*a*	1	0·5	0·875	0·0947
Si	Si3_1	2*a*	1	0·1042	0·9167	0·507

Table 6.11 Continued

Atom	Label	Position	Occ.	x/a	y/b	z/c
Si	Si3_2	2*a*	1	0·1042	0·9167	0·9499
Si	Si3_3	2*a*	1	0·1042	0·5833	0·2855
Si	Si3_4	2*a*	1	0·1042	0·5833	0·7285
Si	Si3_5	2*a*	1	0·8958	0·0833	0·0501
Si	Si3_6	2*a*	1	0·8958	0·0833	0·6072
Si	Si3_7	2*a*	1	0·8958	0·4167	0·3857
Si	Si3_8	2*a*	1	0·8958	0·4167	0·8286

Ca/Si = 1·22; H_2O:Si = 1·06; D_x = 2·494 g cm^{-3}; for the dimeric chain interlayer, av. Ca4–O = 2·405 Å; Ca4–O polyhedral volume = 17·453 Å^3; for the infinite chain interlayer, av. Ca3–O = 2·390 Å; Ca3–O polyhedral volume = 17·852 Å^3.

Ca4 is the interlayer Ca between the dimeric silicate chains and Ca3 is between the chains that are nominally of infinite length (nominally because there are quite possibly tetrahedral bridging sites that are vacant).

Table 6.12 Model structure for a hypothetical C–S–H(I) 'undecamer' (SOF_{BT} = 0·75; MCL = 11) derived from a 14 Å clinotobermorite structure that had staggered silicate chains. The CIF file data block is T11_14sc (continued on next page)

Systematic name:	nonacalcium undecasilicate hexahydroxide heptahydrate
Structural formula:	$Ca_9(Si_{11}O_{28}(OH)_6)(H_2O)_{7{\cdot}25}$
Chemical formula (sum):	$H_{20{\cdot}5}\ Ca_9\ O_{41{\cdot}25}\ Si_{11}$
a	11·35 Å
b	7·3 Å
c	52·7 Å
α	90°
β	95·5°
γ	90°
Z	4
Space group:	No. 1 (setting 2), *C* 1
Symmetry equivalent positions:	(0, 0, 0)+
2*a*:	(1) + *x*, + *y*, + *z* (2) ½+*x*, ½+*y*, *z*

Atom	Label	Position	Occ.	*x/a*	*y/b*	*z/c*
Ca	Ca1A_1	2*a*	1	0·3667	0·125	0·9809
Ca	Ca1A_2	2*a*	1	0·3667	0·375	0·7076

Table 6.12 Continued

Atom	Label	Position	Occ.	x/a	y/b	z/c
Ca	Ca1A_3	2a	1	0·3667	0·625	0·9809
Ca	Ca1A_4	2a	1	0·3667	0·875	0·7076
Ca	Ca1A_5	2a	1	0·6333	0·125	0·7459
Ca	Ca1A_6	2a	1	0·6333	0·375	0·5656
Ca	Ca1A_7	2a	1	0·6333	0·625	0·7459
Ca	Ca1A_8	2a	1	0·6333	0·875	0·5656
Ca	Ca1_1	2a	1	0·3667	0·125	0·5274
Ca	Ca1_2	2a	1	0·3667	0·375	0·2541
Ca	Ca1_3	2a	1	0·3667	0·625	0·5274
Ca	Ca1_4	2a	1	0·3667	0·875	0·2541
Ca	Ca1_5	2a	1	0·6333	0·125	0·2924
Ca	Ca1_6	2a	1	0·6333	0·375	0·0191
Ca	Ca1_7	2a	1	0·6333	0·625	0·2924
Ca	Ca1_8	2a	1	0·6333	0·875	0·0191
Ca	Ca2A_2	2a	0·25	0·5	0·375	0·8634
Ca	Ca2A_4	2a	0·25	0·5	0·875	0·8634
Ca	Ca2_1	2a	0·25	0·5	0·125	0·1366
Ca	Ca2_2	2a	0·25	0·5	0·375	0·4099
Ca	Ca2_3	2a	0·25	0·5	0·625	0·1366
Ca	Ca2_4	2a	0·25	0·5	0·875	0·4099
Ca	Ca3_1	2a	0·25	0·25	0·125	0·638
Ca	Ca3_2	2a	0·25	0·25	0·625	0·638
O	O1A_2	2a	1	0·4167	0·4529	0·7947
O	O1A_4	2a	1	0·4167	0·8077	0·3412
O	O1A_6	2a	1	0·5833	0·3077	0·4785
O	O1A_8	2a	1	0·5833	0·9529	0·932
O	O1_1	2a	1	0·4167	0·0471	0·068
O	O1_2	2a	1	0·4167	0·4529	0·3412
O	O1_3	2a	1	0·4167	0·6923	0·068
O	O1_4	2a	1	0·4167	0·8077	0·7947
O	O1_5	2a	1	0·5833	0·1923	0·2053
O	O1_6	2a	1	0·5833	0·3077	0·932
O	O1_7	2a	1	0·5833	0·5471	0·2053
O	O1_8	2a	1	0·5833	0·9529	0·4785
O	O2A_1	2a	1	0·7308	0·125	0·2493
O	O2A_2	2a	1	0·7308	0·375	0·9761

Table 6.12 Continued

Atom	Label	Position	Occ.	*x/a*	*y/b*	*z/c*
O	O2A_3	2*a*	1	0·7308	0·625	0·7028
O	O2A_4	2*a*	1	0·7308	0·875	0·9761
O	O2A_5	2*a*	1	0·2692	0·125	0·5704
O	O2A_6	2*a*	1	0·2692	0·375	0·7507
O	O2A_7	2*a*	1	0·2692	0·625	0·0239
O	O2A_8	2*a*	1	0·2692	0·875	0·7507
O	O2_1	2*a*	1	0·7308	0·125	0·7028
O	O2_2	2*a*	1	0·7308	0·375	0·5226
O	O2_3	2*a*	1	0·7308	0·625	0·2493
O	O2_4	2*a*	1	0·7308	0·875	0·5226
O	O2_5	2*a*	1	0·2692	0·125	0·0239
O	O2_6	2*a*	1	0·2692	0·375	0·2972
O	O2_7	2*a*	1	0·2692	0·625	0·5704
O	O2_8	2*a*	1	0·2692	0·875	0·2972
O	O3_1	2*a*	1	0·5	0·125	0·0239
O	O3_2	2*a*	1	0·5	0·125	0·2493
O	O3_5	2*a*	1	0·5	0·125	0·5704
O	O3_6	2*a*	1	0·5	0·125	0·7028
O	O3_11	2*a*	1	0·5	0·375	0·2972
O	O3_12	2*a*	1	0·5	0·375	0·5226
O	O3_13	2*a*	1	0·5	0·375	0·7507
O	O3_14	2*a*	1	0·5	0·375	0·9761
O	O3_15	2*a*	1	0·5	0·625	0·0239
O	O3_16	2*a*	1	0·5	0·625	0·2493
O	O3_19	2*a*	1	0·5	0·625	0·5704
O	O3_20	2*a*	1	0·5	0·625	0·7028
O	O3_25	2*a*	1	0·5	0·875	0·2972
O	O3_26	2*a*	1	0·5	0·875	0·5226
O	O3_27	2*a*	1	0·5	0·875	0·7507
O	O3_28	2*a*	1	0·5	0·875	0·9761
O	O4_1	2*a*	1	0·3919	0·125	0·3228
O	O4_2	2*a*	1	0·3919	0·125	0·7763
O	O4_3	2*a*	1	0·3919	0·375	0·0495
O	O4_4	2*a*	1	0·3919	0·375	0·596
O	O4_5	2*a*	1	0·6081	0·625	0·497
O	O4_6	2*a*	1	0·6081	0·625	0·9505

Table 6.12 Continued

Atom	Label	Position	Occ.	*x/a*	*y/b*	*z/c*
O	O4_7	2*a*	1	0·6081	0·875	0·2237
O	O4_8	2*a*	1	0·6081	0·875	0·6772
OH	OH1A_1	2*a*	OH 0·75 O 0·25	0·4167	0·0471	0·6145
OH	OH1A_3	2*a*	OH 0·75 O 0·25	0·4167	0·6923	0·6145
OH	OH1A_5	2*a*	OH 0·75 O 0·25	0·5833	0·1923	0·6588
OH	OH1A_7	2*a*	OH 0·75 O 0·25	0·5833	0·5471	0·6588
OH	OH3_4	2*a*	OH 0·5 O 0·5	0·5	0·125	0·4388
OH	OH3_8	2*a*	OH 0·5 O 0·5	0·5	0·125	0·8923
OH	OH3_10	2*a*	OH 0·5 O 0·5	0·5	0·375	0·1656
OH	OH3_17	2*a*	OH 0·5 O 0·5	0·5	0·625	0·3809
OH	OH3_21	2*a*	OH 0·5 O 0·5	0·5	0·625	0·8344
OH	OH3_23	2*a*	OH 0·5 O 0·5	0·5	0·875	0·1077
OH	OH5A_1	2*a*	1	0·13	0·125	0·7947
OH	OH5A_3	2*a*	1	0·87	0·625	0·932
OH	OH5_1	2*a*	1	0·13	0·125	0·3412
OH	OH5_2	2*a*	1	0·13	0·375	0·068
OH	OH5_3	2*a*	1	0·87	0·625	0·4785
OH	OH5_4	2*a*	1	0·87	0·875	0·2053
Wa	W3_3	2*a*	0·25	0·5	0·125	0·3809
Wa	W3_7	2*a*	0·25	0·5	0·125	0·8344
Wa	W3_9	2*a*	0·25	0·5	0·375	0·1077
Wa	W3_18	2*a*	0·25	0·5	0·625	0·4388
Wa	W3_22	2*a*	0·25	0·5	0·625	0·8923
Wa	W3_24	2*a*	0·25	0·5	0·875	0·1656
Wa	W6_1	2*a*	1	0·145	0·375	0·6145
Wa	W6_2	2*a*	1	0·145	0·875	0·6145
Wa	W6_3	2*a*	1	0·855	0·375	0·6588
Wa	W6_4	2*a*	1	0·855	0·875	0·6588

Table 6.12 Continued

Atom	Label	Position	Occ.	*x/a*	*y/b*	*z/c*
Wa	W7_1	2*a*	1	0·8421	0·125	0·4785
Wa	W7_2	2*a*	1	0·8421	0·125	0·932
Wa	W7_3	2*a*	1	0·8421	0·375	0·2053
Wa	W7_4	2*a*	1	0·1579	0·625	0·3412
Wa	W7_5	2*a*	1	0·1579	0·625	0·7947
Wa	W7_6	2*a*	1	0·1579	0·875	0·068
Wa	W8_1	2*a*	0·25	0·212	0·125	0·1366
Wa	W8_2	2*a*	0·25	0·212	0·375	0·4099
Wa	W8_3	2*a*	0·25	0·212	0·375	0·8634
Wa	W8_4	2*a*	0·25	0·288	0·625	0·1366
Wa	W8_5	2*a*	0·25	0·288	0·875	0·4099
Wa	W8_6	2*a*	0·25	0·288	0·875	0·8634
Wa	W8_7	2*a*	0·25	0·712	0·125	0·1366
Wa	W8_8	2*a*	0·25	0·712	0·375	0·4099
Wa	W8_9	2*a*	0·25	0·712	0·375	0·8634
Wa	W8_10	2*a*	0·25	0·788	0·625	0·1366
Wa	W8_11	2*a*	0·25	0·788	0·875	0·4099
Wa	W8_12	2*a*	0·25	0·788	0·875	0·8634
Si	Si1_1	2*a*	1	0·3929	0·163	0·0408
Si	Si1_2	2*a*	1	0·3929	0·163	0·5873
Si	Si1_3	2*a*	1	0·3929	0·337	0·314
Si	Si1_4	2*a*	1	0·3929	0·337	0·7675
Si	Si1_5	2*a*	1	0·6071	0·663	0·2325
Si	Si1_6	2*a*	1	0·6071	0·663	0·686
Si	Si1_7	2*a*	1	0·6071	0·837	0·5057
Si	Si1_8	2*a*	1	0·6071	0·837	0·9592
Si	Si2_1	2*a*	1	0·5	0·125	0·4694
Si	Si2_2	2*a*	1	0·5	0·125	0·9229
Si	Si2_3	2*a*	1	0·5	0·375	0·1962
Si	Si2_4	2*a*	1	0·5	0·625	0·3503
Si	Si2_5	2*a*	1	0·5	0·625	0·8038
Si	Si2_6	2*a*	1	0·5	0·875	0·0771
Si	Si3_1	2*a*	1	0·3958	0·5833	0·0411
Si	Si3_2	2*a*	1	0·3958	0·5833	0·5876
Si	Si3_3	2*a*	1	0·3958	0·9167	0·3143
Si	Si3_4	2*a*	1	0·3958	0·9167	0·7679

Table 6.12 Continued

Atom	Label	Position	Occ.	*x/a*	*y/b*	*z/c*
Si	Si3_5	2*a*	1	0·6042	0·0833	0·2321
Si	Si3_6	2*a*	1	0·6042	0·0833	0·6857
Si	Si3_7	2*a*	1	0·6042	0·4167	0·5054
Si	Si3_8	2*a*	1	0·6042	0·4167	0·9589

Ca/Si = 0·8181; H_2O:Si = 0·9318; D_x = 2·063 g cm^{-3}; for the dimeric chain interlayer, av. Ca3–O = 2·402 Å; Ca3–O polyhedral volume = 17·44 Å^3; for the infinite chain interlayer, av. Ca2–O = 2·388 Å; Ca2–O polyhedral volume = 17·78 Å^3.
There are two large layer-spacing peaks for this structure that fall between 5 and 10°2θ with Cu Kα radiation and that have significant intensity: d_{003} = 17·486 Å (i.e. (c/3)sin β; $2\theta_{Cu}$ = 5·050°) and d_{004} = 13·114 Å (i.e. (c/4)sin β; $2\theta_{Cu}$ = 6·735°)).

REFERENCES

Alizadeh R, Beaudoin JJ and Raki L (2007) S–H (I) – a nanostructural model for the removal of water from hydrated cement paste? *Journal of the American Ceramic Society* **90(2)**: 670–672.

Andersen AD, Jakobsen HJ and Skibsted J (2004) Characterization of white Portland cement hydration and the C–S–H structure in the presence of sodium aluminate by ^{27}Al and ^{29}Si MAS NMR spectroscopy. *Cement and Concrete Research* **34(5)**: 857–868.

Andersen AD, Jakobsen HJ and Skibsted J (2006) A new aluminium-hydrate species in hydrated Portland cements characterized by ^{27}Al and ^{29}Si MAS NMR spectroscopy. *Cement and Concrete Research* **36(1)**: 3–17.

Atkins M, Glasser FP and Kindness A (1992) Cement hydrate phases: solubility at 25°C. *Cement and Concrete Research* **22(2-3)**: 241–246.

Atkinson A, Hearne JA and Knights CF (1989) Aqueous chemistry and thermodynamic modeling of CaO–SiO_2–H_2O gels. *Journal of the Chemical Society, Dalton Transactions*: 2371–2379.

Barbarulo R (2002) *Comportement des matériaux cimentaires: Actions des sulfates et de la température.* Département de Génie Civil, Université Laval, Québec, Canada. (In French.)

Barby D, Griffiths T, Jacques AR and Pawson D (1977) Soluble silicates and their derivatives. In *The Modern Inorganic Chemicals Industry* (Thompson R (ed.)). The Chemical Society, London, UK, pp. 320–352.

Barret P and Bertrandie D (1986) Fundamental hydration kinetic features of the major cement constituents: Ca_3SiO_5 and βCa_2SiO_4. *Journal de Chimie Physique* **83**: 765–775.

Barret P and Bertrandie D (1988) Comment on 'Aqueous solubility relationships for two types of calcium silicate hydrate'. *Journal of the American Ceramic Society* **71(2)**: 113–115.

Bauer WH (1974) The geometry of polyhedral distortions. Predictive relationships for the phosphate group. *Acta Crystallographica B* **30**: 1195–1215.

Beaudoin JJ, Dramé H, Raki L and Alizadeh R (2009) Formation and properties of C–S-H-PEG nano-structures. *Materials and Structures* **42**(7): 1003–1014.

Bellmann F, Sowoidnich T, Ludwig HM and Damidot D (2015) Dissolution rates during the early hydration of tricalcium silicate. *Cement and Concrete Research* **72**: 108–116.

Bernal JD, Jeffery JW and Taylor HFW (1952) Crystallographic research on the hydration of Portland cement. A first report on investigations in progress. *Magazine of Concrete Research* **4(11)**: 49–54.

Bessey GE (1939) The calcium aluminate and silicate hydrates. *Proceedings of the International Symposium on the Chemistry of Cement*. Ingenjörsvetenskapsakademien, Stockholm, Sweden, pp. 178–215.

Black L, Garbev K and Gee I (2008) Surface carbonation of synthetic C–S–H samples: a comparison between fresh and aged C–S–H using X-ray photoelectron spectroscopy. *Cement and Concrete Research* **38(6)**: 745–750.

Bonaccorsi E, Merlino S and Taylor HFW (2004) The crystal structure of jennite, $Ca_9Si_6O_{18}(OH)_6{\cdot}8H_2O$. *Cement and Concrete Research* **34(9)**: 1481–1488.

Bonaccorsi E, Merlino S and Kampf AR (2005) The crystal structure of tobermorite 14Å (plombierite), a C–S–H phase. *Journal of the American Ceramic Society* **88(3)**: 505–512.

Brough AR, Dobson CM, Richardson IG and Groves GW (1994) In-situ solid-state NMR studies of Ca_3SiO_5 hydration at room temperature and at elevated temperatures using ^{29}Si enrichment. *Journal of Materials Science* **29(15)**: 3926–3940.

Brown PW (1990) The system Na_2O–CaO–SiO_2–H_2O. *Journal of the American Ceramic Society* **73**: 3457–3461.

Brown PW, Franz E, Frohnsdorff G and Taylor HFW (1984) Analysis of the aqueous phase during early C_3S hydration. *Cement and Concrete Research* **14(2)**: 257–262.

Brunauer S (1962) Tobermorite gel – the heart of concrete. *American Scientist* **50**: 210–229.

Brunauer S and Greenberg SA (1960) The hydration of tricalcium silicate and β-dicalcium silicate at room temperature. *Proceedings of the 4th International Symposium on the Chemistry of Cement, Washington, DC, USA* **43(1)**: 135–165.

Brunauer S, Kantro DL and Copeland LE (1958) The stoichiometry of the hydration of beta-dicalcium silicate and tricalcium silicate at room temperature. *Journal of the American Chemical Society* **80(4)**: 761–767.

Brunet F, Bertani P, Charpentier T, Nonat A and Virlet J (2004) Application of ^{29}Si homonuclear and ^{1}H-^{29}Si heteronuclear NMR correlation to structural studies of calcium silicate hydrates. *The Journal of Physical Chemistry B* **108(40)**: 15494–15502.

Bullard JW, Hagedorn J, Ley T, Hu Q, Griffin W and Terrill JE (2018) A critical comparison of 3D experiments and simulations of tricalcium silicate hydration. *Journal of the American Ceramic Society* **101(4)**: 1453–1470.

Busing WR and Levy HA (1957) Neutron diffraction study of calcium hydroxide. *The Journal of Chemical Physics* **26(3)**: 563–568.

Chambers AV, Heath TG, Hunter FMI, Manning MC and Williams SJ (2005) *The Effect of Sodiumchloride on the Dissolution of Calciumsilicate Hydrate Gels*. Serco Assurance, UK, Report SA/ENV-0623.

Chen JJ, Thomas JJ, Taylor HFW and Jennings HM (2004) Solubility and structure of calcium silicate hydrate. *Cement and Concrete Research* **34(9)**: 1499–1519.

Churakov SV and Labbez C (2017) Thermodynamics and molecular mechanism of Al incorporation in calcium silicate hydrates. *The Journal of Physical Chemistry C* **121(8)**: 4412–4419.

Claringbull GF and Hey MH (1952) A re-examination of tobermorite. *Mineralogical Magazine* **29**: 960–962.

Cong X (1994) *^{29}Si and ^{17}O Nuclear Magnetic Resonance Investigation of the Structure of Calcium–Silicate–Hydrate*. PhD thesis, University of Illinois at Urbana-Champaign, IL, USA.

Cong X and Kirkpatrick RJ (1993) ^{17}O and ^{29}SiMAS NMR study of β-C_2S hydration and the structure of calcium silicate hydrates. *Cement and Concrete Research* **23(5)**: 1065–1077.

Cong X and Kirkpatrick RJ (1995) ^{1}H–^{29}Si CPMAS NMR study of the structure of calcium silicate hydrate. *Advances in Cement Research* **7(27)**: 103–111.

Cong X and Kirkpatrick RJ (1996a) ^{29}Si MAS NMR study of the structure of calcium silicate hydrate. *Advanced Cement Based Materials* **3(3-4)**: 144–156.

Cong X and Kirkpatrick RJ (1996b) ^{17}O MAS NMR investigation of the structure of calcium silicate hydrate gel. *Journal of the American Ceramic Society* **79(6)**: 1585–1592.

Copeland LE and Schulz EG (1962) Electron optical investigation of the hydration products of calcium silicates and Portland cement. *Journal of the PCA Research and Development Laboratories* **4**: 2–12.

Copeland LE, Bodor E, Chang TN and Weise CH (1967) Reactions of tobermorite gel with aluminates, ferrites, and sulfates. *Journal of the PCA Research and Development Laboratories* **9**: 61–74.

Courault AC (2000) *Simulation Expérimentale des C–S–H dans les Bétons Modernes: Etude de la Composition et des Propriétés a l'équilibre dans des milieux Complexes*. Laboratoire de Recherches sur la Réactivité des Solides, Université de Bourgogne, Dijon, France. (In French.)

CrystalMaker (2022) Crystal-structure images generated using CrystalMaker®; distortion indices calculated using CrystalMaker®; XRD patterns calculated using CrystalDiffract®. CrystalMaker Software, Oxford, UK. See crystalmaker.com

Damidot D, Nonat A, Barret P, Bertrandie D, Zanni H and Rassem R (1995) C_3S hydration in diluted and stirred suspensions: (III) NMR study of the C–S–H precipitated during the two kinetic steps. *Advances in Cement Research* **7(25)**: 1–8.

Damidot D, Lothenbach B, Herfort D and Glasser FP (2011) Thermodynamic modelling and cement science. *Cement and Concrete Research* **41**: 679–695.

de Jong JGM, Stein HN and Stevels M (1968) Mutual interaction of C_3A and C_3S during hydration. *Proceedings of the 5th International Symposium on the Chemistry of Cement*, Tokyo, Japan, **2**: 311–320.

Diamond S, White JL and Dolch WL (1966) Effects of isomorphous substitution in hydrothermally-synthesized tobermorite. *American Mineralogist* **51**: 388–401.

Divet L and Randriambololona R (1998) Delayed ettringite formation: the effect of temperature and basicity on the interaction of sulfate and C–S–H phase. *Cement and Concrete Research* **28**: 357–363.

Duchesne J and Reardon EJ (1995) Measurement and prediction of portlandite solubility in alkali solutions. *Cement and Concrete Research* **25(5)**: 1043–1053.

Dunitz JD (1979) *X-Ray Analysis and the Structure of Organic Molecules*. Cornell University Press, Ithaca, NY, USA, pp. 235–240.

Duque-Redondo E, Bonnaud PA and Manzano H (2022) A comprehensive review of C–S–H empirical and computational models, their applications, and practical aspects. *Cement and Concrete Research* **156**: 106784.

El-Hemaly SAS, Mohan K and Taylor HFW (1978) Autoclaved lime-quartz materials. II. Thermogravimetry and trimethylsilylation. *Cement and Concrete Research* **8(6)**: 671–676.

Faucon P, Delaye JM and Virlet J (1996) Molecular dynamics simulation of the structure of calcium silicate hydrates I. $Ca_{4+x}Si_6O_{14+2x}(OH)_{4-2x}(H_2O)_2$ $(0 \leq x \leq 1)$. *Journal of Solid State Chemistry* **127(1)**: 92–97.

Faucon P, Delaye JM, Virlet J, Jacquinot JF and Adenot F (1997) Study of the structural properties of the C–S–H(I) by molecular dynamics simulation. *Cement and Concrete Research* **27(10)**: 1581–1590.

Faucon P, Charpentier T, Nonat A and Petit JC (1998a) Triple-quantum two-dimensional ^{27}Al magic angle nuclear magnetic resonance study of the aluminum incorporation in calcium silicate hydrates. *Journal of the American Chemical Society* **120(46)**: 12075–12082.

Faucon P, Petit JC, Minet J and Viallis H (1998b) Iron in the $CaO–SiO_2–H_2O$ system. *Proceedings of the 1998 Goldschmidt Conference, Mineralogical Magazine* **62A**: 438–439.

Faucon P, Petit JC, Charpentier T, Jacquinot JF and Adenot F (1999a) Silicon substitution for aluminum in calcium silicate hydrates. *Journal of the American Ceramic Society* **82(5)**: 1307–1312.

Faucon P, Delagrave A, Petit JC, Richet C, Marchand JM and Zanni H (1999b) Aluminum incorporation in calcium silicate hydrates (C–S–H) depending on their Ca/Si ratio. *The Journal of Physical Chemistry B* **103(37)**: 7796–7802.

Fierens P and Verhaegen JP (1976) Hydration of tricalcium silicate in paste – kinetics of calcium ions dissolution in the aqueous phase. *Cement and Concrete Research* **6(3)**: 337–342.

Flint EP and Wells LS (1934) Study of the system $CaO–SiO_2–H_2O$ at 30°C and of the reaction of water on the anhydrous silicates. *Bureau of Standards Journal of Research* **12(6)**: 751–783.

Fu Y, Xie P, Gu P and Beaudoin JJ (1994) Effect of temperature on sulfate adsorption/desorption by tricalcium silicate hydrates. *Cement and Concrete Research* **24(8)**: 1428–1432.

Fujii K and Kondo W (1968) Hydration of tricalcium silicate. *Proceedings of the 5th International Symposium on the Chemistry of Cement*, Tokyo, Japan **2**: 362–371.

Fujii K and Kondo W (1975) Rate and mechanism of hydration of tricalcium silicate in an early stage. *Yogyo-Kyokai-Shi (Journal of the Ceramic Association, Japan)* **83**: 214–226.

Fujii K and Kondo W (1981) Heterogeneous equilibrium of calcium silicate hydrate in water at 30°C. *Journal of the Chemical Society, Dalton Transactions* **2**: 645–651.

Fujii K and Kondo W (1983) Estimation of thermochemical data for calcium silicate hydrate (C–S–H). *Journal of the American Ceramic Society* **66**: C-220–C-221.

Funk H (1967) Chemical alteration of poorly crystallized calcium silicate hydrates by hydrothermal treatment. In *Autoclaved Calcium Silicate Building Products*. Society of Chemical Industry, London, UK, pp. 122–124.

Funk H and Thilo E (1955) Über Hydrogensilicate IV. Das Calcium-trihydrogenmonosilicat $Ca[OSi(OH)_3]_2$ und seine Umwandlung in das Calcium-tetrahydrogendisilicat $Ca[Si_2O_3(OH)_4]$. *Zeitschrift für anorganische und allgemeine Chemie* **278**: 237–248. (In German.)

Gagné OC and Hawthorne FC (2016) Bond-length distributions for ions bonded to oxygen: alkali and alkaline-earth metals. *Acta Crystallographica B* **72(4)**: 602–625.

Garbev K, Stemmermann P, Black L, Breen C, Yarwood J and Gasharova B (2007) Structural features of C–S–H(I) and its carbonation in air – a Raman spectroscopic study. Part I: Fresh phases. *Journal of the American Ceramic Society* **90(3)**: 900–907.

Garbev K, Beuchle G, Bornefeld M, Black L and Stemmermann P (2008a) Cell dimensions and composition of nanocrystalline calcium silicate hydrate solid solutions. Part 1: Synchrotron-based X-ray diffraction. *Journal of the American Ceramic Society* **91(9)**: 3005–3014.

Garbev K, Bornefeld M, Beuchle G and Stemmermann P (2008b) Cell dimensions and composition of nanocrystalline calcium silicate hydrate solid solutions. Part 2: X-ray and thermogravimetry study. *Journal of the American Ceramic Society* **91(9)**: 3015–3023.

García-Lodeiro I, Fernández-Jiménez A, Sobrados I, Sanz J and Palomo A (2012) C–S–H gels: interpretation of ^{29}Si MAS-NMR spectra. *Journal of the American Ceramic Society* **95(4)**: 1440–1446.

Gard JA and Taylor HFW (1976) Calcium silicate hydrate (II) ('C–S–H(II)'). *Cement and Concrete Research* **6(5)**: 667–677.

Gard JA, Howison JW and Taylor HFW (1959) Synthetic compounds related to tobermorite: an electron-microscope, X-ray and dehydration study. *Magazine of Concrete Research* **11(33)**: 151–158.

Gard JA, Taylor HFW, Cliff G and Lorimer GW (1977) A reexamination of jennite. *American Mineralogist* **62**: 365–368.

Girão AV (2007) *The Nanostructure and Degradation of C–S–H in Portland and Blended Cements.* PhD thesis, University of Leeds, Leeds, UK.

Girão AV, Richardson IG, Porteneuve CB and Brydson RMD (2007a) Composition, morphology and nanostructure of C–S–H in white Portland cement–fly ash blends hydrated at 85°C. *Advances in Applied Ceramics* **106(6)**: 283–293.

Girão AV, Richardson IG, Porteneuve CB and Brydson RMD (2007b) Composition, morphology and nanostructure of C–S–H in white Portland cement pastes hydrated at 55°C. *Cement and Concrete Research* **37(12)**: 1571–1582.

Girão AV, Richardson IG, Taylor R and Brydson RMD (2010) Composition, morphology and nanostructure of C–S–H in 70% white Portland cement–30% fly ash blends hydrated at 55°C. *Cement and Concrete Research* **40(9)**: 1350–1359.

Glasser FP, Tyrer M, Quillin K *et al.* (1999) *The Chemistry of Blended Cement and Backfills Intended for Use in Radioactive Waste Disposal.* Environment Agency, Bristol, UK.

Grangeon S, Claret F, Lerouge C *et al.* (2013a) On the nature of structural disorder in calcium silicate hydrates with calcium/silicon ratio similar to tobermorite. *Cement and Concrete Research* **52**: 31–37.

Grangeon S, Claret F, Linard Y and Chiaberge C (2013b) X-ray diffraction: a powerful tool to probe and understand the structure of nanocrystalline calcium silicate hydrates. *Acta Crystallographica B* **69(5)**: 465–473.

Greenberg SA and Chang TN (1965) Investigation of colloidal hydrated calcium silicates, II. Solubility relationship in the calcium oxide-silica-water system. *The Journal of Physical Chemistry* **69(1)**: 182–188.

Groves GW (1990) A discussion of the paper 'Phase identification during early and middle hydration of tricalciumsilicate (Ca_3SiO_5)' by Melzer R and Eberhard E. *Cement and Concrete Research* **20**: 315–316.

Grudemo Å (1955) An electronographic study of the morphology and crystallization properties of calcium silicate hydrates. *Proceedings of the Swedish Cement and Concrete Research Institute at the Royal Institute of Technology, Stockholm (Handlingar – Svenska Forskningsinstitutet för Cement och Betong vid Kungliga Tekniska Högskolan i Stockholm)* **26**: 1–103.

Grudemo Å (1962) The microstructure of hardened cement paste. In *Chemistry of Cement. Proceedings of the 4th International Symposium, Washington (1960).* National Bureau of Standards, US Department of Commerce, Washington, DC, USA, monograph **43**(2): 615–658.

Grutzeck MW (1999) A new model for the formation of calcium silicate hydrate (C–S–H). *Materials Research Innovations* **3(3)**: 160–170.

Grutzeck M, Benesi A and Fanning B (1989) Silicon-29 magic angle spinning nuclear magnetic resonance study of calcium silicate hydrates. *Journal of the American Ceramic Society* **72(4)**: 665–668.

Gutteridge WA and Parrott LJ (1976) A study of the changes in weight, length and interplanar spacing induced by drying and rewetting synthetic CSH (I). *Cement and Concrete Research* **6(3)**: 357–366.

Haas J (2012) *Etude Expérimentale et Modélisation Thermodynamique du Système CaOSiO$_2$-(Al$_2$O$_3$)-H$_2$O.* Université de Bourgogne, Dijon, France. (In French.)

Haas J and Nonat A (2015) From C–S–H to C–A–S–H- Experimental study and thermodynamic modelling. *Cement and Concrete Research* **68**: 124–138.

Haga K, Shibata M, Hironaga M, Tanaka S and Nagasaki S (2002) Silicate anion structural change in calcium silicate hydrate gel on dissolution of hydrated cement. *Journal of Nuclear Science and Technology* **39(5)**: 540–547.

Halstead PE and Lawrence CD (1960) Kinetics of Reaction in the System $CaO–SiO_2–H_2O$. *Proceedings of the 4th International Symposium on the Chemistry of Cement, Washington, DC, USA* **1**: 321–325.

Hannawayya F (1980) Study of the structure and crystallization properties of ticalcium silicate (C_3S) and dicalcium silicate beta (beta-C_2S) hydrate, and the new calcium silicate hydrate (C–S–H(III) C–S–H (I) and C–S–H (II)). *Proceedings of the 7th International Congress on the Chemistry of Cement* **II**, II-31.

Harris AW, Manning MC, Tearle WM and Tweed CJ (2002) Testing of models of the dissolution of cements – leaching of synthetic CSH gels. *Cement and Concrete Research* **32(5)**: 731–746.

Harris M, Simpson G, Scrivener K and Bowen P (2022) A method for the reliable and reproducible precipitation of phase pure high Ca/Si ratio (>1.5) synthetic calcium silicate hydrates (C–S–H). *Cement and Concrete Research* **151**: 106623.

Hedin R (1945) *Chemical Processes in the Hardening of Portland Cement*. Almquist & Wiksells, Stockholm and Uppsala, Sweden.

Heller L and Taylor HFW (1951) Hydrated calcium silicates. Part II. Hydrothermal reactions: Lime:silica ratio 1:1. *Journal of the Chemical Society (Resumed)*: 2397–2401.

Heller L and Taylor HFW (1956) *Crystallographic Data for the Calcium Silicates*. HMSO, London, UK.

Henderson E and Bailey JE (1986) Structure and morphological aspects of calcium silicate hydrates. *Proceedings of the 8th International Congress on the Chemistry of Cement* **3**: 376–381.

Henocq P (2005) *Modélisation des interactions ioniques à la surface des silicates de calcium hydratés*. PhD thesis, Université Laval, Quebec, Canada. (In French.)

Hirljac J, Wu ZQ and Young JF (1983) Silicate polymerization during the hydration of alite. *Cement and Concrete Research* **13(6)**: 877–886.

Hong SY and Glasser FP (1999) Alkali binding in cement pastes part I. *The C–S–H Phase. Cement and Concrete Research* **29(12)**: 1893–1903.

Hou D, Ma H and Li Z (2015) Morphology of calcium silicate hydrate (C–S–H) gel: a molecular dynamic study. *Advances in Cement Research* **27(3)**: 135–146.

Ishida H, Sasaki K and Mitsuda T (1992a) Highly reactive β-dicalcium silicate: I, hydration behaviour at room temperature. *Journal of the American Ceramic Society* **75(2)**: 353–358.

Ishida H, Okada Y and Mitsuda T (1992b) Highly reactive β-dicalcium silicate: II, hydration behaviour at 25°C followed by ^{29}Si nuclear magnetic resonance. *Journal of the American Ceramic Society* **75(2)**: 359–363.

Ishida H, Sasaki K, Okada Y and Mitsuda T (1992c) Highly reactive β-dicalcium silicate: III, hydration behaviour at 40–80°C. *Journal of the American Ceramic Society* **75**: 2541–2546,

Ishida H, Sasaki K, Mizuno A, Okada Y and Mitsuda T (1992d) Highly reactive β-dicalcium silicate: IV, ball-milling and static hydration at room temperature. *Journal of the American Ceramic Society* **75(10)**: 2779–2784.

Jennings HM (1986) Aqueous solubility relationships for two types of calcium silicate hydrate. *Journal of the American Ceramic Society* **69(8)**: 614–618.

Jennings HM (1988) Reply to a comment on Jennings (1986) by Barret and Bertrandie (1988). *Journal of the American Ceramic Society* **71**: C115–C116.

Jennings HM, Taleb, Frohnsdorff G and Clifton CR (1986) Interpretation of the effects of retarding admixtures on pastes of C_3S, C_3A plus gypsum, and Portland cement. *Proceedings of the 8th International Congress on the Chemistry of Cement* **3**: 239–243.

Kalousek GL (1944) Studies of portions of the quaternary system soda–lime–silica–water at 25°C. *Journal of Research of the National Bureau of Standards* **32(6)**: 285–302.

Kalousek GL (1954) Application of differential thermal analysis in a study of the system lime–silica–water. *Proceedings of the 3rd International Symposium on the Chemistry of Cement, 1952. Cement and Concrete Association*, London, UK, pp. 296–311.

Kalousek GL (1955) Tobermorite and related phases in the system $CaO–SiO_2–H_2O$. *Journal of the American Concrete Institute (Proceedings)* **51**: 989–1011.

Kalousek GL and Prebus AF (1958) Crystal chemistry of hydrous calcium silicates: III, Morphology and other properties of tobermorite and related phases. *Journal of the American Ceramic Society* **41(4)**: 124–132.

Kirkpatrick RJ, Yarger JL, McMillan PF, Ping Y and Cong X (1997a) Raman spectroscopy of C–S–H, tobermorite and jennite. *Advanced Cement Based Materials* **5(3–4)**: 93–99.

Kirkpatrick RJ, Brown GE, Xu N and Cong X (1997b) Ca X-ray absorption spectroscopy of C–S–H and some model compounds. *Advances in Cement Research* **9(33)**: 31–36.

Klimesch DS and Ray A (2002) Evaluation of phases in a hydrothermally treated $CaO-SiO_2-H_2O$ system. *Journal of Thermal Analysis Calorimetry* **70**: 995–1003.

Klur I, Pollet B, Virlet J and Nonat A (1998) C–S–H structure evolution with calcium content by multinuclear NMR. In *Nuclear Magnetic Resonance Spectroscopy of Cement-Based Materials* (Colombet P, Grimmer AR, Zanni H and Sozzani P (eds)). Springer, Berlin, Germany, pp. 119–141.

Komarneni S, Roy R, Roy DM *et al.* (1985) ^{27}Al and ^{29}Si magic angle spinning nuclear magnetic resonance spectroscopy of Al-substituted tobermorites. *Journal of Materials Science* **20(11)**: 4209–4214.

Komarneni S, Breval E, Roy DM and Roy R (1986) Synthesis and characterization of a 12.6 Å calcium silicate hydrate. *Cement and Concrete Research* **16(4)**: 580–586.

Kulik, DA (2011) Improving the structural consistency of C–S–H solid solution thermodynamic models. *Cement and Concrete Research* **41(5)**: 477–495.

Kumar A, Walder BJ, Kunhi Mohamed A *et al.* (2017) The atomic-level structure of cementitious calcium silicate hydrate. *The Journal of Physical Chemistry C* **121(32)**: 17188–17196.

Kunhi Mohamed A, Parker SC, Bowen P and Galmarini S (2018) An atomistic building block description of C–S–H – towards a realistic C–S–H model. *Cement and Concrete Research* **107**: 221–235.

Kunhi Mohamed A, Moutzouri P, Berruyer P *et al.* (2020) The atomic-level structure of cementitious calcium aluminate silicate hydrate. *Journal of the American Chemical Society* **142(25)**: 11060–11071.

Kurczyk HG and Schwiete HE (1960) Electronenmikroskopische und thermochemische Untersuchungen über die Hydratation der Calciumsilikate $3CaO{\cdot}SiO_2$ und β-$2CaO{\cdot}SiO_2$ und den Einfluß von Calciumchlorid und Gips auf den Hydratationsvorgang. *Tonindustrie-Zeitung und Keramische Rundschau* **84**: 585–598. (In German.)

Kurczyk HG and Schwiete HE (1962) Concerning the hydration products of C_3S and β-C_2S, *Proceedings of the 4th International Symposium on the Chemistry of Cement* **1**: 349–358.

Kwan S, LaRosa-Thompson J and Grutzeck MW (1996) Structures and phase relations of aluminum-substituted calcium silicate hydrate. *Journal of the American Ceramic Society* **79(4)**: 967–971.

Lawrence D (1965) The dehydration of calcium silicate hydrate C–S–H (I). *Proceedings of the 7th Conference on Silicate Industry*. Akadémiai Kiado, Budapest, Hungary, pp. 259–263.

Lawrence CD (1966) Changes in composition of the aqueous phase during hydration of cement pastes and suspensions. *Highway Research Board Special Report* **90**: 378–391.

Lequeux N, Morau A, Philippot S and Boch P (1999) Extended X-ray absorption fine structure investigation of calcium silicate hydrates. *Journal of the American Ceramic Society* **82(5)**: 1299–1306.

L'Hôpital E, Lothenbach B, Le Saout G, Kulik D and Scrivener K (2015) Incorporation of aluminium in calcium-silicate-hydrates. *Cement and Concrete Research* **75**: 91–103.

L'Hôpital E, Lothenbach B, Scrivener K and Kulik DA (2016) Alkali uptake in calcium alumina silicate hydrate (C-A-S-H). *Cement and Concrete Research* **85**: 122–136.

Li J, Geng G, Myers R *et al.* (2019) The chemistry and structure of calcium (alumino) silicate hydrate: a study T by XANES, ptychographic imaging, and wide- and small-angle scattering. *Cement and Concrete Research* **115**, 367–378.

Lognot I, Klur I and Nonat A (1998) NMR and infrared spectroscopies of C–S–H and Al substituted C–S–H synthesized in alkaline solution. In *Nuclear Magnetic Resonance Spectroscopy of Cement-Based Materials* (Colombet P, Grimmer AR, Zanni H and Sozzani P (eds)). Springer, Berlin, Germany, pp. 189–196.

Lothenbach B (2010) Thermodynamic equilibrium calculations in cementitious systems. *Materials and Structures* **43(10)**: 1413–1433.

Lothenbach B and Nonat A (2015) Calcium silicate hydrates: Solid and liquid phase composition. *Cement and Concrete Research* **78**: 57–70.

Love CA, Richardson IG, and Brough AR (2007) Composition and structure of C–S–H in white Portland cement–20% metakaolin pastes hydrated at 25°C. *Cement and Concrete Research* **37(2)**: 109–117.

Macphee DE, Luke K, Glasser FP and Lachowski EE (1989) Solubility and aging of calcium silicate hydrates in alkaline solution at 25°C. *Journal of the American Ceramic Society* **72(4)**: 646–654.

Matsuyama H and Young JF (2000) Effects of pH on precipitation of quasi-crystalline calcium silicate hydrates in aqueous solution. *Advances in Cement Research* **12(1)**: 29–33.

Melzer R and Eberhard E (1989) Phase identification during early and middle hydration of tricalciumsilicate (Ca_3SiO_5). *Cement and Concrete Research* **19(3)**: 411–422.

Merlino S, Bonaccorsi E and Armbruster T (1999) Tobermorites: their real structure and order–disorder (OD) character. *American Mineralogist* **84(10)**: 1613–1621.

Merlino S, Bonaccorsi E and Armbruster T (2001) The real structure of tobermorite 11 Å: normal and anomalous forms, OD character and polytypic modifications. *European Journal of Mineralogy* **13(3)**: 577–590.

Michaux M, Ménétrier D and Barret P (1983) Étude par triméthylsilylation du degré de condensation des ions silicate dans l'hydrosilicate de calcium. *Comptes Rendus de l'Académie des Sciences (Series II)* **296**: 1043–1046. (In French.)

Mitsuda T and Taylor HFW (1978) Normal and anomalous tobermorites. *Mineralogical Magazine* **42(322)**: 229–235.

Mitsuda T, Kobayakawa S and Toraya H (1986) Characterization of hydrothermally formed C–S–H. *Proceedings of the 8th International Congress on the Chemistry of Cement* **3**. Abla Gráfica e Editora, Rio de Janeiro, Brazil, pp. 173–178.

Mitsuda T, Toraya H, Okada Y and Shimoda M (1989) Synthesis of tobermorite: NMR spectroscopy and analytical electron microscopy. *Ceramic Transactions* **5**: 206–213.

Moulin I, Stone WEE, Sanz J, Bottero JY, Mosnier F and Haehnel C (1999) Lead and zinc retention during hydration of tri-calcium silicate: a study by sorption isotherms and ^{29}Si nuclear magnetic resonance spectroscopy. *Langmuir* **15(8)**: 2829–2835.

Naber C, Bellmann F, Sowoidnich T, Goetz-Neunhoeffer F and Neubauer J (2019) Alite dissolution and C–S–H precipitation rates during hydration. *Cement and Concrete Research* **115**: 283–293.

Naber C, Bellmann F and Neubauer J (2020) Influence of w/s ratio on alite dissolution and C–S–H precipitation rates T during hydration. *Cement and Concrete Research* **134**: 10608.

Nicoleau L and Nonat A (2016) A new view on the kinetics of tricalcium silicate hydration. *Cement and Concrete Research* **86**: 1–11.

Noma H, Adachi Y, Yamada H, Nishino T, Matsuda Y and Yokoyama T (1998) ^{29}Si MAS NMR spectroscopy of poorly-crystalline calcium silicate hydrates (C–S–H). In *Nuclear Magnetic Resonance Spectroscopy of Cement-Based Materials* (Colombet P, Grimmer AR, Zanni H and Sozzani P (eds)). Springer, Berlin, Germany, pp. 159–167.

Nonat A (2004) The structure and stoichiometry of C–S–H. *Cement and Concrete Research* **34(9)**: 1521–1528.

Nonat A and Lecoq X (1998) The structure, stoichiometry and properties of C–S–H prepared by C_3S hydration under controlled conditions. In *Nuclear Magnetic Resonance Spectroscopy of Cement-Based Materials*. (Colombet P, Grimmer AR, Zanni H and Sozzani P (eds)), Springer, Berlin, Germany, pp. 197–207.

Nonat A, Courault AC and Damidot D (2001) A new model describing the variation of CSH Ca/Si ratio with lime concentration in solution. *Cement Wapno Beton* **5**: 184–191.

Odler I (1981) Interaction between gypsum and the C–S–H phase formed in C_3S hydration. *Proceedings of the 7th International Congress on the Chemistry of Cement*. Editions Septima, Paris, France **4**: 493–495.

Ohashi Y (1984) Polysynthetically-twinned structures of enstatite and wollastonite. *Physics and Chemistry of Minerals* **10(5)**: 217–229.

Okada Y, Ishida H, Sasaki K, Young JF and Mitsuda T (1994a) Characterization of C–S–H from highly reactive β-dicalcium silicate prepared from hillebrandite. *Journal of the American Ceramic Society* **77(5)**: 1313–1318.

Okada Y, Ishida H and Mitsuda T (1994b) ^{29}Si NMR spectroscopy of silicate anions in hydrothermally formed C–S–H. *Journal of the American Ceramic Society* **77(3)**: 765–768.

Pardal X, Pochard I and Nonat A (2009) Experimental study of Si–Al substitution in calcium-silicate-hydrate (C–S–H) prepared under equilibrium conditions. *Cement and Concrete Research* **39(8)**: 637–643.

Pardal X, Brunet F, Charpentier T, Pochard I and Nonat A (2012) ^{27}Al and ^{29}Si solid-state NMR characterization of calcium-aluminosilicate-hydrate. *Inorganic Chemistry* **51(3)**: 1827–1836.

Pellenq RJM (2009/2010) personal communications with Richardson. Note: MIT1 is the version that is published in the supplementary table S7 of Pellenq *et al.* (2009) but with the changes noted in the caption to Table 6.3. The other four files were supplied to Richardson by Pellenq on 30/10/2009 (MIT2); 13/11/2009 (MIT3) and 07/06/2010 (MIT4 and MIT5); MIT3 was supplied in response to a request for the '*definitive version*' of the published model.

Pellenq RJM, Kushima A, Shahsavari R *et al.* (2009) A realistic molecular model of cement hydrates. *Proceedings of the National Academy of Sciences* **106(38)**: 16102–16107.

Plusquellec G (2014) *Analyse in situ de suspensions de silicate de calcium hydraté: application aux interactions ioniques à la surface des particules*. PhD thesis, Université de Bourgogne, Bourgogne, France. (In French.)

Pointeau I (2000) *Etude mécanistique et modélisation de la retention de radionucléides par les silicates de calcium hydrates (CSH) des ciments*. PhD thesis, University of Champagne-Ardenne, Reims, France. (In French.)

Popova A, Geoffroy G, Renou-Gonnord MF, Faucon P and Gartner E (2000) Interactions between polymeric dispersants and calcium silicate hydrates. *Journal of the American Ceramic Society* **83(10)**: 2556–2560.

Pytel Z and Malolepszy J (1997) The role of MgO in the synthesis of C–S–H and tobermorite. *Proceedings of the 10th International Congress on the Chemistry of Cement*. Amarkai AB and Congrex, Göteborg, Sweden, Paper 2ii064, 7 pp.

Qomi MJA, Ulm FJ and Pellenq JM (2012) Evidence on the dual nature of aluminium in the calcium-silicate-hydrates based on atomistic simulations. *Journal of the American Ceramic Society* **95(3)**: 1128–1137.

Radczewski OE, Müller HO and Eitel W (1939) The hydration of tricalcium silicate. *Die Naturwissenschaften* **27(48)**: 807–807.

Rassem R, Zanni-Theveneau H, Vernet C *et al.* (1992a) An NMR investigation of C_3S hydration in pastes and in stirred dilute suspensions. In *Hydration and Setting of Cements* (Nonat A and Mutin JC (eds)). E & FN Spon, London, UK, pp. 77–85.

Rassem R, Zanni-Théveneau H, Vernet C *et al.* (1992b) An NMR investigation of C_3S hydration in pastes and in stirred diluted suspensions. *Proceedings of the 9th International Congress on the Chemistry of Cement.* National Council for Cement and Building Materials, New Delhi, India **4**: 234–240.

Renaudin G, Russias J, Leroux F, Frizon F and Cau-dit-Coumes C (2009a) Structural characterization of C–S–H and C–A–S–H samples – part I: Long-range order investigated by Rietveld analyses. *Journal of Solid State Chemistry* **182(12)**: 3312–3319.

Renaudin G, Russias J, Leroux F, Cau-dit-Coumes C and Frizon F (2009b) Structural characterization of C–S–H and C–A–S–H samples – part II: Local environment investigated by spectroscopic analyses. *Journal of Solid State Chemistry* **182(12)**: 3320–3329.

Reynolds RC Jr. (1986) The Lorentz-polarization factor and preferred orientation in oriented clay aggregates. *Clays and Clay Minerals* **34(4)**: 359–367.

Richardson IG (1999) The nature of C–S–H in hardened cements. *Cement and Concrete Research* **29(8)**: 1131–1147.

Richardson IG (2002) Electron microscopy of cements. In *Structure and Performance of Cements.* (Bensted J and Barnes P (eds)). Spon, London, UK, pp. 500–556.

Richardson IG (2004) Tobermorite/jennite- and tobermorite/calcium hydroxide-based models for the structure of C–S–H: applicability to hardened pastes of tricalcium silicate β-dicalcium silicate, Portland cement and blends of Portland cement with blast-furnace slag, metakaolin, or silica fume. *Cement and Concrete Research* **34(9)**: 1733–1777.

Richardson IG (2008) The calcium silicate hydrates. *Cement and Concrete Research* **38(2)**: 137–158.

Richardson IG (2013) The importance of proper crystal-chemical and geometrical reasoning demonstrated using layered single and double hydroxides. *Acta Crystallographica B* **69(2)**: 150–162.

Richardson IG (2014) Model structures for C–(A)–S–H(I). *Acta Crystallographica B* **70**: 903–923. https://doi.org/10.1107/S2052520614021982. Licensed under CC BY 2.0: https://creativecommons.org/licenses/by/2.0/

Richardson IG and Groves GW (1992a) Models for the composition and structure of calcium silicate hydrate (C–S–H) gel in hardened tricalcium silicate pastes. *Cement and Concrete Research* **22(6)**: 1001–1010.

Richardson IG and Groves GW (1992b) Microstructure and microanalysis of hardened cement pastes involving ground granulated blast-furnace slag. *Journal of Materials Science* **27(22)**: 6204–6212.

Richardson IG and Groves GW (1993a) The incorporation of minor and trace elements into calcium silicate hydrate (C–S–H) gel in hardened cement pastes. *Cement and Concrete Research* **23(1)**: 131–138.

Richardson IG and Groves GW (1993b) Microstructure and microanalysis of hardened ordinary Portland cement pastes. *Journal of Materials Science* **28(1)**: 265–277.

Richardson IG and Groves GW (1997) The structure of the calcium silicate hydrate phases present in hardened pastes of white Portland cement/blast-furnace slag blends. *Journal of Materials Science* **32**: 4793–4802.

Richardson IG, Brough AR, Brydson R, Groves GW and Dobson CM (1993) The location of aluminum in substituted calcium silicate hydrate (C–S–H) gels as determined by ^{29}Si and ^{27}Al NMR and EELS. *Journal of the American Ceramic Society* **76(9)**: 2285–2288.

Richardson IG, Brough AR, Groves GW and Dobson CM (1994) The characterization of hardened alkali-activated blast-furnace slag pastes and the nature of the calcium silicate hydrate (C–S–H) phase. *Cement and Concrete Research* **24(5)**: 813–829.

Richardson IG, Skibsted J, Black L and Kirkpatrick RJ (2010) Characterisation of cement hydrate phases by TEM, NMR and Raman spectroscopy. *Advances in Cement Research* **22(4)**: 233–248.

Richardson IG, Girão AV, Taylor R and Jia S (2016) Hydration of water- and alkali-activated white Portland cement pastes and blends with low-calcium pulverized fuel ash. *Cement and Concrete Research* **83**: 1–18.

Rodger SA (1984) The chemistry of admixture interaction during cement hydration DPhil thesis, University of Oxford, Oxford, UK.

Rodger SA, Groves GW, Clayden NJ and Dobson CM (1987) A study of tricalcium silicate hydration from very early to very late ages. *Materials Research Society Symposium Proceedings* **85**: 13–20.

Rodger SA, Groves GW, Clayden NJ and Dobson CM (1988) The hydration of tricalcium silicate followed by ^{29}Si NMR with cross polarization. *Journal of the American Ceramic Society* **71(2)**: 91–96.

Roller PS and Ervin G (1940) The system CaO–SiO_2–H_2O at 30°C. The association of silicate ion in dilute alkaline solution. *Journal of the American Chemical Society* **62(3)**: 461–471.

Roosz C, Vieillard P, Blanc P *et al.* (2018) Thermodynamic properties of C–S–H, C-A-S-H and M-S-H phases: Results from direct measurements and predictive modelling. *Applied Geochemistry* **92**: 140–156.

Sasaki K, Ishida H, Okada Y and Mitsuda T (1993) Highly reactive β-dicalcium silicate: V, influence of specific surface area on hydration. *Journal of the American Ceramic Society* **76(4)**: 870–874.

Sasaki K, Masuda T, Ishida H and Mitsuda T (1996) Synthesis of calcium silicate hydrate with Ca/Si = 2 by mechanochemical treatment. *Journal of the American Ceramic Society* **80(2)**: 472–476.

Scrivener KL, Matschei T, Georget F, Juilland P and Mohamed AK (2023) Advances in hydration and thermodynamics of cementitious systems. *Cement and Concrete Research* **174**: 107332.

Skibsted J and Andersen MD (2013) The effect of alkali ions on the incorporation of aluminum in the calcium silicate hydrate (C–S–H) phase resulting from Portland cement hydration studied by ^{29}Si MAS NMR. *Journal of the American Ceramic Society* **96(2)**: 651–656.

Smith RH, Bayliss Gamble BR and Mills RH (1972) Crystallographic investigation of C–S–H(I). *Cement and Concrete Research* **2**: 559–566.

Stade H (1980) Zum Aufbau schlecht geordneter Calciumhydrogensilicate. II Über eine aus Poly- und Disilicat bestehende Phase. *Zeitschrift für anorganische und allgemeine Chemie* **470(1)**: 69–83. (In German.)

Stade H (1989) On the reaction of C–S–H(di- poly) with alkali hydroxides. *Cement and Concrete Research* **19(5)**: 802–810.

Stade H and Müller D (1987) On the co-ordination of Al in ill-crystallized C–S–H phases formed by hydration of tricalcium silicate and by precipitation reactions at ambient temperature. *Cement and Concrete Research* **17(4)**: 553–561.

Stade H and Wieker W (1980) Zum Aufbau schlecht geordneter Calciumhydrogensilicate. I Bildung und Eigenschaften einer schlecht geordneten Calciumhydrogendisilicatphase. *Zeitschrift für anorganische und allgemeine Chemie* **466(1)**: 55–70. (In German.)

Stade H and Wieker W (1982) Zum Aufbau schlecht geordneter Calciumhydrogensilicate. III. Zum Einbau von Al^{3+}-ionen in C–S–H(Di, Poly) und Bildung eines instabilen 11 Å-Tobermorits. *Zeitschrift für anorganische und allgemeine Chemie* **494(1)**: 179–188. (In German.)

Stade H, Wieker W and Garzo G (1983) On the structure of ill-crystallized calcium hydrogen silicates. IV. Anion composition of the hydration products of tricalcium silicate. *Zeitschrift für anorganische und allgemeine Chemie*, **500**, 123–131. (In German.)

Stade H, Müller D and Scheler G (1984a) Zum Aufbau schlecht geordneter Calciumhydrogensilicate. V. ^{27}Al-NMR-spektroskopische Untersuchungen zur Koordination des Al in C–S–H(Di, Poly). *Zeitschrift für anorganische und allgemeine Chemie* **510(3)**: 16–24. (In German.).

Stade H, Ulbricht K and Mehner H (1984b) Zum Aufbau schlecht geordneter Calcium hydrogensilicate. VI. Chemische, ESR- und Mössbauer-spektroskopische Untersuchungen zum Einbau von Fe^{3+}-ionen in C–S–H(Di, Poly). *Zeitschrift für anorganische und allgemeine Chemie* **514(7)**: 149–163. (In German.)

Stade H, Grimmer AR, Engelhardt G, Mägi M and Lippmaa E (1985) Zum Aufbau schlecht geordneter Calciumhydrogensilicate. VII ^{29}Si-NMR Untersuchungen an C–S–H(Di, poly). *Zeitschrift für Anorganische und Allgemeine Chemie* **528(9)**: 147–151. (In German.)

Stein HN (1972) Thermodynamic considerations on the hydration mechanisms of Ca_3SiO_5 and $Ca_3Al_2O_6$. *Cement and Concrete Research* **2(2)**: 167–177.

Stein HN and Stevels JM (1964) Influence of silica on the hydration of 3CaO,SiO_2. *Journal of Applied Chemistry* **14(8)**: 338–346.

Steinour HH (1947) The system CaO–SiO_2–H_2O and the hydration of the calcium silicates. *Chemical Reviews* **40(3)**: 391–460.

Steinour HH (1954) The reactions and thermochemistry of cement hydration at ordinary temperature. *Proceedings of the 3rd International Symposium on the Chemistry of Cement*: 261–289.

Sugiyama D (2008) Chemical alteration of calcium silicate hydrate (C–S–H) in sodium chloride solution. *Cement and Concrete Research* **38(11)**: 1270–1275.

Sun GK, Young JF and Kirkpatrick RJ (2006) The role of Al in C–S–H: NMR, XRD, and compositional results for precipitated samples. *Cement and Concrete Research* **36(1)**: 18–29.

Suzuki K, Nishikawa T and Ito S (1985) Formation and carbonation of C–S–H in water. *Cement and Concrete Research* **15(2)**: 213–224.

Swanton SW, Fairbrother HJ and Turner NA (2005) *The Effect of Sodium Chloride on the Dissolution of Calcium Silicate Hydrate Gels. II. Effect of Temperature and Cation Type*. Serco Assurance, UK, Report SA/ENV-0725.

Swanton SW, Heath TG and Clacher A (2016) Leaching behaviour of low Ca:Si ratio CaO–SiO_2–H_2O systems. *Cement and Concrete Research* **88**: 82–95.

Tadros ME, Skalny J and Kalyoncu RS (1976) Early hydration of tricalcium silicate. *Journal of the American Ceramic Society* **59(7-8)**: 344–347.

Tajuelo Rodriguez E, Richardson IG, Black L, Boehm-Courjault E, Nonat A and Skibsted J (2015) Composition, silicate anion structure and morphology of calcium silicate hydrates (C–S–H) synthesised by silica–lime reaction and by controlled hydration of tricalcium silicate (C_3S). *Advances in Applied Ceramics* **114(7)**: 362–371.

Tajuelo Rodriguez E, Garbev K, Merz D, Black L and Richardson IG (2017) Thermal stability of C–S–H phases and applicability of Richardson and Groves' and Richardson C-(A)-S-H(I) models to synthetic C–S–H. *Cement and Concrete Research* **93**: 45–56.

Taylor HFW (1950) Hydrated calcium silicates. Part I. Compound formation at ordinary temperatures. *Journal of the Chemical Society*: 3682–3690.

Taylor HFW (1953) Hydrated calcium silicates. Part V. The water content of calcium silicate hydrate (I). *Journal of the Chemical Society*: 163–171.

Taylor HFW (1962) Hydrothermal reactions in the system $CaO–SiO_2–H_2O$ and the steam curing of cement and cement-silica products. *Proceedings of the 4th International Symposium on the Chemistry of Cement, Washington, DC 1960*. National Bureau of Standards, US Department of Commerce, Washington, DC, USA (1962): monograph **43**(1): 167–190 (specifically, p. 173).

Taylor HFW (1964) The calcium silicate hydrates. In *The Chemistry of Cements*. Academic Press, London, UK, vol. 1.

Taylor HFW (1969) The calcium silicate hydrates. *Proceedings of the 5th International Symposium on the Chemistry of Cement*, Cement Association of Japan, Tokyo **2**: 1–26.

Taylor HFW (1986) Proposed structure for C–S–H gel. *Journal of the American Ceramic Society* **69**: 464–467.

Taylor HFW (1993) Nanostructure of C–S–H: current status. *Advanced Cement Based Materials* **1(1)**: 38–46.

Taylor HFW and Howison JW (1956) Relationships between calcium silicates and clay minerals. *Clay Minerals Bulletin* **3(16)**: 98–111.

Taylor R, Richardson IG and Brydson RMD (2010) Composition and microstructure of 20-year-old ordinary Portland cement–ground granulated blast-furnace slag blends containing 0 to 100% slag. *Cement and Concrete Research* **40(7)**: 971–983.

Thomas NL and Double DD (1981) Calcium and silicon concentrations in solution during the early hydration of Portland cement and tricalcium silicate. *Cement and Concrete Research* **11(5-6)**: 675–687.

Thomas JJ, Chen JJ, Jennings HM and Neumann DA (2003) Ca-OH bonding in the C–S–H gel phase of tricalcium silicate and white Portland cement pastes measured by inelastic neutron scattering. *Chemistry of Materials* **15(20)**: 3813–3817.

Thorvaldson T and Vigfusson VA (1928) The action of water on tricalcium silicate and beta dicalcium silicate. *Transactions of the Royal Society of Canada*, 3rd Series, 22 (Part II, Section III): 423–431.

Tits J, Wieland E, Müller CJ, Landesman C and Bradbury MH (2006) Strontium binding by calcium silicate hydrates. *Journal of Colloid and Interface Science* **300(1)**: 78–87.

Tränkle S, Jahn D, Neumann T, Nicoleau L, Hüsing N and Volkmer D (2013) Conventional and microwave assisted hydrothermal syntheses of 11A tobermorite. *Journal of Materials Chemistry A* **1(35)**: 10318–10326.

Trettin R, Oliew G, Stadelmann C and Wieker W (1991) Very early hydration of dicalcium silicate-polymorphs. *Cement and Concrete Research* **21(5)**: 757–764.

van Bemst A (1955) Contribution to the study of the hydration of pure calcium silicates. *Bulletin des Sociétés Chimiques Belges* **64**: 333–351.

Viallis H, Faucon P, Petit JC and Nonat A (1999) Interaction between salts (NaCl, CsCl) and calcium silicate hydrates (C–S–H). *Journal of Physical Chemistry B* **103(25)**: 5212–5219.

Vinnakota S (2019) *Understanding the Long-Term Evolution of C–S–H Phases Present in Cement Backfills*. PhD thesis, University of Leeds, Leeds, UK.

Vinnakota S, Richardson IG, Brydson RMD, Swanton SW and Shelton A (2025) Synthesis of high Ca/Si ratio calcium silicate hydrate using an alkali-free supersaturated lime solution method. Submitted for publication; data from Vinnakota S (2019).

Walker CS, Savage D, Tyrer M and Ragnarsdottir KV (2007) Non-ideal solid solution aqueous solution modeling of synthetic calcium silicate hydrate. *Cement and Concrete Research* **37(4)**: 502–511.

Walker CS, Sutou S, Oda C, Mihara M and Honda A (2016) Calcium silicate hydrate (C–S–H) gel solubility data and a discrete solid phase model at 25°C based on two binary non-ideal solid solutions. *Cement and Concrete Research* **79**: 1–30.

Yamnova NA, Sarp K, Egorov-Tismenko YK and Pushcharovskii DY (1993) Crystal structure of jaffeite. *Crystallography Reports (c/c of Kristallografiia)* **38**: 464–466.

Yang SY, Yan Y, Lothenbach B and Skibsted J (2021) Incorporation of sodium and aluminum in cementitious calcium-alumino-silicate-hydrate C–(A)–S–H phases studied by ^{23}Na, ^{27}Al, and ^{29}Si MAS NMR spectroscopy. *Journal of Physical Chemistry C* **125(51)**: 27975–27995.

Yu P (1990) *The Behavior of Chloride in Cement Systems*. PhD thesis, University of Illinois at Urbana-Champaign, IL, USA.

Yu P, Kirkpatrick RJ, Poe B, McMillan PF and Cong X (1999) Structure of calcium silicate hydrate (C–S–H): near-, mid-, and far-infrared spectroscopy. *Journal of the American Ceramic Society* **82**: 742–748.

Yuan T, Wang J and Li Z (2010) Measurement and modelling of solubility for calcium sulfate dihydrate and calcium hydroxide in NaOH/KOH solutions. *Fluid Phase Equilibria* **297(1)**: 129–137.

Zhu X and Richardson IG (2023) Morphology-structural change of C–A–S–H gel in blended cements. *Cement and Concrete Research* **168**: 107156.

Ian G Richardson and HFW Taylor
ISBN 978-1-83549-777-7
https://doi.org/10.1108/978-1-83549-776-020253007

Chapter 7
Hydration of the calcium silicate phases

7.1. Introduction
7.1.1 Definitions and general points
In cement chemistry, the term 'hydration' denotes the totality of the changes that occur when an anhydrous cement, or one of its constituent phases, is mixed with water. The chemical reactions taking place are generally more complex than simple conversions of anhydrous compounds into the corresponding hydrates. A mixture of cement and water in such proportions that setting and hardening occur is called a paste, the meaning of this term being extended to include the hardened material. Unless stated otherwise, the water/cement (w/c) or water/solid (w/s) ratio refers to proportions by mass; for a paste, it is typically 0·30–0·60. Setting is stiffening without significant development of compressive strength and typically occurs within a few hours. Hardening is significant development of compressive strength and is normally a slower process. Curing means storage under conditions such that hydration occurs; conditions commonly employed in laboratory studies include storage in moist air initially and in water after the first 24 hours, storage in air of 100% relative humidity and, less favourable for reaction, storage in a sealed container.

Portland cement is a relatively complex mixture and as a consequence many studies that aim to elucidate its hydration chemistry have focused on its constituent phases. For a given particle size distribution and w/s ratio, tricalcium silicate or alite sets and hardens in a manner similar to that of a typical Portland cement. Using XRD or other methods, it can be shown that about 70–80% of the C_3S typically reacts in 28 days and >90% in 1 year, and that the products are calcium hydroxide (CH) and a nearly X-ray amorphous calcium silicate hydrate, called C–S–H, that has the properties of a rigid gel. β–C_2S behaves similarly, but much less CH is formed and reaction is slower, about 30% typically reacting in 28 days and 90% in 1 year. Figure 7.1 shows the XRD diffraction pattern of a 26-year-old paste of C_3S. Patterns of highly reacted β–C_2S pastes are similar, except that the CH peaks are much less intense. The only effects definitely attributable to the C–S–H are the diffuse peak at 2·7–3·1 Å and the somewhat sharper one at 1·82 Å.

For both C_3S and β–C_2S, reaction rates depend on particle size distribution and other factors. Development of compressive strength runs roughly parallel to the course of the chemical reactions, and the strengths at 1 year are comparable to those of Portland cements of the same w/s ratio and cured under the same conditions.

The calcium silicate hydrate formed on paste hydration of C_3S or β–C_2S is a particular variety of C–S–H, which is a generic name for any amorphous or poorly crystalline calcium silicate hydrate. The dashes indicate that no particular composition is implied and are necessary because CSH in cement chemical nomenclature denotes material specifically of composition $CaO·SiO_2·H_2O$ (e.g. suolunite; see Table 5.1(f)). The term 'C–S–H gel' is sometimes used to distinguish the material formed in cement, C_3S or β–C_2S pastes from other varieties of C–S–H.

Figure 7.1 Powder X-ray diffraction pattern of an almost fully reacted C_3S paste

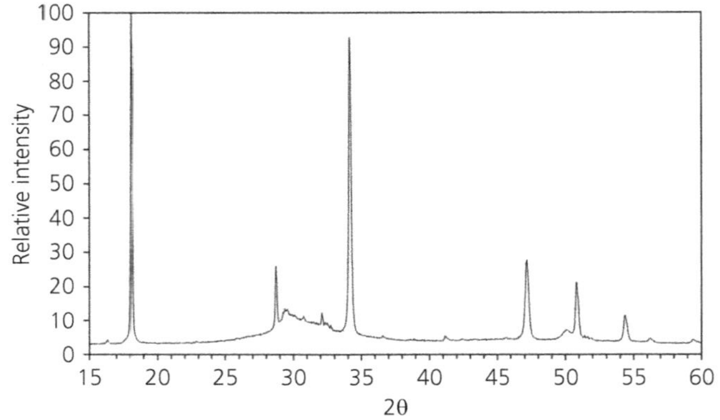

The XRD pattern was collected using a Bruker D2 PHASER (Cu Kα dual-wavelength radiation; $K\alpha_1$ = 1·54060 Å, $K\alpha_2$ = 1·54439 Å; 30 kV; 10 mA) with a step size of 0·01°, a dwell time of 6 s, continuous rotation of 6 rotations/min and angular range of 3–70° 2θ (total number of points = 6647). The powder was backloaded into the sample holder. TEM-EDX and MCL data for the C–S–H in this paste are given in Table 7.3 (the 26-year-old specimen) and §7.4.2, respectively.

This chapter deals primarily with the reactions and products of hydration of C_3S and β–C_2S in pastes at ordinary temperatures. Some closely related studies on cement pastes are also considered. Most of the results for pastes of C_3S or β–C_2S are applicable also to the behaviour of those phases in cement pastes. Except where stated, a temperature of 15–25 °C is assumed. Important background information obtained from studies of C_3S or β–C_2S hydration at higher w/s ratios (i.e. in aqueous suspensions), or on C–S–H prepared in other ways, is considered in Chapter 6.

7.1.2 Experimental considerations; carbonation

C–S–H reacts readily with CO_2 (or CO_3^{2-} ion). $CaCO_3$ is formed and, initially, the Ca/Si ratio of the C–S–H decreases; further reaction destroys the C–S–H, with the formation of hydrous silica. Contamination by atmospheric CO_2 ('carbonation') can probably be more serious in laboratory studies than in practical concrete mixes of lower surface-to-volume ratio. Many studies have been made under, as nearly as possible, CO_2–free conditions. This has the merit of simplifying interpretation, although the extent to which it corresponds to the situation in a real concrete mix is uncertain. To achieve a reasonable approximation to CO_2-free conditions, all operations, such as mixing, curing or drying of pastes, or preparation, filtration or drying of solid phases in suspensions, must be carried out in a CO_2-free atmosphere, either in a dry box or in apparatus otherwise designed to isolate the materials from the atmosphere. Wet or moist materials are especially susceptible to attack, but dry ones are not immune. Water must be deionised and freshly boiled. Because of the alkaline nature of calcium silicate or cement mixes, prolonged contact with glass must also be avoided; wet curing may be carried out in the presence of a small excess of water in a sealed plastic container.

In any laboratory study, therefore, it is usually important to know how much CO_2 is present. This applies both to situations in which it is supposedly absent and to those in which an attempt has been made to include a controlled quantity. Jones's (1940) method, based on decomposition with dilute acid and absorption of the CO_2 in $Ba(OH)_2$ solution, which is then back-titrated with HCl, is simple and accurate. TG has often been used to determine CO_2, but it can be unsatisfactory because the mass loss at 700–900 °C is not wholly due to CO_2 (if available, loss due to CO_2 and H_2O can be distinguished by analysis of the evolved gases using mass spectrometry). Absence of calcite peaks from an XRD pattern is not a proof that CO_2 is absent, as CO_2 can occur in other forms, e.g. as vaterite (μ–$CaCO_3$), in calcium aluminate hydrate phases and possibly within the C–S–H structure. Table 7.1 gives data for the three polymorphs of $CaCO_3$ that can form at ordinary pressures. Calcite is trigonal (rhombohedral), space group $R\bar{3}c$ (No. 167), with Z = 6, and aragonite is orthorhombic, space group $P\,m\,c\,n$ (No. 62, setting **bca**), with Z = 4. Vaterite is a metastable phase (McConnell, 1960) that can occur as an intermediate between an amorphous calcium carbonate (nominally $CaCO_3{\cdot}H_2O$) and calcite or aragonite (Rodriguez-Blanco *et al.*, 2011), which are more stable (Johnston *et al.*, 1916). Many vaterite polytypes have been proposed, e.g. Wang and Becker (2009), Wang *et al.* (2014), Makovicky (2016) and Chakoumakos *et al.* (2016). Christy (2017) reviewed 16 model structures and concluded that the most likely polytypes have space groups $C\,1\,2\,1$ and $P\,3_2\,2\,1$ (Demichelis *et al.*, 2012, 2013), a conclusion supported by ^{43}Ca NMR experiments (Burgess and Bryce, 2015). Small divalent-cation carbonates adopt the calcite-type structure, whereas large divalent-cation carbonates adopt the aragonite-type structure, the switch between the rhombohedral and orthorhombic structures occurring with calcium. This fact has important implications for atomistic modelling of structures for C–S–H; see §6.6.2.

Table 7.1 Crystal and optical data for calcium carbonate polymorphs (continued on next page)

	Calcite[a]	Aragonite	Vaterite[b]
Crystal data			
Crystal system	Trigonal (rhombohedral)	Orthorhombic	Hexagonal
Space group	$R\bar{3}c$ (No. 167)	$P\,m\,c\,n$ (No. 62, setting **bca**)	$P6_3/m\,m\,c$ (No. 194)
a: Å	4·9895 ± 0·0042	4·9614 ± 0·0018	4·13
b: Å		7·9675 ± 0·0031	
c: Å	17·0656 ± 0·0094	5·7416 ± 0·0031	8·49
Angles, where not 90°	120° (γ)		120° (γ)
Z	6	4	2
D_x: g cm^{-3}	2·711	2·929	2·650
Molar volume: cm^3 mol^{-1}	36·93	34·17	37·76
Reference(s)	Av. of 12 structures from: Antao *et al.* (2009); Borodin *et al.* (1979); Graf (1961); Lutterotti *et al.* (1997); Markgraf and Reeder (1985); Maslen *et al.* (1993, 1995); Pilati *et al.* (1998); Sass *et al.* (1957); Sitepu (2009); Wartchow (1989)	Av. of 8 structures from: Antao and Hassan (2009); Caspi *et al.* (2005); dal Negro and Ungaretti (1971); de Villiers (1971); Dickens and Bowen (1971); Jarosch and Heger (1986); Pokroy *et al.* (2007); Ye *et al.* (2012)	Kamhi (1963)

Table 7.1 (Continued)

	Calcite[a]	Aragonite	Vaterite[b]
Optical properties			
Refractive indices (Na_D)	$\omega = 1{\cdot}6584$, $\varepsilon = 1{\cdot}4864$	$\alpha = 1{\cdot}5300$, $\beta = 1{\cdot}681$, $\gamma = 1{\cdot}6854$	$\omega = 1{\cdot}550$, $\varepsilon = 1{\cdot}640$–$1{\cdot}650$
Optic orientation		$X = c$, $Y = a$, $Z = b$	
Optic sign	Negative	Negative (2V = 18°)	Positive
Morphology	Rhombs ($\alpha_R = 101{\cdot}9°$ for morphological cell)	Short columns, needles	Plates, radial aggregates
Reference(s)	Chang *et al.* (1998); Gifford (1902); Winchell and Winchell (1964)	Winchell and Winchell (1964)	Johnston *et al.* (1916); Winchell and Winchell (1964)

[a]Unit cell parameters for hexagonal axes.
[b]Parameters for pseudo-cell; true cell given by Kamhi (1963) as $a = \sqrt{3} \times 4{\cdot}13 = 7{\cdot}15$ Å and $c = 2 \times 8{\cdot}49 = 16{\cdot}98$ Å with Z = 12.

Especially with relatively young pastes, the problem often arises of removing excess water after a specified time of curing to stop the hydration reactions and to render the material less susceptible to carbonation. For this purpose, soaking or grinding in polar organic liquids, such as methanol, propan-2-ol or acetone, has often been employed, but should be used only in situations where it will not invalidate the results of subsequent tests. Organic substances can be completely removed from C_3S or cement pastes only by drastic procedures that seriously alter the material. Their presence is especially undesirable if any form of thermal analysis (e.g. TG or DTA) is to be carried out, since on heating they react with C–S–H to give CO_2 (Day, 1981; Taylor and Turner, 1987). Acetone has especially complex effects, since in the alkaline medium it rapidly undergoes aldol condensations at room temperature, giving mesityl oxide, phorone and other relatively involatile products (Taylor and Turner, 1987). Some workers consider that primary aliphatic alcohols alter the pore structures of C_3S or cement pastes or of synthetic C–S–H preparations; it has also been claimed that methanol reacts with CH at ordinary temperatures, giving calcium methoxide (Beaudoin, 1987).

In choosing a procedure to remove excess water, it is necessary to consider the likely effects on the tests subsequently to be performed. Retention of organic substances is not the only possible complication; unduly drastic drying procedures, such as heating at 105 °C, or equilibration to constant mass in atmospheres of low relative humidity, such as 'D–drying' (§7.2.3), partially dehydrate the C–S–H and, with cement pastes, also partially decompose the hydrated calcium aluminate or aluminate sulfate phases. For some procedures, such as XRD, the presence of organic material may not matter, but even here the possibility of decomposing or altering hydrated aluminate phases must be considered. Zhang and Scherer (2011) and Winnefeld *et al.* (2016) reviewed the literature, compared the different methods of direct drying and solvent exchange and made recommendations for different purposes. Winnefeld *et al.* (2016) suggested that solvent exchange using propan-2-ol (on crushed or cut slices of paste) followed by a few minutes at 40 °C is suitable for many purposes, including XRD, NMR, MIP and SEM. For TGA, they recommended removal of the propan-2-ol with diethyl ether (a less polar solvent with a lower boiling point) prior to the heating although, in apparent contradiction, they also suggested that some calcium aluminate hydrates are affected by even gentle heating. They suggested that short (e.g. 1 hour) freeze-drying is also suitable for TGA, XRD, NMR and SEM. Zhang and Scherer (2011) considered in detail the amount of time that is necessary for the exchange of pore water with propan-2-ol. Their treatment highlights the importance of using powdered paste, unless unavoidable (such as for MIP or microscopy), because the time necessary for exchange can otherwise be very long. For example, using their treatment, 1 mm-diameter granules of pastes with diffusivities (of the propan-2-ol in the cement paste) of $\approx 3 \times 10^{-11}$, $\approx 1 \times 10^{-11}$ and $\approx 3 \times 10^{-12}\,m^2\,s^{-1}$ are estimated to require ≈1·5, 4·5 and 15 hours, respectively, and 2 mm-diameter granules would need 4 times as long – that is, up to 60 hours. A step-by-step guide for the propan-2-ol/diethyl ether procedure is provided by Lothenbach *et al.* (2016). For electron microscopy, including the preparation of polished sections for SEM, freeze-drying has been found to be satisfactory (Jennings *et al.*, 1981; Winnefeld *et al.*, 2016). Kißling *et al.* (2022) preferred freeze-drying (for 3 hours) to solvent exchange with propan-2-ol in a study of early-age (10 to 90 minutes) CEM I samples. For many purposes, including thermal analysis above about 150 °C, pumping with a rotary pump for 1 h is suitable.

All analytical data on hydrated cements or cement constituents should normally be referred to the ignited mass. The original mass includes an arbitrary amount of water, and results referred to it, or to the mass after some arbitrarily chosen drying procedure (e.g. to constant mass at 105 °C), are often impossible to interpret in detail.

7.1.3 Calcium hydroxide

Calcium hydroxide (CH) has a layer structure (Figure 7.2(a); Bernal and Megaw, 1935; Petch, 1961). The calcium atoms are octahedrally, and the oxygen atoms tetrahedrally, coordinated. The interlayer forces are weak, with negligible hydrogen bonding, thus giving good (0001) cleavage. The unit cell is hexagonal, with $a = 3{\cdot}593$ Å, $c = 4{\cdot}909$ Å, space group $P\bar{3}2/m1$ (No. 164), $Z = 1$, $D_x = 2{\cdot}242\,g\,cm^{-1}$; the optical properties are: uniaxial –, $\omega = 1{\cdot}573$, $\varepsilon = 1{\cdot}545$ (Swanson and Tatge, 1953). CH is often described by its mineral name, portlandite.

Under ideal conditions of crystallisation, CH forms hexagonal plates. Admixtures greatly affect the morphology and especially the *c*:*a* aspect ratio (Berger and McGregor, 1972). Euhedral crystals are observed in young pastes and in pores of older pastes, but as hydration proceeds the main deposits of CH become massive and of indeterminate shape, although the good cleavage persists. There does not appear to be any convincing experimental evidence for claims that some of the CH in calcium silicate pastes is amorphous, which is different from the situation in synthetic preparations of C–S–H(I) (as discussed in §6.2.2). TEM studies of ion-thinned sections of C_3S pastes of normal w/s ratio have shown that the CH occurs predominantly as large, imperfect crystals; no finely dispersed, microcrystals were observed, although some crystals were only tens of nanometres thick (Groves *et al.*, 1986; Groves, 1987; Richardson and Groves, 1993a). Cryptocrystalline CH has, however, been found in cement pastes and in pastes made from calcium oxide and silica of high w/s ratio and of blast-furnace slag activated by CH.

Observations that the 0001 peak of CH in XRD powder patterns of calcium silicate or cement pastes is anomalously strong relative to the $10\bar{1}1$ peak (Grudemo, 1984; Zürz and Odler, 1987) have been attributed to overlap with a contribution from C–S–H, but probably arise from preferred orientation, which can be difficult to avoid. Stacking faults can also affect the CH pattern (Spinolo and Tamburini, 1985).

Figure 7.2 Information about calcium hydroxide. (a) Schematic diagram illustrating the structure of a single layer of CH. Medium-grey spheres represent calcium atoms and small dark spheres oxygen atoms; a hydrogen atom, not shown, completes the tetrahedral coordination of each oxygen atom. After Richardson (2004). (b) Solubility of CH measured for solutions made using coarse and fine CH crystals. Data are from Bassett (1934; open circles for large crystals and filled circles for fine) and Hedin (1955; open squares for crystals >1 mm; crosses for crystals that had average size of about 3·2 μm; filled squares for crystals of about 0·5 μm). (c) Thermogravimetric curve of CH (0·2 % CO_2); data are expressed as percentages on the ignited mass. 50 mg sample heated at 10 °C min^{-1} in dry, CO_2-free N_2 flowing at 15 mL min^{-1}

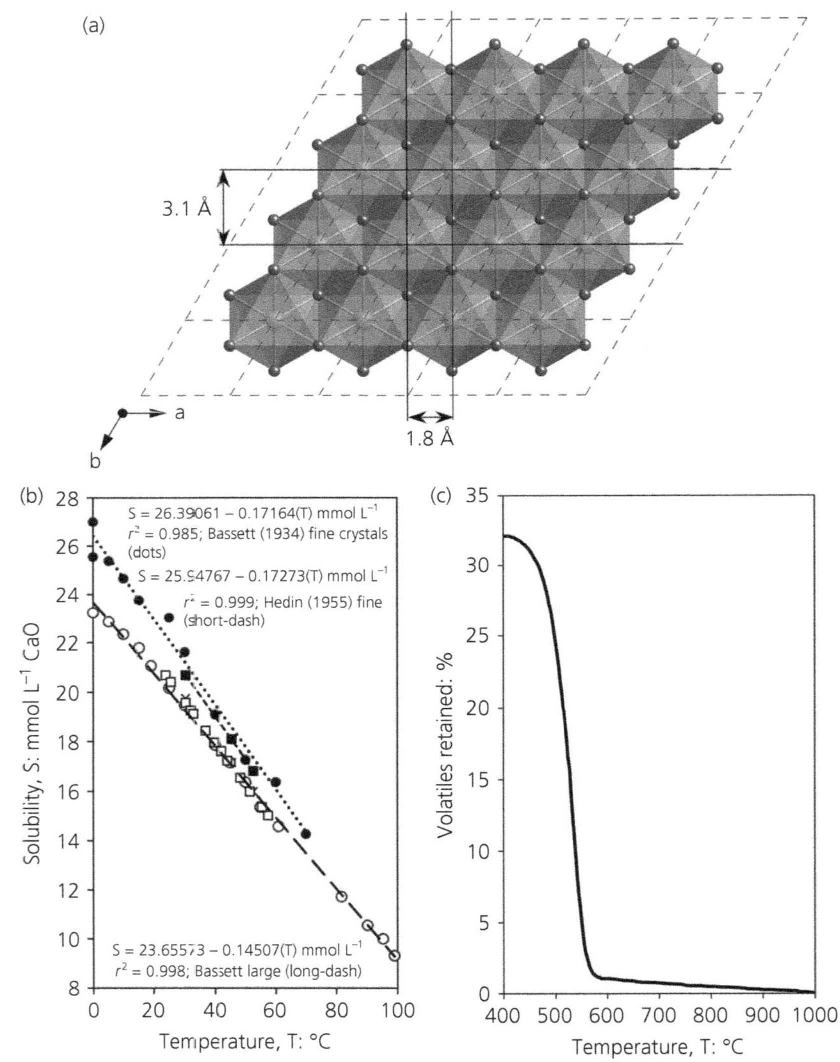

The dissolution of CH in water is exothermic ($\Delta H = -13{\cdot}8\,kJ\,mol^{-1}$ at 25 °C; Greenberg and Copeland, 1960) and the solubility therefore decreases with increasing temperature. The decrease in solubility is essentially linear, see Figure 7.2(b); regression analysis of Bassett's (1934) data for solutions that were made using large crystals gives Equation 7.1, where the solubility, *S*, is in $mmol\,L^{-1}$ CaO and temperature, *T*, is in °C (Richardson *et al.*, 2016).

$$S = 23{\cdot}65573 - 0{\cdot}14507(T)\,(r^2 = 0{\cdot}998) \tag{7.1}$$

$$S = 23{\cdot}39061 - 0{\cdot}17164(T)\,(r^2 = 0{\cdot}985) \tag{7.2}$$

While those data are consistent with the measurements of some other workers – for example, Johnston and Grove (1931) and Bates *et al.* (1956) – it is evident from Figure 7.2(b) that higher apparent solubilities occur readily with finely crystalline material, up to at least 25 and 22 $mmol\,L^{-1}$ at 5 °C and 25 °C, respectively; regression analysis of Bassett's (1934) data for solutions that were made using fine crystals gives Equation 7.2. Inspection of Hedin's (1955) data on Figure 7.2(b) shows that the solubility of crystals that had average size of about 3·2 μm is very similar to those >1 mm, whereas those of about 0·5 μm is greater. This difference has been considered to be due to a high surface energy of CH crystals in water, with the surface energy of fine particles contributing to an overall increase in the Gibbs free energy of formation and thus to an enhancement in its solubility (Duchesne and Reardon, 1995).

Halstead and Moore (1957) reported thermal decomposition equilibria. The decomposition pressure reaches 101 kPa (1 atm) at 512 °C and the mean enthalpy of decomposition at 300–510 °C is 102·9 $kJ\,mol^{-1}$. When heated in a vacuum or in a stream of dry N_2, CH begins to decompose at an easily detectable rate at about 370 °C. Under typical conditions for TG, decomposition is 98 % complete at the end of the step at 580 °C (Figure 7.2(c)). An earlier conclusion that only about 94 % has decomposed at this point (Taylor, 1984) is incorrect. The finely divided CH available as a laboratory or analytical reagent normally contains several per cent of $CaCO_3$, and the curve in Figure 7.2(c) is for coarsely crushed large crystals containing 0·2 % CO_2.

7.2. Composition, density and other data for C–S–H

7.2.1 Calcium hydroxide content

Methods used to determine CH have included TG, DTG, semi-isothermal DTG, thermal evolved gas analysis, DTA, differential scanning calorimetry (DSC), quantitative X-ray diffraction analysis (QXDA), IR spectroscopy, image analysis of backscattered electron images and extraction methods using either aqueous media or organic solvents similar to those used to determine free lime in anhydrous cements. As an example, Figure 7.3 shows the TG and DTA curves of a fully reacted C_3S paste, obtained under the same conditions as the TG curve of CH in Figure 7.2(c). The paste had been prepared at w/s = 0·45 and stored wet in a sealed container for 25 years at 25 °C; it contained 0·8 % of CO_2, referred to the ignited mass. The height of the CH step (mean from seven determinations and estimated as shown) is 8·9 %, referred to the ignited mass, which corresponds to 1·15 mol CH per mole of C_3S reacted. When measuring the CH step (indicated on Figure 7.3 with the grey arrow) allowance should be made for any mass lost by C–S–H at that temperature – that is, simply taking the mass lost between lower and upper cut-off temperatures will result in overestimation of the amount of CH.

Some workers studying pastes of C_3S or cement have found significant differences between the results of thermal methods and QXDA, the latter usually giving lower results for the CH content (Bentur and Berger, 1979; Dalziell and Gutteridge, 1986; Midgley, 1979; Odler and Dörr, 1979a; Ramachandran, 1979), but others have found no differences outside experimental error (Le Sueur *et al.*, 1984; Taylor *et al.*, 1985). There is broad agreement that extraction methods give higher results than either thermal methods or QXDA (Bentur and Berger, 1979; Midgley, 1979; Odler and Dörr, 1979a; Ramachandran, 1979). It has been suggested that QXDA gives low results due to the

Figure 7.3 Thermogravimetric (full line) and differential thermal analysis (dotted line; endothermic down) curves for a fully reacted C_3S paste; experimental conditions as for Figure 7.2(c). TG data are expressed as percentages on the ignited mass. After Taylor and Turner (1987)

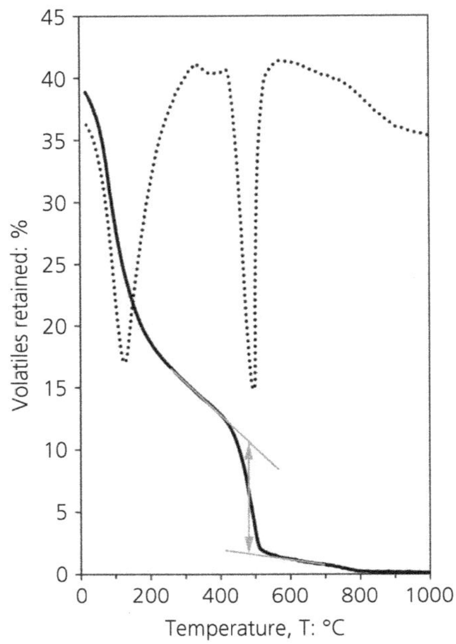

presence of amorphous CH (Ramachandran, 1979). Thermal methods have also been considered to give low results for the same reason (Odler and Dörr, 1979a), or because of CH adsorbed on the C–S–H (Bentur and Berger, 1979) or present as interlayer material (Shebl *et al.*, 1985). Sources of error in CH determination by TG have been discussed (Taylor, 1984). As noted above, large errors can arise from use of thermal methods on material that has been in contact with organic liquids (Taylor and Turner, 1987).

The borderline between CH interlayered with C–S–H and that present as a separate phase is indistinct. The absence of evidence for amorphous or cryptocrystalline CH in C_3S pastes is noted in §7.1.3. The higher results given by extraction methods are probably attributable to partial decomposition of the C–S–H, which is supported by the results on synthetic C–S–H preparations discussed in Chapter 6. Of all the methods, TG or DTG is probably the least open to criticism, although, especially with cement pastes, precision and accuracy are limited by the difficulty of distinguishing sharply between the effects due to CH and to the other phases present. DTA is somewhat suspect because the peak area for CH is reported to depend markedly on particle size (Webb and Heystek, 1957).

7.2.2 Indirect determination of the Ca/Si ratio of C–S–H

C–S–H is the main binding phase in all Portland cement-based concrete, and so knowledge of its chemical composition is important; as a consequence it has been the subject of much study, conducted over many decades. The average Ca/Si ratio of the C–S–H in a fully reacted C_3S or β–C_2S paste can be calculated if the contents of CH and CO_2 are known, although an assumption must be made as to whether the CO_2 occurs in the C–S–H or as $CaCO_3$. Figure 7.3 includes the TG curve for a fully reacted C_3S paste. Assuming that the CH step represents the decomposition of 98% of the CH and that the CO_2 occurs as $CaCO_3$, there is 1·15 mol of CH and 0·04 mol $CaCO_3$ per mole of reacted C_3S; the average Ca/Si ratio of the C–S–H is therefore 1·81 (i.e. 3 – 1·19 = 1·81). In principle, the method can be extended readily to incompletely reacted pastes if the content of unreacted C_3S or β-C_2S is also known; this is evident from Equations 7.3 and 7.4, which summarise the reactions of C_3S or β–C_2S with water (Locher, 1966).

$$C_3S + (3 - x + y)H \rightarrow C_xSH_y + (3 - x)CH \tag{7.3}$$

$$C_2S + (2 - x + y)H \rightarrow C_xSH_y + (2 - x)CH \tag{7.4}$$

In practice, the determination of the Ca/Si ratio of the C–S–H (i.e. x in Equations 7.3 and 7.4) in this way is not straightforward, particularly at low degrees of reaction. The reason for the latter is evident from Figure 7.4(a), which is a plot of % CH (on the basis of ignited mass) against %

Figure 7.4 Plots of % CH (ignited mass) against % hydration for C_3S pastes; experimental details are given in Table 7.2. (a) Data of Dent Glasser *et al.* (1978) compared with 19 grey lines representing Ca/Si ratios between 0·7 and 2·5 in increments of 0·1 (those for 1·4, 1·7 and 2·0 are labelled). (b) Dent Glasser *et al.*'s (1978; filled diamonds) data compared with those of Odler and Dörr (1979a; +), Odler and Becker (1980; ×) and Ogawa *et al.* (1980; open diamonds). (c) Comparison of data from a group at the University of Oxford that used either single-pulse ^{29}Si MAS NMR (Rodger, 1986; filled circles) or QXRD to determine the % hydration (Le Sueur *et al.*, 1984; unfilled circles). (d) Comparison of data from studies that all used ^{29}Si NMR to determine the % hydration: Rodger (1986; filled circles; open squares are for β-C_2S pastes); Clayden *et al.* (1984a; right-half filled circles); Clayden *et al.* (1984b; open circles); Dobson *et al.* (1988; bottom-half filled circles); Goñi *et al.* (2010; open diamonds). (e) Example of the use of image analysis of BE images to determine the % hydration (Costoya Fernandez, 2008); the % CH was either determined by TG (circles) or calculated by mass balance after the Ca/Si ratio of the C–S–H had been measured by SEM-EDX (squares; analyses of large regions of inner product; the % CH calculated by mass balance assuming 0% CH at 0% hydration) (continued on next page)

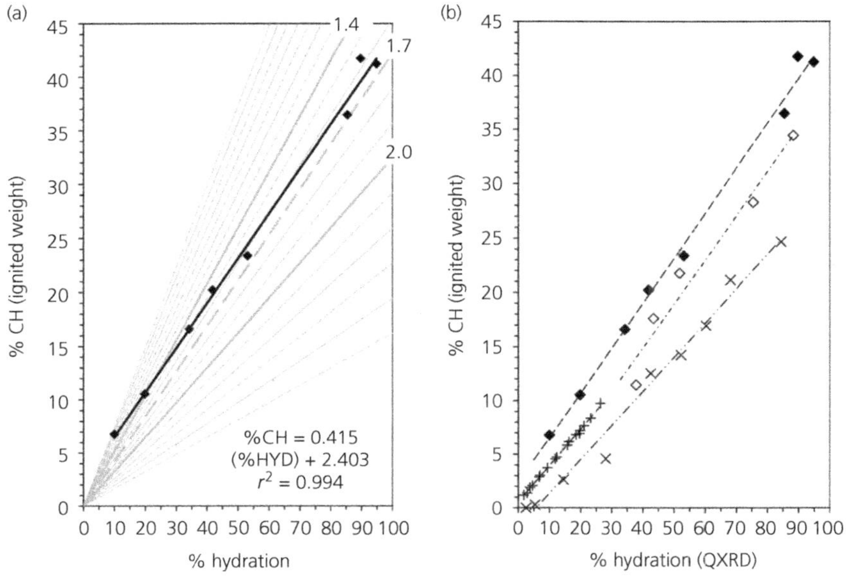

Figure 7.4 Continued

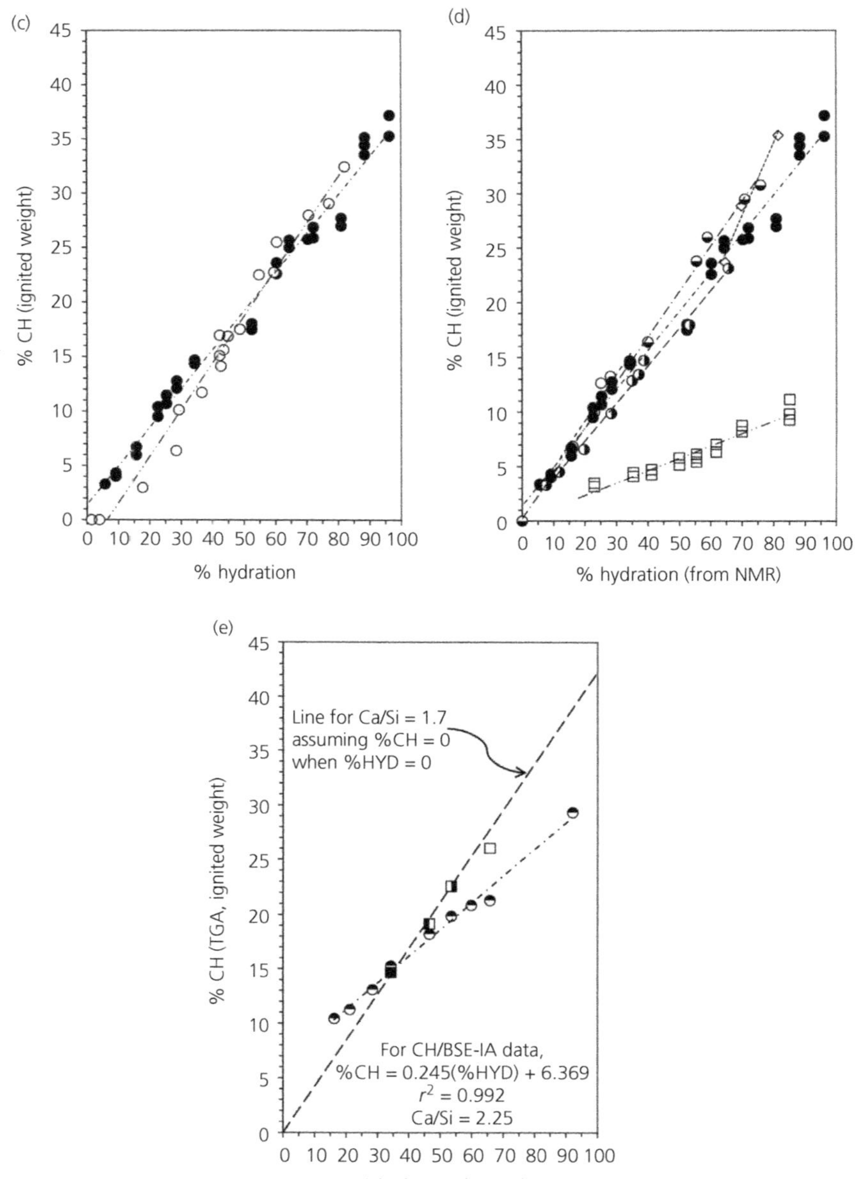

hydration that has 19 lines (in grey) that represent Ca/Si ratios between 0·7 and 2·5 in increments of 0·1 (those for 1·4, 1·7 and 2·0 are labelled). The lines are very close at low degrees of reaction and so even small errors in the determination of % CH or % hydration can have a big impact on the calculated Ca/Si ratio. A consideration of potential errors is complicated by the possibility that the average Ca/Si ratio of the C–S–H is not constant – that it might change as the hydration progresses – which was the interpretation made by Dent Glasser *et al.* (1978). Their data are reproduced on

Figure 7.4(a) (diamonds and full black line); they used thermogravimetry (TG) to measure the quantity of CH and quantitative X-ray diffraction (QXRD) to determine the % hydration. On inspection, it is immediately clear that the data points do indeed correspond to different Ca/Si ratios – that is, the data points are close to different grey lines. However, it is also clear that there is a good linear relationship between the increase in % CH and the increase in % hydration, so a more straightforward interpretation is that the average Ca/Si ratio is constant (for these data, Ca/Si = $1{\cdot}72 \pm 0{\cdot}01$) and that there is a systematic error in the measurement of the % CH or % hydration, or both. The grey lines on Figure 7.4(a) are drawn to the origin, and so the assumption in the figure is that there is no CH present at zero % hydration and that CH is produced as soon as the C_3S starts to react. This is a simplification because CH does not start to form until a small amount of the C_3S has reacted (Rodger, 1986; Rodger *et al.*, 1988); for example, linear regression analysis of the mass-balance data presented in §7.7.3 results in the intercept at 0·3 % hydration. The likelihood that errors (random and systematic) influence the calculated Ca/Si ratio is evident when data from many studies of the reactions of C_3S and β-C_2S with water are compared, as in Figures 7.4(b)–(d).

Figure 7.4(b) compares Dent Glasser *et al.*'s (1978) data with those from three other studies that used the same techniques (i.e. TG and QXRD); some experimental details and the results from re-examination of the data are given in Table 7.2. All four studies show a good linear relationship between % CH and % hydration, but the regression lines are not close to one another, they give different average Ca/Si ratios (the values are given in Table 7.2), they do not all intercept the same axis, and the intercept is generally not close to the origin. The excellent linear relationship given by Odler and Dörr's (1979a) data ($r^2 = 0{\cdot}998$), where each data point is based on 15 separate measurements obtained from three independent samples, strongly supports the view that the average Ca/Si does not change with the extent of hydration (over the range studied). Nevertheless, while their approach certainly resulted in the minimisation of random errors, it is still possible that the calculated Ca/Si ratio could be affected by a systematic error, which could result in an offset and/or affect the gradient, and thus the Ca/Si ratio.

^{29}Si MAS NMR has been used as an alternative to QXRD to measure the percentage hydration of C_3S; it was first demonstrated by Lipmaa *et al.* (1982) but has been used most extensively to study the reaction of C_3S by a group at the University of Oxford during the 1980s and 1990s (Beedle *et al.*, 1989; Brough, 1993; Brough *et al.*, 1994, 1995, 1996; Clayden *et al.*, 1984a, 1984b, 1986; Dobson *et al.*, 1988; Groves *et al.*, 1991; Rodger, 1986; Rodger *et al.*, 1987, 1988). It is not unreasonable to expect that the procedures adopted by members of that group were essentially standardised, so inspection of their data is instructive when considering possible errors: Figure 7.4(c) compares data from Le Sueur *et al.* (1984), who used TG and QXRD, with data from Rodger (1986), who used TG and NMR; Figure 7.4(d) shows four sets of C_3S data that all combined TG with NMR. It is evident from these data (and those on Figure 7.4(b), summarised in Table 7.2) that the Ca/Si ratio determined by indirect techniques is sensitive to experimental errors, either individual or systematic, which is a reasonable explanation for the differences in the values. Nevertheless, it is clear that the average Ca/Si ratio is high: for the nine C_3S studies represented on Figures 7.4(a)–(d) the average Ca/Si = $1{\cdot}82 \pm 0{\cdot}13$. It is also evident that conclusions drawn from the axis intercepts of regression lines should be treated with caution; for example, Le Sueur *et al.* (1984) concluded from the data in Figure 7.4(c) that amorphous CH is present in C_3S pastes (because the regression line intercepts the % hydration axis) and that it is not detected by XRD or TG (because they determined a good correlation between the quantity of CH detected by the two techniques); comparison with the other data on Figures 7.4(c)–(d) suggests strongly that Le Sueur *et al.*'s (1984) data are not sufficiently robust (because of the scatter in the data) to conclude that amorphous CH is present.

Table 7.2 Typical values for the average Ca/Si ratio of C–S–H in hardened C_3S pastes determined by indirect techniques

Reference	System	w/s	°C	Drying method	Technique for % CH	Technique for % hydration	% free lime	Ca/Si	Axis cut by regression (intercept value: %)
Dent Glasser *et al.* (1978)	C_3S 300 $m^2 kg^{-1}$	0·7	25	Evacuation	TG –	QXRD quartz standard	<0·1	1·72	CH (2·4)
Odler and Dörr (1979a)	C_3S 300 $m^2 kg^{-1}$	0·7	20	D-drying	TG –	QXRD quartz standard in original mix	0·2	1·95	CH (0·5)
Odler and Becker (1980)	C_3S 470 $m^2 kg^{-1}$	0·5	20	Acetone / D-drying	TG 5 °C min^{-1}	QXRD	–	2·02	HYD (5·7)
Ogawa *et al.* (1980)	C_3S –	0·4	20	D-drying	TG 10 °C min^{-1}	QXRD MgO standard	<0·3	1·75	HYD (3·3)
Clayden *et al.* (1984a)	C_3S PS ≤50 μm	0·5	–	Propa-2-ol / evacuation	TG –	^{29}Si NMR	–	1·94	CH (0·5)
Clayden *et al.* (1984b)	C_3S PS[a] ≤50 μm	0·5	–	–	TG –	^{29}Si NMR	–	1·69	CH (0·6)

Le Sueur (1984); Le Sueur *et al.* (1984)	C_3S 370 $m^2 kg^{-1}$	0·5	25	Propa-2-ol / evacuation	TG 5 °C min^{-1}	QXRD Rutile standard	0·2	1·68	HYD (6·3)
Rodger (1986)	C_3S PS ≤64 μm	0·5	25	Propa-2-ol / evacuation	TG 5 °C min^{-1}	^{29}Si NMR	0·2	1·90	CH (1·4)
Rodger (1986)	β-C_2S 0·5 wt.% B_2O_5 PS ≤64 μm	0·5	25	Propa-2-ol / evacuation	TG 5 °C min^{-1}	^{29}Si NMR	<0·5	1·74	CH (0·1)
Dobson *et al.* (1988)	C_3S –	0·5	25	Propa-2-ol / evacuation	TG 5 °C min^{-1}	^{29}Si NMR	–	1·72	CH (0·2)
Goñi *et al.* (2010)	C_3S PS ≤40 μm	0·4	–	Acetone / ethanol	TG 4 °C min^{-1}	^{29}Si NMR	Negligible	0·94	HYD (28·2)

[a]PS = particle size.

Goñi *et al.* (2010) also used TG and NMR but claimed that the Ca/Si ratio of the C–S–H can be determined solely from TG data. However, they did so by assuming that 1·3 moles of CH are formed on complete reaction of 1 mole of C_3S, and so a Ca/Si ratio of 1·7 was inevitable; small variations from 1·7 seem to have been due to rounding. Consideration of their NMR and TG data together (presented on Figure 7.4(d) reveals results that are implausible: a Ca/Si ratio for the C–S–H of 0·94 and 28% hydration before the formation of CH, although there are only three data points; unfortunately, experimental details are not given for their NMR spectra, but saturation of NMR peaks is one possible source of systematic error.

Another technique that has been used to determine the degree of reaction of C_3S is the image analysis of backscattered electron micrographs (BEI-IA). Costoya Fernandez (2008) reported results for four different particle size distributions of an alite and combined them with values for % CH determined by TG. Re-analysis of Costoya Fernandez's (2008) data shows that all four data sets show a linear relationship between % CH and % hydration, but in all but one case (for the finest particle size sample, < 6 μm) the regression line intercepts the % CH axis at an implausibly high value (between 3·8 and 6·4%) and the calculated average Ca/Si ratio is in all cases high (between 1·98 and 2·25): an illustrative set of data are shown on Figure 7.4(e), which was chosen because Costoya Fernandez (2008) also measured the Ca/Si ratio for (inner product) C–S–H in that sample directly by SEM-EDX (see discussion in §7.3.4). Costoya Fernandez's (2008) measured values (shown on Figure 7.4(e) with square symbols) are much lower than the value derived indirectly (1·68–1·78 compared with 2·25), which suggests strongly that there is a problem with the gradient from the regression analysis of the CH % hydration data on Figure 7.4(e) as well as the axis intercept. It seems probable that this problem derives largely from the BEI-IA methodology used to determine the % hydration (although it is also possible that there is an additional issue with the TG data), which as a consequence ought to be revisited, particularly given that the technique is being used for neat Portland cement and composite cement pastes, which have more complex microstructures.

7.2.3 Water content

Evidence considered in §7.3 and in Chapters 5 and 6 indicates that the C–S–H of calcium silicate or cement pastes has a layer structure, and that, together with a pore solution, it forms a rigid 'gel' in which the pores range in size from macroscopic to enlarged interlayer spaces of nanometre dimensions. One can therefore define a water content only in relation to a specified drying condition. Three such conditions will be considered.

The most highly hydrated state is that existing in a saturated paste, i.e. one in which the pores are totally filled with water. According to the theory of Powers and Brownyard (1946/7), the C–S–H in a cement paste can form only if sufficient space is available to permit it to be accompanied by a defined proportion of pore space. There is thus a certain w/c ratio below which complete hydration is impossible, and for which a mature paste consists entirely of hydration product, including this essential pore space. The total water content of such a paste in the saturated condition is typically 42–44%, referred to the ignited mass. For nearly saturated C_3S pastes, Young and Hansen (1987) concluded that the composition of the C–S–H was approximately $C_{1.7}SH_4$ – that is, $x = 1{\cdot}7$ and $y = 4{\cdot}0$ in Equation 7.3, which corresponds to a w/c ratio of 0·42 (the number of mols of water per mol of $C_3S = 3 - x + y = 5{\cdot}3$; w/c $= (5{\cdot}3\,\text{mol} \times 18\,\text{g}\,\text{mol}^{-1})/(1\ \text{mol} \times 228\,\text{g}\,\text{mol}^{-1}) = 0{\cdot}42$). This is similar to the values found for cement.

Chemically bound water is most reasonably defined as including that present in interlayer spaces, or more firmly bound, but not that present in pores larger than this. The distinction between interlayer space and micropores is not sharp; water adsorbed on surfaces of pores further blurs the definition. From the experimental standpoint, the determination is complicated by the fact that the amount of water retained at a given RH depends on the previous drying history of the sample and on the rate at which water is removed. An approximate estimate is obtained by equilibrating a sample, not previously dried below saturation, with an atmosphere of 11% RH (Feldman, 1969; Feldman and Ramachandran, 1971, 1974). Saturated aqueous $LiCl \cdot H_2O$ gives the required RH at 25 °C (Young, 1967), which corresponds to a partial pressure of water vapour = 2·7 torr at 25 °C. This is evident on inspection of Figure 7.5, which shows how RH varies with water vapour pressure and temperature. To achieve apparent equilibrium in a reasonable time (several days), the sample must be crushed and the system evacuated; the salt solution should be stirred, at least intermittently. Young and Hansen (1987) found the composition of the C–S–H in C_3S paste thus equilibrated to be $C_{1 \cdot 7}SH_{2 \cdot 1}$.

Figure 7.5 (a) Section of the pressure–temperature phase diagram for water. (b) Part of (a) showing how the vapour pressure of water changes with temperature between 0 and 30 °C. The top curve is the saturated water vapour pressure, and thus represents 100% relative humidity (RH); for example, the equilibrium vapour pressure of water is 6·5 torr at 5 °C while it is 23·8 torr at 25 °C. Lower relative humidities are shown by the other curves in increments of 10% RH

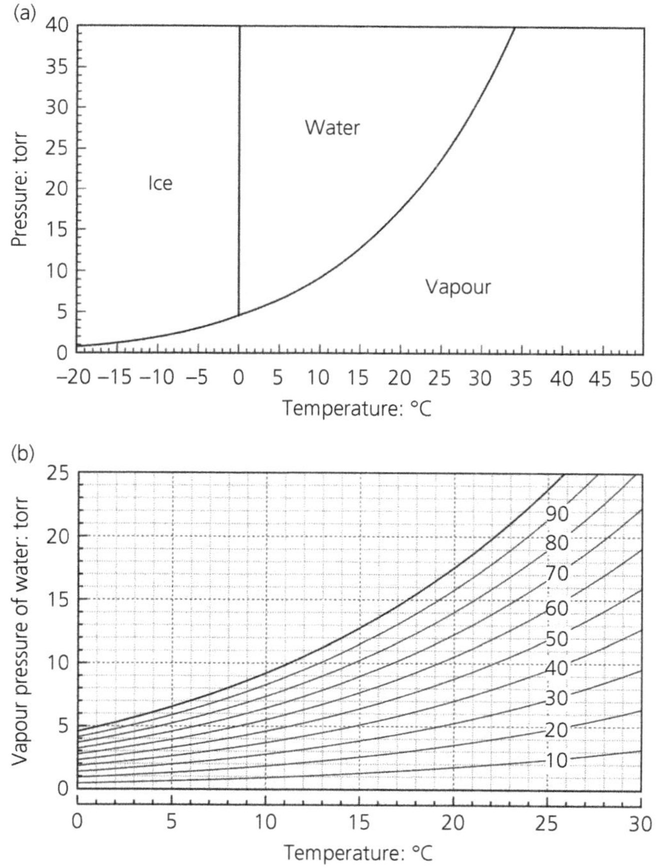

As noted above, treatment of C_3S pastes with methanol leaves some of the latter strongly sorbed. The methanol is removable only on heating at temperatures at which it reacts with the C–S–H, causing carbonation. However, the total mass of volatiles retained in a C_3S paste that has been soaked in methanol and then pumped for about 1 h with a rotary pump is near that obtained on equilibration with saturated $LiCl{\cdot}H_2O$ (Taylor and Turner, 1987). Methanol treatment can thus be used as a rapid, though approximate, method of determining chemically bound water. Other organic liquids could possibly be used in a similar way.

In a procedure known as 'D-drying', the sample is equilibrated with ice at –79 °C by continuous evacuation with a rotary pump through a trap cooled in a mixture of solid CO_2 and ethanol (Copeland and Hayes, 1953). The partial pressure of the water vapour is 5×10^{-4} torr. Heating to constant mass at 105 °C in an atmosphere of uncontrolled humidity, but free from CO_2, reduces the water content to approximately the same value, although it is not a strictly defined drying condition. The corresponding temperature on a TG curve obtained in dry and CO_2-free N_2 at 10 °C min^{-1} is about 145 °C but depends on the rate of gas flow and design of the apparatus. Water retained in pastes subjected to D-drying or equivalent procedures is known as 'non-evaporable' water. Contrary to some statements in the literature, it is not a measure of chemically bound water, as much water is lost from interlayer spaces under these conditions. It can nevertheless be used as an empirical measure of the degree of reaction. Typical values in the literature for the water contents of fully reacted C_3S pastes dried by these procedures are 20·4–22·0% (Odler and Dörr, 1979a), corresponding to C–S–H compositions of $C_{1\cdot7}SH_{1\cdot3-1\cdot5}$.

Powers and Brownyard (1946/7) called the water lost from the C–S–H on passing from the saturated to the D-dry condition 'gel water'. It comprises part of the pore water plus an arbitrarily defined fraction of the interlayer water. Whatever procedure is used to dry or equilibrate a paste, CO_2-free conditions are essential, and the amount of water retained can be obtained accurately only if the ignition loss and CO_2 content are also determined. Calculations based on the initial w/s ratio are often unreliable due to changes in the water content during curing.

7.2.4 Density and infrared spectra

The density of C–S–H depends on the water content. For strongly dried samples, it also depends on how the solid volume is defined, causing observed values to vary with the fluid used in their determination. Widely differing values can be equally meaningful, but for a given application the appropriate one must be used. Since the C–S–H of C_3S or β-C_2S pastes is mixed with CH, it is necessary to correct for this phase ($D_x = 2{\cdot}242\,g\,cm^{-3}$, §7.1.3) or to remove it. Ramachandran (1979) described a way of doing this.

At about 11% RH on first drying, the interlayer spaces are full of water and larger spaces are empty; H_2O/SiO_2 is approximately 2·0 (§7.2.3). Using helium pyknometry, Feldman (1972) obtained values of 2·35–2·36 $g\,cm^{-3}$ for fully reacted C_3S pastes; correction for CH gave 2·43–2·45 $g\,cm^{-3}$ for the C–S–H. Determinations on C–S–H samples made by hydrating C_3S in suspension showed that substitution of methanol or aqueous CH for helium as the fluid had no significant effect on the result. For a fully hydrated C_3S paste with w/s = 0·40 from which the CH had been removed, Hansen (1987) found 2·18 $g\,cm^{-3}$. For all these values, empty pores are excluded from the solid volume.

The C–S–H present in a saturated paste is assumed to include that part of the porosity, and its content of water, without which it cannot be formed. The H_2O/SiO_2 ratio is approximately 4·0. Assuming the water in excess of that present at 11 % RH to have a density of 1·00 $g\,cm^{-3}$, Hansen (1987) calculated the density of the C–S–H to be 1·85–1·90 $g\,cm^{-3}$. This value is near the values of

1·90–2·10 g cm^{-3} for cement pastes under saturated conditions (Powers and Brownyard, 1946/7), if allowance is made for the CH and other hydrated phases that they contain.

Helium pyknometry of D-dried samples has given values of 2·32–2·35 g cm^{-3} for C_3S pastes (Feldman, 1972) and of 2·44 g cm^{-3} for cement pastes (Powers and Brownyard, 1946/7); correction for CH and other phases suggests values of 2·40–2·50 g cm^{-3} for the C–S–H. For a C_3S paste with w/s = 0·40 from which the CH had been removed, Hansen (1987) found 2·39 g cm^{-3}. With D-dried samples, methanol gives results similar to helium, but water or CH solution gives much higher values; Brunauer and Greenberg (1962) obtained values around 2·86 g cm^{-3}, after correction for CH, for the C–S–H of C_3S or β-C_2S pastes. Determinations using water on D-dried cement pastes, based on the difference between the masses of the D-dried and saturated materials, typically give values, uncorrected for CH and other phases, of 2·50–2·60 g cm^{-3}. Relis and Soroka (1980) found these to increase with the degree of hydration, an upper limit of 2·70 g cm^{-3} being found for a specific cement.

Feldman (1972) explained the difference between the values for D-dried samples obtained with aqueous and other fluids. When the material is dried, the layers partially collapse and some empty spaces remain between them. Water penetrates into these spaces, and the density obtained, after correction for CH and any other phases present, is that of the layers themselves. Organic liquids do not penetrate, and helium penetrates only slowly. The empty interlayer space is counted as part of the solid volume, with a consequent decrease in the measured density.

The IR absorption spectra of pastes of C_3S, alite or β–C_2S (Hunt, 1962; Lehmann and Dutz, 1962; Slegers *et al.*, 1977) are relatively diffuse, as would be expected from the poor crystallinity. Si–O stretching is represented by a single, broad band peaking at about 970 cm^{-1}; this is compatible with the existence of condensed silicate anions. Small amounts of carbonate are readily detectable by the absorption near 1400 cm^{-1}.

7.3. Microstructure, microanalysis and electron diffraction

7.3.1 Microstructure

The first effective studies of microstructural development in calcium silicate pastes were made using light microscopy of thin sections (Brown and Carlson, 1936). They showed that the CH grows from relatively few centres in the water-filled space, where it forms isolated masses typically some tens of micrometres in size. Much more CH is observed with C_3S than with β-C_2S. At the same time, a product forms around the anhydrous grains and spreads into the remainder of the water-filled space. The first TEM study of a tricalcium silicate paste was conducted by Radczewski *et al.* (1939) who examined a dispersed specimen. They described the morphology of C–S–H as '*meist nadlige oder spindlige ... deren Bildung in der Weise vor sich zu gehen scheint, das sich die zuerst gebildeten Nadeln zu Spindeln zusammenlegen*' [*mostly acicular or spindle-like ... whose formation seems to proceed in such a way that the needles formed first fold into spindles*; Google Translate]. It is the same morphology (described as cigar-shaped bundles of fibres) as possessed by the semi-crystalline high-Ca C–S–H that Gard and Taylor (1976) called C–S–H(II), as discussed in §6.3.2 and illustrated in Figure 6.10(c). This morphology was confirmed by other early TEM examinations of ground and redispersed material, but those studies did not show where the needles or bundles of fibres came from in the microstructure (Copeland and Schultz, 1962; Grudemo, 1962). Such information was obtained from SEM studies of fracture surfaces (Chatterji and Jeffery, 1966; Diamond, 1976; Goto *et al.*, 1976; Williamson, 1972), which showed that the fibres of C–S–H formed in the originally water-filled space (called 'outer product', Op) radiating from the volume of the original anhydrous grains, whereas the C–S–H formed in situ from the larger anhydrous grains (called 'inner product', Ip) appeared almost structureless. This inner/outer product designation – which is due to Taplin (1959) – is illustrated in Figure 7.6.

Berger *et al.* (1972) studied the morphology of C–S–H visible on fracture surfaces of tricalcium silicate pastes by SEM and TEM of replicas. They designated the fibrillar morphology as Type I (they described it as '*rolled sheet or acicular projections from C_3S grains*'), which they observed for hydration with water and many chemical admixtures. They distinguished another morphological type, designated as Type II, which they described as '*sheets which have numerous concavities or a pseudo "honeycomb" structure*'; Type II was associated with admixtures that cause an acceleration of the hydration. Diamond (1976) extended Berger *et al.*'s (1972) classification, adding Types III and IV. Type III, prominent in older pastes, was more massive and appeared to consist of tightly packed equant grains up to 300 nm across. Type IV, still more featureless and massive, was the inner product, and was also observed in older pastes.

While SEM of fracture surfaces has provided valuable information, its utility is lessened by the inherently unrepresentative nature of a fracture surface (because the fracture follows the weakest path). Further information has come from the use of other techniques, especially examination of ion-thinned sections by TEM or scanning transmission electron microscopy (STEM) (e.g. Dalgleish and Ibe, 1981; Groves *et al.*, 1986; Richardson and Groves, 1993a; Viehland *et al.*, 1996), and backscattered electron imaging (BEI) of polished sections (e.g. Scrivener, 1989;

Figure 7.6 Schematic diagram illustrating Taplin's (1959) inner/outer product reference microstructure, in this case for hardened C_3S pastes. Any hydration product that is within the original boundaries of the C_3S particles is an 'inner product' and any product that is in the originally water-filled space is an 'outer product'

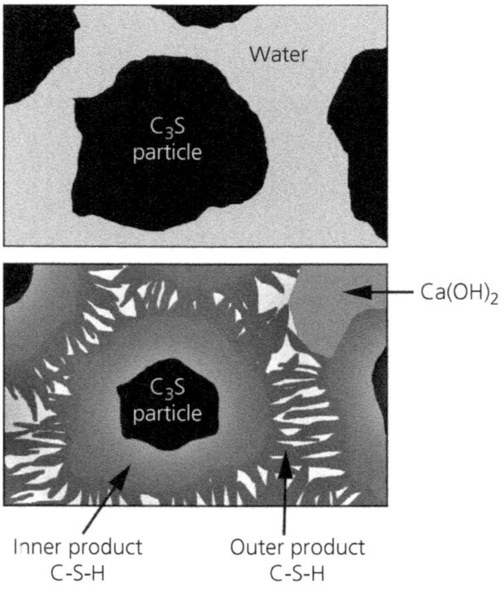

Scrivener and Pratt, 1984a; Scrivener *et al.*, 1987; Kjellsen and Justnes, 2004; Kjellsen and Lagerblad, 2007). Other methods have included X–ray imaging of polished surfaces (e.g. Taylor and Newbury, 1984), SEM of polished surfaces impregnated and etched to reveal porosity (Jennings and Parrott, 1986a, 1986b), micro-X-ray computer tomography and focused ion-beam nanotomography (Kleiner *et al.*, 2022), and soft X-ray transmission microscopy (Juenger *et al.*, 2003, 2005). Jennings *et al.* (1981) introduced another morphological classification system for C–S–H formed in hydrating C_3S pastes based on observations made by TEM. It consists of Types E (or 'early'), O (or 'middle' product, formed between 4 and 24 hours), 1, 3 and 4. Types 1 and 3 were both suggested to form from Type O depending on the amount of space available: Type 1 in open areas $>1\,\mu m$ and occurring as tapered needles radiating perpendicular from C_3S grains with an aspect ratio of about 10, and Type 3 where the inter-particle spacing is less than $1\,\mu m$. Type 4 is the dense inner product (Ip) C–S–H, which they termed 'late' product. Groves *et al.* (1986) found it difficult to make a sharp distinction between the morphologies of Types 1 and 3, and instead characterised all outer product (Op) C–S–H in hardened C_3S (and Portland cement) pastes as 'fibrillar', with the fibrils having a range of aspect ratios, depending on the amount of space available for growth (Richardson, 2002). This 'fibrillar' Op C–S–H observed in TEM studies corresponds to Berger *et al.*'s (1972) Type I morphology, while Op C–S–H described from TEM studies as 'foil-like' corresponds to their Type II morphology.

There has been much discussion about the effects of drying on the observed morphologies of C–S–H, with some workers concluding that fibrous or fibrillar materials may be, at least in part, artefacts of drying (e.g. Jennings *et al.*, 1981; Scrivener and Pratt, 1984c). For example, using a high voltage TEM with an environmental cell, Jennings and Pratt (1979, 1980) claimed to have observed C–S–H 'sheets' roll up on drying, and concluded that the true (delicate) structure is not seen in dried specimens. However, Richardson (2002) noted that their 'wet' C–S–H is very similar to the morphology of coarse Op C–S–H observed in (obviously dry) ion-thinned specimens, and that their 'dry' C–S–H is typical of electron-beam damaged coarse Op C–S–H (their specimen was hydrated at 55 °C; Op C–S–H formed at higher temperatures tends to be coarser).

SEM techniques that have been used to minimise the effects of drying on hydrate structures use either very low temperatures (cryo-SEM) or 'environmental' microscopes or stages. In cryo-SEM, wet samples are rapidly frozen (e.g. by being plunged into a cryogen) and the frozen cement paste is then fractured; its microstructure is gradually revealed by sublimation of the frozen water, e.g. Fylak *et al.* (2006) started sublimation at an N_2 gas pressure of 10–50 Pa (0·075–0·375 torr) and a temperature of –80 to –110 °C. In the environmental SEM (ESEM), a differential pumping system is used to enable the sample chamber to be maintained at relatively low vacuum (around 10 torr). Water vapour is used as the imaging gas and a conductive coating is not needed because the gaseous layer around the specimen becomes ionised and suppresses the accumulation of charge. Water-saturated conditions can be produced by controlling the temperature of the sample. For example, inspection of Figure 7.5(b) shows that with a chamber pressure set at 10 torr, a sample maintained at 10 °C (using a Peltier cooling stage) would be wet; a slight increase in temperature would result in evaporation of water, with the relative humidity reducing to 90 % at about 13 °C. With ESEM, it is important that samples are not dehydrated during the initial pump-down sequence followed by re-wetting at the start of imaging (Donald, 2003). Cryo-SEM, ESEM and other techniques have provided compelling evidence that the true morphology of the high-Ca/Si-ratio Op C–S–H that occurs in hardened C_3S and neat Portland cement pastes is fibrillar, and that the fibrils are not an artefact of drying. The evidence includes the following.

(*a*) Frozen samples of C_3S and Portland cement pastes examined using cryo-SEM show the same needle-like outgrowths of Op C–S–H that are seen in dried pastes (Bailey and Stewart, 1984; Fylak *et al.*, 2006). This is the case for young or old pastes. For example, the fibrous hydration product observed by cryo-SEM in Figure 9 of Zingg *et al.* (2008) appears essentially identical to early-age features observed on conventionally dried samples (e.g. see Figure 3(b) of Le Sueur *et al.*, 1984).

(*b*) Fibrillar Op C–S–H is observed in C_3S pastes by ESEM (e.g. Fonseca and Jennings, 2010; Möser, 2010a, 2010b; Möser and Stark, 2003, 2007).

(*c*) Relicts of AFt observed in ion-thinned specimens of Portland cement pastes in the TEM show very well-defined hexagonal outlines in the surrounding C–S–H that would not be preserved if the C–S–H had changed shape on drying (Richardson and Groves, 1993a).

(*d*) Fibrous Op C–S–H is observed in C_3S pastes by TEM examination when using low electron dose and low temperature (–175 °C) (Rößler *et al.*, 2006).

(*e*) TEM images of replicas of fracture surfaces made using a two-stage method show fibrillar C–S–H morphology that is the same as observed by direct examination of ion-thinned specimens; the two-stage method provides a replica and thus images of the specimen totally without contact with vacuum or electron beam (Groves, 1990a).

(*f*) TEM images of carbonated cement pastes show low-Ca fibre-like features (surrounded by microcrystals of calcium carbonate) that appear strikingly like the fibres of outer product C–S–H, which suggests strongly that the original morphology is preserved as a relict after decalcification; the carbonation would have occurred, of course, prior to any drying associated with preparation of the samples for TEM (e.g. Groves *et al.*, 1991).

(*g*) Fibrillar outer product C–S–H that is intergrown with calcium hydroxide occurs in hardened C_3S pastes (e.g. see Figure 25 of Richardson, 1999). The fibrils of C–S–H could not have formed from the rolling up of sheets on drying because they are encased in calcium hydroxide.

(*h*) Katz *et al.* (2007) examined a hydrating Portland cement paste using a WETSEM™ cell that allows direct observation of wet samples in a conventional SEM. They observed features that are consistent with the fibrillar morphology of Op C–S–H.

(*i*) Images from soft X-ray transmission microscopy of wet samples of early-age C_3S (w/s = 5) show a product with needle-like morphology, which must be Op C–S–H (Juenger *et al.*, 2003, 2005).

There is evidence that the method and rate of drying is important when fracture surfaces are the subject of the examination (Fonseca and Jennings, 2010; Zhang *et al.*, 2018). Zhang *et al.* (2018) concluded that specimens dried using a supercritical drying method or solvent exchange with propan-2-ol were both very close to the natural morphologies but that direct drying by evaporation of the pore solution should not be used.

7.3.2 Stages in microstructural development

The surfaces of C_3S particles examined soon after mixing with water often appear uneven (Henderson and Bailey, 1993; Makar *et al.*, 2007, 2011; Ménétrier *et al.*, 1979, 1980a; Nicoleau and Bertolim, 2016). This has been interpreted as being due to localised dissolution of C_3S

(e.g. Ménétrier *et al.* (1979) referred to 'corroded' surface; Nicoleau and Bertolim (2016) to etch pits) or to holes in an early product (e.g. Makar *et al.*, 2007); it is possible that both exist. Juilland *et al.* (2011) claimed that their Figure 2 (a TEM micrograph referenced to Gallucci *et al.* (2010) but not in that paper) showed '*a clear crystallographic orientation of alite consumption*', but the alignment of striations on each side of the central particle with each other and with those on the edge of a separate particle in the top left of the micrograph suggests that they may have been an artefact of ion-beam milling caused by lack of sufficient rotation of the sample. Ménétrier *et al.* (1979, 1980a) reported that the first detectable product consisted of minute particles. Jennings *et al.* (1981) distinguished three principal stages of product formation in C_3S pastes: an early product formed during the first 4 hours, a middle product formed at 4–24 hours, and a late product formed subsequently. The early product formed by the hydration of C_3S in pastes or compacted powders placed in water has been described variously as foils, flakes, honeycombs or a fibrous gel (Fujii and Kondo, 1969; Groves, 1987; Henderson and Bailey, 1993; Jennings *et al.*, 1981; Ménétrier *et al.*, 1979, 1980a; Stewart and Bailey, 1983).

The middle stage of hydration is characterised by the rapid formation of C–S–H and CH. In fracture surfaces of calcium silicate pastes, Type I C–S–H is prominent, which is the fibrillar material that is seen as the outer product in ion-thinned sections examined in the TEM (Groves *et al.*, 1986). The appearance of the early product in the TEM and the development of fibrillar Op C–S–H are illustrated in Figures 7.7(a)–(g) for pastes aged between 2 hours and 8 years. At 2 hours, the product that seems to be at the surface of the C_3S appears to be nanoporous but otherwise quite featureless (as in Figures 7.7(a) and (b)) but further away it has a more open morphology (as in Figure 7.7(c)), which is quite similar to that shown in Figure 4 of Groves (1983). It occurs both on the surface of the C_3S – which is almost entirely covered – and in clumps. The thickness of the surface product appears to vary quite considerably, between about 10–100 nm, which is the same as the estimate given by Scrivener *et al.* (2015) for the thickness of the early '*hydrated layer*'. It is possible that some clumps of product are formed in the solution some distance from the surface and then deposited on it during TEM sample preparation. Small needles or fibrils appear to grow out from the surface product (examples are evident in Figures 7.7(a) and (b)) and they increase in length with age: at 4 hours (Figure 7.7(d)) they are around 70–150 nm long and at 8 hours (Figure 7.7(e)) they are 150–300 nm. At 4 hours, agglomerations of minute particles are present at the base of the fibrils, as shown in Figure 7.7(g), which is an enlargement of a region of Figure 7.7(d). These particles are around 3–7 nm in diameter and comparison with micrographs at later age (e.g. the bottom half of Figure 7.7(f)) suggests that they constitute Ip C–S–H. At 8 years (Figure 7.7(f)), the fibrils are considerably longer and wider, and they appear to consist of agglomerations of long thin particles that are approximately 3–6 nm in their smallest dimension, with length varying from a few nm to many tens of nm (Richardson, 2004), as illustrated in Figure 7.7(h). The fibrils of Op C–S–H that form on the surface of the original C_3S particle tend to radiate into the outer product region in a fan-like manner from relatively small areas (Richardson and Groves, 1993a). For the 8 hour sample, the % hydration determined by ^{29}Si MAS NMR was 11·2% and the % CH from TG was 4·35% (Li, 2008), which thus gives a Ca/Si ratio for the C–S–H of 1·80. This value is obtained by assuming zero CH at zero hydration and zero Ca and Si in solution. Consideration of the composition of the solution results in a slightly smaller value; for example, it would be 1·77 if the solution had the composition of Naber *et al.*'s (2020) 8 hour sample. The morphology of the Op C–S–H in Figures 7.7(d) and (e) is the same as that observed for synthetic high Ca/Si C–S–H prepared using solutions that are alkali-free and supersaturated with CaO, as discussed in §6.3 (e.g. compare Figures 7.7(e) with the top-right micrograph in Figure 6.8(b)).

Figure 7.7 TEM micrographs showing early product or C–S–H present in hydrating C_3S pastes with w/s = 0·50 hydrated at 25 °C for: (a)–(c) 2 hours (scale bars = 100 nm on (a) and (b)); (d) 4 hours; (e) 8 hours. (f) TEM micrograph showing Ip and Op C–S–H present in a hardened C_3S paste with w/s = 0·40 hydrated at 20 °C for 8 years; modified from Figure 6(a) of Richardson (2004). (g) Enlargement of a region of (d) showing fibrils of Op C–S–H and globular particles of product at the base of the fibrils. The horizontal edge length is 500 nm. (h) TEM micrograph showing fibrils of Op C–S–H present in a hydrating C_3S paste with w/s = 0·40 hydrated at 20 °C for 8 years. An enlargement (×3·75) of the region that is in the black rectangle is inset in the top left-hand corner, with an idealised distribution of the particles in the fibril shown below it (drawn assuming that the darker regions in the fibril are particles of C–S–H and the lighter regions essentially correspond to pore space). The bright-field images in Figures 7.7(a)–(e) and (g) are modified from Li (2008). The image in Figure 7.7(h) was recorded with Figure 7.7(f) (continued on next page)

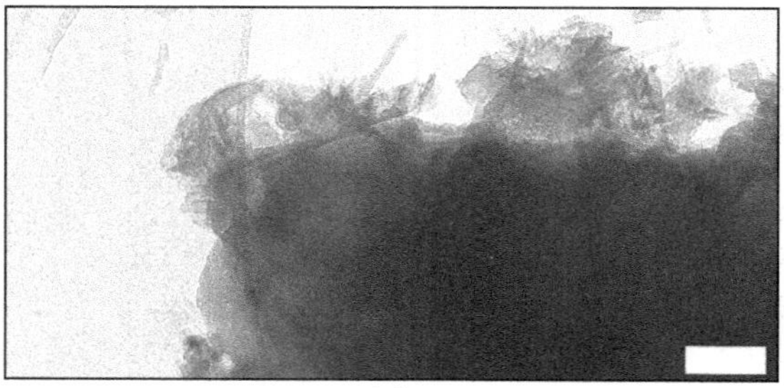

(a)

(b)

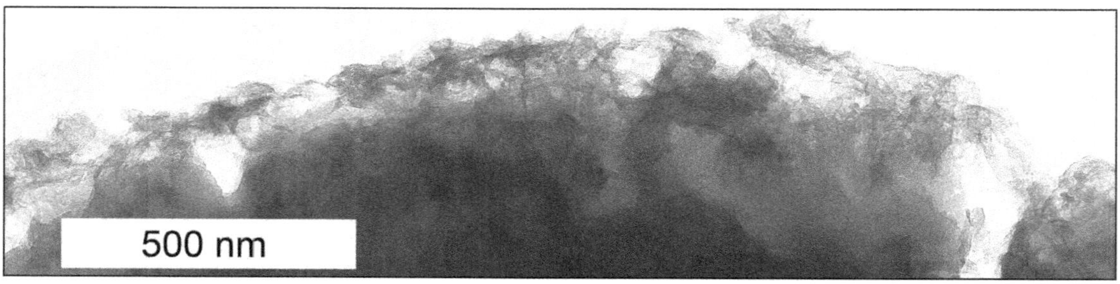

(c)

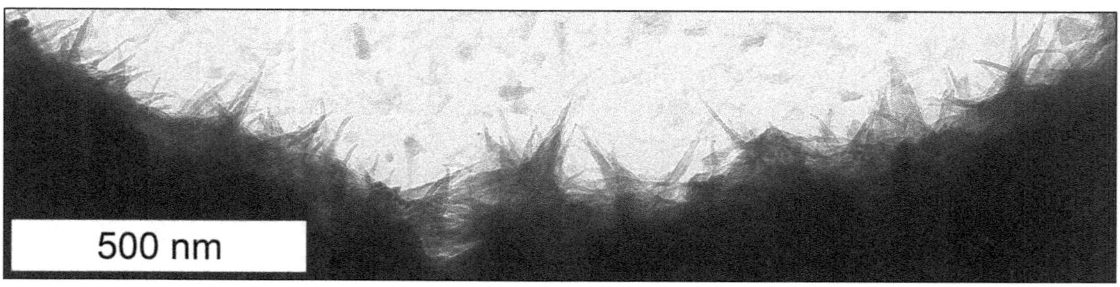

(d)

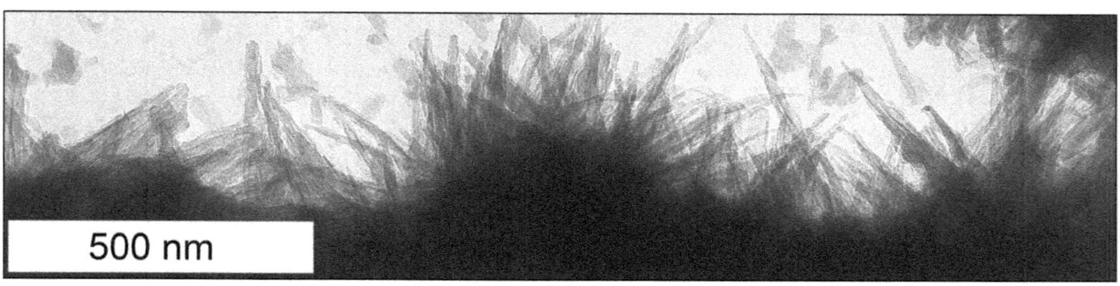

(e)

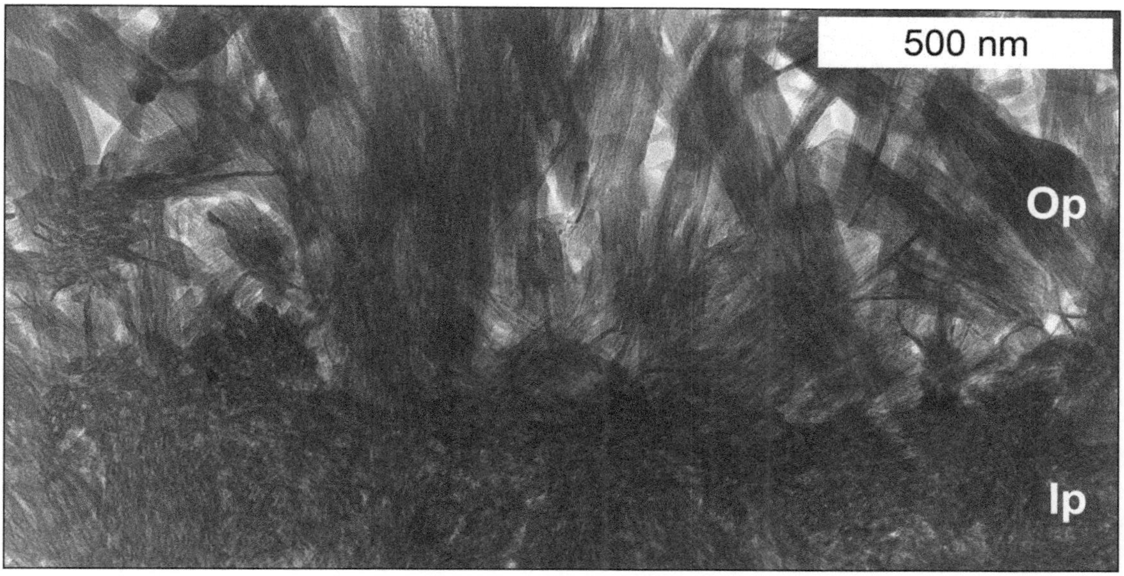

(f)

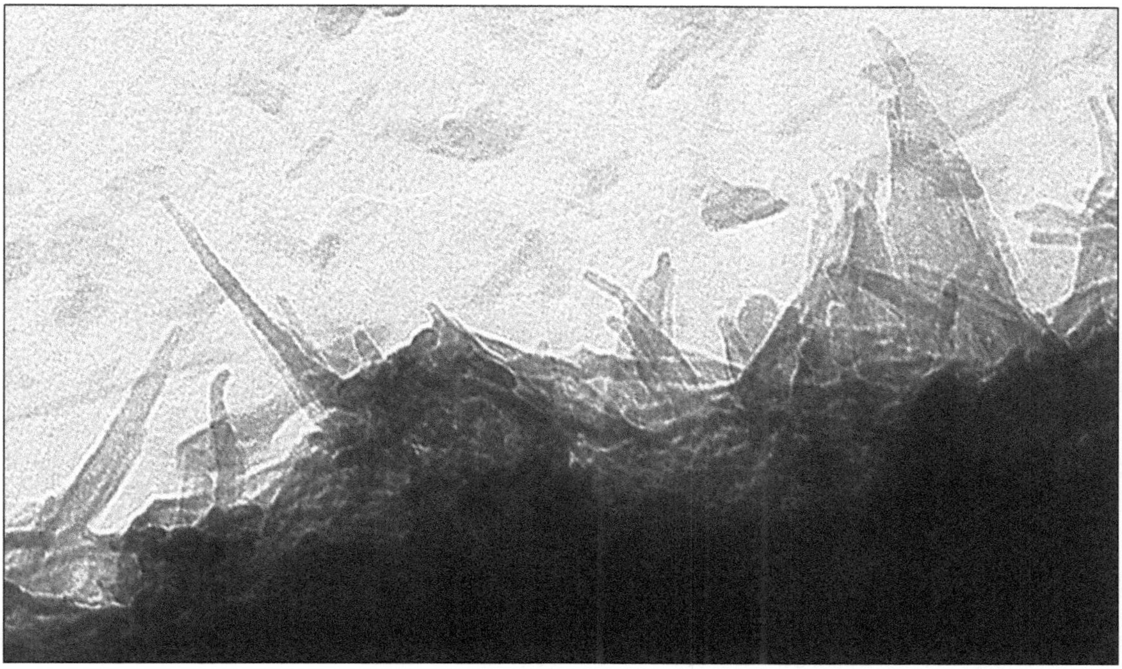

(g)

Figure 7.7 Continued

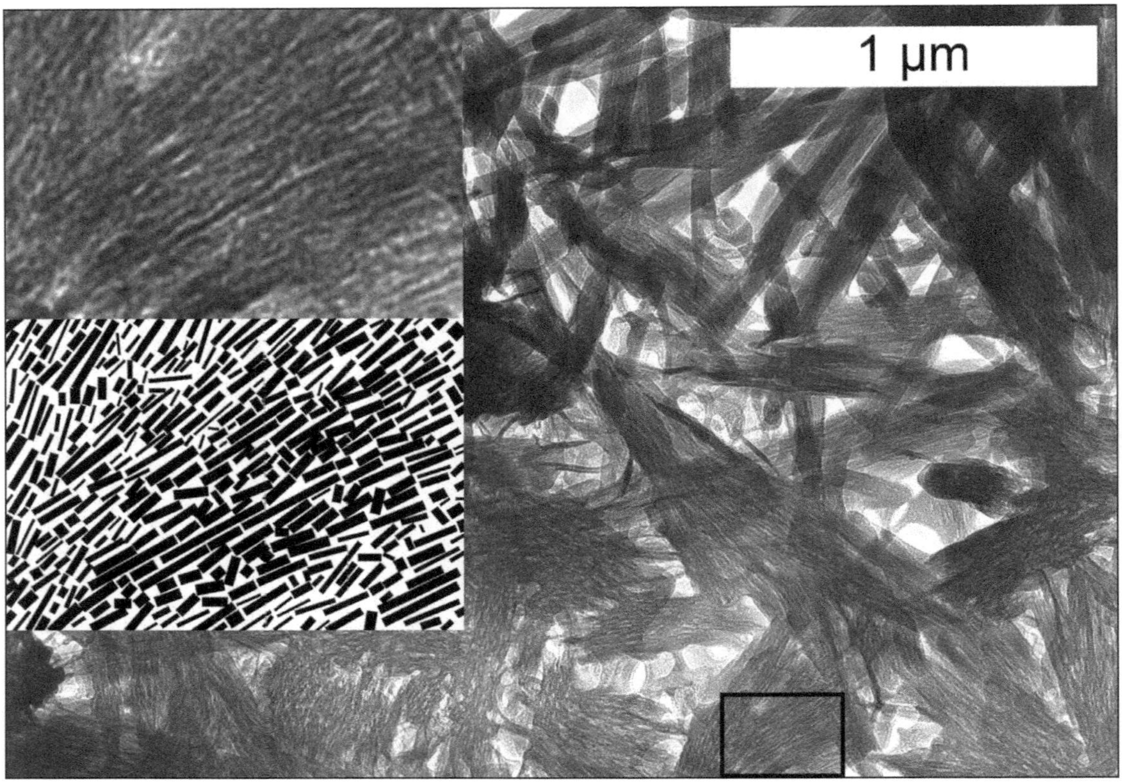

(h)

The distribution of phases within the microstructure of a hardened C_3S paste is most effectively observed by BEI examination of flat polished sections (e.g. Kjellsen and Justnes, 2004; Kjellsen and Lagerblad, 2007). As an example, Figure 7.8 shows backscattered electron images of two regions at different magnifications in a fully reacted 32-year-old C_3S paste (data from Taylor, 2010). The mid-grey regions are Ip C–S–H and they indicate the size and location of the original C_3S particles. The CH is present in very large deposits (labelled on the left-hand-side of the micrograph), which is a common observation (e.g. Kjellsen and Justnes, 2004; Kjellsen and Lagerblad, 2007). The fibrillar morphology of the Op C–S–H is evident on the higher magnification BEI image. For comparison, a TEM micrograph of a region within the same sample that has both Ip and Op C–S–H is inset, although it is at much higher magnification (the relative scales are indicated on the scale bar). Spaces are sometimes observed between the hydrating C_3S particles and the surrounding hydrate (e.g. Kjellsen and Justnes, 2004; Kjellsen and Lagerblad, 2007), which is a feature that is more common in Portland cement pastes.

The early hydration of β-C_2S appears to follow a similar course to that of C_3S, but reaction is slower and begins preferentially at grain boundaries and exsolution lamellae (Groves, 1981; Ménétrier *et al.*, 1980c). The Op C–S–H that forms also has fibrillar morphology (Richardson, 2004; Williamson, 1972), which is evident on comparison of the TEM images shown in Figure 7.7 (for C_3S) and Figure 7.9 (for β-C_2S).

Figure 7.8 Backscattered electron images of regions in a fully reacted 32-year-old C_3S paste

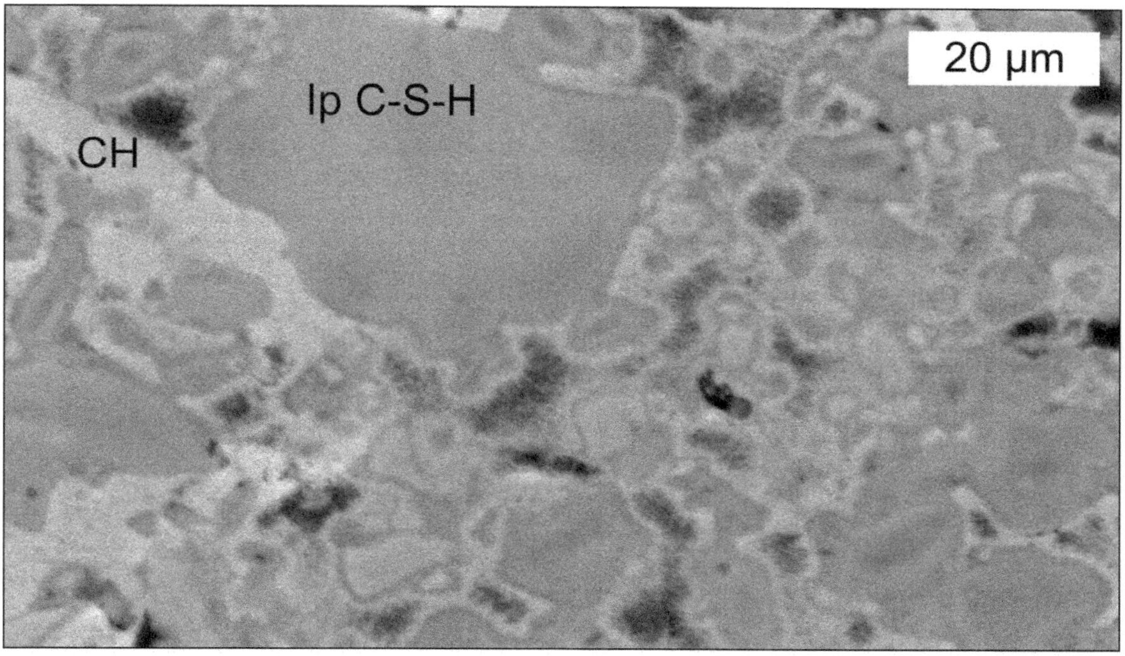

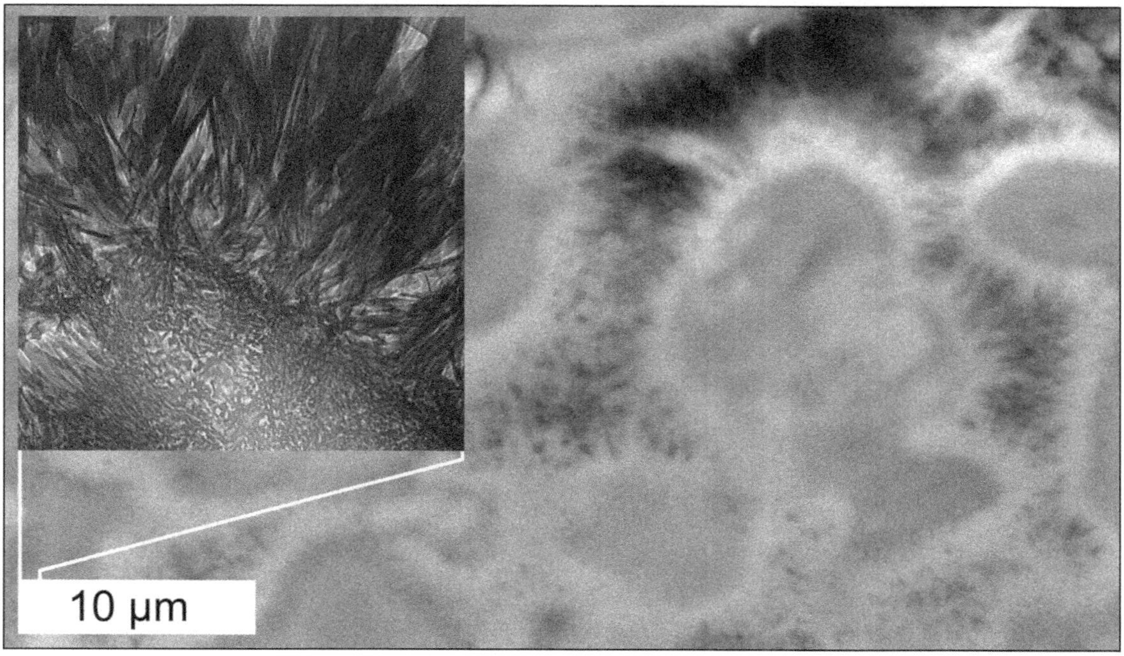

Operating voltage 20 kV, working distance of 12 mm (data from Taylor (2010)). The paste was mixed on 25 July 1978, w/s = 0·50 and the C_3S was supplied by the American Cement Technical Center, 11 July 1969. The free lime content was 0·06 %. The chemical analysis of the C_3S was: CaO = 72·69 %; SiO_2 = 26·06 %; Al_2O_3 = 0·39 %; MgO = 0·46 %; Fe_2O_3 = 0·10 %; ignition loss = 0·18 %. The sample was given to Richardson by Dr J Beaudoin. It was stored sealed in a plastic sleeve together with moist tissue; it was a pale pink colour when opened. Powder X-ray diffraction showed intense peaks for CH, small peaks for calcite and calcium oxalate and broad peaks at about 3·07 and 1·83 Å for C–S–H.

Figure 7.9 Transmission electron micrographs showing coarse Op C–S–H present in hardened β-C_2S pastes: (a) in a region of a carbon replica of a fracture surface of a paste hydrated for 90 days (modified from Figure 40(c) of Williamson (1972; micrograph attributed to RL Berger); w/s and temperature of hydration not given; editing of the image included removal of halftone artefacts); (b) in a region of an ion-thinned paste with w/s = 0·40 hydrated at 20 °C for 3 months (modified from Figure 10 of Richardson, 2004). The horizontal edge length of both images is 1·25 μm. The similarity of the morphology of the Op C–S–H in the two images is striking

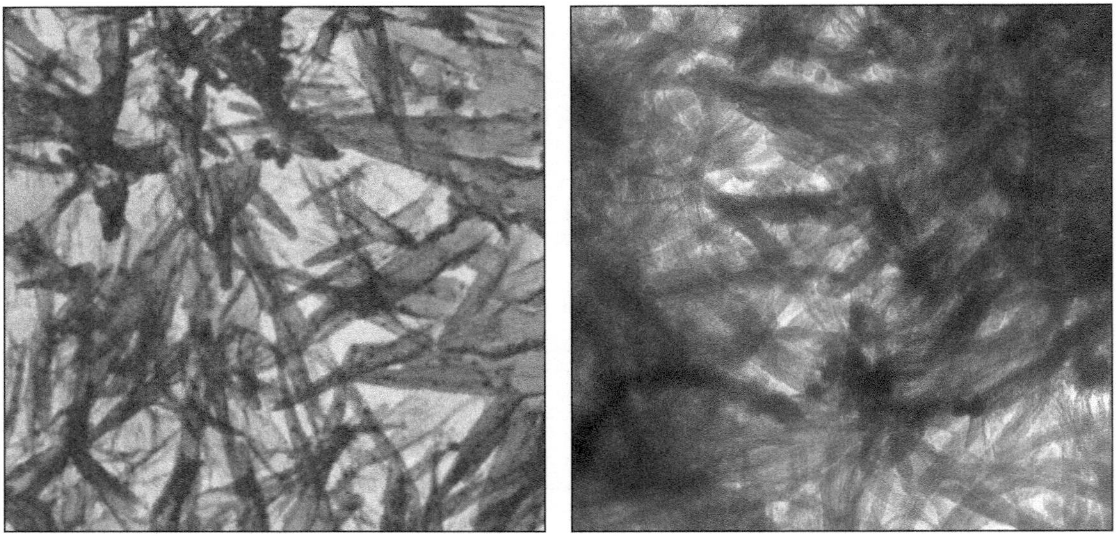

(a) (b)

7.3.3 X-ray microanalysis

Taylor and Newbury (1984) reviewed early X-ray microanalysis (SEM or electron microprobe analysis (EMPA)) studies of polished sections of mature C_3S or β-C_2S pastes and reported new results. As in Figure 7.8, the pastes show well-defined regions of C–S–H and CH. The Ca/Si ratios found for the C–S–H vary irregularly from point to point and increase with the voltage used to accelerate the electrons. As in all analyses of heterogeneous materials by this method, the choice of voltage rests on a compromise: it must be high enough to excite all the elements adequately, but low enough to produce an interaction volume sufficiently small that single phases can be analysed. The mean or median Ca/Si ratios reported for C_3S pastes ranged from 1·4 at 6 kV to around 2·0 at 25 kV. It is questionable whether 6 kV is sufficient to ensure satisfactory analysis for Ca, and at 25 kV the volume analysed, considering the effect of porosity, is several micrometres in each direction, which may be too large to ensure absence of CH or unreacted C_3S. Working at 10 kV, Taylor and Newbury (1984) obtained mean values of 1·72 and 1·78 for mature pastes of C_3S and β–C_2S, respectively. The range of values was approximately 1·5–2·0 in both cases. The mean values are near to those obtained using the indirect methods described in §7.2.2.

The totals obtained in SEM or EMPA microanalyses of C–S–H in C_3S or β-C_2S pastes, after including oxygen equivalent to the calcium and silicon, should in theory fall below 100 % by an amount equal to the water content under the conditions existing during the analysis – that is, high vacuum and electron bombardment. The evidence from thermal dehydration and intensive drying at room temperature (§7.2.3) indicates that this content is unlikely to exceed about 15% on the mass of material analysed, implying an analysis total of at least 85%. The totals are sometimes as high as this, but are more often 70–75%, and sometimes even lower (Taylor and Newbury, 1984). A similar effect is observed with the C–S–H of cement pastes and appears to increase with the porosity of the region analysed. It has not been fully explained, and more than one cause may operate. Possibilities include retention of organic liquids used in preparing the polished section, carbonaceous residues formed under the action of the electron beam, superficial carbonation (Rayment, 1988) and loss of energy of the incident electrons through some ill-understood effect during passage through microporous material (Harrisson *et al.*, 1987).

7.3.4 Analytical electron microscopy

The many attempts to determine the average Ca/Si ratio of C–S–H in C_3S pastes by indirect methods that are discussed in §7.2.2 have been supplemented by others that have attempted direct measurement using the electron-optical techniques of EMPA, SEM-EDX and TEM-EDX. Unfortunately, these techniques are also susceptible to errors. A major issue for the SEM techniques is the question of how finely the phases are mixed together relative to the size of the X-ray generation volume. This has essentially been solved by examination of ion-thinned sections by TEM-EDX, which has shown that CH does not occur in the inner product region of hardened C_3S pastes (Richardson, 1991; Richardson and Groves, 1993a), and so EDX analysis of large areas of Ip observed on flat polished sections in the SEM can be expected to give an accurate chemical composition (see the discussion of data from Costoya Fernandez (2008) in §7.2.2). In contrast, CH can occur intermixed on a fine scale with C–S–H in the Op region and so measurement of the Ca/Si of Op C–S–H using SEM/EMPA should be approached with caution. Its composition can be measured unambiguously (i.e. free of intermixture with other phases) in the TEM because of its higher resolution and the use of very thin specimens – generally by EDX, but it is also possible to do so using electron energy loss spectroscopy (EELS) – but it is important to understand that Ca can migrate away from the analysis region if the electron beam is too finely focused, which results in a Ca/Si ratio that is lower than the true value. Richardson (2002) discussed how this affected the results of Viehland *et al.* (1996) and Zhang *et al.* (2000) for C–S–H in Portland cement pastes, and it seems possible that it also affected the values measured in early TEM-EDX studies of ground and dispersed samples of C_3S and β-C_2S pastes (Gard *et al.*, 1980; Mohan and Taylor, 1981; Lachowski *et al.*, 1981), which are lower than the values determined by EMPA/SEM-EDX of Ip C–S–H or by TEM-EDX of ion-thinned sections. Typical literature values are given in Table 7.3 and the values from TEM-EDX of ion-thinned sections or by EMPA/SEM-EDX are compared on Figure 7.10(a). The dotted and dashed lines are the result of linear regression analysis of the SEM/EMPA and TEM data, respectively. The trends for the two techniques are very similar and, given the scatter in the data and the error bars, it is evident that the Ca/Si ratio is essentially constant for a timescale that extends to many years, which agrees with the data discussed in §7.2.2 for the determination of the Ca/Si ratio using indirect methods. Both curves indicate the possibility of a slight reduction in the average Ca/Si for the oldest samples, which are fully reacted; if the same trends continued to extended ages, the regression equations imply that a Ca/Si of a 1·5 would be reached after about 100 years (i.e. the value for the mineral jennite, §5.4.1), and $\frac{5}{6}$ after about 340 years (i.e. the value for the mineral tobermorite, §5.3). However, the Ca/Si ratio of the Ip C–S–H in a 175-year-old Portland cement paste that was not fully reacted was high ($1{\cdot}74 \pm 0{\cdot}10$, $n = 354$; Owusu and Richardson, 2023; measured by SEM-EDX) and so it is possible that a reduction in Ca/Si ratio with increasing age is associated with the absence of unreacted C_3S. The discussion of

Brown *et al.*'s (1984) data in §6.2.1 (on metastable equilibria in the $CaO–SiO_2–H_2O$ system) is perhaps relevant here, but further study is needed. The distribution of values for the TEM-EDX data of Ip C–S–H in C_3S pastes aged between 1 month and 26 years is shown in Figure 7.10(b); the mean Ca/Si ratio is $1{\cdot}75 \pm 0{\cdot}09$, with 139 of the 143 analyses (97%) between 1·6 and 1·9. Similar average values have also been obtained for the C–S–H in hardened β-C_2S pastes; for example, Taylor and Newbury (1984) obtained 1·78 by EMPA and Richardson (2000) a value of 1·71 by TEM-EDX (see Table 7.3).

Table 7.3 Typical values for the average Ca/Si ratio of C–S–H in hardened C_3S and β-C_2S pastes determined by direct techniques

Reference	System	Technique	w/s	°C	Age	Mean Ca/Si ± SD (*n*)
Grutzeck and Roy (1969)	C_3S	EMPA	0·5	22–28	24 weeks	1·7 ± 0·1
Diamond (1976)	C_3S	SEM-EDX	0·4	Room	3½ years	1·88 ± 0·26 (23)
Stucke and Majumdar (1977)	C_3S	SEM-EDX	0·6	Room	30 days	Ip 1·90 ± 0·05 Op 1·6 ± 0·1
Taylor and Newbury (1984)	C_3S	EMPA	0·45	25	23 years	1·72 ± – (135)
Taylor (2010)	C_3S	SEM-EDX	0·5		40 years	Ip 1·66 ± 0·13 (37)
Rodger (1989)	C_3S	TEM thinned	0·5	25	1 month	Ip 1·71 ± 0·05 (31)
Rodger (1989)	C_3S	TEM thinned	0·5	25	3 months	Ip 1·74 ± 0·19 (18)
Le Sueur (1984)	C_3S	TEM thinned	0·5	25	1 year	Op 1·8 ± 0·3 (30) Ip 1·8 ± 0·3 (16)
Groves *et al.* (1986)	C_3S	TEM thinned	0·5	25	1 year	Op 1·93 ± 0·27 (32) Ip 1·87 ± 0·24 (24) Ip 1·66 ± 0·30 (17) (high voltage electron microscope (HVEM))
Richardson and Groves (1993a)	C_3S	TEM thinned	0·4	20	3½ years	Op 1·74 ± 0·17 (32) Ip 1·74 ± 0·09 (33)
Tajuelo Rodriguez (2015)	C_3S	TEM thinned	0·4	25	5 years	Ip 1·79 ± 0·05 (30)
Rodger (1989)	C_3S	TEM thinned	0·45	25	26 years	Ip 1·75 ± 0·05 (31)
Taylor (2010)	C_3S	TEM thinned	0·5		32 years	Op 1·79 ± 0·18 (48) Ip 1·56 ± 0·08 (19)
Taylor and Newbury (1984)	β-C_2S	EMPA	0·45	25	23 years	1·78 ± – (348)
Richardson (2000)	β-C_2S	TEM thinned	0·4	20	1 month	Op 1·71 ± 0·11 (26)

Figure 7.10 Ca/Si atom ratio of C–S–H in hardened C_3S pastes. (a) Mean Ca/Si ratio of Ip and Op C–S–H for hydration ages between 1 month and 32 years. Data were determined by X-ray analysis in: (i) SEM/EMPA using flat-polished samples (filled squares, data from Grutzeck and Roy (1969), Diamond (1976), Stucke and Majumdar (1977) and Taylor and Newbury (1984); (ii) TEM using ion-milled specimens (unfilled circles, data from Groves *et al.* (1986), Rodger (1989), Richardson and Groves (1993a), Taylor (2010) and Tajuelo Rodriguez (2015)). Some experimental details are given in Table 7.3. (b) Histogram showing the distribution of values of Ca/Si atom ratio for Ip C–S–H in C_3S pastes aged between 1 month and 26 years (TEM-EDX on ion-thinned sections). Data from Rodger (1989), Richardson and Groves (1993a) and Tajuelo Rodriguez (2015)

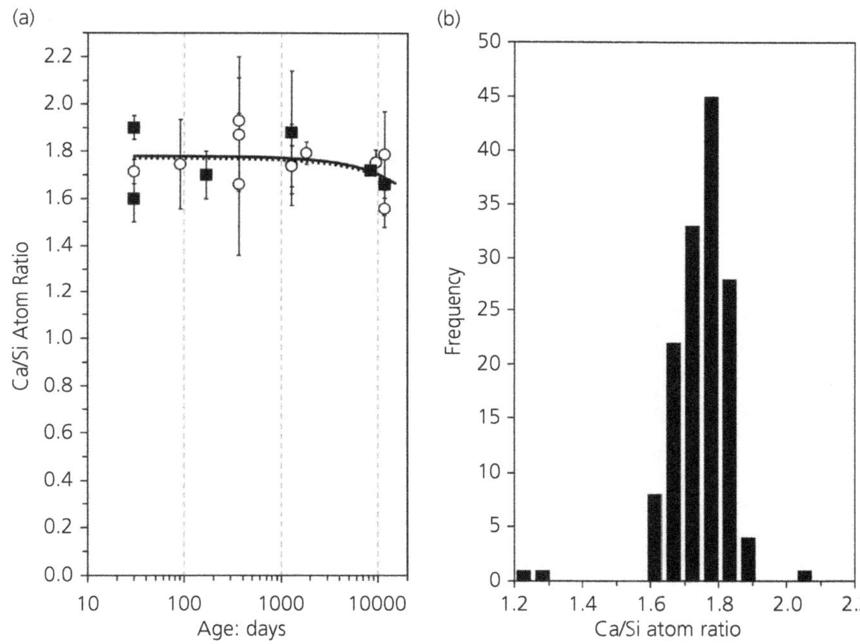

It is clear from the results discussed above and in §7.2.2 that the average Ca/Si ratio of the C–S–H in hardened C_3S and β-C_2S pastes (hydrated for more than a few hours) is about 1·7–1·8, and so this is the value that should be used in Equations 7.2 and 7.3 and in mass balance calculations. An important consequence is that the nanostructure of the C–S–H cannot be based entirely on a tobermorite-like structure that is modified by the omission of bridging tetrahedra and the incorporation of additional interlayer Ca^{2+} ions; it was demonstrated in Chapter 6 that such a structure has a maximum Ca/Si ratio of 1·5, which is for purely dimeric structure.

7.3.5 Electron diffraction

Selected-area electron diffraction of C–S–H in calcium silicate or cement pastes generally yields poorly defined patterns that are consistent with the powder XRD pattern (e.g. Figure 7.1; Copeland *et al.*, 1967; Grudemo, 1962) but have very occasionally resembled patterns from tobermorite and or jennite (Taylor, 1986). These two crystalline calcium silicate hydrate phases are described in Chapter 5. If it is assumed that the particles of C–S–H consist of elements of structure based on tobermorite and/or jennite (§7.5.2), then their dimensions (§7.3.2) are such that they must consist of very few layers of these structures, in some cases as few as two (Richardson, 2004), which would result in poorly defined diffraction patterns. It is also clear that the boundary between interlayer space and gel pore space is likely to be very indistinct (Richardson, 2004). A claim to have identified highly crystalline varieties of C–S–H in C_3S pastes by electron diffraction (Melzer and Eberhard, 1989) was shown to be unsound (Groves, 1990b). Rößler *et al.* (2006) show an electron diffraction pattern (their Figure 4(a)) that was purportedly from a fibrous C–S–H phase of C_3S hydration. However, the pattern is quite poor quality and comparison by way of superposition suggests that it could plausibly be due to calcium hydroxide (superposition through approximate alignment of the spots labelled 3·09 Å and 2·99 Å on Rößler *et al.*'s (2006) Figure 4(a) with five 3·10 Å spots on a simulated pattern for CH viewed along [001]). Comparison of Rößler *et al.*'s (2006) Figure 4(a) with Melzer and Eberhard's (1989) Figure 8(b) – which Groves (1990b) indexed as being mostly due to CH – leads to the same conclusion.

7.4. Silicate anion structure

7.4.1 C–S–H of calcium silicate or cement pastes: chemical methods

Chemical methods for determining the structures of the silicate ions have been based on either the kinetics of the reaction with molybdate (Stade *et al.*, 1983) or the preparation and characterisation of trimethylsilyl (TMS) derivatives. In the TMS method, the sample is treated with a reagent that converts the silicate anions into the corresponding silicic acids, which then react further to replace Si–OH by Si–OSi$(CH_3)_3$. The resulting TMS derivatives can be identified and semi-quantitatively determined by various methods, of which the most widely used have been differential evaporation to isolate the higher molecular weight species, gas liquid chromatography (GLC) and gel permeation chromatography (GPC). It has to be assumed that conversion of the silicic acids into their TMS derivatives is sufficiently rapid to preclude side reactions in which they are altered by hydrolysis or condensation. In practice, such side reactions have never been totally eliminated, although the technique of Tamás *et al.* (1976) was a major advance on that originally used by Lentz (1966). Because of side reactions, and because it is rarely possible to account for more than 80–90% of the total silicon, the absolute accuracy of percentages of individual anionic species is at best a few per cent.

In the remainder of this section, some of the results that are described relate to cement pastes and are in general closely similar to those obtained for C_3S pastes. Chemical formulae of silicate ions are written with negative charges and any attached hydrogen atoms omitted. Most of the chemical investigations on pastes hydrated for more than a few hours at 15–25 °C have shown that any monomer (SiO_4) present is attributable largely to unreacted C_3S; typical results are shown in Figure 7.11(a). The products formed during the first few days contain dimer (Si_2O_7), which is first supplemented and later replaced by larger species, collectively called polymer (Bentur *et al.*, 1979a; Dent Glasser *et al.*, 1981; Hirljac *et al.*, 1983; Lentz, 1966; Massazza and Testolin, 1983; Michaux *et al.*, 1983; Mohan and Taylor, 1982; Richardson, 1999; Tamás *et al.*, 1976; Wu *et al.*, 1983). NMR studies (§7.4.2) have shown that a small amount of monomer is also present in the hydration products at both early and late ages. Figure 7.11(b) gives typical results obtained by the TMS method for C_3S and alite pastes (Mohan and Taylor, 1982). The percentage of the silicon that is present as dimer passes through a maximum after 4 to 6 months, but even after a long time (15–30 years) it is still between 20% and 40%, the exact values varying between studies.

GLC, GPC, mass spectrometry and chemical analyses of TMS derivatives have been used to establish the nature and size distribution of anion sizes within the polymer (Dent Glasser *et al.*, 1981; Hirljac *et al.*, 1983; Mohan and Taylor, 1982; Richardson, 1999; Uchikawa and Furuta, 1981). From chemical analyses, the mean connectivity of the SiO_4 tetrahedra can be calculated – that

Figure 7.11 Selected results from studies that involved trimethylsilylation. (a) Typical results showing the relationship between the % monomer determined by TMS-GPC and the % hydration determined by QXRD. Data from Hirljac *et al.* (1983); a more extensive set of data from the same group are given by Wu *et al.* (1983). (b) Percentages of the total silicon in C_3S or alite pastes recovered as monomer (open symbols), dimer (half-open symbols) and 'polymer' (filled symbols) by a TMS method. Different symbol shapes denote different C_3S or alite specimens. Data from Mohan and Taylor (1982). (c) Percentages of the total silicon in alite pastes recovered as dimer (open circles), pentamer (right half-circles), octamer (left half-circles), > octamer (top half-circles) and total 'polymer' (filled circles) by a TMS method. Data from Hirljac *et al.* (1983)

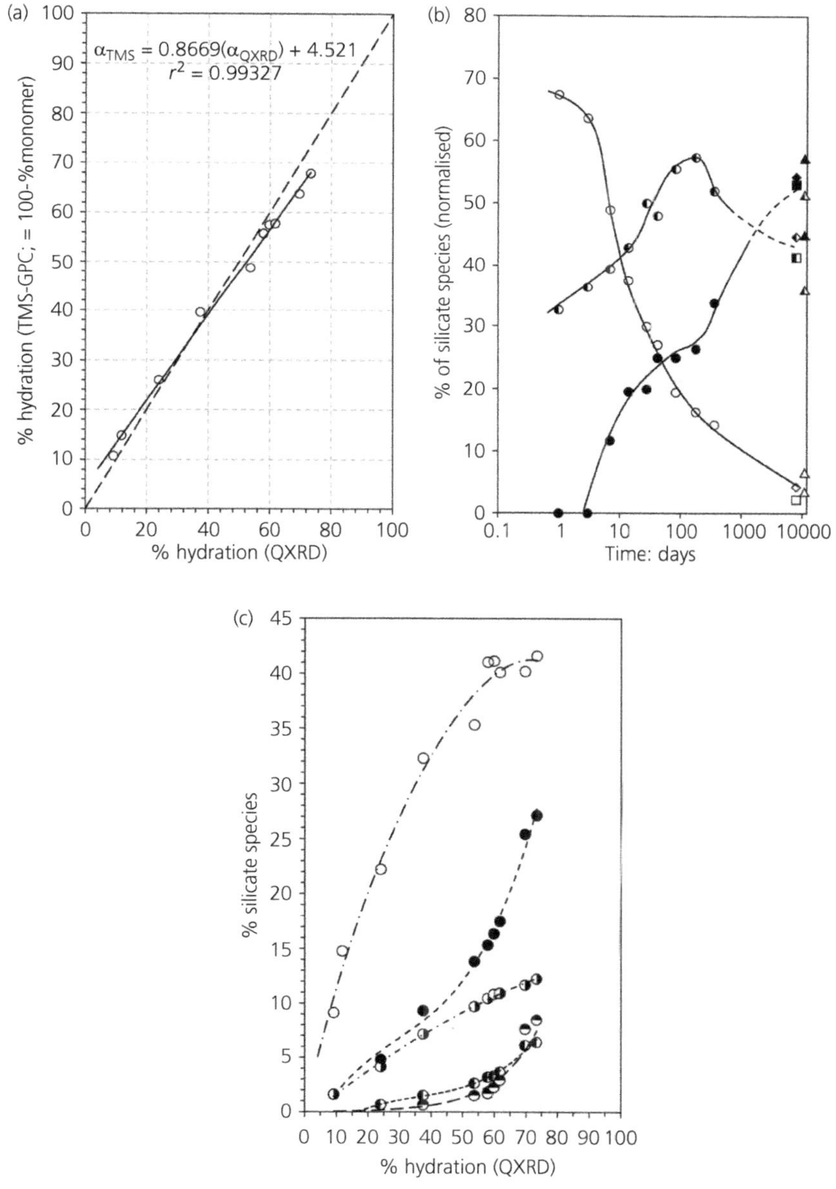

is, the number of other SiO_4 tetrahedra with which each one shares oxygen atoms; typical values are 1·6–2·0 (Massazza and Testolin, 1983; Mohan and Taylor, 1982). The anions are linear, with pentamer (Si_5O_{16}) the most important after dimer (Si_2O_7) at all ages, the next most important being linear octamer (Si_8O_{25}). As the paste ages, the mean size of the anions increases, this process continuing until long after all of the C_3S has reacted. At ages up to 6 months, the polymer fraction consists largely of pentamer and octamer, and even after 20–30 years these species account for a substantial fraction of the total silicon, although larger anions, containing some tens of tetrahedra, are also present (Mohan and Taylor, 1982). These data indicate that the silicate anions in C–S–H follow a 2, 5, 8 ... ($3n-1$) sequence of chain lengths, where $n = 1, 2, 3$ and so on; Figure 7.11(c) shows typical TMS-GPC results plotted against % hydration for alite pastes (data from Hirljac *et al.*, 1983).

Studies at 2–5 °C have given divergent results. Two investigations suggested that the initial products contain substantial proportions of monomer (Bentur *et al.*, 1979a; Stade *et al.*, 1983), but later ones did not confirm this (Hirljac *et al.*, 1983; Parrott and Young, 1981). With increase in temperature above ambient, condensation is increasingly rapid, and anions containing $\geq$100 tetrahedra are reportedly formed (Bentur *et al.*, 1979a; Hirljac *et al.*, 1983; Stade *et al.*, 1983). Early conclusions that admixtures such as $CaCl_2$ affect the degree of polymerisation for a given degree of reaction of the C_3S were not confirmed (Clayden *et al.*, 1986; Hirljac *et al.*, 1983). Early conclusions that the average size of the silicate anions in pastes increases on drying (Bentur *et al.*, 1979b; Milestone, 1980) were also not supported by later work (Parrott and Young, 1981). The behaviour of β-C_2S is similar to that of C_3S, apart from the rate of reaction (Massazza and Testolin, 1983). A conclusion that three-dimensional cluster ions are formed (Currell *et al.*, 1985) is incompatible with observed connectivities and with the NMR evidence discussed below, and can probably be attributed to the occurrence of side reactions.

7.4.2 C–S–H of calcium silicate or cement pastes: nuclear magnetic resonance

^{29}Si magic-angle spinning (MAS) NMR complements the TMS method in several ways. It is a purely physical method, in which there is no possibility of altering the structure through side reactions. Except with very young samples, no preliminary drying is needed. It gives different data, which relate directly not to the fractions of the silicon present in different anionic species, but to the local environments of individual silicon atoms. Quantitatively, it is more reliable than the TMS method. It has been applied mainly to pure calcium silicate pastes or white Portland cements because the substituent ions present in ordinary Portland cements cause the peaks in the spectra to be broadened.

In terms of the Q^n notation for denoting the connectivity of silicate tetrahedra (§5.3.1.7), for C_3S or β-C_2S pastes, the content of Q^0 tetrahedra begins to decrease after a few hours, with the formation of Q^1, later accompanied by Q^2; no Q^3 or Q^4 is detected (Brough, 1993; Brough *et al.*, 1994; Clayden *et al.*, 1984, 1986; Cong and Kirkpatrick, 1993; Geng *et al.*, 2015; Lipmaa *et al.*, 1982; Pustovgar *et al.*, 2016; Rassem *et al.*, 1992; Rodger, 1986; Rodger *et al.*, 1987, 1988). This result is compatible with the formation of dimer, later accompanied by either single chains or rings, but not with that of double chains, clusters or more complex species. Clayden *et al.* (1984) found high correlations between the relative intensities of the Q^0 and Q^1 peaks in the NMR spectrum, the amount of CH indicated by TG and the cumulative heat evolution. Figure 7.12(a) illustrates the development of Q^1 and Q^2 with % hydration at 20 °C; the data, replotted from Brough *et al.* (1994), are particularly good quality because the C_3S was enriched in ^{29}Si to 100 %. The mean length of the

Figure 7.12 Selected results from studies involving solid-state ^{29}Si MAS NMR. (a) Plot of Q^{0H} (circles), Q^1 (squares) and Q^2 (triangles) against % hydration for C_3S pastes (^{29}Si-enriched to 100 %) with w/s = 0·50 hydrated at 20 °C; data with unfilled symbols are from Brough (1993) and Brough *et al.* (1994) and those with filled symbols are from Rodger (1986) and Rodger *et al.* (1988). The dashed line is just a guide for the eye. (b) Plot of MCL (triangles) and Q^{0H} (circles) against % hydration for C_3S pastes (^{29}Si-enriched to 100 %) with w/s = 0·50 hydrated at 20 °C; the MCL data and open circles are from Brough (1993) and Brough *et al.* (1994) and the filled circles are from Rodger (1986) and Rodger *et al.* (1988). The dashed lines are guides for the eye. (c) Plot of MCL against % hydration collated from a number of studies of the hydration of C_3S pastes; the data are from Lipmaa *et al.*, 1982 (half-filled triangles); Clayden *et al.*, 1984 (half-filled squares); Rodger *et al.*, 1987 (open squares); Dobson *et al.*, 1988 (diamonds); Brough *et al.* (1994; open triangles); Pustovgar *et al.* (2016; +); Geng *et al.* (2015; ×) and Bae *et al.* (2020; ×). The dashed line is just a guide for the eye. (d) Estimate of the thickness of early hydration product based on the quantity of Q^{0H} reported by Rodger *et al.* (1988) and based on the method given in Rodger (1986) and outlined in the text. (e) Graph showing the reduction in the surface area of the shrinking core of a smooth-surfaced spherical particle of C_3S that had an initial diameter of 10 μm (continued on next page)

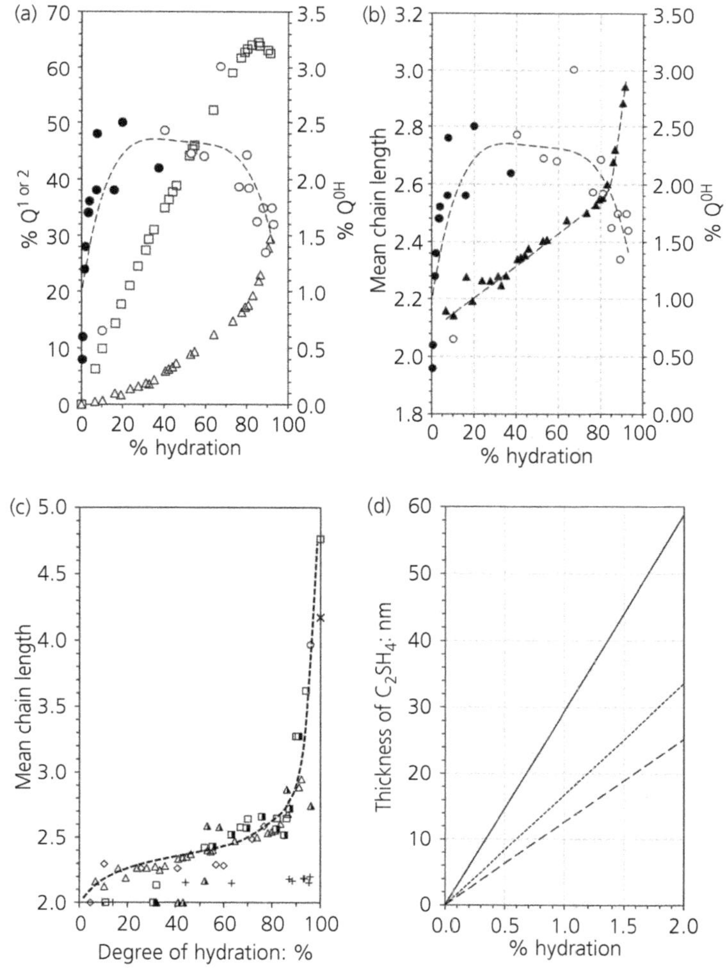

Figure 7.12 Continued

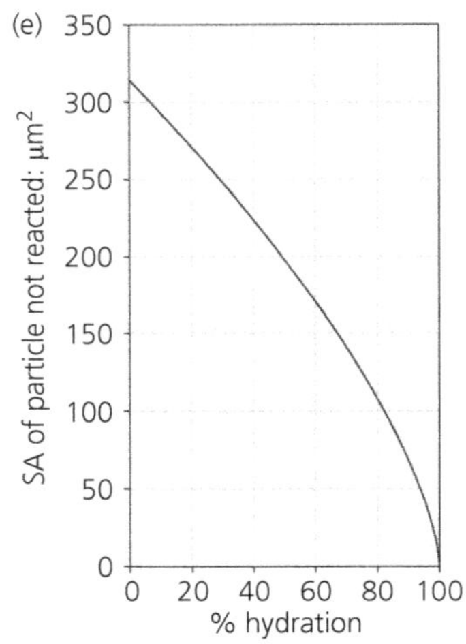

silicate chains (MCL), which is given by $2/q$, where $q = Q^1/(Q^1 + Q^2)$ (Richardson and Groves, 1992), increases only slowly with hydration, from around 2·0 at low degrees of reaction to 2·35 at 50% hydration, 2·5 at 70% and 2·6 at 85%. Degrees of hydration higher than this are associated with increasingly long MCL, which is illustrated in Figure 7.12(b) for Brough *et al.*'s (1994) data, and in Figure 7.12(c) for a compilation of data from a number of sources. It is evident from Figure 7.12(c) that the MCL approaches 5 at extended times: 4·76, 4·18, and 4·17 have been reported for C_3S pastes aged 26 years (Rodger *et al.*, 1987), 50 years (Geng *et al.*, 2015) and 53 years (Bae *et al.*, 2020), respectively. The data of Pustovgar *et al.* (2016; the + symbols on Figure 7.12(c)) are anomalous, showing unusually short silicate chains at high degrees of reaction, which is possibly due to the rapid hydration of their very high surface area C_3S, high w/c ratio (0·80) and the apparent high degree of pre-hydration. Elevated temperatures result in longer MCL for a given degree of hydration: for example, at 75 °C, Brough *et al.* (1994) reported values of approximately 2·6, 3·3 and 3·8 for 50, 80 and 90% hydration, respectively. They also concluded from contents of Q^1 and Q^2 and maximum MCL that the polymerisation of the silicate ions probably occurs through addition of monomer (i.e. separate tetrahedra) to form linkages between dimeric groups, which are thereby converted into pentamer, octamer and longer chains in the $(3n - 1)$ series, consistent with the TMS-GPC data discussed in § 7.4.1.

The TMS method is not sufficiently precise to show whether the hydration products contain a small amount of monomer, and any intensity for that monomer in single-pulse ^{29}Si NMR spectra is masked by unreacted material, except in the unusual case of complete hydration. Examples of spectra for completely reacted cements are shown in Figure 7.13; in (a) for a water-activated Portland cement paste (w/c = 0·50, hydrated for 38 years) and (b) a 5*M* KOH-activated C_3S paste (s/s = 0·60 (by mass), hydrated at 25 °C for 7 weeks). The peaks at chemical shifts of about –85, –79 and –73 ppm are attributable, respectively, to Q^2, Q^1 and hydrated monomer (Rodger *et al.*, 1988). TEM examination of the samples showed that the C–S–H in the water-activated paste had fibrillar morphology and was amorphous while that in the KOH-activated paste had foil-like morphology and was semi-crystalline, which is the reason for the narrower peaks in that spectrum.

Figure 7.13 Single-pulse ^{29}Si MAS NMR spectra for (a) a water-activated Portland cement paste (w/s = 0·5, hydrated for 38 years; the sample was supplied by JJ Beaudoin and corresponds to the first standard sample in Table 1 of Feldman and Swenson (1975); the spectrum was collected by HM Dyson at the University of Leeds) and (b) a 5 *M* KOH-activated C_3S paste (s/s = 0·60 (by mass), hydrated at 25 °C for 7 weeks). Freshly ground samples were packed into zirconia rotors. The spectra were acquired using (a) a Varian InfinityPlus 300 spectrometer (magnetic field 7·05 T; operating frequency of 59·5 MHz for ^{29}Si) with MAS at 6 kHz in a 6 mm rotor, 14595 scans and a pulse recycle delay of 5 s, a pulse width of 4 μs and an acquisition time of 20 ms; (b) a Bruker MSL-200 spectrometer (magnetic field 4·7 T; operating frequency of 39·76 MHz for ^{29}Si) with MAS at 3 kHz in a 7 mm rotor, 3602 scans and a pulse recycle delay of 20 s

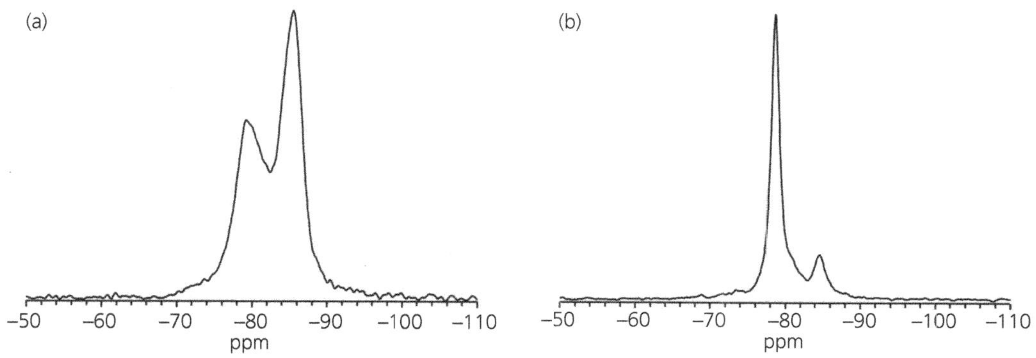

^{1}H-^{29}Si cross-polarisation magic-angle spinning (CPMAS) NMR can distinguish between silicon atoms that are near hydrogen atoms (closer than about 5 Å; Brunet *et al.*, 2004) and ones that are not. This gives a strong, although not always unequivocal, indication as to whether silicon atoms in tetrahedra of a given connectivity carry OH groups. Monomer in a hydration product is almost certain to carry hydrogen and can thus be distinguished from that in the anhydrous compounds. With one exception (Young, 1988), the results have shown that the product formed in C_3S pastes during the first few hours contains only monomeric ions, dimer beginning to form only at the end of the induction period (Brough *et al.*, 1994; Cong and Kirkpatrick, 1993; Rassem *et al.*, 1992; Rodger *et al.*, 1988). As shown on Figure 7.12(b), the quantity of hydrated monomer (Q^{0H}) gradually increases from zero (there is no signal for it on Rodger's (1986) ^{1}H-^{29}Si CPMAS NMR spectrum for anhydrous C_3S, her Figure 5.5a) until it accounts for some 2·0–2·5% of the total silicon after about 20% hydration. It remains at that level until about 80% hydration, after which it decreases. Its formation agrees qualitatively with some of the results at low temperatures mentioned above (Bentur *et al.*, 1979a; Stade *et al.*, 1983). The hydrated monomer has been attributed to superficial hydration or surface hydroxylation of the C_3S (Pustovgar *et al.*, 2016; Rassem *et al.*, 1992), as a distinct phase (Bellmann *et al.*, 2010, 2012; Brough *et al.*, 1994), at defect sites in the C–S–H (Brough *et al.*, 1994) or due to C–S–H particle edge effects (Richardson, 2004). Rodger *et al.* (1988) estimated that a product containing the nearly 2% Q^{0H} that they observed at the end of the induction period would have a thickness of around 40 nm if deposited on the surfaces of the C_3S particles and within the original volume of the particles (6 nm above the original surface of the particle and 34 nm below). The 40 nm thickness was obtained by using a simple calculation that assumed monosized spherical particles of 10 μm diameter, 2% hydration (i.e. 2% Q^{0H}, 98% Q^0) and a complete and evenly distributed product layer of C–S–H with uniform composition C_2SH

and density of $2{\cdot}20\,g\,cm^{-3}$ (Rodger, 1986). The calculated thickness is greater than 40 nm if the value of the density is reduced to account for the presence of some nanoporosity, see §7.3.2. For example, repeating Rodger *et al.*'s (1988) calculation but using a density of $1{\cdot}90\,g\,cm^{-3}$ and H_2O/SiO_2 ratio of 4·0 instead of 1·0 gives a thickness of 59 nm with 25 nm above the original surface and 34 nm below. This is illustrated in Figure 7.12(d), which shows the results of the calculation from 0 to 2% hydration. While some of the product is likely to have formed in the solution away from the surface (Rodger *et al.*, 1988), nevertheless a thickness of 40–59 nm after a few hours' hydration is consistent with the TEM observations described in §7.3.2 when differences in C_3S reactivity between studies are considered. Pustovgar *et al.* (2016) associated the Q^{0H} with surface hydroxylation of C_3S particles, but the quantities involved are far too great: repeating Rodger *et al.*'s (1988) simple calculation with Pustovgar *et al.*'s (2016) data at 1 hour (6 μm diameter particles; 13% Q^{0H}, 87% Q^0) together with the revised values of product density and water content results in an average thickness of 232 nm. Surface hydroxylation only affects one atomic layer at the surface of the C_3S (Barret *et al.*, 1980, 1983; Bellmann *et al.*, 2010; Taylor *et al.*, 1984) – Taylor *et al.* (1984) describe a hydroxylated surface as '*C_3S modified by a collection of point defects*' – and so its thickness is the approximate diameter of an oxygen anion – that is, about 0·28 nm; this would correspond to a degree of hydration of about 0·01% for a 10 μm diameter particle of C_3S. It follows that the observed quantity of Q^{0H} cannot be due to surface hydroxylation and that a layer that is 40–232 nm thick must be mostly due to a distinct phase: 232 nm is over 800× larger than the diameter of an oxygen anion and, as an illustrative example, could accommodate 166 layers of 14 Å tobermorite.

The plateau on Figure 7.12(b) between about 10–20% and 80% hydration is not consistent with attribution of the Q^{0H} to a layer of approximately constant thickness in a shrinking core model of C_3S hydration (i.e. isotropic dissolution of the C_3S) because in that case the quantity of Q^{0H} would reduce as hydration progressed, in proportion to the reduction in the surface area of the unreacted particle, which for a smooth-surface sphere is illustrated in Figure 7.12(e) (initial diameter of 10 μm). Nicoleau and Bertolim (2016) proposed that the reduction in the surface area of the shrinking core is offset by the formation and opening of etch pits (i.e. anisotropic dissolution of the C_3S) and thus accounts for constancy in the quantity of Q^{0H}. However, since their etch pit-based dissolution model approximates as isotropic after about 10% dissolution (Nicoleau and Bertolim, 2016), it cannot account for the form of the curve on Figure 7.12(b); this is discussed further in §7.7.6.

The relative amounts of the Q^n units in the hydration products of a C_3S paste (with a w/c ratio of 0·50 reacted at 20–25 °C) are summarised schematically in Figure 7.14. The figure is based on the combined results of Brough (1993), Brough *et al.* (1994), Rodger (1986) and Rodger *et al.* (1988), some of which are reproduced in Figure 7.12(a). At very low degrees of reaction the small quantity of Q^{0H} constitutes all the silicate in the hydrate but as hydration proceeds, the fraction of Q^{0H} in the hydrate rapidly diminishes. The amount of Q^{0H} at 80% hydration (dashed line) – which corresponds to about 28 days – is around 2·5% of the Si in the hydration products or 2·0% of the total Si in the paste (i.e. 2·5 × 0·8). The curves are extrapolated to 100% hydration from 92%, which is the maximum in Brough *et al.*'s (1994) data; the continued rapid increase in Q^2 is consistent with other work that has data for Q^1 and Q^2 up to 100% hydration (Rodger *et al.*, 1987).

Brough *et al.* (1994) described experiments which showed that rapid quenching of hydration with propan-2-ol followed by evacuation had no detectable effect on spectral line widths or chain lengths and that the high stresses due to spinning had only minimal effect on the reaction.

Figure 7.14 Schematic diagram illustrating the relative amounts of the Q^n units in the hydration products of a C_3S paste (with a w/s ratio of 0·50 reacted at 20–25 °C). The figure is based on the combined results of Brough (1993), Brough *et al.* (1994), Rodger (1986) and Rodger *et al.* (1988), some of which are reproduced in Figure 7.12(a)

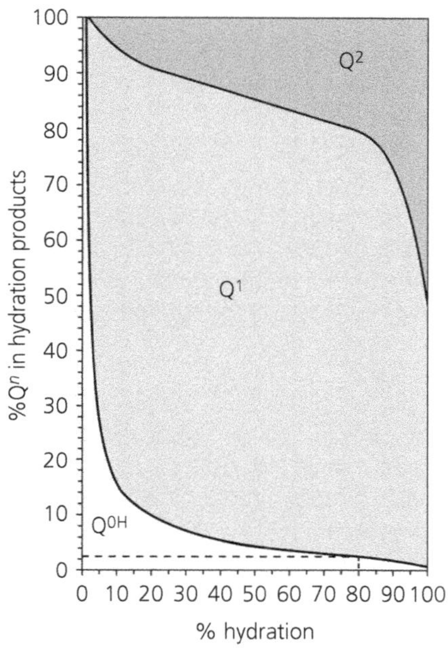

Magic-angle spinning NMR has also been used to study the environments of hydrogen atoms (Rassem *et al.*, 1992) and oxygen atoms (Cong and Kirkpatrick, 1993) in C_3S or β-C_2S pastes. The ^{17}O NMR showed the presence of OH groups attached to calcium and possibly also of OH groups attached to silicon. Rassem *et al.* (1992b) concluded from ^1H combined rotation and multiple pulse spectroscopy (CRAMPS) NMR data that Si–OH is present in C_3S pastes at very early ages, but that by 3 hours Ca–OH is predominant. In subsequent work on a C_3S paste using high speed ^1H MAS NMR, Heidemann and Wieker (1998) could not detect Si–OH at any age. They concluded that any Si–OH formed initially quickly disappears and that increasing quantities of Ca–OH are formed. The method did not distinguish between Ca–OH present in C–S–H and in CH. The presence of Ca–OH bonds in the C–S–H of C_3S pastes was confirmed by Thomas *et al.* (2003) who calculated a (Ca–OH)/Si mole ratio of 0·77 from inelastic neutron scattering experiments. They gave the fraction of Ca that is balanced by OH as 0·23 for C–S–H with Ca/Si = 1·70, with the concentration of Ca–OH bonds decreasing on decalcification by leaching until very little was present at Ca/Si = 1·08. The presence of a substantial concentration of Ca–OH bonds in the C–S–H that is present in hardened tricalcium silicate pastes supports the view that the C–S–H must contain some structure that is based on the mineral jennite and that it cannot be solely based on tobermorite.

7.5. Nanostructures of C–S–H and related materials

7.5.1 Tobermorite-based models for the C–S–H of calcium silicate and cement pastes

Several models based on the tobermorite structure have been suggested for the structure of the C–S–H of calcium silicate and cement pastes or related materials. Kantro *et al.* (1962) proposed that tobermorite-type layers (which will be abbreviated simply as 'T') were interstratified with CH layers, while Stade and Wieker (1980) concluded that both Ca^{2+} and OH^- could be present in the interlayer region of a tobermorite-type structure. Stade (1980) suggested that in 'C–S–H (di,poly)' one surface of a tobermorite-type layer was composed of dimeric and the other surface of polymeric ions, thus accounting for the near constancy that they observed for the dimer/polymer ratio. Structures that have this arrangement are discussed in §6.6.3.4 and illustrated in Figure 6.19 and Figure 6.20. Fujii and Kondo (1983) regarded C–S–H as a solid solution of 14 Å tobermorite and CH. Cong and Kirkpatrick (1996) considered that the C–S–H formed in calcium silicate or cement pastes was essentially similar to C–S–H(I). Richardson and Groves (1992) proposed a generalised model for the C–S–H formed in C_3S or β-C_2S pastes that includes formulations that can be interpreted from the T/CH structural viewpoint but also as T-based structural elements mixed with others derived from jennite – the so-called T/J structural viewpoint, which is discussed in §7.5.2. They started with a general formula for a calcium silicate hydrate that is composed of isolated silicate chains of varying length and with a variable number of –OH groups attached to Si atoms, in 'solid solution' with a variable amount of $Ca(OH)_2$ (Richardson and Groves, 1992).

$$\mathrm{Ca}_X\mathrm{H}_{2(N+1-X)}\mathrm{Si}_N\mathrm{O}_{(3N+1)}\cdot z\mathrm{Ca}(\mathrm{OH})_2\cdot m\mathrm{H}_2\mathrm{O} \tag{7.5}$$

where N is the mean length of the silicate chains (MCL); X is the number of Ca^{2+} ions that are needed to charge-balance the silicate chains; $2(N+1-X)$ is the number of hydroxyl units attached to the chains; z is the number of $Ca(OH)_2$ units in 'solid solution'; m is the number of water molecules that are bound but not present as hydroxyl groups (Richardson, 2004; Richardson and Groves, 1992).

The formula for dimer (i.e. with $N = 2$; dimer is predominant in young C_3S or Portland cement pastes, see §7.4.1 and §7.42) is the same as given by Glasser *et al.* (1987) (their formula 1). Extension of the model to account for the experimentally observed sequence of chain lengths (i.e. $3n - 1$) gave (Richardson and Groves, 1992)

$$\mathrm{Ca}_X\mathrm{H}_{(6N-2X)}\mathrm{Si}_{(3n-1)}\mathrm{O}_{(9n-2)}\cdot z\mathrm{Ca}(\mathrm{OH})_2\cdot m\mathrm{H}_2\mathrm{O} \tag{7.6}$$

Richardson and Groves (1992) gave an alternative formulation as

$$\left\{\mathrm{Ca}_{2n}\mathrm{H}_w\mathrm{Si}_{(3n-1)}\mathrm{O}_{(9n-2)}\right\}\cdot(\mathrm{OH})_{w+n(y-2)}\cdot\mathrm{Ca}_{\frac{n\cdot y}{2}}\cdot m\mathrm{H}_2\mathrm{O} \tag{7.7}$$

where w is the number of silanol groups and w/n is the degree of protonation of the silicate chains (Richardson, 2004). The limiting values of the variables in Formulae 7.6 and 7.7 are set by the requirement to maintain the layer structure and neutrality; they are

$$\begin{array}{lll}\text{For} & 0 \le y \le 2; & n(2-y) \le w \le 2n \\ & 2 \le y \le 4; & 0 \le w \le 2n \\ & 4 \le y \le 6; & 0 \le w \le n(6-y) \\ & X = \frac{1}{2}(6n-w) & z = \frac{1}{2}(w+n(y-2))\end{array}$$

The Ca/Si ratio is given by Equation 7.8

$$\frac{\mathrm{Ca}}{\mathrm{Si}} = \frac{n(4+y)}{2(3n-1)} \tag{7.8}$$

On the T/CH viewpoint, the part of Formula 7.7 that is contained within the braces represents the tobermorite-like core of the structure, which has $2n$ main-layer Ca^{2+} ions. $n-(w/2)$ of the Ca^{2+} ions that are outside the braces are interlayer Ca^{2+} ions that are required for charge balance. The nature of the remainder of the $(n{\cdot}y)/2$ Ca^{2+} ions outside the braces depends on the structural viewpoint: on the T/CH viewpoint they form part of layers of CH-like structure interstratified with tobermorite-like structure; on the T/J viewpoint they form part of the main layer of jennite-based structural units (Richardson, 2004). The model was extended to cement pastes by allowing the possibility of various ionic substitutions, of which the most important were of Si^{4+} by Al^{3+} or Fe^{3+} and of interlayer Ca^{2+} by Na^+ or K^+ (Richardson and Groves, 1993b).

In some cement–based materials, the principal hydration products are forms of C–S–H that are closely similar to the C–S–H(I) that is formed in aqueous suspensions and having Ca/Si ratios ranging from 0·9 to about 1·4. The materials that give such products include alkali-activated slags and autoclaved materials made using cement and siliceous additions. Models based purely on a tobermorite-type structure appear to be appropriate in these cases. The C–S–H that forms in composite cements that contain silica-rich SCMs – while giving only poor X-ray diffraction patterns – are also well-described by purely tobermorite-based models, most of which have been shown to be essentially the same (Richardson, 2004, 2008), as discussed in Chapters 5 and 6; the C–(A–)S–H that is formed in these cements can be represented using Formula 5.6. It is demonstrated in §7.3.4 that the average Ca/Si ratio of the C–S–H that is formed in a hardened C_3S or β-C_2S paste is ≈1·8 (see Figure 7.10) and, as a consequence, purely T-based models are not appropriate.

7.5.2 Mixed tobermorite–jennite models for the C–S–H of calcium silicate and cement pastes

In models that are based strictly on a tobermorite-type structure, it is assumed that any Ca–OH is either present in layers of CH-like structure (e.g. as illustrated in Figure 6.18) or substitutes for silicate ions in a random manner. The latter is reasonable if the content of Ca–OH is relatively small, but scarcely so in materials, or individual regions of them, in which the content of Ca–OH is substantial. As noted in Chapter 5, the conformation of the Ca–O cores of the layers that occur in tobermorite results from the fact that all the O atoms are shared with silicate chains. If only one-half of them are thus shared, the conformation changes to that in jennite. It therefore seems more likely that, if the content of Ca–OH is substantial – which it is in hardened C_3S pastes, §7.4.2 – either a jennite-type conformation and not a tobermorite-type conformation will occur or, alternatively, CH-like structure might be interstratified with T-based structure, as discussed and illustrated (Figure 6.18) in Chapter 6.

Taylor (1986) suggested that the C–S–H that is present in hardened C_3S pastes, assumed to have been equilibrated at 11% RH, contained elements of both 14 Å tobermorite and jennite structure. These were originally considered to form separate layers, but it was later suggested (Taylor, 1993) that the regions could be poorly defined, perhaps merging into each other within individual layers. The material was seen as being highly disordered, and small amounts of monomeric silicate ions could be present. Taylor (1993) assumed that each bridging tetrahedron carries one H atom and that one of the two end-group tetrahedra of each anion also carries an H atom. The observed Ca/Si ratio of about 1·8 can arise in two extreme ways, with intermediate possibilities. One is a mixture of tobermorite-type and jennite-type structures, both with dimeric anions; the other is a purely jennite-type structure, with a mean chain length of five tetrahedra. Initial formation of the first, followed by gradual transition to the second, would explain the observed changes in anion size distribution with time. Since it would be a change towards a more highly ordered structure, it might also explain the direction taken by the process. Several other lines of evidence were cited in support of the hypothesis, based on data from TG, densities and water contents under various drying conditions, electron diffraction patterns and the local variation in composition observed in electron-optical analyses. The model would account for the changes in distribution of local Ca/Si ratio observed for Portland cement pastes (Richardson and Groves, 1993a). The bimodal distributions observed at early ages had ranges typically around 1·2–2·3; the lower value is close to that calculated for a tobermorite-type structure with dimeric ions and the higher one is close to the value of 2·25 calculated for a jennite-type structure with dimeric ions. The observed narrowing of the distribution and change towards unimodal character with a modal value near 1·75 is what would be expected for a change towards a structure purely of jennite type.

The model differs from those assuming a structure based purely on the tobermorite structure in that, in regions having a Ca/Si ratio above 1·5, the conformation of the Ca–O cores of the layers is assumed to be of the jennite and not the tobermorite type. It is also assumed that, in these regions, the silicate ions and OH^- groups are arranged in an ordered way, as in jennite. Taylor (1986) made a particular assumption concerning the content of Si–OH, which he regarded as giving best agreement with the data; however, he later accepted (Taylor, 1993) the possibility of other values that is a feature of the Richardson and Groves (1992) model outlined in §7.5.1. Figure 7.15 shows how the Ca/Si ratio varies with the mean chain length for each type of structure, including the effect of different levels of protonation of the silicate chains. The figure is re-plotted in Figure 7.16 but

Figure 7.15 Variation of the Ca/Si ratio with mean chain length for J- and T-based structural units that are present in Richardson and Groves' (1992) model with three different levels of protonation of the silicate chains

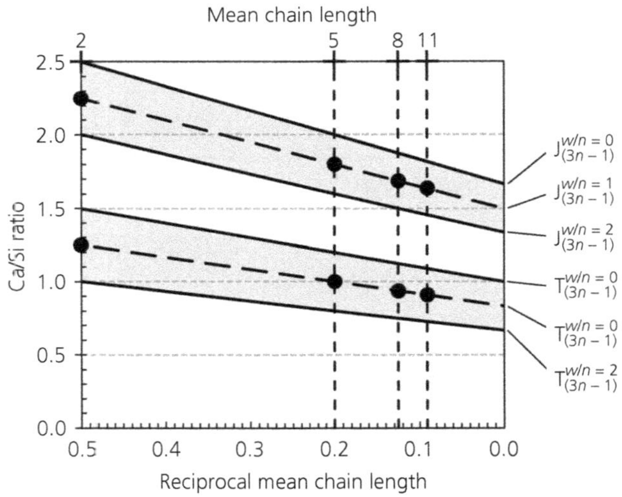

Levels of protonation of the silicate chains: minimum level ($w/n = 0$); intermediate level ($w/n = 1$; sloping dashed lines), which is equivalent to the units in Taylor's (1986) model; maximum level ($w/n = 2$). The figure is based on Figure 3 of Richardson and Groves (1992). The filled circles correspond to the J-based (top) and T-based dimer, pentamer, octamer and undecamer in Taylor's (1986) model; the shaded areas indicate the extra flexibility introduced by Richardson and Groves (1992).

Figure 7.16 Ca/Si ratio frequency histograms of the TEM-EDX analyses of C–S–H present in hardened water-activated C_3S or β-C_2S pastes (data from Richardson and Groves, 1993a) placed against the Ca/Si–MCL plot of Figure 7.15. The data are for C_3S pastes hydrated for (a) 3·5 and (b) 26 years; (c) is the same for a β-C_2S paste hydrated for 3 months. Modified from Figure 44 in Richardson (2004)

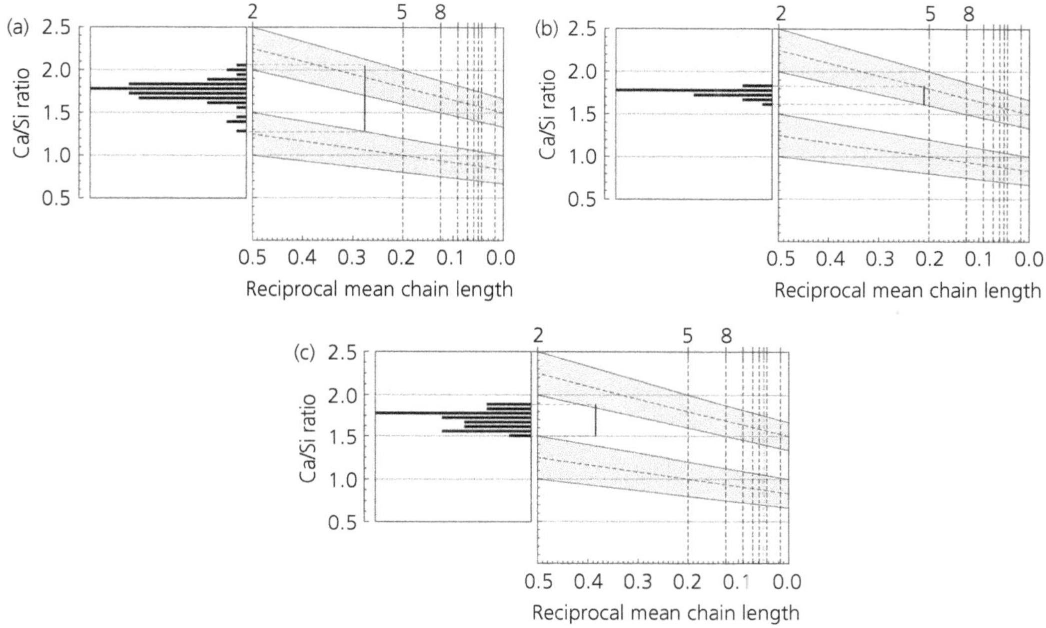

with Ca/Si ratio frequency histograms of the TEM-EDX analyses of C–S–H present in hardened water-activated C_3S or β-C_2S pastes (data from Richardson and Groves, 1993a) placed against it. It is evident that a model that is based entirely on tobermorite-based structure (the bottom three lines) cannot account for the TEM-NMR.

The T- and J-based structural elements that are present in the Taylor and Richardson–Groves models are derived using the same structural variations and charge-balance arguments that are used in Chapter 6 for C–S–H(I) (§6.2.5) and C–S–H(II) (§6.3.3) (structural-chemical formulae for T- and J-based units are given in Formulae 5.6 and 6.9, respectively) and so expressions can be derived easily that express the equivalency of the two approaches, and by extension, also the other models that are included in Table 6.2. Some expressions are given in Tajuelo Rodriguez *et al.* (2017). It follows that the six lines on Figure 7.15 have the same meaning as the six dashed lines on Figure 6.9(a). As an example, the left-hand dotted line on Figure 6.9(a), which represents T-based structure where one Ca^{2+} ion is added to the interlayer region for each bridging site that is vacant (Equation 6.7; Richardson, 2014), is written in terms of the Richardson and Groves (1992) model as

$$\frac{\mathrm{Ca}}{\mathrm{Si}}=\frac{\frac{2}{3}+v}{1-v}=\frac{2n+1}{3n-1}=\frac{1}{\mathrm{MCL}}\times(2n+1) \tag{7.9}$$

Both approaches are useful. For example, the Richardson–Groves (1992) Formula 7.7 includes the MCL (i.e. $3n - 1$), which is easily visualised. However, it can also be misleading because the fraction of bridging sites that need to be occupied to produce a significant increase in MCL is high if starting with dimer, but very small if the chains are already long. The nanostructural changes that occur at later age can thus be easily taken to be much more extensive than is actually the case (for an example see Figure 17 of Richardson *et al.*, 2016), and so such changes are better understood using plots of SOF_{BT} against Ca/Si rather than MCL against Ca/Si. It should be noted that the charge-balance arguments for T-based structure are the same regardless of whether a structural model is derived from an orthotobermorite or clinotobermorite crystal structure.

7.5.3 Other models for the C–S–H of calcium silicate and cement pastes

The mixed tobermorite–jennite and tobermorite–CH models described above account for a range of experimental data including local variability in composition and structure. These and other models are summarised in Table 7.4, including some that have only monomeric silicate (Bernal, 1954; Shpynova *et al.*, 1967; Grudemo, 1984).

While the models that involve monomer are not consistent with the experimentally observed distribution of silicate anions for C–S–H (i.e. $3n - 1$ where n is integer for individual structural units), they are relevant to the product that forms during the induction period, see §7.4.2. For example, Bernal's (1954) formula with $x = 1$ gives a monomer with Ca/Si = 2 (shown in Formula 7.23), a composition that is consistent with the XPS results that are discussed in §7.7.2.

$$Ca\left[SiO_2(OH)_2\right]\left[Ca(OH)_2\right]\left[H_2O\right]_y \tag{7.23}$$

This is also obtained from Formula 7.5 (Richardson and Groves, 1992) with $N = 1$ (i.e. monomer), $X = 1$ and $z = 1$.

A number of models have been proposed that appear to be incompatible with at least some of the experimental evidence. They include an identification with natural tobermorite, based on IR and extraction results (Shebl *et al.*, 1985), one that assumes three-dimensional anionic clusters (Currell *et al.*, 1985), one that consists of intergrowths of sorosilicate-based structure (two forms, one 'active', one 'inactive') and tobermorite-like C–S–H (Grutzeck, 1999; Grutzeck *et al.*, 1999), and the tobermorite-derived structures from atomistic modelling studies that are discussed in §6.6.2.

7.6. Equilibria

7.6.1 Solubility relations

Equilibria in the $CaO–SiO_2–H_2O$ system at ordinary temperatures are discussed in §6.2.1, with part of the water-rich region of the $CaO–SiO_2–H_2O$ phase diagram shown in Figures 6.2(a) and (b). The figures include data points for analytical concentrations of SiO_2 (C_{SiO_2}) and CaO (C_{CaO}) in solution from studies on (a) synthetic C–S–H preparations, and (b) that involve the use of tricalcium silicate. It is concluded in §6.2.1 that curve A on the figures is the solubility curve of C–S–H(I) and that Ca–OH bonds can be present in high Ca/Si C–S–H as part of jennite-like structure, with the highest content where curves B and C converge at CH saturation. Results from studies of highly reactive β-C_2S discussed in §6.3.2 suggest strongly that the part of the C_{SiO_2}–C_{CaO} diagram where curves B and C converge is associated with C–S–H(II), which appears to be a structurally imperfect form of jennite.

Table 7.4 Summary of models for the nanostructure of C–S–H in hardened cements. Modified slightly from Richardson (2008) (continued on next page)

Year(s)	Author(s)	Model based on the structure of:	Silicate anion structure	Formulation or structure-composition equation		Comments
1952	Bernal *et al.*	Tobermorite	Infinite			Infinite-chain models not consistent with experimentally observed distribution of silicate anions; i.e. $3n - 1$ where n is integer for individual structural units
1954	Bernal		Monomer	$Ca\left[SiO_2(OH)_2\right]\left[Ca(OH)_2\right]_x[H_2O]_y$	(7.10)	Monomeric models are not consistent with the experimentally observed distribution of silicate anions for C–S–H (i.e. $3n - 1$ where n is integer for individual structural units) but are relevant to the product that forms during the induction period
1956	Taylor and Howison	Tobermorite	$3n - 1$			Ca/Si ratio raised above 0·83 by the removal of some 'bridging' tetrahedra and replacement by interlayer Ca^{2+} ions

Table 7.4 Continued

Year(s)	Author(s)	Model based on the structure of:	Silicate anion structure	Formulation or structure-composition equation		Comments
1960, 1962	Kurczyk and Schwiete	Tobermorite	Infinite	$Ca\left[Ca_4Si_6O_{16}(OH)_2\right]\left[Ca(OH)_2\right]_{5\to7}\cdot mH_2O$	(7.11)	Ca^{2+} and OH^- ions in an interlayer region, represented in their formula as CH groups, together with water molecules. Formula is consistent with Kantro *et al.*'s (1962) model in which layers of tobermorite are sandwiched between layers of calcium hydroxide
1967	Shpynova *et al.*	CH	Monomer	$Ca_3[SiO_4](OH)_2\cdot 2H_2O$	(7.12)	As Bernal (1954) above
1980–1987	Stade *et al.*	Tobermorite	Dimer and polysilicate	$Ca_{4+y}(OH)_z\left[\left(H_{1+\frac{x}{2}}Si_2O_7\right)\right]_2\cdot mH_2O$	(7.13)	Extension of Kurczyk and Schwiete's (1960, 1962) model by allowing silicate chains that were not infinite, and in particular dimeric. Limits on *y* simply correspond to the compositions studied
1984	Grudemo	CH	Monomer			As Bernal (1954) above
1986	Taylor	14 Å tobermorite and jennite	$3n-1$			Structural elements based on 14 Å tobermorite and jennite (i.e. so-called T/J model). Fixed degree of protonation of the silicate chains

1987	Glasser *et al.*	Tobermorite	Dimer	$Ca_XH_{6-2X}Si_2O_7 \cdot zCa(OH)_2 \cdot mH_2O$	(7.14)	Derived from Stade *et al.*'s dimeric model, Formula 7.13 above (BUT *z* does not have the same meaning in the two equations)
1992, 1993b	Richardson and Groves	Tobermorite, jennite, CH	$3n-1$	$Ca_XH_{(6n-2X)}\left(Si_{1-a}R_a^{[4]}\right)_{(3n-1)}O_{(9n-2)} \cdot I^{c+}_{\frac{a}{c}(3n-1)} \cdot zCa(OH)_2 \cdot mH_2O$	(7.15)	Includes both T/J and T/CH formulations; Formula 7.15 with no substitution (i.e. Equation 7.6) is the polysilicate equivalent of Formula 7.14 (Glasser *et al.*, 1987).
				$\left\{Ca_{2n}H_w\left(Si_{1-a}R_a^{[4]}\right)_{(3n-1)}O_{(9n-2)}\right\} \cdot I^{c+}_{\frac{a}{c}(3n-1)} \cdot (OH)_{w+n(y-2)} \cdot Ca_{\frac{n \cdot y}{2}} \cdot mH_2O$	(7.16)	Variable degree of protonation of silicate chains
				$\frac{Ca}{Si} = \frac{n(4+y)}{2(1-a)(3n-1)}$ $\frac{Ca}{Si} = \frac{n(4+y')+a(3n-1)}{2(1-a)(3n-1)}$	(7.17) and (7.18)	Incorporation of substituent ions, most importantly substitution of Si^{4+} by Al^{3+}. Possibility that T-, J- and CH-based structural elements coexist

Table 7.4 Continued

Year(s)	Author(s)	Model based on the structure of:	Silicate anion structure	Formulation or structure-composition equation		Comments
1993	Taylor	Tobermorite and jennite	$3n-1$	$Ca_4\left[(Si_{6-\Delta}\square_{\Delta}O_{18-2\Delta})H_{2p}Ca_{2-p}\right]_q \cdot (OH)_{8(1-q)} \cdot mH_2O$	(7.19)	Formulation of 1986 model in response to Richardson and Groves' model (1992, 1993b), but extended to incorporate some of their modifications (variable protonation and substituents); as a consequence it is equivalent to Richardson and Groves' (1992, 1993b) T/J formulation
1996	Cong and Kirkpatrick	Tobermorite	$3n-1$	$\frac{C}{S}=\frac{r}{\left(1-\frac{m}{3}\right)} \quad QRT=\frac{2}{\left(\frac{3}{m}-1\right)}$	(7.20) and (7.21)	'Defect tobermorite' model; Equations 7.20 and 7.21 are equivalent to equations in Richardson and Groves' 1992 model; see §4.4.2 in Richardson (2004)
1997–1999	Grutzeck	Tobermorite, sorosilicate	Long-chain, dimer	$Ca_4(OH)_2[SiO_{3.5}]_2 \cdot mH_2O$	(7.22)	Long-chain T-based phase with low Ca/Si and a dimeric sorosilicate-based phase with high Ca/Si

1998	Nonat and Lecoq	Tobermorite	$3n-1$		Two hypotheses: 1st is equivalent to T units with *w/n* = 0 and *y* = 2 in Richardson and Groves' model (1992, 1993b) (Formula 7.7); 2nd involves Ca^{2+} and OH^- ions in an interlayer region, in which respect it is similar to Kurczyk and Schwiete (1960, 1962) and Stade *et al.* (1980–1987)
2004	Chen *et al.*	Tobermorite and jennite	$3n-1$	Chen *et al.*'s Table 5 presented formulae for mean chain lengths of 2, 3, 4, 5 and ∞	Formulae equivalent to T units with *w/n* = 0 and *y* = 2 in Richardson and Groves' model (1992, 1993b) (Formula 7.7)
2004	Nonat	Tobermorite; jennite	$3n-1$		Update of 1998 model; mentions jennite-based structure
T: 2014, J: §6.3.3	Richardson	Tobermorite; jennite	$3n-1$	$T: Ca_4\left[Si_{1-f-v}Al_f \square_v O_{3-2v}\right]_6 H_{2i}Ca_{2-i}(Ca, Na_2, K_2)_{3f} \cdot mH_2O$ $J: Ca_8\left[Si_{1-v} \square_v O_{3-2v}\right]_6 (OH)_8 H_{2i}Ca_{2-i} \cdot mH_2O$	Relationship with Richardson and Groves' model (1992, 1993b) (Formula 7.7) is given in Table 6.2

Many investigators have followed the changes in C_{CaO} and C_{SiO_2} or $[OH^-]$ with time in the course of the reaction of water with C_3S or β-C_2S in suspensions and pastes (e.g. Barret *et al.*, 1980; Brown *et al.*, 1984; Grutzeck and Ramachandran, 1987). Selected data are shown on Figure 7.17. The curves for C_{CaO} or $[OH^-]$ typically show a steep rise during the initial reaction and a more gradual one of uniform slope during the induction period. Significant precipitation of CH begins around the time that maximum concentrations are reached, and is rapid when they decrease, during the acceleratory period (Wu and Young, 1984). Some of the curves reported in the literature show a plateau in the early part of the induction period, followed by a rise (e.g. Barret and Bertrandie, 1986). As noted in §7.6.1, the final values reported for pastes are well above the solubility of CH. The SiO_2 concentration passes through a maximum in the order of 1 mmol L^{-1} during the initial reaction. The data on Figure 7.17 of Bellmann *et al.* (2015), Nicoleau and Nonat (2016) and Naber *et al.* (2019, 2020) all fall along Jenning's curve B while Brown *et al.*'s (1984) data fall along curve A, see Figure 6.2(b).

Figure 7.18(a) is a similar plot to Figure 6.2(b) but it has data from just three studies of C_3S hydration to illustrate how the concentrations in solution develop with time after mixing (Fujii and Kondo, 1969; Naber *et al.*, 2020; Wieker and Trettin, 1983). In the first few hours, the data points essentially follow curve B. Since the product that is formed during this period of time contains only monomeric silicate (see §7.4.2), curve B must be associated with this monomer, and from the XPS data discussed in §7.7.2, the product must have a high Ca/Si of 2·0–2·5. Observations in the TEM show that as the maximum value of C_{CaO} is approached (e.g. the 2-hour sample in §7.3.2)

Figure 7.17 Plots of the analytical concentrations of CaO (C_{CaO}; open symbols) and SiO_2 (C_{SiO_2}; filled symbols) against time for C_3S paste (C_3S 352·5 m^2 kg^{-1}, w/s 0·7, 24°C). Data are from Brown *et al.* (1984: circles; C_3S paste, w/s 0·7, 24°C), Bellmann *et al.* (2015; triangles; 90·7% C_3S, 9·3% β-C_2S, w/s 0·5, 24°C), Nicoleau and Nonat (2016; diamonds; C_3S, hydration started in dilute solution (l/s = 250) and saturated with respect to CH but CH had not precipitated in the experiment), and Naber *et al.* (2019, 2020; squares; C_3S paste, w/s 0·5, 23°C)

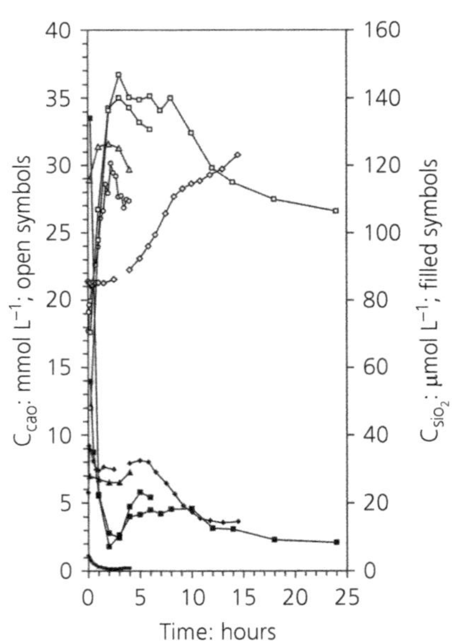

Figure 7.18 Plots of the analytical concentrations of SiO_2 (C_{SiO_2}) and CaO (C_{CaO}) in solution for (a) three studies that involved tricalcium silicate pastes and (b) the data from (a) together with those from Figure 6.9 (b), which are for C–S–H preparations that were synthesised using various methods that involved β-C_2S

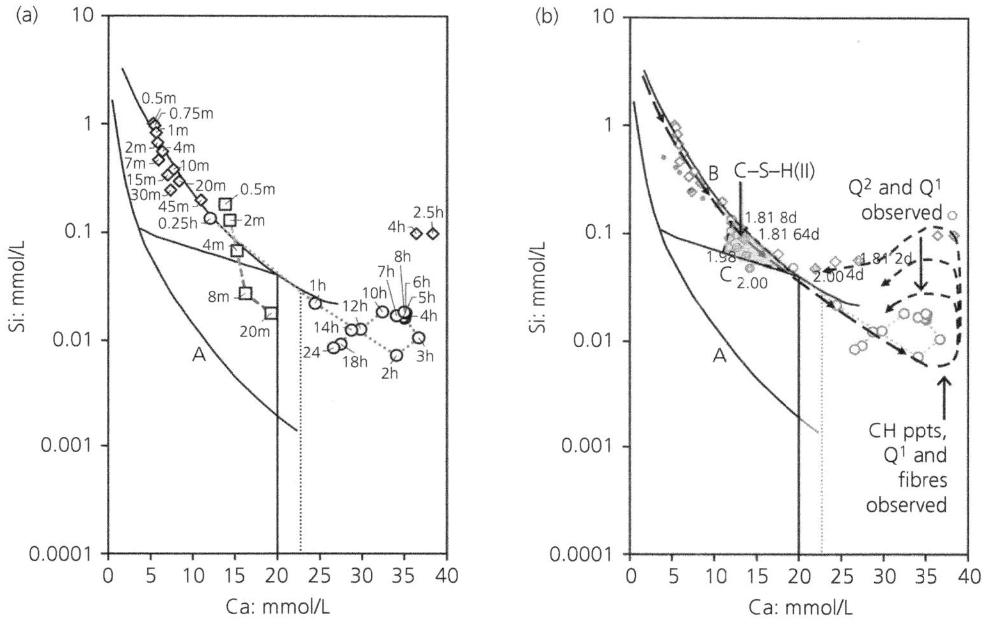

The solid lines and vertical dotted line on both parts have the same meanings as on Figures 6.2(a) and (b). The data that are only on (a) are from Fujii and Kondo (1969) (diamonds; w/s = 0·7; 0·5 minute to 4 hours of hydration), Wieker and Trettin (1983) (squares; w/s = 0·5; 0·5 to 20 minutes) and Naber *et al.* (2020) (circles; w/s = 0·5; 0·25 to 24 hours). The grey large-dotted line curve is an eye guide to show how Naber *et al.*'s (2020) data change with time. The meanings of the other lines on (b) are explained in the text.

the early product appears to have a mixture of nanoporous but otherwise quite featureless morphology close to the surface of the C_3S and a more open morphology further away. Fibrillar morphology is observed as the maximum is reached/exceeded (e.g. the 4- and 8-hour samples in §7.3.2).

To enable direct comparison of the data on Figure 7.18(a) with those for the high Ca/Si C–S–H preparations that were synthesised using various methods that involved β-C_2S, they are replotted on Figure 7.18(b) with those from Figure 6.9(b). In this case, the long-dashed line shows how the composition of the solution develops during the first 2 or 3 hours of C_3S hydration – that is, when the silicate in the product is entirely Q^{0H}. The short-dashed lines show how the composition of the solution develops in the following hours. Three lines are used to simply indicate that there is variability in the data. The part of the diagram that is shaded (where curves B and C converge) was found in §6.3.2 to be associated with C–S–H(II) and closely-related but less well-ordered high Ca/Si ratio C–S–H. The grey numbers on Figure 7.18(b) are Ca/Si ratios of the latter. It is important to note that the Ca/Si ratios correspond unambiguously to single-phase C–S–H (because they were determined by TEM-EDX) and that for the points close to curve B, there was no unreacted β-C_2S

remaining. The evidence presented on this figure therefore suggests that curve B is not associated with a single phase or the presence of unreacted material. Since silicate tetrahedra are present in either *dreierkette* silicate of jennite-based structure (i.e. in C–S–H(II) or related high-Ca/Si C–S–H) or as hydrated monomer (i.e. in the early product) it is plausible that curve B may be associated with calcium silicates that contain Si–O–Ca–OH groups. The grey large-dotted-line curve on the figure is an eye guide to show how Naber *et al.*'s (2020) data change with time. Inspection shows that their data at 0·25 and 1 hour fall on curve B, and then in the next hour or so the curve extends to values of C_{CaO} that appear to be significantly above the solubility of CH (see Equation 7.1), up to about 35 mmol L^{-1}. This has been attributed to the incorporation of SiO_2 in the CH (Wu and Young, 1984), but it might also be due to the very small crystal size of some of this phase (Figure 7.2(b)). At about 3–4 hours, the curve folds back with a decrease in C_{CaO} and initial increase in C_{SiO_2}. As indicated on the figure, this time is associated with the appearance of Q^1 silicate tetrahedra – which initially are present only as dimer – the precipitation of CH, a decrease in the Ca/Si ratio of the silicate-containing product, and – as noted above – the appearance of fibrillar morphology in the outer product – that is, the appearance of Op C–S–H. The trend continues for the next few hours, a period when Q^2 tetrahedra also appear; it remains quite close to the extended curve B even at 24 hours but may be heading down towards curve A – that is, to a curve that is associated with C–S–H (I), which has tobermorite-based structure. It is thus possible that at this stage the C–S–H incorporates elements of both jennite- and tobermorite-like structure, which is consistent with the discussions in §6.2.1 and §6.3.2.

7.6.2 Species in solution

Barby *et al.* (1977) summarised data for silicate species in alkali silicate solutions. Monosilicic acid, H_4SiO_4, has $pK_{a1} = 9{\cdot}46$, $pK_{a2} = 12{\cdot}56$ and pK_{a3} probably about 15. Monomeric species are overwhelmingly predominant at C_{SiO_2} below about 2 mmol L^{-1} at pH values up to 9, and up to higher concentrations at higher pH. The solubility product of CH, in terms of activities, is $9{\cdot}1 \times 10^{-6}$ mol^3 L^{-3} at 25 °C (Greenberg and Copeland, 1960), and the stability constant of the complex $CaOH^+$ is 20 mol L^{-1} (Smith and Martell, 1976). Undissociated CH is not a significant species in saturated CH solution (Gartner and Jennings, 1987). In saturated CH at 25 °C, the activities of Ca^{2+} and $CaOH^+$ are roughly equal and the pH is 12·45. The stability constants of $CaH_2SiO_4^0$ and $CaH_3SiO_4^+$ are reported to be 1240 mol L^{-1} and 2·5 mol L^{-1}, respectively (Smith and Martell, 1976), but some doubt has been cast on these values by the results of thermodynamic calculations using them (Gartner and Jennings, 1987).

Using the equilibrium constants given above, together with standard expressions to calculate ionic strength and activity coefficients, one may calculate the species concentrations and species activities in a system of given bulk composition. For $C_{CaO} = 28{\cdot}0$ mmol L^{-1}, $C_{SiO_2} = 20{\cdot}0$ μmol L^{-1}, and assuming that calcium silicate complexes are not present, the species concentrations (in mol L^{-1}) are Ca^{2+}, $1{\cdot}90 \times 10^{-2}$; $CaOH^+$, $8{\cdot}98 \times 10^{-3}$; OH^-, $4{\cdot}70 \times 10^{-2}$; H_4SiO_4, $4{\cdot}19 \times 10^{-9}$; $H_3SiO_4^-$, $6{\cdot}82 \times 10^{-6}$; $H_2SiO_4^{2-}$, $1{\cdot}30 \times 10^{-5}$; $HSiO_4^{3-}$, $1{\cdot}46 \times 10^{-7}$ (Gartner and Jennings, 1987). The activity of OH^- is $3{\cdot}78 \times 10^{-2}$ mol L^{-1}; this corresponds to a pH of 12·58. If the complexes $CaH_2SiO_4^0$ and $CaH_3SiO_4^+$ are assumed to occur, with stability constants of 2000 and 20 mol L^{-1}, respectively, the concentrations of the SiO_2-containing species become (mol L^{-1}): $CaH_3SiO_4^+$, $2{\cdot}01 \times 10^{-7}$; $CaH_2SiO_4^0$, $1{\cdot}61 \times 10^{-5}$; H_4SiO_4, $7{\cdot}75 \times 10^{-10}$; $H_3SiO_4^-$, $1{\cdot}26 \times 10^{-6}$; $H_2SiO_4^{2-}$, $2{\cdot}41 \times 10^{-6}$; $HSiO_4^{3-}$, $2{\cdot}70 \times 10^{-8}$, the neutral complex $CaH_2SiO_4^0$ thus being the principal SiO_2-containing species. The concentrations of Ca^{2+}, $CaOH^+$ and OH^- are not significantly affected. However, Scrivener *et al.* (2023) noted that there is no general agreement on the correct values of the stability constants for the calcium silicate complexes, which can have a big impact on the results of thermodynamic calculations, as noted in §6.2.1.

7.6.3 Thermochemistry

Using a heat of solution method, Lerch and Bogue (1934) found the enthalpies of hydration of C_3S and β-C_2S in pastes to be $-115\,kJ\,mol^{-1}$ ($-504\,J\,g^{-1}$) and $-45\,kJ\,mol^{-1}$ ($-261\,J\,g^{-1}$), respectively. Similar values have been reported by other investigators; for example, $-118\,kJ\,mol^{-1}$ ($-517\,J\,g^{-1}$) and $-45\,kJ\,mol^{-1}$ ($-261\,J\,g^{-1}$) by Brisi (1954) and $-121\,kJ\,mol^{-1}$ ($-530\,J\,g^{-1}$) and $-46\,kJ\,mol^{-1}$ ($-267\,J\,g^{-1}$) by Fujii and Kondo (1983), who recalculated data obtained by Brunauer *et al.* (1954) for dried products. Brunauer *et al.* (1954) noted that, if the hydration products of C_3S and β-C_2S were identical apart from the quantities of CH formed, the enthalpy of hydration of C_3S should equal the sum of the enthalpies of hydration of β-C_2S, the hydration of CaO ($\Delta H_{298} = -65{\cdot}2\,kJ\,mol^{-1}$; Wagman *et al.*, 1982) and the reaction of β-C_2S and CaO to give C_3S ($\Delta H_{298} = -13{\cdot}4\,kJ\,mol^{-1}$; Wagman *et al.*, 1982). For the moist material, this sum is $(-45) + (-65{\cdot}2) + (-13{\cdot}4) = -123{\cdot}6\,kJ\,mol^{-1}$ ($-541\,J\,g^{-1}$) which is near the observed value. Naber *et al.* (2020) used $-561\,J\,g^{-1}$ (obtained from Jansen *et al.*, 2011) to convert heat flow to reaction rate.

7.7. Kinetics and mechanisms

7.7.1 C_3S: experimental data

The rate of the hydration of C_3S can be followed using a number of experimental techniques. For example, Figure 7.19 is a plot of the fraction of C_3S consumed (α) against time for pastes of w/c = 0·50 hydrated at 20–25 °C as determined by solid-state ^{29}Si NMR (data from Brough *et al.*,

Figure 7.19 General form of curves relating the fraction of C_3S consumed to time in a paste (continued on next page)

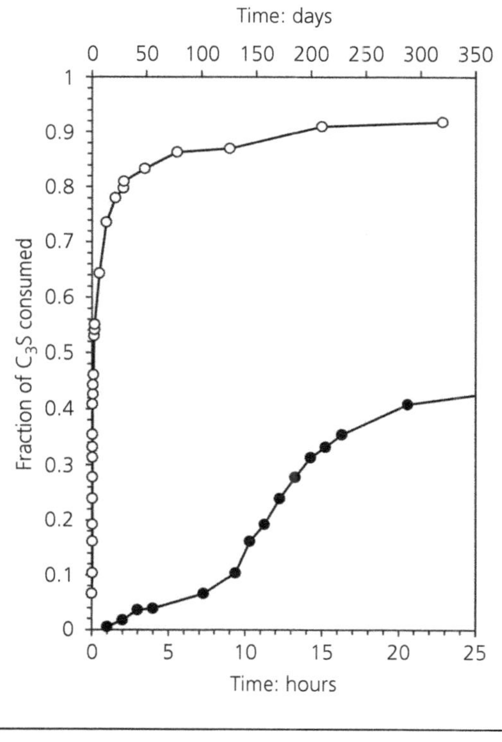

Figure 7.19 Continued

Most data are from Brough *et al.* (1994) measured using ^{29}Si MAS NMR for C_3S pastes (^{29}Si-enriched to 100 %) with w/s = 0·50 hydrated at 20 °C. Filled circles correspond to the bottom horizontal axis. Data for the first 4 hours are from Rodger (1986) and Rodger *et al.* (1988), again measured using ^{29}Si MAS NMR for C_3S pastes (^{29}Si-enriched to 30 %) with w/s = 0·50 hydrated at 25 °C. The slight difference in temperature between the two studies did not greatly affect the kinetics of reaction, with both reaching ≈20 % hydration after ≈12 hours.

1994; Rodger, 1986; Rodger *et al.*, 1988). It shows that in this case about 40 % of the C_3S had reacted by 1 day, and 90 % by about 10 months. Similar curves have been determined using QXDA for unreacted C_3S (e.g. Kondo and Ueda, 1969; Odler and Dörr, 1979a) but there are issues with accuracy and precision with both techniques at low values of α (below about 0·1–0·2). While other methods are used to study early hydration – such as conduction calorimetry (e.g. Prosen *et al.*, 1985), aqueous phase analyses (e.g. Barret *et al.*, 1980; §7.6.1) or determinations of CH content (see §7.2.2) or of non-evaporable water (e.g. Fujii and Kondo, 1969) – at very early ages, it is necessary to allow for the fact that the property determined depends on the nature of the hydration products (e.g. an early product is formed before CH precipitates), which is problematic.

The rate, as followed from that of heat evolution (Figure 7.20), passes through an initial maximum, decreases to a minimum during a so-called induction period, passes through a second maximum

Figure 7.20 Typical calorimetric curve relating the rate of heat liberation to time for a C_3S paste with w/s = 0·40 hydrated at 20 °C. Replotted from Fig. 12 of Richardson *et al.* (1989). The curve was collected using a Calox conduction calorimeter, as described by Le Sueur (1984) and Wilding *et al.* (1984). The C_3S and water were mixed by suction under vacuum after temperature equilibration.

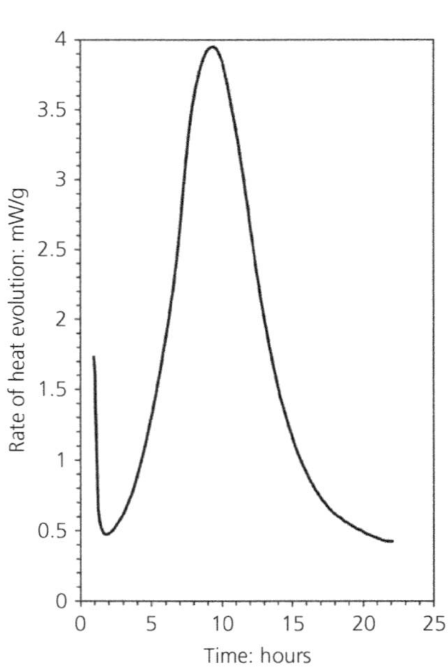

and then gradually declines. Kondo and Ueda (1969) described five periods of reaction, namely (1) the initial reaction, (2) the induction period, (3) the acceleratory period, in which the main reaction first begins to occur rapidly, (4) the deceleratory period and (5) a period of slow, continued reaction. Periods 1 and 2 correspond to the early stage of reaction as defined from microstructural studies (§7.3.2), periods 3 and 4 to the middle stage and period 5 to the late stage. Setting takes place during the acceleratory period.

At early ages, $d\alpha/dt$ increases markedly with w/s ratios above 0·7 (Brown *et al.*, 1984). Moderate variations in specific surface area have little effect on the length of the induction period, but with finer grinding, $d\alpha/dt$ during the acceleratory period increases (Brown *et al.*, 1984; Kondo and Veda, 1969; Odler and Schüppstuhl, 1981). The rate of reaction increases with temperature up to the end of the acceleratory period, but is much less affected thereafter (Kantro *et al.*, 1962), suggesting a change from chemical to diffusion control. Introduction of defects into the C_3S shortens the induction period (Fierens and Verhaegen, 1976; Maycock *et al.*, 1974; Odler and Schüppstuhl, 1981).

The results of experiments at high w/s ratios have been interpreted assuming superficial hydroxylation of the C_3S followed by congruent dissolution and subsequent precipitation (Barret and Bertrandie, 1986; Barret *et al.*, 1980b). Damidot *et al.* (Damidot and Nonat, 1994a, 1994b; Damidot *et al.*, 1990) followed the hydration of C_3S using a technique that allowed simultaneous monitoring of the heat flow and electrical conductivity. The use of high w/s ratios also greatly facilitated the study of the earlier stages of the reaction and made it possible to distinguish endothermic blips on the generally exothermic calorimetric curve, attributable to the onset of precipitation first of C–S–H and later of CH. From their results, the authors concluded that the initial process was one of congruent dissolution of C_3S. When the limiting supersaturation of C–S–H had been reached, this phase began to precipitate, initially by homogeneous nucleation and later by heterogeneous nucleation and growth on a hydroxylated C_3S surface. During this stage, C_{CaO} and the Ca/Si ratio of the C–S–H increased; when the limiting supersaturation of CH had been reached, that phase also precipitated. Reaction in pastes must also occur at least partly by dissolution and precipitation, since some of the product is deposited at a distance from the starting material, and because both C_{CaO} and C_{SiO_2} pass through maxima (Figure 7.17). Topochemical mechanisms, defined as ones in which the material does not pass through a true solution phase, may also occur, especially in the late stage of reaction (Taylor, 1985).

The kinetics up to the middle of the acceleratory period are discussed in the following sections. Those of the later stages have been more thoroughly studied with cement and only some aspects are considered here.

7.7.2 C_3S: the initial reaction

In addition to data on hydration kinetics (§7.7.1) and solution composition (§7.6.1), understanding of the initial reaction of C_3S requires knowledge of the product formed. Measurement of its chemical composition is challenging because of the small quantity, but valuable information has been obtained using the surface analysis techniques of X-ray photoelectron spectroscopy (XPS/ESCA) and secondary neutral mass spectrometry (SNMS). Data from a number of studies are compared on Figure 7.21, which shows the evolution of the Ca/Si ratio of the surface of hydrating compacts of C_3S with w/s = 10 (Figures 7.21 (a)–(c)) or pastes with w/s = 0·50 ((d)–(f)). Comparison of the trends observed in the different studies requires knowledge of how the samples were prepared, and so some experimental details and notes are included in the figure caption.

Figure 7.21 Evolution of the Ca/Si ratio of the surface of hydrating C_3S with w/s = 10 ((a)–(c)) or 0·50 ((d)–(f)) as determined by X-ray photoelectron spectroscopy (XPS/ESCA) or secondary neutral mass spectrometry (SNMS) (continued on next page)

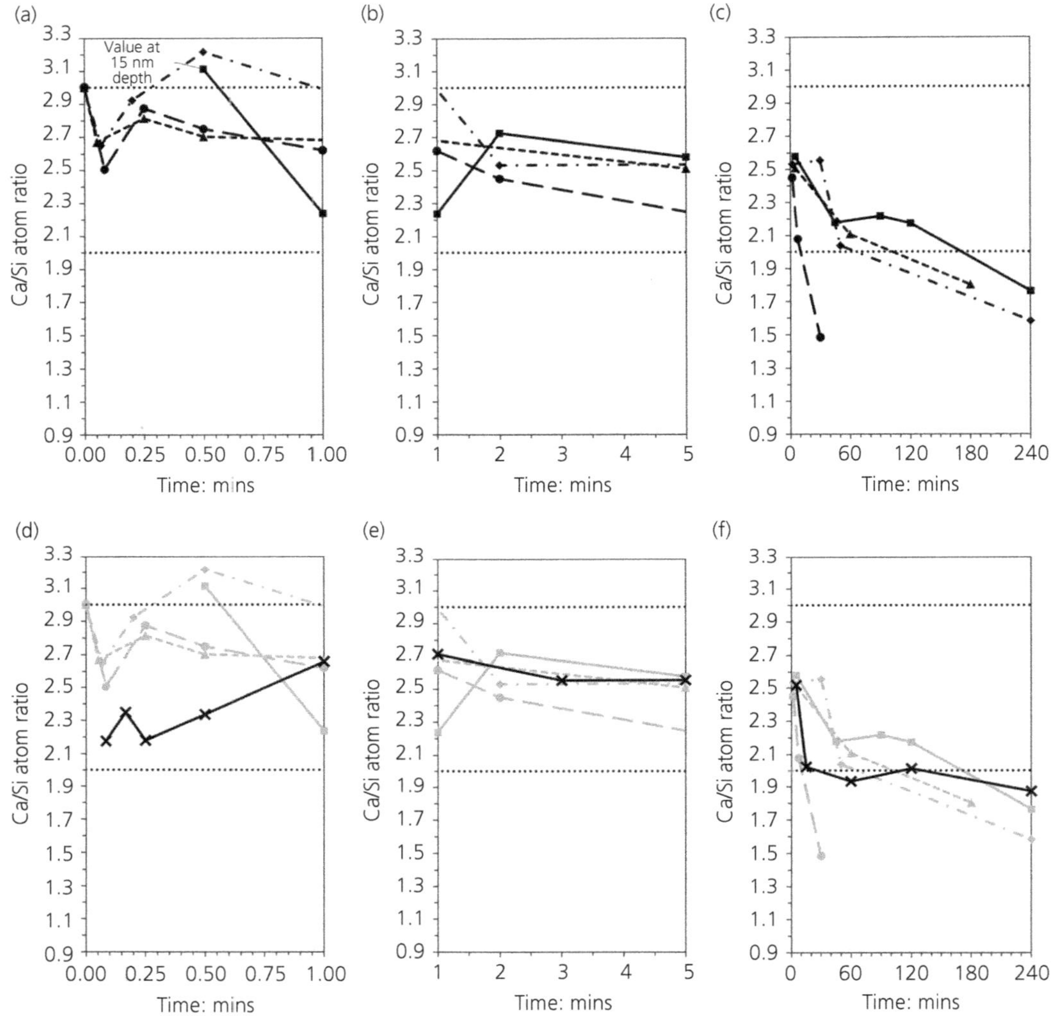

Comparison of the data is facilitated by dotted lines at Ca/Si = 3·0 and 2·0 on all parts of the figure, and by the inclusion of the data from (a)–(c) on (d)–(f) but shaded grey. The data are from: filled square symbols = Odler and Cordes, 2002b (SNMS); filled diamond = Stadelman *et al.*, 1985 (XPS); filled circles = Ménétrier *et al.*, 1979 (XPS); filled triangles = Jawed and Skalny, 1982; × = Regourd *et al.*, 1980 (XPS). The XPS studies of Ménétrier *et al.* (1979), Jawed and Skalny (1982) and Stadelman *et al.* (1985) were all conducted using pellets of pressed C_3S powder with w/s = 10. Some important experimental details of the studies follow. (1) Jawed and Skalny (1982) pressed 200 mg of C_3S powder into pellets at a pressure of 500 kg cm^{-2} using an infrared potassium bromide die. The pellets were placed on a wire holder, immersed in 2 mL of water (w/s = 10) at 25 °C for the desired period of time (2 sec to 3 hr). (2) Ménétrier *et al.* (1979) prepared pellets (0·5 mm thick and 12·5 mm in diameter) by

Figure 7.21 Continued

pressing 200 mg of C_3S powder at 500 kg cm^{-2}. The pellets were sintered at 1000 °C for 1 hour just before the hydration experiment. SEM micrographs showed that the sintered surface was smooth, but the pellet would have contained significant porosity. (3) Odler and Cordes (2002a, 2002b) spread anhydrous or hydrated C_3S powder (fraction 60–90 μm) over a metallic indium foil and pressed it into the metal using a glass plate. The foil measured 7 × 7 mm^2. The area of the surface covered by powder was about 10–15 mm^2. Samples were hydrated in rotating polyethylene bottles at laboratory ambient temperature in deionised water with w/s = 10. It appears that the first SNMS data point considered to be useful is at 15 nm because the data collected in the first seconds of each experiment had to be disregarded (see Odler and Cordes, 2002b; sputtering time of 50 s = 15 nm depth). (4) Stadelman *et al.* (1985) immersed a pellet (300 mg C_3S; pressed powder) in 3 mL of water (w/s = 10; 20 °C) without moving the solution. The authors of the studies consider the probing depth to be between 2 and 10 nm for XPS and '*not greater than 0·5 nm*' for SNMS. Jawed and Skalny were both involved in the work published in Ménétrier *et al.* (1979) and Jawed and Skalny (1982). The descriptions of the C_3S used in the two studies is the same. In both cases, the C_3S was pressed into pellets and the same methods were used for hydration and drying (pellets were placed on a wire holder, immersed in 2 mL of water (w/s = 10; 25 °C) for the desired period of time; then immersed in acetone followed by vacuum drying). The only difference in the experimental part of the two studies is that Ménétrier *et al.* (1979) sintered the C_3S. Inspection of the figure shows that the Ca/Si time behaviour is essentially the same in the first few seconds, but thereafter, while the form of the curves are very similar, Ménétrier *et al.*'s (1979) times need to be multiplied by 9 to get the curves to overlap. This difference in behaviour must have been due to the sintering, which would have reduced the surface area of C_3S that was in contact with the water.

For the C_3S compacts with w/s = 10, the XPS data on Figure 7.21(a) show an initial surface Ca/Si ratio of 3·0, which within the first few seconds falls quite sharply, settling at a value of 2·5–2·7 by 1 minute in two of the studies, with the third study reaching that value by 2 minutes, and a value of about 2·5 is then maintained until at least 5 minutes. The SNMS data show that product is present at a depth of 15 nm after 1 minute, but not at 30 seconds (because of the Ca/Si ratios of about 2·2 and 3, respectively). In addition to their XPS measurements, Jawed and Skalny (1982) also measured the value of C_{CaO} in solution. After 5 mins, it is about 8·8 mmol L^{-1}. If the initial product were C–S–H(I), then this value would be expected to result in a low Ca/Si ratio; using Taylor's (1950) curve (see Figure 6.1), the Ca/Si ratio that corresponds to 8·8 mmol L^{-1} is about 1·25. However, Jawed and Skalny's (1982) measured Ca/Si of the surface layer is very much higher, at about 2·5. Comparison of their XPS data with the SNMS data of Odler and Cordes (2002b) suggests strongly that this high value is not (as they assumed) due to the analysis being affected by the underlying C_3S (because of the XPS probing depth of 2 nm or greater), but real, because it is similar to Odler and Cordes's (2002b) value from SNMS, which has a lower probing depth (about 0·5 nm) and their analysed surface corresponded to about 15 nm below the original, unsputtered surface. The surface analysed by Jawed and Skalny (1982) at 5 mins hydration was thus likely to be entirely product, with no contribution from unreacted C_3S. This conclusion is supported by consideration of the amount of C_3S that is required to have reacted to give the measured solution composition. The data mentioned above from Taylor (1950) fall on curve A of Figure 7.18(a) whereas solution data from hydrating C_3S fall along curve B, and so Jawed and Skalny's (1982) solution after 5 mins with $C_{CaO} \approx 8·8$ mmol L^{-1} would likely have had a value for C_{SiO_2} of ≈ 0·3 mmol L^{-1}. Combining these solution data with the measured Ca/Si of the product (2·509) gives the following mass balance if the H_2O/Si ratio of the product is assumed to be 4

$$0\cdot0376\ C_3S + 126\cdot734916138\ H \rightarrow 0\cdot036915784\ C_{2\cdot509}S\ H_4 + 126\cdot587253000\ H$$

The number of moles of water on the left-hand side of this chemical equation corresponds to the water mixed with 1 mole of C_3S (mass of C_3S = 228·3165 g; mass of H_2O = 126·734916138 mols × 18·01528 g mol^{-1} = 2283·155 g; w/s = 2283·165 g/228·3165 g = 10·00). It has a volume of 2·283165 L. The number of moles of water on the right-hand side is the water remaining after the reaction of 0·0376 moles of C_3S. This water has a volume of 2·280504807 L and so the amount of water consumed (mass %) = 0·12 %. The solid product has a Ca/Si ratio of 2·509, and so it contains 0·092621702 mols CaO (i.e. 0·036915784 × 2·509), 0·036915784 mols SiO_2, and 0·147663136 mols H_2O (i.e. 0·036915784 × 4). The solution thus contains 0·020178298 mols CaO ((0·0376 × 3) – 0·092621702 = 0·020178298), which corresponds to 0·0088 mol L^{-1} (0·020178298 mol / 2·280504807 L), and 0·000684216 mols SiO_2 ((0·0376 × 1) – 0·036915784 = 0·000684216), which corresponds to 0·0003 mol L^{-1} (0·000684216 mol/2·280504807 L). So, at 5 minutes the degree of hydration with w/s = 10 that is required to give a product with Ca/Si = 2·509 together the measured value of C_{CaO} (8·8 mmol L^{-1}) and estimated value of $C_{\mathrm{SiO_2}}$ (0·3 mmol L^{-1}) is 3·76 %. This would give a depth of product that would be very much greater than the XPS probing depth. The degree of hydration in a paste with w/s = 0·50 that is required to give the same compositional values for the product and solution is 3·76/20 = 0·188 % (i.e. 0·00188 C_3S + 6·336745807 H → 0·001845789 $C_{2\cdot509}$ S H_4 + 6·329362650 H). With reference to Figure 7.12(d), this should also result in a product layer that is greater than the depth probed by XPS, with a layer thickness of about 5·5 nm compared with a probing depth of about 2 nm; the XPS data for w/s = 0·50 on Figure 7.21 are consistent with this conclusion.

Allowing for the fact that the region analysed will include unreacted C_3S until the layer of hydration product is sufficiently thick and continuous, the results on Figure 7.21 may be interpreted as follows.

(*a*) On contact with water, an initial hydrate is formed of low Ca/Si ratio, perhaps as low as 1·0 (Regourd *et al.*, 1980; Barret *et al.*, 1983).
(*b*) The Ca/Si ratio of product rapidly increases to about 2·5.
(*c*) This value is maintained for a few minutes, and then falls to about 2·0–2·2 by 15 minutes to 1 hour, depending on whether the sample is a paste or an unsintered compact.
(*d*) The Ca/Si ratio of the product falls to 1·7–1·9 by 4 hours.
(*e*) The average Ca/Si ratio remains within this range for many years, as discussed and illustrated in §7.2.2 and §7.3.4.

The high Ca/Si ratio of the early product determined by XPS is supported by Bellmann *et al.*'s (2010) indirect calculation that combined results from ^{29}Si NMR, thermal analysis and chemical analysis. XPS results also indicate a change in the environment of the silicon atoms at the end of the induction period, possibly associated with the formation of dimeric silicate ions (Regourd *et al.*, 1980). Secondary-ion mass spectrometry (SIMS) is a possible alternative method for studying surface compositions, but it has not yet given information on that of the hydrated material (Gerhard and Nägele, 1983). Electron diffraction shows that an amorphous surface layer can be produced by grinding (Kondo and Ueda, 1969; Ueda *et al.*, 1968); this may explain some of the reported differences in behaviour between apparently similar specimens.

7.7.3 C_3S: the period of slow reaction (often referred to as the 'induction' or 'dormant' period)

Values for α can be calculated by mass balance for the period of slow reaction by combining the Ca/Si ratio of the product with concentrations in the bulk solution. While lack of suitable data mean that it is necessary to use data from different studies – which is not ideal because of varying experimental parameters and reactivities – it is nevertheless useful. As an example, the XPS study of Regourd *et al.* (1980) gave a Ca/Si ratio of about 2 for hydration times of between 15 minutes and 2 hours (for w/s = 0·50; Figure 7.21(f)). The average composition of the solutions for 1-hour (w/s = 0·50; 23 °C) pastes reported by Naber *et al.* (2019, 2020) is $C_{CaO} = 0{\cdot}0256\,mol\,L^{-1}$ and $C_{SiO_2} = 0{\cdot}0000223\,mol\,L^{-1}$. Combining these data gives the following mass balance if the H_2O/Si ratio of the product is assumed to be 4.

$$0{\cdot}00291\,C_3S + 6{\cdot}336745807\,H \rightarrow 0{\cdot}002907464\,C_{2{\cdot}000}S\,H_4 + 6{\cdot}325115950\,H$$

As in §7.7.2, the number of moles of water on the left-hand side of the chemical equation corresponds to the water mixed with 1 mole of C_3S (mass of C_3S = 228·3165 g; mass of H_2O = 6·336745807 mols × 18·01528 g mol^{-1} = 114·15825 g; w/s = 114·15825 g/228·3165 g = 0·50). It has a volume of 0·114158250 L. The number of moles of water on the right-hand side is the water remaining after the reaction of 0·00291 moles of C_3S. This water has a volume of 0·113948735 L and so the amount of water consumed (mass %) = 0·18 %. The solid product has a Ca/Si ratio of 2·000, and so it contains 0·005814928 mols CaO (0·002907464 mols × 2·000), 0·002907464 mols SiO_2, and 0·011629856 mols H_2O (0·002907464 mols × 4). The solution thus contains 0·00291508 mols CaO ((0·00291 × 3) – 0·005814928 = 0·002915072), which corresponds to 0·0256 mol L^{-1} (0·002915072 mol/0·113948735 L), and 0·00000254 mols SiO_2 ((0·00291 × 1) – 0·002907464 = 0·00000254), which corresponds to 0·0000223 mol L^{-1} (0·00000254 mol/0·113948735 L). So, the degree of hydration of C_3S in this paste with w/s = 0·50 required at 1 hour to give a product with Ca/Si = 2·0 together with Naber *et al.*'s (2019, 2020) solution composition is 0·291 %, which is quite similar to the value measured by Rodger *et al.* (1988) using ^1H–^{29}Si CP-NMR with ^{29}Si-enriched C_3S, which was 0·50 %. Figures 7.22 (a) and (b) show the results of repeating this mass balance calculation for the hydration of C_3S paste (w/s = 0·50) up to 6 hours using Ca/Si ratios for the calcium silicate hydrate of 2·0 for times up to and including 3 hours and 1·7 for times > 3 hours (values guided by the results from XPS; see §7.7.2), and CH contents and solution compositions from Naber *et al.* (2019, 2020). Figures 7.22 (a) and (b) show, respectively, the degree of hydration and % CaO in the solution and products. Figure 7.22 (a) also includes data from Naber *et al.* (2020) for the % hydration from XRD (× symbols) and estimated by conduction calorimetry (+). Hydration progresses only slowly for the first three hours, with about two thirds of the CaO in the product and one third in solution. Reaction subsequently starts to accelerate and while the Ca/Si of the silicate part of the product is reduced, the proportion of the CaO that is in the solution decreases steadily as CH is formed. The calculated degree of hydration over the first 6 hours is quite close to that estimated from calorimetry but the values from XRD are much too high. The % hydration at 2 hours of ≈ 0·4 % corresponds to an approximate thickness of early product of ≈ 12 nm (using Rodger's (1986) simple calculation) and thus to just a doubling of the thickness between 5 minutes and 2 hours. The average thickness of the early product throughout the induction period is therefore likely to be on the same order as the thickness of the iron oxy-hydroxide passive film that forms on steel in alkaline solutions, which Angst *et al.* (2017) give as 3–15 nm. These are significant thicknesses because they correspond to between 20–50 times the diameter of an oxygen anion.

Figure 7.22 Results of mass balance calculations for the hydration of C_3S paste (filled circles; w/s = 0·50) using Ca/Si ratios for the calcium silicate hydrate of 2·0 (for times up to and including 3 hours) and 1·7 (for times >3 hours) and solution compositions and CH contents from Naber *et al.* (2019, 2020; the CH values were provided by Sowoidnich, 2022). The solution compositions are shown as squares on Figure 7.17. Figures 7.22 (a) and (b) show, respectively, the degree of hydration and % CaO in the solution and products up to 6 hours; and Figures 7.22 (c) and (d) show the same up to 24 hours. Figures 7.22 (a) and (c) also include data from Naber *et al.* (2020) for the % hydration from XRD (× symbols) and estimated by conduction calorimetry (+) using the enthalpy of hydration of C_3S noted in §7.6.3

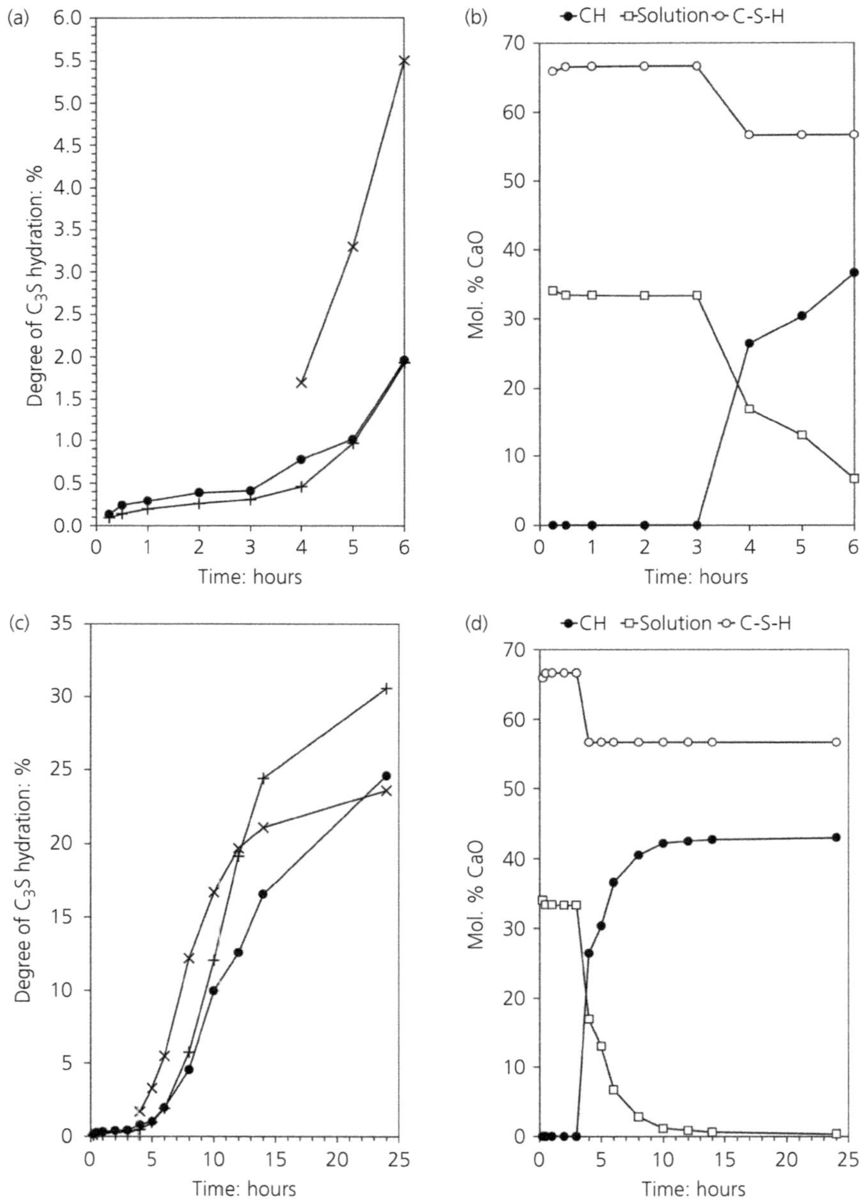

The causes of the period of slow reaction and the reason for the start of the acceleratory period have been the subject of much debate. Hypotheses have been reviewed regularly, for example by Taylor *et al.* (1984), Bullard *et al.* (2011), Scrivener and Nonat (2011), Scrivener *et al.* (2015, 2019, 2023), John and Lothenbach (2023). The main ones are as follows.

1. The product of the initial reaction forms a protective layer on the C_3S particles; the induction period ends when this is destroyed or rendered more permeable by ageing or phase transformation (e.g. Bellmann *et al.*, 2010; Brown *et al.*, 1985; de Jong *et al.*, 1967; Gartner, 2011; Gartner and Gaidis, 1989; Gartner and Jennings, 1987; Grutzeck, 1999; Jennings, 1986; Jennings and Pratt, 1979a; Kondo and Daimon, 1969; Makar *et al.*, 2011; Regourd *et al.*, 1980; Stein and Stevels, 1964).
2. The product of the initial reaction forms a semipermeable membrane which encloses an inner solution and is destroyed by osmotic bursting (e.g. Birchall *et al.*, 1980; Double, 1983; Double *et al.*, 1978).
3. The rate of reaction in the induction and acceleratory periods is controlled by nucleation and growth of the C–S–H formed in the main reaction, the induction period ending when growth begins (e.g. Barret *et al.*, 1980b; Fierens and Verhaegen, 1976; Garrault-Gauffinet and Nonat, 1999; Odler and Dörr, 1979b; Sierra, 1976).
4. The induction period occurs because the CH nuclei are poisoned by SiO_2 and cannot grow, and ends when the level of supersaturation is sufficient to overcome this effect (e.g. Greening, 1972; Tadros *et al.*, 1976; Wu and Young, 1984).
5. Initial dissolution of C_3S is rapid and dominated by the formation of etch pits on surfaces. The induction period begins when dissolution becomes limited to step retreat at pre-existing pits and as a consequence it slows down (Juilland *et al.*, 2010, 2017; Nicoleau and Bertolim, 2016; Nicoleau and Nonat, 2016; Scrivener and Nonat, 2011). The induction period ends when rapid growth of C–S–H commences, which may be related to the formation of a stable form of C–S–H through the polymerisation of silicate monomers (Juilland *et al.*, 2011).

There is no consensus on which of these hypotheses is correct because there are insufficient data to definitively support any of them. Nevertheless, the balance of the evidence reviewed in this chapter favours a combination of hypotheses 1 and 3, as follows.

(*a*) The calculated solubility of C_3S is about 1 molal (Stein, 1972). In the absence of an altered or protective layer, the observed low rate of reaction during the induction period could be explained only if the concentrations close to the C_3S surface were of this order of magnitude. It is doubtful whether the very high concentration gradients that this implies could exist (Gartner and Gaidis, 1989; Jennings, 1986).

(*b*) After an initial step, C_3S dissolves congruently (Barret *et al.*, 1983) and a material is deposited, which will be termed product B. The dissolving C_3S probably has a hydroxylated surface formed by protonation of the O^{2-} and SiO_4^{4-} ions, balanced by loss of Ca^{2+}, as proposed by Barret *et al.* (1980b). Within 30 s the C_3S is almost isolated from the solution, and a metastable equilibrium, represented by curve B of Figure 7.18(a), is established between the solution and product B (Gartner and Jennings, 1987). The silicate in product B is monomeric. The Ca/Si ratio of product B is initially about 2·5, which is maintained for a few minutes, and it then falls to about 2·0–2·2 by 15 minutes to 1 hour. Under typical paste conditions, product B forms a non-uniform overlying surface layer that steadily increases in thickness during the period of slow reaction to about 10–100 nm, similar to a passive layer on a metal. C_{CaO} continues to increase to a high degree of supersaturation.

(c) At a high degree of CaO supersaturation, CH and C–S–H (with jennite-based structure) nucleate and begin to grow, probably as part of the same process because they both have a common nucleus of edge-sharing Ca–OH octahedra (Gartner *et al.*, 2002). This process is represented schematically for formation from solution in Figure 7.23 (steps 1–3) and outlined in §7.7.6. Nevertheless, since product B and the C–S–H are both associated with curve B, it is plausible that the initial sites of nucleation for the C–S–H are defects in particles of product B, which would be consistent with the TEM evidence discussed in §7.3.2. The NMR evidence (§7.4.2) indicates that the hydrated material formed up to the end of the induction period contains only monomeric silicate ions and that the latter persist alongside larger ions when Q^1 appears at the end of the induction period, later followed by Q^2. This suggests that the silicate ions are monomeric in both product B and C–S–H in the early stages of its formation, as illustrated in Figure 7.23. The induction period ends when C–S–H begins to grow. This implies dissolution or transformation of product B, and increased access of solution to the C_3S, which thus dissolves more rapidly. The rate of reaction is then controlled by the growth of C–S–H, the rate of which increases with the amount already formed. The replacement of product B by C–S–H leads to a decrease in the average Ca/Si ratio of silicate-containing product to 1·7–1·9.

7.7.4 The main reaction (C_3S and β-C_2S)

From the kinetic standpoint, there is probably no distinction between the induction period and the early part of the acceleratory period; indeed, as Gartner and Gaidis (1989) noted, there is strictly no

Figure 7.23 Schematic diagram illustrating how particles of C–S–H and CH might form common nuclei

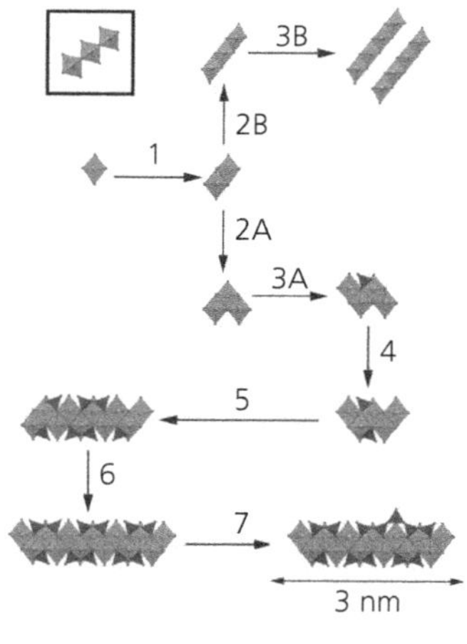

The Ca/Si ratio of the C–S–H decreases as silicate tetrahedra clasp to the central Ca–O core of the structure. For the purpose of straightforward illustration of the *dreierketten* silicate, this schematic diagram is derived from models of tobermorite; however, the initial C–S–H is likely to be based on jennite because of the need for the presence of Ca–OH and its association with curve B on Figure 7.18; as a consequence the Ca/Si ratio would be higher, and consistent with the XPS/SNMS data discussed in §7.7.3.

induction period in the absence of retarding admixtures. This is evident from Figure 7.22(a), which shows a slow but continuous increase in the degree of hydration of C_3S throughout that period. Throughout both stages, growth of silicate-containing product (product B followed by C–S–H) proceeds at an increasing rate, which is reflected in increasingly rapid dissolution of the C_3S (Figure 7.19 and Figure 7.22(c)). The Ca^{2+} and OH^- concentrations in the bulk solution increase steadily; when a sufficient degree of supersaturation has been reached, CH begins to precipitate in quantity, causing these concentrations to fall (Figure 7.17), and the proportion of the CaO that is in solution rather than in product drops dramatically (Figures 7.22(b) and (d)). While the formation of CH is not normally rate-determining, it could become so in the presence of admixtures that retard its growth. The kinetic data for this stage can be fitted by Avrami-type equations, but the assumptions underlying the latter (formation of a product within an initially homogeneous material) do not correspond to the situation existing in the hydration of C_3S or cement.

The smaller grains of C_3S probably react completely, by dissolution and precipitation, during the acceleratory period. In the final, slow periods of hydration, the remaining grains of C_3S, which are relatively large, are gradually replaced by C–S–H through the inward movement of an interface. It is difficult to believe that such a process occurs by dissolution into a true liquid phase followed by precipitation. Following earlier studies (Funk, 1962; Kondo and Ueda, 1969; Kondo *et al.*, 1967; McConnell, 1955), Taylor (1985) suggested that a topochemical reaction may occur with both C_3S and β-C_2S. Comparison of the numbers of atoms of each element in the initial and product phases shows that to convert either C_3S or β-C_2S into C–S–H, relatively little O^{2-} must be gained or lost; much of the Ca^{2+} and smaller proportions of the Si^{4+} must be lost, and equivalent amounts of H^+ gained (Table 7.5). At the interface, a narrow zone could exist, in which the necessary atomic rearrangements take place. It was suggested (Taylor, 1985) that the migrations were literally of Ca^{2+}, Si^{4+}, H^+ and O^{2-}, but outward movement of Ca^{2+} and a silicate species and inward movement of water molecules through the inner product may be considered to be more likely. The Ca^{2+} and silicate expelled must enter the outer product region, which is the reason why the association of the term 'late product' with 'inner product' by Jennings *et al.* (1981) is not appropriate. At the surface, the Ca^{2+} and Si^{4+} enter the solution, and ultimately precipitate, along with the OH^- ions released from the H_2O molecules, as CH and outer product C–S–H. Migration of silicon was suggested as the probable rate-determining step; however, at low w/s ratios, reaction stops because there is no more space in which the outer product can be formed, and in this case, control presumably shifts to some process related to deposition of outer product. Barret (1986) discussed topochemical mechanisms and their relation to dissolution–precipitation processes. More work is needed to establish whether there is any tendency for the inner product C–S–H in cement pastes to be formed topotactically from the alite or belite.

Table 7.5 Numbers of atoms in $(1000/N_A)$ cm^3. Slightly modified from Taylor (1985)

Phase	Density: $g\,cm^{-3}$	Ca	Si	H	O
C_3S	3·120	41·0	13·7		68·3
β-C_2S	3·326	38·6	19·3		77·2
C–S–H (90 % RH)	1·900	14·4	8·2	66·0	63·9

C–S–H composition assumed to be $C_{1\cdot75}SH_{4\cdot0}$.

7.7.5 Early hydration of β-C_2S

The kinetics and mechanism of β-C_2S hydration are similar to those for C_3S, apart from the much lower rate of reaction (Fierens and Tirlocq, 1983a; Fujii and Kondo, 1979; Maycock *et al.*, 1974; Odler and Schüppstuhl, 1982; Tong and Young, 1977) and, as noted earlier, the products are similar apart from the much smaller content of CH. Preparations appear to be more variable in reactivity than those of C_3S; this is partly attributable to differing stabilisers (Fierens and Tirlocq, 1983b), but could also be due to differing amounts or natures of phases in intergranular spaces or exsolution lamellae. Preparations made at low temperatures are especially reactive (Fierens and Tirlocq, 1983a). Because of the low rate of heat evolution, conduction calorimetry is difficult but shows the existence of an induction period similar to that found with C_3S (Maycock *et al.*, 1974), which is also indicated by QXDA (Odler and Schüppstuhl, 1982). XPS, SEM and solution studies show that behaviour during the early period is similar in principle to that of C_3S (Ménétrier *et al.*, 1980). The rate of reaction is increased in the presence of C_3S (Odler and Schüppstuhl, 1982; Tong and Young, 1977). Work on the hydration and products of highly reactive β-C_2S are discussed in §6.3.1 and §6.3.2.

7.7.6 Mechanism of C–S–H formation

Taylor (1964) proposed that the mechanism of the formation of C–S–H initially involves the condensation of mononuclear complexes such as $[Ca(H_2O)_5(OH)]^+$ from solution, which is essentially the same as with CH. These complexes condense repeatedly by the sharing of edges and elimination of water molecules to form slabs of CH-like structure. This process is represented schematically in Figure 7.23, steps 1 and 2; the sharing of edges is emphasised by the alternative orientation of the top-middle slab that is shown in the inset box. There are different ways that the polyhedra can condense: two alternatives are shown in the figure. One alternative leads to formation of particles of C–S–H (route 1–2A–3A) and the second to particles of CH (route 1–2B–3B); the former subsequently involving the attachment of silicate monomers to the Ca–O layer. Taylor (1964) considered that it was not necessary for a whole Ca–O layer to form before silicate tetrahedra attached to it. It is known from ^1H-^{29}Si cross-polarisation NMR studies that the first silicate-containing product to form from the hydration of C_3S contains exclusively monomeric silicate units (Brough *et al.*, 1994; Rodger, 1986; Rodger *et al.*, 1988; §7.4.1) and this is reflected in Figure 7.23, step 3A. The C–S–H nuclei continue to grow (and multiply) slowly through the period of slow reaction (the induction period) and the silicate anions remain monomeric throughout this period; this is illustrated schematically in Figure 7.23, step 4. At the end of the induction period, ^{29}Si NMR data show the formation of Q^1 tetrahedra – that is, dimeric linear silicate chains. This is perhaps because the Ca–O parts of the nuclei are at that point big enough to accommodate the dimerisation and monomers remain as essentially the result of edge effects as the particles grow (steps 5 and 6). This edge effect explanation for the persistence of monomer until high degrees of reaction (§7.4.2) is supported by the observation that the larger amount of Q^{0H} that is formed at elevated temperature (Brough *et al.*, 1994) coincides with smaller particles of Ip C–S–H (Richardson, 2004), and that more Q^{0H} is present when the morphology of the C–S–H is fibrillar rather than foil-like (Richardson, 2004; see Figure 7.13).

The NMR data show that Q^2 tetrahedra are detected quite soon after Q^1 and this corresponds to the detection of pentamers by TMS-GPC (§7.4.1); this is illustrated in Figure 7.23, step 7. As discussed in §7.4.1, continued polymerisation subsequently results in the formation of octamers; the most plausible explanation for this 2, 5, 8, … $(3n - 1)$ sequence of silicate anions (where n is integer for individual anions) is that the linear chains have the *dreierketten* conformation that occurs in a number of phases, including wollastonite, jennite and tobermorite (see Chapter 5).

The reason that the tetrahedra link in this pattern is due to the coordination requirements of the Ca–O layers: the Ca–O layers can exist in isolation (i.e. CH-like structure) but the same is not true for linear silicate chains that have *dreierketten* conformation. The stages in the formation of C–S–H as illustrated in Figure 7.23 are summarised in Table 7.6. In this scenario, polymerisation occurs by the insertion of monomers at bridging sites between the paired tetrahedra of adjacent dimers. This is not the same as occurs with synthetic C–S–H preparations that have tobermorite-based structure (with $Ca/Si < 1{\cdot}5$), where increases in average silicate chain length are more likely achieved by the interstratification of dimeric and long-chain structure (that has some vacant bridging sites), which is described and demonstrated (with model structures) in Richardson (2014) and Chapter 6.

The onset of the formation of dimer at the end of the induction period coincides with the observation of the appearance of needles or fibres of C–S–H growing out from the surface product into the outer product region. These fibres or fibrils, which get both longer and thicker as hydration progresses, consist of agglomerations of long thin particles (§7.3.2). These particles are also observed in the Ip C–S–H, although to a lesser extent. If the silicate chains are aligned along the length of the particles, then the proportion of the particle surface that is chain-end edge reduces as the average length increases. It is plausible that as the extent of hydration increases, the increase in the number of chain-end edges associated with the formation of new particles combined with the reduction in existing particles results in an approximate balance, and thus to the plateau on the Q^{0H} curve on Figure 7.12(b). The reduction in the quantity of Q^{0H} that occurs after 80 % hydration might then occur when the balance is disturbed by an increase in the rate of particle-lengthening (consistent with an increase in polymerisation as indicated by the change in the gradient of the plot of MCL against % hydration, Figure 7.12(c)). A change in the location of Q^{0H} from product B to the edges of C–S–H particles might account for Brough *et al.*'s (1994) report of an upfield shift of the Q^{0H} peak, which they suggested may be due to a change in its chemical environment.

Table 7.6 Summary of the stages of the formation of C–S–H from solution as illustrated schematically in Figure 7.23

S	1	Mononuclear complexes such as $[Ca(H_2O)_5(OH)]^+$ from solution condense by the sharing of edges and elimination of water molecules.
S	2	Repeated condensation forming slabs of Ca–O structure.
S	3A	Monomeric silicate tetrahedra attach to Ca–O slabs.
	3B	Formation of small particles of CH.
S	4	Slow growth of C–S–H particles; silicate anions exclusively monomeric.
A	5	Continued growth of C–S–H particles, which are now large enough to accommodate dimerisation; monomers persist because of edge effects.
A	6	Continued growth of C–S–H particles, the rate of which increases with the amount already formed, and continued dimerisation; monomers remain at edges.
A+	7	Polymerisation: formation of 2, 5, 8, … ($3n - 1$) sequence of linear silicate chains by the insertion of monomers at bridging sites between the paired tetrahedra.

S = period of slow reaction; this is often referred to as the 'dormant' or 'induction' period.
A = period of accelerated reaction.

REFERENCES

Allen AJ, Thomas JJ and Jennings HM (2007) Composition and density of nanoscale calcium-silicate-hydrate in cement. *Nature Materials* **6**: 311–316.

Angst UM, Geiker MR, Michel A *et al.* (2017) The steel–concrete interface. *Materials and Structures* **50**: 143.

Antao SM and Hassan I (2009) The orthorhombic structure of $CaCO_3$, $SrCO_3$, $PbCO_3$ and $BaCO_3$: linear structural trends. *Canadian Mineralogist* **47**: 1245–1255.

Antao SM, Hassan I, Mulder WH, Lee PL and Toby BH (2009) In situ study of the $R\bar{3}c$ to $R\bar{3}m$ orientational disorder in calcite. *Physics and Chemistry of Minerals* **36**: 159–169.

Babushkin VI, Matveyev GM and Mchedlov-Petrossyan OP (1985) *Thermodynamics of Silicates*. Springer, Berlin, Germany.

Bae S, Jee H, Suh H *et al.* (2020) Analysis of atomistic structural deformation characteristics of calcium silicate hydrate in 53-year-old tricalcium silicate paste using atomic pair distribution function. *Construction and Building Materials* **237**: 117714.

Bailey JE and Stewart HR (1984) C_3S hydration products viewed using a cryo stage in SEM. *Journal of Materials Science Letters* **3**: 411–414.

Barby D, Griffiths T, Jacques AR and Pawson D (1977) In *The Modern Inorganic Chemicals Industry* (Thompson R (ed.)). Chemical Society, London, UK, p. 320.

Barret P (1986) Hydration mechanism of calcium silicates (C_3S, C_2S) cement compounds, through the general concepts of the reactivity of solids. *Proceedings of the 8th International Congress on the Chemistry of Cement* **3**: 86–92.

Barret P and Bertrandie D (1986) Fundamental hydration kinetic features of the major cement constituents: Ca_3SiO_5 and βCa_2SiO_4. *Journal de Chimie Physique* **83**: 765–775.

Barret P and Bertrandie D (1988) Comment on 'Aqueous solubility relationships for two types of calcium silicate hydrate'. *Journal of the American Ceramic Society* **71**: C-113–C-115.

Barret P, Bertrandie D and Ménétrier D (1980) Comparatve study of C–S–H formation from super-saturated solutions and C_3S solution mixtures. *Proceedings of the 7th International Congress on the Chemistry of Cement* **2**: II-261-II-266.

Barret P, Ménétrier D and Bertrandie D (1983) Mechanism of C_3S dissolution and problem of congruency in the very initial period and later on. *Cement and Concrete Research* **13**: 728–738.

Bassett H (1934) Notes on the system lime–water, and on the determination of calcium. *Journal of the Chemical Society*: 1270–1275.

Bates RG, Bower VE and Smith ER (1956) Calcium hydroxide as a highly alkaline pH standard. *Journal of Research of the National Institute of Standards and Technology* **56**: 305–312.

Beaudoin JJ (1987) Validity of using methanol for studying the microstructure of cement paste. *Materials and Structures* **20**: 27–31.

Beedle SS, Groves GW, Rodger SA (1989) The effect of fine pozzolanic and other particles on the hydration of C_3S. *Advances in Cement Research* **2**: 3–8.

Bellmann F, Damidot D, Möser B and Skibsted J (2010) Improved evidence for the existence of an intermediate phase during hydration of tricalcium silicate. *Cement and Concrete Research* **40**: 875–884.

Bellmann F, Sowoidnich T, Ludwig HM and Damidot D (2012) Analysis of the surface of tricalcium silicate during the induction period by X-ray photoelectron spectroscopy. *Cement and Concrete Research* **42**: 1189–1198.

Bellmann F, Sowoidnich T, Ludwig HM and Damidot D (2015) Dissolution rates during the early hydration of tricalcium silicate. *Cement and Concrete Research* **72**: 108–116.

Bentur A and Berger RL (1979) Chemical composition of C–S–H gel formed in the hydration of calcium silicate pastes. *Journal of the American Ceramic Society* **62**: 117–120.

Bentur A, Berger RL, Kung JH, Milestone NB and Young JF (1979a) Structural properties of calcium silicate pastes: II, Effect of curing temperature. *Journal of the American Ceramic Society* **62**: 362–366.

Bentur A, Berger RL, Lawrence FV, Milestone NB, Mindess S and Young JF (1979b) Creep and drying shrinkage of calcium silicate pastes III. A hypothesis of irreversible strains. *Cement and Concrete Research* **9**: 83–95.

Berger RL and McGregor JD (1972) Influence of admixtures on the morphology of calcium hydroxide formed during tricalcium silicate hydration. *Cement and Concrete Research* **2**: 43–55.

Berger RL, Young JF and Lawrence FV (1972) Discussion of the paper 'Morphology and surface properties of hydrated tricalcium silicate paste' by Collepardi M and Marchese B. *Cement and Concrete Research* **2**: 633–636.

Bernal JD (1954) The structure of cement hydration compounds. *Proceedings of the 3rd International Symposium on the Chemistry of Cement 1952*. Cement and Concrete Association, London, UK, pp. 216–236.

Bernal JD and Megaw HD (1935) The function of hydrogen in intermolecular forces. *Proceedings of the Royal Society* **151**A: 384–420.

Bernal JD, Jeffery JW and Taylor HFW (1952) Crystallographic research on the hydration of Portland cement. A first report on investigations in progress. *Magazine of Concrete Research* **4**: 49–54.

Bessey GE (1939) The calcium aluminate and silicate hydrates. *Proceedings of the Symposium on the Chemistry of Cement, Ingeniorsventenskapsakademien*, Stockholm, pp. 179–215.

Birchall JD, Howard AJ and Double DD (1980) Some general considerations of a membrane/osmosis model for Portland cement hydration. *Cement and Concrete Research* **10**: 145–155.

Borodin VL, Lyutin VI, Ilyukhin VV and Belov NV (1979) The isomorphous series calcite – otavite. *Doklady Akademii Nauk SSSR* **245**: 1099–1101.

Brisi C (1954) Tonalità termica delle reazioni di idratazione dei costituenti del clinker di Portland. *Scientia Ricerca* **24**: 1436. (In Italian.)

Brough AR (1993) *Solid State NMR Studies of Inorganic Materials*. DPhil thesis, University of Oxford, Oxford, UK.

Brough AR, Dobson CM, Richardson IG and Groves GW (1994) In situ solid state NMR studies of Ca_3SiO_5: hydration at room temperature and at elevated temperatures using ^{29}Si enrichment. *Journal of Materials Science* **29**: 3926–3940.

Brough AR, Dobson CM, Richardson IG, Groves GW (1995) A study of the pozzolanic reaction by solid-state ^{29}Si nuclear magnetic resonance using selective isotopic enrichment. *Journal of Materials Science* **30**: 1671–1678.

Brough AR, Dobson CM, Richardson IG, Groves GW (1996) Alkali activation of reactive silicas in cements: in situ ^{29}Si MAS NMR studies of the kinetics of silicate polymerization. *Journal of Materials Science* **31**: 3365–3373.

Brown LS and Carlson RW (1936) Petrographic studies of hydrated cements. *Proceedings of the ASTM*. **36**: 332–350.

Brown PW (1990) The system Na_2O-CaO-SiO_2-H_2O. *Journal of the American Ceramic Society* **73**: 3457–3461.

Brown PW, Franz E, Frohnsdorff G and Taylor HFW (1984a) Analyses of the aqueous phase during early C_3S hydration. *Cement and Concrete Research* **14**: 257–262.

Brown PW, Galuk K and Frohnsdorff G (1984b) The rate of calcium hydroxide precipitation measured by electrical conductance. *Cement and Concrete Research* **14**: 843–846.

Brown PW, Pommersheim J and Frohnsdorff G (1985) A kinetic model for the hydration of tricalcium silicate. *Cement and Concrete Research* **15**: 35–41.

Brown PW, Harner CL and Prosen EJ (1986) The effect of inorganic salts on tricalcium silicate hydration. *Cement and Concrete Research* **16**: 17–22.

Brunauer S and Greenberg SA (1962) The hydration of tricalcium silicate and β-dicalcium silicate at room temperature. *Proceedings of the 4th International Symposium on the Chemistry of Cement* **1**: 135–165, paper III-1.

Brunauer S, Hayes JC and Hass WE (1954) The heats of hydration of tricalcium silicate and β-dicalcium silicate. *Journal of Physical Chemistry* **58**: 279–285.

Brunet F, Bertani P, Charpentier T, Nonat A and Virlet J (2004) Application of ^{29}Si homonuclear and 1H–^{29}Si heteronuclear NMR correlation to structural studies of calcium silicate hydrates. *Journal of Physical Chemistry B* **108**, 15494–15502.

Bullard JW, Jennings HM, Livingston RA *et al.* (2011) Mechanisms of cement hydration. *Cement and Concrete Research* **41**, 1208–1223.

Burgess KMN and Bryce DL (2015) On the crystal structure of the vaterite polymorph of $CaCO_3$: A calcium-43 solid-state NMR and computational assessment. *Solid State Nuclear Magnetic Resonance* **65**: 75–83.

Carpenter AB, Chalmers RA, Gard JA, Speakman K and Taylor HFW (1966) Jennite, a new mineral. *American Mineralogist* **51**: 56–74.

Caspi EN, Pokroy B, Lee PL, Quintana JP and Zolotoyabko E (2005) On the structure of aragonite. *Acta Crystallographica B* **61**: 129–132.

Chakoumakos BC, Pracheil BM, Koenigs RP, Bruch RM and Feygenson M (2016) Empirically testing vaterite structural models using neutron diffraction and thermal analysis. *Scientific Reports* **6**: 36799, 1–8.

Chang LLY, Howie RA and Zussman J (1998) *Rock-forming Minerals. Non-silicates: Sulphates, Carbonates, Phosphates, Halides*, vol. 5B, 2nd edn. The Geological Society, London, UK.

Chatterji S and Jeffery IW (1966) Three-dimensional arrangement of hydration products in set cement paste. *Nature* **209**: 1233–1234.

Chen JJ, Thomas JJ, Taylor HFW and Jennings HM (2004) Solubility and structure of calcium silicate hydrate. *Cement and Concrete Research* **34**: 1499–1519.

Christy AG (2017) A review of the structures of vaterite: The impossible, the possible, and the likely. *Crystal Growth and Design* **17**: 3567−3578.

Cirilli V (1939) Calorimetric studies of the system SiO_2–CaO–H_2O. *Scientia Ricerca* **10**: 459–461. (In Italian.)

Clayden NJ, Dobson CM, Groves GW, Hayes CJ and Rodger SA (1984a) Solid state NMR studies of cement hydration. *Transactions of the British Ceramic Society* **35**: 55–64.

Clayden NJ, Dobson CM, Hayes CI and Rodger SA (1984b) Hydration of tricalcium silicate followed by solid-state ^{29}Si N.M.R. spectroscopy. *Journal of the Chemical Society, Chemical Communications* 1396–1397.

Clayden NJ, Dobson CM, Groves GW and Rodger SA (1986) The application of solid state nuclear magnetic resonance spectroscopy techniques to the study of the hydration of tricalcium silicate. *Proceedings of the 8th International Congress on the Chemistry of Cement* **3**: 51–56.

Cong X and Kirkpatrick RJ (1993) ^{17}O and ^{29}Si MAS NMR study of β–C_2S hydration and the structure of calcium-silicate hydrates. *Cement and Concrete Research* **23**: 1065–1077.

Cong X and Kirkpatrick RJ (1996) ^{29}Si MAS NMR study of the structure of calcium silicate hydrate. *Advanced Cement-Based Materials* **3**: 144–156.

Copeland LE and Hayes JC (1953) Determination of non-evaporable water in hardened Portland-cement paste. *ASTM Bulletin.* **194**: 70–74.

Copeland LE and Schultz EG (1962) Electron optical investigation of the hydration products of calcium silicates and Portland cement. *Portland Cement Association Research and Development Laboratories Bulletin* **4(1)**: 2–12.

Copeland LE, Bodor E, Chang TN and Weise CH (1967) Reactions of tobermorite gel with aluminates, ferrites, and sulfates. *Portland Cement Association Research and Development Laboratories Bulletin* **9(1)**: 61–74.

Costoya Fernandez MM (2008) *Effect of Particle Size on the Hydration Kinetics and Microstructural Development of Tricalcium Silicate*. PhD thesis No. 4102, École Polytechnique Fédérale de Lausanne, Lausanne, Switzerland.

Currell BR, Midgley HG, Montecinos M and Parsonage JR (1985) A study of Portland cement hydration by trimethylsilylation techniques. *Cement and Concrete Research* **15**: 889–900.

Dal Negro A and Ungaretti L (1971) Refinement of the crystal structure of aragonite. *American Mineralogist* **56**: 768–772.

Dalgleish BJ and Ibe K (1981) Thin-foil studies of hydrated Portland cement. *Cement and Concrete Research* **11**: 729–739.

Dalziell JA and Gutteridge WA (1986) *The Influence of Pulverized-fuel Ash upon the Hydration Characteristics and Certain Physical Properties of a Portland Cement Paste*. Cement and Concrete Association, Slough, UK, Technical Report 560.

Damidot D and Nonat A (1994a) C_3S hydration in diluted and stirred suspensions: (I) study of the two kinetic steps. *Advances in Cement Research* **6**: 27–35.

Damidot D and Nonat A (1994b) C_3S hydration in diluted and stirred suspension: (II) properties of C–S–H precipitated during the two kinetic steps. *Advances in Cement Research* **6**: 83–91.

Damidot D, Nonat A and Barret P (1990) Kinetics of tricalcium silicate hydration in diluted suspensions by microcalorimetric measurements. *Journal of the American Ceramic Society* **73**: 3319–3322.

Damidot D, Nonat A, Barret P, Bertrandie D, Zanni H and Rassem R (1995) C_3S hydration in diluted and stirred suspensions: (III) NMR study of C–S–H precipitated during the two kinetic steps. *Advances in Cement Research* **7**: 1–8.

Davis RW and Young JF (1975) Hydration and strength development in tricalcium silicate pastes seeded with afwillite. *Journal of the American Ceramic Society* **58**: 67–70.

Day RL (1981) Reactions between methanol and Portland cement paste. *Cement and Concrete Research* **11**: 341–349.

de Jong JGM, Stein HN and Stevels JM (1967) Hydration of tricalcium silicate. *Journal of Applied Chemistry* **17**: 246–250.

De Villiers JPR (1971) Crystal structures of aragonite, strontianite, and witherite. *American Mineralogist* **56**: 758–767.

Demichelis R, Raiteri P, Gale JD and Dovesi R (2012) A new structural model for disorder in vaterite from first-principles calculations. *CrystEngComm* **14**: 44–47.

Demichelis R, Raiteri P, Gale JD and Dovesi R (2013) The multiple structures of vaterite. *Crystal Growth and Design* **13**: 2247–2251.

Dent Glasser LS, Lachowski EE, Mohan K and Taylor HFW (1978) A multi-method study of C_3S hydration. *Cement and Concrete Research* **8**: 733–740.

Dent Glasser LS, Lachowski EE, Qureshi MY *et al.* (1981) Identification of some of the polysilicate components of trimethylsilylated cement paste. *Cement and Concrete Research* **11**: 775–780.

Diamond S (1976) Cement paste microstrcuture – an overview at several levels. In *Proceedings of the Conference on Hydraulic Cement Pastes: Their Structure and Properties, Sheffield.* Cement and Concrete Association, Slough, UK, 2–30.

Dickens B and Bowen JS (1971) Refinement of the crystal structure of the aragonite phase of $CaCO_3$. *Journal of Research of the National Bureau of Standards, Section A: Physics and Chemistry* **75**: 27–32.

Dobson CM, Goberdhan DGC, Ramsay JDF and Rodger SA (1988) ^{29}Si MAS NMR study of the hydration of tricalcium silicate in the presence of finely divided silica. *Journal of Materials Science* **23**: 4108–4114.

Donald AM (2003) The use of environmental scanning electron microscopy for imaging wet and insulating materials. *Nature Materials* **2**: 511–516.

Double DD (1983) New developments in understanding the chemistry of cement hydration. *Philosophical Transactions of the Royal Society of London* **A310**: 53–66.

Double DO, Hellawell A and Perry SJ (1978) The hydration of Portland cement. *Proceedings of the Royal Society of London* **359A**: 435–451.

Duchesne J and Reardon EJ (1995) Measurement and prediction of portlandite solubility in alkali solutions. *Cement and Concrete Research* **25**: 1043–1053.

Farmer VC, Jeevaratnam J, Speakman K and Taylor HFW (1966) Thermal decomposition of 14Å tobermorite from Crestmore. *Proceedings of the Symposium on the Structure of Portland Cement Paste and Concrete*, 291–299. Highway Research Board, Washington, DC, USA, Special Report 90.

Feldman RF (1969) Sorption and length-change scanning isotherms of methanol and water on hydrated Portland cement. *Proceedings of the 5th International Symposium on the Chemistry of Cement* **3**: 53–66.

Feldman RF (1972) Helium flow and density measurement of the hydrated tricalcium silicate–water system. *Cement and Concrete Research* **2**: 123–136.

Feldman RF and Ramachandran VS (1971) Differentiation of interlayer and adsorbed water in hydrated Portland cement by thermal analysis. *Cement and Concrete Research* **1**: 607–620.

Feldman RF and Ramachandran VS (1974) A study of the state of water and stoichiometry of bottle-hydrated Ca_3SiO_5. *Cement and Concrete Research* **4**: 155–166.

Feldman RF and Swenson EG (1975) Volume change on first drying of hydrated Portland cement with and without admixtures. *Cement and Concrete Research* **5**: 25–35.

Fierens P and Tirlocq J (1983a) Effect of synthesis temperature and cooling conditions of β-dicalcium silicate on its hydration rate. *Cement and Concrete Research* **13**: 41–48.

Fierens P and Tirlocq J (1983b) Nature and concentration effect of stabilizing elements of beta-dicalcium silicate on its hydration rate. *Cement and Concrete Research* **13**: 267–276.

Fierens P and Verhaegen JP (1976) Hydration of tricalcium silicate in paste – Kinetics of calcium ions dissolution in the aqueous phase. *Cement and Concrete Research* **6**: 337–342.

Flint EP and Wells LS (1934) Study of the system $CaO–SiO_2–H_2O$ at 30°C and the reaction of water on the anhydrous calcium silicates. *Journal of Research of the National Bureau of Standards* **12**: 751–783.

Fonseca PC and Jennings HM (2010) The effect of drying on early-age morphology of C–S–H as observed in environmental SEM. *Cement and Concrete Research* **40**: 1673–1680.

Fujii K and Kondo W (1969) Hydration of tricalcium silicate in a very early stage. *Proceedings of the 5th International Symposium on the Chemistry of Cement* **2**: 362–371.

Fujii K and Kondo W (1979) Rate and mechanism of hydration of β-dicalcium silicate. *Journal of the American Ceramic Society* **62**: 161–167.

Fujii K and Kondo W (1981) Heterogeneous equilibrium of calcium silicate hydrate in water at 30°C. *Journal of the Chemical Society, Dalton Transactions* 645–651.

Fujii K and Kondo W (1983) Estimation of thermochemical data for calcium silicate hydrate. *Journal of the American Ceramic Society* **66**: C-220–C-221.

Funk H (1962) Two different ways of hydration in the reaction of β-Ca_2SiO_4 with water at 25 and 120°C, paper III-S7. *Proceedings of the 4th International Symposium on the Chemistry of Cement* **1**: 291–295.

Fylak M, Göske J, Kachler W, Wenda R and Pöllmann H (2006) Cryotransfer scanning electron microscopy for the study of cementitious systems. *Microscopy and Analysis* July: 9–12.

Gallucci E, Mathur P and Scrivener K (2010) Microstructural development of early age hydration shells around cement grains. *Cement and Concrete Research* **40**: 4–13.

Gard JA and Taylor HFW (1976) Calcium silicate hydrate (II) ('C–S–H(II)'). *Cement and Concrete Research* **6**: 667–677.

Gard JA, Mohan K and Taylor HFW (1980) Analytical electron microscopy of cement pastes: I, Tricalcium silicate pastes. *Journal of the American Ceramic Society* **63**: 336–227.

Gard JA, Mohan K, Taylor HFW and Cliff G (1981) Electron diffraction and microscopy techniques for studying grain-boundary structure. *Journal of the American Ceramic Society* **63**: 336–345.

Gard JA, Taylor HFW, Cliff G and Lorimer GW (1977) A reexamination of jennite. *American Mineralogist* **62**: 365–368.

Garrault-Gauffinet S and Nonat A (1999) Experimental investigation of calcium silicate hydrate (C–S–H) nucleation. *Journal of Crystal Growth* **200**: 565–574.

Gartner EM (1997) A proposed mechanism for the growth of C-S-H during the hydration of tricalcium silicate. *Cement and Concrete Research* **27**: 665–672.

Gartner EM (2011) Discussion of the paper 'Dissolution theory applied to the induction period in alite hydration' by Juilland P *et al.* (2010). *Cement and Concrete Research* **41**: 560–562.

Gartner EM and Gaidis JM (1989) Hydration mechanisms. In *Materials Science of Concrete I* (Skalny JP (ed.)). American Ceramic Society, Westerville, OH, USA, pp. 95–125.

Gartner EM and Jennings HM (1987) Thermodynamics of calcium silicate hydrates and their solutions. *Journal of the American Ceramic Society* **70**: 743–749.

Gartner EM, Kurtis KE and Monteiro PJM (2000) Proposed mechanism of C-S-H growth tested by soft X-ray microscopy. *Cement and Concrete Research* **30**: 817–822.

Gartner EM, Young JF, Damidot DA and Jawed, I (2002) Hydration of Portland cement In *Structure and Performance of Cements* (Barnes P and Bensted J (eds)), 2nd edn. Spon, London, UK, pp. 57–113.

Geng G, Taylor R, Bae S *et al.* (2015) Atomic and nano-scale characterization of a 50-year-old hydrated C_3S paste. *Cement and Concrete Research* **77**: 36–46.

Gerhard W and Nägele E (1983) The hydration of cement studied by secondary ion mass spectrometry (SIMS). *Cement and Concrete Research* **13**: 849–859.

Gifford JW (1902) The refractive indices of fluorite, quartz, and calcite. *Proceedings of the Royal Society of London* **70**: 329–340.

Glasser FP, Lachowski EE and Macphee DE (1987) Compositional model for calcium silicate hydrate (C–S–H) gels, their solubilities, and free energies of formation. *Journal of the American Ceramic Society* **70**: 481–485.

Goñi S, Puertas F, Hernández MS *et al.* (2010) Quantitative study of hydration of C_3S and C_2S by thermal analysis. *Journal of Thermal Analysis and Calorimetry* **102**: 965–973.

Goto S, Daimon M, Hosaka G and Kondo R (1976) Composition and morphology of hydrated tricalcium silicate. *Journal of the American Ceramic Society* **59**: 281–284.

Graf DL (1961) Crystallographic tables for the rhombohedral carbonates. *American Mineralogist* **46**: 1283–1316.

Greenberg SA and Copeland LE (1960) The thermodynamic functions for the solution of calcium hydroxide in water. *Journal of Physical Chemistry* **64**: 1057–1059.

Greening NR (1972) quoted in Young JF (1972) A review of the mechanisms of set-retardation in Portland cement pastes containing organic admixtures. *Cement and Concrete Research* **2**: 415–433.

Groves GW (1981) Portland cement clinker viewed by transmission electron microscopy. *Journal of Materials Science* **16**: 1063–1070.

Groves GW (1983) TEM studies of cement clinker compounds, their hydration and strong cement pastes. *Philosophical Transactions of the Royal Society, London* **A310**: 79–83.

Groves GW (1987) TEM studies of cement hydration. *Materials Research Society Symposium Proceedings* **85**: 3–11.

Groves GW (1990a) Transmission electron microscopy of cements and mortars. *Materials Forum* **14**: 1–8.

Groves GW (1990b) A discussion of the paper 'Phase identification during early and middle hydration of tricalcium silicate' by Melzer R and Eberhard E. *Cement and Concrete Research* **20**: 315–316.

Groves GW, Le Sueur PJ, and Sinclair W (1986) Transmission electron microscopy and microanalytical studies of ion-beam-thinned sections of tricalcium silicate paste. *Journal of the American Ceramic Society* **69**: 353–356.

Groves GW, Brough A, Richardson I G and Dobson CM (1991) Progressive changes in the structure of hardened C_3S cement pastes due to carbonation. *Journal of the American Ceramic Society* **74**: 2891–2896.

Grudemo Å (1955) An electronographic study of the morphology and crystallization properties of calcium silicate hydrates. *Proceedings of the Swedish Cement and Concrete Research Institute at the Royal Institute of Technology, Stockholm, Sweden (Handlingar – Svenska Forskningsinstitutet för Cement och Betong vid Kungliga Tekniska Högskolan i Stockholm)* **26**: 1–103.

Grudemo Å (1962) The microstructure of hardened cement paste. *Proceedings of the 4th International Symposium on the Chemistry of Cement*, paper V-2, **2**: 615–647.

Grudemo Å (1984) Variation with solid-phase concentration of composition, structure and strength of cement pastes at high age. *Cement and Concrete Research* **14**: 123–132.

Grutzeck MW (1999) A new model for the formation of calcium silicate hydrate (C–S–H). *Materials Research Innovations* **3**: 160–170.

Grutzeck MW and Ramachandran AR (1987) An integration of tricalcium silicate hydration models in light of recent data. *Cement and Concrete Research* **17**: 164–170.

Grutzeck MW and Roy DM (1969) Electron microprobe studies of the hydration of $3CaO \cdot SiO_2$. *Nature* **223**: 492–494.

Grutzeck MW, Benesi A and Fanning B (1989) Silicon-29 magic angle spinning nuclear magnetic resonance study of calcium silicate hydrates. *Journal of the American Ceramic Society* **72**: 665–668.

Grutzeck MW, La Rosa-Thompson JL and Kwan S (1997) Characteristics of C-S-H gels. *Proceedings of the 10th International Congress on the Chemistry of Cement*. Amarkai AB and Congrex Göteborg AB, Sweden, Paper 2ii067, 10 pp.

Grutzeck MW, Kwan S, Thompson JL and Benesi A (1999) A sorosilicate model for calcium silicate hydrate (C–S–H). *Journal of Materials Science Letters* **18**: 217–220.

Hadley DW, Dolch WL and Diamond S (2000) On the occurrence of hollow-shell hydration grains in hydrated cement paste. *Cement and Concrete Research* **30**: 1–6.

Halstead PE and Moore AE (1957) The thermal dissociation of calcium hydroxide. *Journal of the Chemical Society* 3873–3875.

Hamid SA (1981) The crystal structure of the 11 Å natural tobermorite $Ca_{2.25}[Si_3O_{7.5}(OH)_{1.5}] \cdot 1H_2O$. *Zeitschrift für Kristallographie* **154**: 189–198.

Hansen W (1987) quoted in Young JF and Hansen W (1987) Volume relationship for C–S–H formation based on hydration stoichiometry. *Materials Research Society Symposium Proceedings* **85**: 313–322.

Hara N and Inoue N (1980) Formation of jennite from fumed silica. *Cement and Concrete Research* **10**: 677–682.

Hara N, Chan CF and Mitsuda T (1978) Formation of 14 Å tobermorite. *Cement and Concrete Research* **8**: 113–115.

Harrisson AM, Winter NB and Taylor HFW (1987) X-ray microanalysis of microporous materials. *Journal of Materials Science Letters* **6**: 1339–1340.

Hedin R (1945) *Chemical Processes in the Hardening of Portland Cement*. Almquist and Wiksells, Stockholm/Uppsala, Sweden, in particular pp. 56–62, 101–102.

Hedin R (1955) Saturation concentration of calcium hydroxide. *Svenska forskningsinstitutet for cement och betong vid kungl. tekniska hogskolan i Stockholm: Handlingar* (Swedish Cement and Concrete Research Institute at the Royal Institute of Technology, Stockholm: Proceedings) **27**: 1–14.

Heidemann D and Wieker W (1998) Characterization of protons in C–S–H phases by means of high-speed ^1H MAS NMR investigations. In *Nuclear Magnetic Resonance Spectroscopy of Cement-Based Materials* (Colombet P, Grimmer A-R, Zanni H, Sozzani P (eds)), Springer, Berlin, Germany, pp. 169–179.

Henderson E and Bailey JE (1993) The compositional and molecular character of the calcium silicate hydrates formed in the paste hydration of $3CaO \cdot SiO_2$. *Journal of Materials Science* **28**: 3681–3691.

Hirljac J, Wu ZQ and Young JF (1983) Silicate polymerization during the hydration of alite. *Cement and Concrete Research* **13**: 877–886.

Hunt CM (1962) Infrared absorption of some compounds in the CaO-SiO_2-H_2O system, paper III-S8. *Proceedings of the 4th International Symposium on the Chemistry of Cement* **1**: 297–304.

Ings JB, Brown PW and Frohnsdorff G (1983) Early hydration of large single crystals of tricalcium silicate. *Cement and Concrete Research* **13**: 843–848.

Jansen D, Bergold ST, Goetz-Neunhoeffer F and Neubauer J (2011) The hydration of alite/a time-resolved quantitative X-ray diffraction approach using the G-factor method compared with heat release. *Journal of Applied Crystallography* **44**: 895–901.

Jarosch D and Heger G (1986) Neutron diffraction refinement of the crystal structure of aragonite. *Tschermaks Mineralogische und Petrographische Mitteilungen* **35**: 127–131.

Jawed I and Skalny J (1982) Surface phenomena during tricalcium silicate hydration. *Journal of Colloid and Interface Science* **85**: 235–243.

Jennings HM (1986) Aqueous solubility relationships for two types of calcium silicate hydrate. *Journal of the American Ceramic Society* **69**: 614–618.

Jennings HM (1988) Reply to a comment on Jennings (1986) by Barret and Bertrandie (1988). *Journal of the American Ceramic Society* **71**: C115–C116.

Jennings HM and Parrott LJ (1986a) Microstructural analysis of hydrated alite paste. Part 1. Porosity. *Journal of Materials Science* **21**: 4048-4052.

Jennings HM and Parrott LJ (1986b) Microstructural analysis of hydrated alite paste. Part 2. Microscopy and reaction products. *Journal of Materials Science* **21**: 4053–4059.

Jennings HM and Pratt PL (1979a) An experimental argument for the existence of a protective membrane surrounding Portland cement during the induction period. *Cement and Concrete Research* **9**: 501–506.

Jennings HM and Pratt PL (1979b) On the hydration of Portland cement. *Proceedings of the British Ceramic Society* **28**: 179–193.

Jennings HM and Pratt PL (1980) The use of a high-voltage electron microscope and gas reaction cell for the microstructural investigation of wet Portland cement. *Journal of Materials Science* **15**: 250–253.

Jennings HM, Dalgleish BJ and Pratt PL (1981) Morphological development of hydrating tricalcium silicate as examined by electron microscopy techniques. *Journal of the American Ceramic Society* **64**: 567–572.

John E and Lothenbach B (2023) Cement hydration mechanisms through time – a review. *Journal of Materials Science* **58**: 9805–9833.

Johnston J and Grove C (1931) The solubility of calcium hydroxide in aqueous salt solutions. *Journal of the American Chemical Society* **53**: 3976–3991.

Johnston J, Merwin HE and Williamson ED (1916) The several forms of calcium carbonate. *American Journal of Science* **s4, 41**: 473–512.

Jones FE (1940) A method for the determination of carbonate in small amounts of materials. *Journal of the Society of Chemical Industry* **59**: Issue Supplement S1, Trans. Commun. 21–23.

Juenger MCG, Lamour VHR, Monteiro EM, and Denbeaux GP (2003) Direct observation of cement hydration by soft X-ray transmission microscopy. *Journal of Materials Science Letters* **22**: 1335–1337.

Juenger MCG, Monteiro EM, Gartner EM and Denbeaux GP (2005) A soft X-ray microscope investigation into the effects of calcium chloride on tricalcium silicate hydration. *Cement and Concrete Research* **35**: 19–25.

Juilland P, Gallucci E, Flatt R and Scrivener K (2010) Dissolution theory applied to the induction period in alite hydration. *Cement and Concrete Research* **40**: 831–844.

Juilland P, Gallucci E, Flatt R and Scrivener K (2011) Reply to the discussion by Makar J, Beaudoin JJ, Sato T, Alizadeh R and Raki L of the paper 'Dissolution theory applied to the induction period in alite hydration'. *Cement and Concrete Research* **41**: 568–569.

Juilland P, Nicoleau L, Arvidson RS and Gallucci E (2017) Advances in dissolution understanding and their implications for cement hydration. *RILEM Technical Letters* **2**: 90–98.

Kalousek GL (1944) Studies of portions of the quaternary system soda–lime–silica–water at 25°C. *Journal of Research of the National Institute of Standards and Technology* **32**: 285–302.

Kalousek GL (1954) Application of differential thermal analysis in a study of the system lime–silica–water. *Proceedings of the 3rd International Symposium on the Chemistry of Cement, London, 1952*. Cement and Concrete Association, London, UK, pp. 296–311.

Kalousek GL and Roy R (1957) Crystal chemistry of hydrous calcium silicates: II, Characterization of interlayer water. *Journal of the American Ceramic Society* **40**: 236–239.

Kamhi SR (1963) On the structure of vaterite, $CaCO_3$. *Acta Crystallographica* **16**: 770–772.

Kantro DL and Weise CH (1979) Hydration of various beta-dicalcium silicate preparations. *Journal of the American Ceramic Society* **62**: 621–626.

Kantro DL, Brunauer S and Weise CH (1959) The ball-mill hydration of tricalcium silicate at room temperature. *Journal of Colloid Science* **14**: 363–376.

Kantro DL, Brunauer S and Weise CH (1962) Development of surface in the hydration of calcium silicates. II. Extension investigations to earlier and later stages of hydration. *Journal of Physical Chemistry* **66**: 1804–1809.

Katz A, Bentur A and Kovler K (2007) A novel system for in-situ observations of early hydration reactions in wet conditions in conventional SEM. *Cement and Concrete Research* **37**: 32–37.

Kißling PA, Lübkemann F, Mundstock A *et al.* (2022) Is freeze-drying an alternative to solvent exchange for the hydration stop of cementitious suspensions? *Cement and Concrete Research* **159**: 106841.

Kjellsen KO and Justnes H (2004) Revisiting the microstructure of hydrated tricalcium silicate – a comparison to Portland cement. *Cement and Concrete Composites* **26**: 947–956.

Kjellsen KO and Lagerblad B (2007) Microstructure of tricalcium silicate and Portland cement systems at middle periods of hydration-development of Hadley grains. *Cement and Concrete Research* **37**: 13–20.

Kleiner F, Rößler C, Vogt F, Osburg A and Ludwig HM (2022) Reconstruction of calcium silicate hydrates using multiple 2D and 3D imaging techniques: light microscopy, μ-CT, SEM, FIB-nT combined with EDX. *Journal of Microscopy* **286**: 102–107.

Komarneni S, Roy DM, Pyfe CA and Kennedy GJ (1987) Naturally occurring 1.4 nm tobermorite and synthetic jennite: characterization by ^{27}Al and ^{29}Si MASNMR spectroscopy and cation exchange properties. *Cement and Concrete Research* **17**: 891–895.

Kondo R and Daimon M (1969) Early hydration of tricalcium silicate: A solid reaction with induction and acceleration periods. *Journal of the American Ceramic Society* **52**: 503–508.

Kondo R and Ueda S (1969) Kinetics and mechanisms of the hydration of cements. *Proceedings of the 5th International Symposium on the Chemistry of Cement* **2**: 203–248.

Kondo R, Ueda S and Kodama M (1967) Reaction process of the hydration of 3 $CaO{\cdot}SiO_2$. *Semento Gijutsu Nenpo* **21**: 83–90.

Kühl H and Mann A (1934) Calcium silicate hydrates. A contribution to the hardening theory of silicate cements. *Tonindustrie-Zeitung und Keramische Rundschau* **58**: 862–865, 896–897, 918–919, 930–932, 944–945, 955–957, 990–991, 1003, 1014–1016. (In German.)

Kurczyk HG and Schwiete HE (1960) Electronenmikroskopische und thermochemische Untersuchungen über die Hydratation der Calciumsilikate $3CaO{\cdot}SiO_2$ und β-$2CaO{\cdot}SiO_2$ und den Einfluß von Calciumchlorid und Gips auf den Hydratationsvorgang. *Tonindustrie-Zeitung und Keramische Rundschau* **84**: 585–598. (In German.)

Kurczyk HG and Schwiete HE (1962) Concerning the hydration products of C_3S and β-C_2S. *Proceedings of the 4th International Symposium on the Chemistry of Cement* **1**: 349–358.

Lachowski EE, Mohan K, Taylor HFW, Lawrence CD and Moore AE (1981) Analytical electron microscopy of cement pastes: III, Pastes hydrated for long times. *Journal of the American Ceramic Society* **64**: 319–321.

Le Sueur PJ (1984) *Studies of the Hydration of Tricalcium Silicate*. DPhil Thesis, University of Oxford, Oxford, UK.

Le Sueur PJ, Double DD and Groves GW (1984) Chemical and morphological studies of the hydration of tricalcium silicate. *Transactions of the British Ceramic Society (Proceedings)* **35**: 177–191.

Lehmann H and Dutz H (1962) Infrared spectroscopy studies on the hydration of clinker minerals and cements, paper IV-S6. *Proceedings of the 4th International Symposium on the Chemistry of Cement* **1**: 513–518.

Lentz CW (1966) The silicate structure analysis of hydrated Portland cement paste. *Proceedings of the Symposium on the Structure of Portland Cement Paste and Concrete*. Highway Research Board, Washington, DC, USA, Special Report 90: 269–283.

Lerch W and Bogue RH (1934) Heat of hydration of Portland cement pastes. *Journal of Research of the National Bureau of Standards* **12**: 645–664.

Li Q (2008) *The Nature of Early Age Hydration Products*. PhD Thesis, University of Leeds, Leeds, UK.

Lipmaa E, Magi M, Tarmak M, Wieker W and Grimmer AR (1982) A high resolution ^{29}Si NMR study of the hydration of tricalcium silicate. *Cement and Concrete Research* **12**: 597–602.

Locher FW (1966) Stoichiometry of tricalcium silicate hydration. *Proceedings of the Symposium on the Structure of Portland Cement Paste and Concrete*. Highway Research Board, Washington, DC, USA, Special Report 90: 300–308.

Long JVP and McConnell JDC (1959) A mineralogical application of X-ray absorption microspectroscopy: The hydration of larnite. *Mineralogical Magazine* **32**: 117–127.

Lothenbach B, Durdzinski P and De Weerdt K (2016) Thermogravimetric analysis. In *A Practical Guide to Microstructural Analysis of Cementitious Materials* (Scrivener K, Snellings R and Lothenbach B (eds)). CRC Press, Boca Raton, FL, USA, pp. 177–211.

Lothenbach B, Kulik DA, Matschei T *et al.* (2019) Cemdata18 – A chemical thermodynamic database for hydrated Portland cements and alkali-activated materials. *Cement and Concrete Research* **115**: 472–506.

Lutterotti L, Matthies S, Wenk HR, Schultz AS and Richardson JW Jr. (1997) Combined texture and structure analysis of deformed limestone from time-of-flight neutron diffraction spectra. *Journal of Applied Physics* **81**: 594–600.

Macphee DE, Lachowski EE and Glasser FP (1988) Polymerization effects in C–S–H: implications for Portland cement hydration. *Advances in Cement Research* **1**: 131–137.

Macphee DE, Luke K, Glasser FP and Lachowski EE (1989) Solubility and aging of calcium silicate hydrates in alkaline solutions at 25°C. *Journal of the American Ceramic Society* **72**: 646–654.

Makar JM, Chan GW and Esseghaier KY (2007) A peak in the hydration reaction at the end of the cement induction period. *Journal of Materials Science* **42**: 1388–1392.

Makar JM, Beaudoin JJ, Sato T, Alizadeh R and Raki L (2011) Discussion of 'Dissolution theory applied to the induction period in alite hydration'. *Cement and Concrete Research* **41**: 565–567.

Makovicky E (2016) Vaterite: Interpretation in terms of OD theory and its next of kin. *American Mineralogist* **101**: 1636–1641.

Markgraf SA and Reeder RJ (1985) High-temperature structure refinements of calcite and magnesite. *American Mineralogist* **70**: 590–600.

Maslen EN, Strel'tsov VA and Strel'tsova NR (1993) X-ray study of the electron density in calcite, $CaCO_3$. *Acta Crystallographica B* **49**: 636–641.

Maslen EN, Strel'tsov VA, Strel'tsova NR and Ishizawa N (1995) Electron density and optical anisotropy in rhombohedral carbonates. III. Synchrotron X-ray studies of $CaCO_3$, $MgCO_3$ and $MnCO_3$. *Acta Crystallographica B* **51**: 929–939.

Massazza F and Testolin M (1983) Trimethylsilylation in the study of pozzolana-containing pastes. *Il Cemento* **80**: 49–62.

Maycock JN, Skalny J and Kalyoncu R (1974) Crystal defects and hydration I. Influence of lattice defects. *Cement and Concrete Research* **4**: 835–847.

McConnell JDC (1955) The hydration of larnite (β-Ca_2SiO_4) and bredigite (α_1-Ca_2SiO_4) and the properties of the resulting gelatinous mineral plombierite. *Mineralogical Magazine* **30**: 672–680.

McConnell JDC (1960) Vaterite from Ballycraigy, Larne, Northern Ireland. *Mineralogical Magazine* **32**: 535–544.

Megaw HD and Kelsey CH (1956) Crystal structure of tobermorite. *Nature* **177**: 390–391.

Melzer R and Eberhard E (1989) Phase identification during early and middle hydration of tricalciumsilicate (Ca_3SiO_5). *Cement and Concrete Research* **19**: 411–422.

Ménétrier D, Jawed I, Sun TS and Skalny J (1979) ESCA and SEM studies on early C_3S hydration. *Cement and Concrete Research* **9**: 473–482.

Ménétrier D, Jawed I and Skalny J (1980a) Hydration of tricalcium silicate – A study by high resolution scanning electron microscopy. *Silicates Industriels* **45**: 243–248.

Ménétrier D, Jawed I and Skalny J (1980b) Effect of gypsum on C_3S hydration. *Cement and Concrete Research* **10**: 697–701.

Ménétrier D, McNamara DK, Jawed I and Skalny J (1980c) Early hydration of β-C_2S: Surface morphology. *Cement and Concrete Research* **10**: 107–110.

Ménétrier D, Jawed I, Sun TS and Skalny J (1980d) Surface studies of hydrated β-C_2S. *Cement and Concrete Research* **10**: 425–432.

Michaux M, Ménétrier D and Barret P (1983) Étude par triméthylsilylation du degré de condensation des ions silicate dans l'hydrosilicate de calcium. *Comptes Rendus de l'Académie des Sciences Paris (Ser. 2)* **296**: 1043–1046. (In French.)

Midgley HG (1979) The determination of calcium hydroxide in set Portland cements. *Cement and Concrete Research* **9**: 77–82.

Milestone NB (1980) Ageing and drying of tricalcium silicate pastes. *Proceedings of the 7th International Congress on the Chemistry of Cement* **3**: VI-61–VI-66.

Mohan K and Taylor HFW (1981) Analytical electron microscopy of cement pastes: IV, β-dicalcium silicate pastes. *Journal of the American Ceramic Society* **64**: 717–719.

Mohan K and Taylor HFW (1982) A trimethylsilylation study of tricalcium silicate pastes. *Cement and Concrete Research* **12**: 25–31.

Möser B (2010a) Progress in building materials analysis (Part 1). *ZKG International* **63**: 54–63.

Möser B (2010b) Progress in building materials analysis (Part 2). *ZKG International* **63**: 63–72.

Möser B and Stark J (2003) High resolution imaging of wet building material samples in their natural state using environmental scanning electron microscope. *Proceedings of the 11th International Congress on the Chemistry of Cement*, Durban, South Africa, 11–16 May 2003.

Möser B and Stark J (2007) Nanoscale characterization of hydration processes by means of high resolution scanning electron microscopy imaging technique. *Proceedings of the 12th International Congress on the Chemistry of Cement*, Montreal, Canada, 8–12 July 2007.

Naber C, Bellmann F, Sowoidnich T, Goetz-Neunhoeffer F and Neubauer J (2019) Alite dissolution and C-S-H precipitation rates during hydration. *Cement and Concrete Research* **115**: 283–293.

Naber C, Bellmann F and Neubauer J (2020) Influence of w/s ratio on alite dissolution and C-S-H precipitation rates T during hydration. *Cement and Concrete Research* **134**: 10608.

Nicoleau L and Bertolim MA (2016) Analytical model for the alite (C_3S) dissolution topography. *Journal of the American Ceramic Society* **99**: 773–786.

Nicoleau L and Nonat A (2016) A new view on the kinetics of tricalcium silicate hydration. *Cement and Concrete Research* **86**: 1–11.

Nonat A (2004) The structure and stoichiometry of C–S–H. *Cement and Concrete Research* **34**: 1521–1528.

Nonat A and Lecoq X (1998) The structure, stoichiometry and properties of C–S–H prepared by C_3S hydration under controlled conditions. In *Nuclear Magnetic Resonance Spectroscopy of Cement-Based Materials* (Colombet P, Grimmer AR, Zanni H and Sozzani P (eds)). Springer, Berlin, Germany, pp. 197–207.

Odler I and Becker T (1980) Effect of some liquefying agents on properties and hydration of Portland cement and tricalcium silicate pastes. *Cement and Concrete Research* **10**: 321–331.

Odler I and Cordes S (2002a) Initial hydration of tricalcium silicate as studied by secondary neutrals mass spectrometry I. Sample preparation and calibration. *Cement and Concrete Research* **32**: 1071–1075.

Odler I and Cordes S (2002b) Initial hydration of tricalcium silicate as studied by secondary neutrals mass spectrometry II. Results and discussion. *Cement and Concrete Research* **32**: 1077–1085.

Odler I and Dörr H (1979a) Early hydration of tricalcium silicate I. Kinetics of the hydration process and the stoichiometry of the hydration products. *Cement and Concrete Research* **9**: 239–248.

Odler I and Dörr H (1979b) Early hydration of tricalcium silicate II. The induction period. *Cement and Concrete Research* **9**: 277–284.

Odler I and Schüppstuhl J (1981) Early hydration of tricalcium silicate III. Control of the induction period. *Cement and Concrete Research* **11**: 765–774.

Odler I and Schüppstuhl J (1982) Combined hydration of tricalcium silicate and β-dicalcium silicate. *Cement and Concrete Research* **12**: 13–20.

Odler I and Stassinopoulos EN (1982) The composition of the pore liquid in hydrated cement pastes. *TIZ-Fachberichte* **106**: 394–401. (In German.)

Ogawa K, Uchikawa H, Takemoto K and Yasui I (1980) The mechanism of the hydration in the system C_3S-pozzolana. *Cement and Concrete Research* **10**: 683–696.

Owusu AB and Richardson IG (2023) Analysis of William Aspdin's 1848 hardened cement paste: compositional and microstructural analysis. *42nd Cement and Concrete Science Conference*, 11–12 September 2023, Imperial College, London, UK.

Parrott LJ and Young JF (1981) Effect of prolonged drying upon the silicate structure of hydrated alite pastes. *Cement and Concrete Research* **11**: 11–17.

Parrott LJ, Patel RG, Killoh DC and Jennings HM (1984) Effect of age on diffusion in hydrated alite cement. *Journal of the American Ceramic Society* **67**: 233–237.

Pellenq RJM, Kushima A, Shahsavari R *et al.* (2009) A realistic molecular model of cement hydrates. *Proceedings of the National Academy of Sciences of the USA* **106**: 16102–16107.

Petch HE (1961) The hydrogen positions in portlandite, $Ca(OH)_2$, as indicated by the electron distribution. *Acta Crystallographica* **14**: 950–957.

Pilati T, Demartin F and Gramaccioli CM (1998) Lattice-dynamical estimation of atomic displacement parameters in carbonates: Calcite and aragonite $CaCO_3$, dolomite $CaMg(CO_3)_2$, and magnesite $MgCO_3$. *Acta Crystallographica B* **54**: 515–523.

Pokroy B, Fieramosca JS, von Dreele RB, Fitch AN, Caspi EN and Zolotoyabko E (2007) Atomic structure of biogenic aragonite. *Chemistry of Materials* **19**: 3244–3251.

Powers TC and Brownyard TL (1946–1947) Studies of the physical properties of hardened Portland cement paste. Parts 1–9. *Journal of the American Concrete Institute* **18**: 101–132, 249–336, 469–504, 549–602, 669–712, 845–864, 865–880, 933–969, 971–992.

Prosen EJ, Brown PW, Frohnsdorff G and Davis F (1985) A multichambered microcalorimeter for the investigation of cement hydration. *Cement and Concrete Research* **15**: 703–710.

Pustovgar E, Sangodkar RP, Andreev AS *et al.* (2016) Understanding silicate hydration from quantitative analyses of hydrating tricalcium silicates. *Nature Communications* **7**: 10952.

Radczewski OE, Müller HO and Eitel W (1939) Zur Hydratation des Trikalziumsilikats (The hydration of tricalcium silicate). *Die Naturwissenschaften* **27**: 807. (In German.).

Ramachandran VS (1979) Differential thermal method of estimating calcium hydroxide in calcium silicate and cement pastes. *Cement and Concrete Research* **9**: 677–684.

Rassem R, Zanni-Theveneau H, Vernet C *et al.* (1992) An NMR investigation of C_3S hydration in pastes and in stirred dilute suspensions. In *Hydration and Setting of Cements* (Nonat A and Mutin JC (eds)). E and FN Spon, London, UK, pp. 77–85.

Regourd M (1983) Microanalytical studies (X-ray photoelectron spectrometry) of surface hydration reactions of cement compounds. *Philosophical Transactions of the Royal Society A* **310** 85–92.

Regourd M, Thomassin JH, Baillif P and Touray JC (1980) Study of the early hydration of Ca_3SiO_5 by X-ray photoelectron spectrometry. *Cement and Concrete Research* **10**: 223–230.

Relis M and Soroka J (1980) Limiting values for density, expansion and intrinsic shrinkage in hydrated Portland cement. *Cement and Concrete Research* **10**: 499–508.

Richardson IG (1991) *The Microstructure and Composition of Hardened Cement Pastes*. DPhil thesis, University of Oxford, Oxford, UK.

Richardson IG (1999) The nature of C–S–H in hardened cements. *Cement and Concrete Research* **29**: 1131–1147.

Richardson IG (2000) The nature of the hydration products in hardened cement pastes. *Cement and Concrete Composites* **22**: 97–113.

Richardson IG (2002) Electron microscopy of cements. In *Structure and Performance of Cements* (Barnes P and Bensted J (eds)), 2nd edn. Spon, London, UK, pp. 500–556.

Richardson IG (2004) Tobermorite/jennite- and tobermorite/calcium hydroxide-based models for the structure of C–S–H: applicability to hardened pastes of tricalcium silicate β-dicalcium silicate, Portland cement and blends of Portland cement with blast-furnace slag, metakaolin, or silica fume. *Cement and Concrete Research* **34**: 1733–1777.

Richardson IG (2008) The calcium silicate hydrates. *Cement and Concrete Research* **38**: 137–158.

Richardson IG (2014) Model structures for C–(A)–S–H(I). *Acta Crystallographica* **B70**: 903–923.

Richardson IG and Groves GW (1992) Models for the composition and structure of calcium silicate hydrate (C–S–H) gel in hardened tricalcium silicate pastes. *Cement and Concrete Research* **22**: 1001–1010.

Richardson IG and Groves GW (1993a) The microstructure and microanalysis of hardened ordinary Portland cement pastes. *Journal of Materials Science* **28**: 265–277.

Richardson IG and Groves GW (1993b) The incorporation of minor and trace elements into calcium silicate hydrate (C–S–H) gel in hardened cement pastes. *Cement and Concrete Research* **23**: 131–138.

Richardson IG, Wilding CR and Dickson MJ (1989) The hydration of blastfurnace slag cements. *Advances in Cement Research* **2**: 147–157.

Richardson IG, Girão AV, Taylor R and Jia S (2016) Hydration of water- and alkali-activated white Portland cement pastes and blends with low-calcium pulverized fuel ash. *Cement and Concrete Research* **83**: 1–18.

Rodger SA (1986) *The Chemistry of Admixture Interaction during Cement Hydration*. DPhil Thesis, University of Oxford, Oxford, UK.

Rodger SA (1989) Unpublished results.

Rodger SA, Groves GW, Clayden NJ and Dobson CM (1987) A study of tricalcium silicate hydration from very early to very late stages. *Materials Research Society Symposium Proceedings* **85**: 13–20.

Rodger SA, Groves GW, Clayden NJ and Dobson CM (1988) Hydration of tricalcium silicate followed by ^{29}Si NMR with cross-polarization. *Journal of the American Ceramic Society* **71**: 91–96.

Rodriguez-Blanco JD, Shaw S and Benning LG (2011) The kinetics and mechanisms of amorphous calcium carbonate (ACC) crystallization to calcite, via vaterite. *Nanoscale* **3**: 265–271.

Roller PS and Ervin G (1940) The system calcium oxide–silica–water at 30°C. The association of silicate ion in dilute alkaline solution. *Journal of the American Chemical Society* **62**: 461–471.

Rößler C, Stark J, Steiniger F and Tichelaar W (2006) Limited-dose electron microscopy reveals the crystallinity of fibrous C–S–H phases. *Journal of the American Ceramic Society* **89**: 627–632.

Sass RL, Vidale R and Donohue J (1957) Interatomic distances and thermal anisotropy in sodium nitrate calcite. *Acta Crystallographica* **10**: 567–570.

Scrivener KL (1984) *Development of Microstructure during the Hydration of Portland Cement*. PhD thesis, University of London, London, UK.

Scrivener KL (1989) The microstructure of concrete. In *Materials Science of Concrete 1* (Skalny JP (ed.)). American Ceramic Society, Westerville, OH, USA, pp. 127–161.

Scrivener KL and Nonat A (2011) Hydration of cementitious materials, present and future. *Cement and Concrete Research* **41**: 651–665.

Scrivener KL and Pratt PL (1984a) Backscattered electron images of polished cement sections in the SEM. *Proceedings of the 6th International Conference on the Cement Microscopy*, pp. 145–155.

Scrivener KL and Pratt PL (1984b) Characterisation of Portland cement hydration by electron optical techniques. *Materials Research Society Symposium Proceedings* **31**: 351–356.

Scrivener KL and Pratt PL (1984c) Microstructural studies of the hydration of C_3A and C_4AF independently and in cement paste. *Proceedings of the British Ceramic Society* **35**: 207–219.

Scrivener KL, Patel HH, Pratt PL and Parrott LJ (1987) Analysis of phases in cement paste using backscattered electron images, methanol adsorption and thermogravimetric analysis. *Materials Research Society Symposium Proceedings* **85**: 67–76.

Scrivener KL, Juilland P and Monteiro PJM (2015) Advances in understanding hydration of Portland cement. *Cement and Concrete Research* **78**: 38–56.

Scrivener KL, Ouzia A, Juilland P and Mohamed AK (2019) Advances in understanding cement hydration mechanisms. *Cement and Concrete Research* **124**: 105823.

Scrivener KL, Matschei T, Georget F, Juilland P and Mohamed AK (2023) Advances in hydration and thermodynamics of cementitious systems. *Cement and Concrete Research* **174**: 107332.

Shebl FA, Helmy FM and Ludwig U (1985) A new approach on the hydration mechanism of tricalcium silicate. *Cement and Concrete Research* **15**: 747–757.

Shibata S, Kishi K, Asaga K, Daimon M and Shrestha PR (1984) Preparation and hydration of β-C_2S without stabilizer. *Cement and Concrete Research* **14**: 323–328.

Shpynova LG, Nabitovich ID and Belov NV (1967) Microstructure of alite cement stone (hydrated tricalcium silicate). *Soviet Physics. Crystallography* **11**(6): 747–751 (English translation of *Krystallografiya*).

Sierra R (1976) Study of the hydration kinetics of tricalcium silicate. *Proceedings of the 6th International Congress on the Chemistry of Cement* **2** Part 1: 138–143. (In Russian.)

Sierra R (1980) Distribution of different forms of water in the structure of pure slurries of C_3S and Portland cement. *Proceedings of the 7th International Congress on the Chemistry of Cement* **3**: VI/201–206.

Sitepu H (2009) Texture and structural refinement using neutron diffraction data from molybdite (MoO_3) and calcite ($CaCO_3$) powders and a Ni-rich $Ni_{50.7}Ti_{49.3}$ alloy. *Powder Diffraction* **24**: 315–326.

Slegers PA, Genet M, Leonard AJ and Fripiat JJ (1977) Structural transformation of tricalcium silicate during hydration. *Journal of Applied Crystallography* **10**: 270–276.

Smith RM and Martell AE (1976) *Critical Stability Constants, 4: Inorganic Complexes*. Plenum Press, New York, NY, USA.

Snellings R (2016) X-ray powder diffraction applied to cement. In *A Practical Guide to Microstructural Analysis of Cementitious Materials* (Scrivener K, Snellings R and Lothenbach B (eds)). CRC Press, Boca Raton, FL, USA, pp. 107–176.

Spinolo G and Tamburini UA (1985) Stacking faults in $Ca(OH)_2$ produced by vapour phase hydration. *Zeitschrift für Naturforschung A* **40**: 73–78.

Stade H (1980) Zum Aufbau schlecht geordneter Calciumhydrogensilicate. II. Über eine aus Poly- und Disilicat bestehende Phase (On the structure of ill-crystallized calcium hydrogen silicates. II. A phase consisting of poly- and disilicate). *Zeitschrift für anorganische und allgemeine Chemie* **470**: 69–83. (In German.)

Stade H and Müller D (1987) On the coordination of Al in ill-crystallized C–S–H phases formed by hydration of tricalcium silicate and by precipitation reactions at ambient temperature. *Cement and Concrete Research* **17**: 553–561.

Stade H and Wieker W (1980) On the structure of ill-crystallized calcium hydrogen silicates. I. Formation and propeties of an ill-crystallized calcium hydrogen disilicate phase, *Zeitschrift für anorganische und allgemeine Chemie*, **466**: 55–70. (In German.)

Stade H, Wieker W and Garzo G (1983) On the structure of ill-crystallized calcium hydrogen silicates. IV. Anion composition of the hydration products of tricalcium silicate. *Zeitschrift für anorganische und allgemeine Chemie* **500**: 123–131. (In German.)

Stade H, Müller D and Scheler G (1984) On the structure of ill-crystallized calcium hydrogen silicates. V. Studies on the coordination of Al in C–S–H(di,poly) by ^{27}Al NMR spectroscopy, *Zeitschrift für anorganische und allgemeine Chemie* **510**: 16–24.

Stade H, Grimmer AR, Engelhardt G, Magi M and Lipmaa E (1985) Zum Aufbau schlecht geordneter Calciumhydrogensilicate. VII. ^{29}Si-NMR-Untersuchungen an CSH(Di, Poly). *Zeitschrift für anorganische und allgemeine Chemie* **528**: 147–151. (In German.)

Stadelman C, Trettin R, Wieker W and Ramm M (1985) Studies of the hydration of Ca_3SiO_5 by ESCA- and Krypton-adsorption-measurements. *Cement and Concrete Research* **15**: 145–150.

Stein HN (1972) Thermodynamic considerations on the hydration mechanisms of Ca_3SiO_5 and $Ca_3Al_2O_6$. *Cement and Concrete Research* **2**: 167–177.

Stein HN and Stevels JM (1964) Influence of silica on the hydration of 3 CaO,SiO_2. *Journal of Applied Chemistry* **14**: 338–346.

Steinour HH (1947) The system CaO-SiO_2-H_2O and the hydration of the calcium silicates. *Chemical Reviews* **40**: 391–460.

Steinour HH (1954) The reactions and thermochemistry of cement hydration at ordinary temperature. *Proceedings of the 3rd International Symposium on the Chemistry of Cement*, pp. 261–289.

Stewart HR and Bailey JE (1983) Microstructural studies of the hydration products of three tricalcium silicate polymorphs. *Journal of Materials Science* **18**: 3686–3695.

Stucke MS and Majumdar AJ (1977) The morphology and composition of an immature C_3S paste. *Cement and Concrete Research* **7**: 711–718.

Swanson HE and Tatge E (1953) *Standard X-ray Powder Diffraction Patterns*, US Department of Commerce, Washington, DC, USA, NBS Circular 539: 1.

Tadros ME, Skalny J and Kalyoncu RS (1976) Early hydration of tricalcium silicate. *Journal of the American Ceramic Society* **59**: 344–347.

Tajuelo Rodriguez (2015) *Relation between composition, structure and morphology in C–S–H.* PhD Thesis, University of Leeds, Leeds, UK.

Tajuelo Rodriguez E, Garbev K, Merz D, Black L and Richardson IG (2017) Thermal stability of C-S-H phases and applicability of Richardson and Groves' and Richardson C–(A)–S–H(I) models to synthetic C-S-H. *Cement and Concrete Research* **93**: 45–56.

Tamás FD, Sarkar AK and Roy DM (1976) Effect of variables upon silylation products of hydrated cements. In *Hydraulic Cement Pastes: Their Structure and Properties*. Cement and Concrete Association, Slough, UK, pp. 55–72.

Taplin JH (1959) A method for following the hydration reaction in Portland cement paste. *Australian Journal of Basic and Applied Sciences* **10**: 329–345.

Taylor HFW (1950) Hydrated calcium silicates. Part I. Compound formation at ordinary temperatures. *Journal of the Chemical Society* 3682–3690.

Taylor HFW (1961) The chemistry of cement hydration. *Progress in Ceramic Science* **1**: 89–145.

Taylor HFW (1964) *The Chemistry of Cements*. Academic Press, New York, NY, USA, vol. 1.

Taylor HFW (1984) Studies on the chemistry and microstructure of cement pastes. *Proceedings of the British Ceramic Society* **35**: 65–82.

Taylor HFW (1985) The reactions of cement compounds with water. *Materials Science Monographs (Reactivity of Solids, Part A)* **28A**: 39–45.

Taylor HFW (1986) Proposed structure for calcium silicate hydrate gel. *Journal of the American Ceramic Society* **69**: 464–467.

Taylor HFW (1992) Tobermorite, jennite, and cement gel. *Zeitschrift für Kristallographie* **202**: 41–50.

Taylor HFW (1993) Nanostructure of C–S–H: current status. *Advanced Cement Based Materials* **1**: 38–46.

Taylor R (2010) *Characterization of C–S–H in early and late age systems containing admixtures*. PhD Thesis, University of Leeds, Leeds, UK.

Taylor HFW and Howison JW (1956) Relationships between calcium silicates and clay minerals. *Clay Minerals Bulletin* **3**: 98–111.

Taylor HFW and Newbury DE (1984) Calcium hydroxide distribution and calcium silicate hydrate composition in tricalcium silicate and β-dicalcium silicate pastes. *Cement and Concrete Research* **14**: 93–98.

Taylor HFW and Turner AB (1987) Reactions of tricalcium silicate paste with organic liquids. *Cement and Concrete Research* **17**: 613–623.

Taylor HFW, Barret P, Brown PW *et al.* (1984) The hydration of tricalcium silicate. *Materials and Structures* **17**: 457–468.

Taylor HFW, Mohan K and Moir GK (1985) Analytical study of pure and extended Portland cement pastes: I, Pure Portland cement pastes. *Journal of the American Ceramic Society* **68**: 680–685.

Thomas JJ, Chen JJ, Jennings HM and Neumann DA (2003) Ca-OH bonding in the C–S–H gel phase of tricalcium silicate and white Portland cement pastes measured by inelastic neutron scattering. *Chemistry of Materials* **15**: 3813–3817.

Thomassin JH, Regourd M, Baillif P and Touray JC (1979) Étude de l'hydratation initiale du silicate tricalcique par spectrometrie de photo-électrons. *Comptes Rendus de l'Académie des Sciences Paris*. **288C**: 93–95. (In French.)

Thorvaldson T and Vigfusson VA (1928) The action of water on tricalcium silicate and beta dicalcium silicate. *Transactions of the Royal Society of Canada* 3rd series 22, **III**: 423–431.

Tong HS and Young JF (1977) Composition of solutions in contact with hydrating β-dicalcium silicate. *Journal of the American Ceramic Society* **60**: 321–323.

Trettin R, Oliew G, Stadelmann C and Wieker W (1991) Very early hydration of dicalcium silicate-polymorphs. *Cement and Concrete Research* **21**: 757–764.

Uchikawa H and Furuta R (1981) Hydration of C_3S-pozzolana paste estimated by trimethylsilylation. *Cement and Concrete Research* **11**: 65–78.

Ueda S, Hashimoto M and Kondo R (1968) Study on the hydration of cement by surface electron diffraction. *Semento Gijutsu Nenpo* **22**: 67–72.

Viehland D, Li JF, Yuan LJ and Xu Z (1996) Mesostructure of calcium silicate hydrate (C–S–H) gels in Portland cement paste: Short-range ordering, nanocrystallinity, and local compositional order. *Journal of the American Ceramic Society* **79**: 1731–1744.

Wagman DD, Evans WH, Parker VB *et al.* (1982) The NBS tables of chemical thermodynamic properties: selected values for inorganic and C1 and C2 organic substances in SI units. *Journal of Physical Chemistry* Ref. Data **11** (Supplement No. 2).

Walker CS, Sutou S, Oda C, Mihara M and Honda A (2015) Calcium silicate hydrate (C–S–H) gel solubility data and a discrete solid phase model at 25°C based on two binary non-ideal solid solutions. *Cement and Concrete Research* **79**: 1–30.

Wang J and Becker U (2009) Structure and carbonate orientation of vaterite ($CaCO_3$). *American Mineralogist* **94**: 380–386.

Wang J, Zhang F, Zhang J, Ewing RC, Becker U and Cai Z (2014) Carbonate orientational order and superlattice structure in vaterite. *Journal of Crystal Growth* **407**: 78–86.

Wartchow R (1989) Datensammlung nach der 'Learnt profile'-Methode(LP) für Calcit und Vergleich mit der 'Background peak background'-Methode (BPB). *Zeitschrift für Kristallographie* **186**: 300–302.

Webb TL and Heystek H (1957) The carbonate minerals. In *The Differential Thermal Investigation of Clays* (Mackenzie RC (ed.)), Mineralogical Society, London, UK, pp. 329–363.

Wieker W (1968) Silikatanionenstruktur des 14 Å Tobermorits von Crestmore und seiner Entwässerungsprodukte. *Zeitschrift für anorganische und allgemeine Chemie* **360**: 307–316. (In German.)

Wieker W and Trettin R (1983) Surface area development in the very early stages of hydration of C_3S. *Proceedings of the Conference on the Characterization and Performance Prediction of Cement and Concrete*, 25–30 July 1982, Henniker, NH, USA. Engineering Foundation, 87–91.

Wieker W, Grimmer AR, Winkler A, Mägi M, Tarmak M and Lipmaa E (1982) Solid-state high-resolution ^{29}Si NMR spectroscopy of synthetic 14 Å, 11 Å and 9 Å tobermorites. *Cement and Concrete Research* **12**: 333–339.

Wilding CR, Walter A, and Double DD (1984) A classification of inorganic and organic admixtures by conduction calorimetry. *Cement and Concrete Research* **14**: 185–194.

Williamson RB (1968) Constitutional supersaturation in Portland cement solidified by hydration. *Journal of Crystal Growth* **3/4**: 787–794.

Williamson RB (1972) Solidification of Portland cement. *Progress in Materials Science* **15**: 189–286.

Winchell AN and Winchell H (1964) *The Microscopical Characters of Artificial Inorganic Solid Substances*. Academic Press, London and New York, NY, USA.

Winnefeld F, Schöler A and Lothenbach B (2016) Sample preparation. In *A Practical Guide to Microstructural Analysis of Cementitious Materials* (Scrivener K, Snellings R and Lothenbach B (eds)), CRC Press, Boca Raton, FL, USA, pp. 1–35.

Wu ZQ and Young JF (1984) Formation of calcium hydroxide from aqueous suspensions of tricalcium silicate. *Journal of the American Ceramic Society* **67**: 48–51.

Wu ZQ, Hriljac J, Hwang CL and Young JF (1983) Orthosilicate analyses: a measure of hydration in pastes of alite and Portland cement. *Journal of the American Ceramic Society* **66**: C-86–C-87.

Ye Y, Smyth JR and Boni P (2012) Crystal structure and thermal expansion of aragonite-group carbonates by single-crystal X-ray diffraction. *American Mineralogist* **97**: 707–712.

Young JF (1967) Humidity control in the laboratory using salt solutions – a review. *Journal of Applied Chemistry* **17**: 241–245.

Young JF (1988) Investigations of calcium silicate hydrate structure using silicon-29 nuclear magnetic resonance spectroscopy. *Journal of the American Ceramic Society* **71**: C-118–C-120.

Young JF and Hansen W (1987) Volume relationship for C–S–H formation based on hydration stoichiometry. *Materials Research Society Symposium Proceedings* **85**: 313–322.

Zhang J and Scherer GW (2011) Comparison of methods for arresting hydration of cement. *Cement and Concrete Research* **41**: 1024–1036.

Zhang X, Chang W, Zhang T and Ong CK (2000) Nanostructure of calcium silicate hydrate gels in cement paste. *Journal of the American Ceramic Society* **83**: 2600–2604.

Zhang Z, Scherer GW and Bauer A (2018) Morphology of cementitious material during early hydration. *Cement and Concrete Research* **107**: 85–100.

Zingg A, Holzer L, Kaech A *et al.* (2008) The microstructure of dispersed and non-dispersed fresh cement pastes – New insight by cryo-microscopy. *Cement and Concrete Research* **38**: 522–529.

Zur Strassen H and Stditling W (1940) Die Reaktion zwischen gebranntem Kaolin und Kalk in wäßriger Lösung. II. Die Reaktionsprodukte im Zusammenhang mit dem System Kalk–Kieselsäure–Tonerde–Wasser. *Zeitschrift für anorganische und allgemeine Chemie* **245**: 267–278. (In German.)

Zürz A and Odler I (1987) XRD studies of portlandite present in hydrated Portland cement paste. *Advances in Cement Research* **1**: 27–30.

Ian G Richardson and HFW Taylor
ISBN 978-1-83549-777-7
https://doi.org/10.1108/978-1-83549-776-020253007

Index

abbreviations 1–2
acceleratory hydration period 418, 423, 428–31
acetone
  reaction with C_3S pastes 374
  use in drying 266, 374, 382–3, 425
afwillite 166, 223, 231, 241
åkermanite 62–3, 71
alite (*see also* tricalcium silicate)
  chromium substitution, defects 149
  crystal structure and data for polymorphs 7–20
  density 10–1, 16
  enthalpy change on formation 93–5
  factors affecting crystal size
  factors affecting reactivity 147–50
  fluoride substitution 78–9
  formation in cement kiln 89–90, 108–9, 119
  hydration, *see* hydration
  in clinkers
    compositions 12–14
    content 1, 91–4, 140–6
    effect of cooling rate 114, 133–4
    polymorphic types 15–6
    zoning 15
  magnesium substitution 8–16, 20
  natural occurrence 20
  optical and thermal data 19–20
  polymorphic transitions in cement kiln 15–6, 110
  polymorphism 7–20
  recrystallisation and crystal growth 15–6, 105, 108–9
  solid solutions or ionic substitutions 9–14, 78–9, 149
  X-ray powder patterns 16–19
alkali and alkali calcium sulfates
  data, equilibria 74–7
  in Portland cement clinkers 102–4, 114–7, 133, 140–1
  optical properties 76
alkali carbonates, in cement clinkers 74, 103, 116
alkalis in C–A–S–H 331–4
  structural-chemical formulae 415, 417
alkalis in C–S–H 324–6
alkalis in major constituent phases of clinker
  alite 12–3, 15
  aluminate 13, 31–4, 36–8
  belite 13, 27–8
  ferrite 13
alkalis in Portland cement making 99–100, 109, 115–8
alkalis in tobermorite 168, 186, 188, 206
alkali silicate
  in laboratory preparation of C–S–H 251, 257, 270–1, 282–3
  silicate species in solutions of 420
alumina ratio (alumina modulus) (AR) 7, 89–90, 94, 97, 107, 109, 111–13, 118, 133, 142
aluminate (phase) 31–8
  alkali substitution 31–3
  crystal structures and data for variants 31–4
  density 36
  enthalpy change on formation 93–4
  equilibria 58–9, 61, 63–9, 72, 77
  factors affecting reactivity 147–50
  formation in cement making 90, 102–3, 111–3
  in clinkers
    compositions and modifications 34–6
    content 1, 91–4, 140–6
  ionic substitutions 31–6
  optical properties 36
  structural variants 31–4
  X-ray powder patterns 35–8
Andreason pipette 128–9
anhydrite
  data, equilibria 74–6
  in cement making or clinker 100–4, 109, 114–17, 131, 140

anorthite 63
anomalous 11 Å tobermorite, *see* tobermorite phases
aphthitalite, *see* alkali and alkali calcium sulfates
aragonite 293, 373–4
arcanite, *see* alkali and alkali calcium sulfates
ash deposition, and light microscopy 133
Aspdin, Joseph 2–5
Aspdin, Thomas 3–5
Aspdin Works 2–4
atom coordinates, conversion between Cartesian and fractional 295–6
atomistic modelling, C–(A–)S–H 291–6, 331, 373, 412
autoclaved materials 160, 206, 334, 409

backscattered electron imaging (BEI) 7, 135–7, 377, 380, 384, 388, 394–5
belite (*see also* dicalcium silicate polymorphs) 20–31
  enthalpy change on formation 93–5
  formation in cement making 89–90, 103–4, 108–9, 114, 119
  hydration, *see* calcium silicate pastes
  in clinkers
    cell parameters 28, 31
    composition 13–4, 27–8, 31
    content 1, 91–4, 140–6
    lamellar textures 25–7
      Type I 26, 28
      Type II 27
    optical properties 28
    polymorphic types 25
    X-ray powder patterns 28
  modulated structures 25
  natural occurrence (larnite) 21, 28, 231
  reactive (from hillebrandite), and product of 256, 274–9, 281–2, 322, 419
  reactivities of different polymorphs 147–8
BET (Brunauer–Emmett–Teller) gas adsorption 130, 232
bicchulite 167
Billhead 4
Blaine method 130
blast-furnace slag 7, 96, 100, 329–30, 375
Bogue calculations
  original, including derivation of equations 91–3, 146
  modified 140–3, 146
  sources of error in 92
  updated to include variable Al/Fe of ferrite 92–3
bond distances 145
  aluminate (phase) 31
  atomistic models for C–S–H 290–5
  CA 59
  CA_2 / CA_6 / $C_{12}A_7$ 60
  C_5A_3 62
  C_3S 8
  calcium silicate hydrates and related phases, including tabulated 234–42
  distortion indices 41
  ferrite (phase) 323–4
  importance of checking after Rietveld refinement / atomistic modelling 145, 290–1, 296
  jaffeite 323–4
  jennite 228, 283, 323–4
  model crystal structures for C–S–H(I) 296–311, 335, 337–8, 344, 346, 351, 356
  tobermorite phases 171-2, 175–7, 179, 182, 191, 193, 196, 200, 269, 283
bottle hydration 277, 279, 281, 315, 317, 425
bound water (chemically bound water)
  calcium silicate pastes 384–6
  definition 384
Bower, John & Joshua 2–5
bredigite 28, 71, 231
bridging tetrahedra (BT), calcium silicate hydrates
  adjacent, in model structure for 14 Å orthotobermorite 192, 194–5, 197
  cross-linked (*see also* cross-linking quotient (CLQ)) 214
  C–S–H(I) 260, 263–5, 267, 270, 286, 290–1
    in model structures 297, 299, 300, 302, 306–8
  C–S–H, calcium silicate pastes 399, 409
  C–S–H, high Ca/Si 278, 280
  definition 172, 212
  double-chain structure, schematic 208–10
  jennite 225–6, 324
  missing 207, 260, 270, 280, 320–1, 324, 413
  occupied by aluminium 328–9, 331–3
  orientation 173
  plombièrite 232
  single-chain structure, schematic 211–2
  tobermorite, 14 Å 269

bridging tetrahedral sites, calcium silicate hydrates
  aluminium at 185, 207–22, 330–3
  mechanism of C–S–H formation 432
  models for C–S–H 407–11
  ^{17}O MAS NMR spectra 267
  fraction occupied (*see also* SOF_{BT}) 212–3
  vacant 186, 264, 306, 315, 344, 346, 351, 411
brownmillerite (*see also* ferrite phase) 45
bultfonteinite 166, 241
burnability (of cement raw mixes), *see* Portland cement clinker
burning conditions of clinker, and light microscopy 133–4

CA
  crystal chemistry 59, 147–9
  density 59
  equilibria 58–9, 63, 65–6
  structure 58
CA_2
  crystal chemistry 60, 148–9
  density 60
  equilibria 58–9, 63
  structure 59–60
CA_6
  crystal chemistry 60, 148–9
  density 60
  equilibria 59, 63
  structure 60
C_2A 38–41, 62, 70
C_2AH_8 328–9
C_2AS, *see* gehlenite
C_3A and substituted modifications, *see* aluminate (phase)
C_3A_2M 71, 73–4
$C_3A_3{\cdot}CaF_2$ 79
C_4AF, *see* ferrite (phase)
C_4A_3 62
$C_4A_3H_3$ 62
$C_4A_3\overline{S}$
  data, structure, preparation, equilibria 75–7
  in cement making 102–4
C_5A_3 60–2, 66, 71, 73, 148–9
C_6A_2F, *see* ferrite (phase)
C_7A_5M 71, 73–4
$C_{11}A_7{\cdot}CaCl_2$ 61
$C_{11}A_7{\cdot}CaF_2$ 7, 60–1, 79
$C_{11}A_7{\cdot}CaS$ 61
$C_{12}A_7$
  absence in Portland cement clinkers 111–3
  data, equilibria, structure, thermal behaviour 58–68, 70–2, 77
  derived structures containing additional ions 60–1
  in cement making 90, 102–3
$C_{20}A_{13}M_3S_3$ ('Phase Q') 71, 73
C–(A–)S–H, *see* calcium (alumino)silicate hydrates, synthetic preparations
CF 65–6
CF_2 65
C_2F (*see also* ferrite phase) 41, 43, 65–6, 72, 111, 149
CS, structures, polymorphism, data 57–8, 104, 147–8, 283–5
C_2S, *see* belite, dicalcium silicate polymorphs
α-C_2SH (dicalcium silicate α-hydrate) 166, 241
C_3S, *see* alite / hatrurite / tricalcium silicate
C_3S pastes, *see* calcium silicate pastes
C_3S_2, *see* kilchoanite
$C_5S_2\overline{S}$ ('silicosulfate')
  data, structure, preparation, equilibria 75–7
  in cement making 102–5
$C_{10}S_3\overline{S}_3H$ (hydroxyl ellestadite) 102
$C_{10}S_4{\cdot}CaF_2$ 78–9, 119
C–S–H in calcium silicate pastes
  carbonation 372–5
  Ca/Si ratio
    determination by indirect methods 378–84
    determination directly by X–ray microanalysis 396–9
  definition 371
  density 386–7
  electron diffraction 399–400
  formation 371, 387–96
  infrared spectrum 387
  kinetics and mechanism of formation 421–33
  morphology, microstructure 387–96
  relation with synthetic C–S–H preparations 334
  silicate anion structure 400–7
  structural models 407–17
  thermogravimetry and DTA 377–86
  water content 384–6
  X-ray powder pattern 371–2
C–S–H(di, poly) 285–6, 327
C–S–H(I), *see* calcium (alumino)silicate hydrates, synthetic preparations
C–S–H(II)
  Ca–OH groups 322–324
  electron diffraction 277, 324

metastable equilibria, $CaO–SiO_2–H_2O$ system, association with Curve B 256, 276, 279–80, 412, 419–20
morphology 277–9, 387
position on plot of SOF_{BT} against the Ca/Si 274
silicate anion structure
^{29}Si MAS NMR 279
molybdate / TMS methods 277
structural-chemical formulae for high-Ca C–S–H phases 280–82, 408, 410–17
water content 317
X-ray powder pattern 278–9
calcium (alumino)silicate hydrates, synthetic preparations (C–(A–)S–H(I))
alkalis in C–A–S–H 331–4
alkalis in C–S–H 324–6
aluminium substitution 326–33
(alumino)silicate anion structure
early studies 260–1
mean chain length 213–4, 221–3, 262–6, 286, 288
molybdate method 261
^{29}Si MAS NMR 261–7, 282–3, 286, 327–34
anion substitution 334
atomistic modelling studies 291–6
average structure 256–60
calcium extraction 286–8
density 315–6
electron diffraction 260–1, 273
elevated temperature preparations of C–S–H
formation, XRD and other data 283–5
nuclear magnetic resonance 285
calcium extraction 286–8
formation in pastes 190, 220, 222
infrared absorption spectroscopy 269–70, 331
iron(III) substitution 333–4
layer spacing 257–60, 263, 265
metastable equilibria, $CaO–SiO_2–H_2O$ system, association with Curve A 251–6, 273, 412, 419–20, 425
model crystal structures 296–312
dimeric silicate chains 259, 263, 274, 288, 295, 297–305, 316, 335–8
density 315–316
'octameric' silicate chains 259, 274, 288, 306, 316–8
'pentameric' silicate chains 259, 263, 274, 288, 306–8, 316–8, 342–6
'trimeric' silicate chains 259, 263, 274, 308–11, 316–8, 347–51
'undecameric' silicate chains 259, 263, 274, 288, 311–12, 316–8, 351–6
morphology
change with absolute concentration of Ca in solution 273, 275, 283, 324–5
foil-like 260–1, 271, 273, 277, 324, 326, 331
nuclear magnetic resonance spectroscopy
^{27}Al MAS NMR 326–34
heteronuclear chemical shift correlation (HETCOR) NMR 267–9
^{17}O, ^{43}Ca and ^{1}H MAS NMR 267–9, 282
^{29}Si MAS NMR 261–7, 282–3, 286, 327–34
useful equations for the interpretation of single-pulse data 212–23
Raman spectroscopy 270
relation to 14 Å tobermorite 169, 214, 251, 256–60
room temperature preparations of C–S–H
Ca/Si ratios 0·7–1·5 251–71, 315
anion substitution and exchange 334
average structure of C–S–H(I) 256–60
conditions of formation and metastable equilibria 251–6
incorporation of alkali metal cations 324–6
incorporation of aluminium 326–31
incorporation of aluminium and alkali 331–33
incorporation of iron(III) and other cations 333–34
infrared absorption spectroscopy 269–70
morphology and electron diffraction 260
^{17}O, ^{43}Ca and ^{1}H MAS NMR 267–9
Raman and Ca X-ray absorption spectra 270
silicate anion structure of C–S–H(I), early studies 260–1
silicate anion structure of C–S–H(I), ^{29}Si MAS NMR 261–7
thermal behaviour and water content of C–S–H(I) 316–322
Ca/Si ratios greater than 1·5 271–82
Ca–OH groups in C–S–H preparations with Ca/Si ratios >1·3–1·5 322–324
preparations of higher crystallinity 277–80
preparations of low crystallinity 271–7

structural-chemical formulae for high-Ca C–S–H phases 280–82
preparations containing monomer or high proportions of dimer 282–3, 322
relation of synthetic C–S–H to the C–S–H in cement pastes 334
vibratory milling of lime–silica slurries 283
structural-chemical formulae, low-Ca C–(A–) S–H phases 174–5, 207–13, 263–5, 288, 319, 408, 410–17
thermal behaviour and water content 316–322
water content 316–22
X-ray diffraction pattern
model structures (calculated) 312–3
observed 256–8, 313
calcium aluminium silicates 62–9, *see* anorthite / garnet / gehlenite
^{43}Ca MAS NMR
of synthetic C–S–H 267–8, 373
of vaterite 373
calcium aluminoferrite, *see* ferrite (phase)
calcium carbonate
as product of carbonation 140, 372–4, 386, 390
data for polymorphs 187, 278, 293, 373–4
thermal decomposition 89–90, 93–7, 100–2, 373
calcium chondrodite 165, 240
calcium extraction from C–S–H 286–8
calcium fluoride 60–1, 77–9, 95, 118–9, 143
calcium hydroxide (portlandite; CH)
cryptocrystalline or amorphous 253, 259–60, 270–1, 288, 320, 375, 378, 381
electron diffraction 400
in calcium silicate pastes 371–2,
morphology and microstructure 375, 390, 394–5, 397
onset of precipitation 418, 423, 427–31
quantity, thermogravimetry 377–84
in C–S–H preparations 259, 263, 270–1, 274–5,
in models for high-Ca C–S–H 295, 304, 408–9, 412, 414–5
in nature 231, 375
solubility 252–5, 272–3, 275–6, 376–7, 418–20
reduction with alkali 324–6
structure, crystal chemistry 172, 294–5, 303, 375–6
thermal behaviour 376–8
XRD powder pattern 278, 372
calcium langbeinite, *see* alkali and alkali calcium sulfates
calcium magnesium aluminates 71, 73
calcium magnesium silicates (*see also* åkermanite, bredigite, diopside, merwinite, monticellite) 71
calcium oxide (*see also* free lime),
structure, data 57
calcium potassium sulfate, *see* alkali and alkali calcium sulfates
hydrated, *see* syngenite
calcium silicate carbonates (*see also* spurrite, tilleyite) 79
calcium salts 251, 257, 270–1, 282–3, 322, 332
calcium silicate fluorides 77–9
calcium silicate hydrates (*see also* individual phases / C–S–H in calcium silicate pastes)
crystal chemistry, general aspects 233–42
natural minerals 159–242
crystal chemical data, tabulated 238–42
crystal structure data, tabulated 161–7
synthetic preparations (*see also* C–S–H(II) / calcium (alumino)silicate hydrates) 251–356
calcium silicate pastes 371–433
C–S–H, *see* C–S–H in calcium silicate pastes
calcium hydroxide, *see* calcium hydroxide
calorimetric curve / data 421–3, 427–8, 432
carbonation 372–5, 386, 390, 397
density 386–7
early hydration reactions 402–7, 418–33
early hydration product
composition 423–7, 429
morphology 391–2, 419
silicate anions (monomer) 282, 291, 334, 403, 405–7, 419, 426–31, 432–3
thickness 391–2, 403, 405–6, 426–7, 429
hydration kinetics and mechanisms, *see* hydration
microstructures 387–96
phase compositions 371, 377–84, 396–99
silicate anion structures 387, 400–17, 430, 432–3
thermogravimetry and DTA
curves for a fully reacted C_3S paste 377–8
results from 377–84, 386, 391, 402, 410
water content 384–6
X-ray diffraction, powder
pattern for 26-year-old C_3S paste 371–2
results from 371–2, 379–83, 399, 401, 427–8
calcium sulfate phases (*see also* anhydrite, gypsum, hemihydrate) 1–2, 74–7

calcium sulfide, in reduced clinkers 110
calorimetry (isothermal conduction) 421–2, 427–8, 432
calorimetry (differential scanning), *see* differential scanning calorimetry (DSC)
calcium X-ray absorption spectra 187, 200, 270
carbonation 140, 372–4, 386, 390, 397
  avoidance in laboratory experiments 372
Cartesian coordinates, converting to/from fractional coordinates 295–6
cement pastes, *see* calcium silicate pastes
chegemite 165, 240
chemical nomenclature, cement 6–7, 371
chemically bound water, *see* bound water
chloride
  alkali 6
  binding in C–S–H 334
  calcium, in the preparation of C–(A–)S–H 282, 285–6
  phases containing 61, 100–4, 109
chromatography
  gas liquid 7, 400
  gel permeation 7, 400–2, 404, 432
  paper 185
chromium substitution in clinker phases 119, 149
CIF, *see* crystallographic information files
clay minerals
  reactions in cement making 89–90, 93–5, 100–4, 107
  TEM of 277
clinker formation in pure systems
  CaO–Al_2O_3–Fe_2O_3–SiO_2 68–70
  CaO–Al_2O_3–SiO_2 64–5
clinker, Portland cement, *see* Portland cement clinker
clinotobermorite, *see* tobermorite phases
CLQ, *see* cross-linking quotient
colour of clinker, *see* Portland cement clinker
combined rotation and multiple pulse spectroscopy (CRAMPS) NMR
  C_3S pastes 407
  C–S–H, synthetic 268
compressive strength
  effect of minor components of clinker 117–9
  in definition of hardening 371
  in specifications for Portland cement 5–6
  of mortars prepared with fluoride-containing C_3S 78–9
  prediction of, *see* Ono's method
  relationship with amount of cement reacted 131
  size of alite crystals 109, 134
coolers, in Portland cement manufacture 99–100
cooling of Portland cement clinker
  reactions during 111–114
  studied by light microscopy 133
coordination number
  for aluminium (to oxygen)
    in atomistic models for C–A–S–H 331
    in calcium aluminates 8, 59–60
    in calcium silicates 326–33
  for calcium (to oxygen)
    in atomistic models for C–S–H 291–3, 295
    in calcium aluminates 31, 59–60,
    in calcium carbonates 293, 373
    in calcium silicates 8, 25, 147, 191–2, 196, 233–42, 269, 290, 296, 299–301, 308
  for oxygen in calcium aluminates / hydroxide / silicates 8, 59–60, 147–9, 172, 376
CRAMPS, *see* combined rotation and multiple pulse spectroscopy
cross-linking quotient (CLQ) 220–3
cross-polarisation magic-angle spinning (CPMAS) NMR, *see* ^{1}H-^{29}Si CPMAS NMR
cryo-SEM 389–90
crystallisation
  alite 15, 108, 112, 134
  anhydrous calcium aluminosilicates 63
  calcium hydroxide 271, 375
  interstitial phases 33, 89, 111, 113
  sulfate phases 114
  tobermorite 168
  wollastonite 201, 203, 224, 283–4, 320
crystallographic information files / datablock names (CIF) 296–7
  clinotobermorite, 11 Å 181–2
  clinotobermorite, 14 Å 197–8, 339–40
  dimeric C–S–H(I) 295, 298, 301, 303–5, 335–7
  orthotobermorite, 11 Å 178
  'pentameric' C–S–H(I) 307, 309, 342, 344
  'trimeric' C–S–H(I) 310, 347
  'undecameric' C–S–H(I) 312, 351
cristobalite 58, 63, 102–3
curing, definition 371
cuspidine 79, 166, 241, 322

d-block elements 119
D-drying, definition 386
deceleratory period 423

decomposition
calcium carbonate 95, 119
calcium hydroxide 377–9
C–S–H 232, 378
during clinker formation 89, 93–5, 100–3, 110, 114
sulfate phases 74, 109
tobermorite phases 189, 202
tricalcium silicate 33, 65, 68, 118, 134
defects, and reactivity of clinker phases 149–50
dellaite 166, 238, 241
density (*see also* under individual phases)
Portland cement clinker 127
deposits, in cement kilns or preheaters 99–100, 104–5, 108, 117, 119, 132
DI, *see* distortion indices
dicalcium ferrite, *see* C_2F
dicalcium silicate polymorphs (*see also* belite / calcium silicate pastes)
data, structures, thermal behaviour 20–27
dusting 20, 114
enthalpy change on formation 93–5
enthalpy change on hydration 421
equilibria 57–8, 63–5, 66–72
modulated structures in 25
preparation 79–80
stabilization of β-polymorph 20–5, 79
thermal behaviour 20
differential scanning calorimetry (DSC) 205, 224, 232, 377
differential thermal analysis (DTA) 7
alite 20
calcium silicate hydrates 278, 283, 320
calcium silicate paste 374, 377–8
tobermorite phases 201, 203
white cement 110
dimeric silicate chains
as elements in models for C–S–H with longer average silicate chains 305–312, 342–56
in ($3n-1$) chain length sequence, C–S–H 260, 288, 291, 402, 404, 410, 432–3
in atomistic model for C–S–H 291
in dreierkette-based models for C–(A–)S–H 212–4, 265, 288, 317–9
in model crystal structures for C–S–H 262–3, 274, 295, 297–305, 311–20, 335–7
in structural-chemical formulae for C–(A–)S–H 210, 214, 408–17
observed in calcium silicate pastes 233, 261, 280, 400–2, 404–5, 420, 426, 432–3
observed in synthetic C–(A–)S–H preparations 266–7, 277, 282–3, 285, 289, 296, 322–4, 328–9
schematic illustration 212, 430
diopside 71
dissolution
of C_3S 256, 390–1, 406, 423, 429, 431
of calcium hydroxide 376
of calcium sulfate 117
of product B 430
distillation method, for preparation of high Ca/Si C–S–H 271–5
distortion indices (DI) 293–5
dolomite 101, 133
dormant hydration period, *see* tricalcium silicate, hydration, period of slow reaction
double quantum homonuclear CP/MAS correlation NMR, of synthetic C–S–H 266–7, 306
dreierkette conformation of (alumino)silicate chains
C–S–H(I) 267
connectivity of tetrahedra 185
density of phases that have dreierketten silicate 235–7
double chains, *dreierdoppelketten* (*see also* cross-linking quotient)
clinotobermorite, 11 Å 179–82
schematic diagrams 208–10
tobermorite, 11 Å 171–7
xonotlite 186
single chains, *dreiereinfachketten*
calcium silicate pastes, C–S–H 420, 430, 432–3
clinotobermorite, 9 Å 202–3
clinotobermorite, 11 Å 182–5
high Ca/Si C–S–H, synthetic 275
jennite 225–6, 228–9, 283, 323
models for C–S–H 283, 288, 297, 430, 432–3
net charge per Si atom, z_{Sil} 262, 264
relationship with ($3n-1$) sequence 225, 260, 263, 265, 269, 288, 413, 415–7, 432
schematic diagrams 191, 211–2
tobermorite, 9 Å 201–2
tobermorite, 11 Å 178–9
tobermorite, 12 Å 205
tobermorite, 14 Å 190–9
symmetry 174

dry process 96–7, 127
DSC, *see* differential scanning calorimetry
DTA, *see* differential thermal analysis
dust, formation in cement kiln 99–100, 103–4, 108

early product
  in C_3S hydration 391–2, 419–29
  in the preparation of synthetic C–S–H 288
EDX, *see* Energy dispersive X-ray spectroscopy
electron diffraction
  alite 16
  belite 25
  C_3A 33
  C–S–H, from hydration of C_3S or β-C_2S 277, 399–400, 410
  C–S–H, high Ca/Si 277, 324
  C–S–H, synthetic 260–1, 273
  calcium hydroxide 400
  jennite 223
  phase Q 73
  plombièrite 231–2
  tacharanite 229
  tobermorite, anomalous 160
electron microprobe analysis, *see* electron probe microanalysis
electron microscop(e,y), *see* scanning electron microscop(e,y) / transmission electron microscop(e,y)
electron probe microanalysis (EPMA), *see* X-ray microanalysis
electron spectroscopy for chemical analysis (ESCA), *see* X-ray photoelectron spectroscopy (XPS)
ellestadite, hydroxyl 102
endothermic reactions on heating
  α'_H to α transition of belite 28, 110
  β to α'_L transition of belite 20, 28
  decomposition of calcite 93, 95, 100
  dehydration of calcium hydroxide 278, 378
  dehydration of high-Ca synthetic C–S–H 278
  dehydration of jennite 224
  dehydration of plombièrite 231
  dehydration of tobermorite 201, 203, 205, 320
  melting 110
  convention on DTA curves 378
Energy dispersive X-ray spectroscopy (EDX), *see* X-ray microanalysis
enstatite 71
enthalpy
  changes in clinker formation 93–5
  change on hydration of β-C_2S 421
  change on hydration of C_3S 421, 428
  of decomposition of calcium carbonate 93, 95
  of decomposition of calcium hydroxide 377
EPMA, *see* X-ray microanalysis
equilibria, *see under* system or phases concerned
ESCA, *see* X-ray photoelectron spectroscopy (XPS)
etching, of clinker for microscopy 133
ettringite 116, 139, 326, 331, 390
evaporation
  differential, in the TMS method 400
  in the cryo-SEM 389
  in the drying of cement paste 374–5, 390
  in the preparation of high Ca/Si C–S–H, *see* distillation method
  of cyclohexane / propan-2-ol in the separation of clinker phases 138–9
  of volatiles during the production of clinker 105, 108–10
exit gases 97–8
exothermic reactions on heating
  crystallisation of β-C_2S 110, 278
  crystallisation of wollastonite 201, 203, 224, 283–4, 320
experimental considerations, *see* carbonation / high-temperature phases / hydration
extraction methods
  for calcium silicate pastes 377–8, 412
  for unreacted cement or clinker 34, 116, 138–9, 146
  of Ca from C–S–H 286–8

fedorite 164, 240
ferrite (phase) 38–45
  cell parameters 38–40, 43–45
  colour 45, 111
  composition in clinkers 42–3
  content in clinkers 1, 91–4, 140–6
  data, structure 38–42
  density 39, 41
  electrical, magnetic and optical properties 45
  enthalpy change on formation 93–4
  equilibria 65–70, 72–3
  factors affecting reactivity 149–50
  formation in cement making 90, 102–3, 113
  ionic substitutions 42–5
  magnetic and optical properties 45
  X-ray powder patterns 44–5
  zoning 113, 143, 150

flyash 6, 7, 96, 329–30
fluoride phases in Portland clinker formation 77–9, 105, 118–19
fluorosilicates, as fluxes in cement making 79
foshagite 161, 171, 238
fractional coordinates, converting to/from Cartesian coordinates 295–6
free lime (*see also* calcium oxide)
  data, structure 57
  determination 139
  in cement making 5, 65, 70, 89–91, 102–3, 105–8, 111–2, 140, 142
  microscopic observation in clinker 132–3
fuels, in cement making 95–7, 99–100, 102, 109–10, 117
fukalite 167, 229–30, 242

garnet 63
gas–liquid chromatography (GLC), *see* chromatography
gehlenite 27, 62–3, 71, 73, 102–3
gel permeation chromatography, *see* chromatography
GGBS, *see* blast-furnace slag
glass content of clinker, *see* Portland cement clinker
glass works 2–5
glaserite 20, 74
GLC, *see* chromatography
GPC, *see* chromatography
grinding / ground clinker, *see* Portland cement clinker
grossular 63
ground granulated blast-furnace slag (GGBS), *see* blast-furnace slag
gypsum 1, 100
  behaviour on grinding and storage of cement 117, 131
  impurities in 140–1
  peaks in XRD pattern of typical Portland cement 137–8
  thermal analysis 139
gyrolite 164, 240

hardened cement pastes (HCP) (*see also* calcium silicate pastes) 7, 135, 330–1
hardening 1, 6
  definition 371
  calcium silicate pastes 371
hatrurite 20
heavy metals, and Portland cement making 95, 105
helium pyknometry 386–7
hematite 65
hemihydrate
  behaviour on grinding and storage of cement 103, 117, 131
  thermal analysis 139
heteronuclear chemical shift correlation (HETCOR) NMR 267–9
higher crystallinity preparations of high Ca/Si C–S–H 277–82
high-resolution transmission electron microscopy (HRTEM)
  tobermorite 168, 190
high-temperature phases, laboratory preparation 79–80
hillebrandite 161, 171, 238, 275–9, 281–2
HRTEM, *see* high-resolution transmission electron microscopy
hydration, calcium silicates 371–433
  acceleratory period 418, 421–3, 428–31
  concentrations of CaO and SiO_2 in solution 412, 418–20, 423
  definition 371
  dormant period, *see* period of slow reaction below
  enthalpy change 421, 428
  experimental considerations 372–87, 397, 400, 404–6
  induction period, *see* period of slow reaction below
  kinetics and mechanisms 421–33
  period of slow reaction 421–3, 426–31
    monomer in early product 282, 334, 403–7, 412–3, 429–30, 432–3
  reactions and products, *see* calcium silicate pastes
  species in solution 420
  termination of 374–5
hydrogarnet, *see* katoite
hydroxyl-ellestadite 102
^{1}H-^{29}Si CPMAS NMR
  of C_3S pastes 402–6, 427, 432
  of plombièrite 232
  of synthetic C–S–H 405, 266–7, 286
^{1}H MAS NMR 196, 225, 267–9, 407

image analysis
  calcium hydroxide in pastes 377
  degree of reaction of C_3S 379–80, 384
  Portland cement or clinker 137, 143, 146

induction period, *see* tricalcium silicate, hydration, period of slow reaction
infrared (IR) absorption spectra / spectroscopy
  of calcium silicate paste 377, 386–7, 412
  of clinker phases 113, 140, 143
  of (meta)jennite / high Ca/Si C–S–H 223–5, 277, 280
  of synthetic C–S–H 269–70, 283, 331
  of tobermorite 186–7, 200-1, 203, 205, 224, 228–9
inner product
  definition 388
  calcium silicate / cement pastes 233, 379, 384, 389, 391, 394, 397–9, 431–3
interstitial material in Portland cement clinkers (*see also* aluminate (phase) / ferrite (phase))
  composition 64, 113, 133
  prismatic, dark 34
  effects of cooling rate 64, 113–4, 134
  effect on grindability 117
  microscopy 133–7, 143, 146
invention of Portland cement 2–5
ion exchange 188, 206, 324–5
ionic substitutions, and reactivities of clinker phases 149–50,
IR, *see* infrared
iron, metallic, in reduced clinkers 110
iron (II) oxide, *see* wüstite
iron(III) substitution in C–S–H 333–4

jaffeite 166, 225–9, 241, 281, 322–4
jennite
  and C–S–H 159, 251, 256, 269, 277–8, 283, 322, 407–12, 420, 430, 432
  crystal chemistry 239, 294–5, 323
  data, structure 163, 171, 223–9, 323–4
  density 223, 236
  electron microscopy and diffraction 223–4, 399
  infrared absorption spectroscopy 224–5
  nuclear magnetic resonance spectroscopy 225, 228
  optical properties 223
  Raman and Ca X-ray absorption spectroscopy 270
  Structural-chemical formulae 280–2, 408, 415–7
  thermal dehydration 224
  X-ray diffraction 223–5, 228

$KC_{23}S_{12}$, non-existence of 27
K-phase 164, 240
kalsilite 63
katoite 167, 329
kilchoanite 57, 147–8, 165, 240
killalaite 166, 241
kiln
  bottle 2–4
  rotary 99
    clinker from (*see also* Portland cement clinker) 127, 132
    conditions / reactions in 64–5, 76–9, 90, 99–111, 117–19, 133
    gases 97–9
    rings and deposits 104, 117, 119
kinetics of hydration of calcium silicates 421–3, 432
Klein's compound, *see* $C_4A_3\overline{S}$
KOSH reagent (KOH and sucrose) 139

larnite (*see also* dicalcium silicate polymorphs) 231–2
laser granulometry 128–30
late product 391, 431
Lea and Nurse method 130
Lepol process 100
light microscopy
  alite 15
  aluminate 33–4
  belite 25
  calcium silicate pastes 387
  Portland cement clinker 104, 111–2, 132–4, 143, 145
lime, free, *see* free lime
lime saturation factor (LSF) 7, 65, 70, 89–91, 94, 97, 107–9, 118, 133
limestone (*see also* calcium carbonate)
  as raw material for making cement 1, 89, 95, 102, 133
  nature of 95, 107, 118
liquid, in formation of Portland cement clinker
  composition 105–6, 113
  physical properties 108, 117–9
  quantity 105–6, 118
  solidification 111–2, 114
litre weight 127
low heat cement (Type IV) 6
LSF, *see* lime saturation factor

magic-angle spinning nuclear magnetic resonance (MAS NMR), *see* nuclear magnetic resonance spectroscopy
magnesium compounds, anhydrous, of cement-chemical interest 71
magnesium oxide, *see* periclase
mass balance calculations
  clinkers 141–2
  hydration of C_3S 379, 399, 426–8
mayenite 62
MD, *see* molecular dynamics
mean (alumino)silicate chain length (MCL)
  calcium silicate pastes, C–S–H 233, 372, 403–4, 408–12, 433
  calculation 213–14, 263–4, 404
  C–S–H(I), synthetic 256, 262–8, 282, 286
  high Ca/Si C–S–H, synthetic 256, 274–5, 279
  of model structures for C–S–H(I) 288, 347, 351
  tobermorite-based structure 185, 220–3, 262–6, 281, 286, 288, 417
melilite (*see also* åkermanite and gehlenite) 62–3, 71, 73–4
mercury intrusion porosimetry (MIP) 7
  clinker 127
  drying cement paste for 374–5
merwinite 71
metajennite 163, 223–5, 228
metakaolin 7, 102, 330
metastable equilibria, $CaO–SiO_2–H_2O$ system 251–6, 273, 276, 279–80, 412, 419–20, 425
methanol, reactions with C_3S or cement pastes 374, 386
MgO (as oxide component)
  in compounds, *see under* name
  in Portland cement making 105–6, 111–4, 118
microanalysis, *see* X-ray microanalysis
microstructure
  calcium silicate pastes 387–96
  Portland cement, ground 135–7
  Portland cement clinkers 131–7, 149
middle product, calcium silicate pastes 391
mineralizers, in Portland cement making 79, 103, 118–9
MIP, *see* mercury intrusion porosimetry
modified Bogue calculation 140–2, 146
molecular dynamics (MD)
  simulations of C–S–H structures 291–4, 329
molybdate method 185, 196, 201, 261, 277, 282, 285, 327, 400
monocalcium aluminate, *see* CA
monomeric silicate (in early hydration product) 282, 291, 334, 403, 405–7, 419, 426–31, 432–3
monticellite 71
morphology
  aluminate (phase) 36
  calcium carbonates 374
  calcium hydroxide 375, 390, 394–5, 397
  C–S–H, change with absolute concentration of Ca in solution 273, 275, 283, 324–5
  C–S–H, synthetic with low Ca/Si 260–1, 271, 324, 326, 331
  C–S–H, synthetic with high Ca/Si 273, 275, 277
  C–S–H, in calcium silicate pastes 271, 279, 387–96, 404, 420, 432
    see also inner product
    see also outer product
  early product, in calcium silicate pastes 391–2, 419
  pleochroite 73
  tobermorite 189, 206

NC_8A_3, non-existence of 33, 36
nagelschmidtite 28
nekoite 161, 238
nepheline 63
net charge per Si atom in C–S–H, z_{Sil} 262, 264
nitrates
  aluminium, in the preparation of C–A–S–H 332
  calcium, in the preparation of C–(A–)S–H 251, 257, 332
nodulisation, in clinker formation 1, 89, 99–100, 105, 108–11, 117, 132–4
nomenclature, cement chemical 6–7, 371
non-evaporable water
  definition 386
  in calcium silicate pastes 386, 422
nuclear magnetic resonance spectroscopy (NMR)
  ^{27}Al
    alite 12
    belite 27
    C–A–S–H 327–33
    tobermorite, 11 Å 170, 185
    tobermorite, 12·6 Å 205
    tobermorite, 14 Å 196

^{1}H (proton) 196, 205, 407
^{1}H–^{29}Si cross polarisation 225, 232, 266–7, 282, 286, 402–6, 427, 432
^{43}Ca
synthetic C–S–H 267–8, 373
vaterite 373
^{17}O 187, 196, 200, 207, 225, 267–8, 282, 407
^{29}Si
C–(A–)S–H 261–8, 282–3, 286, 327–34
C–S–H(II) / high-Ca/Si C–S–H 273–5, 279, 286
calcium silicate pastes 233, 291, 379–84, 391, 402–7, 421–2, 426–7, 432
jennite 225
plombièrite 230
Portland cement 146
tobermorite, 11 Å 187, 196, 219–20
tobermorite, 12·6 Å 205
tobermorite, 14 Å 189, 203
useful equations for the interpretation of single-pulse ^{29}Si MAS NMR data 212–23, 264–5

octameric silicate chains
in $(3n-1)$ chain length sequence, C–S–H 260, 288, 291, 402, 404, 410, 432–3
detection by TMS-GPC 401–2, 404, 432
models for C–(A–)S–H 263, 274, 285, 288, 306, 311–20
okenite 161, 238
olivine, structure type 21, 25, 71
Ono's method 134
OPC, *see* ordinary Portland cement
optical data / properties, *see* individual phases
ordinary Portland cement (OPC) (*see also* Portland cement / clinker) 5, 13, 42, 73, 127, 130–1, 140–1, 402
orthotobermorites, *see* tobermorite phases
outer product
calcium silicate pastes 325, 389–96, 420, 431, 433
definition, 123 387–8
oxygen atom environments in calcium silicates / aluminates 147–9
oyelite 162, 189, 196, 204, 207

P_2O_5, as oxide component in cement making 95, 104, 118–9
paired tetrahedra (PT), calcium silicate hydrates (*see also* dimeric silicate chains)
aluminium 328–9, 332–3
C–S–H(I) 261, 265–9
in model structures 296–312
C–S–H, calcium silicate pastes 409, 433
definition 172, 212
double-chain structure, schematic 208–10
jennite / jaffeite 226–9, 323-4
single-chain structure, schematic 211–2, 430
tobermorite, 14 Å 269
paired tetrahedral sites, calcium silicate hydrates
fraction occupied 212–3, 218, 265, 289, 319
paper chromatography, *see* chromatography
parawollastonite 57, 147–8
particle size distribution
in studies of hydration 131, 371, 384
of cement 127–31
raw materials 107–8
paste (*see also* calcium silicate pastes)
definition 371
p-block elements 119
pectolite 161, 171, 238
pentameric silicate chains
bridging-site vacancy 214
C–S–H(I) 286
detection by TMS-GPC 401–2, 404, 432
in $(3n-1)$ chain length sequence, C–S–H 260, 288, 291, 402, 404, 410, 432–3
models for C–(A–)S–H 212, 263, 274, 285, 288, 306–9, 311–20, 342–6, 410–1
schematic illustration 212, 430
periclase (magnesium oxide)
data, structure, equilibria 70–3
in Portland cement clinkers 2, 91, 101, 103, 113–4, 118, 133, 141
phase composition, potential 91
phase equilibria, high temperature 57–80
phase Q ($C_{20}A_{13}M_3S_3$ approx.; *see also* pleochroite) 71, 73
phase T ($C_{1\cdot7}M_{0\cdot3}S$ approx.) 71
phosphates, *see* P_2O_5
pipette, Andreason, *see* Andreason pipette
pleochroite 73–4
plombièrite
as amorphous and quasi-crystalline calcium silicate hydrates 160, 207, 230–3, 257
as defined as 14 Å tobermorite 162
recrystallisation to form 14 Å tobermorite 231
poldervaartite (*see also* α-C_2SH) 166, 241
pore size distributions, in clinkers 127

pore solution, C_3S pastes 256, 384, 390
Portland cement (*see also* Portland cement clinker)
  chemical analysis 132
  constituent phases 1
  depth of reaction as function of time 131
  high early strength 6
  infrared and Raman spectra 140
  invention of 2–5
  low heat 6
  moderate heat of hardening 6
  particle size distribution 127–130
  phase compositions of different size fractions 131
  phase composition within grains 135–7
  rapid hardening 6
  reactions during storage 117
  specific surface area 130–1
  specifications, standard 5
  sulfate resisting 6
  thermogravimetry 139–40
  types of 5–6
  white, *see* white cements
  X-ray powder pattern, unhydrated (observed) 137–8
Portland cement clinker
  calculation of phase composition, *see* Bogue calculations
  colour 6, 45, 110–1, 119, 127
  density 127
  effects of minor components 117–9
  general description 127
  glass content 111–12
  grinding 6, 100, 117, 127-31, 133
  light microscopy 132–4
  manufacture 95–100
    effects of reducing conditions 110–11, 134
    enthalpy changes during 93–5
    fuels 74, 95–7, 99, 100, 102, 109–10, 117
  raw materials 6, 70, 74, 77–8, 89–119, 133
    burnability 106–8
  reactions during formation 89–90, 100–117
  physical properties, pore structure 127
  quantitative phase composition 91–4, 140–6
  reactivities of individual phases 147–50
  scanning electron microscopy 135–7
  typical compositions of constituent phases 13–4
  X-ray powder pattern (observed, portion of patterns) 35
portlandite, *see* calcium hydroxide
potassium aluminate, in cement clinkers 116
potassium calcium sulfates, *see* alkali and alkali calcium sulfates
potassium iron sulfide, in reduced clinkers 110
potassium sulfate (arcanite), *see* alkali and alkali calcium sulfates
precalciners, in cement making 97–9
preheaters 97–100, 102–4
preparation, laboratory
  high-temperature phases 79–80
  hydrated phases in general 372–5
primary air 99
product B, *see* calcium silicate pastes, early hydration product
propan-2-ol, use in drying cement paste 374–5, 406
protected phases 66, 80
proto C_3A 34
pseudowollastonite 57
pulverised fuel ash, *see* flyash
pyknometry, helium 386–7
pyrosilicate groups (*see also* dimeric silicate) 280–1

quantitative phase composition, clinker 91–4, 140–6
quantitative X-ray diffraction analysis (QXRD / QXDA) 7
  clinker / anhydrous cement 112, 137–9, 143–6
  calcium silicate pastes 377–83, 400–1, 422, 432
quartz
  in synthesis of tobermorite 168, 185, 206
  reactions in cement making 89–90, 95, 103–4, 107–8
  thermal changes 102
quasi-crystalline calcium silicate hydrates 160, 162, 207, 230–3, 257

Raman spectroscopy
  of synthetic C–S–H 270
  of tobermorites 187, 200, 204
  of unhydrated cements 140
rankinite 57, 147–8
rapid-hardening cements 6, 130
raw materials, *see* Portland cement clinker
reactivity of clinker phases, factors affecting 147–50
recrystallisation
  alite / belite 105, 108–9, 134
  plombièrite, to form 14 Å tobermorite 231
reducing conditions, in cement kiln 6, 71, 99, 103, 110–11, 114, 134, 140
reinhardbraunsite, *see* calcium chondrodite

relative humidity (RH) 7, 117, 232–3, 316, 329, 371, 374, 384–6, 389, 409, 431
reyerite 164, 240
Rietveld analysis 43, 112, 144–6, 270, 289, 315
rosenhahnite 166, 241
Rosin–Rammler function 129, 131
rotary kilns 97, 99–102
rustumite 167, 242
Ryley, John 4–5

salicylic acid in methanol (SAM) 138–40
sampling of clinker 102
s-block elements 118
scanning electron microscop(e,y) (SEM)
  abbreviation 7
  aluminate (phase) 34
  C–(A–)S–H, synthetic 251, 334
  calcium silicate paste, C–S–H
    backscattered electron imaging (BEI), polished surface 384, 388–9, 394–5
    cryo-SEM 389–90
    environmental SEM (ESEM) 389–90
    secondary electron imaging (SEI), fracture surface 387–391, 425, 432
    X-ray microanalysis 396–9
  Portland cement / clinker 112, 129, 135–7, 146
  tobermorite, 12 Å 205
scanning transmission electron microscop(e,y) (STEM) 7, 388
scawtite 167, 189, 242
secondary air 99
secondary ion mass spectroscopy (SIMS) 426
secondary neutral mass spectrometry (SNMS) 423–5, 430
sedigraph, X-ray, *see* X-ray sedigraph
sedimentation, *see* Andreason pipette
selected area electron diffraction (SED), *see* electron diffraction
SEM, *see* scanning electron microscop(e,y) (SEM)
separation of phases in Portland cement 138–9
setting 5, 110–11, 114, 119, 423
  definition 371
shales, as raw materials for cement making 89, 95
sieve analysis 127–31
silica ratio (silica modulus) 7, 89–90, 94, 97, 107–9, 133
silicate anion species in solution 420
'silicosulfate' ($C_5S_2\overline{S}$)
  data, equilibria, preparation, structure 75–7
  in cement making 102–5
site occupancy factor, bridging, *see* SOF_{BT}
slag, see blast-furnace slag
SNMS, *see* secondary neutral mass spectrometry
sodalite 62, 76
sodium sulfate, *see* alkali and alkali calcium sulfates
SOF_{BT} 216, 262–265, 273–5, 280–1, 285–6, 288–9, 296, 306–12, 315, 347, 351, 412
soft X-ray transmission microscopy 389–90
solubility
  calcium hydroxide 252–5, 272–3, 275–6, 376–7, 418–20
    reduction with alkali 324–6
  C–S–H 255–6
  C–S–H(I) 251–6, 412
  C–S–H(II) / jennite-like C–S–H 256, 276, 279–80, 412
  C_3S 255, 429
species in solution 420
specific surface area 5, 129–131, 232, 423
spinel 71, 102–3, 231
spurrite
  data, structure, equilibria 76, 79, 230
  formation in cement kiln 103–5
STEM, *see* scanning transmission electron microscop(e,y)
strätlingite 167, 329
structural-chemical formulae / arguments
  for tobermorite-related phases 174–5, 207–13, 263–5, 288, 319, 408, 410–17
  for jennite / C–S–H(II) / high Ca/Si C–S–H 280–2, 408, 410–17
sulfate phases in Portland cement clinker (*see* also alkali and alkali calcium sulfates)
  estimation from bulk analysis 115–7
  formation 102–5, 115–8
  light microscopy 132–4
sulfate-resisting Portland cements 6, 13, 43, 90, 142
sulfate, substitution in C–S–H 334
sulfide ion in reduced clinkers 110
'sulfospurrite', *see* 'silicosulfate'
suolunite 166, 236, 241, 371
suspension preheaters 97–100
sylvine 103
syngenite 117–8, 140

systems
  $CaO–Al_2O_3$ 58–61
  $CaO–Al_2O_3–CaF_2$ 79
  $CaO–Al_2O_3–Fe_2O_3$ 65–6
  $CaO–Al_2O_3–Fe_2O_3–SiO_2$ 66–70
  $CaO–Al_2O_3–SiO_2$ 62–5
  $CaO–C_2S–C_4AF$ 67–8
  $CaO–C_2S –C_{12}A_7–C_2F–MgO$ 72–3
  $CaO–C_2S–C_{12}A_7–C_4AF$ 66–8
  $CaO–C_2S–C_3P$ 119
  CaO–'FeO'–SiO_2 71
  $CaO–Fe_2O_3$ 65
  $CaO–MgO–Al_2O_3–SiO_2$ 71–3
  $CaO–SiO_2$ 57–8
    plus alkalis 77
    plus CaF_2 77–9
    plus CO_2 or $CaCO_3$ 79
  including SO_3 74–7
  $CaO–SiO_2–H_2O$ 251–6, 412, 418–21
    plus Na_2O or K_2O 325
synthetic preparations
  C–(A–)S–H, *see* calcium (alumino)silicate hydrates
  ferrite 39
  high-temperature phases 79–80
  hydrated phases in general 372–5
  tobermorite phases 170, 188, 196, 206, 225

tacharanite 205, 229–30
TEM, *see* transmission electron microscop(e,y)
ternasite, *see* 'silicosulfate'
tertiary air 100
tetracalcium aluminoferrite, *see* ferrite (phase)
TG, *see* thermogravimetry
thermal analysis,
  see differential scanning calorimetry
  see differential thermal analysis
  see thermogravimetry
thermochemistry
  calcium hydroxide dissolution and dehydroxylation 376–7
  calcium silicate hydration 421
  clinker formation 93–4
thermodynamics, $CaO-SiO_2-H_2O$ system 231, 233, 251, 255–6, 325, 420
thermogravimetry (TG) 7
  calcium hydroxide 376
  calcium silicate hydrates 188, 201, 224, 232, 278
  calcium silicate pastes 373–84, 391, 402
  unhydrated cements 139–40
tilleyite 79, 166, 230, 241, 323
tobermorite phases
  chemical compositions and cell contents 170–1
  clinotobermorite, 11 Å 162, 168–170, 174, 179–185, 206
    clinotobermorite, definition 168
    model crystal structure 179–85
  conditions of formation 168
  conversion to wollastonite 202–3
  crystal data 162, 168–173
  infrared spectra 186–7
  ionic substitutions 188
  morphology 206
  orthotobermorite, 11 Å 162, 168–179
    model crystal structure 175–9
    orthotobermorite, definition 168
  Raman spectra, ^{17}oxygen NMR and Ca X-ray absorption spectra 187
  structural-chemical formulae 207–13
  tobermorite, 9 Å 162, 201–3
    silicate anion structure (9 Å) 202
  tobermorite, 10 Å; oyelite 162, 204
  tobermorites, 11 Å, thermal behaviour
    anomalous variety 160, 162, 168–179, 185-8, 204–6, 239
    behaviour 160
    determinants and mechanisms of normal and anomalous thermal behaviour 160, 205–6
    normal variety 160, 162, 168–179, 186, 200–1, 204, 239
    silicate anion structure (11 Å) 185–6
  tobermorite, 12 Å 205
    silicate anion structure (12 Å) 202
  tobermorite, 14 Å
    model crystal structures for 14 Å orthotobermorites 190–195
    model crystal structures for theoretical 14 Å clinotobermorites 195–9
    natural occurrence, synthesis and crystal data 162, 188–9
    spectroscopic investigations 196, 200
      silicate anion structure (14 Å) 196
    thermal behaviour 200–1
    transmission electron microscopy, diffraction and microanalysis (Ca/Si ratio) 189–90
  useful equations for the interpretation of single-pulse data 212–23

trace components, effects on clinker 117–19
transition elements, and Portland cement making 119
transmission electron microscop(e,y) (*see also* electron diffraction)
  abbreviation 7
  alite 16
  alkali-activated cement 304
  C–(A–)S–H, synthetic 251, 254–5, 260–1, 271, 320, 326, 334
  C–(A–)S–H, hardened cement 329–31, 404
  calcium silicate pastes
    early hydration product 391–2, 418–9, 430
    CH 375
    C–S–H
      morphology 387–96, 406, 430
      X-ray microanalysis 372, 397–9, 411, 419
  high-Ca C–S–H, synthetic 273, 275–9, 281, 322, 419
  jennite 224
  ferrite (phase) 43
  tobermorite, 10 Å 204
  tobermorite, 11 Å 168
  tobermorite, 12 Å 205
  tobermorite, 14 Å 189–90
  vibratory milled lime–silica slurries 283
tricalcium aluminate and substituted modifications, *see* aluminate (phase)
tricalcium silicate, *see also* alite
  enthalpy change on hydration 421
  equilibria 57–8, 62–5, 67–70, 72, 78–9
  hydration 277, 371–433
    composition of solutions 254–5, 273, 418–9, 426–8
    initial reaction 423–6
    main reaction 430–1
    period of slow reaction 427–9
    preparation 79–80
  solubility 255, 429
  thermal stability 57, 112, 114
  X-ray powder patterns 7, 16–19
tricalcium silicate hydrate, *see* jaffeite
tricalcium silicate pastes, *see* calcium silicate pastes
tridymite 58–9, 102
'trimeric' silicate chains
  absence from $(3n-1)$ chain length sequence, C–S–H 291
  models for C–(A–)S–H 263, 274, 306, 308, 310–1, 347–51
trimethylsilyl(ation) 7, 139, 277, 282, 285, 400–402
truscottite 164

undecameric silicate chains
  in $(3n-1)$ chain length sequence, C–S–H 260, 288, 291, 402, 404, 410, 432–3
  models for C–(A–)S–H 263, 274, 285, 306, 311–20, 351–6
units of pressure or stress 7

vaterite (*see also* calcium carbonates) 373–4
vibrational spectroscopy, *see* infrared (IR) absorption spectroscopy, Raman spectroscopy
vibratory milling, in synthesising C–S–H 283
volatiles, circulation in cement manufacture 99–100, 103–5, 108–9

Wagner turbidimeter 130
Wakefield cement works 2–5
waste material utilisation in making clinker or cement 95–6, 100
water, categories of in calcium silicate pastes 384–6
water/cement (w/c) ratio 131, 276–81, 315, 371, 384, 404–6, 421
water content, *see* bound water / non-evaporable water
water, pressure–temperature phase diagram 385
water/solid (w/s) ratio 266, 289, 371–2, 375–7, 382–7, 390–6, 398, 403–7, 418–28, 431
water sorption 232
wet process 96–7, 100, 119, 127
white cements
  alite in 12
  aluminate in 13, 35
  alumina ratio 90, 107
  belite in 27
  Bogue calculation 142
  ferrite in 113
  glass in 112
  manufacture 6, 64, 99
  MAS NMR 402
  thermal analysis 110

White's reagent 133
wollastonite
  crystal structure 161
  density 236–7
  equilibria 57–8, 79
  occurrence 103–4, 201–5, 224–5, 283–4, 320
  crystal chemistry 147–8, 238
wüstite 71, 110
WYAS Wakefield Deeds 2–5

XPS, *see* X-ray photoelectron spectroscopy
X-ray absorption spectroscopy, *see* calcium X-ray absorption spectra
X-ray diffraction (XRD) powder patterns (*see also* quantitative X-ray diffraction analysis)
  alite, M_3 (calculated) 18–9
  aluminate (phase), cubic (calculated) 37–8
  aluminate (phase), orthorhombic (calculated) 36–7
  belite (calculated) 29–31
  C_3S paste, almost fully reacted (observed) 372
  C–S–H(I) model structures (calculated) 313
  C–S–H(I) (observed) 256–8, 313
  C–S–H, high-Ca, relatively high crystallinity (observed) 278
  C–S–H, high-Ca formed from highly-reactive β-C_2S (observed) 278
  C–S–H preparations of Nonat and Lecoq (1998) (observed, low-angle section) 314
  calcium hydroxide 278, 372
  clinotobermorite, 11 Å (calculated) 169
  dicalcium silicate polymorphs (calculated, portions of patterns) 26
  ferrite (phase) (calculated) 44–5
  orthotobermorite, anomalous 11 Å (calculated) 169
  Portland cement, unhydrated (observed) 137–8
  Portland cement clinker (observed, portion of patterns) 35
  tricalcium silicate polymorphs (calculated) 17
X-ray fibre rotation
  jennite 226
  plombièrite 231
X-ray fluorescence spectroscopy (XRF) 7, 132, 141
X-ray microanalysis
  alite 12–4, 137
  aluminate 13–4, 34–6
  belite 13–4, 27, 137
  C–A–S–H, in hardened cements 329–31, 334
  clinker 12–4, 102, 112–3, 135, 141
  C–S–H, in C_3S or alite pastes 372, 379–80, 384, 396–9, 411
  C–S–H, synthetic 255, 260–1, 273–6, 279, 419
  ferrite 13–4, 43–4
  jennite 224
  phase Q 73
  tobermorite, 10 Å 204
  tobermorite, 12 Å 205
  tobermorite, 14 Å 189–90
X-ray photoelectron spectroscopy (XPS) 7, 256, 412, 418, 423–7, 430, 432
X-ray sedigraph 129–30
xonotlite 161, 168, 186–7, 204, 230, 235, 237–8
XRD, *see* X-ray diffraction
XRF, *see* X-ray fluorescence spectroscopy

ye'elimite, *see* $C_4A_3\overline{S}$

zinc salts in cement making 105
zinc substitution in alites 149–50
zoning
  in alite 15
  in ferrite 113, 143, 150
Z-phase 164, 240
z_{SiI}, *see* net charge per Si atom in C–S–H